Active Metals

Edited by
Alois Fürstner

Further Titles from VCH:

C. Elschenbroich, A. Salzer
Organometallics – A Concise Introduction
XIII, 495 pages, 57 tables
2nd ed. 1992. Hardcover ISBN 3-527-28165-7
2nd ed. 1991. Softcover ISBN 3-527-28164-9

E. Shustorovich
Metal Surface Reaction Energetics
Theory and Applications to Heterogeneous Catalysis, Chemisorption, and Surface Diffusion
XII, 232 pages, 56 figures, 43 tables
1991. Hardcover ISBN 3-527-27938-5

J. Mulzer, et al.
Organic Synthesis Highlights
1991. XII, 410 pages, 45 figures, 3 tables
Hardcover ISBN 3-527-27355-5

H. Waldmann (ed.)
Organic Synthesis Highlights II
1995. XIII, 407 pages, 318 figures, 3 tables
Hardcover ISBN 3-527-29200-4

K. C. Nicolaou, E. J. Sorensen
Classics in Total Synthesis
1996. ca. 600 pages
Hardcover ISBN 3-527-29284-5
Softcover ISBN 3-527-29231-4

Distribution:
VCH, P. O. Box 101161, D-69451 Weinheim, Federal Republic of Germany
Switzerland: VCH, P. O. Box, CH-4020 Basel, Switzerland
United Kingdom and Ireland: VCH, 8 Wellington Court, Cambridge CB1 1HZ, United Kingdom
USA and Canada: VCH, 220 East 23rd Street, New York, NY 10010–4606, USA
Japan: VCH, Eikow Building, 10-9 Hongo 1-chome, Bunkyo-ku, Tokyo 113, Japan

ISBN 3-527-29207-1

Active Metals

- Preparation
- Characterization
- Applications

Edited by
Alois Fürstner

Weinheim · New York
Basel · Cambridge · Tokyo

Dr. Alois Fürstner
Max-Planck-Institut
für Kohlenforschung
Kaiser-Wilhelm-Platz 1
D-45470 Mülheim

Published jointly by
VCH Verlagsgesellschaft, Weinheim (Federal Republic of Germany)
VCH Publishers, New York, NY (USA)

Editorial Director: Dr. Ute Anton
Assistant Editor: Eva Schweikart
Production Manager: Claudia Grössl

Cover illustration: (top) STM image of a 4.1 nm Pd colloid stabilized by $N(nC_8H_{17})_4Br$ (see Chapter 7). M. T. Reetz, W. Helbig, S. A. Quaiser, U. Stimming, N. Breuer, R. Vogel *Science* **1995**, *267*, 367–369. Reproduced with the kind permission of the American Association for the Advancement of Science; (bottom) taxol precursor synthesized by a McMurry coupling using low-valent titanium (see Chapter 3).

Library of Congress Card No. applied for

A catalogue record for this book is available from the British Library

Die Deutsche Bibliothek – CIP-Einheitsaufnahme
Active metals : preparation, characterization, applications / ed. by Alois Fürstner. – Weinheim ; New York ; Basel ; Cambridge ; Tokyo : VCH, 1995
ISBN 3-527-29207-1
NE: Fürstner, Alois [Hrsg.]

Printed on acid-free and low-chlorine paper

Composition, Printing and Bookbinding: Druckhaus „Thomas Müntzer“, D-99947 Bad Langensalza
Printed in the Federal Republic of Germany

Preface

Many preparative chemists have bad experiences with metal-induced reactions in the early stages of their careers, for example when they try for the first time to start a Grignard reaction. And as the first impression usually persists, many metal-promoted transformations have a bad reputation as being tricky and highly "co-worker dependent".

Indeed, experimental skill is an important prerequisite for many such reactions when they are carried out in a conventional set-up. Since the early recommendations of Grignard himself to use a "crystal of iodine" as initiation agent, many preparative tricks for inducing the reaction have been put forward, evaluated, rejected, and re-introduced. Many other metal-induced transformations share these problems of reproducibility with magnesium chemistry. And even more seriously, only a few metals will (more or less spontaneously) react at all with organic molecules under conventional conditions. However, the various examples described in this book provide clear evidence that this is often not an intrinsic problem of a given metal but rather the consequence of an unsuitable physical form.

The use of activated metals offers convenient solutions to many of the shortcomings mentioned above. This monograph therefore aims to provide a fairly comprehensive overview of those metal activating procedures that are of greatest importance for preparative chemistry and catalysis. In most cases the significant advances achieved result from a substantial alteration of the particle size and texture of the metals employed. All the relevant information on the scope and limitations of the different activation procedures is brought together, and many practical aspects, such as the preparation and handling of the reagents and the specific requirements for chemicals and equipment, are illustrated by the representative laboratory procedures incorporated into the text. This should help the chemist at the bench to find the best solution for a given problem.

Despite these considerable accomplishments, metal activation must remain an immature science until a better insight into the origin of the chemical reactivity becomes possible. The reader will notice that even among different "active" forms of the same metal there may be great differences in the physical properties and chemical performance. We are still far from understanding the intricate phenomenon of activation and the complex processes taking place at a metal–liquid interface. Even the detailed physico-chemical characterization of reactive metal samples using the arsenal of modern solid-state analytical chemistry is only just getting started. Therefore, special emphasis is given in this book to the recent fascinating achievements in this direction. One effect of these has been to teach us important lessons about the discord between generally accepted anticipations and physical reality. For example, some of the activated "metals" have turned out (and others may well do so) not to be zero-valent elements but rather low-valent, polyphasic materials.

We should also keep in mind that finely dispersed (metal) particles deserve some attention as new materials in their own right. Although an in-depth treatment of all these aspects is beyond the scope of this monograph, a lot of information on this rapidly growing area is included. Metal activation is likely to become an important technique not only for preparative organic chemists but also for inorganic chemists and materials scientists. It is beginning to conquer domains that were previously the preserve of metallurgy.

Finally, I should like to take this opportunity to express my sincere gratitude to all the fellow chemists who agreed to write a contribution for this volume. I was lucky to succeed in persuading leading experts to participate in the project. It is their expertise that will familiarize the reader with the essence of the topic.

Mülheim, October 1995 Alois Fürstner

Contents

Preface V

List of Contributors XVII

1 Rieke Metals: Highly Reactive Metal Powders Prepared by Alkali Metal Reduction of Metal Salts

R. D. Rieke, M. S. Sell, W. R. Klein, T. Chen, J. D. Brown, and M. V. Hanson

1.1 Introduction 1
1.1.1 Physical Characteristics of Highly Reactive Metal Powders 3

1.2 Rieke Magnesium, Calcium, Strontium, and Barium 3
1.2.1 Formation of Rieke Magnesium 4
1.2.2 Formation of Rieke Calcium, Strontium, and Barium 4
1.2.3 Grignard Reactions Using Rieke Metals 5
1.2.4 1,3-Diene-Magnesium Reagents 7
1.2.4.1 Preparation 7
1.2.4.2 Regioselectivity 7
1.2.4.3 Carbocyclization of (1,4-Diphenyl-2-butene-1,4-diyl)magnesium with Organic Dihalides 8
1.2.4.4 1,2-Dimethylenecycloalkane-Magnesium Reagents 11
1.2.4.5 Synthesis of Fused Carbocycles, β,γ-Unsaturated Ketones and 3-Cyclopentenols from Conjugated Diene-Magnesium Reagents 12
1.2.4.6 Synthesis of Spiro γ-Lactones and Spiro δ-Lactones from 1,3-Diene-Magnesium Reagents 16
1.2.4.7 Synthesis of γ-Lactams from Conjugated Diene-Magnesium Reagents 20

1.3 Rieke Zinc 24
1.3.1 The Preparation of Rieke Zinc 24
1.3.2 Direct Oxidative Addition of Functionalized Alkyl and Aryl Halides 24
1.3.3 Reactions of Organozinc Reagents with Acid Chlorides 25
1.3.4 Reactions of Organozinc Reagents with α,β-Unsaturated Ketones 25
1.3.5 Reactions with Allylic and Alkynyl Halides 28
1.3.6 Cross-Coupling of Vinyl and Aryl Organozinc Reagents Using a Palladium Catalyst 28
1.3.7 Intramolecular Cyclizations and Conjugate Additions Mediated by Rieke Zinc 29

1.3.8 Formation of Tertiary and Secondary Alkylzinc Bromides 32
1.3.9 Cyanide-Based Rieke Zinc 32

1.4 Organocopper Reagents Utilizing Rieke Copper 33
1.4.1 Introduction 33
1.4.2 Background to the Development of Rieke Copper 33
1.4.3 Phosphine-Based Copper 34
1.4.4 Lithium 2-Thienylcyanocuprate-Based Copper 37
1.4.5 Copper Cyanide-Based Active Copper 41
1.4.6 Two-Equivalent Reduction of Copper(I) Complexes: A Formal Copper Anion 44

1.5 Rieke Aluminum, Indium, and Nickel 48
1.5.1 Aluminum 48
1.5.2 Indium 48
1.5.3 Nickel 49

1.6 Synthesis of Specialized Polymers and New Materials via Rieke Metals 50
1.6.1 Formation of Polyarylenes Mediated by Rieke Zinc 50
1.6.2 Regiocontrolled Synthesis of Poly(3-alkylthiophenes) and Related Polymers Mediated by Rieke Zinc 51
1.6.3 Synthesis of Poly(phenylcarbyne) Mediated by Rieke Calcium, Strontium, or Barium 52
1.6.4 Chemical Modification of Halogenated Polystyrenes Using Rieke Calcium or Copper 53
1.6.5 Polymer Supported Rieke Metal Reagents and their Applications in Organic Synthesis 55

2 Allylic Barium Reagents

A. Yanagisawa and H. Yamamoto

2.1 Introduction 61

2.2 Preparation of Stereochemically Homogeneous Allylic Barium Reagents 61
2.2.1 Direct Insertion Method Using Reactive Barium 61
2.2.2 Stereochemical Stability 62
2.2.3 Silylation of Stereochemically Homogeneous Allylic Barium Reagents . . 64
2.2.3.1 Procedure for Generation of Reactive Barium (Ba*) 65
2.2.3.2 Procedure for Protonation of the Geranyl Barium Reagent 65
2.2.3.3 Silylation of (*E*)-2-Decenylbarium Chloride 66

2.3 Allylic Barium Reagents in Organic Synthesis 66
2.3.1 α-Selective and Stereospecific Allylation of Carbonyl Compounds 66
2.3.1.1 Metal Effects on α/γ-Selectivity 66
2.3.1.2 Generality of α-Selectivity and Stereospecificity 68
2.3.1.3 Secondary Allylic Barium Compounds 69
2.3.1.4 Mechanistic Considerations 70
2.3.1.5 Typical Procedure for the Allylation of a Carbonyl Compound with an Allylic Barium Reagent: Synthesis of (*E*)-4,8-Dimethyl-1-phenyl-3,7-nonadien-1-ol 71

2.3.2 Regioselective and Stereospecific Synthesis of β,γ-Unsaturated Carboxylic Acids . . . 71
2.3.2.1 Typical Procedure for Carboxylation of Allylic Barium Reagents: Synthesis of (*E*)-4,8-Dimethyl-3,7-nonadienoic Acid . . . 72
2.3.3 Highly α,α'-Selective Homocoupling and Cross-Coupling Reactions of Allylic Halides . . . 73
2.3.3.1 Typical Procedure for Homocoupling Reactions of Allylic Halides Using Reactive Barium: Synthesis of Squalene . . . 77
2.3.4 Michael Addition Reaction . . . 77
2.3.4.1 Typical Procedure for One-Pot Double Alkylations of α,β-Unsaturated Ketones: Synthesis of *trans*-2-(3-Methyl-2-butenyl)-3-(2-propenyl)cyclopentanone . . . 80
2.3.5 Other Reactions . . . 82

2.4 Summary and Conclusions . . . 83

3 The McMurry Reaction

T. Lectka

3.1 Introduction . . . 85

3.2 Historical Perspective . . . 86

3.3 Reaction Mechanism and Theory . . . 87
3.3.1 Introduction . . . 87
3.3.2 Prelude to the Reaction: Formation of the Active Reagent . . . 87
3.3.3 Step I: Formation of the Ketyl Radical . . . 88
3.3.4 Steps II and III: Organization and Coupling . . . 88
3.3.5 Step IV: Deoxygenation . . . 89

3.4 Optimized Procedures for the Coupling Reaction . . . 90
3.4.1 The Reductant . . . 90
3.4.2 The Titanium Source . . . 91
3.4.3 Practical Comments about Couplings . . . 92

3.5 The McMurry Alkene Synthesis . . . 93
3.5.1 Intermolecular Couplings . . . 93
3.5.1.1 Prototype Couplings . . . 93
3.5.1.2 Synthesis of Other Tetrasubstituted Olefins . . . 95
3.5.1.3 Mixed Coupling Reactions . . . 95
3.5.2 The Intramolecular Coupling Reaction . . . 99
3.5.2.1 Basic Couplings . . . 99
3.5.2.2 Synthesis of Natural Products . . . 100
3.5.2.3 The Potassium–Graphite Modification . . . 103
3.5.2.4 Synthesis of Strained Rings and Non-Natural Products . . . 104
3.5.2.5 The Acyl Silane Coupling . . . 108

3.6 The McMurry Pinacol Reaction . . . 109
3.6.1 Mechanistic Considerations and Prototype Couplings . . . 109
3.6.2 Synthesis of Natural Products . . . 111
3.6.2.1 Coupling on a Highly Oxygenated System . . . 112

3.6.2.2 The Synthesis of Periplanone C . . . 114
3.6.2.3 The McMurry Coupling in the Total Synthesis of Taxol . . . 115
3.6.3 The Intermolecular Pinacol Reaction . . . 115
3.6.4 Other Pinacol Methodology: The Pederson Modifications . . . 117
3.7 The Keto Ester Coupling . . . 121
3.8 Tandem Couplings . . . 124
3.9 The Allyl/Benzyl Alcohol McMurry Coupling . . . 125
3.10 The Reaction of Other Functional Groups . . . 126
3.11 The McMurry Reaction in Polymer Synthesis . . . 128
3.12 Conclusion and Future Directions . . . 129

4 Ultrasound-Induced Activation of Metals: Principles and Applications in Organic Synthesis

J.-L. Luche and P. Cintas

4.1 Introduction . . . 133
4.2 The Physical and Chemical Effects of Ultrasound . . . 133
4.2.1 Dynamics of the Cavitation Bubble. Transient and Stable Cavitation . . . 134
4.2.2 Cavitation in Heterogeneous Solid-Liquid Systems . . . 135
4.2.2.1 The Physical Nature of Activation . . . 135
4.2.2.2 The Chemical Component of the Sonochemical Activation . . . 137
4.2.3 The Relationship between Sonochemical and Mechanochemical Activation 138
4.2.4 Sonochemical Reactivity, a General Approach . . . 139
4.3 Sonochemical Reactions of Inorganic Compounds . . . 139
4.3.1 Activation of Metals . . . 139
4.3.2 Activation by Cementation . . . 140
4.3.3 Reduction and Sonolysis of Metallic Compounds . . . 140
4.4 Metals in Organic Sonochemistry: Preliminary Remarks . . . 142
4.5 Electron Transfer to Conjugated Hydrocarbons . . . 142
4.5.1 Aromatic Radical Anions . . . 142
4.5.2 Radical Anions from Dienes . . . 145
4.6 Electron Transfers to Multiple Bonds . . . 145
4.6.1 Reductions of Carbon–Carbon Double Bonds . . . 146
4.6.2 Reductions of Carbonyl Groups . . . 147
4.7 Electron Transfers to Single Bonds (Excluding Carbon–Halogen Bonds) 151
4.7.1 Cleavage of the C–H Bond . . . 151
4.7.2 Reduction of Carbon–Carbon and Carbon–Heteroatom Single Bonds . . . 152
4.7.3 Reduction of Various Bonds Involving Heteroatoms . . . 154
4.8 The Carbon–Halogen Bond. Formation of Organometallics . . . 154
4.8.1 Formation of Organoalkali and Grignard Reagents . . . 155
4.8.2 Transmetallation with Sonochemically Prepared Organometallic Reagents 156
4.8.3 Direct Access to Organozinc Reagents . . . 157
4.8.4 Other Reactions . . . 158

4.9 Organic Reactions Using Sonochemically in-situ Generated Organometallics 160
4.9.1 Deprotonations by in-situ Generated Organoalkali-Metal Reagents 160
4.9.2 Wurtz- and Ullmann-Type Coupling Reactions Forming Carbon–Carbon Bonds 161
4.9.3 Coupling Reactions Involving Silicon, Tin, and Germanium 162
4.9.4 Addition of Organozinc Reagents to Multiple Bonds 165
4.9.5 In-situ Preparation of Dichloroketene 167

4.10 Additions of Organic Halides to Aldehydes and Ketones in the Presence of Metals 168
4.10.1 The Barbier Reaction in the Presence of Lithium or Magnesium 168
4.10.2 Barbier Reactions in the Presence of Zinc 171
4.10.3 Extensions of the Barbier Reaction. The Use of Aqueous Media 172
4.10.4 Additions to Trivalent Functionalities 176
4.10.5 Reactions from α-Halocarbonyl Compounds 177

4.11 Reactions Using Activated Metal Catalysts 179
4.11.1 Nickel-Catalyzed Hydrogenations and Hydrogenolyses 180
4.11.2 Hydrogenations, Hydrogenolyses, and Hydrosilylations Using Various Transition Metal Catalysts 181

4.12 Experimental Sonochemistry 182
4.12.1 Technical Aspects 182
4.12.1.1 Ultrasonic Cleaning Baths 182
4.12.1.2 Horn Generators 183
4.12.2 Typical Sonochemical Syntheses 183
4.12.2.1 Dieckman Condensation with Ultrasonically Dispersed Potassium 183
4.12.2.2 Lithium Di-isopropylamide 184
4.12.2.3 Barbier Reaction with 1-Bromo-4-chlorobutane 184
4.12.2.4 Conjugate Addition of 6-Bromo-1-hexene to 1-Buten-3-one 184
4.12.2.5 Hydrogenation of 4,4-Dimethylcyclopent-2-en-1-one 184

4.13 Conclusions 184

5 Preparation and Applications of Functionalized Organozinc Reagents

P. Knochel

5.1 Preparation of Functionalized Organozinc Reagents 191
5.1.1 Introduction 191
5.1.2 The Direct Insertion of Zinc Metal into Organic Substrates 191
5.1.3 The Halide–Zinc Exchange Reaction 195
5.1.4 The Boron–Zinc Exchange Reaction 202
5.1.5 The Lithium–Zinc Transmetallation 206
5.1.6 Carbon–Carbon Bond-Forming Reactions for the Preparation of Organozinc Compounds 207
5.1.6.1 Homologation of Zinc–Copper Compounds with Zinc Carbenoids 207
5.1.6.2 The Use of Bimetallic Reagents for the Preparation of Organozinc Compounds 209

5.2 Applications of Organozinc Compounds in Organic Synthesis 211
5.2.1 The Reactivity of Organozinc Derivatives 211
5.2.2 Copper(I)-Mediated Reactions of Organozinc Compounds 212
5.2.2.1 Substitution Reactions 212
5.2.2.2 Addition Reactions 216
5.2.3 Palladium(0)-Catalyzed Reactions of Organozinc Compounds 224
5.2.4 Titanium(IV)-Catalyzed Reactions of Organozinc Compounds and their Applications in Asymmetric Synthesis 226

6 Metal Atom/Vapor Approaches to Active Metal Clusters/Particles

K. J. Klabunde and G. Cardenas-Trivino

6.1 Atoms, Clusters, and Nanoscale Particles 237
6.1.1 Introduction 237
6.1.2 Gas-Phase Metal Clusters 237
6.1.3 Metal Clusters/Particles Formed in Low-Temperature Matrices 238

6.2 Experimental Metal Atom/Vapor Chemistry. The Solvated Metal Atom Dispersion (SMAD) Method to Ultrafine Powders 239
6.2.1 Metal Vaporization Methods 239
6.2.2 Preparation of Ultrafine Nickel in Pentane, Ni–FOP (Nickel–Fragments of Pentane) 241
6.2.3 Preparation of Nickel Carbide (Ni_3C) from Ultrafine Nickel Particles (Ni–FOP) 242
6.2.4 Preparation of Non-Solvated Ethylcadmium Iodide (EtCdI) from Ultrafine Cadmium Particles 242
6.2.5 Preparation of Methyltin Iodides [CH_3SnI_3, $(CH_3)_2SnI_2$, $(CH_3)_3SnI$] from Ultrafine Tin Particles 242
6.2.6 Coupling Reactions of Organohalides Using Ultrafine Nickel Particles 243
6.2.7 Grignard Reagent Preparation Using Solvated Metal Atom Dispersed Magnesium (SMAD-Mg) 243
6.2.7.1 Active Magnesium Preparation 243
6.2.7.2 Manipulation of the Active Magnesium under Inert Atmosphere 243
6.2.7.3 Grignard Reactions with Active Magnesium 244

6.3 The Matrix Clustering Process. Ultrafine Powders from Solvated Metal Atoms 244
6.3.1 Controlling Features 244
6.3.2 Nickel with Various Solvents (Example of a Monometallic System) 245
6.3.3 Gold–Tin with Various Solvents (Example of a Bimetallic System) 246
6.3.4 Summary of the Clustering Process 248

6.4 Reaction Chemistry of Solvated Metal Atom Dispersed (SMAD) Powders 249
6.4.1 Introduction 249
6.4.2 Reactions with Organic Halides 249
6.4.2.1 Mechanistic Considerations 249
6.4.2.2 Syntheses 250
6.4.3 Reactions with Hydrogen 254
6.4.3.1 Magnesium 254
6.4.4 Catalysis by SMAD Metals (without Catalyst Supports) 255

6.4.4.1 Introduction 255
6.4.4.2 Nickel 255
6.4.4.3 Other Metals 259
6.4.5 Formation of Carbides 259
6.4.5.1 Introduction 259
6.4.5.2 Nickel and Palladium 260
6.4.6 Colloidal Metals in Solvents and Polymers 263
6.4.6.1 Initial Studies 263
6.4.6.2 Particle Growth and Solvent Studies 266
6.4.6.3 Films from Non-Aqueous Metal Colloids 267
6.4.6.4 Metal Colloids in Polymers 269
6.4.7 Magnetic Properties of SMAD Metal Particles (Mono- and Bimetallic) .. 270

6.5 Conclusions 275

7 Electrochemical Methods in the Synthesis of Nanostructured Transition Metal Clusters

M. T. Reetz, W. Helbig and S. A. Quaiser

7.1 Introduction 279

7.2 Conventional Syntheses of Transition Metal Clusters Stabilized by Ammonium Salts 280

7.3 Electrochemical Synthesis of Metal Clusters Stabilized by Tetraalkylammonium Salts 281
7.3.1 Syntheses Based on Sacrificial Metal Anodes 281
7.3.1.1 Palladium Clusters 281
7.3.1.2 Other Metals 285
7.3.1.3 Bimetallic Clusters 285
7.3.2 Synthesis Based on the Reduction of Added Transition Metal Salts 286
7.3.3 Properties of Electrochemically Prepared Metal Clusters Stabilized by Tetraalkylammonium Salts 286
7.3.3.1 Solubility 286
7.3.3.2 Geometrical Parameters as Determined by Combined TEM/STM Studies 287
7.3.3.3 UV-Vis Spectra 291
7.3.3.4 Electrochemical Properties as Determined by Cyclic Voltammetry 292
7.3.3.5 Superparamagnetism 294
7.3.4 Conclusions 295

8 The Magnesium Route to Active Metals and Intermetallics

L. E. Aleandri and B. Bogdanović

8.1 Introduction 299

8.2 Magnesium–Anthracene Systems in Organic, Organometallic, and Inorganic Synthesis 300
8.2.1 MgA as a Source of *Soluble* Zerovalent Magnesium 300
8.2.2 Magnesium–Anthracene–THF and Magnesium–Anthracene–$MgCl_2$–THF Systems as Phase-Transfer Catalysts for Metallic Magnesium 302

8.2.2.1 Catalytic Synthesis of Magnesium Hydrides 302
8.2.2.2 Anthracene-Activated Magnesium for the Preparation of Finely-Divided Metal Powders and Transition Metal Complexes 303
8.2.2.3 Catalytic Synthesis of Grignard Compounds 306

8.3 Active Magnesium (Mg*) Powder from MgH_2^* and MgA in Synthetic Chemistry 307
8.3.1 Preparation and Characterization of Active Magnesium (Mg*) Powder... 307
8.3.2 Applications of Mg* Powder in Organic and Organometallic Synthesis .. 310

8.4 Wet-Chemical Routes to Magnesium Intermetallics and/or their Hydrides and Carbides 315
8.4.1 The First Route: Reaction of MgH_2^* or Et_2Mg with Bis(allyl)-Metal Complexes 315
8.4.2 The Second Route: Reaction of Metal Halides with Excess Amounts of Magnesium Hydrides, Organomagnesium Compounds, or Magnesium ... 319
8.4.2.1 $[HTiCl(THF)_{\sim 0.5}]_x$ – a Highly Reactive Titanium Hydride and an Active Species in a McMurry Reaction 320
8.4.2.2 Mg–Pt Intermetallics and their Hydrides from $PtCl_2$ and MgH_2' or MgH_2^* 323
8.4.2.3 Magnesium Intermetallics from Metal Halides and MgA or Mg* 325

8.5 Inorganic Grignard Reagents and their Applications in Inorganic Synthesis 326
8.5.1 Preparation and Characterization of Novel Inorganic Grignard Reagents.. 327
8.5.2 Reactive Alloys, Intermetallics, and Metals via Inorganic Grignard Reagents 330
8.5.3 Conclusions and Outlook 333

8.6 Experimental Procedures for Selected Examples 334
8.6.1 Magnesium Anthracene ·3 THF and the Synthesis of Allenylmagnesium Chloride 334
8.6.2 The Novel Magnesium–Platinum Intermetallic $t\text{-}MgPtC_xH_y$ 334
8.6.3 The Titanium Grignard Reagent, $[Ti(MgCl)_2(THF)_x]_2$ 335
8.6.4 Application of the Platinum Grignard Reagent to the Synthesis of PtFe .. 335

9 Catalytically Active Metal Powders and Colloids

H. Bönnemann and W. Brijoux

9.1 Introduction 339

9.2 Metal and Alloy Powders 340
9.2.1 Finely Divided Metals and Alloy Powders from Metal Salts 340
9.2.2 Metal and Alloy Powders from Metal Oxides 348

9.3 Formation of Colloidal Transition Metals and Alloys in Organic Solvents 352
9.3.1 Ether-Stabilized Ti^0, Zr^0, V^0, Nb^0, and Mn^0 352
9.3.1.1 Hydrogenation of Ti and Zr Sponge with $[Ti^0 \cdot 0.5\ THF]_x$ 353
9.3.2 Colloidal Transition Metals and Alloys Protected by Tetraalkylammonium Halides via Reduction of Metal Salts with Tetraalkylammonium Hydrotriorganoborates 355
9.3.3 General Approach to NR_4^+-Stabilized Metal and Alloy Colloids Using Conventional Reducing Agents 361

9.3.4 Synthesis of $PEt(octyl)_3^+$- and $SMe(octyl)_2^+$-Stabilized Metal and Alloy Colloids 363
9.4 Nanoscale Metal Powders via Metal Colloids 366
9.5 Catalytic Applications 368
9.5.1 Free and Supported Metal and Alloy Colloid Catalysts in Liquid-Phase Hydrogenation 368
9.6 Conclusions 376

10 Supported Metals

A. Fürstner

10.1 Introduction 381
10.2 General Features of Supported Metals 381
10.3 Applications of Supported Metals in Synthesis 384
10.3.1 "High Surface Area Alkali Metals" 384
10.3.2 Activated Metals by Reduction of Metal Salts with "High Surface Area Sodium" 388
10.3.3 Potassium–Graphite Laminate (C_8K) 389
10.3.4 Metal–Graphite Surface Compounds 401
10.3.4.1 Zinc– and Zinc/Silver–Graphite 401
10.3.4.2 Titanium–Graphite 410
10.3.4.3 Magnesium–Graphite 415
10.3.4.4 Group-10-Metal–Graphite Combinations as Catalysts 416
10.3.4.5 Other Metal–Graphite Combinations 417
10.3.5 Graphimets 418
10.3.6 Supported Rieke Metals 419
10.3.7 Miscellaneous Applications 419
10.4 Some Representative Procedures 420
10.4.1 "High Surface Area Sodium" on NaCl 420
10.4.2 Example of a [Ti]/Al_2O_3-Induced Macrocyclization of a Diketone: 1,2-Diphenyl(2,26)[36]paracyclophane 420
10.4.3 Large-Scale Preparation of Potassium–Graphite Laminate (C_8K) 421
10.4.4 C_8K-Induced Synthesis of a Furanoid Glycal from a Thioglycoside: 1,4-Anhydro-3-*O*-(*tert*-butyldiphenylsilyl)-2-deoxy-5,6-*O*-isopropylidene-D-arabino-hex-1-enitol 421
10.4.5 Zn(Ag)–Graphite-Induced Reductive Ring-Opening of a Deoxyiodo Sugar: (2*R*,4*S*)-Bis(benzyloxy)hex-5-enal 421
10.4.6 Titanium–Graphite-Induced Cyclization of Oxoamides: Synthesis of 2-(2′-Furanyl)-3-phenylindole 422

11 Morphological Considerations on Metal–Graphite Combinations

F. Hofer

11.1 Introduction 427

11.2 Characterization Methods 428
11.2.1 X-Ray Methods 428
11.2.2 Electron Microscopy 428
11.2.3 Surface-Sensitive Methods 429
11.2.4 In-situ Investigations 429

11.3 Intercalation in Graphite and the Structure of C_8K 430

11.4 The Structures of the Metal–Graphite Combinations 431
11.4.1 The Graphite Support 431
11.4.2 Platinum– and Palladium–Graphites 432
11.4.3 Silver– and Copper–Graphites 435
11.4.4 Zinc–Graphite and Zinc/Metal–Graphites 436
11.4.5 Nickel–Graphite and Nickel/Metal–Graphites 439
11.4.6 Titanium–Graphite Combinations 441

11.5 The Structure of Graphimets 442

11.6 Conclusions and Outlook 444

Experimental Procedures 447

Subject Index 449

List of Contributors

L. E. Aleandri
Max-Planck-Institut
für Kohlenforschung
Kaiser-Wilhelm-Platz 1
D-45470 Mülheim/Ruhr
Germany

Borislav Bogdanović
Max-Planck-Institut
für Kohlenforschung
Kaiser-Wilhelm-Platz 1
D-45470 Mülheim/Ruhr
Germany

Helmut Bönnemann
Max-Planck-Institut
für Kohlenforschung
Kaiser-Wilhelm-Platz 1
D-45470 Mülheim/Ruhr
Germany

Werner Brijoux
Max-Planck-Institut
für Kohlenforschung
Kaiser-Wilhelm-Platz 1
D-45470 Mülheim/Ruhr
Germany

Jeffrey D. Brown
Department of Chemistry
University of Nebraska-Lincoln
P.O. Box 880 304
Lincoln, NE 68588-0304
USA

Galo Cardenas-Trivino
Department of Chemistry
University of Concepcion
Concepcion
Chile

Tian-an Chen
Department of Chemistry
University of Nebraska-Lincoln
P.O. Box 880 304
Lincoln, NE 68588-0304
USA

Pedro Cintas
Department of Organic Chemistry
Faculty of Sciences
University of Badajoz
E-06071 Badajoz
Spain

Alois Fürstner
Max-Planck-Institut
für Kohlenforschung
Kaiser-Wilhelm-Platz 1
D-45470 Mülheim/Ruhr
Germany

Mark V. Hanson
Department of Chemistry
University of Nebraska-Lincoln
P.O. Box 880 304
Lincoln, NE 68588-0304
USA

Wolfgang Helbig
Max-Planck-Institut
für Kohlenforschung
Kaiser-Wilhelm-Platz 1
D-45470 Mülheim/Ruhr
Germany

Ferdinand Hofer
Research Institute
for Electron Microscopy
Graz University of Technology
A-8010 Graz
Austria

Kenneth J. Klabunde
Department of Chemistry, Willard Hall
Kansas State University
Mannhattan, KA 66506-3701
USA

Walter R. Klein
Department of Chemistry
University of Nebraska-Lincoln
P.O. Box 880 304
Lincoln, NE 68588-0304
USA

Paul Knochel
Fachbereich Chemie
der Universität
D-35032 Marburg
Germany

Thomas Lectka
Department of Chemistry
The Johns Hopkins University
3400 N. Charles Street
Baltimore, MD 21218
USA

Jean-Louis Luche
Université Paul Sabatier
Bâtiment de Chimie
118 Route de Narbonne
F-31062 Toulouse
France

Stefan A. Quaiser
Max-Planck-Institut
für Kohlenforschung
Kaiser-Wilhelm-Platz 1
D-45470 Mülheim/Ruhr
Germany

Manfred T. Reetz
Max-Planck-Institut
für Kohlenforschung
Kaiser-Wilhelm-Platz 1
D-45470 Mülheim/Ruhr
Germany

Reuben D. Rieke
Department of Chemistry
University of Nebraska-Lincoln
P.O. Box 880 304
Lincoln, NE 68588-0304
USA

Matthew S. Sell
Department of Chemistry
University of Nebraska-Lincoln
P.O. Box 880 304
Lincoln, NE 68588-0304
USA

Hisashi Yamamoto
Department of Applied Chemistry
Nagoya University
Chikusa
Nagoya 464-01
Japan

Akira Yanagisawa
Department of Applied Chemistry
Nagoya University
Chikusa
Nagoya 464-01
Japan

1 Rieke Metals: Highly Reactive Metal Powders Prepared by Alkali Metal Reduction of Metal Salts

Reuben D. Rieke, Matthew S. Sell, Walter R. Klein, Tian-an Chen, Jeffrey D. Brown, and Mark V. Hanson

1.1 Introduction

In 1972 we reported a general approach for preparing highly reactive metal powders by reducing metal salts in ethereal or hydrocarbon solvents using alkali metals as reducing agents [1]. Several basic approaches are possible and each has its own particular advantages. For some metals, all approaches lead to metal powders of identical reactivity. However, for other metals one method can lead to far superior reactivity. High reactivity, for the most part, refers to oxidative addition reactions. Since our initial report, several other reduction methods have been reported including metal–graphite compounds, a magnesium–anthracene complex, and dissolved alkalides [2].

Although our initial entry into this area of study involved the reduction of $MgCl_2$ with potassium biphenylide, our early work concentrated on reductions without the use of electron carriers. In this approach, reductions are conveniently carried out with an alkali metal and a solvent whose boiling point exceeds the melting point of the alkali metal. The metal salt to be reduced must also be partially soluble in the solvent, and the reductions are carried out under an argon atmosphere. Equation (1.1) shows the reduction of metal salts using potassium as the reducing agent.

$$MX_n + nK \rightarrow M^* + nKX \tag{1.1}$$

The reductions are exothermic and are generally completed within a few hours. In addition to the metal powder, one or more moles of alkali salt are generated. Convenient systems of reducing agents and solvents include potassium and THF, sodium and 1,2-dimethoxyethane (DME), and sodium or potassium with benzene or toluene. For many metal salts, solubility considerations restrict reductions to ethereal solvents. Also, for some metal salts, reductive cleavage of the ethereal solvents requires reductions in hydrocarbon solvents such as benzene or toluene. This is the case for Al, In, and Cr. When reductions are carried out in hydrocarbon solvents, solubility of the metal salts may become a serious problem. In the case of Cr [3], this was solved by using $CrCl_3 \cdot 3\,THF$.

A second general approach is to use an alkali metal in conjunction with an electron carrier such as naphthalene. The electron carrier is normally used in less than stoichiometric proportions, generally 5 to 10% by mole based on the metal salt being reduced. This procedure allows reductions to be carried out at ambient temperature or at least at lower temperatures compared with the previous approach, which requires refluxing. A convenient reducing metal is lithium. Not only is the procedure much safer when lithium is used rather than sodium or potassium, but in many cases the reactivity of the metal powders is greater.

A third approach is to use a stoichiometric amount of preformed lithium naphthalenide. This approach allows for very rapid generation of the metal powders in that the reductions are diffusion controlled. Very low to ambient temperatures can be used for the reduction. In some cases the reductions are slower at low temperatures because of the low solubility of the metal salts. This approach frequently generates the most active metals, as the relatively short reduction times at low temperatures restrict the sintering (or growth) of the metal particles. This approach has been particularly important for preparing active copper. Fujita et al. have shown that lithium naphthalenide in toluene can be prepared by sonicating lithium, naphthalene, and *N,N,N′,N′*-tetramethylethylenediamine (TMEDA) in toluene [4]. This allows reductions of metal salts in hydrocarbon solvents. This proved to be especially beneficial with cadmium [5]. An extension of this approach is to use the solid dilithium salt of the dianion of naphthalene. Use of this reducing agent in a hydrocarbon solvent is essential in the preparation of highly reactive uranium [6].

For many of the metals generated by one of the above three general methods, the finely divided black metals will settle after standing for a few hours, leaving a clear, and in most cases colorless, solution. This allows the solvent to be removed via a cannula. Thus the metal powder can be washed to remove the electron carrier as well as the alkali salt, especially if it is a lithium salt. Moreover, a different solvent may be added at this point providing versatility in solvent choice for subsequent reactions.

The wide range of reducing agents under a variety of conditions can result in dramatic differences in the reactivity of the metal. For some metals, essentially the same reactivity is found no matter what reducing agent or reduction conditions are used. In addition to the reducing conditions, the anion of the metal salt can have a profound effect on the resulting reactivity. These effects are discussed separately for each metal. However, for the majority of metals lithium is by far the preferred reducing agent. First, it is much safer to carry out reductions with lithium. Second, for many metals (magnesium, zinc, nickel, etc.), the resulting metal powders are much more reactive if they have been generated by lithium reduction.

An important aspect of the highly reactive metal powders is their convenient preparation. The apparatus required is very inexpensive and simple. The reductions are usually carried out in a two-neck flask equipped with a condenser (if necessary), septum, heating mantle (if necessary), magnetic stirrer, and argon atmosphere. A critical aspect of the procedure is that *anhydrous* metal salts must be used. Alternatively, anhydrous salts can sometimes be easily prepared as, for example, $MgBr_2$ from Mg turnings and 1,2-dibromoethane. In some cases, anhydrous salts can be prepared by drying the hydrated salts at high temperatures in vacuum. This approach must be used with caution as many hydrated salts are very difficult to dry completely by this method or lead to mixtures of metal oxides and hydroxides. This is the most common cause when metal powders of low reactivity are obtained. The introduction of the metal salt and reducing agent into the reaction vessel is best done in a dry box or glove bag; however, very nonhygroscopic salts can be weighed out in the air and then introduced into the reaction vessel. Solvents, freshly distilled from suitable drying agents under argon, are then added to the flask with a syringe. While it varies from metal to metal, the reactivity will diminish with time and the metals are best reacted within a few days of preparation.

We have never had a fire or explosion caused by the activated metals; however, extreme caution should be exercised when working with these materials. Until one becomes familiar with the characteristics of the metal powder involved, careful consideration should be taken at every step. To date, no metal powder we have generated will spontaneously ignite if removed from the reaction vessel while wet with solvent. They do, however, react rapidly with oxygen and with moisture in the air. Accordingly, they should be handled under an argon atmosphere. If the metal powders are dried before being exposed to the air, many will begin to smoke and/or ignite. Perhaps the most dangerous step in the preparation of the active metals is the handling of sodium or potassium. This can be avoided for most metals by using lithium as the reducing agent. In rare cases, heat generated during the reduction process can cause the solvent to reflux excessively. For example, reductions of $ZnCl_2$ or $FeCl_3$ in THF with potassium are quite exothermic. This is generally only observed when the metal salts are very soluble and the molten alkali metal approach (method one) is used. Sodium–potassium alloy is very reactive and difficult to use as a reducing agent; it is used only as a last resort in special cases.

1.1.1 Physical Characteristics of Highly Reactive Metal Powders

The reduction generates a finely divided black powder. Particle size analyses indicate a range of sizes varying from 1–2 μm to submicron dimensions depending on the metal and, more importantly, on the method of preparation. In cases such as nickel and copper, black colloidal suspensions are obtained that do not settle and cannot be filtered. In some cases even centrifugation is not successful. It should be pointed out that the particle size analyses as well as surface area studies have been done on samples that have been collected, dried, and sent off for analysis, and are thus likely to have experienced considerable sintering. Scanning electron microscopy (SEM) photographs reveal a range from sponge-like material to polycrystalline material [7]. Results from X-ray powder diffraction studies range from those for metals such as Al and In, which show diffraction lines for both the metal and the alkali salt, to those for Mg and Co, which only show lines for the alkali salt. This result suggests that the metal in this latter case is either amorphous or has a particle size less than 0.1 micrometer. In the case of Co, a sample heated to 300 °C under argon and then reexamined showed diffraction lines due to Co, suggesting the small crystallites had sintered upon heating [8].

ESCA (XPS) and Auger spectroscopy studies have been carried out on several metals and in all cases the metal has been shown to be in the zerovalent state. Bulk analysis also clearly shows that the metal powders are complex materials containing in many cases significant quantities of carbon, hydrogen, oxygen, halogens, and alkali metal.

A BET surface area measurement [9] was carried out on the activated Ni powder showing it to have a specific surface area of 32.7 m^2 g^{-1}. Thus, it is clear that the highly reactive metals have very high surface areas which, when initially prepared, are probably relatively free of oxide coatings. The metals contain many dislocations and imperfections which probably add to their reactivity.

1.2 Rieke Magnesium, Calcium, Strontium, and Barium

The significance of the Grignard reaction is clearly expressed in the following quotation: "... every chemist has carried out the Grignard reaction at least once in his lifetime ..." [10]. In the light of this ubiquity, the generation of new Grignard reagents is indeed a worthwhile endeavor. Although the formation of Grignard reagents is commonly thought to be completely general, there are many organic halides that do not form Grignard reagents. The preparation of Grignard reagents from primary, secondary, and tertiary halides as well as aryl iodides and bromides using ordinary metallic magnesium turnings can be accomplished in a low boiling solvent such as diethyl ether. However, many organic halides including bicyclic bridgehead halides, aryl chlorides, and vinyl chlorides exhibit a low propensity for the formation of their corresponding Grignard reagents via ordinary magnesium turnings. Also, there are many alkyl and aryl bromides which appear to be likely candidates for Grignard formation but fail because of the presence of certain functional groups, such as ethers. Moreover, organic fluorides are generally regarded as inert towards ordinary magnesium. Also, in some cases the Grignard reagents are not stable under the reaction conditions, which commonly involve refluxing in tetrahydrofuran (THF). Clearly, any technique which could make these more difficult organic halides accessible to Grignard reagent formation is desirable. To this end, if ordinary magnesium could be somehow "activated" (either physically or chemically), some of these difficultly accessible Grignard reagents could be prepared. The Rieke method of metal activation has greatly expanded the range of organic halides that are amenable to Grignard reagent preparation.

1.2.1 Formation of Rieke Magnesium

The most common approach for the generation of Rieke magnesium involves a one pot procedure using lithium in conjunction with a catalytic amount of naphthalene (10% by mole based on lithium) in THF under an argon atmosphere. The Rieke magnesium is obtained after vigorous stirring for 3 h. Although the stoichiometric lithium naphthalenide procedure affords Rieke magnesium which is of identical reactivity, the catalytic method is the method of choice because it facilitates work-up.

In an oven-dried, 50-mL, two-necked, round-bottomed flask is placed anhydrous magnesium chloride (0.402 g, 4.22 mmol), lithium (cut foil, 0.061 g, 8.86 mmol) and naphthalene (0.170 g, 1.33 mmol) in a Vacuum Atmospheres Company dry-box under argon. The flask is then connected through an adapter to a double manifold providing vacuum and purified argon, and the flask is evacuated and back-filled with argon three times. Freshly distilled THF (15 mL) from Na/K alloy under argon is added via a syringe. The mixture is stirred vigorously at room temperature for 3 h. The newly formed Rieke magnesium slurry is allowed to settle for 3 h and the supernatant is removed via a cannula. Freshly distilled THF (10 mL) is added to the Rieke magnesium (black powder) followed by the appropriate substrate.

1.2.2 Formation of Rieke Calcium, Strontium, and Barium

The current methodology for the production of Rieke calcium, strontium, and barium is similar to that of magnesium but involves reduction of these metal salts using a stoichiometric amount of lithium biphenylide.

Lithium (9.0 mmol) and biphenyl (9.8 mmol) are stirred in freshly distilled THF (20 mL) under argon until the lithium is completely consumed (ca. 2 h). To a well-suspended anhydrous metal halides (4.4 mmol) in freshly distilled THF (20 mL) solution, the preformed lithium biphenylide is transferred via a cannula at room temperature and stirred for 1 h. The Rieke calcium, strontium, or barium is then ready for use.

1.2.3 Grignard Reactions Using Rieke Metals

The reactions of Rieke magnesium and calcium with various dihaloarenes, some of which react only with difficulty under the conditions of normal Grignard preparations, were recently investigated [11]. It was found that a one-pot, low temperature formation of mono-organometallic intermediates derived from the incorporation of (1:1) Rieke magnesium or Rieke calcium with *meta-* and *para-*dihalobenzenes and 2,5-dibromothiophene is possible in very good isolated yields. The ability of Rieke magnesium and calcium to undergo formal oxidative addition reactions with aryl-chlorides, bromides, and fluorides at low temperature is possible owing to the high reactivity of the metal. For example, reaction of 1,4-dibromobenzene and Rieke magnesium (1:1) at –78 °C for 30 min formed the mono-organomagnesium intermediate, which when reacted with carbon dioxide followed by acidic hydrolysis, afforded 4-bromobenzoic acid in 99% isolated yield. Similarly, reaction of 1,4-dichlorobenzene and Rieke magnesium (1:1) under reflux for 30 min followed by carbon dioxide and acidic hydrolysis afforded 4-chlorobenzoic acid in 95% isolated yield. This is in stark contrast to the reaction of 1,4-dibromobenzene with ordinary magnesium, which affords a mixture of the mono- and di-Grignards, as well as unreacted starting material [12]. Remarkably, reaction of 1,4-difluorobenzene with Rieke calcium (1:1) at –78 °C for 30 min, followed by addition of benzaldehyde and acidic hydrolysis afforded 4-fluorobenzhydrol in 93% isolated yield. This one-pot, high yield preparation of mono-organometallic intermediates via Rieke metals could find application in the synthesis of many types of conducting polymers as well as many asymmetrical aryl molecules.

Previous efforts to prepare di-magnesium derivatives of benzene have been successful only with dibromo- or bromoiodobenzene, required forcing conditions, and typically resulted in the mono-magnesium derivative as the major product. This work has been surveyed by Ioffe and Nesmayanov [13]. Using the $MgCl_2$–KI–K–THF system, it was shown that the preparation of the di-Grignard of 1,4-dibromobenzene in 100% yield in 15 min at room temperature is possible. The yield was determined by gas chromatography after hydrolysis and was based on the disappearance of starting material and bromobenzene and the formation of benzene. In earlier work, only one halogen atom of dichloro derivatives of benzene and naphthalene reacted with magnesium [14], and the chlorine atom of *p*-chlorobromobenzene was found to be completely unreactive [15].

We have reported Grignard-type reactions of organocalcium reagents with cyclohexanone [16]. It was found that reactions of aryl halides with Rieke calcium required slightly higher temperatures, up to –30 °C for aryl bromides and up to –20 °C for aryl chlorides. Reactions of *m*-bromotoluene, *m*-bromoanisole, and *p*-chlorotoluene with Rieke calcium gave the corresponding arylcalcium reagents in quantitative yields based on the GC analyses of reaction quenches. As expected, 1,2-addition of these arylcalcium compounds with cyclohexanone gave the alcohols in excellent yields (76%, 79%, and 86%, respectively). Surprisingly, Rieke calcium reacted readily with fluorobenzene at room temperature to form the corresponding organometallic compound which underwent an addition reaction with cyclohexanone to give 1-phenylcyclohexanol in 85% yield. Except for Rieke magnesium prepared by the reduction of magnesium salts [17], few metals undergo oxidative addition with aryl fluorides to form organometallic compounds [18].

A simple example of the advantage of generating Grignard reagents at low temperatures is the reaction of 3-halophenoxypropanes [19]. Although the Grignard reagent is easy to generate at room temperature or above, it eliminates the phenoxy group by an

S_N2 reaction, generating cyclopropane. This reaction is, in fact, a standard way to prepare cyclopropanes and the cyclization cannot be stopped; however, by using Rieke magnesium, the Grignard reagent can be prepared at –78 °C and it does not cyclize at these low temperatures. It then can be added to a variety of other substrates in standard Grignard reactions.

When benzonitrile was reacted with Rieke magnesium in refluxing glyme, a deep red color developed after 1 h. Gas chromatography revealed most of the starting material to be unconsumed. After refluxing overnight, the reaction mixture turned brown, and most of the starting material was consumed. An aliquot injected directly into the gas chromatograph did not reveal benzil formation. Work-up showed two main products, 2,4,6-triphenyl-1,3,5-triazine and 2,4,5-triphenylimidazole, in 26% and 27% yields, respectively. These products are shown in Scheme 1-1. The imidazole was quite unexpected but nevertheless present as the most abundant product. A product with a TLC R_f value equivalent to benzil was also present in less than 5 mol % yield, as were numerous other unidentified products in low yields. The imidazole was shown to arise, at least in part, from the triazine. 2,4,6-Triphenyl-1,3,5-triazine reacted with Rieke magnesium in refluxing THF to afford 2,4,5-triphenylimidazole in 27% yield; 65% was unreacted starting material; 8% was unaccounted for.

2,4,6-triphenyl-1,3,5-triazin 2,4,5-triphenylimidazole

Scheme 1-1. Two products formed from reaction of benzonitrile with Rieke magnesium in refluxing glyme.

Giordano and Belli have formed imidazoles from 2,4,6-triaryl-4*H*-1,3,5-thiadiazines via a base-catalyzed extrusion of sulfur [20]. Schmidt et al. also refer to an analogous extrusion of oxygen from the 4*H*-1,3-oxazine [21]. Scheme 1-2 depicts two extrusion reactions used to form imidazoles. Radzisewski has prepared imidazoles by the extrusion of nitrogen from triazines using Zn in refluxing acetic acid [22].

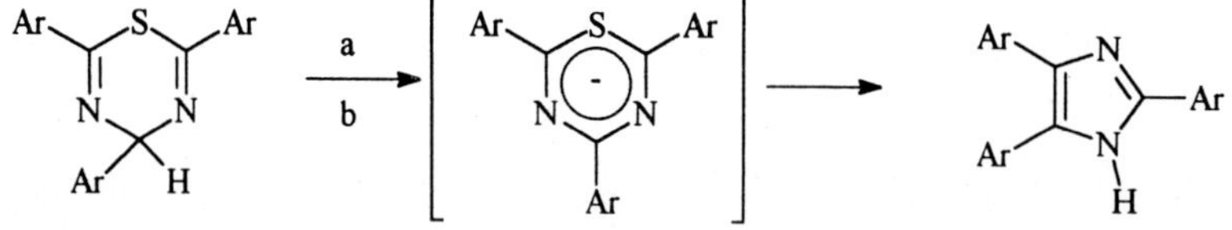

Scheme 1-2. Two extrusion reactions used to form imidazoles. Conditions: (a) aliphatic amines/benzene; (b) *n*-butyllithium.

The trimerization of aromatic nitriles to give symmetrical triazines is not unknown, but generally the reactions must be catalyzed by strong acids or weak bases and carried out under extremely high pressure. The action of *preformed* Grignard reagents also gives symmetrical triazines. Organoalkalis are also known to give trimers but not symmetrical triazines. This direct reaction of magnesium to afford symmetrical triazines is unprecedented in the literature.

1.2.4 1,3-Diene-Magnesium Reagents

1.2.4.1 Preparation

Metallic magnesium is known to react with certain 1,3-dienes yielding halide-free organomagnesium compounds [23]; however, there is a primary problem associated with the preparation of these reagents. The reaction of ordinary magnesium with 1,3-dienes such as 1,3-butadiene or isoprene is usually accompanied by dimerization, trimerization, and oligomerization. This reaction may be catalyzed by alkyl halides or transition metal salts but is generally accompanied by a variety of by-products and is very time-consuming. Consequently, the utilization of these reagents in organic synthesis has been quite limited [24], except for 1,3-butadiene-magnesium, which has found considerable application in organometallic synthesis [25].

In contrast, Rieke magnesium reacts with (*E,E*)-1,4-diphenyl-1,3-butadiene, 2,3-dimethyl-1,3-butadiene, isoprene, myrcene, or 2-phenyl-1,3-butadiene in THF at ambient temperature to afford the corresponding substituted (2-butene-1,4-diyl)magnesium complexes in near quantitative yields. The structures of these complexes have not been determined to date except for (1,4-diphenyl-2-butene-1,4-diyl)magnesium, which has been shown to be a five-membered ring metallacycle [26]. Accordingly, the most likely structures for these complexes are five-membered metallacycles or oligomers. It is also possible that an equilibrium exists between these various forms.

1.2.4.2 Regioselectivity

The reactions discussed below illustrate the unusual reactivity of these magnesium-diene complexes. While one might expect the complexes to react as bis-Grignard reagents, it is clear that they are much more powerful nucleophiles. Remarkably, they react with alkyl bromides and chlorides at temperatures as low as $-80°C$. They appear to react via a standard S_N2 mechanism. However, they can serve as electron transfer agents with certain reagents such as metal salts. With most electrophiles accommodating good leaving groups, clean S_N2 chemistry is observed. Soft electrophiles such as organic halides demonstrate complete regioselectivity for the 2-position as shown in Scheme 1-3. The resulting primary organomagnesium intermediate can then be reacted with a wide range of second electrophiles. Harder electrophiles such as monohalosilanes, monohalostannanes etc. react completely regioselectively in the 1-position as shown in Scheme 1-4. Addition of a second electrophile occurs

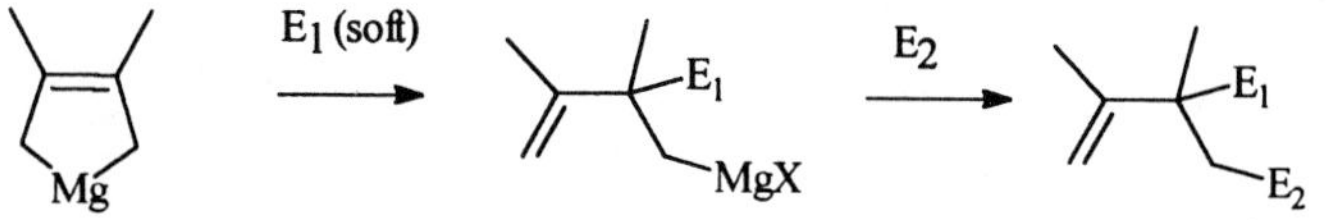

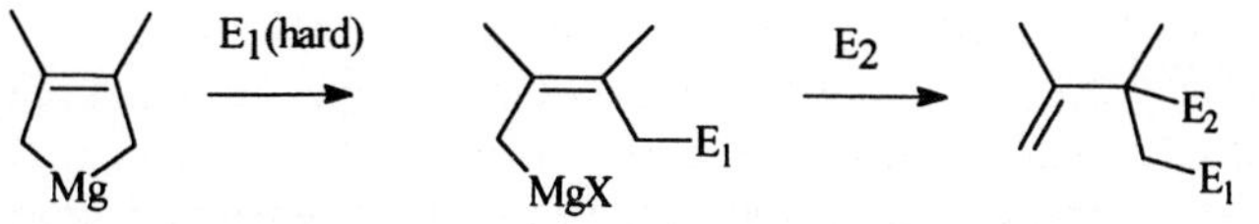

Scheme 1-3. Reaction of a soft electrophile with (2,3-dimethyl-2-butene-1,4-diyl)magnesium.

Scheme 1-4. Reaction of a hard electrophile with (2,3-dimethyl-2-butene-1,4-diyl)magnesium.

Table 1-1. Stepwise reactions of (2,3-Dimethyl-2-butene-1,4-diyl)magnesium.

Entry	E_1	E_2	Product	Yield (%)
1	$Br(CH_2)_4Br$	MeCOCl	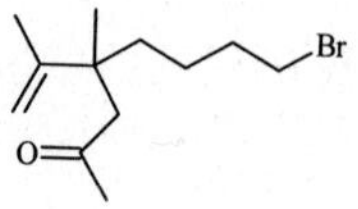 O, Br	62
2	$Br(CH_2)_4Br$	PhCOCl	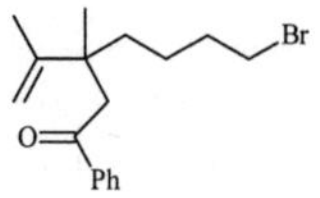 O, Ph, Br	60
3	$Me(CH_2)_3Br$	MeCOCl	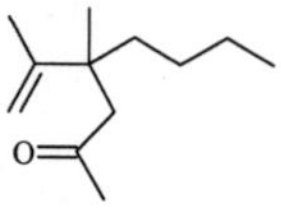 O	61
4	$Me(CH_2)_3Br$	PhCOCl	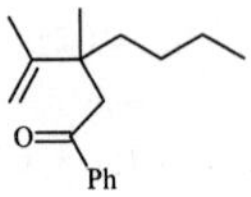 O, Ph	82
5	Me_3SiCl	MeCOCl	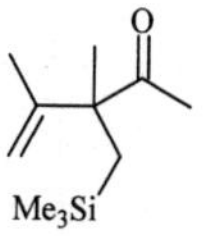 O, Me_3Si	73
6	Me_3SiCl	PhCOCl	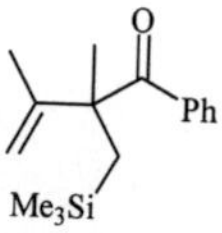 O, Ph, Me_3Si	79
7	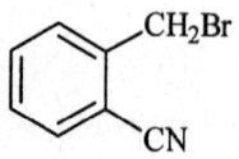 CH_2Br, CN	H_3O^+	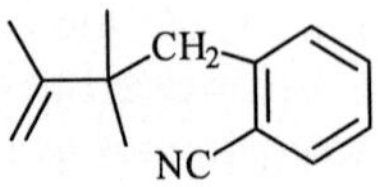 CH_2, NC	35
8	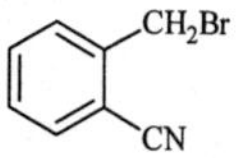 CH_2Br, CN	–	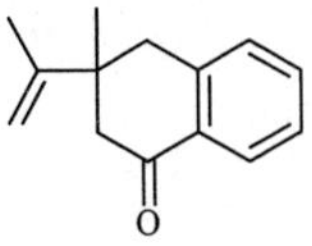 O	30
9	$Br(CH_2)_3CN$	H_3O^+	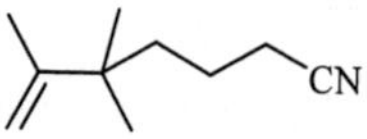 CN	58

primarily in the 2-position. The overall result is that one can effect a net 2,1-addition or a 1,2-addition by choosing appropriate electrophiles. Complex, highly functionalized molecules have been prepared using this approach and are shown in Table 1-1 [27].

1.2.4.3 Carbocyclization of (1,4-Diphenyl-2-butene-1,4-diyl)magnesium with Organic Dihalides

(*E,E*)-1,4-Diphenyl-1,3-butadiene has been found to be one of the most reactive conjugated dienes towards metallic magnesium. Reaction of this diene with Rieke magnesium affords (1,4-diphenyl-2-butene-1,4-diyl)magnesium (**2**) which can be treated with various electrophiles (Table 1-2). Reactions of **2** with 1,*n*-dibromoalkanes resulted in either cyclization or reduction of the electrophile, depending on the initial dibromide. For example, cyclization proceeded rapidly at −78 °C in the reaction of **2** with 1,3-dibromopropane or 1,3-dichloropropane, yielding a single product, *trans*-1-phenyl-2-((*E*)-2-phenylethenyl)cyclopentane (**4**) in 65% or 81% isolated yield, respectively (Scheme 1-5). Similar cyclizations were obtained with (2,3-dimethyl-2-buten-1,4-diyl)-magnesium (Table 1-3).

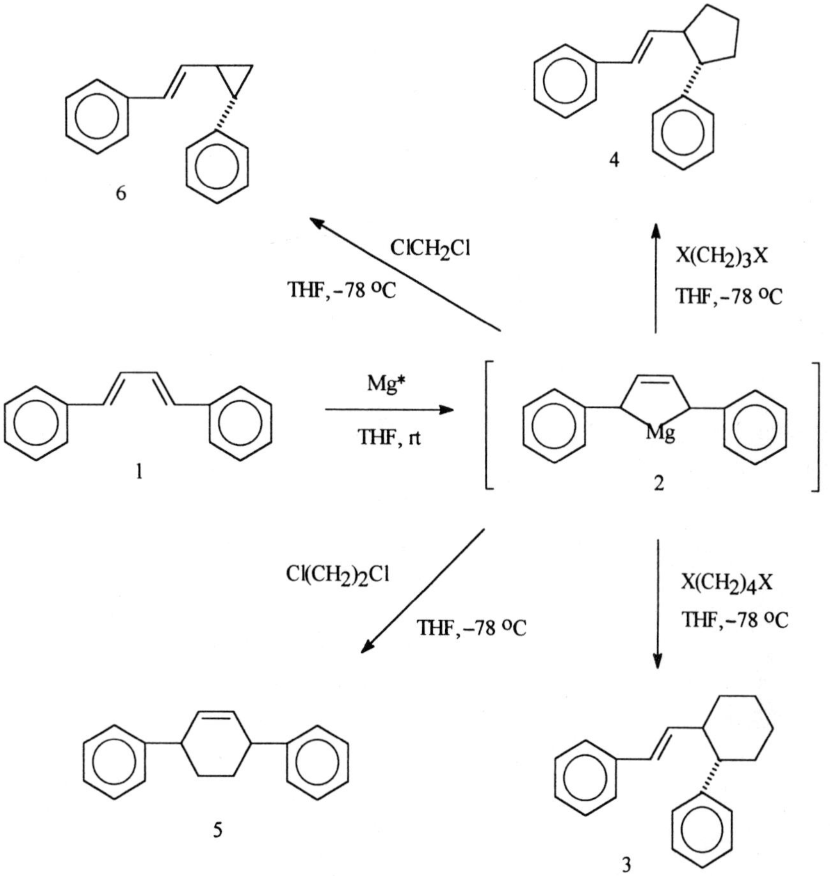

Scheme 1-5. Carbocyclization of (1,4-diphenyl-2-butene-1,4-diyl)magnesium with organic dihalides [27].

(1,4-Diphenyl-2-butene-1,4-diyl)magnesium can be easily prepared using Rieke magnesium. Treatment of the diene-magnesium reagent with 1,*n*-dihaloalkanes provides a convenient method for the generation of substituted three-, five-, and six-membered carbocycles. Significantly, the cyclizations are always stereoselective and completely regioselective.

It should be noted that the (1,4-diphenyl-2-butene-1,4-diyl)barium complex has recently been prepared in our laboratories [28] and exhibits higher reactivity than its magnesium-diene counterpart.

Table 1-2. Reaction of (1,4-diphenyl-2-butene-1,4-diyl)magnesium with electrophiles.

Entry	Electrophile	Product	Yield (%)
1	$Br(CH_2)_4Br$	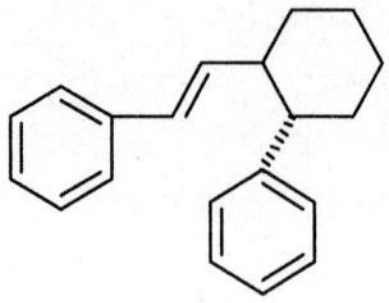	40
2	$Cl(CH_2)_4Cl$	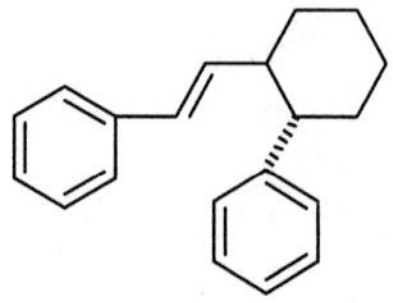	51
3	$Br(CH_2)_3Br$	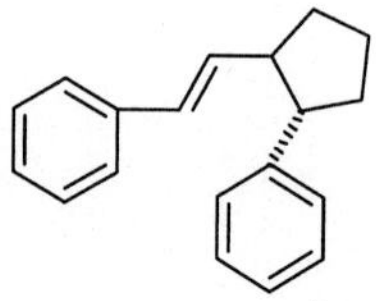	65
4	$Cl(CH_2)_3Cl$	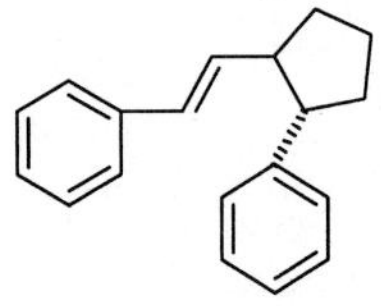	81
5	$Br(CH_2)_2Br$	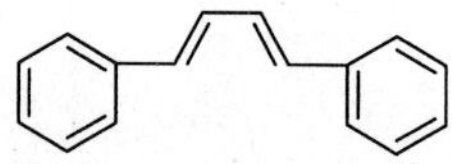	–
6	$Cl(CH_2)_2Cl$	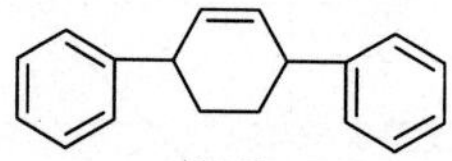	59
7	$BrCH_2Br$	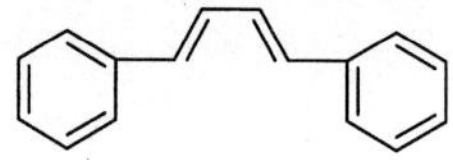	–
8	$ClCH_2Cl$	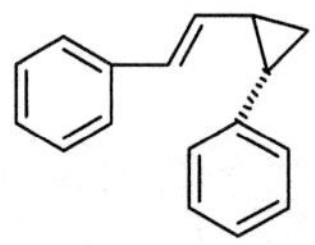	76
9	$CH_3(CH_2)_3Br$	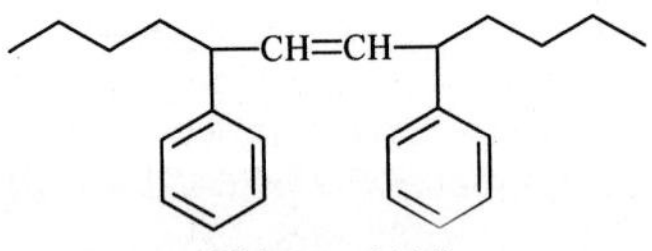 –CH=CH– (cis/trans = 56:44)	93
10	$CH_3(CH_2)_3Cl$	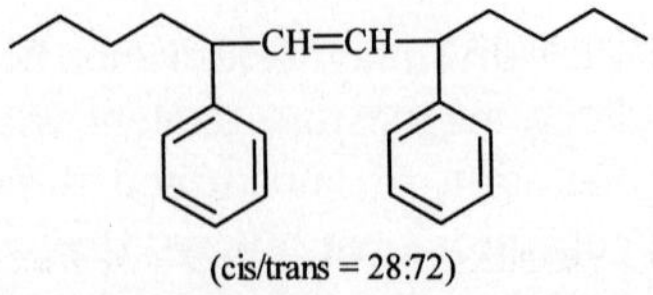 –CH=CH– (cis/trans = 28:72)	87
11	Me_2SiCl_2	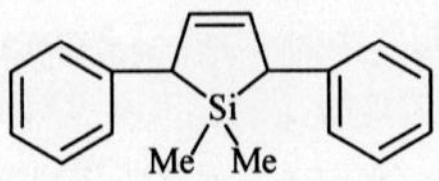 Si, Me, Me	66

Table 1-3. Reactions of (2,3-dimethyl-2-butene-1,4-diyl)magnesium with organodihalides.

Entry	Dihalide	Product	Yield (%)[a]
1	$Br(CH_2)_4Br$	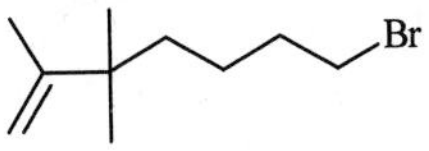 Br	79
2	$Br(CH_2)_4Br$	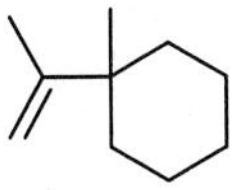	53 (69)
3	$Br(CH_2)_3Br$	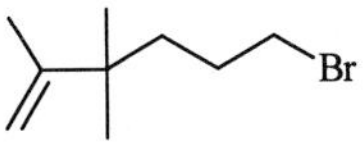 Br	72
4	$Br(CH_2)_3Br$	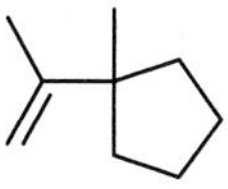	– (75)
5	$Cl(CH_2)_3Cl$	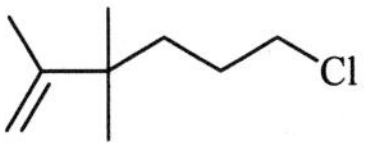 Cl	81
6	$Br(CH_2)_2Br$	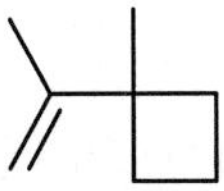	– (49)
7	$Cl(CH_2)_2Cl$	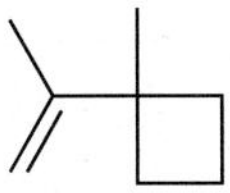	– (61)
8	*o*-$BrC_6H_4CH_2Br$	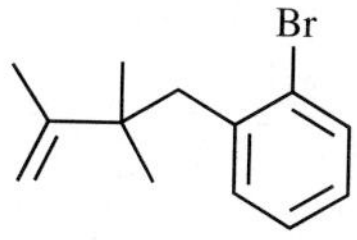 Br	62
9	*o*$BrC_6H_4CH_2Br$	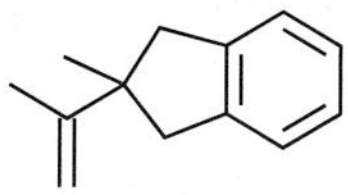	~30
10	Ph_2SiCl_2	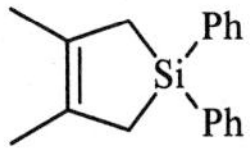 Si Ph Ph	65

a) Isolated yield. GC yields in parentheses.

1.2.4.4 1,2-Dimethylenecycloalkane-Magnesium Reagents

Based upon the bis-nucleophilicity of 1,3-diene-magnesium intermediates, reactions of these intermediates with bis-electrophiles can lead to spiro or fused bicyclic molecules, depending upon the regioselectivity of the cyclization. It has been shown that treatment of

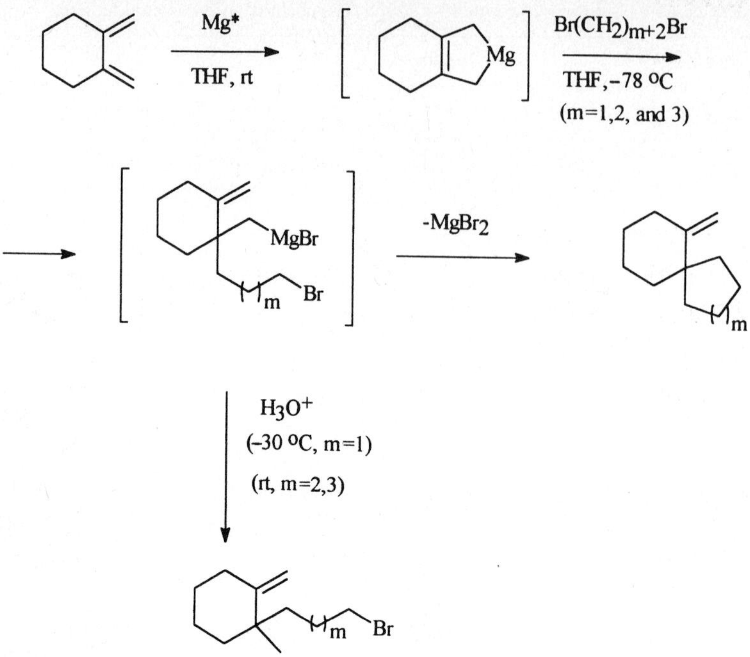

Scheme 1-6. Synthesis of spiro-olefins.

magnesium complexes of 1,2-bis(methylene)cyclohexane with 1,*n*-dibromoalkanes results in overall 1,2-cyclizations of the original dienes, affording spirocarbocycles in good to excellent yields [29]. Scheme 1-6 depicts a general route for the spiro-olefin synthesis. Typically, a 1,*n*-dibromoalkane was added to the THF solution of the diene-magnesium complex at −78 °C, producing an organomagnesium intermediate accommodating a bromo group. The intermediate cyclized upon warming, affording the corresponding spirocarbocycle containing an exocyclic double bond.

The spiroannulation approach described in Scheme 1-6 has been easily extended to other 1,2-dimethylenecycloalkanes. A wide variety of spirocycles were synthesized from the reactions of magnesium complexes of 1,2-dimethylenecyclopentane and 1,2-dimethylenecycloheptane with 1,*n*-dibromoalkanes as shown in Table 1-4. Therefore, this spiroannulation method provides a very general approach to a large number of spirocarbocycles with different combinations of ring sizes.

1.2.4.5 Synthesis of Fused Carbocycles, β,γ-Unsaturated Ketones and 3-Cyclopentenols from Conjugated Diene-Magnesium Reagents

With the aid of Rieke magnesium, fused carbocyclic enols can be effortlessly synthesized from 1,3-diene-magnesium complexes by reacting with carboxylic esters [30]. By controlling the reaction temperature, β,γ-unsaturated ketones can also be produced.

As shown in Scheme 1-7, reaction of the 1,2-bis(methylene)cycloalkane-magnesium, **1**, with ethyl acetate at −78 °C, followed by acidic quenching at −10 °C, afforded the β,γ-unsaturated ketone, (2-methyl-1-cyclohexenyl)propan-2-one (**6**). Alternatively, upon refluxing the initially formed intermediate, **2** followed by acidic work-up, the fused bicyc-

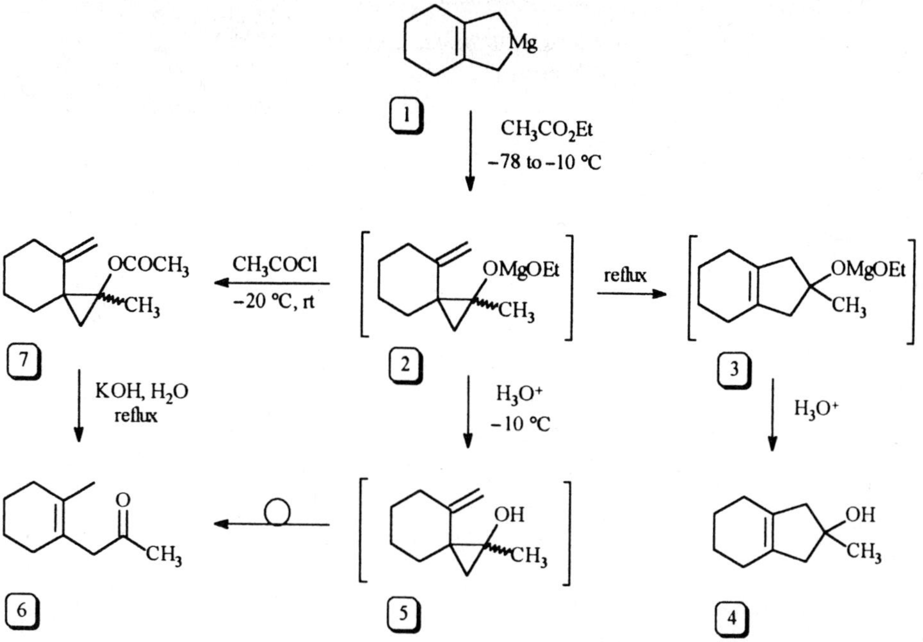

Scheme 1-7. Reaction of the 1,2-bis(methylene)cyclohexane-magnesium complex with an ester.

Table 1-4. Reactions of magnesium complexes of 1,2-dimethylenecycloalkanes with bis-electrophiles.

Entry	Diene[a]	Electrophile	Product	Yield[b] (%)	Reaction condition[c]
1	1a	$Br(CH_2)_5Br$	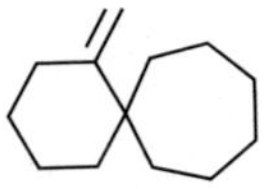	45	A
2	1a	$Br(CH_2)_5Br$	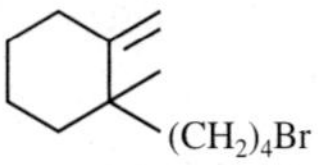 $(CH_2)_4Br$	79	B
3	1a	$Br(CH_2)_4Br$	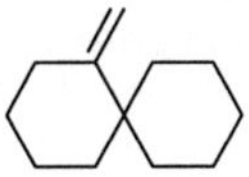	75(81)	C
4	1a	$Br(CH_2)_4Br$	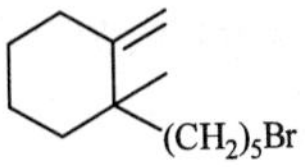 $(CH_2)_5Br$	81	B
5	1a	$Br(CH_2)_3Br$	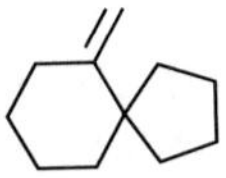	75(87)	D
6	1a	$Br(CH_2)_3Br$	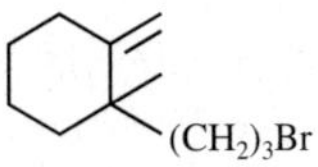 $(CH_2)_3Br$	78	E

Table 1-4. (continued).

Entry	Diene[a]	Electrophile	Product	Yield[b] (%)	Reaction condition[c]
7	1a	$Cl(CH_2)_3Cl$	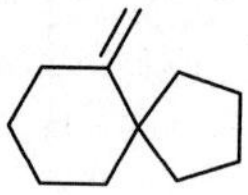	–(78)	F
8	1a	$Br(CH_2)_2Br$	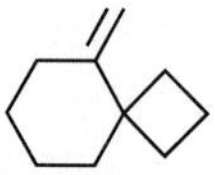	–(15)	B
9	1a	$Cl(CH_2)_2Cl$	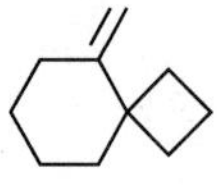	–(40)	B
10	1a	$TsO(CH_2)_2OTs$	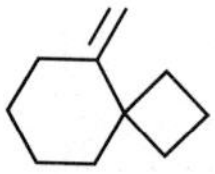	52(67)	B
11	1b	$Br(CH_2)_3Br$	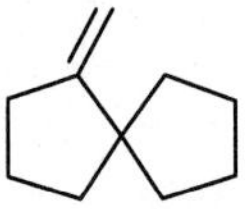	60(70)	D
12	1c	$Br(CH_2)_4Br$	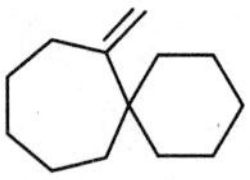	73	C
13	1c	$Br(CH_2)_3Br$	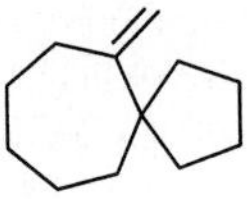	77(86)	D
14	1c	$TsO(CH_2)_2OTs$		46(59)	B
15	1a	Ph_2SiCl_2	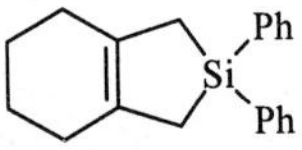	89	B

a) (1a) 1,2-bis(methylene)cyclohexane; (1b) 1,2-bis(methylene)cyclopentane; (1c) 1,2-bis(methylene)cycloheptane. b) Isolated overall yields were based on 1,2-bis(methylene)cycloalkanes. GC yields are shown in parentheses. c) Bis-electrophiles were added to the THF solution of the diene-magnesium reagent at −78 °C. The reaction mixture was then stirred at −78 °C for 1 h prior to warming to the specified temperature. (A) reflux (15 h); (B) room temperature (30 min); (C) reflux (10 h); (D) room temperature (10 h); (E) −30 °C; (F) room temperature (30 h).

lic enol 2,3,4,5,6,7-hexahydro-2-methyl-1*H*-inden-2-ol (**4**) was obtained. It was found that the initial product generated upon treating the 1,2-bis(methylene)cycloalkane-magnesium, **1**, with ethyl acetate was **2**, a magnesium salt of a spiroenol containing a cyclopropane ring. This intermediate underwent ring expansion upon warming, to produce the fused carbocyclic product **4**. Alternatively, protonation of the intermediate **2** at −10 °C generated the corresponding spiroenol **5**, which underwent an in-situ rearrangement to afford the β,γ-unsaturated ketone **6**. The identity of the initially formed versatile intermediate, **2**, was validated by trapping with acetyl chloride to give 1-methyl-4-methylenespiro[2.5]oct-1-yl acetate, **7**. Upon basic work-up, the spiroacetate **7** also afforded the β,γ-unsaturated ketone product, **6**.

It is important to note that the ring enlargement from **2** to **3** involves a vinylcyclopropane-cyclopentene ring expansion which has also been observed for the lithium salts of 2-vinylcyclopropanol systems [31]. On the other hand, the rearrangement of **5** to **6** is formally a 2-vinylcyclopropanol ring opening with a proton transfer. To our knowledge this is the first report of such a rearrangement, although 1-vinylcyclopropanol-cyclobutanone rearrangements have been well documented [32].

The absence of the formation of a five-membered ring at low temperature suggests that the initial attack of ethyl acetate occurred at the 2-position of the 1,2-bis(methylene)-cycloalkane-magnesium (**1**), followed by an intramolecular cyclization, generating the spiro intermediate (**2**).

This new methodology is quite general and can be applied to the magnesium complexes of 1,2-dimethylenecyclopentane, 1,2-dimethylenecycloheptane, and 2-methyl-3-phenyl-1,3-butadiene. Likewise, other carboxylic esters, such as butyl and ethyl benzoates, can also be used to make various fused carbocyclic enols or β,γ-unsaturated ketone products. Significantly, the overall synthetic process to form fused carbocyclic enols from the corresponding 1,2-bis(methylene)cycloalkanes represents a formal [4+1] annulation.

Table 1-5 shows representative examples of the high temperature pathway affording cyclopentenols, and Table 1-6 depicts the low temperature pathway which generates β,γ-unsaturated ketones from the versatile intermediate **2**.

We have recently extended this chemistry, incorporating a lactone as the electrophile, and generated β,γ-unsaturated ketone alcohols, as well as cyclopentenols containing both 3° and 1° alcohols within the same molecule depending on the reaction temperature [33]. We have also broadened this chemistry to included magnesium-diene intermediates derived from isoprene, myrcene, and 2,3-dimethyl-1,3-butadiene in addition to the 1,2-bis(methylene)cyclohexane complex.

Table 1-5. Reactions of diene-magnesium reagents with carboxylic esters. Formation of cyclopentenols.

Entry	Diene	Ester	Product	Yield[a)]
1	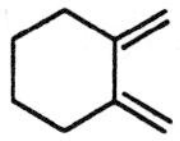	CH_3COOEt	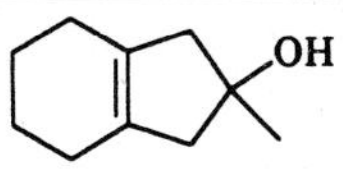 OH	91
2	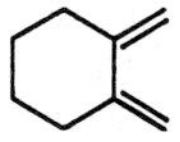	$CH_3(CH_2)_2COOEt$	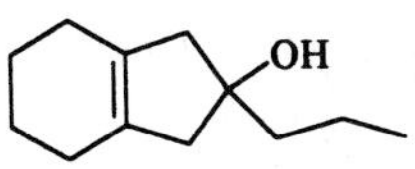 OH	96
3	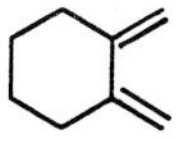	PhCOOEt	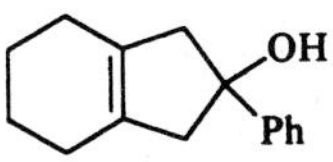 OH Ph	55
4	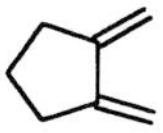	$CH_3(CH_2)_2COOEt$	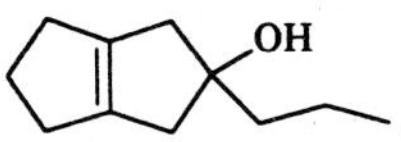 OH	59
5	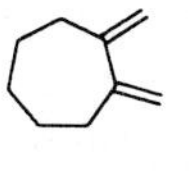	$CH_3(CH_2)_2COOEt$	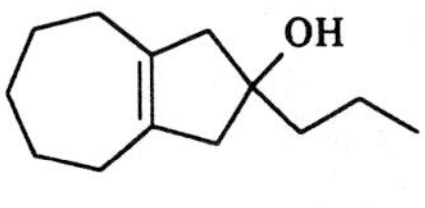 OH	74

a) Fused bicyclic product was obtained under reflux.

Table 1-6. Reactions of diene-magnesium reagents with carboxylic esters. Formation of β,γ-unsaturated ketones.

Entry	Diene	Ester	Product	Yield[a] (%)
1	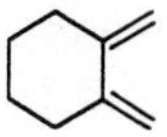	CH_3COOEt	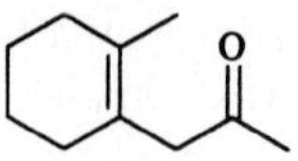	72
2	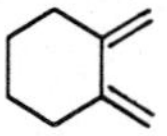	$CH_3(CH_2)_2COOEt$	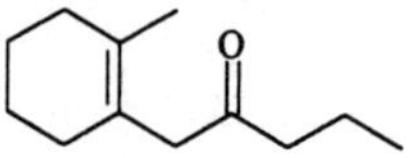	81
3	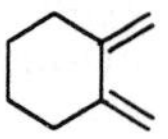	PhCOOEt	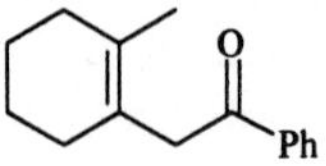	62
4	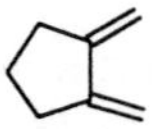	$CH_3(CH_2)_2COOEt$	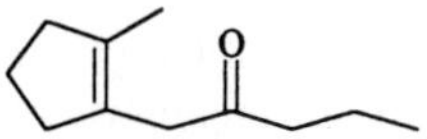	76
5	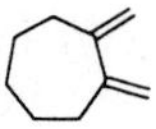	$CH_3(CH_2)_2COOEt$	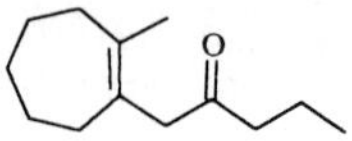	84

a) Quenching the reaction at –10 °C gave the β,γ-unsaturated ketone.

1.2.4.6 Synthesis of Spiro γ-Lactones and Spiro δ-Lactones from 1,3-Diene-Magnesium Reagents

The generation of a quaternary carbon center, and the introduction of functional groups, which are present in the process of lactonization, are some of the difficulties associated with the formation of spirolactones. There have been many reports in the literature regarding the development of synthetic strategies to overcome these difficulties [34]. However, many require the use of complex reagents and multiple synthetic steps to achieve the overall lactonization. One of the more effective methods is the reaction of bis(bromomagnesio)alkanes with dicarboxylic anhydrides [35].

We have recently developed a direct synthetic method for the one-pot synthesis of lactones, spirolactones, di-spirolactones, tertiary alcohols, and even 1,2-diols from the corresponding conjugated diene-magnesium reagents mediated by Rieke magnesium [36]. The overall lactonization procedure can be considered as a molecular assembling process in which three distinct individual species, a conjugated diene, a ketone, and carbon dioxide, are used to build a complex organic molecule in a well-controlled manner. Likewise, the generation of a quaternary center and the incorporation of both a hydroxyl and a carboxyl group required for lactonization are achieved in one synthetic procedure.

An example of the formation of a spiro γ-lactone from the corresponding 1,2-bis(methylene)cyclohexane-magnesium complex is shown in Scheme 1-8. The reaction between 1,2-bis(methylene)cyclohexane-magnesium (**1**) and acetone at –78 °C yielded a 1,2-adduct, **2**, resulting from the incorporation of one molecule of acetone with the diene complex. Acidic work-up of the 1,2-adduct at –78 °C gave a tertiary alcohol, **3**, containing both a quaternary center and a vinyl group at the β-position. Upon warming, the 1,2-adduct can further undergo a nucleophilic addition to a second electrophile. Carbon dioxide introduced into the reaction mixture at 0 °C as the second electrophile reacted immediately with **2**, presumably giving a magnesium salt of a γ-hydroxy acid (**4**). After

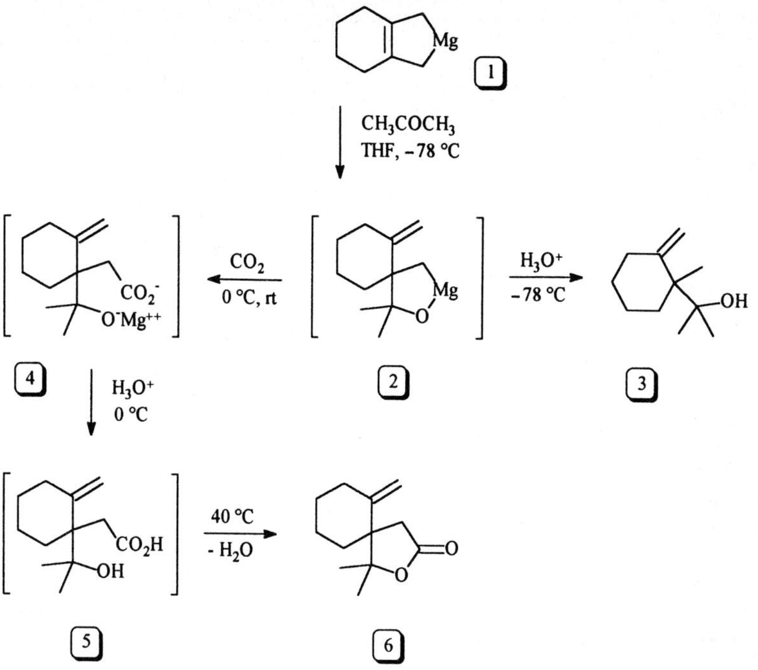

Scheme 1-8. Synthesis of spiro-γ-lactones.

acidic hydrolysis and mild warming the spiro-γ-lactone product, 4,4-dimethyl-6-methylene-3-oxaspiro[4.5]decan-2-one (**6**) was obtained. Table 1-7 shows representative spiro-γ-lactones that have been prepared in this way.

We have recently reported an extension of this methodology in the production of spiro-δ-lactones [37]. Much work has been done to elucidate novel synthetic routes for these types of molecules [38]. In particular, δ-substituted δ-lactones have recently attracted considerable attention, mainly because molecules of this class include many natural products that exhibit significant biological activity [39].

Scheme 1-9 illustrates a route for the synthesis of spiro δ-lactones from the magnesium complex of 1,2-bis(methylene)cyclohexane **1** [40]. Initially, treatment of the 1,2-bis(methylene)cyclohexane-magnesium reagent **2** with an excess of ethylene oxide at −78 °C resulted in the formation of the 1,2-adduct **3** by the incorporation of one equivalent of epoxide with the diene complex. Significantly, the bis-organomagnesium reagent **2** reacted with only one mole of epoxide with 100% regioselectivity in the 2-position to give the intermediate **3**. Upon warming to 0 °C, **3** reacted with CO_2 to yield the magnesium salt of a δ-hydroxy acid (**4**). Acidic hydrolysis followed by warming to 40 °C generated the spiro δ-lactone 7-methylene-3-oxaspiro[5.5]undecan-2-one (**6**) in 69% yield (Table 1-8, entry 1).

Importantly, the approach described in Scheme 1-9 can be used to prepare both bicyclic and tricyclic spiro δ-lactones, and representative examples are listed in Table 1-8 (entries 3 and 4, respectively). For example, 1,2-bis(methylene)cyclohexane-magnesium reagent was treated with cyclohexene oxide at −78 °C followed by the introduction of CO_2 at 0 °C and acidic hydrolysis with subsequent warming to 40 °C. Work-up afforded the tricyclic spiro δ-lactone, hexahydro-2′-methylene-spiro[4*H*-1-benzopyran-4,1′-cyclohexan]-2(3*H*)-one, in 63% isolated yield as a 1:1 mixture of diastereomers (Table 1-8, entry 4).

Table 1-7. Synthesis of spiro-γ-lactones from conjugated dienes, ketones and CO_2.

Entry	Diene	Ketone	Product	Yield[a]
1	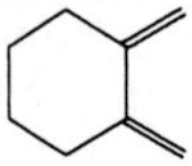	Acetone	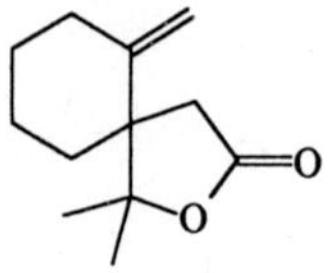 O	68
2	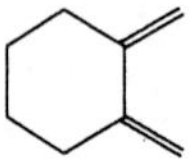	Cyclopentanone	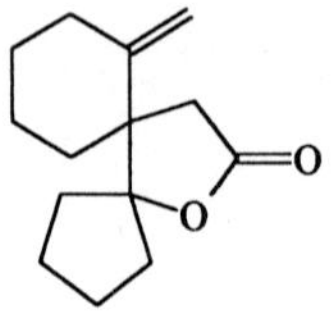 O	66
3	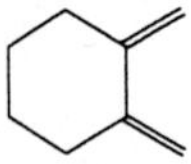	Cyclohexanone	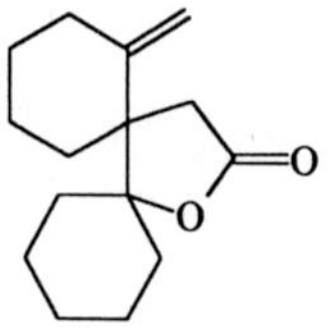 O	60
4		Cyclopentanone	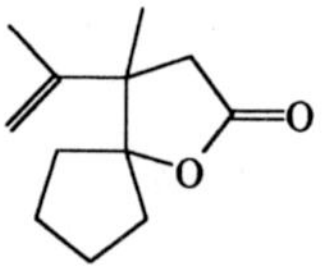 O	68
5	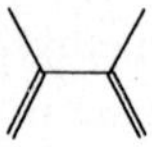	Cyclohexanone	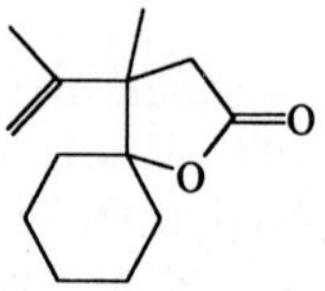 O	61

a) Isolated yields.

1 → Mg*, THF, rt → [2, Mg] → epoxide (O), THF, -78 °C → [3, Mg, O] → CO_2, 0 °C to rt

[4, O^-, CO_2^- Mg^{++}] → H_3O^+, 0 °C → [5, OH, CO_2H] → 40 °C, $-H_2O$ → 6 (O, O)

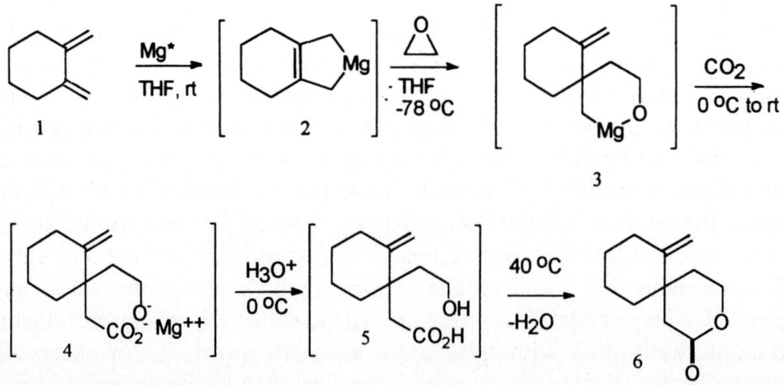

Scheme 1-9. Synthesis of spiro-δ-lactones.

Table 1-8. Reactions of conjugated diene-magnesium reagents with epoxides followed by carbon dioxide.

Entry	Diene	Epoxide[a)]	Product[b)]	Yield[c)] (%)
1	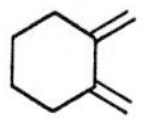		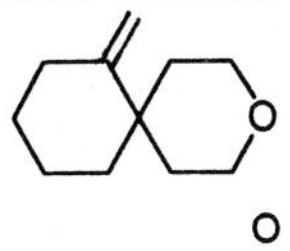	69
2			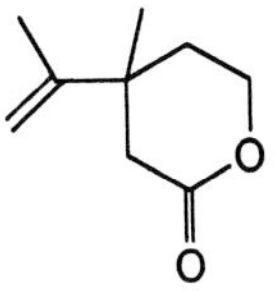	39[d)]
3		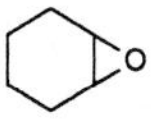	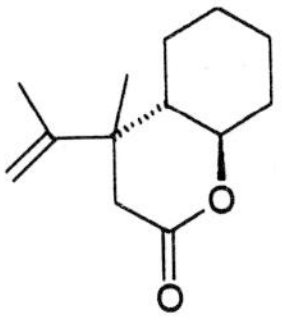	69[e)]
4	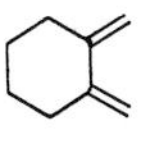	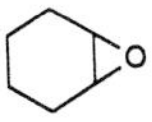	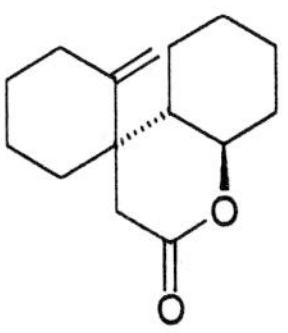	63[e)]
5		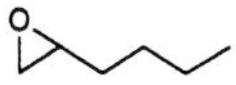	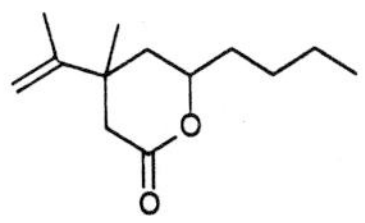	72[e)]

a) The epoxide was added to the diene-magnesium complex at −78 °C and the reaction mixture was stirred at −78 °C for 30 min, then gradually warmed to 0 °C followed by the bubbling of CO_2. b) Elemental analysis, mass spectra, ^{1}H NMR, ^{13}C NMR, and FTIR were all consistent with the indicated formulation. c) Isolated yields. d) Yield was based on amount of active magnesium. e) A 1:1 mixture of diastereomers as determined by ^{1}H NMR.

This method also exhibited good regioselectivity when an unsymmetrical epoxide was used as the primary electrophile. The formation of 6-butyltetrahydro-4-methyl-4-(1-methylethenyl)-2*H*-pyran-2-one (Table 1-8, entry 5) demonstrates that the attack of the asymmetric epoxide occurs at the less sterically hindered carbon.

The generation of a quaternary carbon center is not a trivial undertaking in organic synthesis, and multiple synthetic steps are often required. In work related to our studies of the synthesis of spiro-δ-lactones [41], the preparation of alcohols and *vic*-diols has also been realized utilizing this one-pot methodology [42]. In the process the formation of a quaternary carbon center is achieved. Scheme 1-10 shows that the initial adduct **3**, when treated with dilute acid at low temperature, affords the alcohol **4** containing a quaternary carbon center. It is important to note that in all cases where asymmetric epoxides were used, the bis-organomagnesium reagent **2** reacted with good regioselectivity at the least hindered carbon atom (Table 1-9, entries 3–6). Also, the organomagnesium reagent **2** reacted with only one mole of epoxide in cases in which an excess of epoxide was used

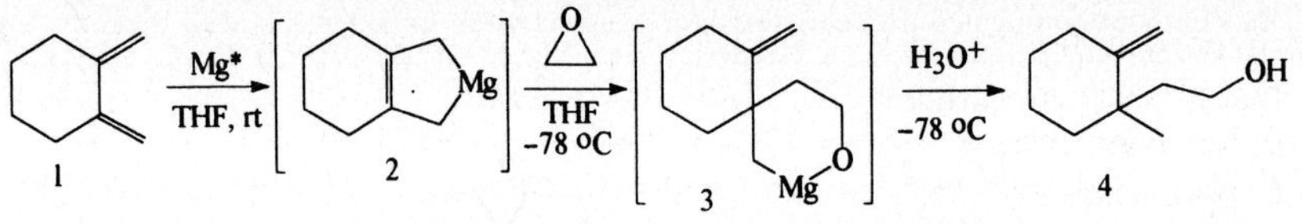

Scheme 1-10. Preparation of alcohols containing a quaternary carbon center.

(Table 1-9, entries 1 and 2). Low temperature protonation of the initial adduct **3** afforded a primary alcohol containing a quaternary center, 2-(β-hydroxyethyl)-2-methyl-1-methylene-cyclohexane (**4**) in excellent yield (Table 1-9, entry 1). An asymmetric chiral epoxide was utilized as the primary electrophile, followed by protonation which afforded a 1,2-diol. It was hoped that the attack of the organomagnesium intermediate **2** could be selectively induced to produce only one diastereomer. Unfortunately, the chiral epoxide did not exhibit any influence on the diastereoselectivity of the attack (Table 1-9, entry 6). However, the reactions proved that the organomagnesium reagent **2** will attack the epoxide even in the presence of the unprotected proximal hydroxyl functional group with no epimerization of the chiral center present in the epoxide.

The process presented here is a facile means for the preparation of alcohols and *vic*-diols in one pot in high yields.

Table 1-9. Reactions of conjugated diene-magnesium reagents with epoxides followed by acidic hydrolysis.

Entry	Diene	Epoxide	Product[a]	Yield[b] (%)
1	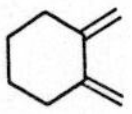		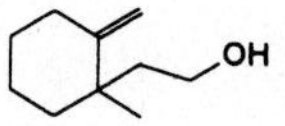	86
2				60[c]
3	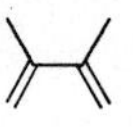	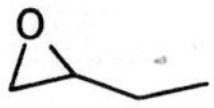	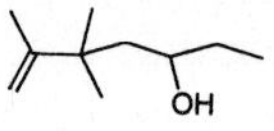	64
4		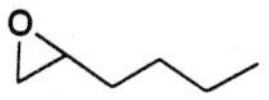	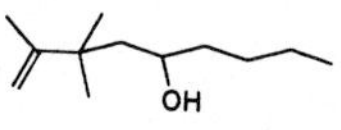	90
5		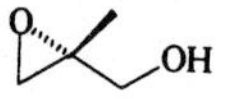	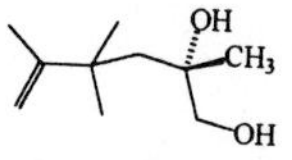	47
6	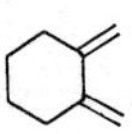		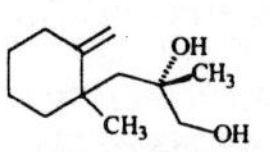	62[d]

a) Elemental analysis, mass spectra, ^{1}H NMR, ^{13}C NMR, and FTIR were all consistent with the indicated formulation. b) Isolated yields. c) Yield was based on amount of active magnesium. d) A 1:1 mixture of diastereomers as determined by ^{1}H NMR.

1.2.4.7 Synthesis of γ-Lactams from Conjugated Diene-Magnesium Reagents

γ-Lactams are important intermediates in synthetic routes to five-membered heterocyclic compounds. Moreover, tetramic acids and 3-pyrrolin-2-ones represent a diverse and profoundly important family of biologically active secondary metabolites, many of which have potential uses in medicine and agriculture [43]. Synthetic interest in this class of molecules has been intense, particularly in the past decade [44].

Most approaches to γ-lactams have relied on cyclization via acyl-nitrogen bond formation [45]. Cyclization involving carbon-carbon bond formation is an alternative route; however, until recently this approach has received little attention. Mori and co-workers reported a palladium-catalyzed cyclization of *N*-allyl iodoacetamides, in which the intramolecular addition reaction of the carbon-iodine bond to an olefinic linkage is a key step [46].

Itoh and co-workers recently developed a new route to γ-lactams by the ruthenium-catalyzed cyclization of *N*-allyltrichloroacetamides [47]. Also, Stork and Mah have reported on the radical cyclization of *N*-protected haloacetamides to yield *N*-protected lac-

Table 1-10. Lactamization of conjugated diene-magnesium reagents with imines and CO_2.

Entry	Diene	Imine	Product	Yield[a] (%)
1		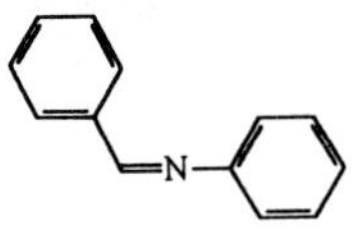	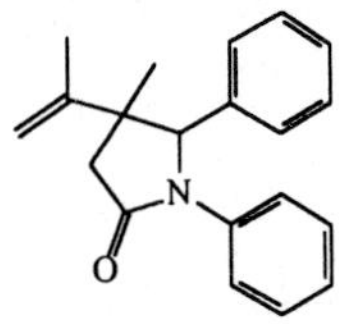	47 (80 : 20)
2		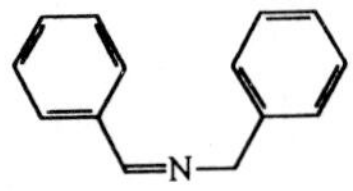	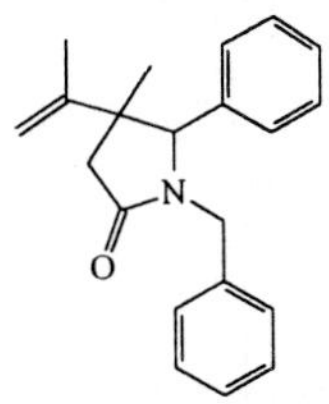	67 (75 : 25)
3		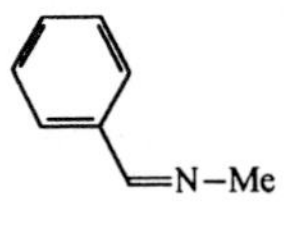	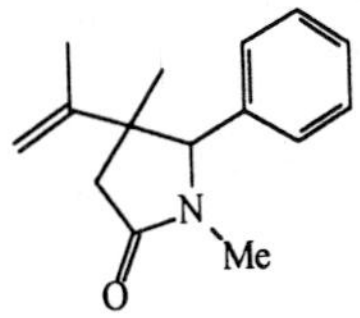	58 (70 : 30)
4	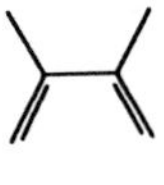	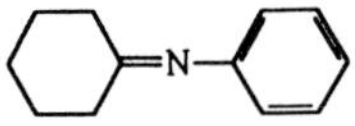	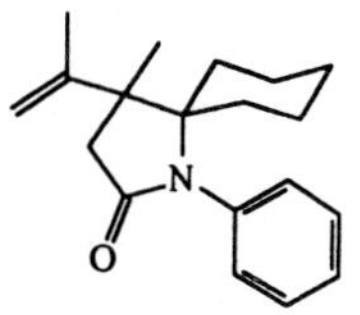	39
5	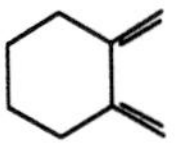	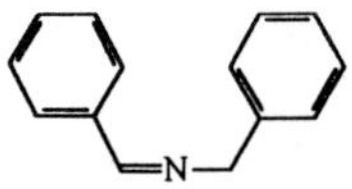	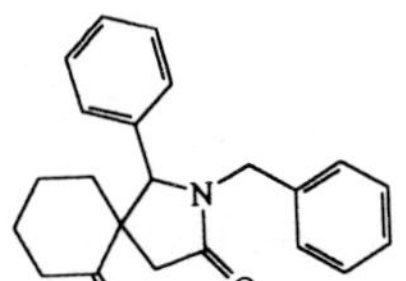	36 (57 : 43)

a) Isolated yields based on the imine. Diasteriomeric ratios in parenthesis determined by ^{1}H NMR or by GC analysis.

tams. The protecting groups can then be easily removed under a variety of conditions [48]. This efficient radical cyclization route to *cis*-fused pyrrolidones and piperidones is interesting because of the widespread occurrence of related systems in natural products [49]. We have recently developed a direct method for the one-pot synthesis of γ-lactams and secondary amines utilizing conjugated diene-magnesium complexes [50].

Scheme 1-11 illustrates a route for the synthesis of a γ-lactam from the (2,3-dimethyl-2-butene-1,4-diyl)magnesium intermediate (**2**). Initially, treatment of **2** with *N*-benzylidene-benzylamine at –78 °C resulted in the formation of the 1,2-adduct (**3**). Significantly, the bis-organomagnesium reagent (**2**) reacted with 100% regioselectivity in the 2-position to give the intermediate **3**. Upon warming to 0 °C, **3** reacted with CO_2 to yield the magne-

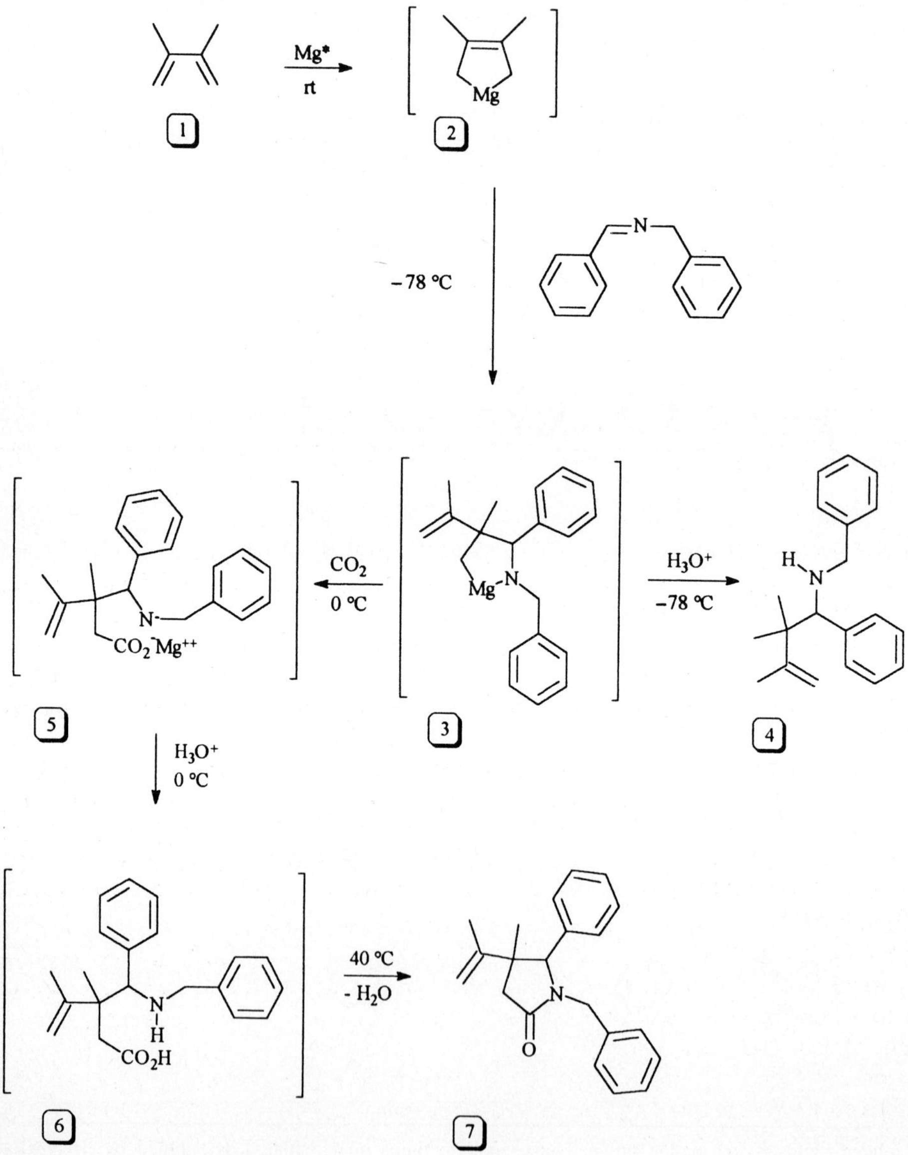

Scheme 1-11. Formation of γ-lactams.

Table 1-11. Reactions of conjugated diene-magnesium reagents with imines followed by acidic hydrolysis.

Entry	Diene	Imine	Product	Yield[a] (%)
1		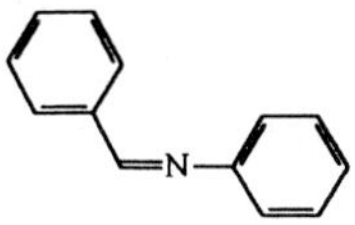	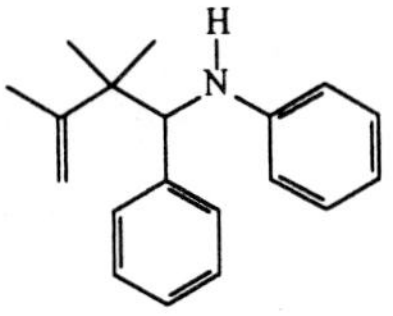	87
2		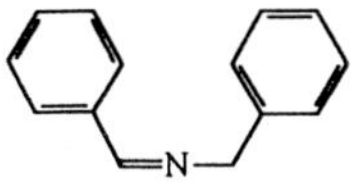	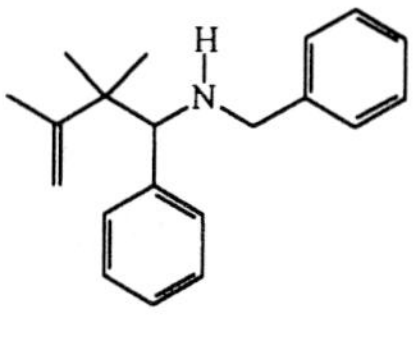	92
3		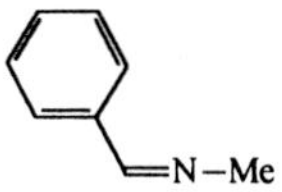 =N–Me	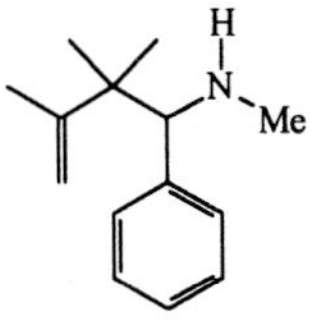	96
4		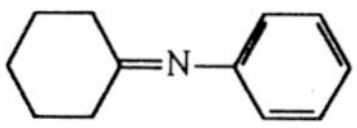	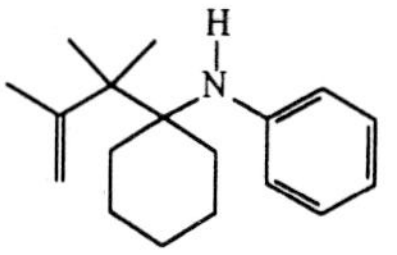	85

a) Isolated yields based on the imine.

sium salt of a γ-amino acid (**5**). Acidic hydrolysis followed by warming to 40 °C generated β-(1-methylethenyl)-β-methyl-γ-phenyl-*N*-benzyllactam (**7**) in 67% isolated yield as a 75 : 25 mixture of diastereomers (Table 1-10, entry 2). Importantly, the generation of a highly substituted γ-lactam was accomplished in a one-pot process and in good overall chemical yield. This approach was equally applicable to 1,2-bis(methylene)cyclohexane and provided a facile route to a spiro γ-lactam, β-((2-methylene)cyclohexyl)-γ-phenyl-*N*-benzyllactam in 36% isolated yield (Table 1-10, entry 5).

Scheme 1-11 also illustrates a facile route to secondary amines containing a quaternary carbon center and a vinyl group in the β-position. When the magnesium-diene intermediate (**2**) was treated with *N*-benzylidenebenzylamine at –78 °C, the initially formed adduct (**3**) could be easily hydrolyzed at 0 °C to afford *N*-(1-phenyl-2,2,3-trimethylbut-3-en-1-yl)-benzylamine (**4**) in 92% isolated yield (Table 1-11, entry 2). Other secondary amines are shown in Table 1-11 and all were prepared in high isolated yields.

This facile one-pot transformation of a 1,3-diene-magnesium intermediate provides a direct route to the formation of both secondary amines and β,γ,*N*-trisubstituted γ-lactams in good to high isolated yields. The overall procedure of the γ-lactam synthesis can be thought of as a molecular assembling process in which three independent species, namely a conjugated diene, an imine, and carbon dioxide, mediated by Rieke magnesium, are transformed into a complex organic molecule in a well-controlled fashion. In the process, the construction of a quaternary carbon center and the introduction of both the amino and carboxyl groups required for lactamization are achieved in one synthetic step.

1.3 Rieke Zinc

1.3.1 The Preparation of Rieke Zinc

In 1973 we reported the formation of Rieke zinc, which was shown to add oxidatively to alkyl and aryl bromides [51]. The Rieke zinc used in those reactions was prepared using anhydrous zinc bromide and potassium or sodium metal in refluxing tetrahydrofuran (THF) or 1,2-dimethoxyethane (DME) for 4 h. This was the same method as used for preparing active magnesium, which had been reported the previous year [52].

Currently, Rieke zinc is prepared by placing lithium metal (10 mmol), a catalytic amount of naphthalene (1 mmol), and 12–15 mL tetrahydrofuran (THF) in one flask. Once this mixture has stirred for about 20–40 sec, it will turn dark green, indicating the formation of lithium naphthalenide. Zinc chloride dissolved previously in 12–15 mL of THF is then cannulated dropwise (ca. 3 sec per drop) into the lithium naphthalenide and stirring is continued for 15 min after the transfer is complete. This method is not only safer due to the use of lithium metal rather than sodium or potassium but also yields a more reactive zinc [53]. A second method sometimes employed to prepare Rieke zinc uses a stoichiometric amount of naphthalene with respect to lithium. Both methods yield Rieke zinc with the same reactivity. It should also be noted that the reactivities are similar regardless of the choice of solvent (THF or DME) and that of the halide salt. Further, the electron carrier is not limited to naphthalene. Other carriers such as biphenyl and anthracene have also been used.

1.3.2 Direct Oxidative Addition of Functionalized Alkyl and Aryl Halides

Prior to the discovery of Rieke zinc it was not possible to react alkyl, aryl, and vinyl bromides or chlorides directly with zinc. The one exception was the Reformatsky reaction, which is used to produce β-hydroxy esters as shown in Scheme 1-12. The reaction was typically done in refluxing solvent and gave only modest yields. However, using Rieke zinc it was possible to react ethyl α-bromoacetate at –5 °C in THF or diethyl ether. Ethyl α-chloroacetate will also react but at a slightly higher temperature (10 °C). Diethyl ether was found to be the best solvent for these reactions. The zinc is prepared as usual in THF, which is then stripped off and replaced with dry diethyl ether. A one-to-one mixture of aldehyde or ketone and α-bromoester is then added at –5 °C. After stirring for one hour at room temperature the reaction is followed by the normal work-up procedure [54].

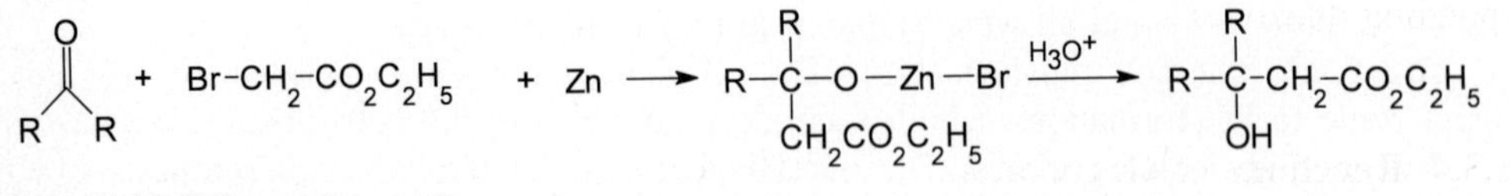

$$R_2C{=}O + Br\text{-}CH_2\text{-}CO_2C_2H_5 + Zn \longrightarrow R\text{-}C(R)(CH_2CO_2C_2H_5)\text{-}O\text{-}Zn\text{-}Br \xrightarrow{H_3O^+} R\text{-}C(R)(OH)\text{-}CH_2\text{-}CO_2C_2H_5$$

Scheme 1-12. Reformatsky reaction.

The formation of organozincs from alkyl, aryl, and vinyl bromides or chlorides was formerly possible only by a metathesis reaction of a zinc halide salt with a preformed organolithium or Grignard reagent. Unfortunately, the organolithium and Grignard re-

Table 1-12. Preparation of organozinc compounds.

$$RX + Zn^* \xrightarrow[\text{THF}]{\text{temp., time}} RZnX$$

Entry	Organic halide	Zn*:RX ratio	Temp., (°C)	Time (h)	Yield (%)
1	$Br(CH_2)_6Cl$	1.2:1	23	4	100
2	$Br(CH_2)_7CH_3$	1.2:1	23	6	100
3	$Br(CH_2)_3CO_2Et$	1:1	23	3	100
4	*p*-IC_6H_4Cl	2:1	23	3	100
5	*p*-BrC_6H_4CN	2:1	reflux	3	90
6	*p*-BrC_6H_4CN	3:1	reflux	3	100
7	*p*-$BrC_6H_4CO_2Et$	2:1	reflux	2	100
8	*o*-$BrC_6H_4CO_2Et$	2:1	reflux	2	100
9	*m*-$BrC_6H_4CO_2Et$	3:1	reflux	4	100
10	$Cl(CH_2)_3CO_2Et$	3:1	reflux	4	100

agents are not compatible with many types of functionality, which limit their utility. Rieke zinc, on the other hand, will react directly with these substrates while at the same time tolerating a wide variety of functional groups such as chlorides, nitriles, esters, and ketones. Also of significance is that aryl halides showed no scrambling of position when *ortho-*, *meta-*, or *para-*substituted substrates were used. It was also found that Rieke zinc will react with alkyl chlorides in the presence of potassium iodide with refluxing. Most likely the zinc reacts with an alkyl iodide formed in situ after halogen exchange with the potassium iodide [55]. Table 1-12 shows the preparation of organozinc intermediates from organohalides containing both ester and nitrile functionalities.

1.3.3 Reactions of Organozinc Reagents with Acid Chlorides

Once the organozinc reagent is formed, it easily transmetallates to copper under mild conditions, yielding a highly functionalized cuprate. For example, the organozinc reagent can be added to CuCN · LiBr at –35 °C giving RCu(CN)ZnCl. When reacted with various acid chlorides, this organocopper species gives highly functionalized asymmetrical ketones in high yields. Representative examples of this type of transformation are shown in Table 1-13. Aryl substrates require excess zinc and occasional refluxing to complete the oxidative addition, owing to their reduced reactivity. Di-organozinc bromides and iodides were also prepared and reacted with two equivalents of acid chloride affording the corresponding diketones.

1.3.4 Reactions of Organozinc Reagents with α,β-Unsaturated Ketones

These highly functionalized organocopper reagents, derived from the corresponding organozinc, have also been used in 1,4-conjugate additions [56]. Several methods were explored in attempting to optimize the 1,4-conjugate addition reaction. We found that a zinc cuprate formed from CuCN · LiBr in THF at –35 °C resulted in the highest yields. This

Table 1-13. Reactions of organozinc halides mediated by copper with acid chlorides.

$$RX + Zn^* \longrightarrow [RZnX] \xrightarrow{CuCN \cdot 2LiBr} [RCu(CN)ZnX] \xrightarrow{R'COCl} RCOR'$$

Entry	RX	R′COCl	Zn*:RX:R′COCl	Product	Yield (%)
1	$Br(CH_2)_7CH_3$	PhCOCl	1.5:1.0:0.9	$PhCO(CH_2)_7CH_3$	92
2	$Br(CH_2)_6CN$	PhCOCl	1.1:1.0:0.8	$PhCO(CH_2)_6CN$	94
3	$Br(CH_2)_6Cl$	PhCOCl	1.0:1.0:1.0	$PhCO(CH_2)_6Cl$	85
4	$BrCH_2CH_2Ph$	PhCOCl	1.2:1.0:0.9	$PhCOCH_2CH_2Ph$	97
5	$Br(CH_2)_3CO_2Et$	$CH_3(CH_2)_3COCl$	1.0:1.0:0.9	$CH_3(CH_2)_3CO(CH_2)_3CO_2Et$	91
6	$Br(CH_2)_3CO_2Et$	PhCOCl	1.0:1.0:0.9	$PhCO(CH_2)_3CO_2Et$	95
6′	$Cl(CH_2)_3CO_2Et$	PhCOCl	1.0:1.0:0.9	$PhCO(CH_2)_3CO_2Et$	91
7	*p*-BrC_6H_4Me	$CH_3(CH_2)_3COCl$	3.5:1.0:0.9	$CH_3(CH_2)_3COC_6H_4$-*p*-Me	86
8	*p*-$BrC_6H_4CO_2Et$	$CH_3(CH_2)_3COCl$	2.0:1.0:1.0	$CH_3(CH_2)_3COC_6H_4$-*p*-(CO_2Et)	83
9	*p*-IC_6H_4Cl	$CH_3(CH_2)_3COCl$	1.5:1.0:1.0	$CH_3(CH_2)_3COC_6H_4$-*p*-Cl	90
10	*p*-$IC_6H_4CO_2Et$	PhCOCl	2.0:1.0:1.0	$PhCOC_6H_4$-*p*-(CO_2Et)	88
11	*m*-$BrC_6H_4CO_2Et$	PhCOCl	4.0:1.0:0.9	$PhCOC_6H_4$-*m*-(CO_2Et)	83
12	*o*-$BrC_6H_4CO_2Et$	PhCOCl	2.0:1.0:0.9	$PhCOC_6H_4$-*o*-(CO_2Et)	92
13	*o*-$BrC_6H_4CO_2Et$	$CH_3(CH_2)_3COCl$	2.0:1.0:0.9	$CH_3(CH_2)_3COC_6H_4$-*o*-(CO_2Et)	94
14	*p*-BrC_6H_4CN	$CH_3(CH_2)_3COCl$	2.5:1.0:1.0	$CH_3(CH_2)_3COC_6H_4$-*p*-CN	71
15	*p*-BrC_6H_4CN	PhCOCl	3.0:1.0:0.9	$PhCOC_6H_4$-*p*-CN	73
16	*o*-BrC_6H_4CN	PhCOCl	3.0:1.0:0.9	$PhCOC_6H_4$-*o*-CN	98
17	*o*-BrC_6H_4CN	$CH_3(CH_2)_3COCl$	3.0:1.0:0.9	$CH_3(CH_2)_3COC_6H_4$-*o*-CN	97
18	*p*-$BrC_6H_4COCH_3$	$CH_3(CH_2)_3COCl$	2.0:1.0:0.9	$CH_3(CH_2)_3COC_6H_4$-*p*-$(COCH_3)$	80
19	$PhCH{=}CHBr$	$CH_3(CH_2)_3COCl$	3.0:1.0:0.9	$CH_3(CH_2)_3COCH{=}CHPh$	82
20	$PhCH_2Cl$	PhCOCl	1.5:1.0:0.9	$PhCOCH_2Ph$	81
21	$Br(CH_2)_4Br$	PhCOCl	3.0:1.0:2.0	$PhCO(CH_2)_4COPh$	78
22	*p*-IC_6H_4I	CH_3COCl	3.0:1.0:2.0	$CH_3COC_6H_4$-*p*-$(COCH_3)$	76

includes reactions with "higher order" cuprates as developed by Lipshutz, using 2-thienylcyanocuprate and our highly functionalized organozinc reagents [57]. Equation (1.2) shows the preparation of a higher order cuprate.

$$\mathrm{LiCu(CN)(2\text{-}th) + RZnX \rightarrow R(2\text{-}th)Cu(CN)ZnX} \quad (1.2)$$
$$(\mathrm{th = thienyl,\ C_4H_3S})$$

Unfortunately, we obtained a slightly lower yield together with a small amount (ca. 9% by GC yield) of the product resulting from the 1,4-addition of the 2-thienyl group. The reaction conditions were similar to those using CuCN · LiBr but without boron trifluoride etherate ($BF_3 \cdot Et_2O$). When $BF_3 \cdot Et_2O$ or trimethylsilyl chloride was used, even more of the 2-thienyl adduct was observed. Representative copper-mediated conjugate additions of organozinc halides with α,β-unsaturated ketones are shown in Table 1-14.

Table 1-14. Copper-mediated conjugate additions of organozinc halides with α,β-unsaturated ketones.

Entry	Organic halide	Enone	Cu^I-complex[a)]	Additives	Products	Yield (%)
1	$Br(CH_2)_3CO_2Et$	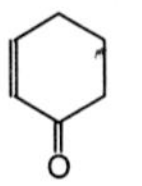 (I)	A	$BF_3 \cdot OEt_3$ TMSCl	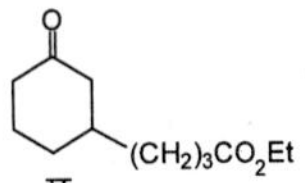 $(CH_2)_3CO_2Et$ II	92
2	$Br(CH_2)_3CO_2Et$	I	A	TMSCl	II	75
3	$Br(CH_2)_3CO_2Et$	I	B	TMSCl	II	76
					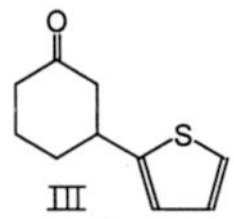 S III	
4	$Br(CH_2)_3CO_2Et$	I	B	$BF_3 \cdot OEt_2$	III II	59 10
5	$Br(CH_2)_6Cl$	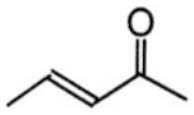	A	$BF_3 \cdot OEt_2$ TMSCl	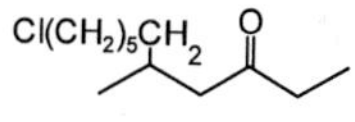 $Cl(CH_2)_5CH_2$	72
6	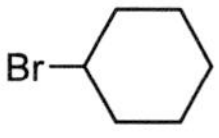 Br–	I	A	$BF_3 \cdot OEt_2$ TMSCl	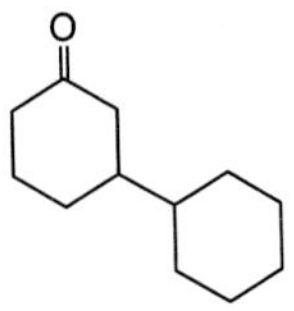	58
7	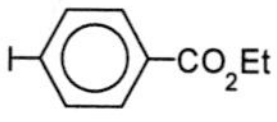 I–C$_6$H$_4$–CO_2Et	I	A	$BF_3 \cdot OEt_2$ TMSCl	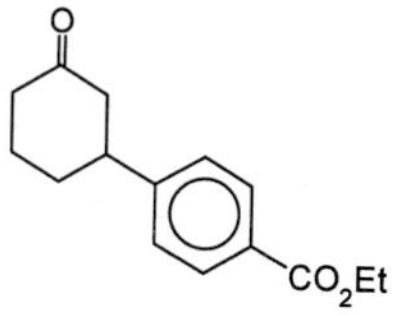 CO_2Et	68

a) A = CuCN · LiBr; B = LiCu(CN) · 2th (2th = 2-thienyl).

1.3.5 Reactions with Allylic and Alkynyl Halides

Our cuprates, derived from the highly functionalized organozincs, reacted with allylic halides with high regioselectively giving S_N2' products in high yield. This includes geminal disubstituted alkenes. Table 1-15 shows examples of reactions of organozincs with allylic halides mediated by CuCN · 2 LiBr.

We also found it possible to make 2,3-disubstituted-1,3-butadienes using CuCN · LiBr and cross-coupling with 1,4-dichloro-2-butyne. This reaction was very regiospecific, giving S_N2' products. Unfortunately, arylzinc compounds did not yield the S_N2' product when reacted with 1,4-dichloro-2-butyne. However, when reacted with 1,4-ditosyloxy-2-butyne, aryl and alkyl organozincs gave excellent yields of the S_N2' products. Table 1-16 shows examples of this type of transformation. The use of Rieke zinc allows a wide variety of functional groups to be included [58].

This same catalyst can be used to make 2-halo-olefins incorporating a high degree of functionality by reacting with 2,3-dibromopropene [59]. Examples of this type of reaction are shown in Table 1-17.

1.3.6 Cross-Coupling of Vinyl and Aryl Organozinc Reagents Using a Palladium Catalyst

The organozinc can be transmetallated to palladium and the resulting organopalladium can subsequently be reacted with vinyl or aryl halides giving the cross-coupled products.

Table 1-15. Reactions of RZnX with allylic halides mediated by CuCN · 2 LiBr.

$$RX + Zn^* \longrightarrow RZnX \xrightarrow{CuCN\ 2LiBr} [RCu(CN)ZnX]$$

$$[RCu(CN)ZnX] + R^1R^2C{=}CH{-}CH_2Y \longrightarrow \underset{S_N2'}{R^1R^2C(R){-}CH{=}CH_2} + \underset{S_N2}{R^1R^2C{=}CH{-}CH_2R}$$

Entry	RX	Allylic halide	Products S_N2' : S_N2	Yield (%)
1	$Br(CH_2)_3CO_2Et$	H_3C–CH=CH–CH_2–Cl	96:4	83
2	$Br(CH_2)_6CN$		97:3	91
3	$Br(CH_2)_6Cl$		98:2	94
4	*p*-$BrC_6H_4CO_2Et$		80:20	86
5	$Br(CH_2)_3CO_2Et$	CH_3–CH(Cl)–CH=CH_2	100:0	87
6	$Br(CH_2)_6CN$		97:3	87
7	*p*-$IC_6H_4CO_2Et$		100:0	93
8	$Br(CH_2)_3CO_2Et$		97:3	86
9	$Br(CH_2)_6CN$	Ph–CH=CH–CH_2–Cl	98:2	88
10	$Br(CH_2)_2Ph$		95:5	94
11	$Br(CH_2)_3CO_2Et$	H_3C–C(CH_3)=CH–CH_2–Br	98:2	88
12	$Br(CH_2)_6CN$		95:5	90
13	$Br(CH_2)_2Ph$		95:5	89

Table 1-16. Reactions of organozinc halides with $Y–CH_2–C\equiv C–CH_2–Y$.

$$2RZnBr + Y\text{-}CH_2\text{-}C\equiv C\text{-}CH_2\text{-}Y \xrightarrow[\text{CuCN 2LiBr}]{\text{THF}}$$
(Y = Cl or TsO)

Entry	Y	R	Yield (%)	Entry	Y	R	Yield (%)
1	Cl	$–(CH_2)_7CH_3$	95	8	Cl	$–CH_2C_6H_4$-*p*-CN	93
2	Cl	$–(CH_2)_6Cl$	92	9	TsO	$–(CH_2)_7CH_3$	(98)
3	Cl	$–(CH_2)_3CO_2Et$	95	10	TsO	$–(CH_2)_3CO_2Et$	(82)
4	Cl	$–(CH_2)_3CN$	84	11	TsO	$–C_6H_5$	88
5	Cl	$–CH_2$–cyclohexyl	87	12	TsO	$–CH_2C_6H_5$	95
6	Cl	–cyclohexyl	82	13	TsO	$–C_6H_4$-*p*-COMe	93
7	Cl	$–CH_2C_6H_5$	91	14	TsO	$–C_6H_4$-*p*-CN	97

Table 1-17. Preparation of 2-bromo-1-alkenes.

$$RX + Zn^* \xrightarrow[\text{Temp.}]{\text{Time}} RZnX \xrightarrow{\text{CuCN 2LiBr}}$$

Entry	RX	Y	RX/Zn*	Temp.	Time (h)	Yield (%)
1	*p*-BrC_6H_4CN	Br	1:2.5	reflux	1	85
2	*p*-$BrC_6H_4COCH_3$	Br	1:2.0	reflux	1	81
3	*p*-$IC_6H_4CO_2Et$	Br	1:1.5	reflux	0.3	90
4	*p*-BrC_6H_4Cl	Br	1:1.7	reflux	1.5	82
5	$Br(CH_2)_3CO_2Et$	Br	1:1.2	rt	2	96
6	$Br(CH_2)_3CN$	Br	1:1.5	reflux	0.5	94
7	$Br(CH_2)_6Cl$	Br	1:1.2	rt	2	88
8	$Br(CH_2)_3CO_2Et$	Cl	1:1.2	rt	2	95
9	$Br(CH_2)_3CN$	Cl	1:1.5	rt	1	78
10	*m*-$IC_6H_4CO_2Et$	Cl	1:1.5	reflux	1	83

This method can be used to yield a variety of highly functionalized compounds that would be difficult to prepare using other methodologies. In conjunction, it is also an important route to many substituted 1,3-dienes. Table 1-18 shows coupling reactions of organozinc and aryl and vinyl halides. It should be noted that organozinc reagents can also be transmetallated to nickel, allowing for the full range of chemistry available to that metal as well.

1.3.7 Intramolecular Cyclizations and Conjugate Additions Mediated by Rieke Zinc

Rieke zinc is superior to other forms of activated zinc when applied to intramolecular cyclizations and conjugate additions. The reaction of alkyl iodides with Rieke zinc proceeds instantly at room temperature, and the intramolecular cyclization of enones to form five- and six-membered rings requires 50 min to 48 h at room temperature [60]. An en-

Table 1-18. Coupling reactions of RZnX with aryl and vinyl halides catalyzed by $Pd(PPh_3)_4$.

$$RZnX + R'Y \xrightarrow[\text{THF}]{\text{5 mol \% } Pd(PPh_3)_4} RR'$$

Entry	RZnX	R′Y	Product	Yield (%)
1	$EtO_2C(CH_2)_3ZnBr$	*p*-$BrC_6H_4COCH_3$	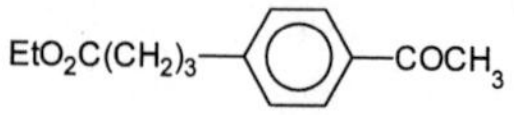 $EtO_2C(CH_2)_3$–C_6H_4–$COCH_3$	86
2	$EtO_2C(CH_2)_3ZnBr$	*p*-BrC_6H_4CN	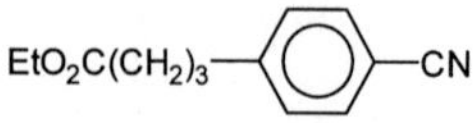 $EtO_2C(CH_2)_3$–C_6H_4–CN	93
3	$EtO_2C(CH_2)_3ZnBr$	*p*-$BrC_6H_4NO_2$	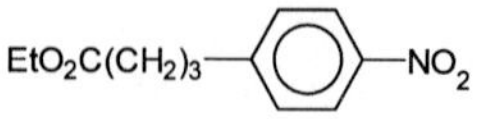 $EtO_2C(CH_2)_3$–C_6H_4–NO_2	90
4	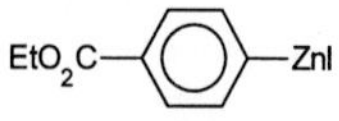 EtO_2C–C_6H_4–ZnI	*p*-BrC_6H_4CN	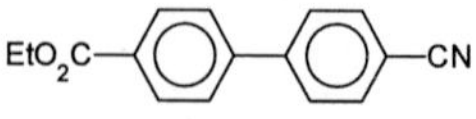 EtO_2C–C_6H_4–C_6H_4–CN	80
5	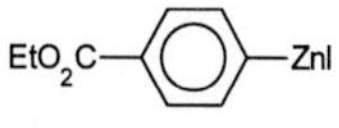 EtO_2C–C_6H_4–ZnI	*p*-$IC_6H_4CO_2Et$	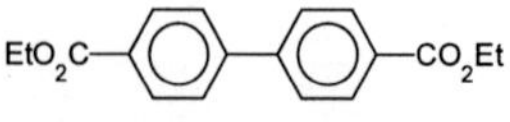 EtO_2C–C_6H_4–C_6H_4–CO_2Et	94
6	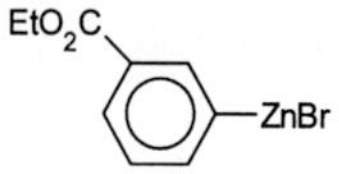 EtO_2C, ZnBr	*p*-BrC_6H_4CN	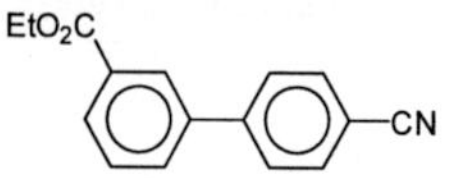 EtO_2C, CN	82
7	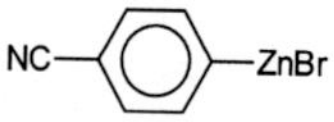 NC–C_6H_4–ZnBr	*p*-BrC_6H_4CN	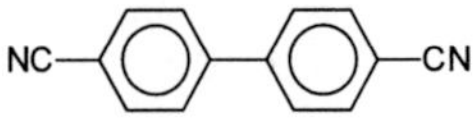 NC–C_6H_4–C_6H_4–CN	95
8	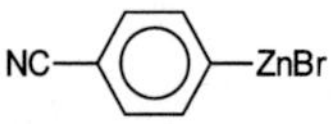 NC–C_6H_4–ZnBr	*p*-$IC_6H_4CO_2Et$	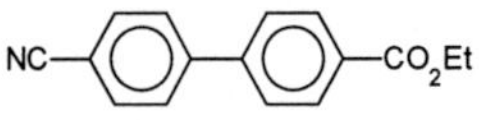 NC–C_6H_4–C_6H_4–CO_2Et	82
9	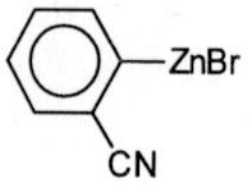 ZnBr, CN	*m*-$BrC_6H_4CO_2Et$	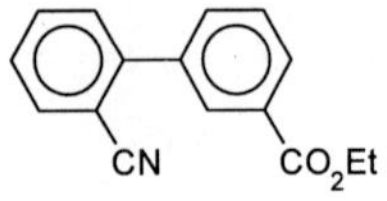 CN, CO_2Et	93
10	ZnBr	CH_3, Br	CH_3	85 (91)
11				86
12	ZnBr, CO_2Et	H, Br	H, CO_2Et	95
13				93

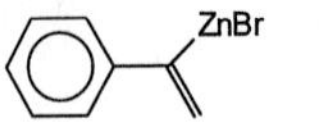

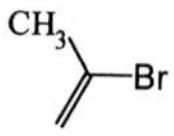

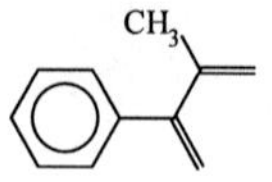

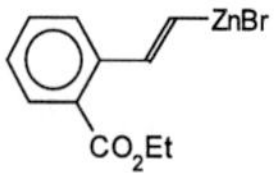

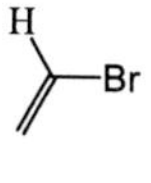

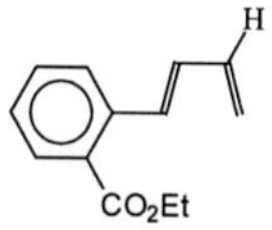

hancement of rate and yield was noted with trimethylsilyl chloride as an additive in the cyclization step. This methodology requires no subsequent metal metathesis to form a more reactive metal reagent, and is a good method for ring closure and the formation of spiro compounds. The direct insertion of Rieke zinc into the carbon–iodine bond of 6-iodo-3-functionalized-1-hexenes forms the primary alkylzinc iodide reagents, which then undergo an intramolecular insertion of the olefinic π-bond into the zinc–carbon bond to form functionalized methylcyclopropanes [61]. This is a significant finding in that this is a regiospecific 5-*exo*-trig cyclization which can occur in the presence of functional groups, with the possibility of further functionalization with various electrophiles. This methodology can be extended to ω-alkenyl-*sec*-alkylzinc reagents as well as the tertiary organozinc iodide reagents [62].

Rieke zinc can be quite useful in the generation of heteroarylzinc halide reagents. Pyridinylzinc halides can be formed from bromo- or iodopyridines and Rieke zinc at room temperature [63]. These reagents can be cross-coupled with iodobenzene, benzoyl chloride, or benzoic anhydride. The oxidative insertion of Rieke zinc proceeds at room temperature for 2- and 3-indolyl iodides [64]. This application towards direct formation of these indolylzinc iodide reagents is important since the indolyl system failed for the transmetallation of 3-indolyllithium with zinc chloride, whereas this is readily achieved via Rieke zinc.

Table 1-19. Formation and coupling reactions of *sec*- and *tert*-alkylzinc bromides.

Entry	Alkyl bromide	Electrophile	Product	Yield (%)
1	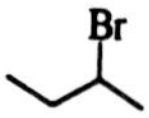	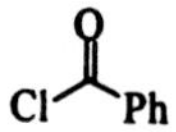	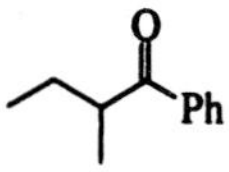	95
2	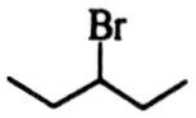	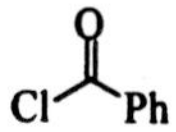	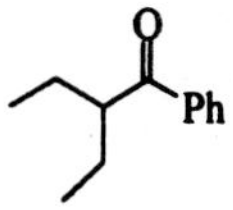	87
3	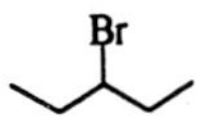	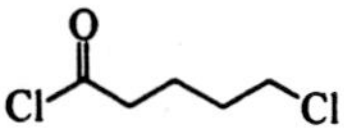	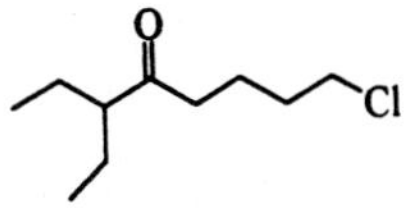	62
4	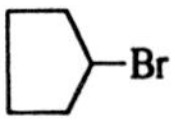	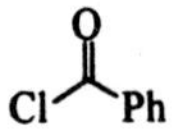	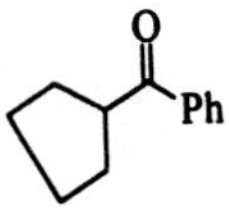	84
5	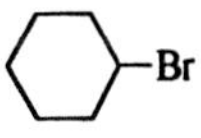	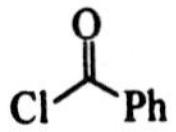	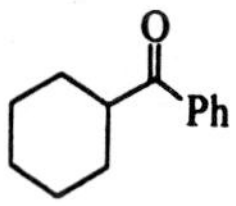	99
6	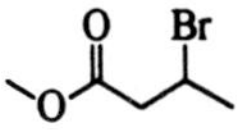	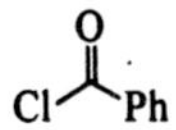	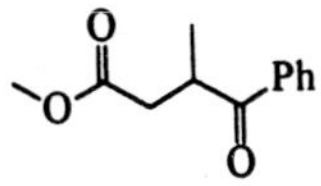	75
7	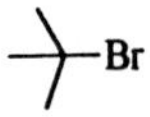	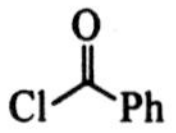	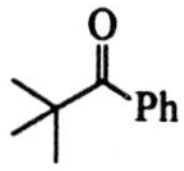	75
8		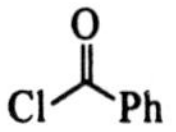	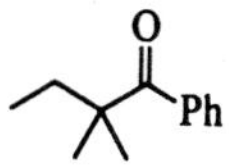	86
9	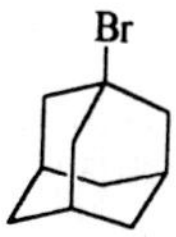	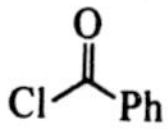	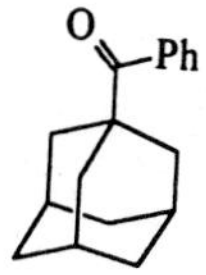	72

1.3.8 Formation of Tertiary and Secondary Alkylzinc Bromides

Tertiary and secondary alkylzinc bromides can be formed directly from tertiary alkyl bromides at room temperature (secondary alkyl bromides at reflux) with Rieke zinc [65]. This is a significant result, especially for the formation of tertiary alkylzinc bromide reagents, which would be difficult or impossible by other methods, which do not have the capacity to tolerate functionality. These tertiary and secondary alkylzinc bromide reagents cross-couple with acid chlorides to form ketones, mediated through Cu^{I} catalysis [66], more slowly than the primary alkylzinc bromides. Table 1-19 shows representative coupling reactions of secondary and tertiary alkyl bromides. Generally these reactions take 4–8 h at 0 °C for tertiary alkylzinc reagents and 2–4 h in the range –35 to 0 °C for secondary alkylzinc reagents.

1.3.9 Cyanide-Based Rieke Zinc

The lithium reduction of zinc(II) cyanide using napthalene or biphenyl as a catalytic electron carrier yields a more reactive form of Rieke zinc. This new form of Rieke zinc is able to undergo direct oxidative addition to alkyl chlorides under mild conditions and tolerates the presence of nitriles and bulky tertiary amides [67]. Table 1-20 shows representative reactions of alkylzinc chloride reagents with benzoyl chloride. The activation of the zinc surface could originate by the adsorption of the Lewis base cyanide ion on the metal surface. The adsorbed cyanide ion can affect the metal's reactivity in two possible ways. One possible mode of activation would be the reduction of the metal's work function in the vicinity of the adsorbed cyanide ion, and the second could be that the cyanide ion is acting as a conduction path for the transfer of the metal's electrons to the alkyl chloride. One or both processes could account for the observed enhanced chemical reactivity.

Table 1-20. Reactions of alkylzinc chloride reagents with benzoyl chloride.

Entry	Organohalide	Zn* : R–Cl	PhCOCl	Product	Yield (%)
1	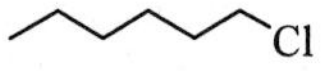 Cl	3 : 1	0.8	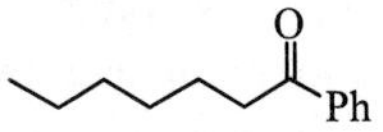 O, Ph	74
2	 Cl	4 : 1	0.8	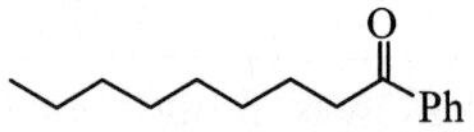 O, Ph	65
3	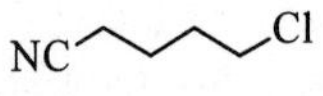 NC, Cl	2 : 1	0.7	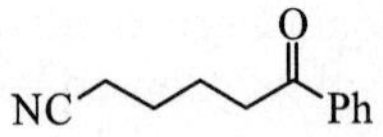 NC, O, Ph	41
4	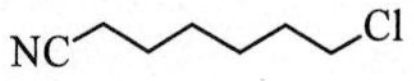 NC, Cl	2.5 : 1	0.9	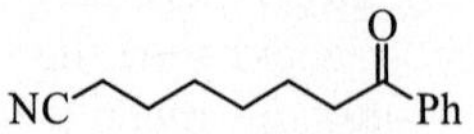 NC, O, Ph	68
5	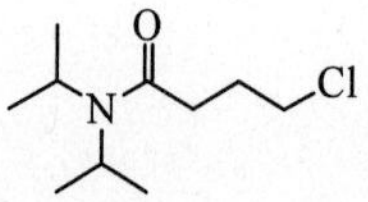 O, N, Cl	3 : 1	0.9	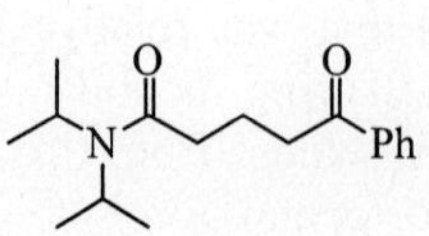 O, N, O, Ph	61

1.4 Organocopper Reagents Utilizing Rieke Copper

1.4.1 Introduction

The use of organocopper reagents in organic synthesis has grown dramatically over the past decade. Both their ease of preparation and their ability to react with other substrates have led to their widespread use and acceptance. However, ordinary copper metal is not sufficiently reactive to add oxidatively to organic halides. This is evidenced by the Ullman biaryl synthesis which involves the reaction of copper bronze with aryl iodides to form biaryl compounds [68]. Typically, these reactions are carried out in sealed tubes at temperatures of 100 to 300 °C for long periods of time ranging from hours to days. Of much more synthetic utility are the organocopper, lithium di-organocuprate, and various heterocuprate reagents developed over the past few years. The vast majority of organocopper reagents are prepared by a transmetallation reaction involving an organometallic reagent and a copper(I) salt. A variety of organometallic reagents, derived from metals more electropositive than copper, have been utilized in the preparation of these organocopper reagents. The most common organometallic precursors have been organomagnesium and organolithium reagents. However, this approach severely limits the functionalities that may be incorporated into the organocopper reagent. The use of the traditional organolithium or Grignard precursors can be circumvented by using a highly reactive zero-valent copper which undergoes direct oxidative addition to organic halides [69]. Significantly, the organocopper reagents prepared in our laboratories with this form of active copper may incorporate a wide variety of functionalities, such as allyl, nitrile, chloride, fluoride, epoxide, ketone, and ester moieties. These functionalized organocopper reagents undergo many of the same reactions as other organocopper species, including cross-coupling reactions with acid chlorides, 1,4-conjugate additions with α,β-unsaturated carbonyl compounds, and intermolecular and intramolecular epoxide-opening reactions.

1.4.2 Background to the Development of Rieke Copper

Initial attempts to prepare highly reactive copper in our laboratories were only partially successful due to rapid sintering of the finely divided copper [70]. This method involved the reduction of cuprous iodide with a stoichiometric amount of potassium metal, along with a catalytic amount of naphthalene (10 mole%) as an electron carrier, under refluxing conditions in 1,2-dimethoxyethane (DME) under argon. The reductions are usually complete within 8–12 h at room temperature. While the heterogeneous copper metal formed by this process was found to be more reactive than ordinary copper metal, the active copper could not undergo oxidative addition at temperatures low enough to form stable organocopper reagents. Consequently, the active copper formed by this method was prone to sinter, giving larger particles upon extended stirring or prolonged periods of standing. As a result, the reactivity of the copper metal was reduced. It was also found that substitution of the potassium metal by lithium resulted in much longer reduction times, giving a copper metal that was relatively unreactive. The reactivity of the copper metal was determined by measuring its ability to homocouple organic halides, analogous to the Ullman reaction. It was evident from this initial work that copper could be activated by the Rieke method. Indeed, pentafluoroiodobenzene (C_6F_5I) easily formed the homocoupled product, $C_{12}F_{10}$, under very mild conditions in high yield when compared to the classical Ullman reaction. Similarly, allyl iodide also homocoupled in high yield. However, the copper was not reactive enough to give good yields of homocoupled products for C_6F_5Br, p-$NO_2C_6H_4Br$ or n-butyliodide. Although this copper was found to be more reactive than the ordinary copper bronze used in the traditional Ullman reaction, it lacked the reactivity necessary to react with the majority of organic halides.

When compared with other metals produced by the Rieke method of metal activation, copper suffers from the disadvantage of sintering into larger particles, thereby reducing its reactivity. It was apparent that the main difficulty associated with the copper metal sintering was the exceptionally long reduction time. Previous reports from our laboratory have shown a correlation between particle size and reactivity [71]. It was anticipated that the reduction of a soluble copper complex would result in a short reduction time, thereby giving a finely divided copper metal. It was later discovered that a highly reactive zerovalent copper solution could be prepared by the reduction of a soluble copper(I) salt complex ($CuI \cdot PEt_3$) by a stoichiometric amount of preformed lithium naphthalenide in an ethereal solvent under inert, ambient conditions [72]. By using this method the reduction time was substantially shortened. The formation of lithium naphthalenide required 2 h and the subsequent reduction of the copper(I) salt complex was usually completed within 5 min, giving an active copper that was much more reactive than previous forms of copper. The resulting copper solutions appeared homogeneous and could be stirred for prolonged periods without sintering. Moreover, these new active copper species could undergo oxidative addition with functionalized alkyl and aryl organic halides to form the corresponding functionalized organocopper reagents at temperatures conducive to the formation of organocopper compounds. Since then many other copper(I) salt complexes have been reduced by preformed lithium naphthalenide. The most common include copper(I) halide phosphine complexes, copper cyanide-lithium halide complexes, and lithium 2-thienylcyanocuprate. The choice of ligand used in preparing the active copper is a critical factor influencing the reactivity of the resulting active copper species. Although all of the complexes used to form active copper are capable of undergoing the transformations previously mentioned, each has its inherent advantages and disadvantages.

1.4.3 Phosphine-Based Copper

The choice of ligand used in preparing the active copper is a critical factor affecting the reactivity of the active copper species. The use of phosphine ligands significantly enhanced the reactivity of active copper. In general, the more electron donating the phosphine the more reactive towards oxidative addition was the copper. Also, the resulting organocopper reagent was generally more nucleophilic. The general trend of reactivities for phosphine-based copper is $P(NMe_2)_3 > PEt_3 > P(CH_2NMe_2)_3 > P(cyclohexyl)_3 > PBu_3 > PPh_3 >$ Diphos $> P(OEt)_3$ as shown in Table 1-21. Phosphites were of little use, but nitrogen-containing phosphine ligands had the advantage that phosphine-containing impurities could be removed by a simple dilute acid work-up. The active copper prepared utilizing trialkylphosphines, such as PEt_3, PBu_3 or $P(Cy)_3$, gave 10–30% of homocoupled products when using alkyl bromides ($2\,RBr \rightarrow R\text{–}R$). Higher proportions of the homocoupled product were found when using alkyl iodides. The tendency for homocoupling was highly dependent upon the ratio of copper to alkyl halide used, the temperature of addition, and the manner in which the alkyl halide was added. In general, rapid low-temperature addition of the alkyl bromide to the active copper gives the best yield of the organocopper species while minimizing the formation of homocoupled products. Conver-

Table 1-21. Reactions of activated copper with alkyl halides.

Equivalent octyl halide	Copper complex	Temp. (°C)	Time (min)	Product yields (%)		
				R–H	R–X	R–R
0.45 I	CuI-PEt_3	0	2	46	0	49
0.45 I	CuI-PBu_3	0	2	68	0	27
0.50 I	CuI-PBu_3	–70	15	35	0	64
0.50 I	CuI-PBu_3	–50	20	44	0	52
0.45 I	CuI-$P(OEt)_3$	0	140	0	78	0
0.40 I	CuI-PPh_3	–78	30	67	9	23
0.40 I	CuI-$P(CH_2NMe_2)_3$	–78	75	64	1	34
0.43 I	CuI-DIPHOS	–78	10	49	13	23
0.50 I	CuBr-PBu_3	0	5	58	0	42
0.50 I	$CuBrSMe_2$	0	60	0	99	0
0.50 I	CuCN	0	60	7	86	8
0.50 I	CuCN-PEt_3	–78	30	18	56	14
0.50 I	CuCN-PEt_3	0	60	30	31	16
0.46 I	CuCN-PBu_3	0	10	40	8	27
0.45 I	CuCN-$P(OEt)_3$	0	60	30	28	2
0.45 I	CuCN-Bipy	0	60	40	2	2
0.40 Br	CuI-PBu_3	–50	110	71	3	30
0.40 Br	CuI-PBu_3	–78	20	65	5	25
0.50 Br	CuI-PBu_3	–50	110	58	24	30
0.50 Br	CuI-PBu_3	–50	105	44	22	35
0.50 Br	CuI-PBu_3	–50	75	60	27	13
0.40 Br	CuI-PBu_3	–78	120	65	3	26
0.40 Br	CuI-PPh_3	–78	60	53	42	1
0.40 Br	CuI-PPh_3	–30	150	67	31	1
0.41 Br	Cu-(NMe_2)	–78	15	68	6	21
0.41 Br	CuI-PCy_3	–78	20	70	0	25
0.44 Br	CuI-$PCy_2(CH_2)_2NMe_2$	–78	65	37	31	26
0.40 Br	CuI-$P(CH_2NMe_2)_3$	–78	130	64	28	8
0.41 Br	CuILiSCN	24	170	0	28	0
0.37 Br	CuCN-PBu_3	0	5	19	50	3
0.30 Br	CuCNLi(2-thienyl)	–60	30	91	0	5
0.35 Br	$CuCN(LiBr)_2$	–78	5	70	25	5
0.35 Br	$CuCN(LiBr)_2$	–78	5	84	3	2
0.37 Br	$CuSCN(LiBr)_2$	–35	10	29	60	0
0.50 Cl	CuI-PBu_3	–78	100	19	76	0
0.50 Cl	CuI-PBu_3	–50	80	23	71	0
0.50 Cl	CuI-PBu_3	–30	50	26	70	0
0.50 Cl	CuI-PBu_3	24	35	32	61	0

sely, dropwise addition of the alkyl bromide to active copper at low temperature is not desirable and usually forms considerable amounts of homocoupled products. The active copper prepared utilizing triphenylphosphine gave less than 1% of homocoupled product with alkyl bromides. However, the resulting organocopper reagent was less nucleophilic and not efficient in reactions that require a higher degree of nucleophilicity such as intermolecular epoxide ring opening reactions.

Although the choice of phosphine ligand was found to be crucial for the formation and subsequent reactivity of the organocopper species, the presence of malodorous phosphine ligands was often found to interfere with product isolation. In our search for a more reactive copper that did not require the use of phosphorus ligands, we discovered that the lithium napthalenide reduction of a solution of a commercially available lithium 2-thienylcyanocuprate or CuCN · LiCl complex produced a highly reactive zero valent copper complex.

Alkylcopper reagents derived from copper(I) iodide trialkylphosphine complexes are very reactive in conjugate addition reactions. Ligands such as PBu_3, PEt_3, PCy_3, and HMPT have been used quite effectively in preparing organocopper species that undergo 1,4-conjugate additions. The 1,4-conjugate addition of these alkylcopper species with 2-cyclohexenone proceeds readily at –78 °C with the enone being consumed upon warming to –50 °C. In contrast, organocopper species derived using triphenylphosphine gave little or no conjugate addition products. The amount of phosphine ligands present in the reaction mixture has a pronounced effect upon the yield of the conjugate adduct in these reactions. These results are in agreement with those reported by Noyori, in which a significant enhancement in the yields of the conjugate adducts was observed when using excess PBu_3 with the organocopper reagents [73]. Chlorotrimethylsilane is also compatible with the reaction conditions [74]. The low-temperature addition of chlorotrimethylsilane to the organocopper reagents, followed by addition of the enone and subsequent acidic work-up, results in the formation of the 3-alkylated cyclohexanone products in good yields.

The triphenylphosphine-based copper reacts rapidly with functionalized alkyl bromides to give stable alkylcopper reagents, followed by subsequent trapping with acid chlorides to give good yields of the functionalized ketone products. An excess of the acid chloride must be used in order to trap the organocopper reagent [75], since otherwise the unreacted copper species will further react with the acid chloride; for example, the use of benzoyl chloride gives *cis*-α,α'-stilbenediol dibenzoate. Homocoupling of primary alkyl bromides does not occur when using triphenylphosphine-based active copper, but some homocoupled products are formed when using primary alkyl iodides. Although triphenylphosphine alkylcopper reagents are less nucleophilic, they give higher yields in cross-coupling reactions with acid chlorides since none of the primary alkyl bromide is lost as the homocoupled product. In the aryl cases, the choice of the phosphine ligand is not crucial in cross-coupling reactions with acid chlorides because aryl halides do not homocouple to the same extent as alkyl halides. Trialkylphosphine-based copper has also been used to form ketone-functionalized, remote ester-functionalized, stable *ortho*-halophenylcopper reagents and in the cyclization of α,ω-dihaloalkanes [76].

The functionalized organocopper reagents prepared from organic halides and the trialkylphosphine derived copper species also undergo epoxide-opening reactions. The reactions of primary alkylcopper reagents with 1,2-epoxybutane gave a single regioisomer, with the alkylation taking place at the least hindered position of the epoxide, proceeding to completion at –15 °C or lower. The epoxides are found to be reasonably stable in the presence of activated copper under these conditions, otherwise the alkylcopper reagents decompose above –10 °C. Aryl halides form arylcopper compounds at 25 °C or lower. Arylcopper compounds also undergo epoxide-opening reactions with 1,2-epoxybutane to form a single regioisomer in good yield at room temperature or with moderate heating.

Molecules containing both an epoxide and a nucleophile which can undergo an intramolecular cyclization are synthetically useful and have increasingly been reported [77]. Epoxyalkylcopper species, using trialkylphosphine-based copper, are able to undergo intramolecular cyclization via an epoxide-opening process upon warming, thereby generating new carbocycles. It should be noted that triphenylphosphine-based organocopper reagents may also be used for these intramolecular epoxide-opening reactions. In general, the regioselectivity of intramolecular epoxide opening reactions usually follows

Table 1-22. Intramolecular epoxide-opening reactions of epoxy aryl halides using phosphine-based active copper.

Entry	R_1	R_2	R_3	R_4	PR_3	Solvent	Temp. (°C)	*Exo*:*Endo*	Yield (%)
1	H	H	H	H	PBu_3	THF	$0 \rightarrow$ rt	82: 18	61
2	H	H	H	H	PPh_3	THF	$0 \rightarrow$ rt	88: 12	63
3	H	H	H	H	PPh_3	Toluene	$0 \rightarrow$ rt	58: 42	29
4	H	H	H	H	PBu_3	THF/DMF	$0 \rightarrow$ rt	92: 8	25
5	H	H	H	H	PPh_3	THF/DMF	$0 \rightarrow$ rt	100: 0	5
6	CO_2Et	H	H	H	PBu_3	THF	0	87: 13	80
7	Cl	H	H	H	PBu_3	THF	$0 \rightarrow$ rt	83: 17	53
8	Cl	H	H	H	PPh_3	THF	$0 \rightarrow$ rt	89: 11	63
9	Me	H	H	H	PBu_3	THF	$0 \rightarrow$ rt	82: 18	59
10	Me	H	H	H	PPh_3	THF	$0 \rightarrow$ rt	87: 13	56
11	H	Me	H	H	PBu_3	THF	0	100: 0	86
12	H	Me	H	H	PPh_3	THF	0	100: 0	85
13	CO_2Et	Me	H	H	PBu_3	THF	0	100: 0	74
14	CO_2Et	Me	H	H	PPh_3	THF	0	100: 0	68
15	Cl	Me	H	H	PBu_3	THF	0	100: 0	96
16	Cl	Ph	H	H	PBu_3	THF	0	84: 16	86
17	Cl	H	H	Me	PBu_3	THF	0	0: 100	53
18	CO_2Et	Me	Me	H	PBu_3	THF	0	100: 0	79

Baldwin's rules [78]. *Exo*-mode ring formations are favored in medium size ring cyclizations when both termini of the epoxide are equally substituted. For a shorter connecting chain, such as four-carbon bromoepoxides, reactions always prefer the *exo*-mode ring closures.

Parham cyclialkylations [79] and cycliacylations [80] involving an anionic cyclization of *ortho* functionalized aryllithium compounds have been found to be synthetically useful [81]. The intramolecular cyclizations of epoxyarylcopper compounds, mediated by Rieke copper, undergo *exo*-mode ring closures to form 2,3-dihydrobenzofuran and *endo* ring closures to give 3-chromanol, as shown in Table 1-22. Both trialkylphosphine- and triphenylphosphine-based copper may be used for the intramolecular epoxide-opening reactions of these arylcopper reagents. The regioselectivity of these cyclizations is also affected by the substitution pattern, reaction solvent, and the $CuI \cdot PR_3$ complex used to generate the active copper. The *exo*-mode ring closure is usually preferred for these reactions.

1.4.4 Lithium 2-Thienylcyanocuprate-Based Copper

The reduction of lithium 2-thienylcyanocuprate with lithium naphthalenide at –78 °C generates a highly reactive form of active copper [82]. This clearly demonstrates that highly reactive copper can also be produced from copper salts that do not contain phosphines. As a result, product isolation and purification is greatly facilitated by avoiding the use of malodorous phosphine ligands. The resulting zero-valent copper species will react oxidatively with alkyl halides under very mild conditions at –78 °C to form stable alkylcopper compounds in high yields. The formation of the alkylcopper reagent is generally accompanied by trace amounts (< 2%) of eliminated and homocoupled by-products. A minor amount of the thienyl ligand transfer product has been found. However, these thienyl transfer products only affected product isolation in a few cases.

The ability to directly form various organocopper reagents utilizing this thienyl-based active copper solution allows easy preparation of functionalized organocopper reagents containing ester, nitrile, chloride, fluoride, epoxide, amine, and ketone moieties. In turn, these stable alkylcopper reagents are able to cross-couple with benzoyl chloride at –78 °C to generate ketone products in excellent isolated yields, as shown in Table 1-23. Significantly, this thienyl-based active copper adds oxidatively to allyl chlorides and acetates at –78 °C to allow the direct formation of allylic organocopper reagents, accompanied by less than 10% of the Wurtz-type homocoupling products. In turn, these allylic organocopper reagents are capable of undergoing cross-coupling reactions with electrophiles such as benzoyl chloride and benzaldehyde to give the corresponding ketone and alcohol products respectively in good isolated yields, as shown in Table 1-24. Another common, although tedious, route to allylic organocopper reagents is the transmetallation of allylic stannanes with an appropriate organocopper reagent, itself derived from a transmetallation of an organolithium or Grignard precursor.

Conjugate additions of thienyl-based organocopper reagents to α,β-unsaturated ketones proceed in excellent yields, and these are the organocopper reagents of choice. The addition of TMSCl to the organocopper reagent prior to the addition of the enone allows the facile formation of the 1,4-conjugate product at –78 °C in excellent isolated yields, as shown in Table 1-25. Competitive 1,2-addition products were not observed for these conjugate addition reactions. The addition of TMSCl significantly increased product formation as much as five-fold and is considered essential. The addition of Lewis acids, such as $BF_3 \cdot Et_2O$, did not have as dramatic an effect upon product formation as did TMSCl. These thienyl-based organocopper reagents are very efficient for conjugate addition reactions, the ideal ratio of organocopper reagent to enone being as low as 2 : 1. The use of larger excesses of the organometallic reagent (10 equiv) has been reported in the literature

Table 1-23. Cross-coupling reactions of thienyl-based organocopper reagents with acid chlorides.

Entry	Halide	Acid chloride	Product	Yield (%)
1	$Br(CH_2)_7CH_3$	PhCOCl	$PhCO(CH_2)_7CH_3$	73
2	$Br(CH_3)_3CO_2Et$	PhCOCl	$PhCO(CH_2)_3CO_2Et$	47
3	$Br(CH_2)_3CN$	PhCOCl	$PhCO(CH_2)_3CN$	61
4	$Br(CH_2)_6Cl$	PhCOCl	$PhCO(CH_2)_6Cl$	42
5	$Br(CH_2)_6\overset{O}{\overbrace{CH\text{-}CH_2}}$	PhCOCl	$PhCO(CH_2)_6\overset{O}{\overbrace{CH\text{-}CH_2}}$	65
6	$BrC_6H_4(p\text{-}CH_3)$	PhCOCl	$PhCOC_6H_4(p\text{-}CH_3)$	86
7	$BrC_6H_4(p\text{-}OCH_3)$	PhCOCl	$PhCOC_6H_4(p\text{-}OCH_3)$	87
8	$IC_6H_4(p\text{-}OCH_3)$	*n*-BuCOCl	$n\text{-}BuCOC_6H_4(p\text{-}OCH_3)$	63
9	$BrC_6H_4(p\text{-}CN)$	PhCOCl	$PhCOC_6H_4(p\text{-}CN)$	75
10	$BrC_6H_4(m\text{-}CN)$	PhCOCl	$PhCOC_6H_4(m\text{-}CN)$	62
11	$BrC_6H_4(p\text{-}NMe_2)$	PhCOCl	$PhCOC_6H_4(p\text{-}NMe_2)$	81
12	$IC_6H_4(p\text{-}Cl)$	PhCOCl	$PhCOC_6H_4(p\text{-}Cl)$	90
13	$BrC_6H_4(p\text{-}F)$	PhCOCl	$PhCOC_6H_4(p\text{-}F)$	93
14	$BrC_6H_4(p\text{-}COPh)$	PhCOCl	$PhCOC_6H_4(p\text{-}COPh)$	44

Table 1-24. Reactions of thienyl-based allylic organocopper reagents with electrophiles.

Entry	Allylic precursor	Electrophile	Product	Yield (%)
1	Cl	PhCHO	OH	80
2	Cl	PhCOCl	O	56
3	Cl	PhCHO	OH	70
4	Cl	PhCOCl	O	40
5	Cl	PhCHO	OH	78
6	Cl	PhCOCl	O	55
7	Cl	PhCHO	OH	61
8	OAc	PhCHO	OH	26

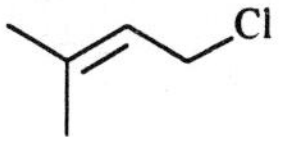

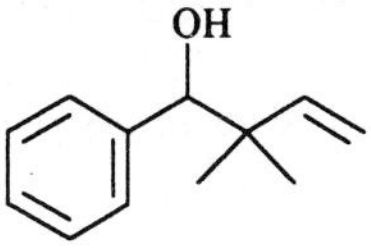

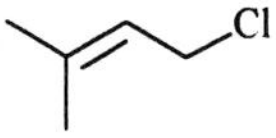

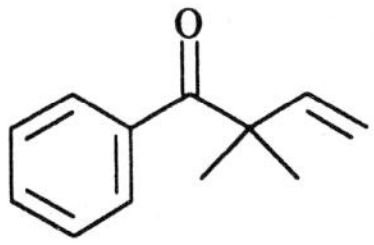

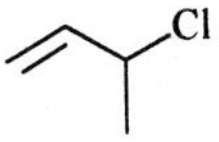

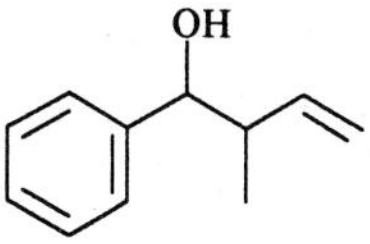

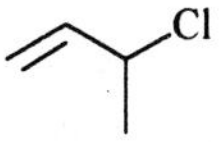

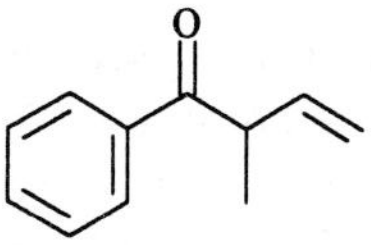

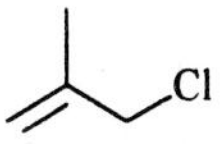

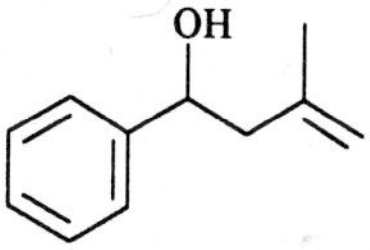

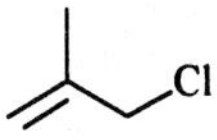

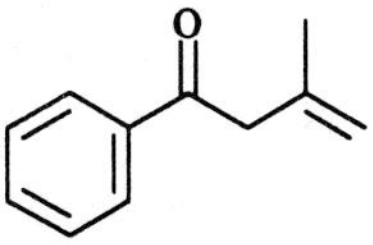

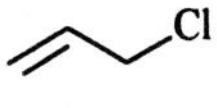

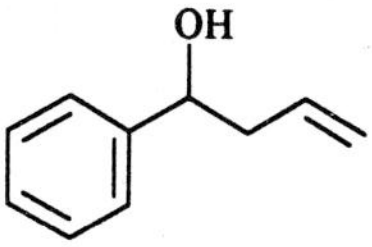

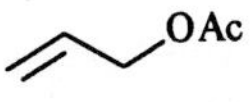

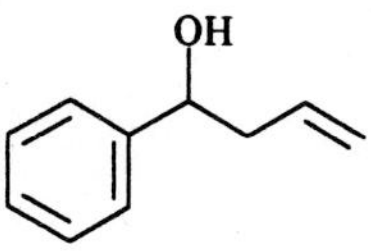

Table 1-25. 1,4-Conjugate addition reactions with thienyl-based organocopper reagents.

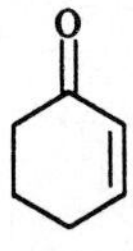

A 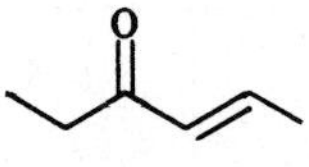 B

Entry	Eq. halide	Eq. additive	Eq. enone	Product	Yield (%)
1	0.5 $Br(CH_2)_7CH_3$	2.0 TMSCl	0.25 A	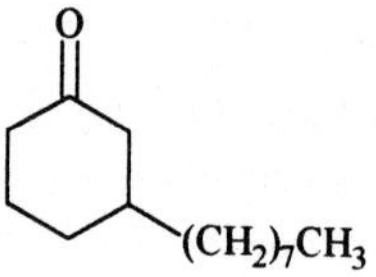 $(CH_2)_7CH_3$	91
2	0.5 $Br(CH_2)_7CH_3$	1.0 TMSCl	0.125 A	"	97
3	0.5 $Br(CH_2)_7CH_3$	none	0.25 A	"	20
4	0.5 $Cl(CH_2)_7CH_3$	2.0 TMSCl	0.25 A	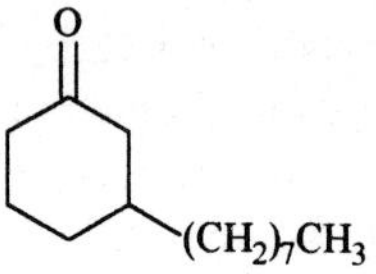 $(CH_2)_7CH_3$	71
5	0.5 $Cl(CH_2)_7CH_3$	none	0.25 A	"	16
6	0.5 $Cl(CH_2)_7CH_3$	1.0 TMSCl	0.167 A	"	87
7	0.5 $Br(CH_2)_7CH_3$	1.0 TMSCl	0.125 B	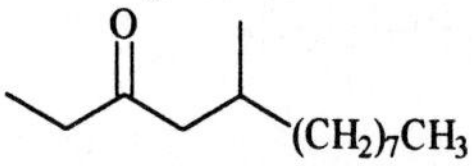 $(CH_2)_7CH_3$	88
8	0.5 $Cl(CH_2)_7CH_3$	1.0 TMSCl	0.125 B	"	87
9	0.5 $Br(CH_2)_6Cl$	1.0 TMSCl	0.25 A	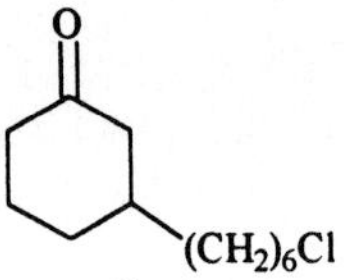 $(CH_2)_6Cl$	80
10	0.5 $Br(CH_2)_6Cl$	1.0 TMSCl	0.125 B	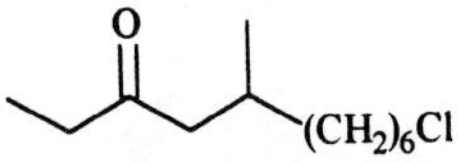 $(CH_2)_6Cl$	83
11	0.5 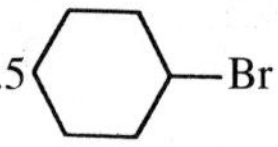 cyclohexyl–Br	1.0 TMSCl	0.125 A	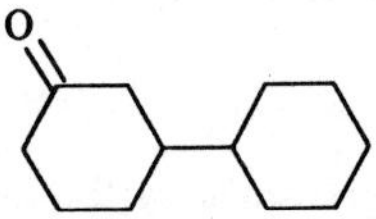	72
12	0.5 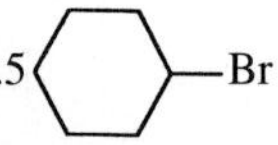 cyclohexyl–Br	1.0 TMSCl	0.125 B	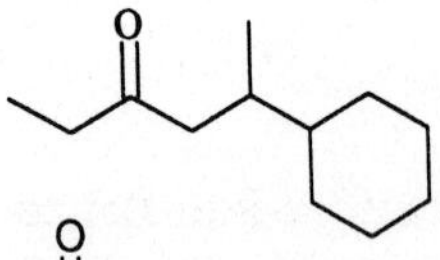	80
13	0.5 $Br(CH_2)_3CO_2Et$	1.0 TMSCl	0.125 A	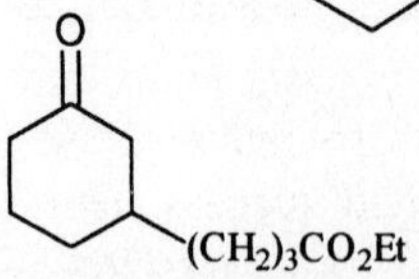 $(CH_2)_3CO_2Et$	79
14	0.5 $Br(CH_2)_3CO_2Et$	1.0 TMSCl	0.125 B	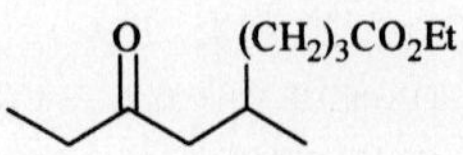 $(CH_2)_3CO_2Et$	81

Table 1-26. Reactions of thienyl-based organocopper reagents with 1,2-epoxybutane.

Entry	Halide	Product	Yield (%)
1	$Br(CH_2)_7CH_3$	$CH_3CH_2CH(OH)(CH_2)_8CH_3$	71
2	$Br(CH_2)_5OPh$	$CH_3CH_2CH(OH)(CH_2)_6OPh$	60
3	$Br(CH_2)_6Cl$	$CH_3CH_2CH(OH)(CH_2)_7Cl$	68
4	$IC_6H_4(p\text{-}CH_3)$	$CH_3CH_2CH(OH)CH_2C_6H_4(p\text{-}CH_3)$	64
5	$IC_6H_4(p\text{-}OCH_3)$	$CH_3CH_2CH(OH)CH_2C_6H_4(p\text{-}OCH_3)$	78
6	$IC_6H_4(p\text{-}Cl)$	$CH_3CH_2CH(OH)CH_2C_6H_4(p\text{-}Cl)$	62

[83]. Both cyclic and acyclic enones worked well with this formulation of active copper. However, more sterically hindered enones, such as isophorone, carvone, and 2,4,4-trimethyl-2-cyclohexen-1-one, were not amenable for these 1,4-conjugate additions.

The thienyl-based organocopper reagents were also found to be nucleophilic enough to undergo intermolecular epoxide-opening reactions with 1,2-epoxybutane to form single regioisomers in good isolated yields, as shown in Table 1-26. We have attempted to activate the epoxide using Lewis acids to accelerate the organometallic epoxide-opening reaction, but to no avail. Analogously, intramolecular cyclizations via an epoxide cleavage process can also be realized. For example, treatment of 6-bromo-1,2-epoxyhexane with thienyl-based copper resulted in intramolecular cyclization; subsequent trapping of the intermediate with benzoyl chloride gave a 64 : 36 mixture of cyclohexyl benzoate and cyclopentylmethyl benzoate in 52% isolated yield as shown in Scheme 1-13.

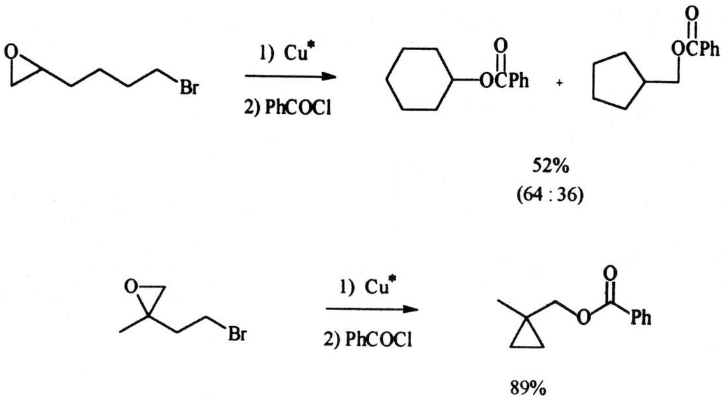

Scheme 1-13. Intramolecular cyclizations via epoxide cleavage.

1.4.5 Copper Cyanide-Based Active Copper

Initial attempts to make active copper from $CuCN \cdot nLiX$ complexes (X = Br, Cl) met with only limited success. However, exploiting the low-temperature reduction effects seen with other Cu^I complexes led to a successful approach for the formation of active copper from $CuCN \cdot nLiX$ [84]. Lowering the reduction temperature to −110 °C allowed the active copper solution to react with organic iodides, bromides, and to some extent chlorides, to produce organocopper species in high yields. Like the triphenylphosphine and thienyl-based copper species, active copper derived from CuCN · *n*LiX produced very little of the homocoupled products when using alkyl bromides. Better product yields were obtained when two equivalents of the lithium salt were used to solubilize CuCN. Both LiBr and LiCl gave comparable results. Also, the manner in which the CuCN · *n*LiX and lithium naphthalenide are added together affects the reactivity of the resulting active copper solution. The best results were obtained when the CuCN · *n*LiX was added into the preformed lithium naphthalenide. It should be noted that the phosphine-based active copper is more reactive than the active copper derived from CuCN · *n*LiX. However, active copper derived from CuCN · *n*LiX offers several advantages. First, CuCN is an inexpensive and stable source of Cu^I which can be used as received without further purification. Both active copper chemistry and traditional organocopper chemistry involving Cu^I have been shown to be effected by the purity of the Cu^I used [85]. Second, product isolation is facile and product purity is greatly improved using this phosphine-free source of Cu^I. The lithium salts are easily removed during aqueous work-up. However, this form of active copper is not as nucleophilic, as evidenced by its inability to undergo epoxide-opening.

Table 1-27 shows the results of reacting various functionalized organocopper reagents, produced from both alkyl and aryl bromides, with benzoyl chloride to form the corresponding functionalized ketone products, which were obtained in good isolated yields. Chloride, nitrile, and ester functionalities can be incorporated into the organocopper reagent, but if the functional group is proximal to the carbon–bromine bond the yields are lower.

The essential addition of TMSCl to organocopper reagents made from CuCN · 2 LiBr-based active copper allowed 1,4-conjugate additions to occur at –78 °C in good isolated yields, as shown in Table 1-28. Both cyclic and acyclic enones can be used, the ideal organocopper to enone ratio being 2.5 : 1. As with other species of active copper, competitive 1,2-addition products were not observed

The reactions of several non-functionalized allyl chlorides and acetates with active copper followed by cross-coupling with various electrophiles at –100 °C gave ketone and alcohol products in good isolated yields. Although this species of active copper is not nucleophilic enough to undergo intramolecular or intermolecular epoxide-opening reactions, the addition of MeLi enhances the nucleophilicity of the allylic organocopper reagents and allows substitution reactions with epoxides. The MeLi is believed to act as a "dummy" non-transferable ligand, presumably forming a higher-order cuprate.

Primary and secondary allyl chlorides containing diverse functionalities are tolerated by the CuCN-based copper. Allyl organocopper reagents containing ketone, α,β-unsaturated ketone, epoxide, nitrile, alkyl acetate, ester, alkyl chloride, and carbamate functional-

Table 1-27. Cross-coupling of benzoyl chloride with organocopper reagents derived from CuCN · 2 LiBr-based active copper.

Entry	Halide (equiv.)	Product	Yield (%)
1	$Br(CH_2)_7CH_3$ (0.25)	$PhCO(CH_2)_7CH_3$	82
2	$Br(CH_2)_6Cl$ (0.25)	$PhCO(CH_2)_6Cl$	80
3	$Br(CH_2)_3CO_2Et$ (0.25)	$PhCO(CH_2)_3CO_2Et$	81
4	$Br(CH_2)_2CO_2Et$ (0.25)	$PhCO(CH_2)_2CO_2Et$	43
5	$Br(CH_2)_3CN$ (0.25)	$PhCO(CH_2)_3CN$	86
6	Bromobenzene (0.20)	PhCOPh	87
7	*p*-BrC_6H_4CN (0.20)	*p*-NCC_6H_4COPh	60
8	*o*-BrC_6H_4CN (0.20)	*o*-NCC_6H_4COPh	74
9	*o*-$BrC_6H_4CO_2Et$ (0.20)	*o*-$EtO_2CC_6H_4COPh$	51
10	*p*-BrC_6H_4Cl (0.20)	*p*-ClC_6H_4COPh	83

Table 1-28. Conjugate additions with organocopper reagents derived from $CuCN \cdot 2\,LiBr$-based active copper.

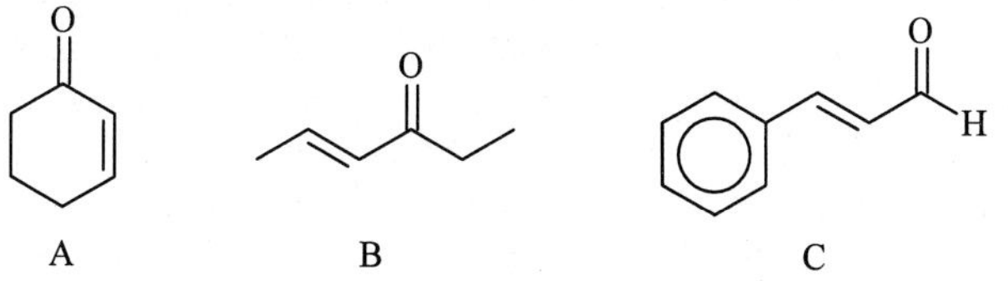

Entry	Halide	Enone (equiv.)	Yield (%)
1	$Br(CH_2)_7CH_3$	A (0.17)	92
2	$Cl(CH_2)_7CH_3$	A (0.16)	42
3	$Br(CH_2)_3CO_2Et$	A (0.17)	70
4	$Br(CH_2)_3CO_2Et$	A (0.12)	90
5	$Br(CH_2)_3CO_2Et$	B (0.11)	94
6	$Br(CH_2)_3CO_2Et$	C (0.11)	87
7	$Br(CH_2)_3CN$	A (0.12)	87
8	$Br(CH_2)_3CN$	B (0.11)	92
9	$Br(CH_2)_6Cl$	A (0.12)	82
10	BrC_6H_{11}	A (0.12)	80
11	BrC_6H_5	A (0.11)	45
12	$ClCH_2CH{=}C(CH_3)_2$	A (0.10)	81

ities have been prepared. The ability of this CuCN-based active copper to tolerate a wide variety of functionalities allows the facile formation of highly functionalized homoallylic alcohols, β,γ-unsaturated ketones, and amines in excellent isolated yields. The organocopper reagents derived from primary allyl chlorides showed a remarkable thermostability with little decomposition at 0 °C, unlike secondary allyl organocopper reagents, which decompose at a significant rate. When MeLi was added to these functionalized organocopper species derived from allyl chlorides, an intramolecular opening of the epoxide produced a bicyclic product, as shown in Scheme 1-14.

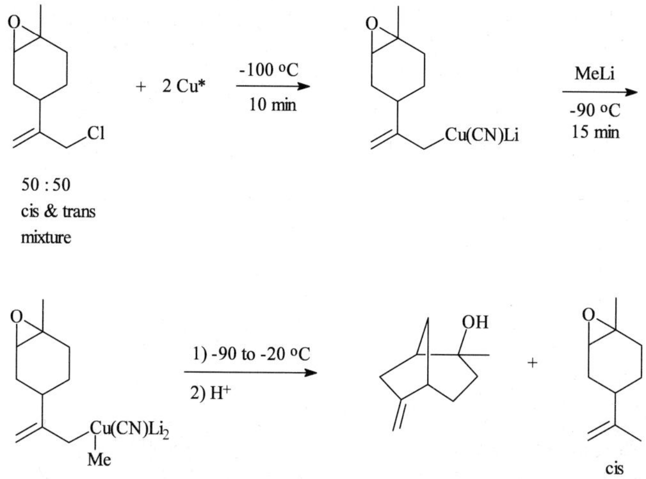

Scheme 1-14 Intramolecular epoxide opening to afford a bicyclic product.

Table 1-29. Reaction of CuCN · 2 LiBr-derived copper with 2,3-dichloropropene.

Entry	E1	E2	Product	Yield (%)
1	PhCHO	H^+	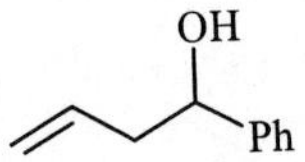 OH, Ph	88
2	PhCHO	I_2	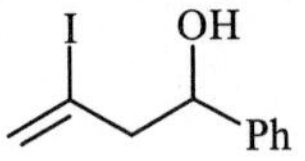 I, OH, Ph	69
3	PhCHO	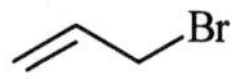 Br	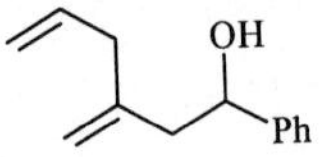 OH, Ph	71
4	PhCHO	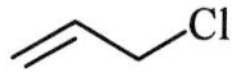 Cl	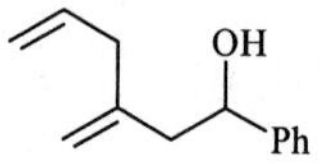 OH, Ph	83
5	PhCHO	MeI	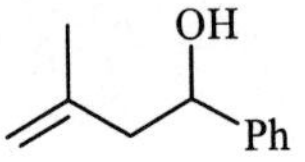 OH, Ph	63
6	$PhCH_2COCH_3$	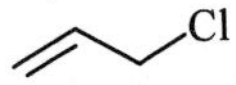 Cl	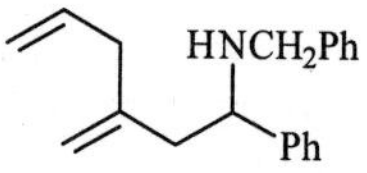 $HNCH_2Ph$, Ph	22
7	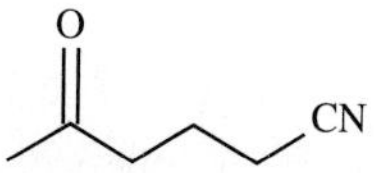 O, CN	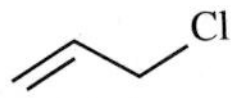 Cl	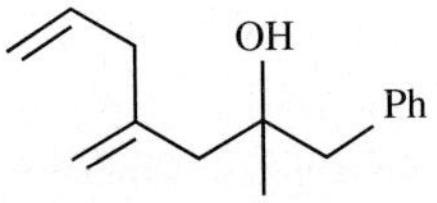 OH, Ph	54
8	NCH_2Ph ‖ PhCH	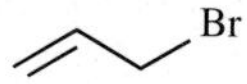 Br	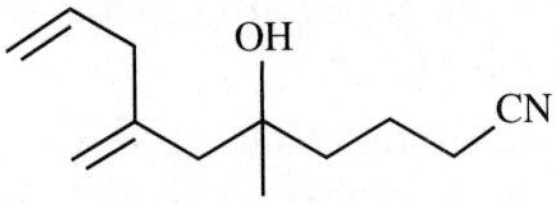 OH, CN	79

This highly reactive copper derived from CuCN · 2 LiCl reacts directly with 2,3-dichloropropene to yield a new bis-organocopper species which contains both a nucleophilic allylic and a vinylic moiety [86]. This novel bis-organocopper reagent undergoes a selective one-pot addition to two different electrophiles in good to excellent yields. The more reactive allyl carbon-copper bond adds to the first electrophile, followed by the incorporation of the second electrophile into the vinyl carbon-copper bond. Table 1-29 shows the one-pot reactions of the bis-organocopper reagent with various combinations of electrophiles. Aldehydes, ketones, and imines are all suitable electrophiles for reaction with the allylic terminus, whereas allyl chlorides, allyl bromides, alkyl iodides, and iodine add to the vinyl-copper bond.

1.4.6 Two-Equivalent Reduction of Copper(I) Complexes: A Formal Copper Anion

A new, highly reactive copper species made from the reduction of copper(I) complexes with two equivalents of preformed lithium naphthalenide leads to a reagent that behaves chemically as a formal copper anion solution [87]. Lithium naphthalenide provides a suitable reducing agent for a variety of metals and offers unique advantages over other techniques used in metal activation, such as metal vaporization. It was believed that an exten-

Table 1-30. Reaction of copper anion with organohalides and subsequent cross-coupling reactions with benzoyl chloride.

$$Cu^{I}\text{ Complex} + 2\text{ LiNp} \xrightarrow[\text{THF}]{^{\circ}\text{C}} \xrightarrow[-35\,^{\circ}\text{C}]{\text{RX}} \xrightarrow[-35\,^{\circ}\text{C}]{\text{PhCOCl}} \text{RCOPh}$$

Entry	Halide	Complex	Temp. (°C)	Yields (%)		
				RCOPh	R–X	R–R
1	$CH_3(CH_2)_7Cl$	CuI · PPh_3	0	40	39	0
2	$CH_3(CH_2)_7Cl$	CuI · PPh_3	– 78	77	12	0
3	$CH_3(CH_2)_7Cl$	CuI · PPh_3	–107	96	0	0
4	$CH_3(CH_2)_7Br$	CuI · PPh_3	–107	92	0	0
5	$CH_3(CH_2)_5CHBrCH_3$	CuI · PPh_3	–107	33	–	–
6	$C_6H_{11}Cl$	CuI · PPh_3	–107	99	–	–
7	$C_6H_{11}Cl$	CuI · PPh_3	–107	90	–	–
8	PhCl	CuI · PPh_3	–107	93	–	–
9	PhBr	CuI · PPh_3	–107	80	–	–
10	PhF	CuI · PPh_3	–107	38	–	–
11	$CH_3(CH_2)_7Cl$	CuCN · 2 LiBr	– 78	61	19	8
12	$CH_3(CH_2)_7Cl$	CuCN · 2 LiBr	–107	93	0	4

Table 1-31. 1,4-Conjugate addition reactions of copper-anion-based organocopper reagents with 2-cyclohexen-1-one.

Entry	Halide	Product	Yield (%)
1	$CH_3(CH_2)_7Cl$	O, $(CH_2)_7CH_3$	73
2	$CH_3(CH_2)_7Br$	O, $(CH_2)_7CH_3$	81
3	PhBr	O, Ph	42
4	$Br(CH_2)_6Cl$	O, $(CH_2)_6Cl$	30
5	$Br(CH_2)_3CN$	O, $(CH_2)_3CN$	18

sion of the investigations on zero-valent active copper could include a further equivalent reduction of Cu^0 to Cu^{-1} [$4s^1 3d^{10} + e^- = 4s^2 3d^{10}$] to yield a closed shell anion. This work initially involved the two-equivalent reduction of $CuI \cdot PPh_3$ with lithium naphthalenide in THF at –108 °C. The resulting copper anion solutions are homogeneous and, like the zero-valent copper, readily undergo oxidative addition to carbon-halogen bonds. However, the copper anion is a two-equivalent electron reagent, and one equivalent of halide reacts with one equivalent of copper anion. Reaction of 2 mmol of 1-bromooctane with 2 mmol of copper anion at –108 °C followed by warming to –35 °C results in the total consumption of the bromide. Similar results were obtained with 1-chlorooctane. The resulting organocopper species can be cross-coupled with benzoyl chloride to form ketones in high yields as shown in Table 1-30. The copper anion solution displayed a dramatic increase in reactivity with halides otherwise not reactive with zero-valent active copper, undergoing oxidative addition to alkyl and aryl chlorides and to some extent even aryl fluorides. The significant increase in reactivity and stoichiometry provides evidence of a new copper species, either a copper monoanion or possibly a polyatomic anionic copper complex. The resulting organocopper species derived from copper anion also undergo 1,4-conjugate additions with 2-cyclohexen-1-one in moderate yields, as shown in Table 1-31. However, copper anion is less suitable for 1,4-conjugate additions when compared with the thienyl-based organocopper species.

Table 1-32. Formation and reactions of copper-anion-based higher-order allyl cyanocuprates.

Entry	Allyl chloride	Electrophile	Product	Yield (%)
1	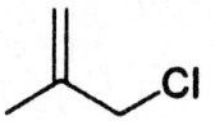 Cl	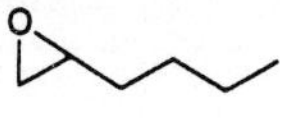 O	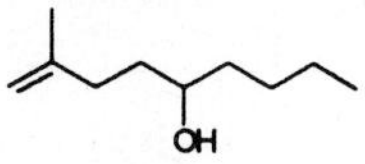 OH	98
2	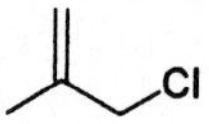 Cl	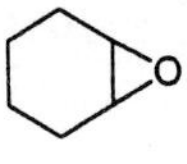 O	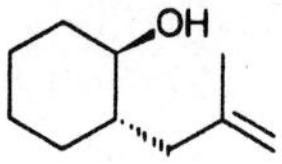 OH	98
3	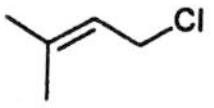 Cl	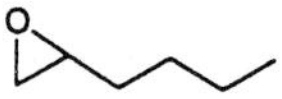 O	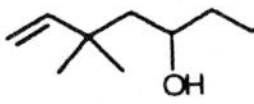 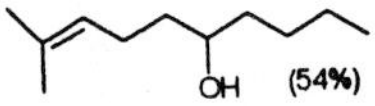 OH (46%); OH (54%)	97
4	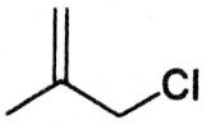 Cl	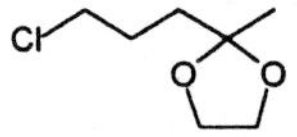 Cl, O O	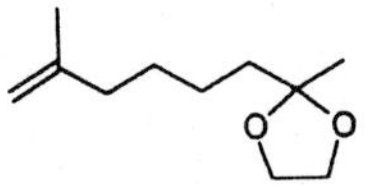 O O	82
5	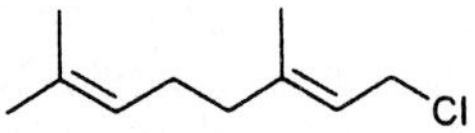 Cl	$PhO(CH_2)_4Cl$	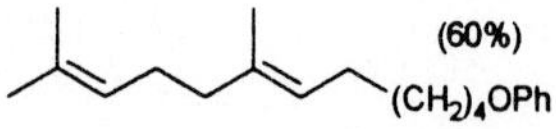 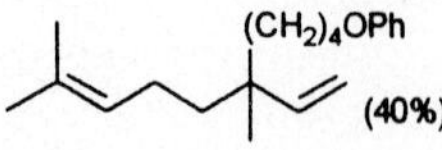 $(CH_2)_4OPh$ (60%); $(CH_2)_4OPh$ (40%)	81
6	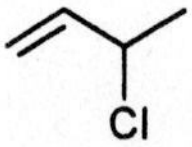 Cl	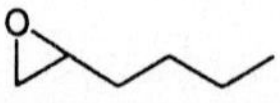 O	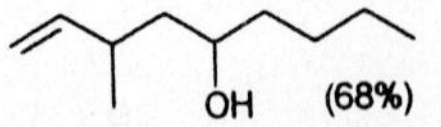 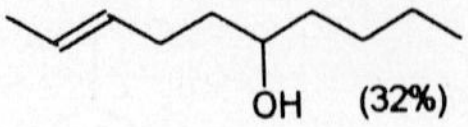 OH (68%); OH (32%)	86

Table 1-33. 1,4-Conjugate additions with copper-anion-based allyl organocopper reagents.

Entry	Allyl chloride	Enone	Product	Yield (%)
1	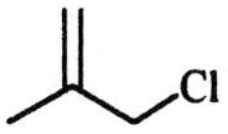 Cl	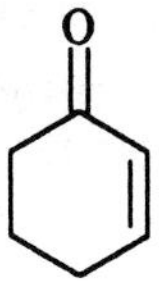 O	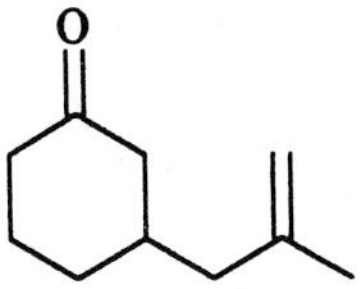 O	91
2	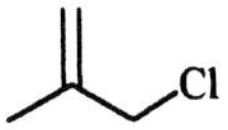 Cl	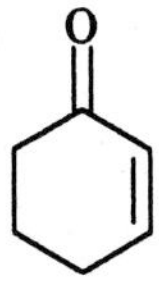 O	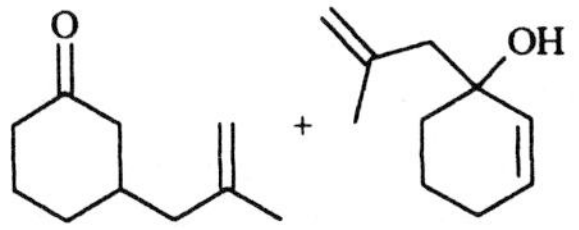 O + OH (12%) (31%)	43
3	 Cl	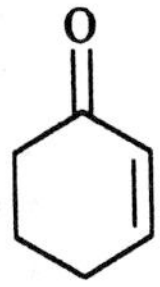 O	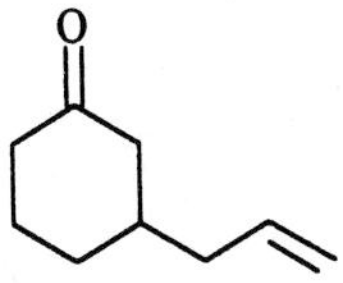 O	72
4	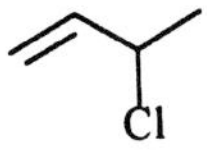 Cl	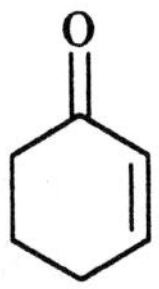 O	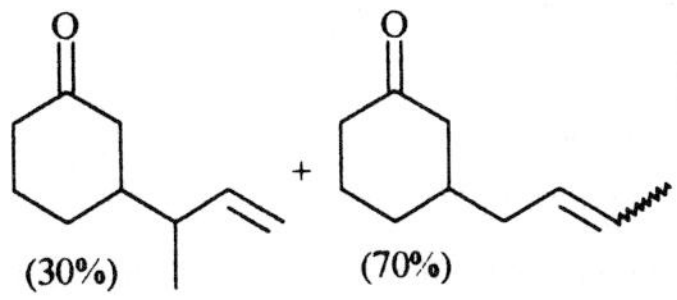 O + O (30%) (70%)	24
5	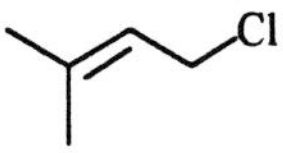 Cl	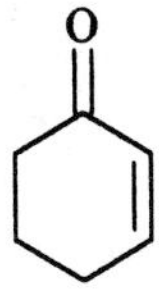 O	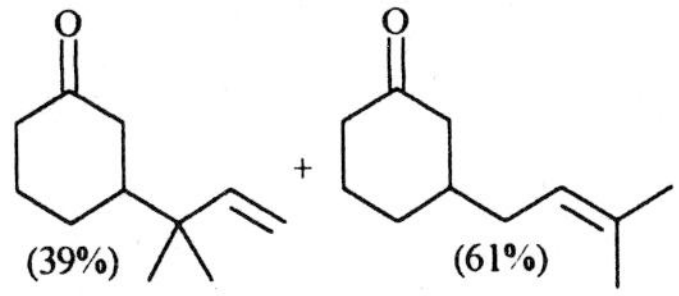 O + O (39%) (61%)	98
6	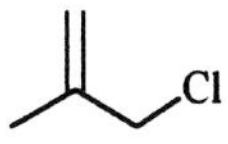 Cl	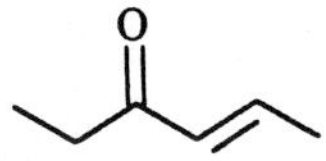 O	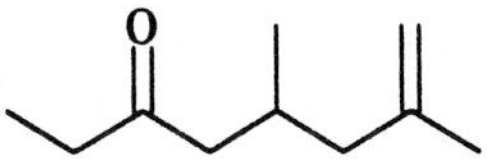 O	39

While both the thienyl-based and CuCN-based forms of active copper allow the straightforward production of allyl organocopper compounds, the type of cuprates generated possess reactivities comparable to those of the lower-order cyanocuprates. As an aside, Lipshutz et al. have shown that higher order allyl cyanocuprates possess remarkable reactivities and undergo substitution with alkyl chlorides and epoxides at –78 °C [88]. Furthermore, in order to add an allyl moiety in a 1,4-manner across an α,β-unsaturated enone, Lipshutz developed a less reactive "non-ate" allyl copper reagent [89]. A straightforward preparation of both higher-order and "non-ate" allyl copper reagents directly from allyl chlorides utilizing a formal copper anion solution is now possible. Table 1-32 shows the results of producing higher order allyl cyanocuprates with copper anion. The reaction between the allyl chloride and the copper anion is similar to those previously discussed, but in order to form the higher order cuprate, methyllithium is added to act as a non-transferable "dummy" ligand. The resulting copper solution shows reactivity comparable to that of the higher order allylcyanocuprates. The regioselectivity of the asymmetrical allyl chlorides proved a disappointment, as substantial yields of products from α and γ

attack were observed. Since higher-order allyl cyanocuprates produced via transmetallation of allyl stannanes have shown regioselective addition to various electrophiles [90], the nature of these higher-order allyl cyanocuprates must be associated with their method of preparation.

In order to facilitate 1,4-conjugate addition to α,β-unsaturated enones, the production of a less reactive "non-ate" organocopper reagent was sought, since otherwise the 1,2-addition product was formed. Since the cyano ligand is known to remain fixed in organocuprates produced from cuprous cyanide salts [91], the THF-soluble $CuCN \cdot LiCl$ copper(I) complex was chosen as the copper anion precursor. The results with various allyl moieties in their transfer to α,β-unsaturated enones are shown in Table 1-33. As with many less reactive organocopper reagents, such as entry 2 of Table 1-33, the use of TMSCl is essential in these reactions [92]. The use of Lewis acids such as $BF_3 \cdot Et_2O$ did not enhance the conjugate product yields to the same extent as with TMSCl.

1.5 Rieke Aluminum, Indium, and Nickel

1.5.1 Aluminum

Highly reactive aluminum has been prepared by reducing anhydrous aluminum halides in refluxing organic solvents such as THF, xylene, or triethylamine with potassium or sodium metal under an inert atmosphere [93]. The most convenient combinations are $AlCl_3$/K/THF, $AlCl_3$/K/xylene, and $AlCl_3$/Na/xylene. Once formed, Rieke aluminum reacts with iodobenzene and bromobenzene in refluxing xylene to give phenylaluminum halides ($Ph_3Al_2I_3$) in quantitative yields (100%). Slightly lower yields of the phenylaluminum halide product were observed with chlorobenzene. In comparison, Grosse [94] prepared $Ph_3Al_2I_3$ by reacting aluminum shavings with iodobenzene at 100 °C for 44 h to give phenylaluminum halide in only 84% yield; chloro- or bromobenzene failed to react under these conditions. Similarly, Wittenburg [95] attempted to activate aluminum metal by grinding with aluminum salts, subsequent reaction with bromobenzene in xylene at 140 °C for 5 h gave only a 79% yield of the phenylaluminum halide product. Thus, the reactivity of Rieke aluminum appears to be vastly superior to that found using other reported methods for the activation of aluminum.

1.5.2 Indium

The use of indium metal in organic or organometallic synthesis is quite limited. Schumb and Crane reported that attempts to react several alkyl halides with metallic indium were unsuccessful, but a reaction did occur with methylene iodide [96]. Deacon reported that C_6F_5I is reactive with indium metal [97]. Similarly, Gynane et al. reported successful reactions of methyl, ethyl, and propyl bromides and iodides with metallic indium [98]. However, these reactions were very slow, requiring 1–5 days for completion, and afforded a mixture of R_2InX and $RInX_2$.

We have found that the reduction of anhydrous $InCl_3$ with potassium metal in refluxing xylene produces a highly reactive indium metal [99]. Ethereal solvents do not work as there appears to be extensive reductive cleavage of the solvent. The reduction is usually complete within 4 to 6 h. The reaction of Rieke indium with alkyl iodides was found to proceed very rapidly, and afforded nearly quantitative yields of the single product dialkylindium iodide [100]. Furthermore, Rieke indium reacted with iodobenzene to give high yields of diphenylindium iodide. In fact, even bromobenzene will react with active indium, although slowly. Additional studies showed that Rieke indium reacted readily with α-haloesters to give the corresponding organoindium products in high yields. This new Reformatsky reagent was found to add to the carbonyl group of ketones and aldehydes to give β-hydroxyesters in high yields. These reactions are best carried out in nonpolar solvents and require an excess of the ketone or aldehyde substrate. It should be noted that the yields for alkyl aldehydes are generally poor, on the order of 25%. Rieke indium also reacts with diorganomercury compounds yielding the corresponding triaryl and trialkyl indium products in essentially quantitative yields. These reactions are usually completed within 3 h. By comparison, Dennis [101] attempted to prepare Me_3In from indium metal and Me_2Hg by heating at 100 °C for eight days without success. However, Rieke indium reacted with Me_2Hg within 3 h at 100 °C to give quantitative yields of Me_3In. Similarly, the preparation of triaryl indium compounds is accomplished in quantitative yields using Rieke indium.

1.5.3 Nickel

Initial attempts to prepare active nickel by the reduction of nickel salts with potassium in refluxing THF met with only limited success. Although finely divided particles of black nickel powders were obtained, they showed limited reactivity towards oxidative addition with carbon-halogen bonds. It should be noted that similar results were obtained for both palladium and platinum. However, when the reductions were performed at room temperature for 12 h in DME, with lithium metal and a catalytic amount of naphthalene (10 mole %) as an electron carrier, a highly reactive form of finely divided nickel powder was obtained [102]. The resulting black slurry settled after stirring ceased, allowing removal of the solvent and naphthalene via a cannula. The resulting Rieke nickel slurries were found to undergo oxidative addition with a wide variety of aryl, vinyl, and many alkyl carbon-halogen bonds. A variety of iodobenzenes and bromobenzenes reacted with Rieke nickel at 85 °C to give the corresponding biphenyl products in good to high yields. Substituents such as methoxy, chloro, cyano, and acetyl groups in the *para* position were all compatible with the reaction conditions employed. Zero-valent nickel complexes such as bis(1,5-cyclooctadiene)nickel or tetrakis(triphenylphosphine)nickel have been shown to be acceptable coupling reagents; however, these complexes are unstable and difficult to prepare. The use of Rieke nickel is an attractive and viable alternative for carrying out aryl coupling reactions, thereby eliminating the need for the arduous preparation of zero-valent nickel complexes. The homocoupling reaction of benzylic halides by Rieke nickel proceeded at room temperature. Thus 1,2-diarylethanes accommodating a variety of functional groups were easily prepared in good to high yields. Conversely, benzylic polyhalides were converted to the corresponding olefins via a vicinal dihalide intermediate. Rieke nickel was also shown to be useful for the dehalogenation of vicinal dihalides. Cross-coupling reactions of benzylic halides and acyl halides produced the corresponding ketones in good to high yields. In comparison, the corresponding Grignard reagent does not work well, since benzylmagnesium halides readily undergo dimerization. Moreover, the Grignard approach has the disadvantage of forming by-products from the addition of the Grignard reagent to the newly formed ketones. Hence, with the aid of Rieke nickel, the yields of benzyl ketones are good and the reaction will tolerate a wide variety of functionalities. When alkyl oxalyl chlorides were employed, symmetrical dibenzyl ketones were obtained in good yields. Usually, transition metal carbonyls or metal salts/carbon monoxide have been used for the synthesis of dibenzyl ketones. With the aid of Rieke nickel, the use of toxic nickel carbonyl or carbon monoxide/transition metal complexes can be avoided in a laboratory-scale preparation. Benzylic nickel halides prepared in situ by the oxidative addition of benzyl halides to Rieke nickel were found to give the corresponding α-hydroxyketones on reaction with 1,2-diketones in excellent yields. The reaction is general and will tolerate a wide range of functionalities.

Additional cross-coupling reactions mediated by Rieke nickel have been developed. 3-Arylpropanenitriles can be easily prepared in high yields from benzyl halide, bromoacetonitrile, and Rieke nickel. This reaction will tolerate ester, ketone, nitrile, and halogen functionalities in the benzylic halides. A Reformatsky-type reaction can also be carried out with α-haloacetonitriles. Rieke nickel reacts with α-haloacetonitriles and the resulting organonickel species adds to aryl and alkyl aldehydes to give the corresponding β-hydroxynitrile products in good yields upon aqueous work-up. While this organonickel species adds to aldehydes, it does not react with ketones, which may prove to be quite useful in the elaboration of complex organic molecules. In conclusion, Rieke nickel also reacts with α,α-dibromo-*o*-xylene at room temperature to generate *o*-xylylene, which can then undergo a Diels-Alder reaction with an electron-deficient olefin such as acrylonitrile, methyl acrylate, dialkyl maleate/fumarate, or maleic anhydride to give the corresponding tetrahydronaphthalene derivative in good to high yield.

1.6 Synthesis of Specialized Polymers and New Materials via Rieke Metals

1.6.1 Formation of Polyarylenes Mediated by Rieke Zinc

Polyarylenes, such as poly(*para*-phenylenes) (PPP) and polythiophenes (PTh), are a class of conjugated polymers with electrical conductivity and other interesting physical properties [103]. The synthetic methodology for the formation of polyarylenes includes electrochemical polymerization of aromatic compounds, oxidative polymerization of benzene or thiophene with an oxidant such as $FeCl_3$, and catalytic cross-coupling polymerization of Grignard reagents. A facile synthesis for polyarylenes was recently achieved in our laboratories utilizing Rieke zinc (Zn*) [104].

Rieke zinc was found to react chemoselectively with dihaloarenes (**1**) to yield quantitatively monoorganozinc arenes (**2**) as shown in Scheme 1-15. Significantly, no bis(halo-

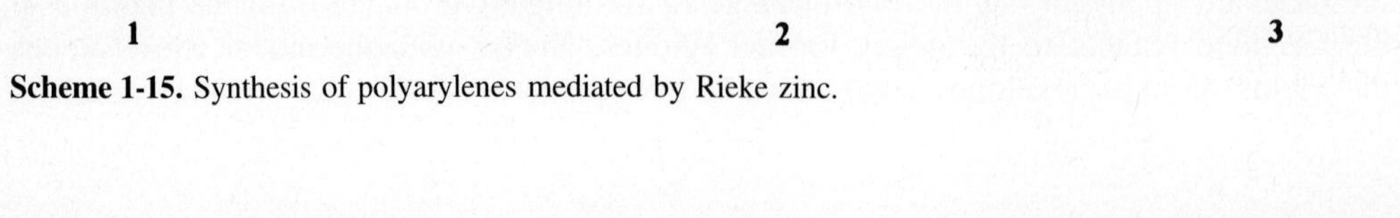

I–C6H4–I + 1.1 Zn* —(THF, Ar. / r.t. 30 min)→ IZn–C6H4–I —(Pd(0), 0.2 mol-% / Reflux 4 h)→ –(C6H4)–n

Br–C4H2S–Br + 1.1 Zn* —(THF, Ar. / r.t. 20 min)→ BrZn–C4H2S–Br —(Pd(0), 0.2 mol-% / Reflux 4 h)→ –(C4H2S)–n

1 **2** **3**

Scheme 1-15. Synthesis of polyarylenes mediated by Rieke zinc.

zincio)arene was formed. The organozinc intermediate **2** was polymerized upon the addition of a catalytic amount of $Pd(PPh_3)_4$ (0.2 mol-%) to give a quantitative yield of a polyarylene **3**, such as PPP or PTh. The method has been extended to synthesize a series of polymeric 3-substituted thiophenes such as poly(3-methylthiophene), poly(3-phenylthiophene), and poly(3-cyclopentylthiophene) as shown in Scheme 1-16. The advantages of this method for the synthesis of polyarylenes include the quantitative selectivity of Zn* for dihaloarenes, the use of only a very small amount of catalyst, which makes the separation of the polymer easier, mild reaction conditions, and high yields of the resulting polymers.

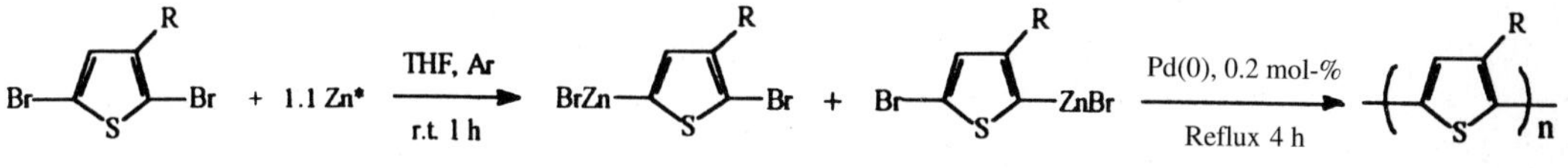

R	Regioselectivity (%)		Yield (%)
CH_3	**80**	**20**	**98**
Cyclopentyl	**71**	**29**	**96**
Phenyl	**66**	**34**	**98**

Scheme 1-16. Synthesis of poly(3-substituted thiophenes) mediated by Rieke zinc.

1.6.2 Regiocontrolled Synthesis of Poly(3-alkylthiophenes) and Related Polymers Mediated by Rieke Zinc

Regioregular head-to-tail (HT) poly(3-alkylthiophenes) (P3ATs) have been found to be one of the most highly conducting groups of polymers, and have recently been synthesized in our laboratories [105] and elsewhere [106]. Our synthesis involves a catalytic cross-coupling polymerization of a 2-bromo-5-(bromozincio)-3-alkylthiophene (**5**), which is generated by the regioselective oxidative addition of Zn* to the 2,5-dibromo-3-alkylthiophene (**4**) as shown in Scheme 1-17. The synthesis has been extended to a fully regiocontrolled synthesis, mediated by Rieke zinc, for a series of poly(3-alkylthiophenes) with different percentages of head-to-tail linkages in the polymer chain [107]. The organozinc intermediate (**5**) is polymerized either by $Ni(DPPE)Cl_2$ to afford a completely regioregular HT P3AT (**6**) or by $Pd(PPh_3)_4$ to afford a totally regiorandom HT-HH P3AT (**7**). The regioregular HT P3AT has a longer electronic absorption wavelength [108], a smaller band gap (1.7 eV), and a much higher iodine-doped film electrical-conductivity (~10^3 S cm^{-1}) than that of the regiorandom P3AT (band gap: 2.1 eV, iodine-doped film conductivity: ~5 S cm^{-1}). The cast film of regioregular P3AT is a flexible, crystalline, self-ordered, bronze-colored film with a metallic luster, whereas that of the regiorandom polymer is an amorphous, orange transparent film.

Scheme 1-18 depicts the regiocontrolled synthesis of poly(3-alkylthio-thiophenes) using Rieke zinc [109]. This novel regiocontrolled synthesis of poly(3-alkylthio-thiophenes) mediated by Rieke zinc is one of very few examples of a regiospecific synthesis for this class of polymers.

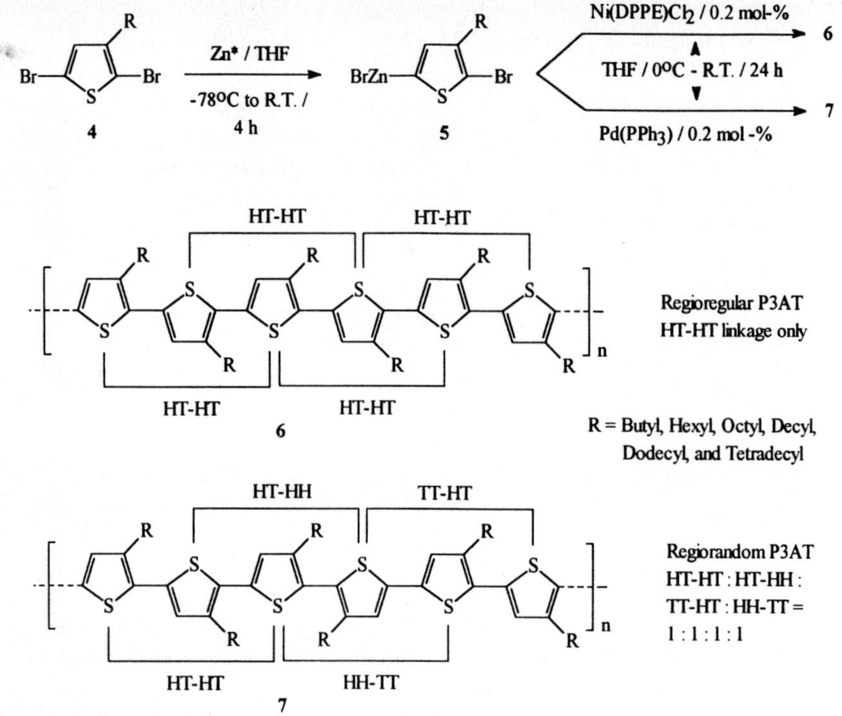

Scheme 1-17. Regiocontrolled synthesis of poly(3-alkylthiophenes) mediated by Rieke zinc.

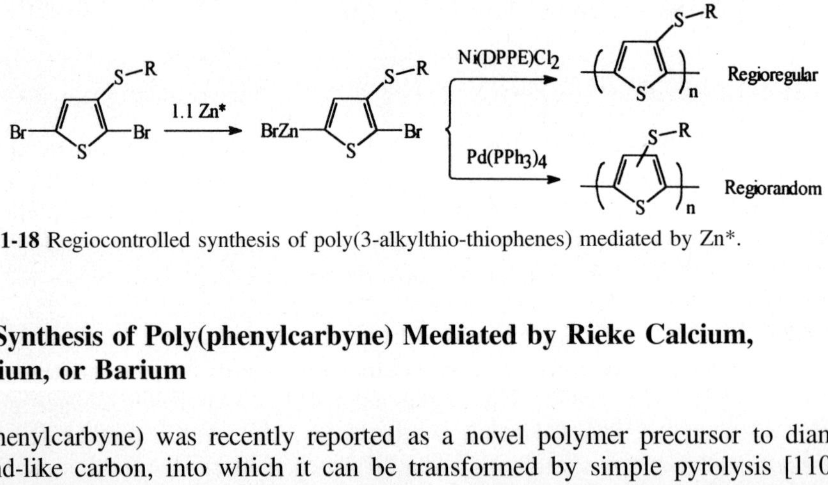

Scheme 1-18 Regiocontrolled synthesis of poly(3-alkylthio-thiophenes) mediated by Zn*.

1.6.3 Synthesis of Poly(phenylcarbyne) Mediated by Rieke Calcium, Strontium, or Barium

Poly(phenylcarbyne) was recently reported as a novel polymer precursor to diamond or diamond-like carbon, into which it can be transformed by simple pyrolysis [110]. "The discovery implies there is a chemical means of forming diamond without hydrogen or plasma. The discovery also implies a chemistry-based layer-by-layer growth of diamond is feasible" [111]. However, the original synthesis for poly(phenylcarbyne) requires high-intensity ultrasonic immersion equipment and the yield is only 25%, as shown in Scheme 1-19 (top).

A facile synthesis for poly(phenylcarbyne) was recently developed in our laboratories using Rieke metals. The synthesis was quite straightforward (Scheme 1-19, bottom). The reaction of α,α,α,-trichlorotoluene with Rieke calcium, strontium, or barium in THF at

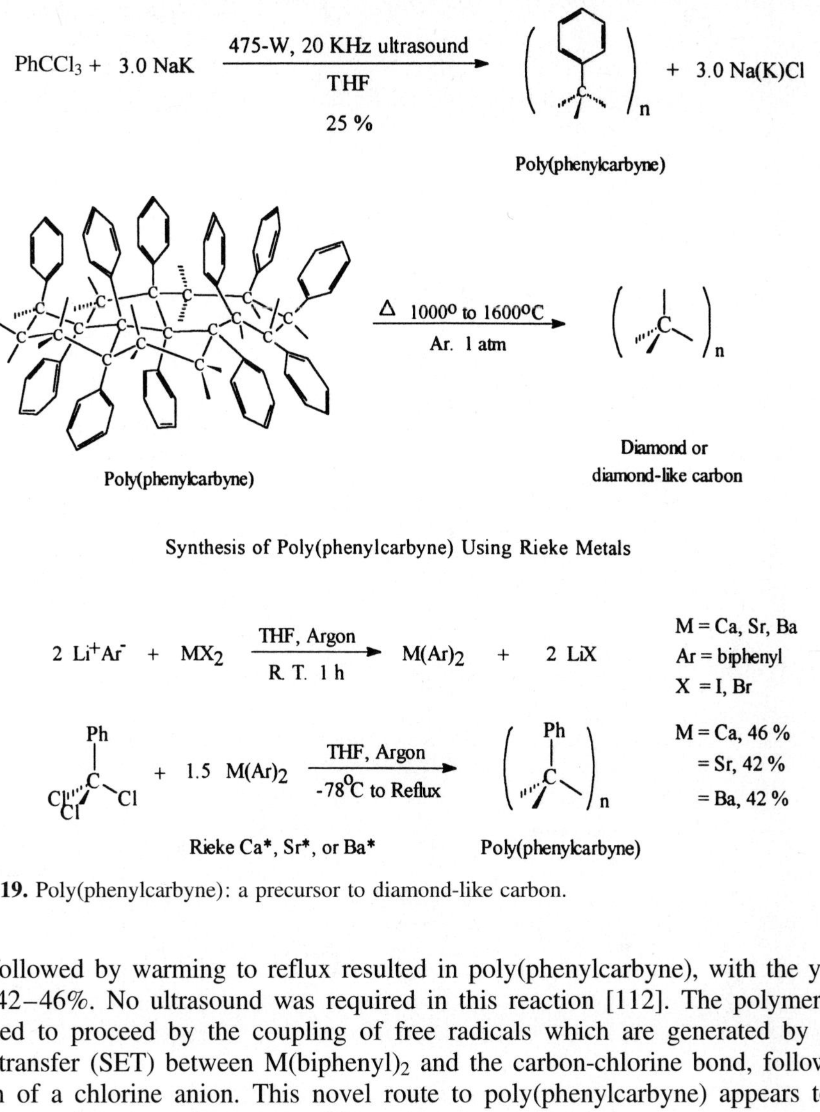

PhCCl$_3$ + 3.0 NaK —(475-W, 20 KHz ultrasound; THF; 25 %)→ Poly(phenylcarbyne) + 3.0 Na(K)Cl

Poly(phenylcarbyne) —(Δ 1000° to 1600°C; Ar. 1 atm)→ Diamond or diamond-like carbon

Synthesis of Poly(phenylcarbyne) Using Rieke Metals

2 Li$^+$Ar$^-$ + MX$_2$ —(THF, Argon; R. T. 1 h)→ M(Ar)$_2$ + 2 LiX M = Ca, Sr, Ba; Ar = biphenyl; X = I, Br

PhCCl$_3$ + 1.5 M(Ar)$_2$ —(THF, Argon; -78°C to Reflux)→ Poly(phenylcarbyne) M = Ca, 46 %; = Sr, 42 %; = Ba, 42 %

Rieke Ca*, Sr*, or Ba*

Scheme 1-19. Poly(phenylcarbyne): a precursor to diamond-like carbon.

–78 °C followed by warming to reflux resulted in poly(phenylcarbyne), with the yield as high as 42–46%. No ultrasound was required in this reaction [112]. The polymerization is assumed to proceed by the coupling of free radicals which are generated by single-electron transfer (SET) between M(biphenyl)$_2$ and the carbon-chlorine bond, followed by exclusion of a chlorine anion. This novel route to poly(phenylcarbyne) appears to be a general approach to several new materials.

1.6.4 Chemical Modification of Halogenated Polystyrenes Using Rieke Calcium or Copper

The chemical modification of cross-linked polymers has received considerable interest since the discovery of Merrifield's resin and its use in peptide synthesis. Cross-linked polystyrene resins have been mainstays in solid-phase synthesis and organic synthesis. Few successful examples of functionalizing cross-linked polystyrene resins have been reported, as the insoluble cross-linked polymers are particularly resistant to reagents. Rieke metals have been found to be an effective tool for the functionalization of cross-linked polymers.

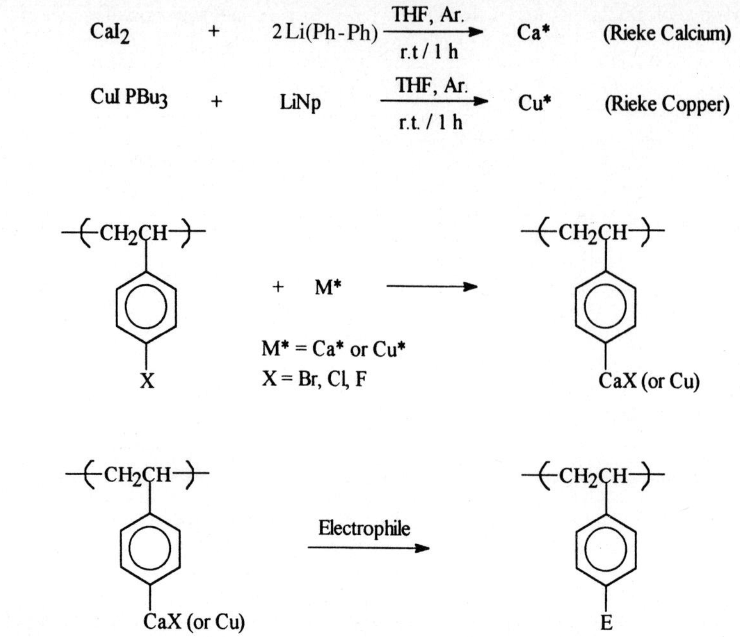

$$CaI_2 + 2\,Li(Ph\text{-}Ph) \xrightarrow[r.t/1\,h]{THF,\ Ar.} Ca^* \quad \text{(Rieke Calcium)}$$

$$CuI\,PBu_3 + LiNp \xrightarrow[r.t./1\,h]{THF,\ Ar.} Cu^* \quad \text{(Rieke Copper)}$$

$-(CH_2CH)-$ (aryl–X) + M* → $-(CH_2CH)-$ (aryl–CaX (or Cu))

M* = Ca* or Cu*
X = Br, Cl, F

$-(CH_2CH)-$ (aryl–CaX (or Cu)) —Electrophile→ $-(CH_2CH)-$ (aryl–E)

Scheme 1-20. Modifications of halogenated polystyrene resin using Rieke calcium or copper.

$-(CH_2CH)-$ (aryl–CH_2Cl) —Ca*→ $-(CH_2CH)-$ (aryl–CH_2CaCl) —CuCN·2LiBr→ $-(CH_2CH)-$ (aryl–$CH_2Cu(CN)CaCl$)

CH_2CaCl —Electrophile→ $-(CH_2CH)-$ (aryl–CH_2E)

$CH_2Cu(CN)CaCl$ —Electrophile→ $-(CH_2CH)-$ (aryl–CH_2E)

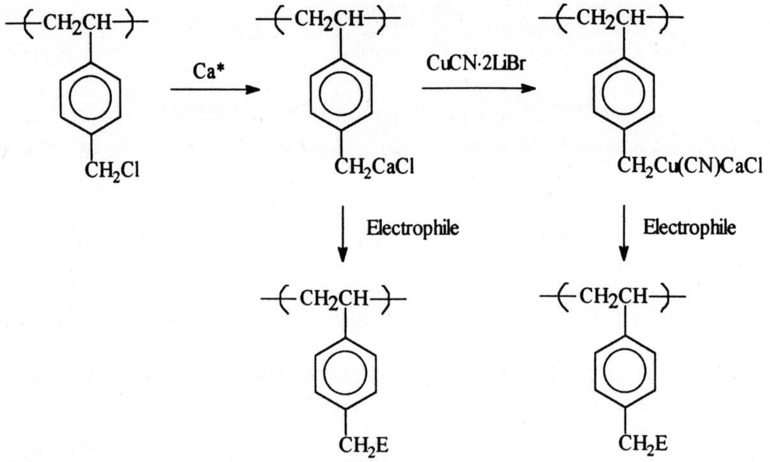

Scheme 1-21. Reactions of organocalcium and calcium cuprate reagents prepared from chloromethylated polystyrene and Rieke calcium.

Rieke calcium or copper undergoes direct oxidative addition to halogenated cross-linked polystyrene resins as shown in Schemes 1-20 and 1-21 [113]. The resulting insoluble polymeric aryl-calcium or -copper reagents can be used to carry out a number of useful transformations on polymers by reacting them with various electrophiles as shown in Table 1-34 [114].

Rieke calcium is highly reactive. Thus, fluorinated, chlorinated, brominated, and chloromethylated cross-linked polystyrene resins are all successfully converted to the corre-

Table 1-34. Reactions of organocalcium reagents prepared from *p*-halopolystyrenes and Rieke calcium.

Polystyrene polymer	Electrophile	Polystyrene product	Yield (%)
(P)-Br	CO_2	(P)-COOH	83
(P)-Br	$ClSiPh_3$	(P)-$SiPh_3$	72
(P)-Br	$ClPPh_2$	(P)-PPh_2	71
(P)-Br	C_6H_5CHO	(P)-$CH(OH)C_6H_5$	100
(P)-Cl	H_2O	(P)-H	100
(P)-Cl	CO_2	(P)-COOH	82
(P)-Cl	C_6H_5COCl	(P)-COC_6H_5	26
(P)-Cl	CH_3COCl	(P)-$COCH_3$	28
(P)-F	CO_2	(P)-COOH	78
(P)-F	$ClSiMe_3$	(P)-$SiMe_3$	26
(P)-F	CH_3COCl	(P)-$COCH_3$	59
(P)-F	C_6H_5CHO	(P)-$CH(OH)C_6H_5$	25

Table 1-35. Reactions of calcium cuprate reagents from chloromethylated polystyrene and Rieke calcium with various electrophiles.

Polystyrene polymer	Electrophile	Polystyrene product	Yield (%)
(P)-CH_2Cl	C_6H_5COCl	(P)-$CH_2COC_6H_5$	100
(P)-CH_2Cl	CH_3COCl	(P)-CH_2COCH_3	100
(P)-CH_2Cl	$Br(CH_2)_3CN$	(P)-$(CH_2)_4CN$	37
(P)-CH_2Cl	$Br(CH_2)_3CO_2Et$	(P)-$(CH_2)_4CO_2Et$	80
(P)-CH_2Cl	C_6H_8O	(P)-$CH_2C_6H_9O$	100

sponding calcium reagents. It should be noted that few metals undergo oxidative addition to aryl fluorides. Thus, it is noteworthy that Ca* reacts with *p*-fluoropolystyrene at room temperature to give the polymeric aryl-calcium reagent [115]. The polymeric calcium intermediates are converted to functionalized polystyrene resins by reacting with different electrophiles or converted to polymeric cuprate reagents, which then afford the ketone-containing polymers shown in Table 1-35.

1.6.5 Polymer Supported Rieke Metal Reagents and their Applications in Organic Synthesis

Rieke metals can be attached to polymers to form polymer supported Rieke metal reagents. Polymer supported Rieke copper, for example, can be prepared by attachment to a phosphine-bound polystyrene resin [116]. These polymer-supported Rieke copper reagents can be used as normal copper reagents in organic synthesis to afford organic compounds with various functional groups [117]. This methodology is significant, since it allows the stabilization, isolation, and long-term storage of Rieke metal reagents and organometallic compounds. The Rieke-metal-bound polymer materials themselves have some novel physical properties and potential applications, which are currently under investigation in our laboratories.

The basic approach for preparing polymer-bound triphenylphosphine Rieke copper reagent involves using lithium naphthalenide to reduce phosphine-bound polymer loaded

Table 1-36. Reactions of polymer-supported Rieke copper reagents with various alkyl/aryl halides and electrophiles.

RX	Electrophile	Product	Yield (%)
$Br(CH_2)_7CH_3$	PhCOCl	$PhCO(CH)_7CH_3$	78
$Br(CH_2)_3COOEt$	PhCOCl	$PhCO(CH_2)_3CO_2Et$	63
$Br(CH_2)_3CN$	PhCOCl	$PhCO(CH_2)_3CN$	73
$Br(CH_2)_6Cl$	PhCOCl	$PhCO(CH_2)_6Cl$	65
$Br(CH_2)_3COOEt$	ClCOOEt	$EtO_2C(CH_2)_3CO_2Et$	43
$Br(CH_2)_7CH_3$	PhCHO	$PhCH(OH)\text{-}C_8H_{17}$	45
$Br(CH_2)_7CH_3$	1,2-Epoxybutane	3-Dodecanol	79
p-IC_6H_4COOEt	PhCOCl	*p*-$EtO_2CC_6H_4COPh$	45
p-$BrC_6H_4CH_3$	PhCOCl	*p*-$CH_3C_6H_4COPh$	56

with CuI. The polymer-supported Rieke copper reagent easily adds to alkyl halides to afford the corresponding polymer supported organocopper species, which are readily attacked by different electrophiles to yield the target organic compounds as shown in Scheme 1-22 and Table 1-36.

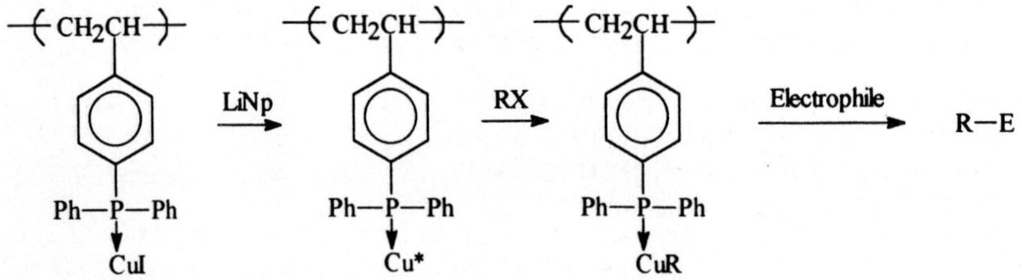

Scheme 1-22. Reactions of polymer-supported Rieke copper reagent.

In conclusion, Rieke metals have been found to be useful tools in the synthesis and modification of specialized polymers and new materials. The exploration of practical and potential applications of Rieke metals in this area is currently under investigation in our laboratories.

References

[1] R. D. Rieke, *CRC Critical Reviews in Surface Chemistry* **1991**, *1*, 131–166, and references therein.

[2] (a) R. Csuk, B. L. Glänzer, A. Fürstner, *Adv. Organomet. Chem.* **1988**, *28*, 85–137; (b) D. Savoia, C. Trombini, A. Umani-Ronchi, *Pure Appl. Chem.* **1985**, *57*, 1887–1896; (c) B. Bogdanovic, *Acc. Chem. Res.* **1988**, *21*, 261–267; (d) P. Marceau, L. Gautreau, F. Beguin, *J. Organomet. Chem.* **1991**, *403*, 21–27; (e) K. L. Tsai, J. Dye, *J. Am. Chem. Soc.* **1991**, *113*, 1650–1652.

[3] R. D. Rieke, K. Öfele, E. O. Fischer, *J. Organomet. Chem.* **1974**, *76*, C19-C21.

[4] T. Fujita, S. Watanaba, K. Suga, K. Sugahara, K. Tsuchimoto, *Chem. Ind.* (London) **1983**, *4*, 167–168.

[5] E. Burkhardt, R. D. Rieke, *J. Org. Chem.* **1985**, *50*, 416–417.

[6] (a) B. E. Kahn, R. D. Rieke, *Organometallics* **1988**, *7*, 463–469; (b) B. E. Kahn, R. D. Rieke, *J. Organomet. Chem.* **1988**, *346*, C45-C48.

[7] (a) R. D. Rieke, *Top. Curr. Chem.* **1975**, *59*, 1–31; (b) R. D. Rieke, *Acc. Chem. Res.* **1977**, *10*, 301–306.

[8] G. L. Rochfort, R. D. Rieke, *Inorg. Chem.* **1986**, *25*, 348–355.

[9] A. V. Kavaliunas, A. Taylor, R. D. Rieke, *Organometallics* **1983**, *2*, 377–383.
[10] M. S. Kharasch, O. Reinmuth, *Grignard Reactions of Non-Metallic Substances*, Prentice-Hall, New York, **1954**.
[11] M. S. Sell, M. V. Hanson, R. D. Rieke, *Synth. Commun.* **1994**, *24*, 2379–2386.
[12] T. Yamamoto, A. Yamamoto, *Chem. Lett.* **1977**, 353–356.
[13] S. T. Ioffe, A. N. Nesmayanov, *Methods of Elemento-Organic Chemistry, Vol. 2*, North Holland, Amsterdam, **1967**.
[14] (a) M. Gomberg, L. H. Cove, *Ber.* **1906**, *39*, 3274–3297. (b) H. S. Pink, *J. Chem. Soc.* **1923**, *123*, 3418–3420. (c) E. St. John, N. St. John, *Recl. Trav. Chim. Pays-Bas* **1936**, *55*, 585–588. (d) H. Normant, *C. R. Acad. Sci.* **1954**, *239*, 1510.
[15] E. Krause, K. Weinberg, *Ber.* **1929**, *62*, 2235–2241.
[16] T. C. Wu, H. Xiong, R. D. Rieke, *J. Org. Chem.* **1990**, *55*, 5045–5051.
[17] R. D. Rieke, S. E. Bales, *J. Am. Chem. Soc.* **1974**, *96*, 1775–1781.
[18] S. H. Yu, E. C. Ashby, *J. Org. Chem.* **1971**, *36*, 2123–2128, and references cited therein.
[19] T. P. Burns, R. D. Rieke, *J. Org. Chem.* **1983**, *48*, 4141–4143.
[20] C. Giordano, A. Belli, *Synthesis* **1975**, 167–168.
[21] R. R. Schmidt, W. J. W. Mayer, H. U. Wagner, *Liebigs Ann. Chem.* **1973**, 2010–2024.
[22] B. Radzisewski, *Ber.* **1892**, *15*, 1493–1497.
[23] (a) K. Fujita, Y. Ohnuma, H. Yasuda, H. Tani, *J. Organomet. Chem.* **1976**, *113*, 201–213. (b) M. Yang, K. Yamamoto, N. Otake, M. Ando, K. Takase, *Tetrahedron Lett.* **1970**, *44*, 3843–3846. (c) Y. Nakano, K. Natsukawa, K. Yasuda, H. Tani, *Tetrahedron Lett.* **1972**, *28*, 2833–2836.
[24] G. E. Herberich, W. Boveleth, B. Hessner, M. Hostalek, D. P. J. Koeffer, H. Ohst, D. Soehnen, *Ber.* **1986**, *119*, 420–433.
[25] For relevant review articles, see (a) G. Erker, C. Krüger, G. Müller, *Adv. Organomet. Chem.* **1985**, *24*, 1–39. (b) H. Yasuda, K. Tasumi, A. Nakamura, *Acc. Chem. Res.* **1985**, *18*, 120–126.
[26] Y. Kai, N. Kanehisa, K. Miki, N. Kasai, K. Mashima, H. Yasuda, A. Nakamura, *Chem. Lett.* **1982**, 1277–1280.
[27] R. D. Rieke, H. Xiong, *J. Org. Chem.* **1991**, *56*, 3109–3118.
[28] M. S. Sell, R. D. Rieke, unpublished results.
[29] R. D. Rieke, H. Xiong, *J. Org. Chem.* **1992**, *57*, 6560–6565.
[30] H. Xiong, R. D. Rieke, *J. Am. Chem. Soc.* **1992**, *114*, 4415–4417.
[31] (a) R. L. Danheiser, C. Martinex-Davila, J. M. Morin, Jr., *J. Org. Chem.* **1980**, *45*, 1340–1341. (b) R. L. Danheiser, C. Martinez-Davila, R. J. Auchus, J. T. Kadonaga, *J. Am. Chem. Soc.* **1981**, *103*, 2443–2446.
[32] For a review, see Z. Goldschmidt, B. Crammer, *Chem. Soc. Rev.* **1988**, *17*, 229–267. For representative references, see (a) H. H. Wasserman, R. E. Cochoy, M. S. Baird, *J. Am. Chem. Soc.* **1969**, *91*, 2375–2376. (b) B. M.. Trost, D. C. Lee, *J. Am. Chem. Soc.* **1988**, *110*, 6556–6558. (c) J. Ollivier, J. Salaun, *Tetrahedron Lett.* **1984**, *25*, 1269–1272. (d) B. M. Trost, M. K. Mao, *J. Am. Chem. Soc.* **1983**, *105*, 6753–6755.
[33] R. D. Rieke, M. S. Sell, H. Xiong, *J. Am. Chem. Soc.* **1995**, *117*, 5429–5437.
[34] (a) D. Alonso, J. Font, R. M. Ortuno, *J. Org. Chem.* **1991**, *56*, 5567–5572. (b) B. Mudryk, C. A. Shook, T. Cohen, *J. Am. Chem. Soc.* **1990**, *112*, 6389–6391. (c) W. E. Fristad, S. Hershberger, *J. Org. Chem.* **1985**, *50*, 1026–1031. (d) B. M. Trost, M. K. Mao, *J. Am. Chem. Soc.* **1983**, *105*, 6753–6755.
[35] P. Canonne, D. Belanger, G. Lemay, *J. Org. Chem.* **1982**, *47*, 3953–3959.
[36] (a) R. D. Rieke, M. S. Sell, H. Xiong, *J. Org. Chem.* **1995**, *60*, 5143–5149. (b) M. S. Sell, H. Xiong, R. D. Rieke, *Tetrahedron Lett.* **1993**, *34*, 6007–6010; *Tetrahedron Lett.* **1993**, *34*, 6011–6012. (c) H. Xiong, R. D. Rieke, *J. Org. Chem.* **1992**, *57*, 7007–7008.
[37] M. S. Sell, H. Xiong, R. D. Rieke, *Tetrahedron Lett.* **1993**, *34*, 6007–6010.
[38] L. Set, D. Cheshire, D. L. Clive, *J. Chem. Soc. Chem. Commun.* **1985**, 1205- 1207.
[39] (a) N. C. Barua, R. R. Schmidt, *Synthesis* **1986**, 1067–1070. (b) C. M. Thompson, *Tetrahedron Lett.* **1987**, *28*, 4243–4246; (c) P. Canonne, D. Belanger, G. Lemay, G. B. Foscolos, *J. Org. Chem.* **1981**, *46*, 3091–3097.
[40] N. Le, M. Jones, F. Bickelhaupt, W. H. deWolf, *J. Am. Chem. Soc.* **1989**, *111*, 8691–8698.
[41] M. S. Sell, H. Xiong, R. D. Rieke, *Tetrahedron Lett.* **1993**, *34*, 6007–6010.
[42] M. S. Sell, H. Xiong, R. D. Rieke, *Tetrahedron Lett.* **1993**, *34*, 6011–6012.
[43] A. I. Laskin, H. A. Lechevalier, *CRC Handbook of Microbiology, 2nd Edition, Vol. 5,* CRC, New York **1984**, and references therein.
[44] L. A. Paquette, D. MacDonald, L. G. Anderson, J. Wright, *J. Am. Chem. Soc.* **1989**, *111*, 8037–8039, and references therein.

[45] R. L. Frank, W. R. Schmitz, B. Zeidman, *Org. Synth.* **1955**, *3*, 328–329.
[46] M. Mori, I. Oda, Y. Ban, *Tetrahedron Lett.* **1982**, *23*, 5315–5318.
[47] H. Nagashima, H. Wakamatsu, N. Ozaki, T. Ishii, M. Watanabe, T. Tajima, K. Itoh, *J. Org. Chem.* **1992**, *57*, 1682–1689.
[48] G. Stork, T. Mah, *Heterocycles* **1989**, *28*, 723–727.
[49] (a) M. D. Bachi, F. Frolow, C. Hornaert, *J. Org. Chem.* **1983**, *48*, 1841–1849. (b) J. K. Choi, D. J. Hart, *Tetrahedron* **1985**, *41*, 3959–3971. (c) S. J. Danishefsky, J. S. Panek, *J. Am. Chem. Soc.* **1987**, *109*, 917–918.
[50] M. S. Sell, W. R. Klein, R. D. Rieke, *J. Org. Chem.* **1995**, *60*, 1077–1080.
[51] R. D. Rieke, S. J. Sung, P. M. Hudnall, *J. Chem. Soc. Chem. Comm.*, **1973**, 269–270.
[52] R. D. Rieke, P. M. Hudnall, *J. Am. Chem. Soc.* **1972**, 7178–7179.
[53] R. D. Rieke, P. T. Li, T. P. Burns, S. T. Uhm, *J. Org. Chem.* **1981**, *46*, 4323–4324.
[54] R. D. Rieke, S. J. Uhm, *Synthesis*, **1975**, 452–453.
[55] L. Zhu, R. M. Wehmeyer, R. D. Rieke, *J. Org. Chem.* **1991**, *56*, 1445–1453.
[56] (a) P. Knochel, M. C. P. Yeh, S. C. Berk, J. Talbert, *J. Org. Chem.* **1988**, *53*, 2390–2392; (b) P. Knochel, M. C. P. Yeh, S. C. Berk, J. Talbert, *Tetrahedron Lett.* **1988**, *29*, 2395–2396; (c) P. Knochel, M. C. P. Yeh, W. M. Butler, S. C. Berk, *Tetrahedron Lett.* **1988**, *29*, 6693–6696.
[57] B. H. Lipshutz, D. A. Parker, S. L. Nguyen, K. E. McCarthy, J. C. Barton, S. E. Whitney, H. Kotsuki, *Tetrahedron* **1986**, *42*, 2873–2879.
[58] L. Zhu, R. D. Rieke, *Tetrahedron Lett.* **1991**, *32*, 2865–2866.
[59] L. Zhu, K. H. Shaughnessy, R. D. Rieke, *Synth. Commun.* **1993**, *23*, 525–529.
[60] B. S. Bronk, S. J. Lippard, R. L. Danheiser, *Organometallics* **1993**, *12*, 3340–3349.
[61] C. Meyer, I. Marek, G. Courtemanche, J. F. Normant, *Synlett*, **1993**, 266–268.
[62] C. Meyer, I. Marek, G. Courtemanche, J. F. Normant, *Tetrahedron Lett.* **1993**, *34*, 6053–6056.
[63] T. Sakamoto, Y. Kondo, N. Murata, H. Yamanaka, *Tetrahedron Lett.* **1992**, *33*, 5373–5374.
[64] T. Sakamoto, Y. Kondo, N. Takazawa, H. Yamanaka, *Tetrahedron Lett.* **1993**, *34*, 5955–5956.
[65] M. V. Hanson, J. D. Brown, Q. J. Niu, R. D. Rieke, *Tetrahedron Lett.* **1994**, *35*, 7205–7208.
[66] P. Knochel, M. C. P. Yeh, S. C. Berk, J. Talbert, *J. Org. Chem.* **1988**, *53*, 2390–2392.
[67] M. Hanson, R. D. Rieke, *Synth. Commun.* **1995**, *25*, 101–104.
[68] (a) F. Ullman, *Ann.* **1904**, *332*, 38; (b) P. E. Fanta, *Chem. Rev.* **1964**, *64*, 613; (c) R. G. Bacon, H. A. Hill, *Proc. Chem. Soc.* **1962**, 113.
[69] (a) R. D. Rieke, R. M. Wehmeyer, T. C. Wu, G. W. Ebert, *Tetrahedron* **1989**, *45*, 443–454; (b) G. W. Ebert, R. D. Rieke, *J. Org. Chem.* **1988**, *53*, 4482–4488; (c) R. M. Wehmeyer, R. D. Rieke, *Tetrahedron Lett.* **1988**, 29, 4513–4516; (d) T. C. Wu, R. D. Rieke, *Tetrahedron Lett.* **1988**, *29*, 6753–6756; (e) T. C. Wu, R. M. Wehmeyer, R. D. Rieke, *J. Org. Chem.* **1987**, *52*, 5057–5059; (f) R. M. Wehmeyer, R. D. Rieke, *J. Org. Chem.* **1987**, *52*, 5056–5057; (g) G. W. Ebert, R. D. Rieke, *J. Org. Chem.* **1984**, *49*, 5280–5282; (h) R. D. Rieke, L. D. Rhyne, *J. Org. Chem.* **1979**, *44*, 3445–3446; (i) R. D. Rieke, A. V. Kavaliunas, L. D. Rhyne, D. J. Frazier, *J. Am. Chem. Soc.* **1979**, *101*, 246–248; (j) R. D. Rieke, *CRC Critical Reviews in Surface Chemistry* **1991**, *1*, 131–136; *Science* **1989**, 1260–1264.
[70] R. D. Rieke, L. D. Rhyne, *J. Org. Chem.* **1979**, *44*, 3445–3446.
[71] R. D. Rieke, *CRC Critical Reviews in Surface Chemistry* **1991**, *1*, 131–136; *Science* **1989**, 1260–1264.
[72] G. W. Ebert, R. D. Rieke, *J. Org. Chem.* **1984**, *49*, 5280–5282.
[73] M. Suzuki, Y. Suzuki, T. Kawagishi, R. Noyori, *Tetrahedron Lett.* **1980**, 1247.
[74] (a) E. J. Corey, N. W. Boaz, *Tetrahedron Lett.* **1985**, *26*, 6019–6022; (b) A. Alexakis, J. Berlan, Y. Besace, *Tetrahedron Lett.* **1986**, *27*, 1047–1050.
[75] T. C. Wu, R. D. Rieke, *J. Org. Chem.* **1988**, *53*, 2381–2383.
[76] (a) F. O. Ginah, T. A. Donovan, S. D. Suchan, D. R. Pfening, G. W. Ebert, *J. Org. Chem.* **1990**, *55*, 584–589; (b) G. W. Ebert, W. R. Klein, *J. Org. Chem.* **1991**, *56*, 4744–4747; (c) G. W. Ebert, J. W. Cheasty, S. S. Tehrani, E. Aoud, *Organometallics* **1992**, *11*, 1560–1564; (d) G. W. Ebert, D. R. Pfening, S. D. Suchan, T. A. Donovan, *Tetrahedron Lett.* **1993**, *34*, 2279–2282.
[77] (a) J. G. Smith, *Synthesis* **1984**, 629–656; (b) G. Stork, J. F. Cohen, *J. Am. Chem. Soc.* **1974**, *96*, 5270–5272; (c) G. Stork, L. D. Cama, D. R. Coulson, *J. Am. Chem. Soc.* **1974**, *96*, 5268–5270.
[78] J. E. Baldwin, *J. Chem. Soc. Chem. Commun.* **1976**, 734–736.
[79] W. E. Parham, L. D. Jones, Y. A. Sayed, *J. Org. Chem.* **1976**, *41*, 1184–1186.
[80] W. E. Parham, L. D. Jones, Y. A. Sayed, *J. Org. Chem.* **1975**, *40*, 2394–2399.
[81] (a) R. J. Boatman, B. J. Whitlock, H. W. Whitlock Jr., *J. Am. Chem. Soc.* **1978**, *100*, 2935; *J. Am. Chem. Soc.* **1977**, *99*, 4822–4824; (b) P. D. Brewer, J. Tagat, C. A. Hergruetor, P. Helquist, *Tetrahedron Lett.* **1977**, 4573–4574.

[82] (a) R. D. Rieke, W. R. Klein, T. C. Wu, *J. Org. Chem.* **1993**, *58*, 2492–2500; (b) W. R. Klein, R. D. Rieke, *Synth. Commun.* **1992**, *18*, 2635–2644; (c) R. D. Rieke, T. C. Wu, D. E. Stinn, R. M. Wehmeyer, *Synth. Commun.* **1989**, *19*, 1833–1840.
[83] (a) W. Oppolzer, T. Stevenson, T. Godel, *Helv. Chim. Acta.* **1985**, *68*, 212–215; (b) W. Oppolzer, H. J. Löher, *Helv. Chim. Acta.* **1981**, *64*, 2808–2811.
[84] D. E. Stack, B. T. Dawson, R. D. Rieke, *J. Am. Chem. Soc.* **1992**, *114*, 5110–5116; *J. Am. Chem. Soc.* **1991**, *113*, 4672–4673.
[85] (a) G. M. Whitesides, W. F. Fischer, J. SanFilippo, C. M. Basche, H. O. House, *J. Am. Chem. Soc.* **1969**, *91*, 4871–4882; (b) B. H. Lipshutz, S. Whitney, J. A. Kozolowski, C. M. Breneman, *Tetrahedron Lett.* **1986**, *27*, 4273–4276.
[86] D. E. Stack, R. D. Rieke, *Tetrahedron Lett.* **1992**, *33*, 6575–6578.
[87] (a) D. E. Stack, W. R. Klein, R. D. Rieke, *Tetrahedron Lett.* **1993**, *34*, 3063–3066; (b) R. D. Rieke, B. T. Dawson, D. E. Stack, D. E. Stinn, *Synth. Commun.* **1990**, *20*, 2711–2721.
[88] (a) B. H. Lipshutz, R. Crow, S. H. Dimock, E. L. Ellsworth, *J. Am. Chem. Soc.* **1990**, *112*, 4063–4064; (b) B. H. Lipshutz, T. R. Elworthy, *J. Org. Chem.* **1990**, *55*, 1695–1696; (c) B. H. Lipshutz, E. L. Ellsworth, S. H. Dimock, *J. Org. Chem.* **1989**, *54*, 4977–4979; (d) P. Knochel, *J. Am. Chem. Soc.* **1989**, *111*, 6474–6476.
[89] B. H. Lipshutz, E. L. Ellsworth, S. H. Dimock, R. A. J. Smith, *J. Am. Chem. Soc.* **1990**, *112*, 4404–4410.
[90] (a) B. H. Lipshutz, R. Crow, S. H. Dimock, E. L. Ellsworth, *J. Am. Chem. Soc.* **1990**, *112*, 4063–4064; (b) B. H. Lipshutz, T. R. Elworthy, *J. Org. Chem.* **1990**, *55*, 1695–1696; (c) B. H. Lipshutz, E. L. Ellsworth, S. H. Dimock, *J. Org. Chem.* **1989**, *54*, 4977–4979.
[91] S. H. Bertz, *J. Am. Chem. Soc.* **1990**, *112*, 4031–4032.
[92] (a) A. Alexakis, J. Berlan, Y. Besace, *Tetrahedron Lett.* **1986**, *27*, 1047–1050; (b) E. J. Corey, N. W. Boaz, *Tetrahedron Lett.* **1985**, *26*, 6019–6022.
[93] R. D. Rieke, L. C. Chao, *Syn. React. Inorg. Metal-Org. Chem.* **1974**, *4*, 101–105.
[94] A. V. Grosse, J. M. Mavity, *J. Org. Chem.* **1940**, *5*, 106–121.
[95] D. Wittenberg, *Ann.* **1962**, *654*, 23–26.
[96] W. C. Schumb, H. I. Crane, *J. Am. Chem. Soc.* **1938**, *60*, 306–308.
[97] G. B. Deacon, J. C. Parrot, *Austral. J. Chem.* **1971**, *24*, 1771–1779.
[98] M. J. S. Gynane, L. G. Waterworth, I. J. Worrall, *J. Organomet. Chem.* **1972**, *40*, C9–C10.
[99] L. I. Chao, R. D. Rieke, *Syn. React. Inorg. Metal-Org. Chem.* **1975**, *5*, 165–173; *Syn. React. Inorg. Metal-Org. Chem.* **1974**, *4*, 373–378; *J. Org. Chem.* **1975**, *40*, 2253–2255.
[100] L. I. Chao, R. D. Rieke, *Syn. React, Inorg. Metal-Org. Chem.* **1974**, *4*, 101–105.
[101] L. M. Dennis, R. W. Work, E. G. Rochow, *J. Am. Chem. Soc.* **1934**, *56*, 1047–1049.
[102] (a) S. Inaba, H. Matsumoto, R. D. Rieke, *Tetrahedron Lett.* **1982**, *23*, 4215–4216; (b) A. V. Kavaliunas, A. Taylor, R. D. Rieke, *Organometallics* **1983**, *2*, 377–383; (c) H. Matsumoto, S. Inaba, R. D. Rieke, *J. Org. Chem.* **1983**, *48*, 840–843; (d) S. Inaba, R. D. Rieke, *Tetrahedron Lett.* **1983**, *24*, 2451–2452; (e) S. Inaba, H. Matsumoto, R. D. Rieke, *J. Org. Chem.* **1984**, *49*, 2093–2098; (f) S. Inaba, R. D. Rieke, *Chem. Lett.* **1984**, 25–28; *Synthesis*, **1984**, 842–843; *Synthesis*, **1984**, 844–845; *J. Org. Chem.* **1985**, *50*, 1373–1381; *Tetrahedron Lett.* **1985**, *26*, 155–156.
[103] (a) T. A. Skotheim (eds.), *Handbook of Conducting Polymers,* Marcel Dekker, New York, **1986**; (b) M. G. Kanatzidis, *Chem. Eng. News* **1990**, *(December 3)*, 36–54; (c) P. Kovacic, M. B. Jones, *Chem. Rev.* **1987**, *87*, 357–379.
[104] T.-A. Chen, R. A. O'Brien, R. D. Rieke, *Macromolecules* **1993**, *26*, 3462–3463.
[105] (a) T.-A. Chen, R. D. Rieke, *J. Am. Chem. Soc.* **1992**, *114*, 10087–10088; (b) T.-A. Chen, R. D. Rieke, *Synth. Met.* **1993**, *60*, 175–177; (c) T.-A. Chen, R. D. Rieke, *Polym. Prepr.* **1993**, *34*, 426–427; (d) R. Menon, C. O. Yoon, T.-A. Chen, D. Moses, R. D. Rieke, A. J. Heeger, *Proceedings of International Conference on Synthetic Metals,* July 1994.
[106] R. D. McCullough, R. D. Lowe, M. Jayaraman, D. L. Anderson, *J. Org. Chem.* **1993**, *58*, 904–912.
[107] R. D. Rieke, T.-A. Chen, X. Wu, to be published.
[108] D. J. Sandman, *Trends in Polym. Sci.* **1994**, *2*, 44–55.
[109] X. Wu, T.-A. Chen, R. D. Rieke, *Tetrahedron Lett.* **1994**, *in press.*
[110] G. T. Visscher, D. C. Nesting, J. V. Badding, P. A. Bianconi, *Science* **1993**, *260*, 1496–1499.
[111] *Diamond Depositions, Sci. Technol.* **1993**, *3*, 1–19.
[112] R. D. Rieke, T.-A. Chen, *Chem. Mater.* **1994**, *6*, 576–577.
[113] R. A. O'Brien, T.-A. Chen, R. D. Rieke, *J. Inorg. Organomet. Polym.* **1992**, *2*, 345–356.
[114] R. A. O'Brien, T.-A. Chen, R. D. Rieke, *J. Org. Chem.* **1992**, *57*, 2667–2677.
[115] C. U. Pittman, Jr., *Polym. News* **1993**, *18*, 79–80.
[116] R. A. O'Brien, A. K. Gupta, R. D. Rieke, R. K. Shoemaker, *Magn. Reson. Chem.* **1992**, *30*, 398–402.
[117] R. A. O'Brien, R. D. Rieke, *J. Org. Chem.* **1990**, *55*, 788–790.

2 Allylic Barium Reagents

Akira Yanagisawa and Hisashi Yamamoto

2.1 Introduction

Allylic metal compounds are favorable reagents for the formation of carbon-carbon bonds [1]. Although a large number of allylic organometallics have been developed for selective allylation reactions, the allylic organometallic compounds of heavier alkaline earth metals have found little application in organic synthesis. Actually, they do not offer any particular advantages over simple Grignard reagents [2]. We have been interested in the use of barium or strontium reagents with the anticipation that such species should exhibit stereochemical stability markedly different from that of the ordinary magnesium reagent [3]. Described herein are methods of generating allylic barium reagents stereospecifically and performing subsequent reactions of these anions with a variety of electrophiles in regio- and stereoselective ways (Scheme 2-1).

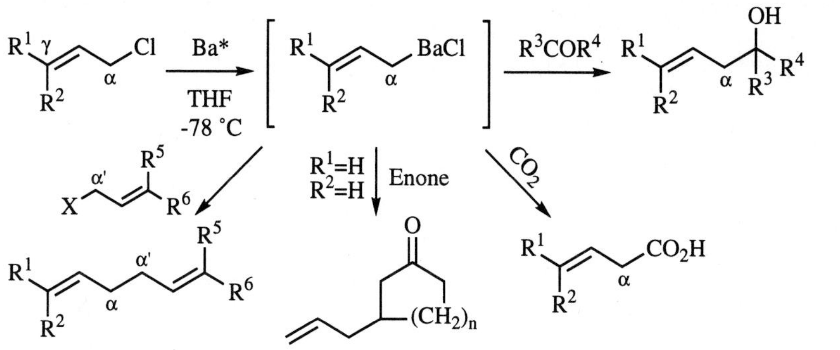

R^1 γ Cl R^2 α — Ba*, THF, -78 °C → [R^1 BaCl R^2 α] — R^3COR^4 → R^1 R^2 α OH R^3 R^4

X α' R^5 R^6 → R^1 R^2 α α' R^5 R^6

R^1=H, R^2=H, Enone → O, $(CH_2)_n$

CO_2 → R^1 R^2 α CO_2H

Scheme 2-1. Various allylation reactions using allylic barium reagents.

2.2 Preparation of Stereochemically Homogeneous Allylic Barium Reagents

2.2.1 Direct Insertion Method Using Reactive Barium

Reactive alkaline-earth metals (Mg and Ca) are often prepared according to Rieke's method, in which the corresponding anhydrous metal halide is reduced by lithium naphthalenide or biphenylide at room temperature [4, 5]. Treatment of allyl chloride with these reactive metals affords allylmagnesium and calcium reagents without any trouble. Allylbarium reagents can also be prepared from reactive barium by a procedure similar to that for calcium [5, 6]. Reactive barium is readily generated by the reduction of anhydrous barium iodide with two equivalents of lithium biphenylide in dry THF at room temperature for 30 min [Eq. (2.1)]. The dark brown suspension thus obtained is exposed to an allylic chloride at –78 °C. A slightly exothermic reaction takes place immediately to give a dark red solution of allylic barium reagent that can be used directly for the succeeding reaction [Eq. (2.2)].

$$BaI_2 + 2\,Li^+\,[Ph\text{-}Ph]^{\bar{\cdot}} \xrightarrow[\text{25 °C, 30 min}]{\text{THF}} Ba^* \qquad (2.1)$$

$$R^1(R^2)C{=}CH{-}CH_2Cl + Ba^* \xrightarrow[-78\text{ °C}]{\text{THF}} R^1(R^2)C{=}CH{-}CH_2BaCl \qquad (2.2)$$

2.2.2 Stereochemical Stability

In the realm of stereoselectivity, one great challenge not previously met was the preparation of stereochemically homogeneous allylic metals directly from allylic halides [7]. This is not a simple problem, since γ-substituted allylmetals are known to isomerize rapidly between the *E*- and *Z*-isomers, probably via metallotropic 1,3-rearrangements (Scheme 2.2) [8]. The facility of the isomerization makes it difficult to control the regio- and *E*/*Z*-stereochemistry of the subsequent reaction. We initially assumed that such metallotropic rearrangements of allylic metals were temperature-dependent. Thus, a γ-substituted allylic chloride was transformed into the corresponding barium reagent at –75, –55, –30, and 0 °C. The mixture was stirred for 30 min at each temperature and quenched with methanol to give a mixture of *E*- and *Z*-hydrocarbons. The rate of isomerization was measured by analyzing this hydrocarbon mixture (Scheme 2-3).

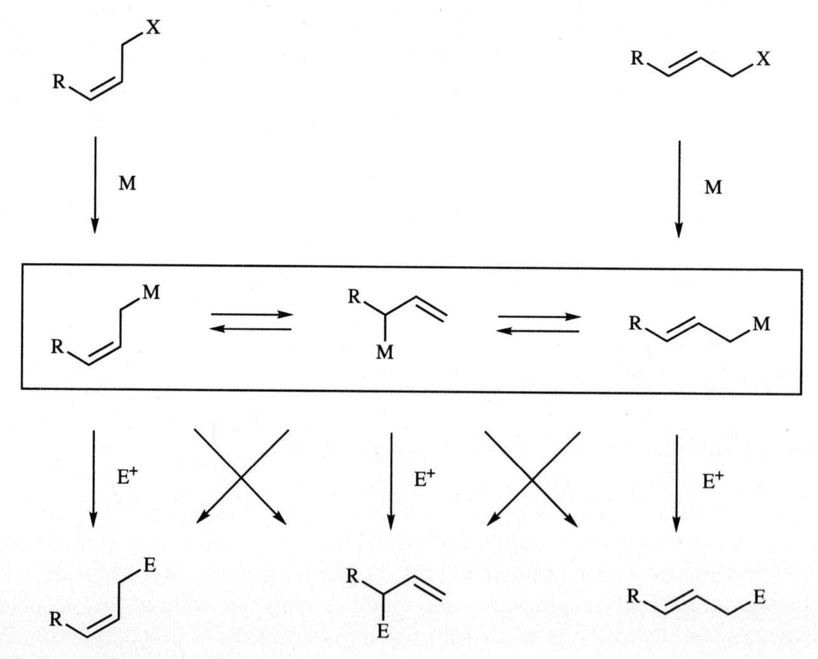

Scheme 2-2. Stereomutation of allylic metal reagents.

1) Ba*/THF
2) MeOH
-75 to 0 °C

Scheme 2-3 *E/Z* ratio

The results with geranyl and neryl barium compounds are shown in Fig. 2-1. Similarly, magnesium [4] and lithium [9] derivatives were prepared and quenched as above. The implications of the figure are apparent. There are two experimental variables (the temperature of the system and the choice of metal) and three consequences (the *E*/*Z* ratio of the olefins produced, the yield (%), and the α/γ ratio of the protonation products). Although there is no remarkable *E*/*Z* selectivity obtained by protonation of magnesium derivatives above –60 °C, extremely high stereoretention is observed below –95 °C [3]. In contrast, the double-bond geometry of the allylic barium compounds is retained even at –50 °C, a temperature higher than the corresponding lithium compounds [10]. The superiority of the barium reagent for stereoselectivity is thus apparent. It should be further

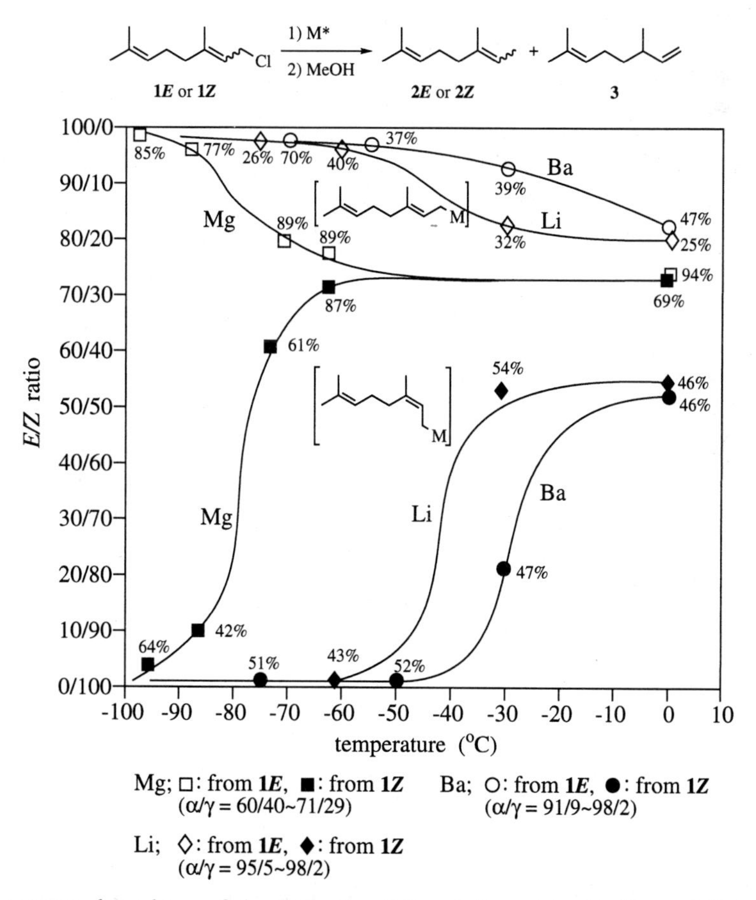

Fig. 2-1. Temperature dependence of the *E*/*Z* ratio of the allylic metals (Li, Mg, and Ba) derived from geranyl chloride (**1*E***; *E*/*Z* > 99/1) and neryl chloride (**1*Z***; *E*/*Z* < 1/99). Numbers refer to combined yields of the products **2** and **3**.

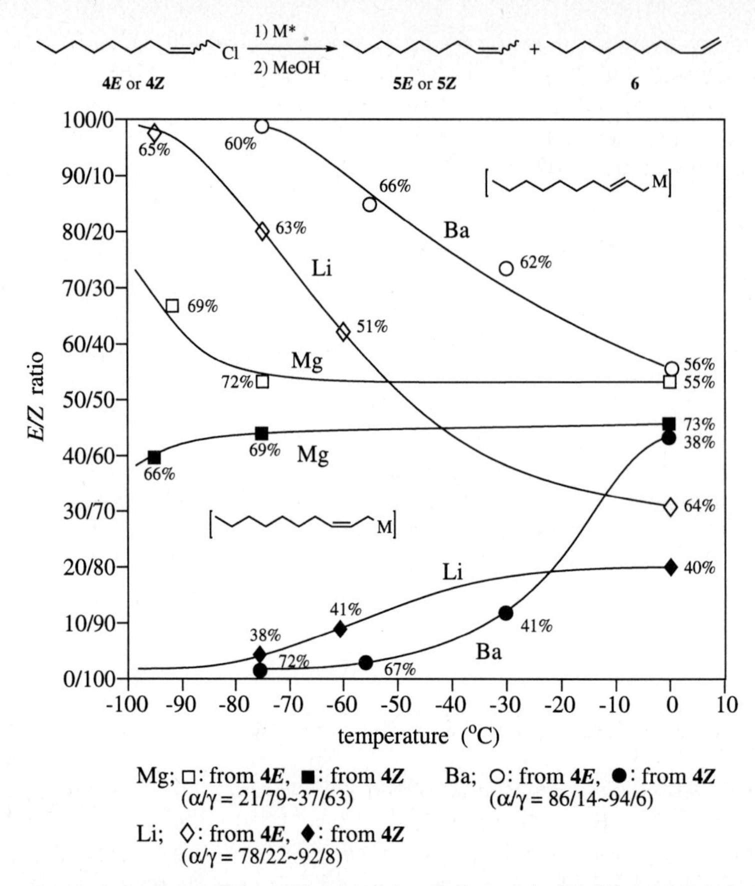

Fig. 2-2. Temperature dependence of the *E/Z* ratio of the allylic metals (Li, Mg, and Ba) derived from (*E*)-2-decenyl chloride (**4*E***; *E/Z* > 99/1) and (*Z*)-2-decenyl chloride (**4Z**; *E/Z* = 2/98). Numbers refer to combined yields of the products **5** and **6**.

noted that the combined yields of the derived olefins **2** and **3** are sufficiently high for practical purposes.

The temperature dependence of the *E/Z* ratio of 2-decenylmetals was also investigated, and the results are summarized in Fig. 2-2 [10]. In contrast to the γ-disubstituted allylmetals, a significant enhancement of the isomerization rate is observed for these γ-monosubstituted allylmetals, and rapid stereoisomerization of magnesium derivatives is found even at –100 °C [3]. Although the cause of this enhancement is not immediately apparent, it does indicate that barium at below –70 °C should be chosen rather than magnesium or lithium for the effective generation of configurationally homogeneous γ-monosubstituted allylmetals.

2.2.3 Silylation of Stereochemically Homogeneous Allylic Barium Reagents

The versatility of stereochemically homogeneous mono- and disubstituted allylmetals [11] in synthesis is noteworthy, as is their complementary relationship to other key functional groups. Stereochemically pure allylic silanes can be prepared readily from the corresponding barium derivatives [12]. Silylation of (*E*)-2-decenylbarium chloride (**7**), generated from reactive barium and (*E*)-2-decenyl chloride (**4*E***) at –78 °C, with chlorotrimethylsilane provides (*E*)-2-decenyl trimethylsilane (**8**) with an *E*/*Z* ratio of 97/3 (Scheme 2-4). The *cis*-isomer, (*Z*)-2-decenyl trimethylsilane (**9**), is also stereoselectively synthesized from (*Z*)-2-decenyl chloride (**4*Z***) by a similar experimental procedure (Scheme 2-5).

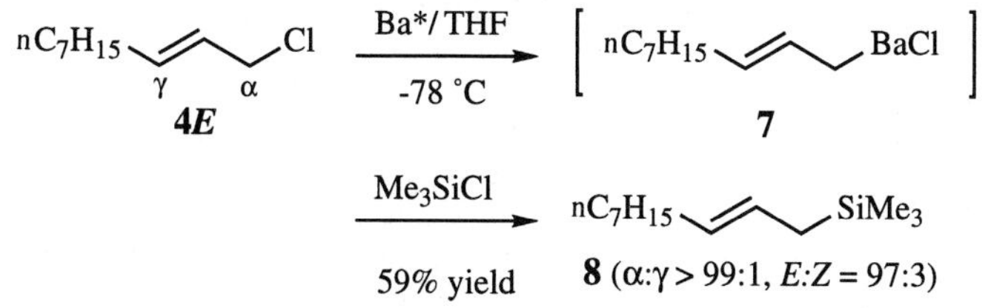

nC_7H_{15}–CH=CH–CH$_2$Cl (γ, α) **4*E*** —[Ba*/THF, -78 °C]→ [nC_7H_{15}–CH=CH–CH$_2$BaCl] **7** —[Me_3SiCl, 59% yield]→ nC_7H_{15}–CH=CH–CH$_2SiMe_3$ **8** (α:γ > 99:1, *E*:*Z* = 97:3)

Scheme 2-4

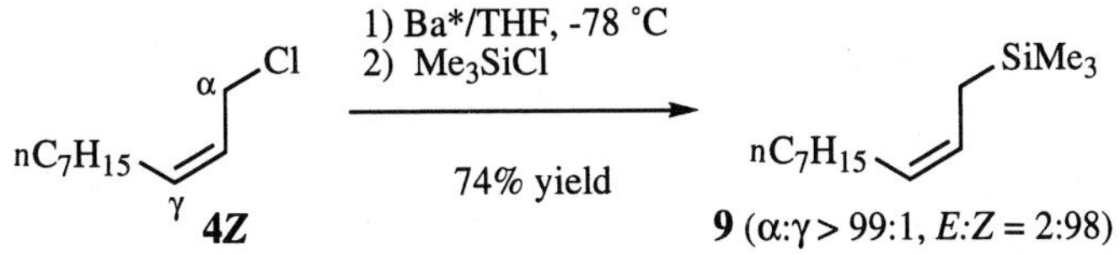

4*Z* (γ, α; Cl, nC_7H_{15}) —[1) Ba*/THF, -78 °C; 2) Me_3SiCl; 74% yield]→ **9** ($SiMe_3$, nC_7H_{15}) (α:γ > 99:1, *E*:*Z* = 2:98)

Scheme 2-5

2.2.3.1 Procedure for Generation of Reactive Barium (Ba*)

An oven-dried, 20-mL Schlenk tube equipped with a Teflon-coated magnetic stirring bar is flushed with argon. Freshly cut lithium (15 mg, 2.2 mmol) and biphenyl (350 mg, 2.3 mmol) are put into the apparatus and covered with dry THF (5 mL), and the mixture is stirred for 2 h at 20–25 °C (lithium was completely consumed) [9]. Into a separate oven-dried 50-mL Schlenk tube equipped with a Teflon-coated magnetic stirring bar is placed anhydrous BaI_2 (430 mg, 1.1 mmol) under argon atmosphere; this is covered with dry THF (5 mL) and stirred for 5 min at room temperature. To the resulting yellowish solution of BaI_2 in THF is added at room temperature a solution of the lithium biphenylide through a stainless steel cannula under an argon stream. The reaction mixture is stirred for 30 min at room temperature, and the resulting dark brown suspension of reactive barium thus prepared is ready to use.

2.2.3.2 Procedure for Protonation of the Geranyl Barium Reagent

To the suspension of reactive barium (1.1 mmol) in THF (10 mL) is slowly added a solution of geranyl chloride (**1*E***, 170 mg, 0.98 mmol) in dry THF (1.5 mL) at the specified temperature. A digital thermometer (Model HH81, OMEGA Engineering, Inc.) is used to measure the internal reaction temperature by immersing the thermocouple sensor into the reaction mixture. The reaction mixture is stirred for 30 min at this temperature and quenched by slow addition of MeOH (0.5 mL). After stirring for 5 min, 1 N HCl (10 mL) is added, and the aqueous layer is extracted with pentane (10 mL). The combined organic extracts are washed with 1 N sodium thiosulfate solution (20 mL), dried over anhydrous $MgSO_4$, and concentrated in vacuo after filtration. The residual oil is purified by flash-column chromatography on silica gel (hexane as eluent) to afford a mixture of olefins **2** and **3** (see Fig. 2-1); the α/γ and *E*/*Z* ratios are determined by GLC analysis. Typical results of the protonation are as follows (temperature, yield, α/γ ratio, *E*/*Z* ratio): 0 °C, 47%, 96/4, 82/18; –30 °C, 39%, 91/9, 93/7; –55 °C, 37%, 93/7, 97/3; –75 °C, 70%, 97/3, 98/2.

2.2.3.3 Silylation of (*E*)-2-Decenylbarium Chloride (7)

To the suspension of reactive barium (1.8 mmol) in dry THF (15 mL) is slowly added a solution of (*E*)-2-decenyl chloride (**4*E***, 280 mg, 1.6 mmol) in dry THF (2 mL) at –78 °C. After being stirred for 20 min, the mixture is treated with chlorotrimethylsilane (0.2 mL, 1.58 mmol) at –78 °C and stirred for another 20 min at this temperature. To the mixture is added a saturated NH_4Cl aqueous solution, and the organic material is extracted with pentane. The combined organic extracts are washed with 1 N sodium thiosulfate solution, dried over anhydrous $MgSO_4$, and concentrated in vacuo after filtration. The crude product is purified by flash-column chromatography on silica gel (hexane as eluent) to give (*E*)-2-decenyl trimethylsilane (**8**, 197 mg, 59% yield): the α/γ and *E*/*Z* ratios are determined to be >99/1 and 97/3, respectively, by GLC analysis (see Scheme 2-4).

2.3 Allylic Barium Reagents in Organic Synthesis

2.3.1. α-Selective and Stereospecific Allylation of Carbonyl Compounds

2.3.1.1 Metal Effects on α/γ-Selectivity

The key problem in utilization of an allylmetal reagent for the preparation of homoallylic compounds is the control of regio- and stereochemistry [13]. In general, γ-monosubstituted allylmetals react with carbonyl compounds exclusively at the γ-position, bulky ketones being the only exception [1a,14]. The α-regioselective synthesis of homoallylic alcohols from the corresponding allylic organometallics has remained a challenge in organic synthesis [15]. In contrast to the ordinary metal reagents, allylic barium reagents react with carbonyl compounds with high α-selectivity and stereospecificity (Scheme 2-6) [10, 16]. The α/γ-selectivity in the reaction of allylic organometallic reagents with carbonyl compounds depends on the choice of metal. Table 2-1 shows the results of the reaction of various geranylmetals with benzaldehyde [10]. It is well established that the allylic magnesium or calcium reagent gives the γ-substituted product **11** predominantly (entries 3 and 4) while the allylation with the lithium or potassium reagent is less selective (entries 1 and 2). In marked contrast, however, the barium reagent reacts with a remarkable α-selectivity ($\alpha/\gamma = 92/8$) and retention of the configuration of the starting halide **1*E*** (*E*/*Z* = 98/2, entry 6). Geranyl cerium reagent also shows a moderate α-selectivity (α/γ = 72/28, entry 8) [15b].

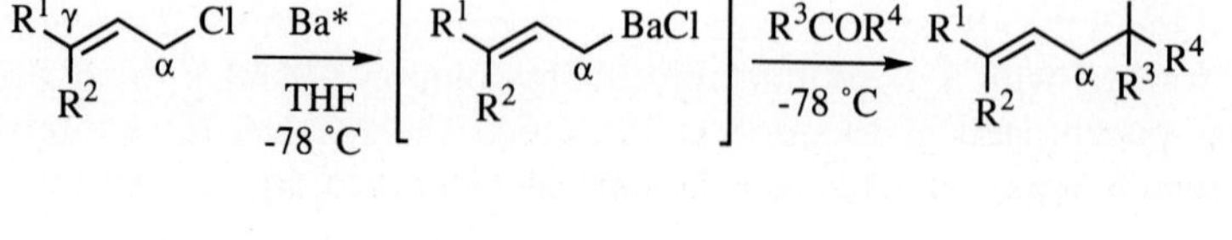

Scheme 2-6

Table 2-1. Reactions of various geranylmetals with benzaldehyde.

1*E* → (1) M*/THF, 2) PhCHO, −78 °C) → **10** (α-product) + **11** (γ-product)

Entry	M*	Combined yield, %[a]	α-product α : γ[b]	E : Z[b]
1	$Li^+[Ph\text{-}Ph]^{\cdot-}$	36	47 : 53	> 99 : 1
2	$K^+[Ph\text{-}Ph]^{\cdot-}$	35	67 : 33	98 : 2
3	Mg	99	< 1 : 99	–
4	Ca	70	12 : 88	98 : 2
5	Sr	89	54 : 46	97 : 3
6	Ba	90	92 : 8	98 : 2
7	Cu[c]	39	45 : 55	97 : 3
8	Ce	52	72 : 28	> 99 : 1

a) Isolated yield. b) Determined by GLC analysis. c) Prepared by reduction of $CuI \cdot PBu_3$ with lithium naphthalenide.

Table 2-2. α-Selective allylation of carbonyl compounds with γ-monosubstituted allylbarium reagents.

R^1-CH=CH-CH_2Cl → (1) Ba*, 2) R^2COR^3, THF, −78 °C) → α-product + γ-product

Entry	Allylic chloride	Carbonyl compound	Combined yield, %[a]	α-Product α : γ[b]	E : Z[b]
1	(*E*)-$nC_7H_{15}CH{=}CHCH_2Cl$	PhCHO	80	97 : 3	> 99 : 1
2		$nC_5H_{11}CHO$	82	98 : 2	97 : 3
3		(*E*)-PhCH=CHCHO	73[c]	94 : 6	98 : 2
4		Cyclohexanone	95	99 : 1	99 : 1
5		$PhCOCH_3$	94	96 : 4	99 : 1
6	(*Z*)-$nC_7H_{15}CH{=}CHCH_2Cl$	PhCHO	98	73 : 27	2 : 98
7		$nC_5H_{11}CHO$	75	86 : 14	2 : 98
8		Cyclohexanone	89	75 : 25	2 : 98
9		*i*PrCO*i*Pr	99	82 : 18	2 : 98
10		*t*BuCO*t*Bu	99	> 99 : 1	< 1 : 99
11		$nC_5H_{11}COSiMe_3$	89[d]	99 : 1	< 1 : 99
12	(*E*)-$CH_3CH{=}CHCH_2Cl$	$nC_5H_{11}CHO$	65	83 : 17	97 : 3
13	(*Z*)-$CH_3CH{=}CHCH_2Cl$	$nC_5H_{11}CHO$	56	77 : 23	1 : 99

a) Isolated yield. b) Determined by GLC analysis. c) 1,4-Adduct was also obtained in 14% yield.
d) The intermediate α-hydroxysilanes were desilylated with ^nBuNF in DMF.

2.3.1.2 Generality of α-Selectivity and Stereospecificity

The generality of the α-selectivity has been studied using various allylic barium reagents. Table 2-2 summarizes the results obtained for the reaction of a variety of carbonyl compounds with barium reagents generated from (*E*)- or (*Z*)-γ-monosubstituted allyl chlorides in THF at −78 °C [10, 16]. All reactions result in high yields with remarkable α-selectivities, not only with aldehydes but also with ketones. The double bond geometry of the starting allylic chloride is completely retained in each case. In the reaction with an α,β-unsaturated aldehyde, 1,2-addition proceeds preferentially (entry 3). (*Z*)-γ-Monosubstituted allylbarium shows relatively low α-selectivities in reactions with carbonyl compounds (entries 6–9 and 13). However, the condensation with bulky carbonyl compounds, 2,2,4,4-tetramethyl-3-pentanone [14], and *n*-hexanoyltrimethylsilane [15d, e] produces the α-product exclusively (entries 10 and 11).

The high α-regioselectivities and stereoselectivities are still observed for the reaction of γ,γ- and β,γ-disubstituted allylbarium reagents (Table 2-3) [10, 16]. The alkyl substitu-

Table 2-3. α-Selective allylation of carbonyl compounds with γ,γ- and β,γ-disubstituted allylbarium reagents.

1) Ba*; 2) R^4COR^5; THF, −78 °C → α-product + γ-product

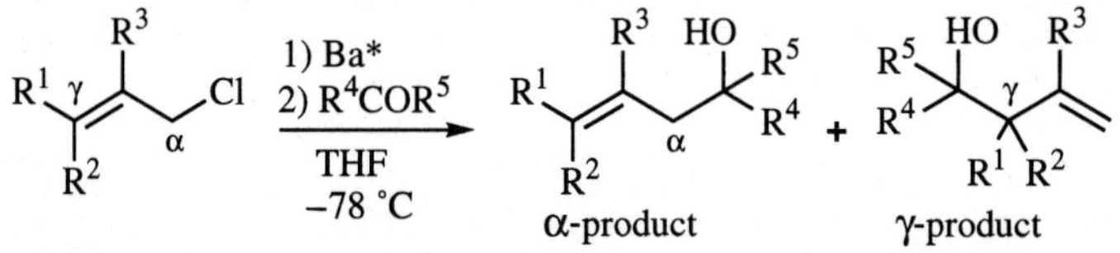

Entry	Allylic chloride	Carbonyl compound	Combined yield, %[a]	α-Product α : γ[b]	E : Z[b]
1	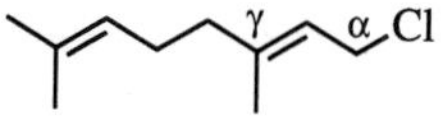	PhCHO	90	92 : 8	98 : 2
2		$nC_5H_{11}CHO$	90	94 : 6	> 99 : 1
3		Cyclohexanone	98	89 : 11	> 99 : 1
4		$PhCOCH_3$	96	93 : 7	99 : 1
5		PhCHO	89	94 : 6	2 : 98
6	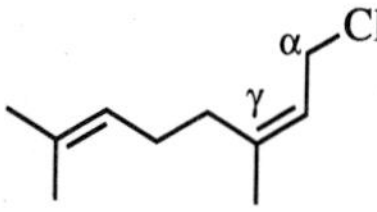	$nC_5H_{11}CHO$	73	96 : 4	< 1 : 99
7		Cyclohexanone	98	91 : 9	< 1 : 99
8		$PhCOCH_3$	84	94 : 6	< 1 : 99
9		$nC_5H_{11}CHO$	64	94 : 6	> 99 : 1
10	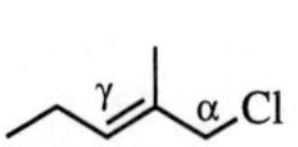	Cyclohexanone	92	96 : 4	99 : 1
11	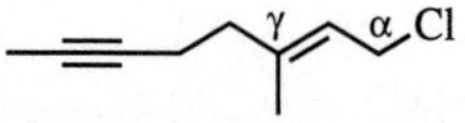	PhCHO	98	91 : 9	> 99 : 1
12	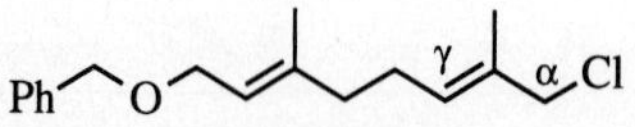	C_2H_5CHO	74	97 : 3	> 99 : 1

a) Isolated yield. b) Determined by GLC analysis.

ent at the β-position of allylic barium has no effect on the regioselectivity (entries 10 and 12). Existence of a triple bond or benzyl ether group in the allylic barium reagent does not interrupt its preparation or the course of the reaction (entries 11 and 12).

2.3.1.3 Secondary Allylic Barium Compounds

It is more difficult to control regio- and stereochemistries of secondary allylic metals than those of the corresponding primary allylic metals because of their rapid stereoisomerization. Generation of a stereochemically pure secondary allylic barium reagent was attempted using various secondary allylic chlorides [10]. Reaction of benzaldehyde with allylic barium reagent **13** derived from 3-chloro-1-butene (**12**) at –78 °C gives a 40:60 mixture of the α-product **14** and γ-product **15** in 98% yield (Scheme 2-7). At lower temperature (–95 °C), a slight increase of α-selectivity is observed ($\alpha/\gamma = 50/50$). Existence of two methyl groups at the γ-position of the secondary allylic chloride **16** proves effective for obtaining a higher regioselectivity ($\alpha/\gamma = 56/44$) (Scheme 2-8). The highest α-selectivity ($\alpha/\gamma = 91/9$) is gained using 2-chloro-4,8-dimethylnona-3,7-diene (**20**), which possesses a long alkyl chain at the γ-position (Scheme 2-9). A similar regioselectivity ($\alpha/\gamma = 92/8$) is observed in the condensation of the same barium reagent **21** with acetone, and the double bond geometry of the starting allylic chloride **20** is completely retained throughout the reaction ($E/Z > 99/1$).

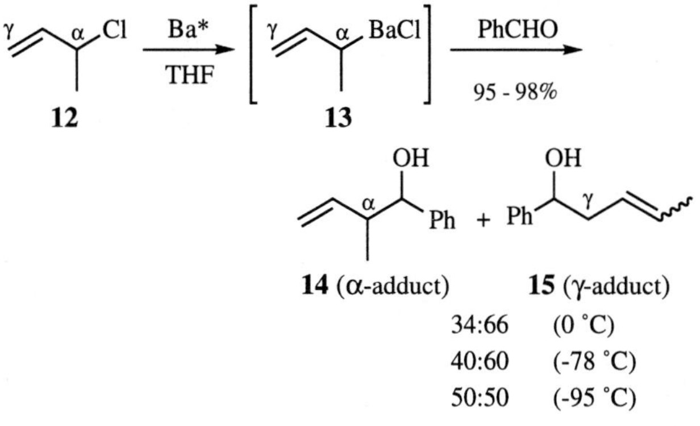

Scheme 2-7

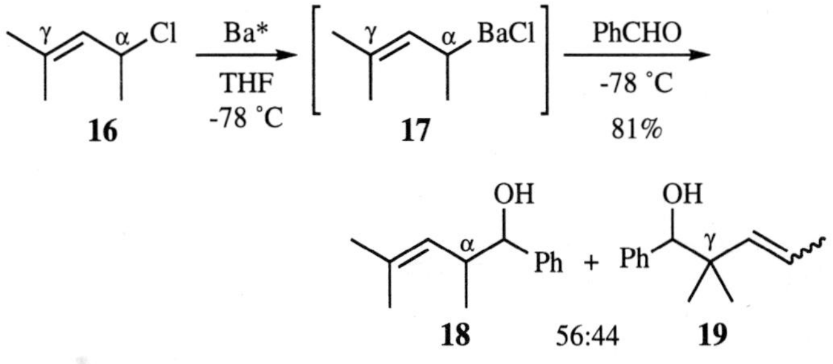

Scheme 2-8

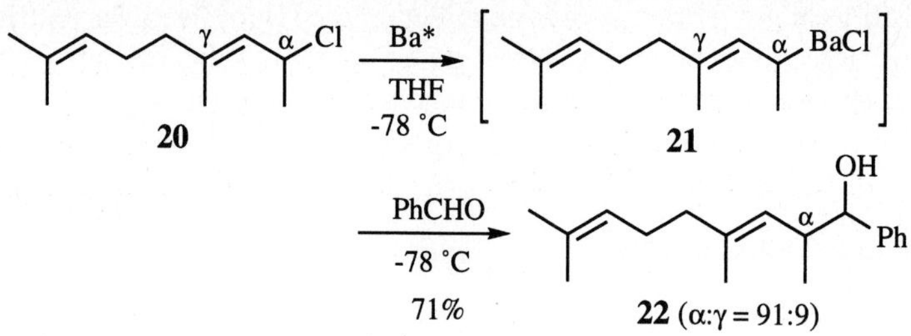

γ α Cl
Ba*
THF
-78 °C
20
γ α BaCl
21
PhCHO
-78 °C
71%
OH
α
Ph
22 (α:γ = 91:9)

Scheme 2-9

2.3.1.4 Mechanistic Considerations

Scheme 2-10 provides a graphical interpretation of the reaction pathway of primary allylic barium reagents with carbonyl compounds. Selective formation of the *E*-isomer of the α-product **F** from the (*E*)-allylic chloride **A** is readily accounted for by the oxidative addition of barium metal to the halide **A** to generate a primary (*E*)-allylic barium compound (**A** $\rightarrow$ **C**), followed by its condensation at the α-carbon with the carbonyl compound (**C** $\rightarrow$ **F**). Since stereoisomerization of the barium reagent (**C** $\rightarrow$ **D** $\rightarrow$ **E**) does not occur at -78 °C, the *Z*-isomer **H** of the α-product is not formed from the (*E*)-allylic chloride **A** at this temperature. The formation of the *Z*-isomer **H** of the α-product from the (*Z*)-allylic chloride **B** proceeds by a similar pathway, **B** $\rightarrow$ **E** $\rightarrow$ **H**. The minor γ-product **G** may arise from **C** or **E** via a six-membered cyclic transition structure. On the other hand, the secondary allylic barium reagent seems to isomerize rapidly between the α- and γ-forms even at low temperatures. The presence of a long alkyl substituent at the γ-position prevents such a metallotropic rearrangement.

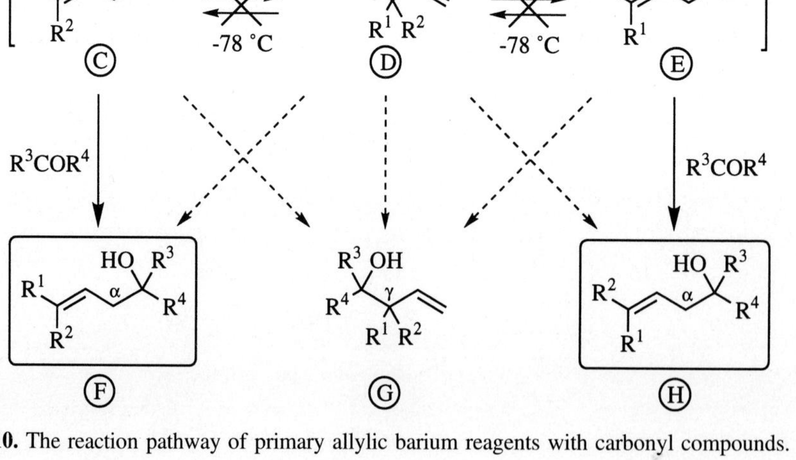

R^1 γ α Cl
R^2
A ($R^1 > R^2$)
R^2 γ α Cl
R^1
B ($R^1 > R^2$)
Ba*
Ba*
R^1 α BaCl
R^2
C
-78 °C
ClBa γ
R^1 R^2
D
-78 °C
R^2 α BaCl
R^1
E
R^3COR^4
R^3COR^4
HO R^3
R^1 α R^4
R^2
F
R^3 OH
R^4 γ
R^1 R^2
G
HO R^3
R^2 α R^4
R^1
H

Scheme 2-10. The reaction pathway of primary allylic barium reagents with carbonyl compounds.

Exactly what causes an allylic barium compound to react selectively at the α-carbon with a carbonyl compound is not yet clear; however, it is conceivable that the unusually long barium-carbon bond (2.76–2.88 Å) [17] might prevent the formation of a six-membered cyclic transition structure leading to the γ-product. A four-membered cyclic transition structure containing Ba–C and C=O bonds is one possible transition-state model for the α-selective allylation.

2.3.1.5 Typical Procedure for the Allylation of a Carbonyl Compound with an Allylic Barium Reagent: Synthesis of (*E*)-4,8-Dimethyl-1-phenyl-3,7-nonadien-1-ol

To the suspension of reactive barium (1.1 mmol) in THF (10 mL) is slowly added a solution of geranyl chloride (**1E**, 170 mg, 0.98 mmol) in THF (1.5 mL) at –78 °C. After being stirred for 20 min, the mixture is treated with a solution of benzaldehyde (40 µL, 0.39 mmol) in THF (1 mL) at –78 °C and stirred for another 20 min at this temperature. To the mixture is added 1 N HCl (10 mL), and the aqueous layer is extracted with ether (10 mL). The combined organic extracts are washed with 1 N sodium thiosulfate solution (20 mL), dried over anhydrous $MgSO_4$, and concentrated in vacuo after filtration. The crude product is purified by flash-column chromatography on silica gel (1:5 ethyl acetate/hexane as eluent) to afford a mixture of the homoallylic alcohols **10** and **11** (86 mg, 90% combined yield): the α/γ and *E*/*Z* ratios are determined to be 92/8 and 98/2, respectively, by GLC analysis (see Table 2-1, entry 6).

2.3.2 Regioselective and Stereospecific Synthesis of β,γ-Unsaturated Carboxylic Acids

β,γ-Unsaturated carboxylic acids and their derivatives are valuable intermediates in the synthesis of various natural products. The two most commonly used multi-stage processes for the synthesis of β,γ-unsaturated acids are (1) the Knoevenagel reaction/isomerization with base [18] and (2) allylic cyanide/hydrolysis [19]. Other methods have been developed [20–23], but most of these suffer from the problem of inadequate *E*/*Z* stereoselectivity. One straightforward way to obtain β,γ-unsaturated acids is by the carboxylation of an allylmetal reagent. In the substituted allylic series, the reaction usually occurs at the more sterically hindered terminus [1a, 24]. However, carboxylation of allylic barium reagent

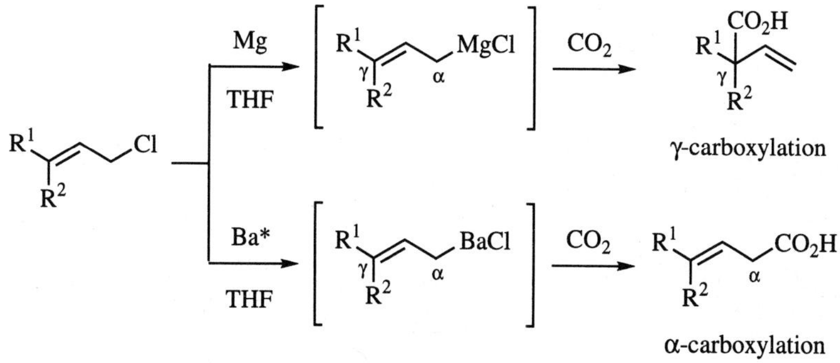

Scheme 2-11. α- and γ-carboxylations of allylic metal compounds.

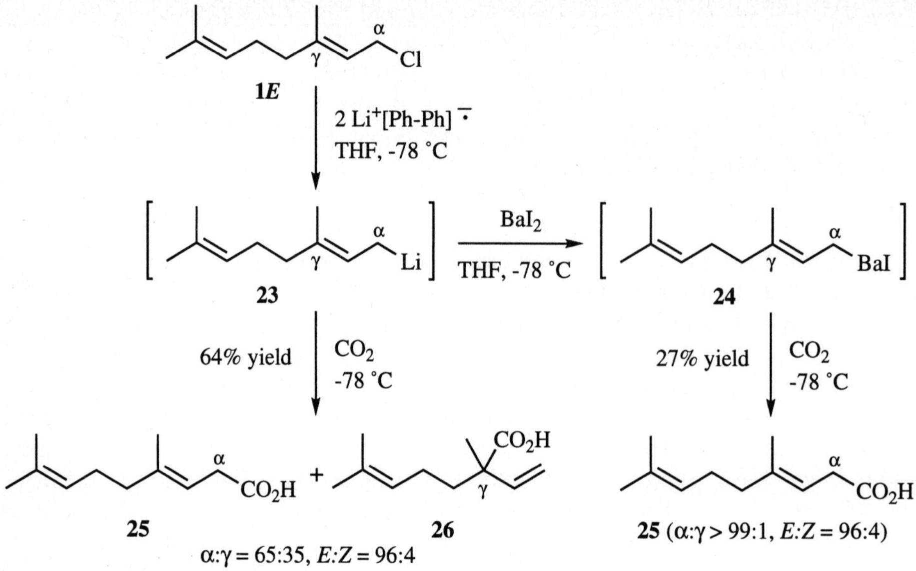

1*E* (Cl)

2 $Li^+[Ph\text{-}Ph]^{\overline{\cdot}}$, THF, -78 °C

[**23** (Li)] — BaI_2, THF, -78 °C → [**24** (BaI)]

64% yield; CO_2, -78 °C — 27% yield; CO_2, -78 °C

25 (CO_2H) + **26** (CO_2H) — α:γ = 65:35, *E*:*Z* = 96:4

25 (α:γ > 99:1, *E*:*Z* = 96:4)

Scheme 2-12. Carboxylation of geranylbarium reagent prepared by transmetallation.

shows α-selectivity without loss of the double bond geometry (Scheme 2-11) [10, 25]. Some results of carboxylation of allylic magnesium and barium reagents are summarized in Table 2-4. Allylic barium reagents generated from a variety of allylic chlorides show high α-selectivities, while in contrast the corresponding allylic magnesium reagents show γ-selectivities. The double-bond geometry of the allylbarium is completely retained in each case. The alkyl substituent at the β-position of allylic barium has no effect on the regioselectivity (entries 5 and 12). Use of geranyl barium **24**, prepared by transmetallation of geranyllithium **23** [3] with BaI_2, also results in the exclusive formation of the α-product **25** (Scheme 2-12). Carboxylation of geranyllithium **23** affords a lower α-selectivity (α/γ = 65/35).

2.3.2.1 Typical Procedure for Carboxylation of Allylic Barium Reagents: Synthesis of (*E*)-4,8-Dimethyl-3,7-nonadienoic Acid (25)

To the suspension of reactive barium (1.1 mmol) in THF (10 mL) is slowly added a solution of geranyl chloride (**1E**, 168 mg, 0.97 mmol) in THF (2 mL) at –78 °C. After stirring for 30 min, carbon dioxide is introduced into the solution through a stainless steel cannula under an argon stream for 30 min at –78 °C. 1 N HCl (10 mL) is added to the mixture at –78 °C, and the aqueous layer is extracted twice with ethyl acetate (2×20 mL). The combined organic extracts are washed with 1 N sodium thiosulfate solution (20 mL), dried over anhydrous $MgSO_4$, and concentrated in vacuo. The crude acid is purified by column chromatography on silica gel (ethyl acetate as eluent) to afford the β,γ-unsaturated carboxylic acid **25** (154 mg, 87% yield): the α/γ and *E*/*Z* ratios are determined to be >99/1 and 98/2 respectively by GLC analysis after conversion to the methyl ester with diazomethane in ether.

Table 2-4. Regio- and stereoselective carboxylation of allylic magnesium and barium reagents.

Entry	Allylic chloride	Metal	Combined yield, %[a]	α-Product	
				$\alpha : \gamma$[b]	$E : Z$[b]
1	(E)-n$C_7H_{15}\overset{\gamma}{C}H{=}CH\overset{\alpha}{C}H_2Cl$	Mg	99	< 1 : 99	
2		Ba	82	98 : 2	99 : 1
3	(Z)-n$C_7H_{15}\overset{\gamma}{C}H{=}CH\overset{\alpha}{C}H_2Cl$	Ba	65	82 : 18	1 : 99
4		Mg	32	< 1 : 99	
5		Ba	58	95 : 5	99 : 1
6	$(CH_3)_2\overset{\gamma}{C}{=}CH\overset{\alpha}{C}H_2Cl$	Mg	53	< 1 : 99	
7		Ba	59	> 99 : 1	
8		Mg	59	< 1 : 99	
9		Ba	87	> 99 : 1	98 : 2
10		Ba	51	> 99 : 1	< 1 : 99
11		Mg	65	< 1 : 99	
12		Ba	79	98 : 2	

a) Isolated yield. b) Determined by GLC analysis after conversion to methyl ester.

2.3.3 Highly *α,α′*-Selective Homocoupling and Cross-Coupling Reactions of Allylic Halides

Homo- and cross-coupling reactions of allylic halides are among the most basic carbon-carbon bond-forming methods in organic synthesis [1a, 26]. Highly α,α'-selective and stereocontrolled homocoupling and cross-coupling reactions of allylic halides have been achieved using barium reagent (Scheme 2-13) [27].

Various metal reagents have been examined for homocoupling of geranyl bromide (**27**) at low temperature (Table 2-5). Alkali metal naphthalenide (1.5 equiv.) or reactive alkaline-earth metal (0.7 equiv.) in THF is exposed to the bromide **27** (1 equiv.) at –95 °C or –78 °C. Among these metals, barium is found to be unique for the α,α-selective homocoupling reaction ($\alpha,\alpha'/\alpha,\gamma' = 97/3$, Table 2-5, entry 7). Furthermore, the configurational purity of the α,α' product (**28**/**29** = 96/1) indicates that configurational isomerization (*trans* to *cis*) of the allylic barium can be kept minimal during the coupling reaction.

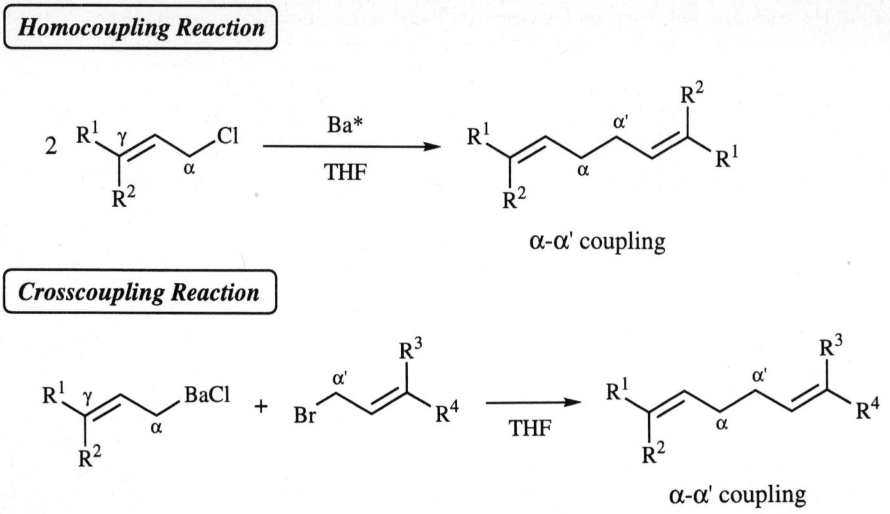

Homocoupling Reaction

2 R^1 γ α Cl —Ba*, THF→ R^1 R^2 α α' R^2 R^1

α-α' coupling

Crosscoupling Reaction

R^1 γ α BaCl R^2 + Br α' R^3 R^4 —THF→ R^1 R^2 α α' R^3 R^4

α-α' coupling

Scheme 2-13. Regio- and stereoselective 1,5-diene synthesis using barium metal.

Table 2-5. Effect of metals on the regio- and stereoselectivities of dimerization reaction of geranyl bromide.

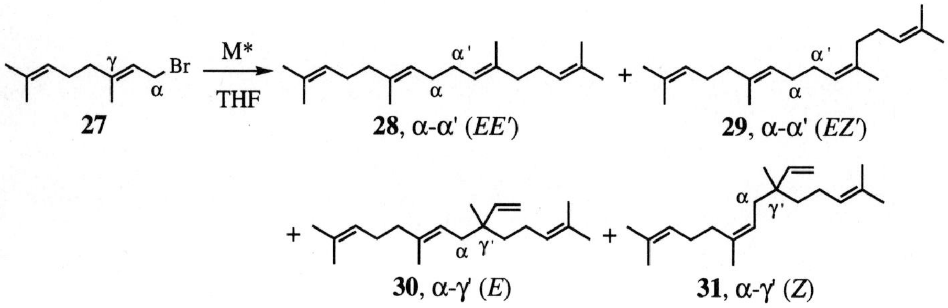

27 —M*, THF→ **28**, α-α' (*EE'*) + **29**, α-α' (*EZ'*) + **30**, α-γ' (*E*) + **31**, α-γ' (*Z*)

Entry	M*	*T*, °C	Combined yield, %[a]	Ratio (α-α'/α-γ)	Ratio of isomers[b]			
					28	**29**	**30**	**31**
1	Li-Np[c]	−95	62	69 : 31	67	2	31	0
2	Na-Np[c]	−95	86	61 : 39	60	1	39	0
3	K-Np[c]	−95	99	78 : 22	76	2	22	0
4	Cs-Np[c]	−95	27	65 : 35	65	0	35	0
5	Mg	−95	61	63 : 37	60	3	34	3
6	Ca	−78	58	62 : 38	58	4	38	0
7	Ba	−78	47	97 : 3	96	1	3	0
8	Cr	−40	85	74 : 26	68	6	26	0
9	Mn	−40	60	74 : 26	62	12	26	0

a) Isolated yield. b) Determined by GLC analysis. c) Np = naphthalene.

Table 2-6 summarizes the results obtained for the reactions of a variety of allylic halides **32** with reactive barium in THF at −78 °C. Reaction of -γ-mono- or disubstituted allyl halides results in >90% α,α'-selectivities, except with (*Z*)-2-decenyl chloride and

Table 2-6. Regio- and stereoselective dimerization reaction of various allylic halides using barium metal.

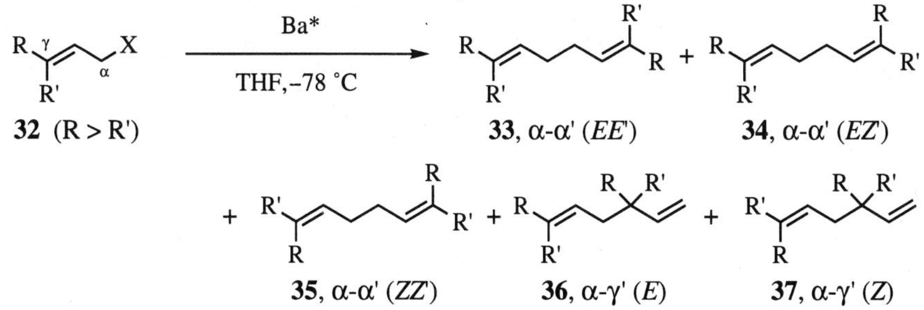

32 (R > R') —[Ba*, THF, –78 °C]→ **33**, α-α' (*EE*) + **34**, α-α' (*EZ*) + **35**, α-α' (*ZZ*) + **36**, α-γ' (*E*) + **37**, α-γ' (*Z*)

Entry	Allylic halide **32**	Combined yield, %[a]	Ratio (α-α'/α-γ')	Ratio of isomers[b] **33**	**34**	**35**	**36**	**37**
1	 nC_7H_{15} (E)-allyl Cl	86	95 : 5	95	0	0	5	0
2	 nC_7H_{15} (E)-allyl Br	68	92 : 8	92	0	0	8	0
3	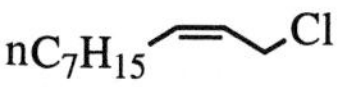 nC_7H_{15} (Z)-allyl Cl	88	51 : 49	0	0	51	0	49
4	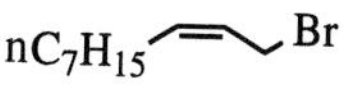 nC_7H_{15} (Z)-allyl Br	50	77 : 23	0	0	77	0	23
5	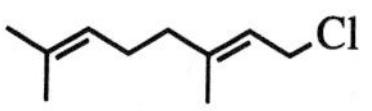 geranyl Cl	70	91 : 9	89	2	0	9	0
6	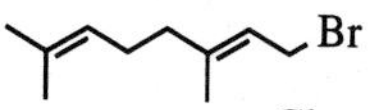 geranyl Br	47	97 : 3	96	1	0	3	0
7	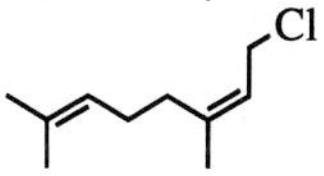 neryl Cl	44	92 : 8	0	2	90	0	8
8	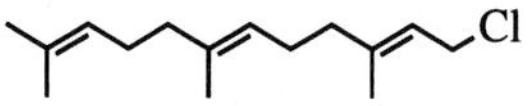 farnesyl Cl	68	94 : 6	94	0	0	6	0
9	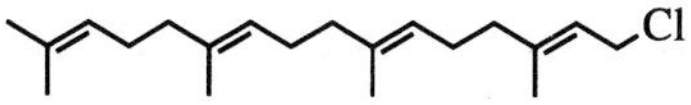 geranylgeranyl Cl	64	93 : 7	92	1	0	7	0

a) Isolated yield. b) Determined by GLC analysis.

bromide (entries 3 and 4). Both allylic chlorides and bromides can be used for the reaction with equal efficiency. The double-bond geometry of the starting allylic halide is completely retained in each case. (*E,E*)-Farnesyl chloride is stereospecifically converted to squalene in 64% yield (entry 8). *α,α′* cross-coupling products can also be prepared stereospecifically and regioselectively by this method. Treatment of (*E*)-2-decenylbarium reagent (**38**) with (*E*)-2-decenyl bromide and (*Z*)-2-decenyl bromide affords the (*E,E*)-diene **39** and the (*E,Z*)-diene **40** respectively in high yield (Schemes 2-14 and 2-15). The benzyl

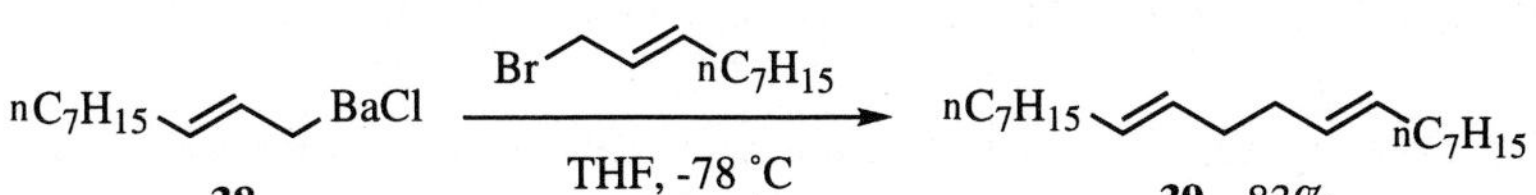

nC_7H_{15}–CH=CH–CH$_2$BaCl (**38**) —[Br–CH$_2$CH=CH–nC_7H_{15}, THF, -78 °C]→ **39**, 83%

Scheme 2-14

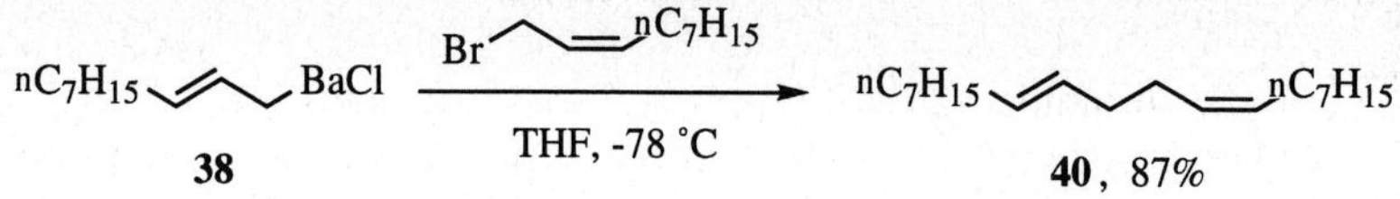

Br nC7H15; nC7H15 BaCl 38 → THF, -78 °C → nC7H15 nC7H15 40, 87%

Scheme 2-15

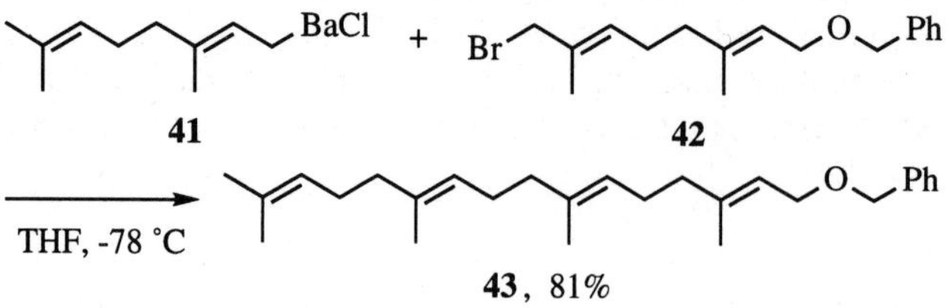

BaCl 41 + Br O Ph 42 → THF, -78 °C → O Ph 43, 81%

Scheme 2-16

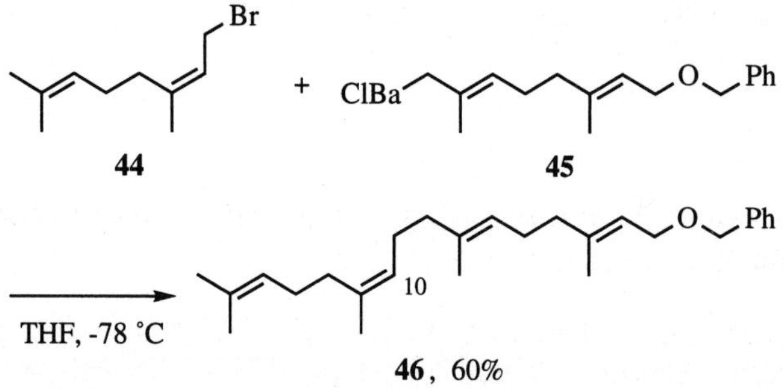

Br 44 + ClBa O Ph 45 → THF, -78 °C → O Ph 10 46, 60%

Scheme 2-17

Cl Br 47; BaCl 48, THF, -78 °C, 55% → Cl 49; Ba*, THF, -78°C → BaCl 50; Me Br O 51, THF, -78 °C, 66% → O 52

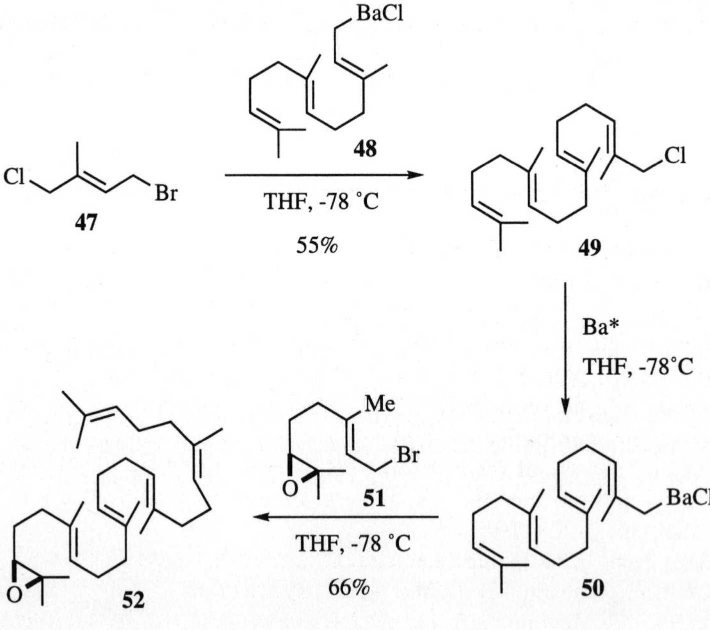

Scheme 2-18. Synthesis of (*3S*)-2,3-oxidosqualene.

ether of geranylgeraniol (**43**) is obtained almost exclusively in 81% yield on treatment of the primary allylic bromide **42** with geranylbarium reagent **41** in THF at –78 °C (Scheme 2-16). The functionalized allylic barium reagent **45** is readily prepared, and when allowed to react with neryl bromide (**44**) yields a (10*Z*)-isomer of benzyl geranylgeranyl ether, **46** (Scheme 2-17).

Corey and coworkers successfully applied this cross-coupling method to the synthesis of (3*S*)-2,3-oxidosqualene (**52**, Scheme 2-18) [28]. The utility of allylic barium reagents for nucleophilic substitution reactions was further demonstrated by the synthesis of cembrol A (**54**), in which ring closure of the epoxy barium compound occurred regioselectively (Scheme 2-19) [29].

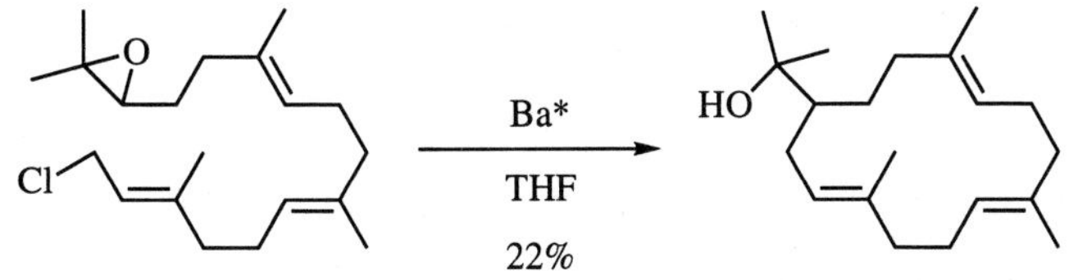

O
Cl
Ba*
THF
22%
HO

Scheme 2-19 **53** Cembrol A (**54**)

2.3.3.1 Typical Procedure for Homocoupling Reactions of Allylic Halides Using Reactive Barium: Synthesis of Squalene

To the suspension of reactive barium (1.1 mmol) in THF (8 mL) is slowly added a solution of (*E*,*E*)-farnesyl chloride (378 mg, 1.6 mmol) in THF (1.5 mL) at –78 °C. The reaction mixture is stirred for 1 h at this temperature. 2 N HCl (10 mL) is added to the mixture at –78 °C, and the aqueous layer is extracted with ether. The combined organic extracts are washed with 1 N sodium thiosulfate solution, dried over anhydrous $MgSO_4$, and concentrated in vacuo. The crude product is purified by column chromatography on silica gel (hexane as eluent) to afford a mixture of squalene and its regioisomer (220 mg, 68% combined yield): the $\alpha,\alpha'/\alpha,\gamma'$ ratio is determined to be 94/6 by GLC analysis.

2.3.4 Michael Addition Reaction

The conjugate addition of a carbanion to an α,β-enone is an extremely useful process for the introduction of a β-functionalized substituent [30]. However, allylic copper reagents are unstable and do not always give satisfactory results [31]. In 1977, Hosomi and Sakurai reported the smooth reaction of an allylsilane with an α,β-enone preferentially in a conjugate mode in the presence of titanium chloride as an activator of the enone, leading to a δ,ε-enone by simple protonolysis [32]. Although the process is exceedingly useful, sequential regio- and stereoselective alkylation is not possible under Lewis acidic conditions. By changing the metal from magnesium (or other ordinary metals) [1a, 30, 33] to barium, the dominant course of the reaction can be transformed from a 1,2- to a 1,4-addition reaction (Scheme 2-20) [10].

1,4/1,2 Regioselectivity in the reactions of various allylmetals **59** with 2-cyclopentenone (**55**) has been investigated (Table 2-7). Treatment of the allylmetal **59**, generated from reactive alkali or alkaline-earth metal (2 equiv.) and allyl chloride (**58**, 2 equiv.), with 2-cyclopentenone (**55**, 1 equiv.) at –78 °C in THF affords a mixture of the 1,4-adduct **57**

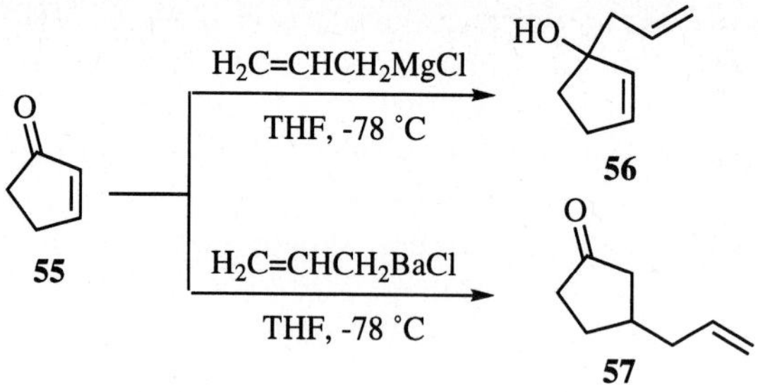

$H_2C{=}CHCH_2MgCl$ / THF, -78 °C; $H_2C{=}CHCH_2BaCl$ / THF, -78 °C

55 → **56**, **57**

Scheme 2-20. 1,2- and 1,4-addition reactions using allylic metals.

Table 2-7. Reaction of various allylmetals with 2-cyclopentenone.

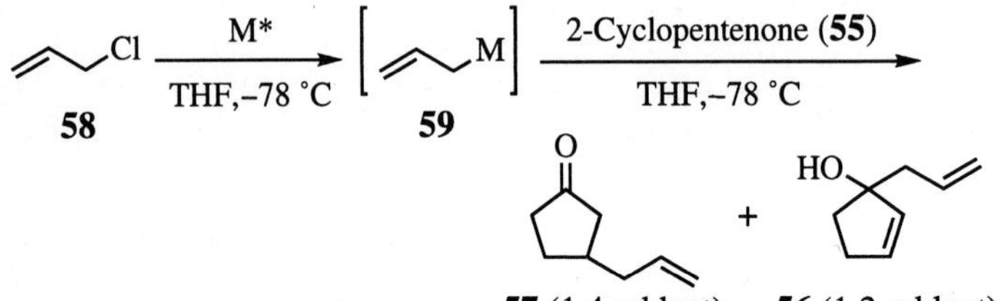

58 → (M*, THF, −78 °C) → [**59**] → (2-Cyclopentenone (**55**), THF, −78 °C) → **57** (1,4-adduct) + **56** (1,2-adduct)

Entry	M*	Combined yield %[a]	1,4 : 1,2[b]
1	$Li^+[Ph\text{-}Ph]^{\cdot-}$	79	< 1 : 99
2	$K^+Ph\text{-}Ph]^{\cdot-}$	38	62 : 38
3	Mg	81	5 : 95
4	Ca	65	28 : 72
5	Sr	63	67 : 33
6	Ba	94	> 99 : 1

a) Isolated yield. b) Determined by GLC analysis.

and the 1,2-adduct **56**. Among these allylmetals **59**, the allylbarium reagent is found to be uniquely suitable for the Michael addition reaction (1,4/1,2 > 99/1, entry 6). In contrast, allyllithium and allylmagnesium reagents show nearly exclusive 1,2-selectivities (entries 1 and 3). Reactions with allylpotassium and allylstrontium reagents result in moderate 1,4-selectivities (entries 2 and 5).

Some results of this reaction between allylic barium reagents and α,β-unsaturated ketones are listed in Table 2-8. Substituted allylbarium and benzylbarium reagents can be used for the reaction with 2-cyclopentenone with equal effectiveness (1,4/1,2 > 99/1, entries 1–5). The presence of a substituent at the C-2, C-3, or C-4 position of 2-cyclopentenone has no effect on the course of the reaction (entries 6–8). In the reaction of 2-cyclohexenone with allylbarium reagent, a lower temperature (< −95 °C) is necessary for obtaining a higher regioselectivity (1,4/1,2 = 89/11, entries 9 and 10). The presence of methyl groups at the C-6 position of 2-cyclohexenone raises the 1,4-selectivities (entries 14 and 15). Not all competitive 1,2-addition reactions can be effectively controlled with barium reagents. Thus, use of 2-cycloheptenone and acyclic enones as electrophiles resulted in relatively lower 1,4-selectivities (entries 16–21).

The in situ generated barium enolate has adequate nucleophilicity. Selected results of a double alkylation process [34] are summarized in Table 2-9: direct alkylation of the en-

Table 2-8. Conjugate addition reaction of allylic barium reagents to enones.

$R^1C(O)C(R^2)=CR^3R^4$ + R^5BaCl → (THF, –78 °C) 1,4-adduct + 1,2-adduct

Entry	Enone	Barium reagent	Yield, %[a]	1,4 : 1,2[b]
1	2-cyclopentenone	allyl-BaCl	94	> 99 : 1
2		2-methylallyl-BaCl	96	> 99 : 1
3		nC_7H_{15}CH=CHCH$_2$BaCl	69[c]	> 99 : 1
4		prenyl-BaCl	92[d]	> 99 : 1
5		$PhCH_2BaCl$	63	96 : 4
6	2-methyl-2-cyclopentenone	allyl-BaCl	75[e]	> 99 : 1
7	3-methyl-2-cyclopentenone	allyl-BaCl	86	> 99 : 1
8	4-(tBuMe$_2$SiO)-2-cyclopentenone	allyl-BaCl	96[f]	97 : 3
9	2-cyclohexenone	allyl-BaCl	93	79 : 21
10[g]		2-methylallyl-BaCl	99	89 : 11
11[g]	2-cyclohexenone		99	78 : 22
12[g]		prenyl-BaCl	91[h]	91 : 9
13[g]		$PhCH_2BaCl$	57	95 : 5
14[g]	6-methyl-2-cyclohexenone	allyl-BaCl	92[i]	92 : 8
15[g]	6,6-dimethyl-2-cyclohexenone	allyl-BaCl	97	99 : 1

Table 2-8. (Continued)

Entry	Enone	Barium reagent	Yield, %[a]	1,4 : 1,2[b]
16[g]	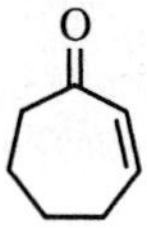 (cycloheptenone)	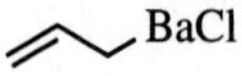 allyl-BaCl	99	42 : 58
17[g]	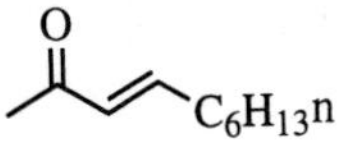 (enone, $C_6H_{13}n$)	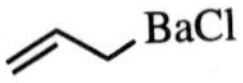 allyl-BaCl	99	40 : 60
18[g]	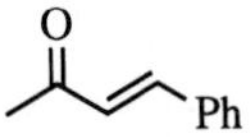 (enone, Ph)	 allyl-BaCl	93	57 : 43
19[g]	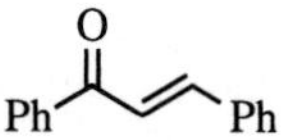 Ph (enone) Ph	 allyl-BaCl	95	46 : 54
20[g]	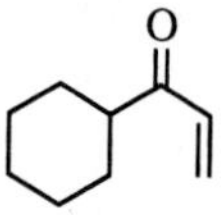 (cyclohexyl vinyl ketone)	 allyl-BaCl	92	62 : 38
21[g]	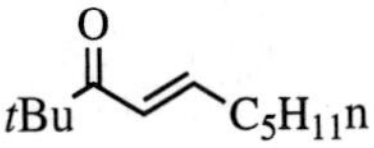 *t*Bu (enone) $C_5H_{11}n$	 allyl-BaCl	98	76 : 24

a) Isolated yield. b) Determined by GLC analysis. c) The α/γ and *E*/*Z* ratios were 73/27 and > 99/1, respectively. d) The α/γ ratio was 36/64. e) The trans/cis ratio was 62/38. f) The trans/cis ratio of 1,4-adduct was > 99/1. g) Run at −95 °C. h) The α/γ ratio of 1,4-adduct and 1,2-adduct were 44/56 and 68/32, respectively. i) Ratio of two diastereomers was 65/35.

olate is achieved by treating it with an excess of prenyl bromide or 2-octynyl bromide (entries 1, 2, 5, and 6) [35]. The aldol condensation and acylation reactions also proceed smoothly with the same facility (entries 3 and 4).

The hypothesis that the barium alkoxide **60** is formed as an intermediate, followed by an anionic oxy-Cope rearrangement, is refuted by the following experiments (Scheme 2-21) [31b]. Upon heating to 0 °C, no rearrangement product **57** is obtained from barium alkoxide **60** which is generated from the corresponding *tert*-alcohol **56**. Thus, the allylbarium reagent reacts directly at the *β*-carbon of enones, which reveals an additional unique feature of barium reagents.

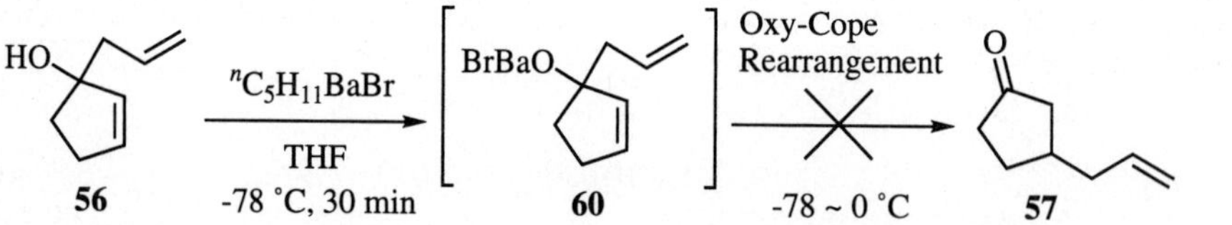

Scheme 2-21

Table 2-9. One-pot double alkylation of α, β-unsaturated ketones.

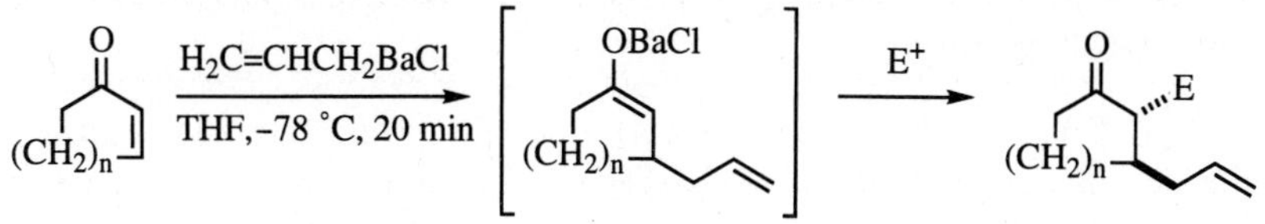

Scheme: enone with $(CH_2)_n$ ring + $H_2C{=}CHCH_2BaCl$, THF, −78 °C, 20 min → [OBaCl enolate, $(CH_2)_n$] → E^+ → product with E and allyl, $(CH_2)_n$

Entry	Enone	Electrophile	Conditions Temp. °C	Conditions Time, h	Product	Yield, %[a]
1	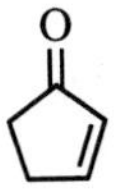 cyclopentenone	$(CH_3)_2C{=}CHCH_2Br$	−30	1.5	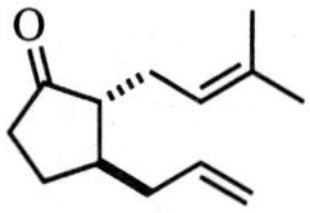	81[b]
2		$nC_5H_{11}C{\equiv}CCH_2Br$	−25	4	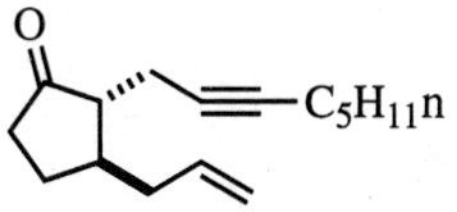 $C_5H_{11}n$	50[b]
3		$nC_5H_{11}CHO$	−78	0.3	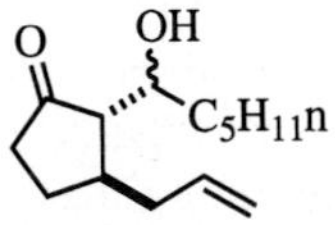 OH, $C_5H_{11}n$	85[b]
4		CH_3COCl	−78	0.3	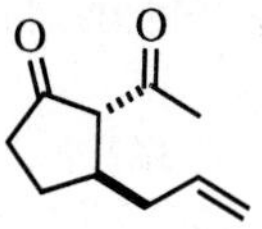	87[b]
5	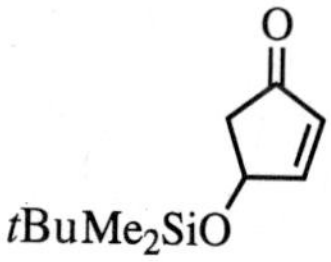 tBuMe$_2$SiO	$(CH_3)_2C{=}CHCH_2Br$	−30	1	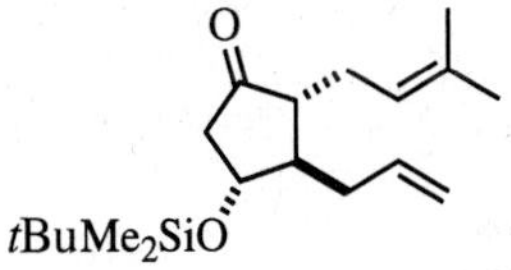 tBuMe$_2$SiO	46[b]
6[b]	 cyclohexenone	$(CH_3)_2C{=}CHCH_2Br$	−30	1.5	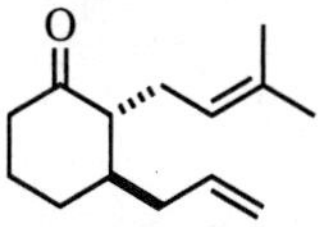	74[b]

a) Isolated yield. b) The trans/cis ratio was > 99/1. c) The threo/erythro ratio was 83/17. d) A mixture of keto and enol compounds. e) The conjugate addition reaction was performed at −95 °C. f) The trans/cis ratio was 98/2.

2.3.4.1 Typical Procedure for One-Pot Double Alkylations of *α,β*-Unsaturated Ketones: Synthesis of *trans*-2-(3-Methyl-2-butenyl)-3-(2-propenyl)cyclopentanone

To a suspension of barium powder (1.2 mmol) in THF (10 mL) is slowly added a solution of allyl chloride (**58**, 75 mg, 0.98 mmol) in THF (1.5 mL) at –78 °C. The reaction mixture is stirred for 20 min at this temperature. A solution of 2-cyclopentenone (**55**, 48 mg, 0.58 mmol) in THF (1 mL) is introduced into the mixture at –78 °C (see Table 2-7). After being stirred for 20 min, the mixture is treated with prenyl bromide (250 mg, 1.7 mmol) at –78 °C and warmed to –30 °C over a period of 30 min. The stirring is continued for another 1 h at this temperature. To the mixture is added saturated aqueous NH_4Cl solution (10 mL), and the organic material is extracted with ether (10 mL). The combined organic extracts are washed with 1 N sodium thiosulfate solution (20 mL), dried over anhydrous $MgSO_4$, and concentrated in vacuo after filtration. The crude product is purified by flash-column chromatography on silica gel (1:50 to 1:10 ether/hexane as eluent) to afford the *α,β*-diallylated cyclopentanone (90 mg, 81% yield); the purity is checked by GLC analysis (>99/1).

2.3.5 Other Reactions

Heteroatom-substituted allylic metal reagents are valuable homoenolate anion equivalents and are widely used in organic synthesis [36]. Siloxyallylbarium reagents **62**, prepared from anhydrous BaI_2 and siloxyallyllithiums, react with carbonyl compounds and alkyl halides highly selectively at the least substituted allyl terminus (*γ*-position) in high yields (Scheme 2-22) [37].

The method used to prepare stereochemically homogeneous allylic metal reagents (Mg and Ba) from the corresponding allylic chlorides [3, 13] is applied to a diastereoselective *γ*-allylation reaction of aldehydes using *B*-OMe-9-BBN as an additive. *Threo*-homoallylic alcohols are obtained selectively from (*E*)-allylic barium reagents, and *erythro*-isomers from (*Z*)-allylic barium reagents, in high yields (Scheme 2-23) [38].

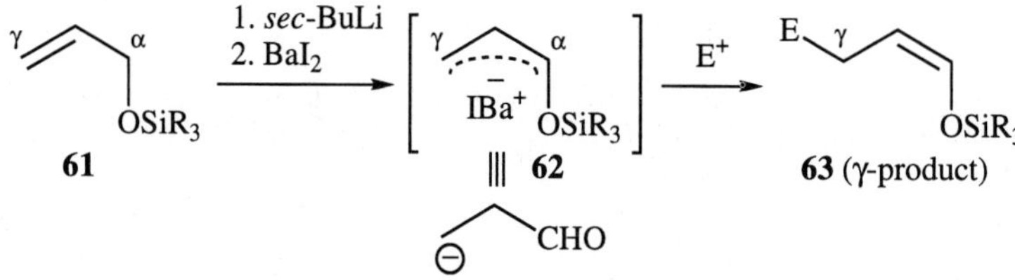

Scheme 2-22

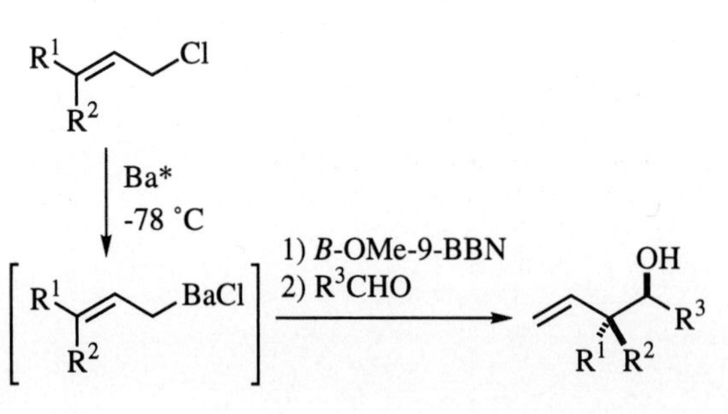

Scheme 2-23

2.4 Summary and Conclusions

Described herein are practical methods for generating allylic barium reagents stereospecifically and performing subsequent reactions of these anions with a variety of electrophiles in regio- and stereoselective ways. The main features of the present scheme are: (1) allylic barium reagents are readily prepared by treatment of the corresponding allylic chlorides with reactive barium; (2) no stereoisomerization of the primary allylic barium compound is observed below –70 °C; (3) the barium reagent reacts selectively at the α-position with aldehydes, ketones, carbon dioxide, and allylic halides with complete retention of the stereochemistry of the starting halide; (4) selective 1,4-addition reaction occurs with α,β-unsaturated ketones. The extraordinary α-selectivity and stereospecificity of the reactions of allylic barium reagents provide unprecedented routes to homoallylic alcohols, β,γ-unsaturated carboxylic acids, and 1,5-dienes and are broadly applicable in organic synthesis.

Acknowledgements. We are deeply indebted to our coworkers, Dr. Shigeki Habaue, and Messrs. Katsutaka Yasue, Hiroaki Hibino, and Yoshiyuki Hisada. Financial support from the Ministry of Education, Science and Culture of the Japanese Government, and the Asahi Glass Foundation is gratefully acknowledged.

References

[1] Reviews: (a) G. Courtois, L. Miginiac, *J. Organomet. Chem.* **1974**, *69*, 1; (b) W. R. Roush in *Comprehensive Organic Synthesis, Vol. 2* (eds.: B. M. Trost, I. Fleming, C. H. Heathcock), Pergamon, Oxford, **1991**, p. 1; (c) Y. Yamamoto, N. Asao, *Chem. Rev.* **1993**, *93*, 2207.

[2] Reviews: (a) S. T. Ioffe, A. N. Nesmeyanov, *The Organic Compounds of Magnesium, Beryllium, Calcium, Strontium and Barium*, North-Holland, Amsterdam, **1967**; (b) K. Nützel, in *Houben-Weyl: Methoden der Organischen Chemie, Vol. 13/2a*, (ed.: E. Müller), Thieme, Stuttgart, **1973**, p. 529; (c) B. G. Gowenlock, W. E. Lindsell, in *Journal of Organometallic Chemistry Library: 3 Organometallic Chemistry Reviews* (eds.: D. Seyferth, A. G. Davies, E. O. Fischer, J. F. Normant, O. A. Reutov), Elsevier, Amsterdam, **1977**, p. 1; (d) W. E. Lindsell, in *Comprehensive Organometallic Chemistry, Vol. 1*, (eds.: G. Wilkinson, F. G. A. Stone, E. W. Abel), Pergamon, Oxford, **1982**, Ch. 4, p. 223; (e) B. J. Wakefield, in *Comprehensive Organometallic Chemistry, Vol. 7*, (eds.: G. Wilkinson, F. G. A. Stone, E. W. Abel), Pergamon, Oxford, **1982**, Ch. 44.

[3] A. Yanagisawa, S. Habaue, H. Yamamoto, *J. Am. Chem. Soc.* **1991**, *113*, 5893.

[4] T. P. Burns, R. D. Rieke, *J. Org. Chem.* **1987**, *52*, 3674.

[5] T.-C. Wu, H. Xiong, R. D. Rieke, *J. Org. Chem.* **1990**, *55*, 5045.

[6] Diallylbarium has been prepared by transmetallation with diallylmercury or tetraallyltin in THF: a) P. West, M. C. Woodville, Ger. Offen. 2, 132, 955, **1972**; b) P. West, M. C. Woodville, U.S. Pat. 3, 766, 281, **1973**.

[7] J. L. Wardell, in *Comprehensive Organometallic Chemistry, Vol. 1*, (eds.: G. Wilkinson, F. G. A. Stone, E. W. Abel), Pergamon, Oxford, **1982**, Ch. 2.

[8] D. A. Hutchinson, K. R. Beck, R. A. Benkeser, J. B. Grutzner, *J. Am. Chem. Soc.* **1973**, *95*, 7075.

[9] Prepared using lithium biphenylide: (a) N. L. Holy, *Chem. Rev.* **1974**, *74*, 243; (b) T. Cohen, M. Bhupathy, *Acc. Chem. Res.* **1989**, *22*, 152.

[10] A. Yanagisawa, S. Habaue, K. Yasue, H. Yamamoto, *J. Am. Chem. Soc.* **1994**, *116*, 6130.

[11] (a) M. Schlosser, J. Hartmann, V. David, *Helv. Chim. Acta* **1974**, *57*, 1567; (b) M. Stähle, J. Hartmann, M. Schlosser, *Helv. Chim. Acta* **1977**, *60*, 1730.

[12] A. Yanagisawa, S. Habaue, H. Yamamoto, unpublished results.

[13] Reviews: (a) Y. Yamamoto, K. Maruyama, *Heterocycles* **1982**, *18*, 357; (b) R. W. Hoffmann, *Angew. Chem.* **1982**, *94*, 569; *Angew. Chem. Int. Ed. Engl.* **1982**, *21*, 555.

[14] R. A. Benkeser, M. P. Siklosi, E. C. Mozdzen, *J. Am. Chem. Soc.* **1978**, *100*, 2134.
[15] (a) Y. Yamamoto, K. Maruyama, *J. Org. Chem.* **1983**, *48*, 1564; (b) B.-S. Guo, W. Doubleday, T. Cohen, *J. Am. Chem. Soc.* **1987**, *109*, 4710. Indirect methods using acylsilanes: (c) S. R. Wilson, M. S. Hague, R. N. Misra, *J. Org. Chem.* **1982**, *47*, 747; (d) A. Yanagisawa, S. Habaue, H. Yamamoto, *J. Org. Chem.* **1989**, *54*, 5198; (e) A. Yanagisawa, S. Habaue, H. Yamamoto, *Tetrahedron* **1992**, *48*, 1969; (f) M. Suzuki, Y. Morita, R. Noyori, *J. Org. Chem.* **1990**, *55*, 441.
[16] A. Yanagisawa, S. Habaue, H. Yamamoto, *J. Am. Chem. Soc.* **1991**, *113*, 8955.
[17] M. Kaupp, P. v. R. Schleyer, *J. Am. Chem. Soc.* **1992**, *114*, 491.
[18] N. Ragoussis, *Tetrahedron Lett.* **1987**, *28*, 93.
[19] P. Gosselin, F. Rouessac, *C. R. Séances Acad. Sci., Ser. 2* **1982**, *295*, 469.
[20] Ene reaction of diethyl oxomalonate: M. F. Salomon, S. N. Pardo, R. G. Salomon, *J. Am. Chem. Soc.* **1984**, *106*, 3797.
[21] β-Vinyl-β-propiolactone/organocopper reagent: M. Kawashima, T. Sato, T. Fujisawa, *Bull. Chem. Soc. Jpn.* **1988**, *61*, 3255.
[22] Transition-metal-catalyzed carbonylation: (a) N. Satyanarayana, H. Alper, I. Amer, *Organometallics* **1990**, *9*, 284; (b) Y. Imada, O. Shibata, S.-I. Murahashi, *J. Organomet. Chem.* **1993**, *451*, 183.
[23] Review: E. W. Colvin, E. W. in *Comprehensive Organic Chemistry, Vol. 2*, (eds.: D. H. R. Barton, W. D. Ollis), Pergamon, Oxford, **1979**, p. 620.
[24] A stereospecific route for the synthesis of homogeranic acid and homoneric acid by carboxylation of the lithiated allylic sulfone has been reported: P. Gosselin, C. Maignan, F. Rouessac, *Synthesis* **1984**, 876.
[25] A. Yanagisawa, K. Yasue, H. Yamamoto, *Synlett* **1992**, 593.
[26] Reviews: D. C. Billington, in *Comprehensive Organic Synthesis, Vol. 3*, (eds.: B. M. Trost, I. Fleming, G. Pattenden), Pergamon, Oxford, **1991**, p. 413.
[27] A. Yanagisawa, H. Hibino, S. Habaue, Y. Hisada, H. Yamamoto, *J. Org. Chem.* **1992**, *57*, 6386.
[28] E. J. Corey, M. C. Noe, W.-C. Shieh, *Tetrahedron Lett.* **1993**, *34*, 5995.
[29] E. J. Corey, W.-C. Shieh, *Tetrahedron Lett.* **1992**, *33*, 6435.
[30] Reviews: (a) *Comprehensive Organic Synthesis*, Vol. 4, (eds.: B. M. Trost, I. Fleming, H. F. Semmelhack), Pergamon, Oxford, **1991**, pp. 1–268; (b) P. Perlmutter, *Conjugate Addition Reactions in Organic Synthesis*, Pergamon, Oxford, **1992**.
[31] (a) H. O. House, W. F. Fischer Jr., *J. Org. Chem.* **1969**, *34*, 3615; (b) H. O. House, T. S. B. Sayer, C.-C. Yau, *J. Org. Chem.* **1978**, *43*, 2153; (c) H. O. House, J. M. Wilkins, *J. Org. Chem.* **1978**, *43*, 2443.
[32] A. Hosomi, H. Sakurai, *J. Am. Chem. Soc.* **1977**, *99*, 1673.
[33] (a) W. D. K. Macrosson, J. Martin, W. Parker, A. B. Penrose, *J. Chem. Soc.(C)* **1968**, 2323; (b) S. Wolff, W. C. Agosta, *Tetrahedron Lett.* **1979**, 2845; (c) R. A. Benkeser, *Synthesis* **1971**, 347.
[34] M. J. Chapdelaine, M. Hulce, *Org. React.* **1990**, *38*, 225.
[35] A. E. Greene, P. Crabbé, *Tetrahedron Lett.* **1976**, 4867.
[36] Y. Yamamoto in *Comprehensive Organic Synthesis, Vol. 2*, (eds.: B. M. Trost, I. Fleming, C. H. Heathcock), Pergamon, Oxford, **1991**, p. 55, and references therein.
[37] A. Yanagisawa, K. Yasue, H. Yamamoto, *Synlett* **1993**, 686.
[38] S. Habaue, K. Yasue, A. Yanagisawa, H. Yamamoto, *Synlett* **1993**, 788.

3 The McMurry Reaction

Thomas Lectka

3.1 Introduction

Some twenty years ago McMurry [1–3], Tyrlik [4], and Mukaiyama [5] made the simultaneous and independent discovery that carbonyl compounds, including ketones and aldehydes, undergo reductive coupling in the presence of low-valent titanium reagents to form olefins (Scheme 3-1). The Mukaiyama and Tyrlik procedures applied only to aryl ketones, whereas the McMurry discovery was originally made on an aliphatic ketone substrate. In subsequent years, the development of the reaction has occurred primarily in McMurry's laboratories [6]. At the time the immense synthetic utility of the process was immediately apparent; high yields of dimerized product were generally obtained, and most importantly, the intramolecular variant also occurred to form cycloalkenes with similarly high yields for all manner of ring sizes. In the intervening time, the reaction has been fruitfully applied to the synthesis of complex natural products, theoretically interesting strained rings, and non-natural systems, and in the past few years, has been extended to include keto ester coupling modifications [7]. In many cases, molecules have been synthesized for which no other conceivable methods exist.

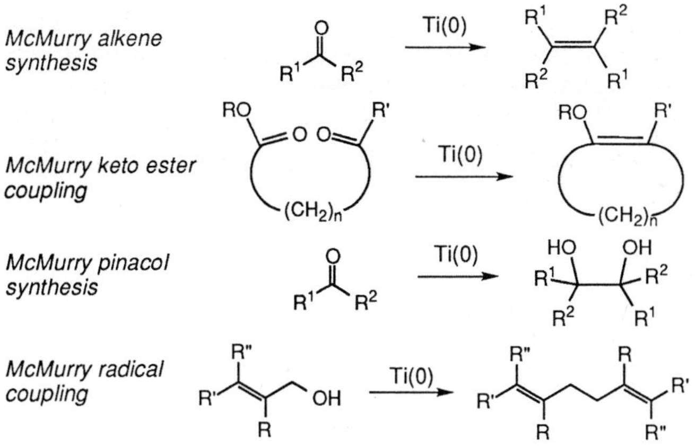

McMurry alkene synthesis

McMurry keto ester coupling

McMurry pinacol synthesis

McMurry radical coupling

Scheme 3-1

For purposes of illustration, the McMurry reaction has been divided into four subgroups, each of which will be discussed in turn: (a) the McMurry alkene synthesis; (b) the McMurry keto ester coupling; (c) the McMurry pinacol synthesis; and (d) the McMurry allyl/benzyl radical coupling. In recent years the most important development has been the McMurry pinacol synthesis [8], in which reaction temperatures are lowered to permit high-yield isolation of diol products.

3.2 Historical Perspective

During the course of a natural product synthesis, McMurry needed to convert the α,β-unsaturated ketone **1** into the alkene **2**, preferably in one high-yielding step (Scheme 3-2) [9]. The idea was to use a hydride-based reducing agent in the presence of an oxophilic transition metal Lewis acid; initial reduction would presumably afford an alkoxide salt possessing strong binding to the metal, opening the possibility that a second equivalent of hydride would be delivered, displacing the metal alkoxide in an S_N2 reaction.

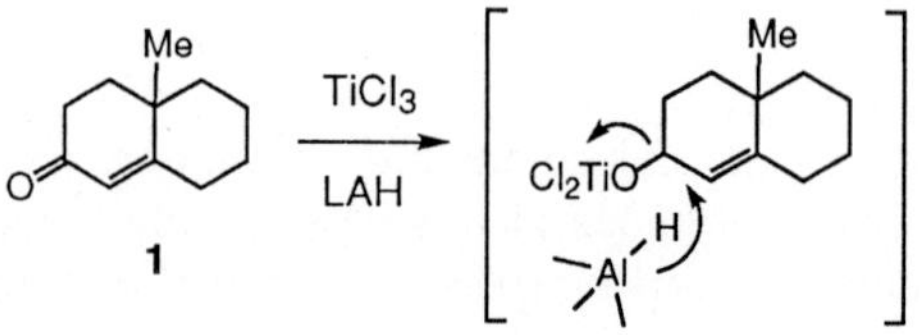

Me TiCl3 LAH 1 Me Cl2TiO H Al LAH Me 2

Scheme 3-2

Lithium aluminum hydride (LAH) was chosen as the hydride source, and $TiCl_3$ as the transition metal salt. When a reagent was prepared by mixing 0.5 equiv. of LAH and 1 equiv. of $TiCl_3$ in anhydrous tetrahydrofuran (THF) and the starting ketone was added, instead of isolating alkene **2**, the reaction afforded the dimeric hydrocarbon **3** as the major product (Scheme 3-3). At the time the reaction was unknown, and McMurry realized that the observed result was of much greater significance than the initially desired outcome. It was quickly discovered that the reaction was successful for all kinds of alkyl and aryl ketones and aldehydes. Eventually the original $TiCl_3$/LAH tandem was replaced by $TiCl_3$/Zn–Cu [10], as a safe, highly effective reagent. Although numerous other transition metal salts were screened as reagents in the reaction, titanium was found by far to be the most effective [11] [12, 13]. However, a large number of different titanium-based reagent systems have since proliferated in the literature; certain modifications including $TiCl_3$/K/graphite [14] have proven useful for some substrates including keto esters, and will be discussed further below.

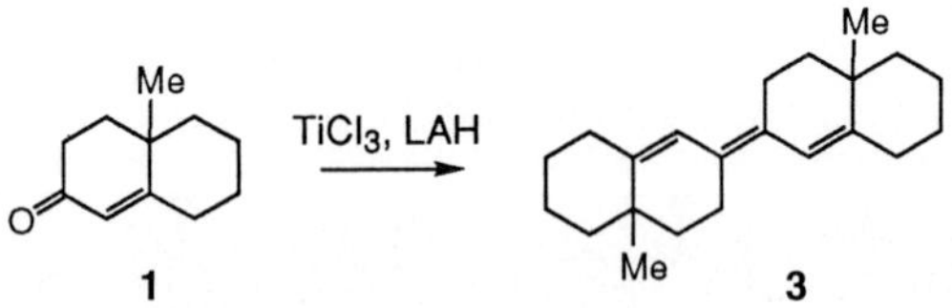

Me TiCl3, LAH 1 Me Me 3

Scheme 3-3

Given the detailed review published by McMurry in 1989 [2(a)], this account seeks to bring the reader up-to-date on the important new developments in the McMurry reaction. Emphasis will be placed on the communication of practical knowledge about running a reaction, and representative experimental procedures are provided. The primary aim of this review is to leave the reader with the knowledge necessary to perform a coupling using the most modern techniques.

3.3 Reaction Mechanism and Theory

3.3.1 Introduction

Scheme 3-4 depicts the generally accepted mechanism of the reaction, which involves four discrete steps after the formation of the active reagent, namely (I) reduction to form the ketyl radical; (II) organization of two ketyl radicals on the metal surface in the proximity of each other; (III) coupling; and (IV) deoxygenation.

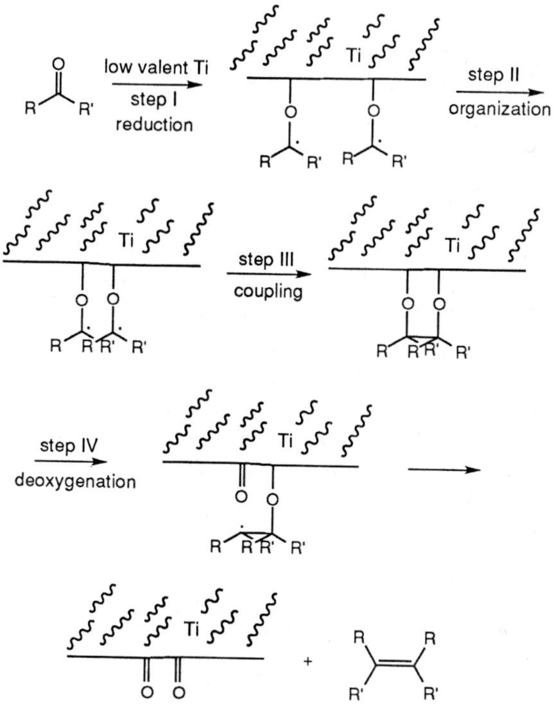

low valent Ti
step I
reduction
step II
organization
step III
coupling
step IV
deoxygenation

Scheme 3-4

3.3.2 Prelude to the Reaction: Formation of the Active Reagent

Reaction of $TiCl_3$ with a suitable reducing agent was thought to form a finely dispersed suspension of $Ti^{(0)}$ [15], which was believed to be the actual coupling agent. Evidence was mainly based on ESR spectroscopic studies [16]. The titanium reagent derived from $TiCl_3$ and Li has also been examined by scanning electron microscopy [6], which revealed that the active material forms a highly porous sponge with an extremely high surface-to-mass ratio. However, comprehensive recent studies by Bogdanovic et al. have demonstrated that $Ti^{(0)}$ need not be the actual coupling reagent [91]. For example, reduction of $TiCl_3(THF)_3$ with $LiAlH_4$ in THF leads to the hydride $[HTiCl(THF)_{0.5}]_n$, not $Ti^{(0)}$. Similarly, the use of Mg as the reducing agent – contrary to previous assumptions – does not afford $Ti^{(0)}$ but leads to the formation of highly reactive Ti–Mg intermetallic species [91]. For a comprehensive treatise of this system see Chapter 8, Section 8.5.1.

The metal dispersion of the coupling reagents is not necessarily homogeneous; traces of different oxidation states of titanium may be present, and when Zn–Cu is used as the reducing agent, Lewis acid zinc and copper salts and unreacted Zn–Cu as well. In any case the low-valent titanium surface serves as the crucial site of reaction, and imparts to the reaction its characteristic efficacy in the formation of large rings. The solvent of choice in the reaction is usually dimethoxyethane (DME).

3.3.3 Step I: Formation of the Ketyl Radical

After formation of the reagent, the first step of the reaction is a one-electron reduction of the carbonyl substrate to an intermediate ketyl radical, the standard first step in a classical pinacol reaction [17, 18]. Geise has observed ESR-active ketyl radicals bound to the titanium metal surface [16], and in a very thorough study Wolczanski and coworkers have spectroscopically characterized a number of titanium-bound ketyl radicals generated through the reaction of the sterically encumbered $Ti^{(III)}$ complex $(silox)_3Ti$ (**4**) and appropriate ketones (Scheme 3-5) [19]. For example, an orange solution of **4** and di-*t*-butyl ketone **5** became ink-blue when cooled below 0 °C, indicating formation of ketyl radical **6**. The ESR spectrum showed a signal attributable to the ketyl species, centered at g = 1.9985 with an isotropic titanium ($^{47,49}Ti$) coupling of 4 G.

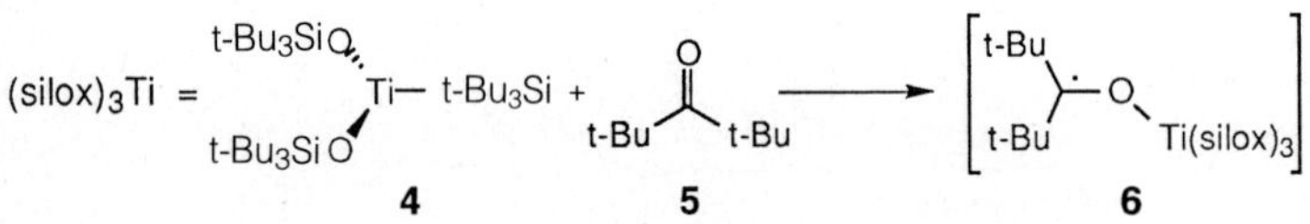

$(silox)_3Ti$ = $(t\text{-}Bu_3SiO)_3Ti$ **4** + t-Bu–CO–t-Bu **5** → [$(t\text{-}Bu)_2\dot{C}$–O–$Ti(silox)_3$] **6**

Scheme 3-5

When **4** was allowed to react with benzophenone at –78 °C, the solution became bright red upon warming to room temperature then faded to yellow, indicating formation of the dimeric species **9**, which is in equilibrium with the corresponding ketyl **8** as indicated by trapping experiments conducted at this temperature (Scheme 3-6).

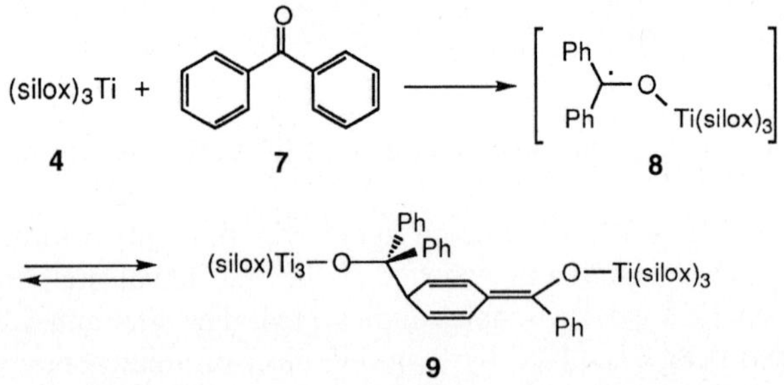

$(silox)_3Ti$ **4** + benzophenone **7** → [$Ph_2\dot{C}$–O–$Ti(silox)_3$] **8** ⇌ **9**

Scheme 3-6

3.3.4 Steps II and III: Organization and Coupling

The ketyl radical species then reacts with another similarly formed radical, from either another molecule or through another site in the molecule if an intramolecular coupling is performed. As mentioned, this step represents a classical pinacol coupling which occurs with a number of other reducing metals. The titanium metal surface itself may act to bring the reacting radical fragments within proximity to one another in a template effect, which is especially notable in the formation of large rings (Scheme 3-7).

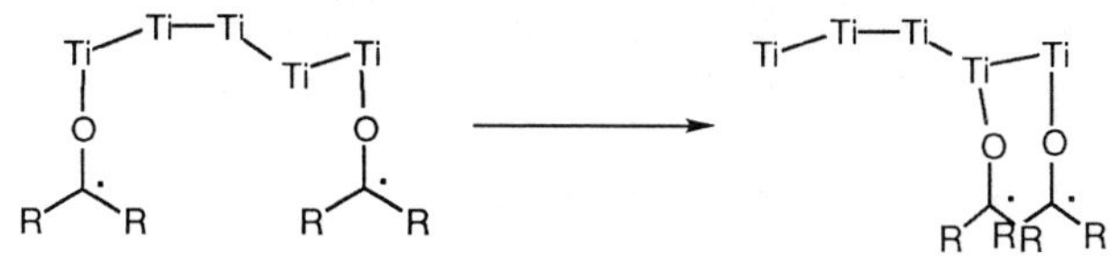

Scheme 3-7

If the reaction is carried out at sufficiently low temperatures, the pinacols themselves may be isolated in high yields after mild hydrolysis of the titanium pinacolate mixture. In related work, it was found that the mononuclear titanium complex **10** reacts with carbonyl compounds to afford discrete titanium pinacol complexes such as **12** (Scheme 3-8) [20]. The difference when the reaction happens on a metal surface is that more than one titanium atom may be involved in the formation of the pinacolate intermediate.

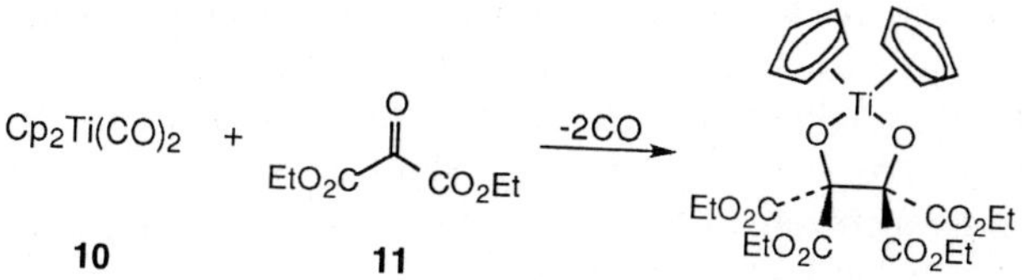

Scheme 3-8

3.3.5 Step IV: Deoxygenation

In many cases, especially for couplings in which little strain energy is added to the framework of the molecule, deoxygenation is the rate-determining step in the process. In fact, when pinacol products initially isolated at low temperatures are resubmitted to the reaction at elevated temperatures, efficient deoxygenation occurs to afford the corresponding alkenes [6]. Deoxygenation occurs in a stepwise fashion; for instance when isomerically pure *meso*-5,6-decanediol **13** is submitted to the conditions of the reaction (Scheme 3-9a), a mixture of *cis* and *trans* olefin isomers in the ratio 60:40 is afforded. Were the deoxygenation step concerted, pure *cis* olefin would be expected. Similarly, *d,l*-5,6-decanediol **16** affords a mixture of *cis* and *trans* olefin products, but here in the ratio 91:9 (Scheme 3-9b).

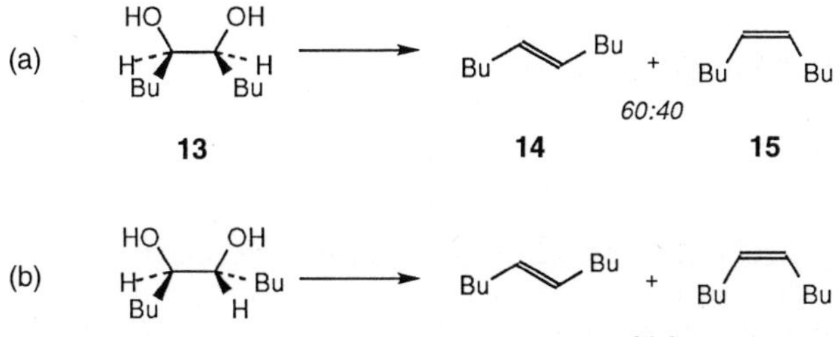

Scheme 3-9

Interestingly enough, however, both oxygens must be simultaneously accessible to the titanium surface for the deoxygenation reaction to occur. Consequently *cis*-decalindiol **17** undergoes smooth deoxygenation, whereas **19** does not (Scheme 3-10).

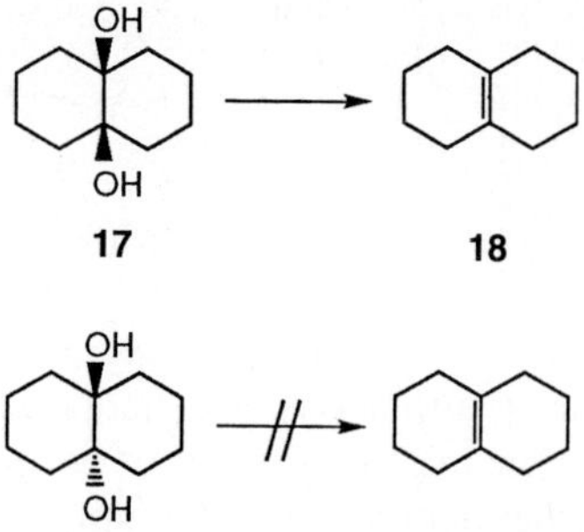

Scheme 3-10

How many titanium atoms on the metal surface are involved in the actual deoxygenation? This question was answered in an elegant experiment performed by McMurry to probe the nature of the metal surface [2a, 6]. Were only one titanium atom involved in the deoxygenation, we would anticipate that diol **20** would react (Scheme 3-11a), but **22** could not (Scheme 3-11b), due to the fact that the single Ti atom would have to be part of a *trans*-fused seven membered ring. In fact, both diols react smoothly under the conditions of the coupling to afford olefin **21**, indicating that more than one atom can be involved in the actual deoxygenation step.

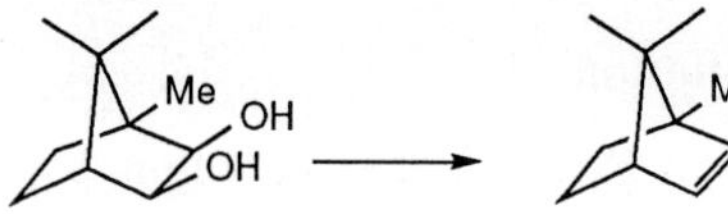

Scheme 3-11

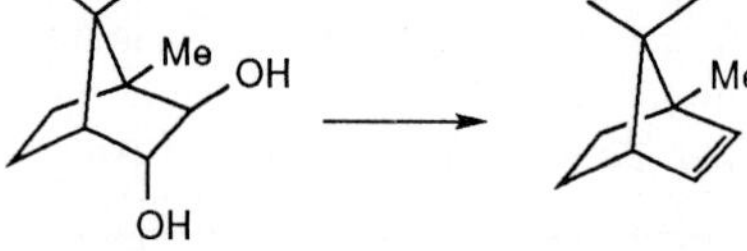

3.4 Optimized Procedures for the Coupling Reaction

3.4.1 The Reductant

There exist a large number of different recipes for the execution of the McMurry reaction, most varying the nature of the reductant while retaining the use of $TiCl_3$ (Table 3-1). Unfortunately the bewilderingly large number of published alternatives has led to confusion about which procedure to use, for example procedures which were abandoned by the

Table 3-1. Experimental procedures for the McMurry reaction.

Metal salt	Reducing agent	Use	Ref.
$TiCl_3$	Zn–Cu	Standard general coupling reagent	[6]
$TiCl_3(DME)_{1.5}$	Zn–Cu	Optimized general coupling reagent	[27]
$TiCl_3$	K	Replaced by $TiCl_3$, Zn–Cu	[15]
$TiCl_3$	Li	Aryl ketones, replaced by $TiCl_3$, Zn–Cu	[24]
$TiCl_3$	C_8K	A method of choice for keto ester, and keto acyl couplings	[22, 23]
$TiCl_3$	LAH	Keto ester couplings, original coupling reagent, replaced by $TiCl_3$, Zn–Cu	[1, 21]
$TiCl_4$	Zn	Mukaiyama pinacol coupling reagent	[5, 25, 26]
$TiCl_4$	Zn, pyridine	Alkene coupling reagent	[26]
$TiCl_4$	Mg	Tyrlik's original reagent	[4]

McMurry group years ago as inferior are still frequently employed for routine couplings. Although complete generalizations cannot be made, we have found the $TiCl_3$/Zn–Cu reagent system in dimethoxyethane (DME) solvent to be the most effective for the widest variety of couplings [6, 27], and we recommend this as the standard coupling reagent for the synthesis of alkenes and pinacol rings. However for the keto ester coupling, a more problematic reaction, $TiCl_3$/LAH [21] and the Fürstner reagent, $TiCl_3/C_8K$ [22–23], are preferable to $TiCl_3$/Zn–Cu. The cross-coupling of aryl ketones may also benefit from the use of $TiCl_3$/Li [24], although $TiCl_3$/Zn-Cu should work well in this case too. It should be emphasized that $TiCl_3$/LAH [1, 21] and $TiCl_3$/K [15] are no longer recommended as general coupling reagents, both for reasons of safety and efficacy.

Recommended Preparation of Zinc–Copper Couple [6, 19]. Zn–Cu couple can be prepared by adding zinc dust (9.8 g, 150 mmol) to 40 mL of nitrogen or argon purged water and purging the resultant gray slurry for 15 additional min, then adding $CuSO_4$ (0.75 g, 4.7 mmol). The black slurry is filtered under nitrogen or argon, then washed with deoxygenated water, acetone, and finally ether, and then dried under high vacuum. We have found fully effective Zn–Cu couple to be black or dark gray in color. The presence of light gray hues from our experience signals poor results in couplings. The couple can then be stored in a Schlenk tube under an inert atmosphere indefinitely.

3.4.2 The Titanium Source

Most versions of the McMurry reaction employ $TiCl_3$ as the titanium source; however some procedures, especially those for the pinacol coupling, may employ $TiCl_4$ [25, 26], including Mukaiyama's original procedure [5]. The quality of the $TiCl_3$ is of paramount importance to the success of the reaction; unfortunately, depending on the supplier or even the batch, results can vary. For example, during the course of a synthesis of a strained ring system we encountered difficulty with the coupling reaction. Yields were variable and often unsatisfactory, and sometimes the reaction did not work at all. The problem was traced to inferior batches of $TiCl_3$. In general, the scope and reproducibility of the reaction seemed to be limited by the origin, age, and history of the $TiCl_3$ starting material, so consequently attempts at purification were made. Converting metal salts into their corresponding solvate complexes is one time-honored method of purification. As an added advantage, solvates are usually more stable and less prone to hydrolysis than the free salts. As a first try we made a $TiCl_3$–THF solvate simply by boiling purple $TiCl_3$ in THF, cooling the reaction mixture and filtering the bright blue product. Unfortunately we found this to be an inferior coupling reagent; one possible reason is that THF, being a fair donor of $H^{\bullet}$, serves to quench a proportion of the intermediate ketyl radicals before they can couple (Scheme 3-12). For this reason, DME is strongly recommended over THF as solvent in the coupling reaction.

Scheme 3-12

However, the sky-blue solvate complex of $TiCl_3$ and DME ($TiCl_3(DME)_{1.5}$) is among the most effective sources of titanium for the reactions discovered to date [27]. As an example of the reagent's efficacy, the coupling of diisopropylketone was originally found to take place in 12% yield using $TiCl_3$/LAH [28]. In contrast, the reaction takes place in 94% yield using the $TiCl_3(DME)_{1.5}$/Zn–Cu reagent combination in DME solvent. The coupling of cyclohexanone results in an essentially quantitative yield of cyclohexylidene-cyclohexane (97%). Intramolecular variants of the coupling reaction using the modified reagent also take place very efficiently, for example the coupling of tetradecanedial takes place in 80% yield, and as a representative diketone coupling, 2,15-hexadecanedione affords 1,2-dimethyltetradecane in 82% yield.

Preparation of $TiCl_3(DME)_{1.5}$ [27]. $TiCl_3$ (25 g, 0.162 mol) was suspended in 350 mL dry DME (doubly distilled over potassium metal), and the mixture was refluxed for two days under argon. The reaction mixture was then cooled to room temperature, filtered under argon, washed with distilled, degassed pentane, and dried under high vacuum to afford fluffy, light blue crystalline $TiCl_3(DME)_{1.5}$ (32 g, 80%) which was used in the coupling reaction. The solvate is air sensitive, but can be stored indefinitely under a rigorously oxygen- and moisture-free atmosphere.

3.4.3 Practical Comments about Couplings

The reader should bear in mind that the McMurry coupling does not always give high yields at the first, or even the second attempt, but that with some experience, reproducibly high yields can be attained. Especially when a valuable substrate is involved, we strongly recommend that the reader perform several test couplings on cyclohexanone before venturing a coupling on the more complex material. Isolated yields of cyclohexylidenecyclohexane should be above 90%; if they are not, optimal results will probably not be attained in the actual reaction. Inferior results are often attributable to one or more of the following: (1) poor quality Zn–Cu couple, (2) inadequately purified DME (freshly doubly distilled over potassium metal is highly recommended), (3) poor quality $TiCl_3$, and (4) intrusion of air and/or water into the reaction vessel during formation of the reagent and/or the actual coupling. Although some workers have employed rigorous inert atmosphere techniques including the use of glove boxes in conducting a coupling, we have found this to be unnecessary, and recommend standard Schlenkware on the benchtop. In regard to the quality of Zn–Cu couple, it is important that the reagent be either black or a dark shade of gray. Light gray batches generally give unsatisfactory results for difficult couplings. The $TiCl_3$ should be a very fine grained, fluid purple or lavender powder; light purple $TiCl_3$ containing splotches of white usually indicates the presence of titanium oxides which can be deleterious to the efficacy of the reaction. One way to overcome an inferior batch of $TiCl_3$ is to employ $TiCl_3(DME)_{1.5}$ as the titanium source. However for batches of $TiCl_3$ which have been sitting on the shelf for years, new $TiCl_3$ is recommended. The low-valent titanium reagent itself should be black or very dark gray; the presence of green hues is also a favorable sign. Once again, light gray signals inferior material.

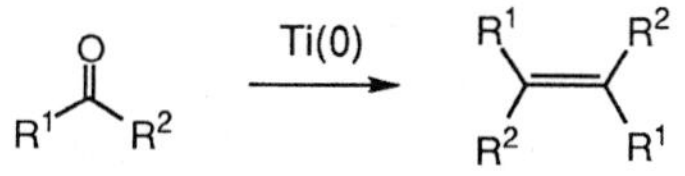

Ti(0)

R1 R2 R1 R2 R2 R1

Scheme 3-13

3.5 The McMurry Alkene Synthesis

3.5.1 Intermolecular Couplings

3.5.1.1 Prototype Couplings

These are the most straightforward types of couplings to perform (Scheme 3-14), and have most notably been used to prepare theoretically interesting substituted olefins. It was realized very early in the development of the reaction that strained products could be formed in high yields, since the thermodynamic driving force of Ti=O bond formation permits a large amount of strain to be built into products. Several sample procedures for performing an intermolecular coupling are provided below; as a general rule of thumb, the more strain to be built into the product, the larger the excess of reagent that should be employed.

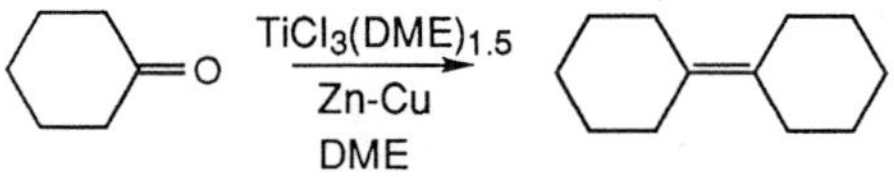

$TiCl_3(DME)_{1.5}$ Zn-Cu DME

23 **24**

Scheme 3-14

Example of an Intermolecular Coupling: Cyclohexylidenecyclohexane [27].
$TiCl_3(DME)_{1.5}$ (5.2 g, 17.9 mmol) and Zn–Cu (4.9 g, 69 mmol) were transferred under argon to a flask containing 100 mL of DME, and the resulting mixture was refluxed to afford a dark gray to black suspension. Distilled cyclohexanone (0.44 g, 4.5 mmol) in 10 mL DME was then added, and the mixture was refluxed an additional 8 h. After being cooled to room temperature, the reaction mixture was diluted with pentane (100 mL), filtered through a pad of Florisil, and concentrated to yield cyclohexylidenecyclohexane (0.36 g, 97%) as white crystals, m.p. 52.5–53.5 °C. If a 3:1 ratio of $TiCl_3(DME)_{1.5}$ to cyclohexanone is employed in the reaction instead of 4:1, the yield decreases slightly to 94%; if a 2:1 ratio is used the yield falls to 75%.

Employing the optimized reagent for the coupling of diisopropylketone yielded tetraisopropylethylene (**26**) in 94% yield (Scheme 3-15). Interestingly enough, olefin **26** is stable on the benchtop in air over a period of several years.

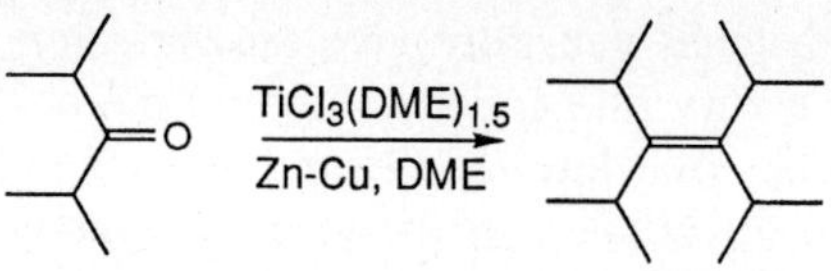

$TiCl_3(DME)_{1.5}$
Zn-Cu, DME

Scheme 3-15 **25** **26**

Example of a Strained Intermolecular Coupling: Tetraisopropylethylene [27]. The procedure for cyclohexylidenecyclohexane was followed except that the following proportions were used: diisopropylketone (0.5 g, 4.5 mmol), $TiCl_3(DME)_{1.5}$ (9.0 g, 31 mmol), and Zn–Cu (8.0 g, 110 mmol). The reaction mixture was refluxed for 12 h and then worked up in the standard fashion to afford the product in 94% yield.

Shown in Table 3-2 are a number of prototype intermolecular couplings. It should be noted that most of the examples were published before the advent of the $TiCl_3$/Zn–Cu protocol [6] which is recommended for most couplings, although for biaryl couplings $TiCl_3$–Li works well.

Table 3-2. Intermolecular couplings of carbonyl compounds.

Entry	Substrate	Product	Yield (%)	Procedure	Ref.
A	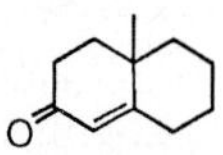	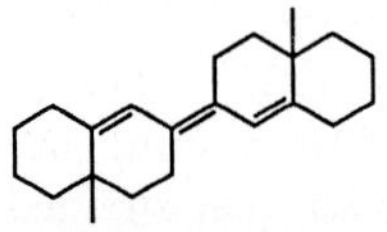	80	$TiCl_3$–LAH	2b
B	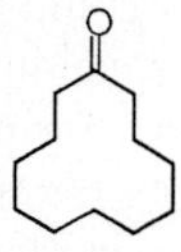	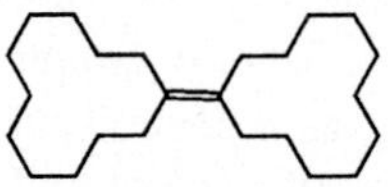	65	$TiCl_3$–Li	6
C	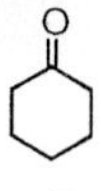	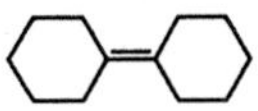	98	$TiCl_3(DME)_{1.5}$ Zn–Cu	27
D	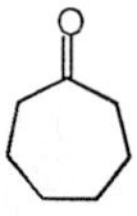	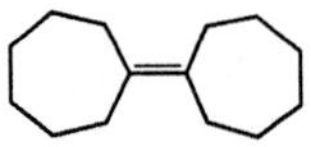	85	$TiCl_3$–Li	6
E	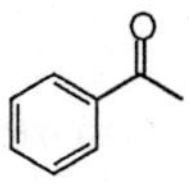	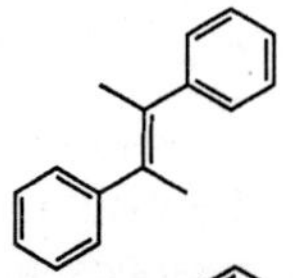	94	$TiCl_3$–Li	6
F	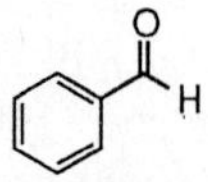	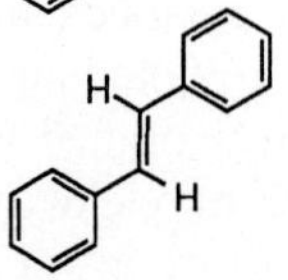	97	$TiCl_3$–Li	6
G	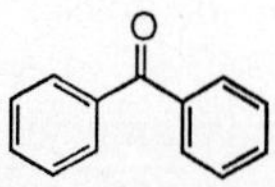	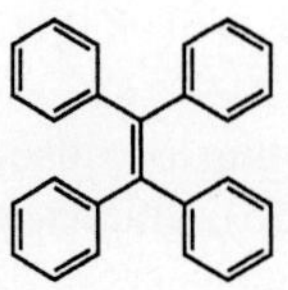	96	$TiCl_3$–Li	6

3.5.1.2 Synthesis of Other Tetrasubstituted Olefins

A number of theoretically interesting tetrasubstituted olefins have been made by the intermolecular McMurry approach. For example, Lenoir has synthesized tetracyclopropylethylene **28** [26] from dicyclopropyl ketone, albeit in modest yield (25%, Scheme 3-16a), which may be attributable to destruction of the starting material through opening of the cyclopropylketyl radical **29** (Scheme 3-16b).

(a) **27** $\xrightarrow[\text{pyridine, THF}]{TiCl_4,\ Zn}$ **28**

(b) **29** ⟶ **30**

Scheme 3-16

Several unsuccessful attempts have been made to use the intermolecular coupling to synthesize the long sought after tetrakis(*tert*-butyl)ethylene molecule **32** (Scheme 3-17) [29]. Although mass spectrometry showed a minor species corresponding to the correct molecular mass, this substance could not be isolated or characterized further.

31 $\xrightarrow{TiCl_3,\ Zn\text{-}Cu}$ (blocked) **32**

Scheme 3-17

Table 3-3 shows a number of other tetrasubstituted olefins of theoretical interest synthesized by the McMurry reaction [30a–k]. In general for these reactions the yields of isolated product are good, even though in many cases considerable strain energy is being added to the products formed.

Aldehydes are among the easiest substrates to couple. One of the most useful applications of the intermolecular aldehyde coupling to date is the synthesis of *β*-carotene (**34**)

33 (CHO) $\xrightarrow{TiCl_3,\ LAH}$ *β-carotene* **34**

Scheme 3-18

Table 3-3. Synthesis of tetrasubstituted olefins.

Entry	Substrate	Product	Ref.
A	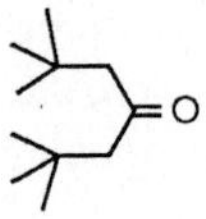	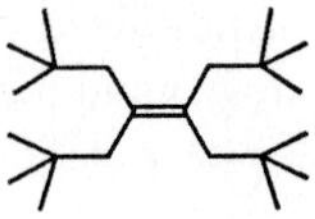	30a
B	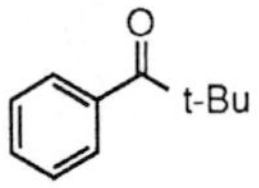	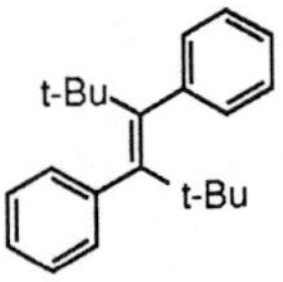	30b
C	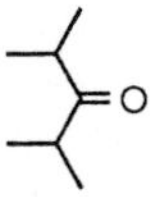	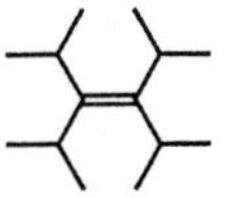	27
D	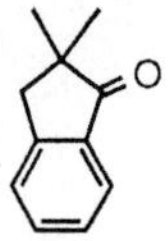	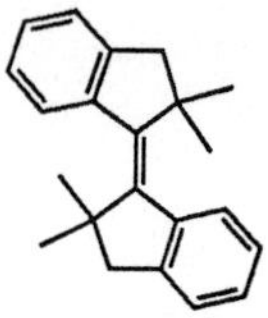	30c
E	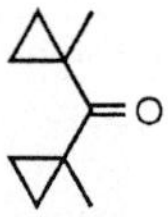	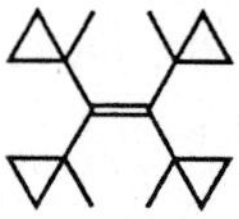	30d
F	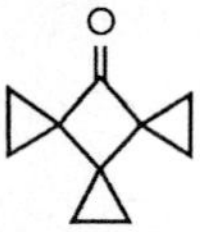	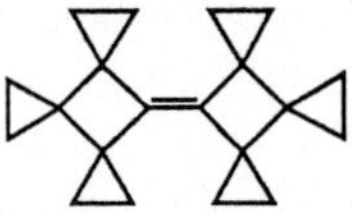	30e
G	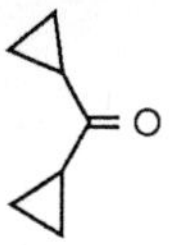	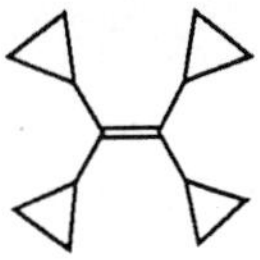	26
H	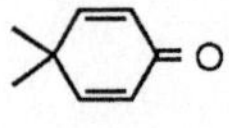	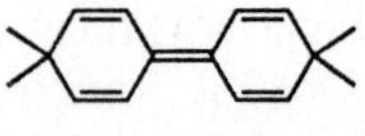	30f
I	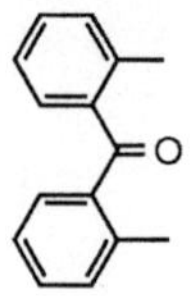	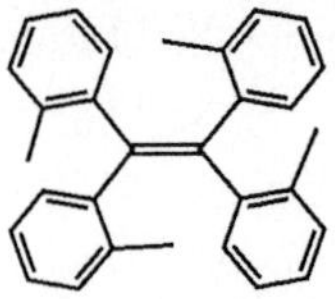	30g
J	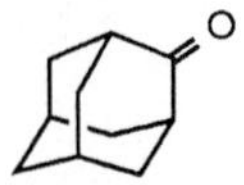	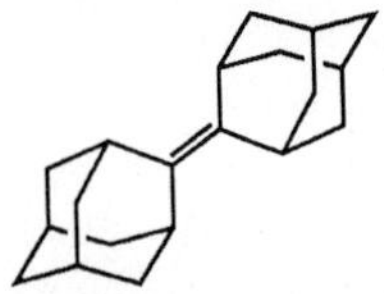	30h
K	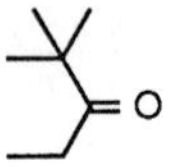	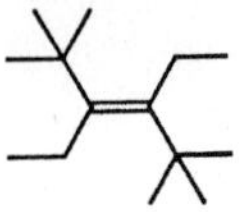	30i
L	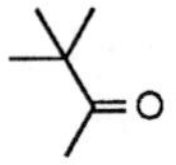	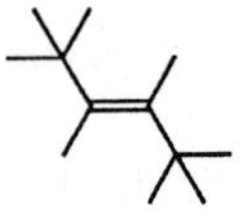	30j
M	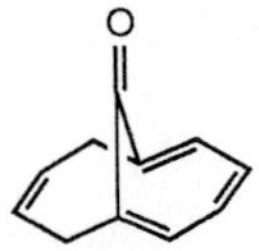		30k

from retinal (**33**) in 84% yield using $TiCl_3$/LAH [1]; the yield with $TiCl_3(DME)_{1.5}$ as the titanium source should be even higher (Scheme 3-18). The result clearly demonstrates that a significant degree of unsaturation is compatible with the reaction conditions.

3.5.1.3 Mixed Coupling Reactions

A mixed coupling reaction is one where two different substrates are combined in an intermolecular process. Unfortunately the utility of these reactions is somewhat limited by the fact that a statistical mixture of products can usually be anticipated. However, in some instances, especially when one relatively inexpensive component is used in excess, the mixed coupling can be biased in favor of the desired product [24]. In many cases, acetone has been used as the excess reagent, with the 2,3-dimethyl-2-butene formed as a by-product easily removed by distillation upon work-up. A good example is the reaction of 3-cholestanone (**35**) with acetone, present in a fourfold excess, which yields isopropylidenecholestane **36** in 54% yield along with 29% of the homodimer **37** (Scheme 3-19) [24].

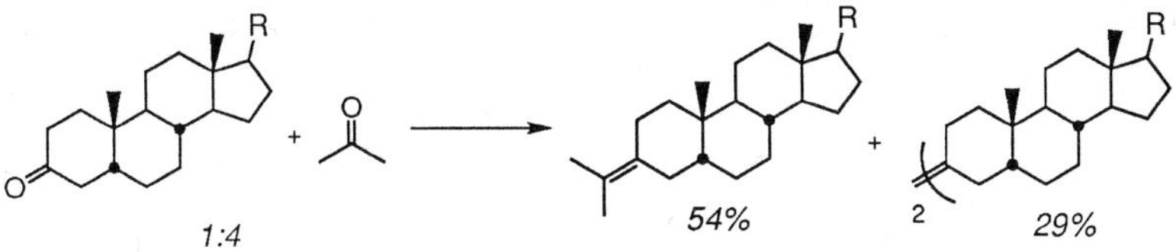

Scheme 3-19

When one of the reactants is an aryl ketone, higher yields from the mixed coupling reaction can be obtained. For example, reaction of a 1:1 ratio of cyclohexanone and benzophenone yields primarily the mixed product **39**. Mechanistically this result can be accounted for by the proposal that the diaryl ketone component undergoes a fast two-electron reduction yielding a dianion which then nucleophically attacks the other component ketone before it can be reduced (Scheme 3-20) [24].

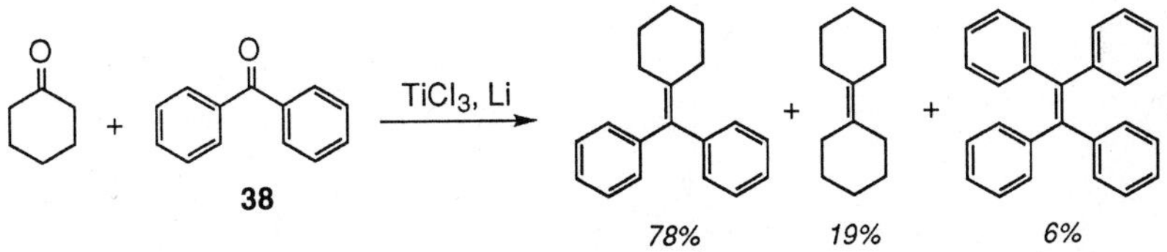

Scheme 3-20

The best published results in the mixed coupling reaction with aryl carbonyl compounds to date have been obtained with the $TiCl_3$/Li reagent combination, possibly because the reaction has not been reinvestigated in recent years; for example, the new $TiCl_3(DME)_{1.5}$ complex has not been systematically screened.

General Procedure for the Mixed Coupling Reaction using $TiCl_3$/Li [24]. Lithium wire (0.45 g, 65 mmol) was added to a stirred suspension of $TiCl_3$ (2.87 g, 18.6 mmol) in 30 mL DME under an argon atmosphere, and the mixture was refluxed for 1 h. The black slurry was then cooled to room temperature, and the two carbonyl components (4.65 mmol total) in 5 mL DME were added. The mixture was stirred for 2 h at room temperature and then refluxed for 16 h. After cooling back to room temperature, the reaction was diluted with pentane and filtered through a small pad of Florisil. Evaporation of solvent from the filtrate gave the crude product. In this manner, the following results, also listed in Table 3-4, were obtained: entry A, adamantanone with 4 equiv. acetone gave 63% isopropylideneadamantane; entry B, cycloheptanone with 4 equiv. acetone gave 50% isopropylidenecyclohexane; entry C, acetophenone with 4 equiv. acetone gave 2-methyl-3-phenyl-2-butene in 65%; entry D, benzophenone with 1 equiv. acetone gave 1,1-diphenyl-2-methyl-1-propene in 94% yield; entry E, benzophenone with 1 equiv. hexanal gave 1,1-diphenyl-1-heptene in 84%; entry F, fluorenone with 1 equiv. cycloheptanone gave 77% cycloheptylidenefluorene.

Table 3-4. Mixed coupling reactions [24].

Entry	Component A	Component B	Product	Yield (%)
A		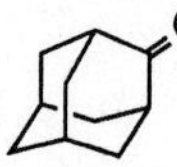	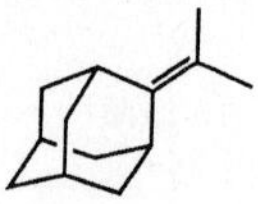	63
B		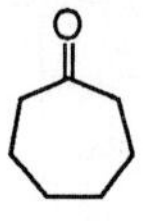	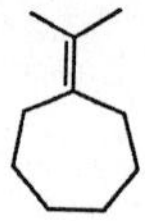	50
C		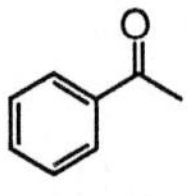	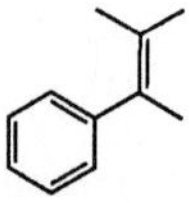	65
D		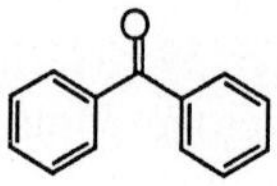	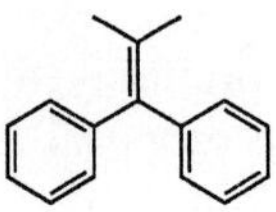	94
E	$Me(CH_2)_4CHO$	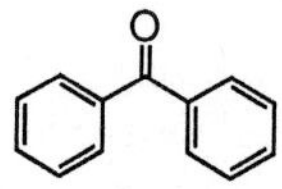	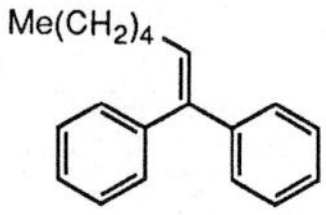 $Me(CH_2)_4$	84
F	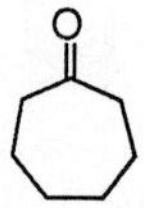	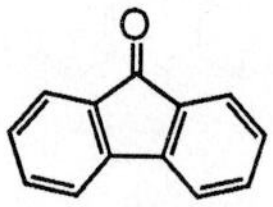	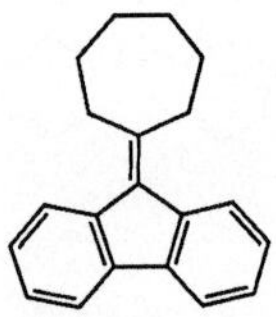	77

In an especially interesting case Paquette has reported the mixed coupling of two similar components in 60% yield (Scheme 3-21) [31]. Both the $TiCl_3/Li$ and $TiCl_3/Zn–Cu$ procedures were implemented, and gave comparable results.

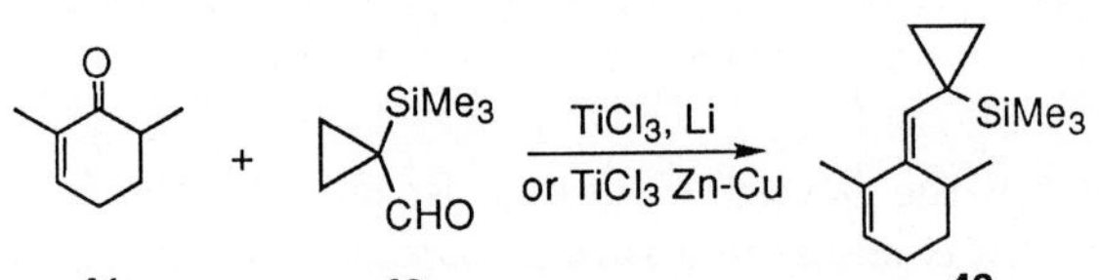

SiMe3 + TiCl3, Li or TiCl3 Zn-Cu CHO SiMe3

Scheme 3-21 **41** **42** **43**

The aryl-alkyl coupling strategy has been employed in a nice synthesis of the anti-tumor agent tamoxifen (**46**) in 88% yield (Scheme 3-22) [32–33]. In this case $TiCl_4/Zn$ was used to produce an exceptional result.

44 Me + TiCl4, Zn Me **45** NMe2 NMe2

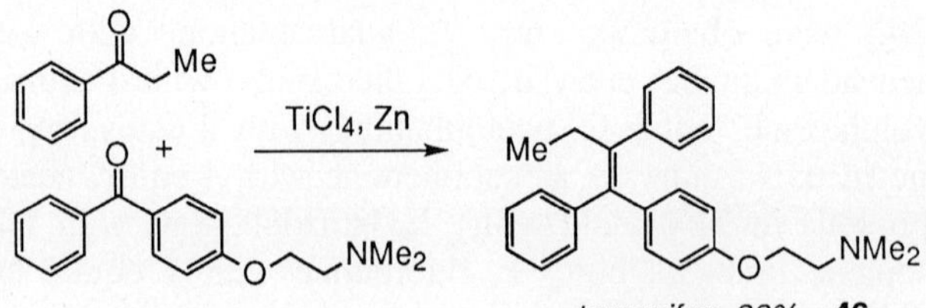

Scheme 3-22 *tamoxifen, 88%* **46**

3.5.2 The Intramolecular Coupling Reaction

3.5.2.1 Basic Couplings

The most notable feature of the McMurry reaction is its ability to produce large ring systems efficiently (Scheme 3-23) [2a, 34]. Virtually all ring sizes from three upwards afford products in high yields; however, for larger ring systems a mixture of *E*- and *Z*-isomers may be obtained. Table 3-5 shows some representative examples using the standard $TiCl_3$/Zn–Cu reagent [34]; yields are consistently good for all ring sizes attempted. Using the optimized procedure with $TiCl_3(DME)_{1.5}$/Zn–Cu, isolated yields should be somewhat higher. Even a 36-membered ring has been formed in good yield (see below), and there is also no reason to expect that much larger ring systems would not be formed.

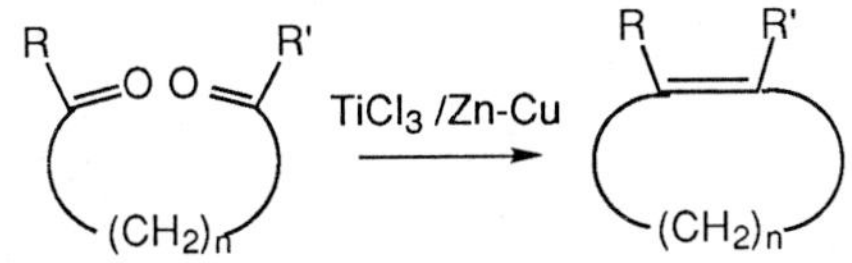

R, R', O O, $(CH_2)_n$ → $TiCl_3$ /Zn-Cu → R, R', $(CH_2)_n$

Scheme 3-23

For the successful synthesis of large rings, high-dilution protocols should be followed. The author recommends the use of a good quality syringe pump which adds the substrate smoothly to the reaction mixture over a 30–40 h interval. Because intramolecular couplings are operationally more elaborate than the intermolecular variants, several archetypi-

Table 3-5. Intramolecular couplings of ring sizes 4–16 [34].

Substrate	Product	Yield (%)	Substrate	Product	Yield (%)
$PhCO(CH_2)_2COPh$	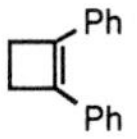 Ph, Ph	87	$BuCO(CH_2)_8COBu$	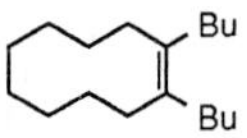 Bu, Bu	75
$MeCO(CH_2)_3COPh$	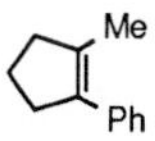 Me, Ph	70	$BuCO(CH_2)_9COBu$	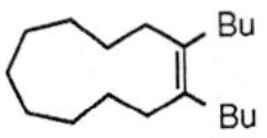 Bu, Bu	76
$PhCO(CH_2)_4COPh$	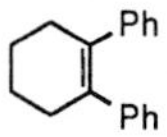 Ph, Ph	95	$OHC(CH_2)_{10}CHO$	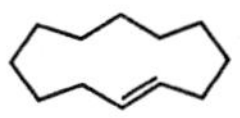	76
$Ph(CH_2)_2CO(CH_2)_4COPh$	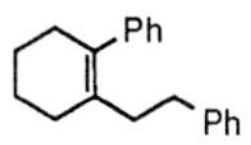 Ph, Ph	50	$OHC(CH_2)_{11}CHO$	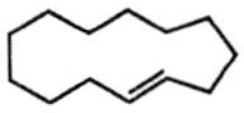	52
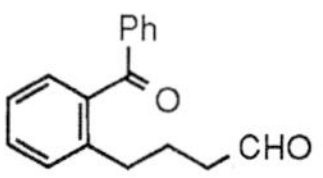 Ph, O, CHO	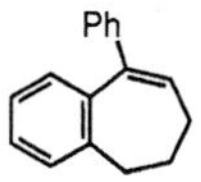 Ph	80	$OHC(CH_2)_{12}CHO$	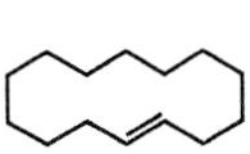	71
$BuCO(CH_2)_6COBu$	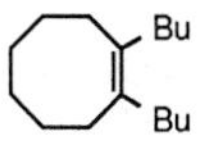 Bu, Bu	67	$PhCO(CH_2)_{13}CHO$	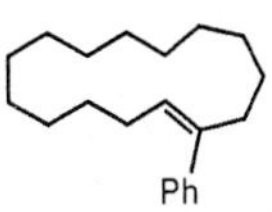 Ph	80
$BuCO(CH_2)_7COBu$	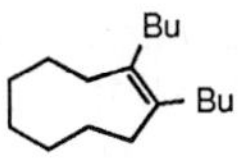 Bu, Bu	68	$OHC(CH_2)_{14}CHO$	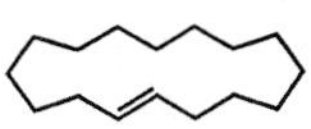	85

cal experimental procedures for performing a successful coupling are provided below. For example, the synthesis of cyclotetradecene (**48**) employing the $TiCl_3(DME)_{1.5}$ reagent results in an increase in yield of 10% over $TiCl_3$ (Scheme 3-24).

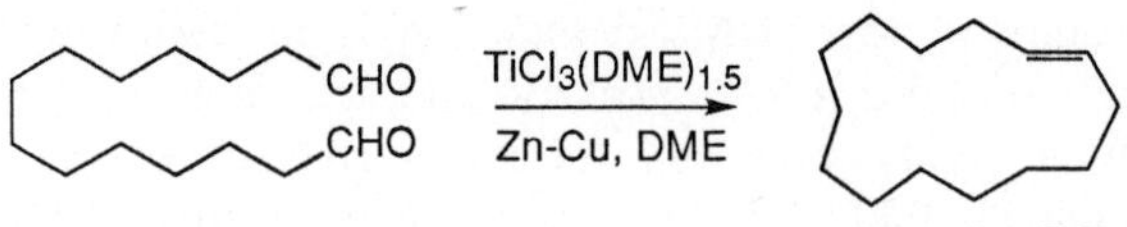

CHO CHO 47 → $TiCl_3(DME)_{1.5}$, Zn-Cu, DME → 48 9:1 *E*:*Z*

Scheme 3-24

Standard Conditions for an Intramolecular Coupling: Cyclotetradecene [27]. $TiCl_3(DME)_{1.5}$ (5.2 g, 17.9 mmol) and Zn–Cu (3.8 g, 58 mmol) were added to a dry, argon-purged flask and were stirred vigorously while DME (150 mL) was added by syringe. After the mixture was heated at 80 °C for 4 h to form the active titanium coupling reagent, freshly-prepared tetradecanedial (**47**) (0.50 g, 2.20 mmol) in 50 mL DME was added via syringe pump over a period of 35 h. The reaction was heated for a further 6 h after addition was complete and was then cooled to room temperature. Pentane (160 mL) was added to the reaction flask, and the slurry was filtered through a small (5 cm × 10 cm) pad of Florisil to remove metal salts. After washing the Florisil pad with an additional 100 mL pentane, the filtrates were combined and concentrated at 0 °C under reduced pressure to give cyclotetradecene (**48**, 340 mg, 80%) as a colorless oil. NMR and capillary GC analysis indicated that the product was a 9:1 mixture of *cis* and *trans* isomers.

Synthesis of Very Large Rings. Recently Kakinuma et al. reported a McMurry coupling to form a 36-membered ring in 53% yield, making this the largest ring to date formed by a titanium coupling reaction (Scheme 3-25) [35]. The syringe pump addition time in this case was extended to 100 h.

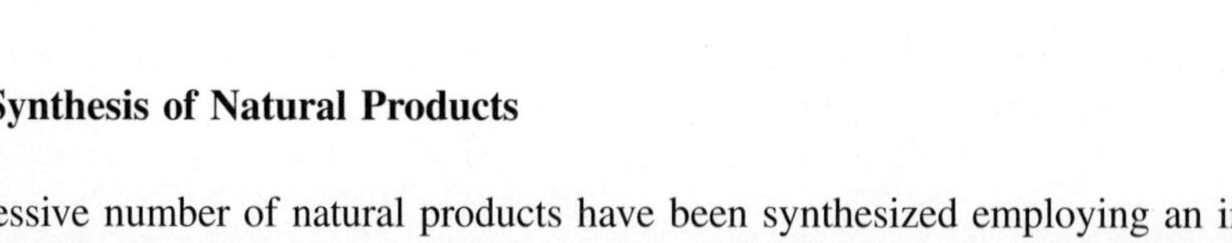

OHC $(CH_2)_{14}$ O OBn H OHC $(CH_2)_{14}$ O 49 → $TiCl_3$, Zn-Cu, DME → $(CH_2)_{14}$ O OBn H $(CH_2)_{14}$ O 50

Scheme 3-25

3.5.2.2 Synthesis of Natural Products

An impressive number of natural products have been synthesized employing an intramolecular McMurry coupling as a key step. In fact, the McMurry cyclization is one of the methods of choice for the synthesis of large-ring terpenes. A simple, elegant example of a natural product synthesis involving an intramolecular coupling is the synthesis of helminthogermacrene (**54**) and β-elemene (**56**) by McMurry and Kočovsky (Scheme 3-26) [36]. The synthesis starts with commercially available geranylacetone **51**, which can undergo selective allylic oxidation in the terminal position, followed by transetherification and Claisen rearrangement to afford the keto aldehyde precursor **53**, which is cyclized using standard conditions to afford **54** and the intermediate **55**, which undergoes a Cope

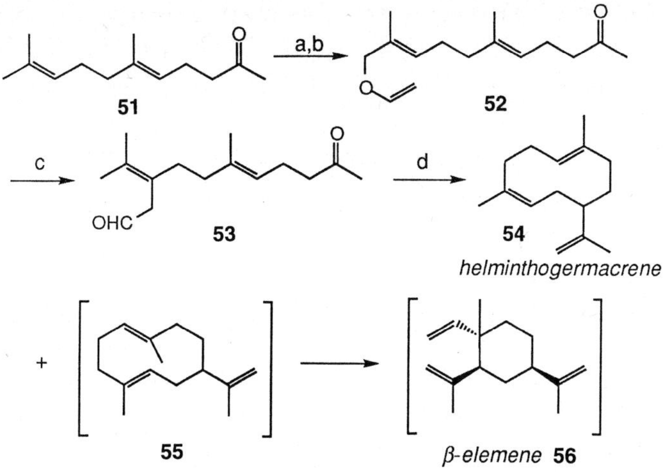

Scheme 3-26

rearrangement to yield β-elemene (**56**). The synthesis by McMurry and Bosch [37] of casbene, (**58**) a 14-membered ring diterpene isolated from seedlings of the castor bean [38], represents a more complex example (Scheme 3-27).

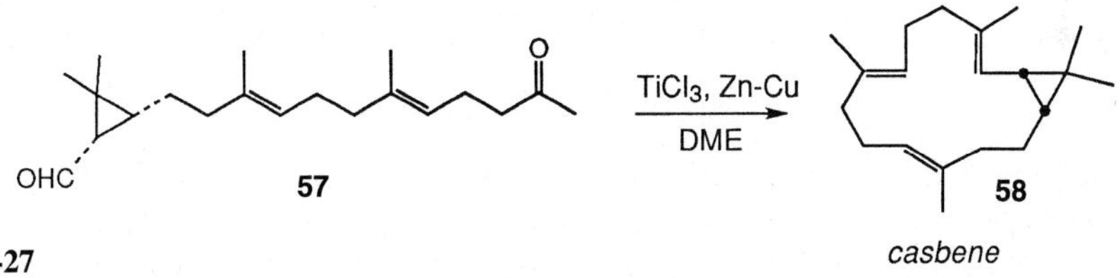

Scheme 3-27

The Synthesis of (+)-Casbene (58). Standard conditions were followed for reagent preparation with $TiCl_3$ (3.57 g, 20.6 mmol), Zn–Cu (4.17 g, 63.8 mmol), DME (65 mL). Keto aldehyde **57** in 50 mL DME was added over 38 h via syringe pump, to yield 57% (+)-casbene (**58**), in addition to 38% isomeric hydrocarbons. Purification was accomplished by reverse-phase HPLC (C18 column), pure MeCN eluent.

Several natural product syntheses employing the McMurry reaction as a key step are shown in Table 3-6, including the syntheses of verticillene [39], bicyclogermacrene [40], lepidozene, humulene [41, 42] and flexibilene [43]. With the exception of verticillene, all the couplings are of the keto aldehyde type.

In other natural products work, Ben et al. have implemented a clever strategy for the synthesis of cannithrene II (**61**) using a McMurry coupling as a key step (Scheme 3-28)

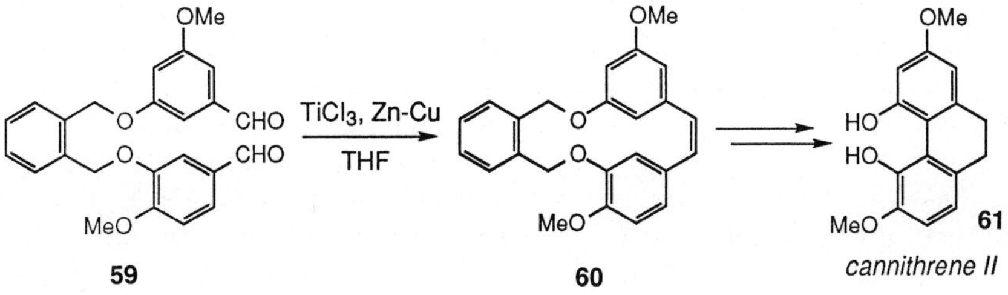

Scheme 3-28

Table 3-6. Syntheses of some natural products using the McMurry alkene reaction.

Entry	Substrate	Product		Ref.
A	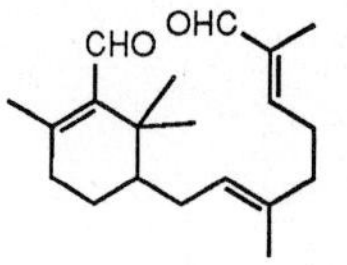 CHO OHC	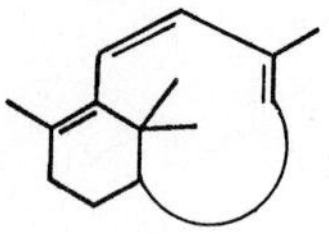	verticillene	39
B	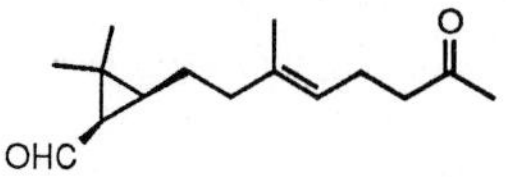 OHC	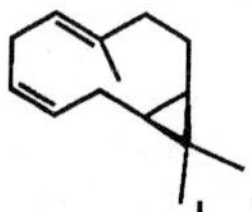	bicyclogermacrene	40
C	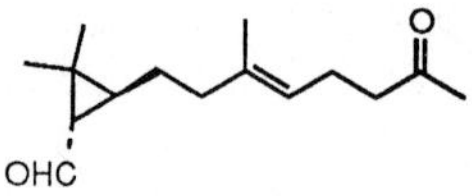 OHC	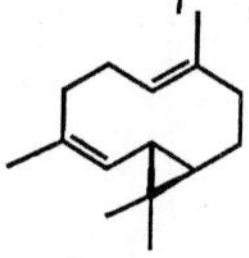	lepidozene	41, 42
D	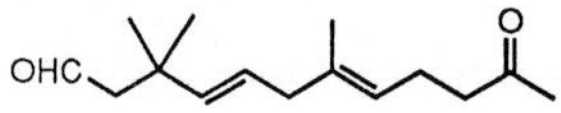 OHC	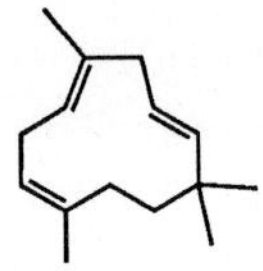	humulene	41, 42
E	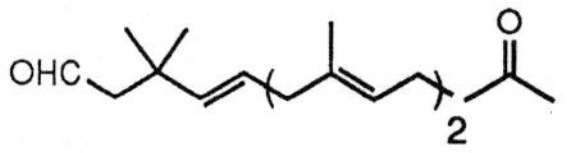 OHC 2	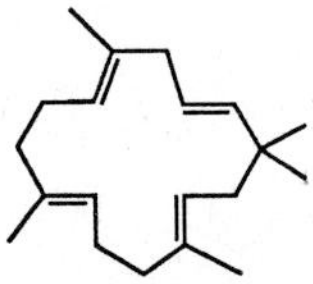	flexibilene	43

[44]. A dibenzylether linker connects to aryl fragments, a fact which renders the coupling intramolecular, thereby controlling the chemoselectivity and stereochemistry of the coupling. Removal of the linker, biaryl coupling and reduction in the subsequent steps then affords the natural product.

In the synthesis of hirsutene (**67**) by Weedon (Scheme 3-29) [45, 46], the diketone precursor **65** is produced by a photocycloaddition followed by a DeMayo reaction. The coupling step was performed by $TiCl_3$/K in THF, which may account for the somewhat low yield (30%).

The synthesis of isokhusimone (**69**) is interesting from the point of view that the precursor **68** contains three carbonyls which can react instead of two [47]. The reaction occurs preferentially to form the six-membered ring product rather than the three-membered.

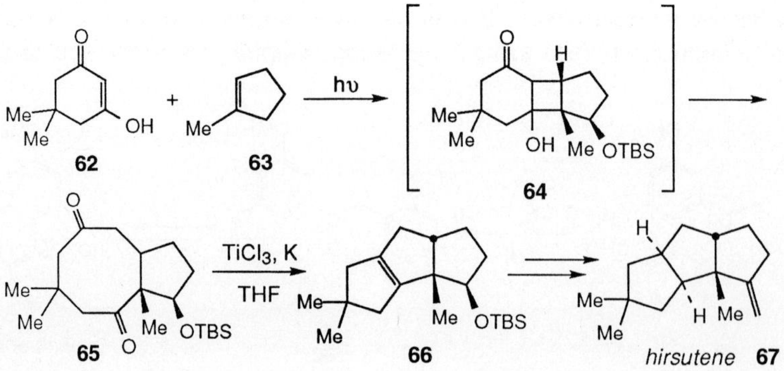

Me Me OH **62** + Me **63** hυ Me Me OH Me OTBS H **64**

Me Me **65** Me OTBS $TiCl_3$, K THF Me Me **66** Me OTBS Me Me H H Me *hirsutene* **67**

Scheme 3-29

Once the coupling occurs, given the high dilution of the reaction, no competing intermolecular coupling of the remaining carbonyls is observed (Scheme 3-30).

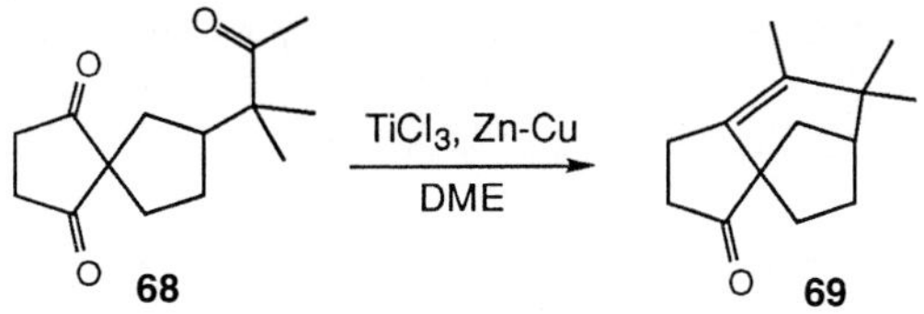

$TiCl_3$, Zn-Cu
DME

68 69

Scheme 3-30

The synthesis of strigol (**72**, Scheme 3-31) also contains a choice of potentially reacting carbonyls in the precursor **70** [48]. As expected, the reaction forms the five rather than the four-membered ring, which is also disfavored by the fact that a deactivated ester group is present.

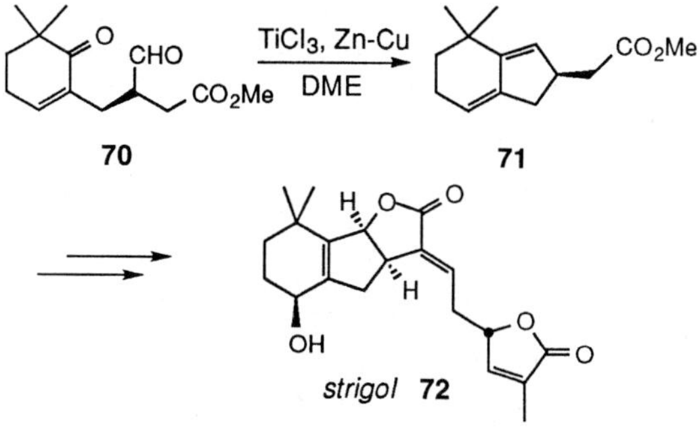

CHO CO_2Me $TiCl_3$, Zn-Cu DME CO_2Me

70 71

OH

strigol **72**

Scheme 3-31

3.5.2.3 The Potassium–Graphite Modification

Fürstner and Weidmann have reported the use of $TiCl_3/C_8K$ as an alternative to the $TiCl_3$/Zn–Cu system for the coupling of ketones and aldehydes [22, 23]. Potassium–Graphite (C_8K) is an easy-to-handle, surprisingly mild reagent in contrast to potassium metal itself, which is no longer recommended for use, since among other reasons it presents a safety hazard.

For certain highly oxygenated compounds, especially those containing ethylene ketals, use of a reagent derived from a $TiCl_3/C_8K$ combination may be beneficial, and seems to be a less Lewis acidic alternative to the standard $TiCl_3$/Zn–Cu combination [49]. The modification has been applied by Clive in the total synthesis of mevinolin, wherein the keto aldehyde **73** was coupled to yield the product **74** (85%, Scheme 3-32) [50].

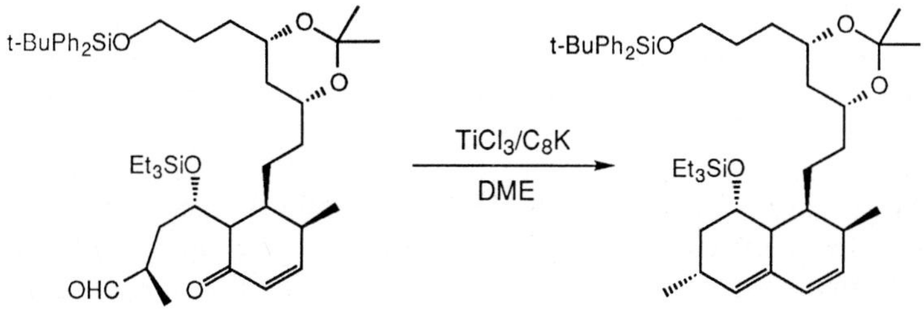

t-BuPh₂SiO t-BuPh₂SiO

$TiCl_3/C_8K$
DME

Et_3SiO Et_3SiO

OHC O

73 74

Scheme 3-32

The Clive Procedure for Titanium-Mediated Dicarbonyl Coupling [49, 50]. Freshly prepared potassium graphite (C_8K, 364 mg, 2.693 mmol) and $TiCl_3$ (192.6 mg, 1249 mmol) were weighed under argon in a glove bag and transferred successively to a 50-mL three-necked flask containing dry DME (15 mL). The mixture was stirred and refluxed for 2 h under argon and then cooled to room temperature. Enone aldehyde **73** (55.2 mg, 0.0737 mmol) in dry DME (5 mL) was added by syringe pump over 10 h to the stirred slurry of titanium reagent. Stirring was continued for an additional 5 h. The mixture was then refluxed for 4 h, cooled to room temperature, and filtered under a blanket of argon through a pad of Florisil (3.5 cm × 6 cm) contained in a sintered funnel equipped with an argon inlet near the top. The column was washed with ether (3 × 50 mL). Evaporation of the combined filtrates and flash chromatography of the residue over silica (1 × 15 cm) with 1:9 ether-petroleum ether gave **74** (45.6 mg, 86%) as an apparently homogeneous (TLC, silica, 1:9 ether-petroleum ether) oil.

3.5.2.4 Synthesis of Strained Rings and Non-Natural Products

Of the many positive attributes of the McMurry reaction, none stands out more than its ability to produce strained products. This notable characteristic has permitted the synthesis of a sizable number of theoretically interesting non-natural olefins. As a general rule of thumb, if a coupling reaction introduces less than 19 kcal mol^{-1} of strain energy into a potential product, the coupling can be expected to occur, all other things being equal. One of the keys to success in all intramolecular strained couplings is the use of carefully optimized syringe-pump additions. The time of substrate addition represents a compromise between that needed to suppress dimer formation and that needed to ensure that an active reagent remains after the specified time of addition.

One interesting application of the alkene coupling is the synthesis of the bicyclo[4.4.4]-1-tetradecene ring system (Scheme 3-33) [51, 52]. The coupling of the keto

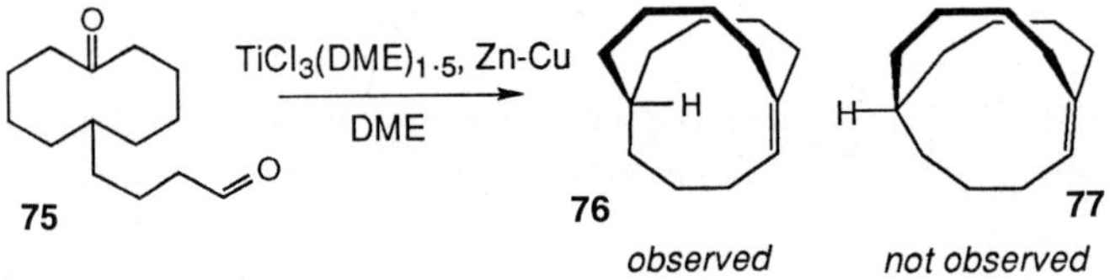

Scheme 3-33

aldehyde **75** takes place in 30% yield using $TiCl_3(DME)_{1.5}$/Zn–Cu using a 35 h addition time [52]. Upon protonation with any number of Brönsted acids, the alkene **76** yields the remarkably stable bicyclo [4.4.4]-1-tetradecyl cation **78**, a molecule possessing a three-center, two-electron C–H–C bond (Scheme 3-34) [53]. Molecular mechanics calculations indicate that the desired isomer, with the bridgehead hydrogen pointing *in*, is more stable than the corresponding *out* isomer **77** by some 12 kcal mol^{-1}. In fact, only the *in* isomer is produced under the conditions of the reaction, in 25% isolated yield after prep-GC purification. Hydrogenation of alkene **76** affords the strained alkane **79**, which upon treatment with mild Brönsted acids regenerates cation **78** and one equivalent of hydrogen gas, representing the first documented case of stoichiometric hydrogen evolution in an alkane protonolysis reaction [54].

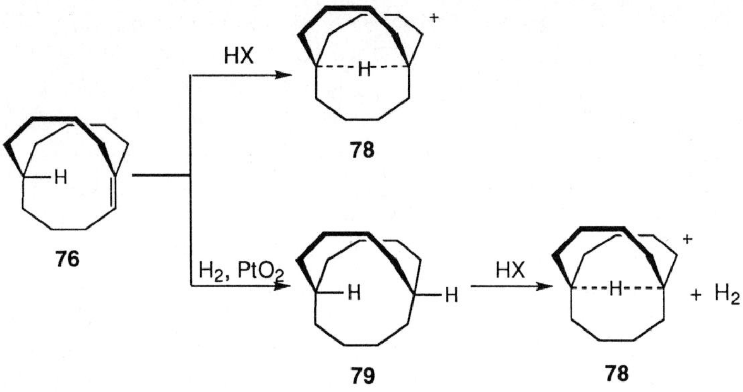

Scheme 3-34

The carbonyl coupling reaction was also employed as the key step in the synthesis of other bicyclic olefins, namely the bicyclo[5.4.4], -[6.3.3] and -[6.4.2] systems shown in Scheme 3-35, each of which similarly leads to cations containing three-center bonds upon protonation [55].

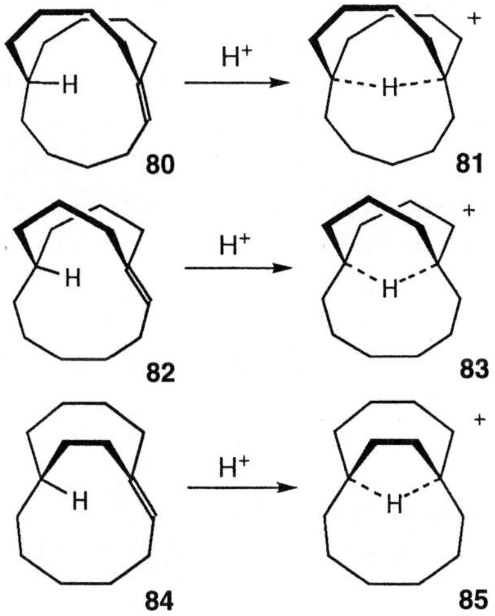

Scheme 3-35

Example of a Strained Intramolecular Coupling Reaction: *In*-Bicyclo[4.4.4]-1-tetradecene (76) [52]. $TiCl_3(DME)_{1.5}$ (7.0 g, 20.9 mmol see [27]) and Zn–Cu couple (4.0 g, 61.9 mmol) were added to 250 mL DME in a 500 mL two-necked flask through a Schlenk tube under argon. The mixture was stirred vigorously at reflux for 3 h to yield a green-black to black homogeneous slurry. Keto aldehyde **75** (50 mg, 0.223 mmol) in 50 mL DME was added to the stirred mixture through a syringe pump over 35 h. The mixture was then cooled to room temperature and diluted with distilled, degassed pentane (300 mL), and filtered through Florisil under argon. The solvent was distilled off at atmospheric pressure and the residue was purified by prep-GC (OV-101 on Supelcoport at 170 °C) to yield 13 mg olefin **76** (30%) as a waxy, white solid.

A synthesis of the stable enol **88** was attempted in unpublished work by McMurry and Lectka (Scheme 3-36) [56]. In one of the more strained couplings performed to date, the keto

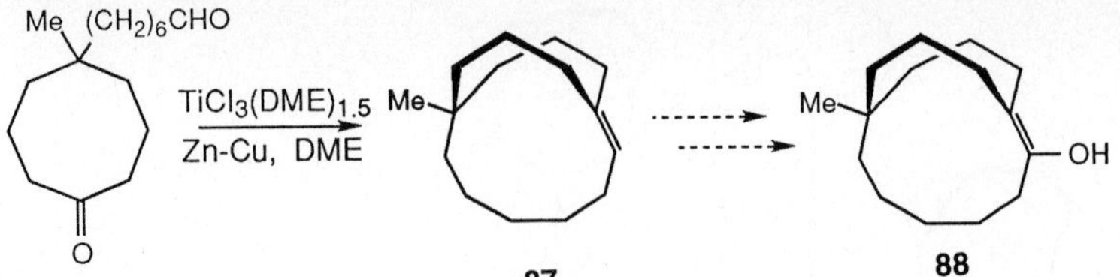

Scheme 3-36

aldehyde **86** was coupled using standard conditions to afford **87**. A bridgehead methyl group is present in the product, forcing it to point outward. Molecular mechanics calculations suggested that approximately 18 kcal of strain energy would be incorporated into the product, close to the limit of the reaction. Consequently, the yield was low (8%), and insufficient material was available to fully characterize the product as being the keto-enol target.

Sorensen has synthesized a series of stable tricyclic cations **91** and **92** through McMurry coupling of keto aldehyde precursors **89**, followed by protonation (Scheme 3-37) [57]. Sorensen found that for the large-ring cation **92** ($n = 8$), spectroscopic data consistent with a rapidly equilibrating set of isomers linked through hydride shifts was obtained. As the ring size was shortened ($n = 8 \rightarrow 5$) the cations became progressively more non-classical. For cation **91** ($n = 5$), a distinctly non-classical structure containing a three-center, two-electron bond was observed.

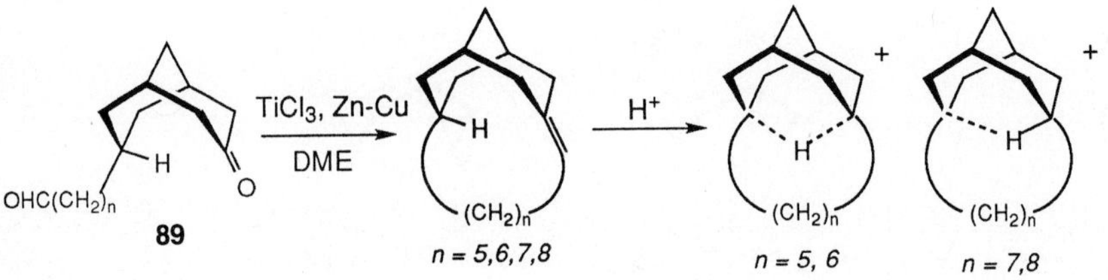

Scheme 3-37

McMurry has used the intramolecular coupling of the diketone **93** to synthesize the triene **94**, a theoretically interesting polyene in which the three double bonds are projected towards the center of the cyclic array and consequently subject to through-space interactions (Scheme 3-38) [58, 59]. Although the molecule is not neutrally homoaromatic [60], the proximity of the double bonds to each other results in a substantial splitting of the π-orbitals as determined by photoelectron spectroscopy. In addition, coupling of the keto aldehyde **95** afforded the tetraene **96**, which was investigated as a tetradentate ligand for transition metals [61]. X-ray crystallographic analysis of **96** indicated a distance of 5.11 Å between the double bonds across the ring, almost perfect to accommodate an Ag^+ ion within the cavity; Ag^+ requires a metal-to-carbon bond length of 2.5 Å. Reaction of **96** with $AgBF_4$ results in a stable complex in which the Ag^+ ion possesses a square-planar geometry, the first such case documented for a d^{10} metal.

The coupling of diketone **93** was expected to be difficult from the outset, since in order for coupling to occur the three cyclohexane rings would necessarily have to attain the boat conformation **98**. In the event, the coupling took place in 24% yield after purification. In contrast, the cyclization of the less-strained four-cyclohexane-ring system **95** occurred in 94% yield.

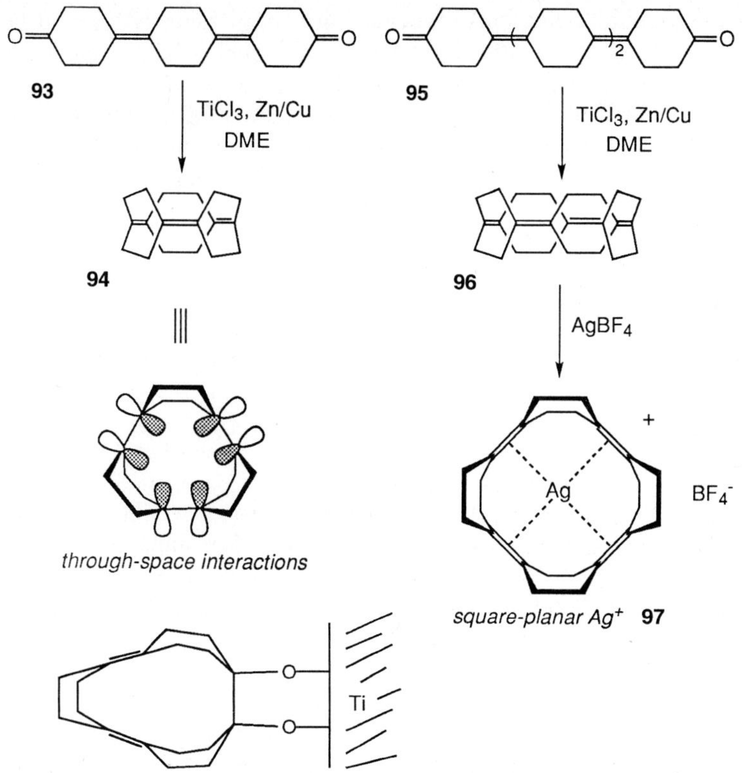

93 TiCl$_3$, Zn/Cu DME 94 95 TiCl$_3$, Zn/Cu DME 96 AgBF$_4$

through-space interactions

square-planar Ag^+ 97 BF_4^-

Ti

Scheme 3-38 98

The precursor diketone (**93**) was synthesized via the unusual route shown in Scheme 3-39, involving a key Barton olefin synthesis [62] to make the first two carbon–carbon double bonds. Other methods of olefin synthesis failed because these bonds proved to be extremely prone to isomerization.

In an attempt to synthesize the diene **102**, in which the two double bonds would interact perpendicularly across the ring system, McMurry and Swenson subjected diketone

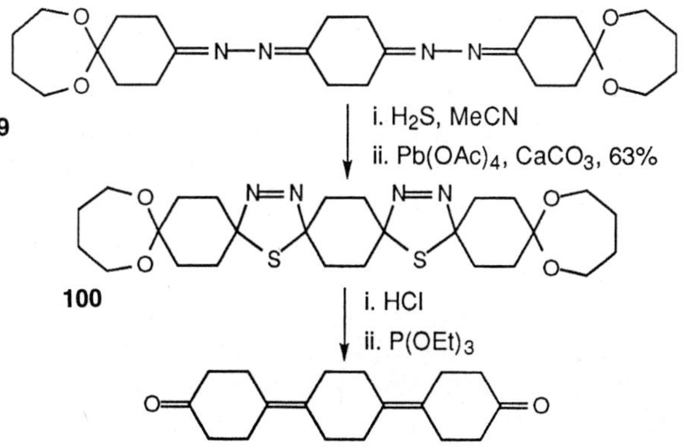

99 i. H_2S, MeCN ii. $Pb(OAc)_4$, $CaCO_3$, 63%

100 i. HCl ii. $P(OEt)_3$

Scheme 3-39 93

101 to standard coupling conditions (Scheme 3-40) [63]. Diene **102** would be the perfect complement to Wiberg's diene **104**, in which the two double bonds interact in a cofacial, parallel fashion. Instead of the desired compound, however, the tricyclic diene **103** was isolated in good yield. Unfortunately under the conditions of the reaction diene **102** was probably formed, but underwent a fast Cope rearrangement due to the geometrical alignment of the double bonds. One possible solution to the rearrangement dilemma would be to synthesize the pinacol **105** at low temperature, then subject it to a mild deoxygenation protocol to generate **102**.

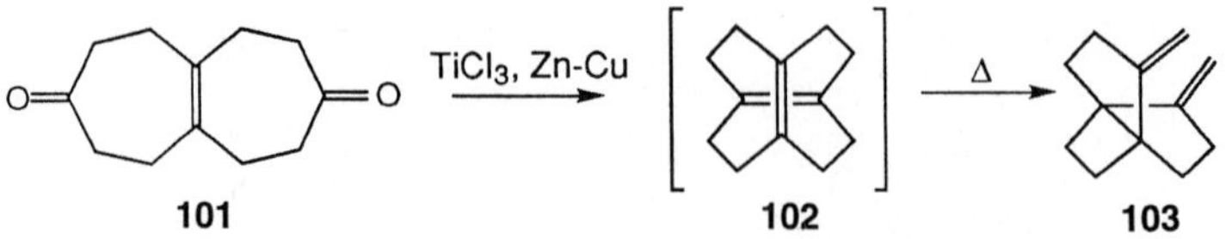

$TiCl_3$, Zn-Cu → [**102**] —Δ→ **103**

101 **102** **103**

Scheme 3-40

What about an instance in which the deoxygenation step would produce an extraordinarily strained molecule? In a coupling of the diketone **106** under standard conditions in refluxing DME, pinacol **107** was isolated instead of olefin **108**, which would possess a severely distorted double bond (Scheme 3-41) [64].

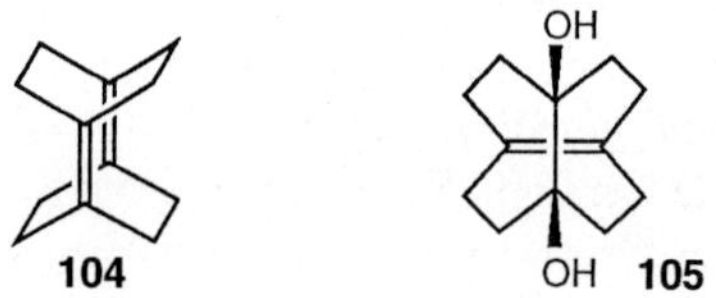

104 OH OH **105**

106 —$TiCl_3$, Zn-Cu→ HO HO **107** —//→ **108**

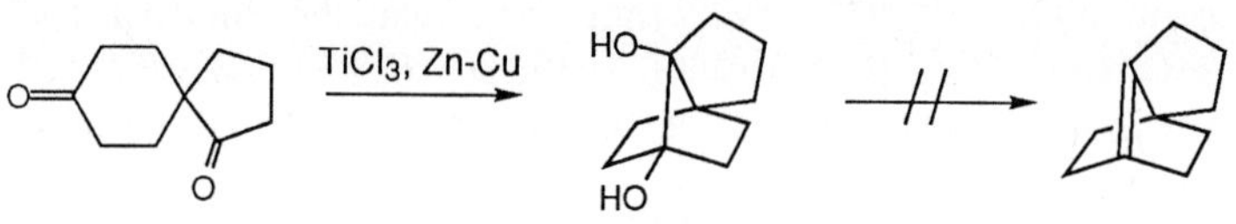

Scheme 3-41

3.5.2.5 The Acyl Silane Coupling

Another interesting strategy for intramolecular coupling would be the use of an acyl silane in place of a ketone. The resulting vinyl silanes would be valuable synthetic intermediates at the oxidation state of a ketone or enol ether. Although the reaction works in acceptable yields for small rings, for larger rings the yields of product are unfortunately poor (Scheme 3-42) [65].

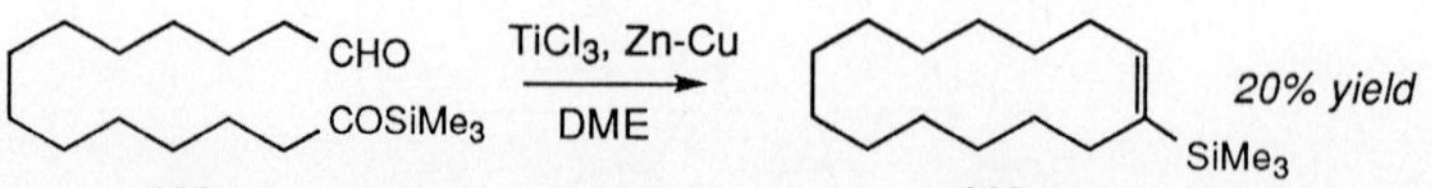

109 CHO $COSiMe_3$ —$TiCl_3$, Zn-Cu, DME→ **110** $SiMe_3$ *20% yield*

Scheme 3-42

3.6 The McMurry Pinacol Reaction

Over the past several years the development of the McMurry reaction has centered around the pinacol variant (Scheme 3-43) [8]. Once again, products derived from both intermolecular and intramolecular couplings are obtained in high chemical yields. The primary challenge of the pinacol reaction is to predict the stereochemistry of product formation, and to design highly diastereoselective reaction protocols.

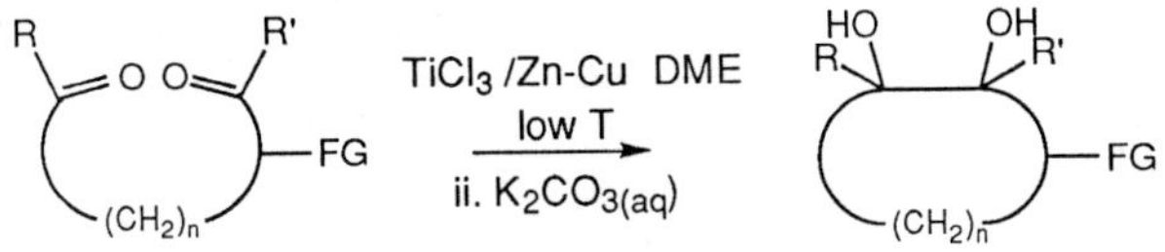

R R' O O FG $(CH_2)_n$ $TiCl_3$ /Zn-Cu DME low T ii. $K_2CO_{3(aq)}$ HO OH R R' FG $(CH_2)_n$

Scheme 3-43

3.6.1 Mechanistic Considerations and Prototype Couplings

For many, especially unstrained couplings, the rate determining step in the McMurry alkene synthesis is presumed to be deoxygenation under standard conditions of the reaction. Consequently as the reaction temperature is lowered, for instance from refluxing DME to room temperature or below, the deoxygenation step can be effectively suppressed and the intermediate pinacols can be isolated in high yields. In analogy to the intramolecular alkene coupling, high dilution procedures are followed. After the completion of substrate addition the pinacol products remain tightly bound to the titanium metal surface, but can be easily liberated through basic hydrolysis (Scheme 3-44). As in the alkene synthesis, for large enough ring systems either or both of the *cis* and *trans* isomers can be produced in the coupling.

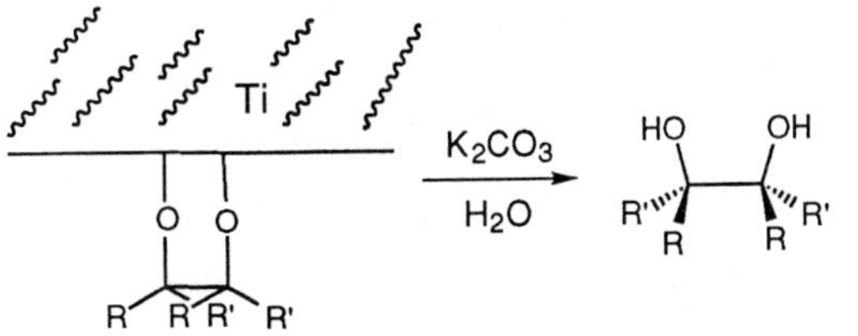

Ti O O R R R' R' K_2CO_3 H_2O HO OH R' R' R R

Scheme 3-44

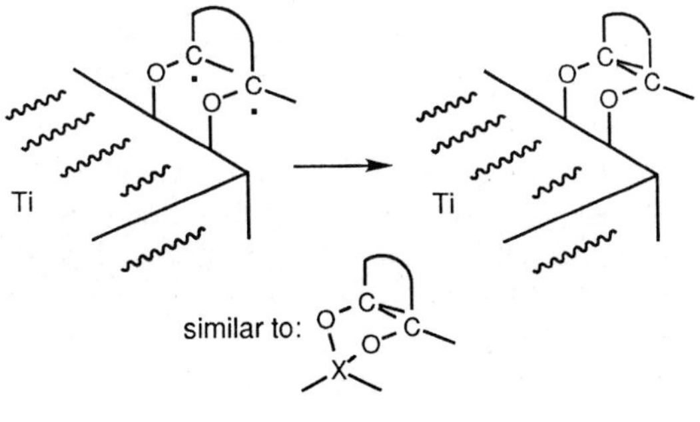

O C O C Ti → O C O C Ti similar to: O C O C X

Scheme 3-45

Table 3-7. Calculated and observed product ratios for the intramolecular pinacol reaction.

Entry	Dialdehyde	Diol	Observed ratio *cis/trans*	Calculated ratio *cis/trans* 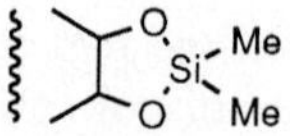
1	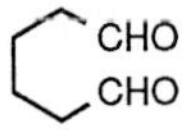	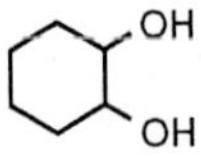	99:1	83:17
2	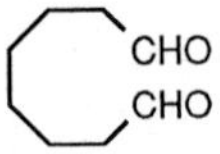	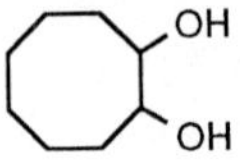	70:30	41:59
3	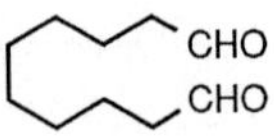	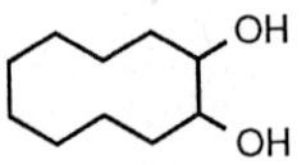	25:75	11:89
4	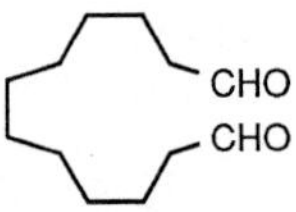	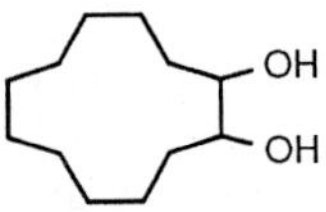	25:75	5:95
5	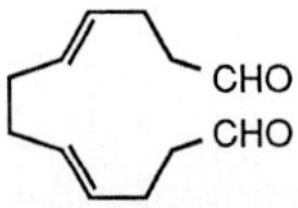	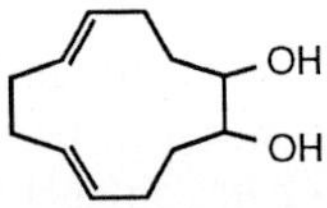	5:95	1:99
6	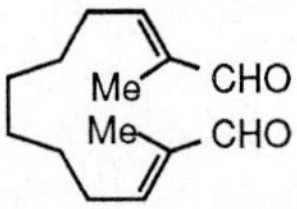	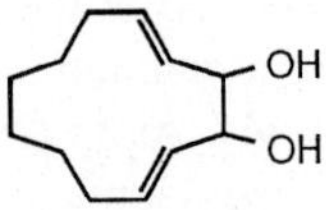	5:95	1:99
7	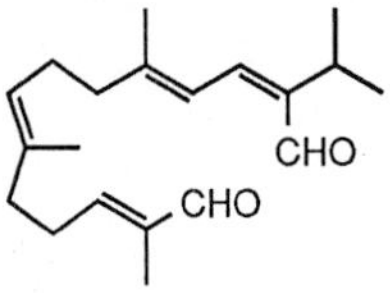	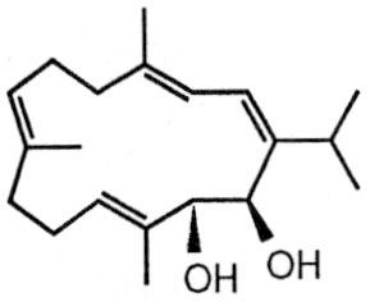	2:98	2:98

McMurry and Siemers have proposed that the diradical coupling step occurs on an edge of the titanium metal surface, so that steric interactions are minimized; thus it may be reasonable to assume that the product stereochemistry is controlled by steric interactions in the forming ring. Consequently, by performing molecular mechanics calculations on bicyclic model acetals in which the acetal atom X substitutes for the metal surface, some prediction of reaction diastereoselectivity can be made (Scheme 3-45) [66].

McMurry and Siemers have found that the relative energies of the *cis* and *trans* isomers of dimethylsilyl acetals, as calculated by Still's Macromodel program using the MM2 force field [67], correlate fairly well with the observed product ratios obtained in the pinacol reaction, as shown in Table 3-7 [66]. Worth comment is the fact that the model performs well for large and complex ring systems. For example, entries 5 and 6 both have a calculated *cis/trans* ratio of 1:99 compared with an observed ratio of 5:95.

Shown in Scheme 3-46 is a representative experimental procedure for the synthesis of 1,2-cyclotetradecanediol (**112**) (Scheme 3-46) [8]. In order to liberate the diol from the low-valent titanium reagent, a mildly basic hydrolysis agent is recommended, such as aqueous potassium carbonate.

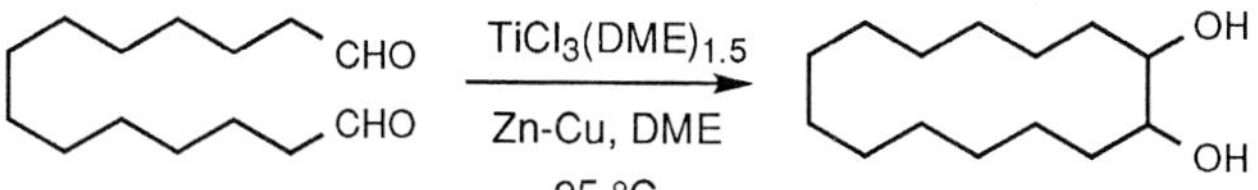

CHO CHO $TiCl_3(DME)_{1.5}$ Zn-Cu, DME 25 °C OH OH

Scheme 3-46 **111** **112**

Representative Procedure for Pinacol Coupling [8]. $TiCl_3(DME)_{1.5}$ (4.38 g, 13.1 mmol) and Zn–Cu couple (2.6 g, 40.0 mmol) were placed under argon in a flame-dried flask via a Schlenk tube. DME was added, the mixture was refluxed for 5 h, and tetradecanedial **111** (0.20 g, 0.88 mmol) in 50 mL DME was added at 25 °C over a period of 30 h using a syringe pump. After the mixture was stirred an additional 8 h, the reaction was hydrolyzed with 70 mL of 20% aqueous K_2CO_3 for 3 h, and then extracted with 1:1 ether/ethyl acetate. The combined organic extracts were washed with brine, dried, and concentrated to afford pinacol product **112** in 90% yield. Analysis of the corresponding acetonide by capillary GC indicated a *cis/trans* ratio of 30:70.

3.6.2 Synthesis of Natural Products

In the first application of pinacol coupling technology the synthesis of sarcophytol B (**116**), an anticancer cembranoid, was undertaken (Scheme 3-47) [68]. Dialdehyde **115** is efficiently available in four steps from farnesal through a Wittig homologation followed by a reduction/oxidation sequence. In the key step, dialdehyde **115** was coupled to form the product in 46% yield, with the desired *trans* stereochemistry. Interestingly enough, the optimal reaction temperature was found to be –40 °C. Although many dialdehydes afford good yields of pinacol at room temperature, with substrates prone to deoxygenation lower reaction temperatures are recommended, and each case should be optimized individually.

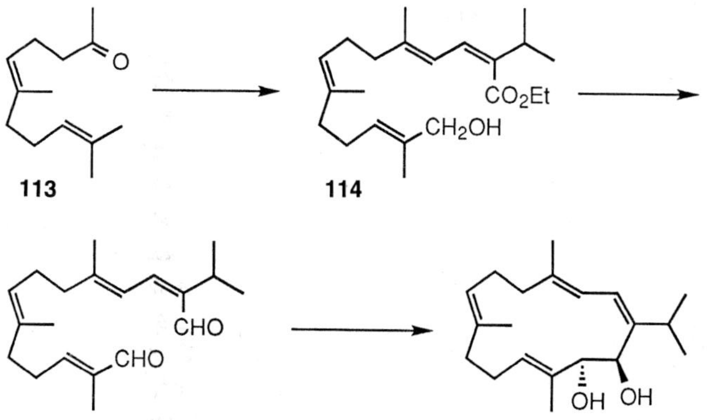

O CO_2Et CH_2OH

113 **114**

CHO CHO OH OH

Scheme 3-47 **115** **116** *Sarcophytol B*

Synthetic Protocol for Sarcophytol B [68]. The dialdehyde **115** in DME was added via syringe pump over 30 h at –40 °C to a stirred slurry of reagent prepared by reduction of $TiCl_3(DME)_{1.5}$/Zn-Cu in DME. The product was hydrolyzed with aqueous K_2CO_3 and purified to afford **116**, (46%).

3.6.2.1 Coupling on a Highly Oxygenated System

One of the major recent developments in coupling methodology involves reaction on highly oxygenated substrates. The total synthesis of the cembranoids isolobophytolide (**117**) and crassin (**119**) are important examples [69, 70]. The keto aldehyde precursors **118** and **120** to these two natural products differ solely in the relative stereochemistry at the β-position of the lactone ring (Scheme 3-48).

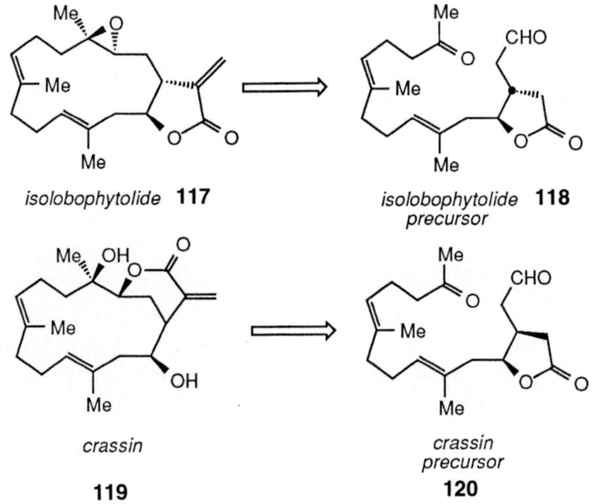

Scheme 3-48

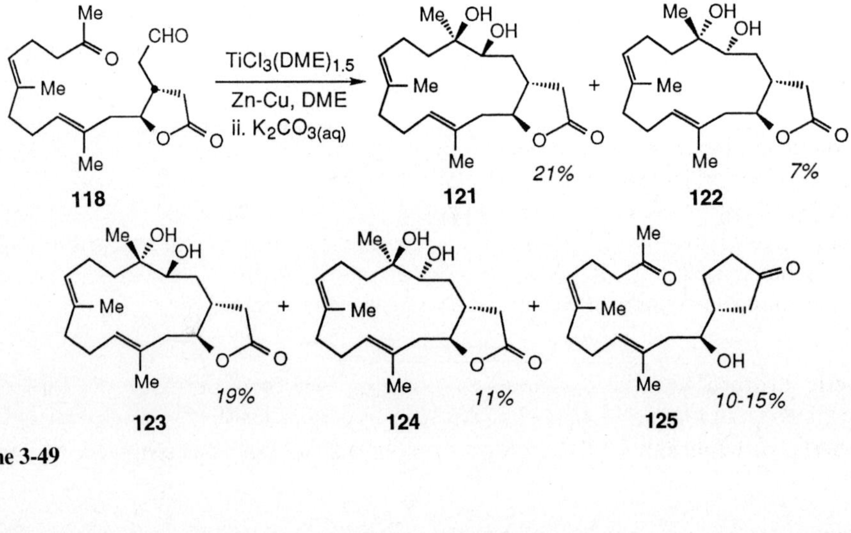

Scheme 3-49

In the case of isolobophytolide, coupling of the keto aldehyde **118** using standard pinacol coupling conditions at room temperature yields a mixture of the diol isomers **121–124** and diketone **125** (Scheme 3-49). Fortunately the desired diol stereoisomer **121** that serves as the direct precursor to isolobophytolide was the major product (21% yield). The synthesis was completed by formation of the epoxide, followed by installation of the *exo*-methylene group to afford isolobophytolide **117** (Scheme 3-50).

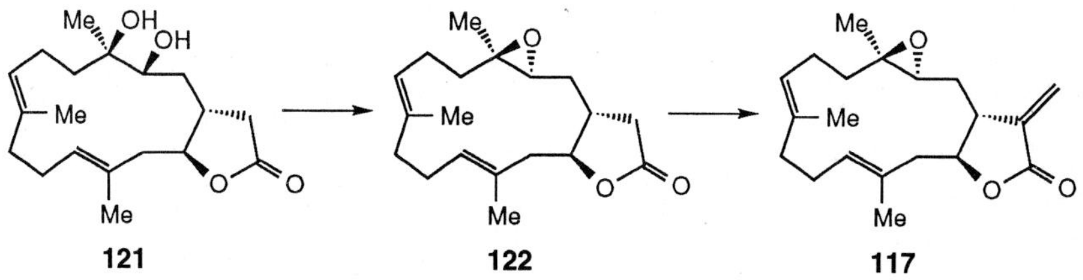

Me OH OH Me Me 121 → Me O Me Me 122 → Me O Me Me 117

Scheme 3-50

When the crassin precursor **120** was subjected to the McMurry coupling at room temperature over 45 h, a mixture of four isomeric diols **126a–d** was also obtained (Scheme 3-51). In contrast to the isolobophytolide synthesis, the diol product **126d** with the stereochemistry corresponding to crassin was obtained in only 1% yield.

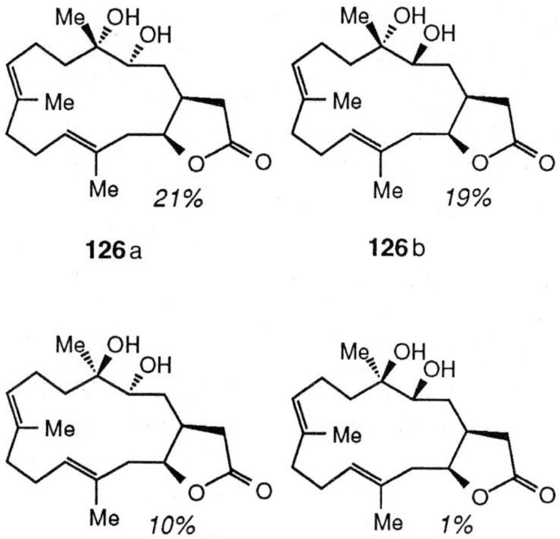

Me OH OH Me Me 21% **126**a

Me OH OH Me Me 19% **126**b

Me OH OH Me Me 10% **126**c

Me OH OH Me Me 1% **126**d

Scheme 3-51

In order to circumvent the dilemma and complete the crassin synthesis, McMurry and Dushin took the major component **126a** and cleverly reversed the diol stereochemistry to provide the direct precursor by first converting **126a** to the epoxide, and then performing a selective hydrolysis to yield **128**, the diol precursor with the correct stereochemistry for crassin (Scheme 3-52). The synthesis of crassin (**119**) was then accomplished by translactonization followed by installation of the *exo*-methylene group.

Synthetic Protocol for the Synthesis of Crassin Intermediates 126 [69, 70]. Zn–Cu couple (0.96 g, 14 mmol, 45 equiv.) was added to a suspension of $TiCl_3(DME)_{1.5}$ (1.56 g, 4.68 mmol, 15 equiv.) in DME (35 mL), and the mixture was refluxed with efficient stir-

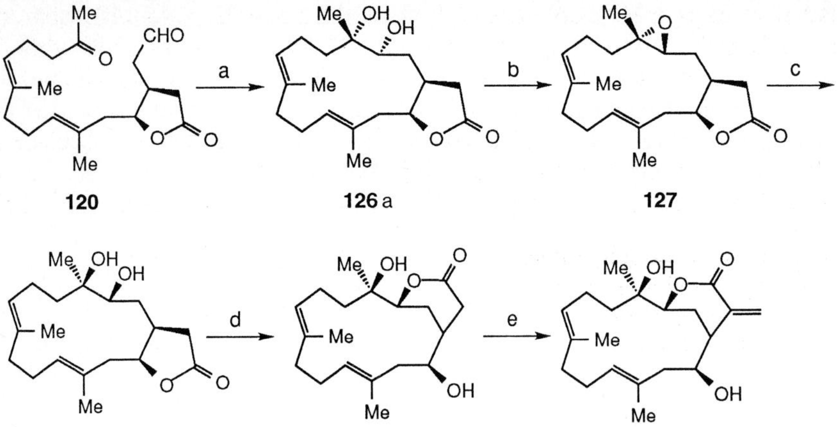

120 **126a** **127**

128 **129** **119**

(a) $TiCl_3(DME)_{1.5}$, Zn-Cu, DME (b) MsCl, pyridine; then $BnMe_3N^+OH^-$, benzene; 68%; (c) H_2O, $HClO_4$, THF; 77%; (d) NaOH, H_2O, EtOH; then H_3O^+, 0 °C, 1 h; 95%; (e) LDA, then CH_2O; then methanesulfonyl chloride, pyridine; then DBU, benzene; 53%.

Scheme 3-52

ring under an argon atmosphere for 1.5 h. Upon cooling to room temperature, keto aldehyde **120** (100 mg, 0.31 mmol) in DME (100 mL) was added via syringe pump over a period of 45 h. After an additional 5 h, aqueous K_2CO_3 (20%, 25 mL) was added, and the mixture was allowed to stir at room temperature for 6 h. The resulting suspension was then extracted with ethyl acetate (4×75 mL). The pooled extracts were washed with cold HCl (2%, 100 mL), water (75 mL), then brine (75 mL) and dried ($MgSO_4$) and concentrated. The resulting oil was passed through a short pad of silica gel and then subjected to HPLC separation. This provided the macrocyclic diol **126b** (19 mg, 19%), and an inseparable mixture of the remaining isomers. From the remaining mother liquor was isolated by careful fractional crystallization diol **126c** (10 mg, 10%), and the mother liquor enriched in diol **126a**. Repeated crystallization of this liquor (20% ethyl acetate in hexanes) then gave clean **126a** as colorless needles (diol **126a** was carried on through the crassin synthesis). The remaining trace isomer **126d** could not be purified from the remaining mother liquors.

3.6.2.2 The Synthesis of Periplanone C

The crassin synthesis pointed out the need for a rational prediction of product stereochemistry for complex pinacol coupling cases. In very recent work, McMurry and Siemers have completed a synthesis of the cockroach sex pheromone periplanone C (**138**) through a key pinacol coupling (Scheme 3-53) which addresses this question [71]. The precursor

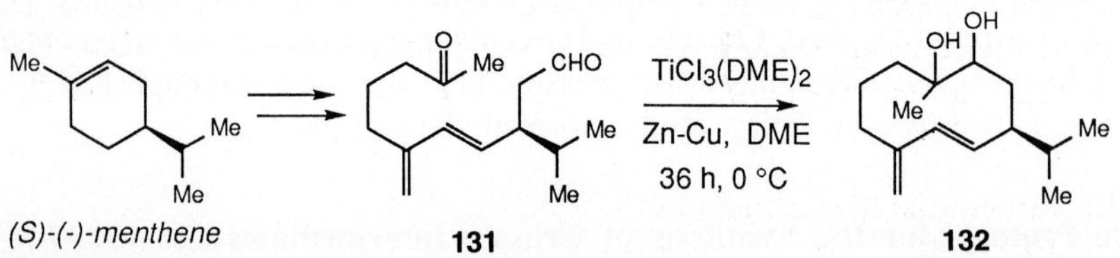

(S)-(-)-menthene **131** **132**

130

Scheme 3-53

keto aldehyde **131** was prepared from commercially available *(S)*-(–)-menthene (**130**) in 12 steps. The authors employed Macromodel (MM2) calculations on the derived dimethylsilyl acetals to predict the two major product diastereomers **133** and **134** of the cyclization reaction; the two minor products **135** and **136** predicted were in the event present in only trace amounts (Scheme 3-54). The major product was elaborated to periplanone C (**138**, Scheme 3-55) by conversion of diol **134** to epoxide **137**, followed by base-induced ring opening and oxidation of the resulting allylic alcohol.

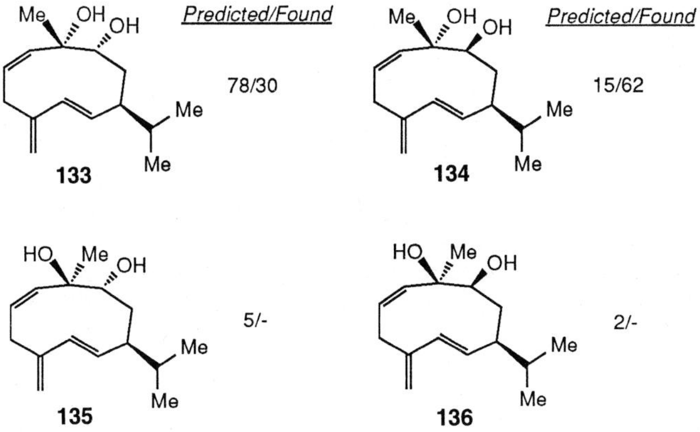

Compound	*Predicted/Found*
133	78/30
134	15/62
135	5/-
136	2/-

Scheme 3-54

134 →(a,b) 137 →(c,d) 138 *periplanone C*

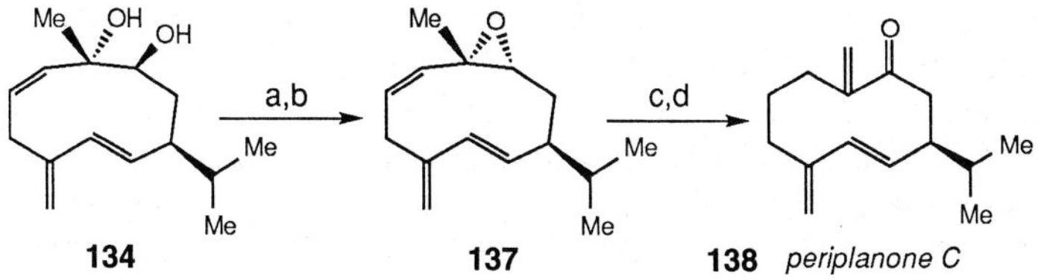

(a) MsCl, ether/pyridine; (b) Bu_4NOH, THF; (c) LDA, THF, 38% for three steps; (d) Dess-Martin periodinane, 84%.

Scheme 3-55

3.6.2.3 The McMurry Coupling in the Total Synthesis of Taxol

Nicoloau and co-workers have recently reported a total synthesis of taxol (**139**) in which the McMurry pinacol coupling reaction figures as a key step [72, 73]. Precursor dialdehyde **140** underwent coupling using the modified titanium reagent to afford **141**, which was then elaborated to natural taxol (Scheme 3-56). The conditions of this pinacol coupling are unusual in that the reaction temperature was 70 °C for 1 h. Presumably the potential alkene product is strained enough so that deoxygenation is inhibited. Interesting features of this particular reaction include the presence of potentially sensitive functional groups including a cyclic carbonate, a benzyl allyl ether, and a cyclic acetal. Although the reaction goes in fairly modest yield (23%), given the immense complexity of the substrate the result is remarkable nonetheless.

3.6.3 The Intermolecular Pinacol Reaction

Until recently, the intermolecular variant of the McMurry pinacol reaction was virtually unexplored (Scheme 3-57). The challeging problem is one of diastereoselectivity. One

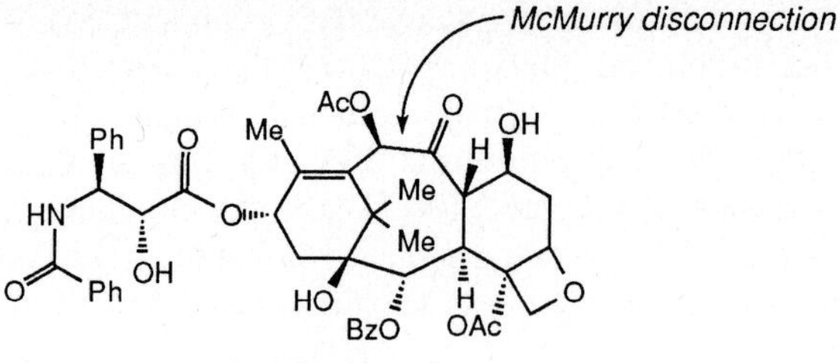

McMurry disconnection

139 *Taxol*

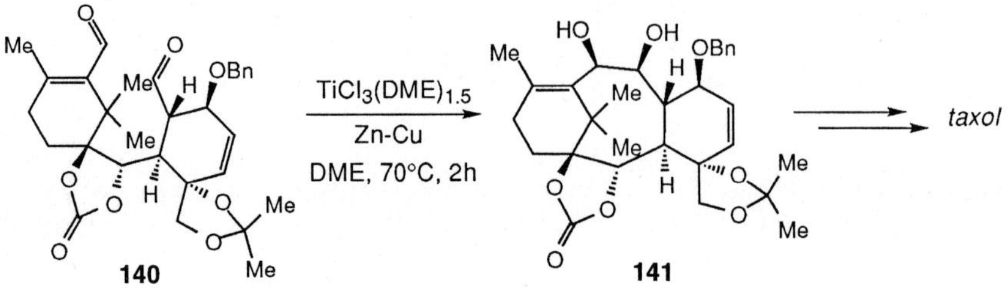

$TiCl_3(DME)_{1.5}$
Zn-Cu
DME, 70°C, 2h

140 141 taxol

Scheme 3-56

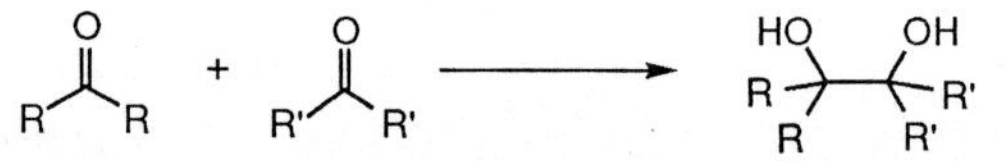

Scheme 3-57

would think that with the help of adjacent stereocenters a diastereoselective bias may be imparted to the reaction. However, work by Kempf on the synthesis of HIV protease inhibitors has recently shown only modest results [74]. He subjected Boc-L-phenylalaninal (**142**) to standard coupling conditions to form the corresponding dimeric molecules **143**, **144** and **145** with 2:1:1 diastereoselectivity (Scheme 3-58).

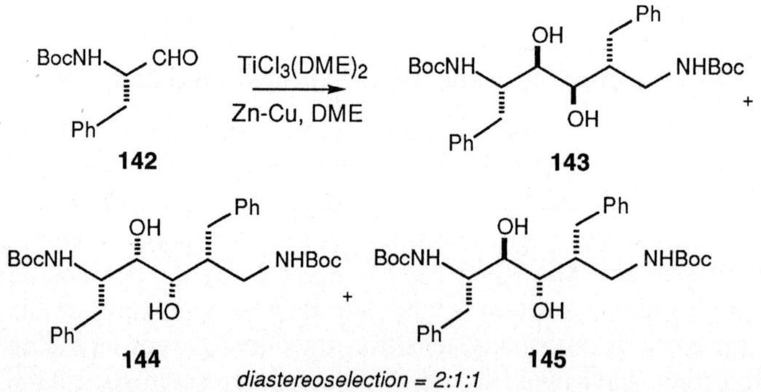

$TiCl_3(DME)_2$
Zn-Cu, DME

142 143 144 145

diastereoselection = 2:1:1

Scheme 3-58

Example of an Intermolecular Pinacol Dimerization: Coupling of Boc-L-phenylalaninal (142) [74]. A suspension of 27 g $TiCl_3(DME)_{1.5}$ in 200 mL DME was treated in portions with 20 g of Zn–Cu couple under positive argon pressure with vigorous stirring. After the addition, stirring was continued while the mixture was heated to 85 °C for 2.5 h. The resulting mixture was cooled to 0 °C and treated via cannula with a solution of

Boc-L-phenylalaninal (20 mmol) in 20 mL DME. After 1 h, the reaction mixture was filtered through Celite, and the residue was washed with ethyl acetate. The filtrate was treated with aqueous $NaHCO_3$, and air was bubbled through the suspension until it became white. The layers were separated, and the organic layer was washed with saturated brine, dried over $MgSO_4$, and concentrated to give 3.7 g of a light yellow solid. Flash chromatography (hexane/ethyl acetate followed by $CHCl_3$/ethyl acetate) provided pure samples of **143**, **144**, and **145** in a ratio of 2:1:1.

Mukaiyama has recently reported an interesting system for cross-coupling reactions using benzyl glyoxylate **146** and a number of aldehydes [75] affording products with modest *syn/anti* diastereoselectivity (Scheme 3-59). In analogy to diaryl-aryl ketone alkene couplings where the diaryl substrate in all likelihood undergoes a fast two-electron reduction before reaction, here the glyoxylate forms the ester-stabilized dianion **149** as the reactive species in what in actuality may be an aldol cross-coupling.

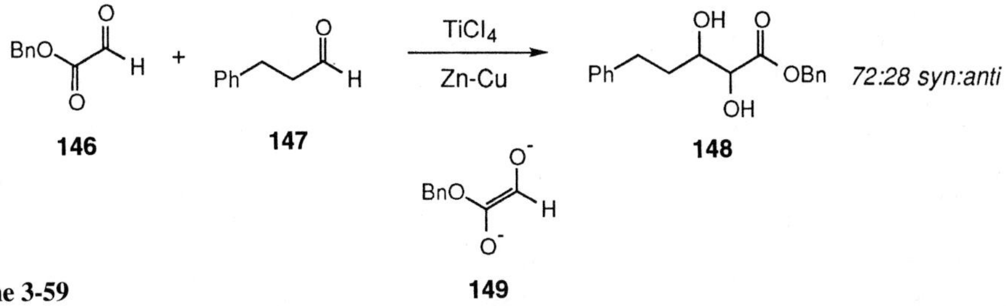

146 + 147 → ($TiCl_4$, Zn-Cu) → 148 (72:28 *syn:anti*)

149

Scheme 3-59

Clerici and Porta have used aqueous $TiCl_3$ as a reagent for the synthesis of asymmetrical 1,2-diols (**152**), in which one of the reacting partners possesses an electron withdrawing group [76]. This system is analogous to Mukaiyama's, in which the electronically stabilized component undergoes a fast two-electron reduction to yield an intermediate dianion which then reacts with the other component (Scheme 3-60).

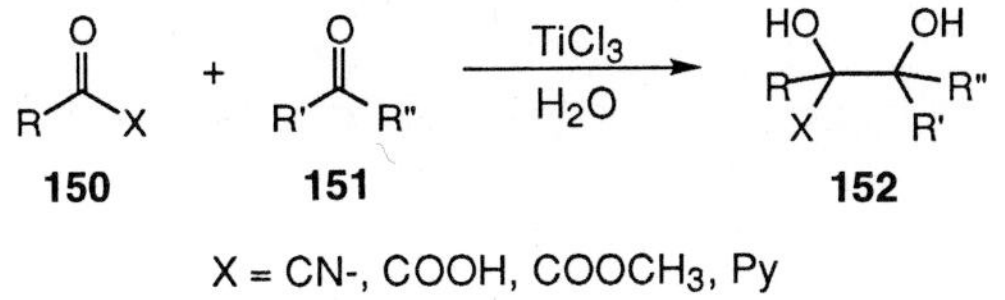

150 + 151 → ($TiCl_3$, H_2O) → 152

X = CN-, COOH, $COOCH_3$, Py

Scheme 3-60

3.6.4 Other Pinacol Methodology: The Pederson Modifications

Over the past few years Pederson and co-workers have published some very elegant reductive couplings which, like the McMurry pinacol reaction, utilize early transition metal chlorides in reducing oxidation states. Shown in Scheme 3-61 is the use of a niobium(IV)

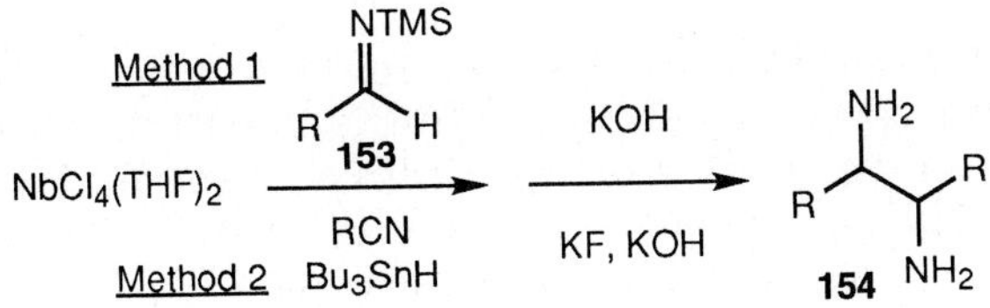

$NbCl_4(THF)_2$ → (Method 1: RCH=NTMS **153**; Method 2: RCN, Bu_3SnH) → (KOH; KF, KOH) → **154**

Scheme 3-61

Table 3-8. Coupling of imines and nitriles to vicinal diamines [77].

Entry	Substrate	Method	Yield (%)	*d,l*:*meso*	Product diamine
A	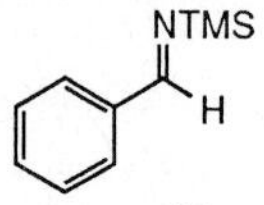 PhCH=NTMS	1	69	19:1	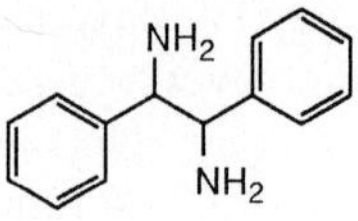 PhCH(NH₂)CH(NH₂)Ph
B	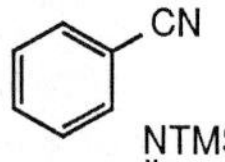 PhCN	2	70	8:1	PhCH(NH₂)CH(NH₂)Ph
C	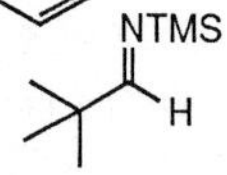 tBuCH=NTMS	1	62	2:1	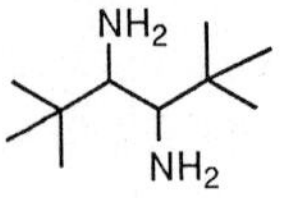 tBuCH(NH₂)CH(NH₂)tBu
D	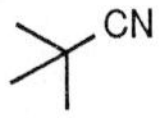 tBuCN	2	52	2:1	tBuCH(NH₂)CH(NH₂)tBu
E	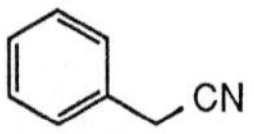 PhCH₂CN	2	45	1.4:1	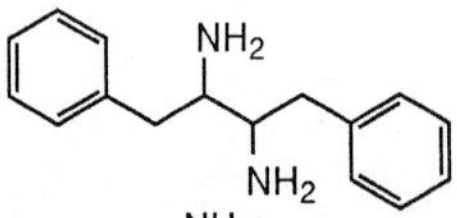 PhCH₂CH(NH₂)CH(NH₂)CH₂Ph
F	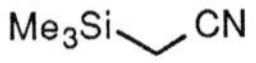 Me₃SiCH₂CN	2	50	3:1	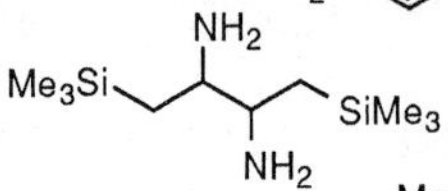 Me₃SiCH₂CH(NH₂)CH(NH₂)CH₂SiMe₃
G	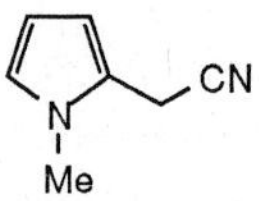 N-methylpyrrol-2-yl-CH₂CN	2	63	1:1	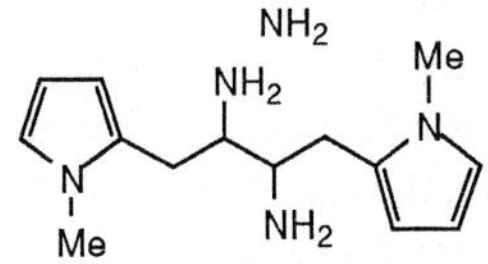 (N-Me-pyrrol-2-yl)CH₂CH(NH₂)CH(NH₂)CH₂(N-Me-pyrrol-2-yl)

solvate complex, $NbCl_4(THF)_2$, in the reductive coupling of nitriles and *N*-trimethylsilylimines to afford synthetically useful vicinal diamines [77]. This complex is conveniently available in large scale from niobium pentachloride by Manzer's procedure [78]. Reaction diastereoselectivities (Table 3-8) range from modest to excellent, depending on the substrate. For example, the imine derived from benzaldehyde (entry A, method 1) couples to afford product selectively (19:1 *d,l/meso*). Presumably the reaction mechanism is analogous to the McMurry pinacol coupling [8], where a delocalized radical species dimerizes to afford the product after hydrolytic workup (Scheme 3-62).

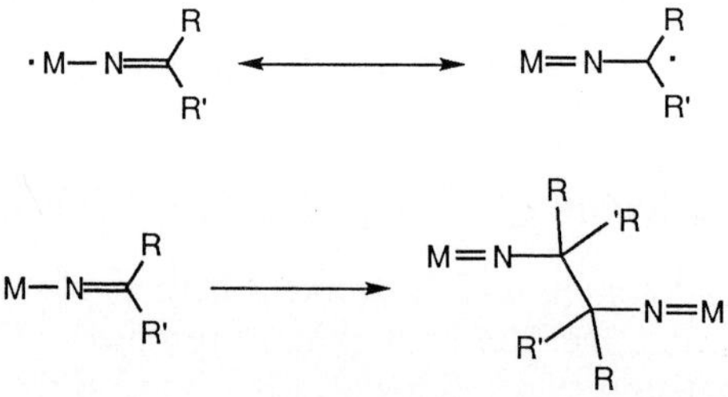

Scheme 3-62

Pederson and Roskamp have also devised some interesting methodology that addresses the problem of pinacol cross-coupling [79]. They have shown that a DME solvate of niobium(III) chloride [$NbCl_3(DME)$] is an effective reagent for the coupling of imines (**155**) with aldehydes and ketones to afford a range of vicinal amino alcohols **156** in good yield with diastereoselectivities in the range of 3:1 to 83:1 (*threo*/*erythro*) (Scheme 3-63).

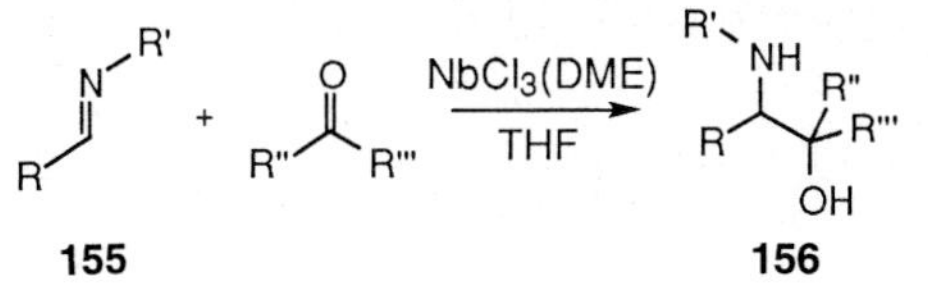

R'
N
R + O R'' R''' NbCl3(DME) THF R' NH R'' R''' R OH

Scheme 3-63 **155** **156**

The first step in the process is the reaction of the imine component **157** with the niobium reagent, with the resulting metallaaziridine **158** (Scheme 3-64) then reacting with the introduced carbonyl compound to form the product amino alcohol. Table 3-9 shows a range of amino alcohol products obtained in the reaction. In entry A, when using a benzaldehyde imine starting material, the diastereoselectivity is especially good (83:1).

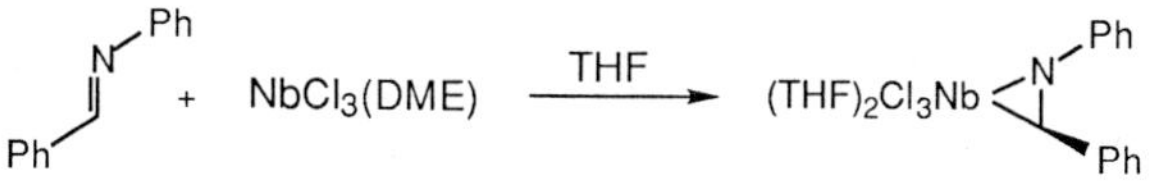

Ph N Ph + NbCl3(DME) THF (THF)2Cl3Nb N Ph Ph

Scheme 3-64 **157** **158**

Synthesis of $NbCl_3(DME)$ [79]. A dry, 1L, three-necked flask was fitted with an overhead mechanical stirrer and a nitrogen inlet adapter and charged with DME (600 mL) and tributyltin hydride (111 g, 0.38 mol). A 100 mL round-bottomed flask containing niobium pentachloride (50 g, 0.19 mol) was attached to the set-up via a 30 cm piece of rubber tubing fitted with two 24/40 male joints on each end. The reaction was then cooled to –78 °C, and the solution was vigorously stirred while the niobium pentachloride was

Table 3-9. Pinacol cross-coupling leading to vicinal amino alcohols.

Entry	Product amino alcohol	Yield (%)	Diastereoselectivity
A	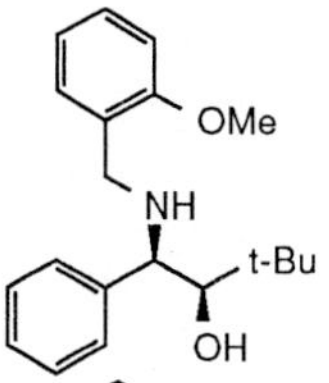 OMe NH t-Bu OH	79	83:1
B	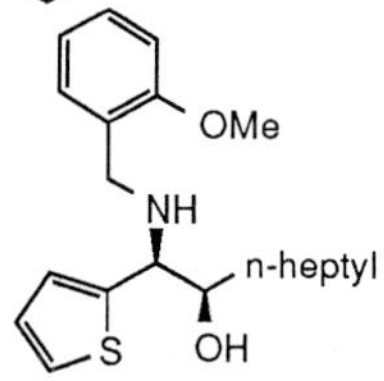 OMe NH n-heptyl S OH	33	6:1
C	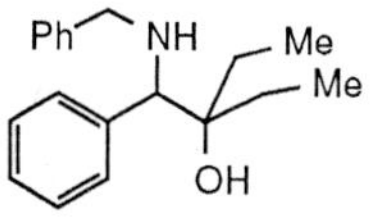 Ph NH Me Me OH	97	–
D	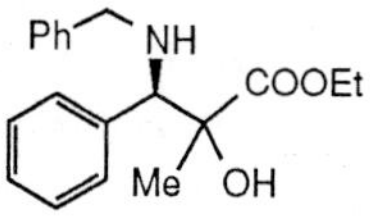 Ph NH COOEt Me OH	82	3:1

added over a 35–45 min period, maintaining the bath temperature between –78 °C and –65 °C. After the addition was complete, the reaction was allowed to warm slowly to between 0 and 20 °C (3–5 h) at which point the apparatus was brought into a drybox. The solid was isolated by filtration and washed with DME (2×150 mL) and pentane (150 mL) followed by drying in vacuo (45 min). A brick-red solid (50.6 g, 92%) was obtained.

2-Amino Alcohol Synthesis [79]. A dry 250 mL flask was charged with $NbCl_3$(DME) (2.0 g, 6.9 mmol) and THF (80 mL). A THF solution (4 mL) of the imine **155** (6.9 mmol) was then added dropwise (via syringe) over 30 s to the stirred mixture. When the solution became a homogeneous dark green (or yellow-green) color (ca. 1 min) the aldehyde or ketone (4.6 mmol) in THF (ca. 2 mL) was added dropwise over 30 s. After stirring for 30 min the reaction was poured into a separatory funnel and treated with KOH (10% w/v, 75 mL) and extracted with ether (2×150 mL). The combined ether layers were dried over $MgSO_4$ and filtered. The ether was removed in vacuo yielding the crude product **156** as an oil which was purified by flash chromatography (silica gel, hexane/ethyl acetate).

Pederson has developed a unique process for the selective intermolecular cross-coupling of electronically similar aldehydes using a dimeric vanadium complex, $[V_2Cl_3(THF)_6]_2$ $[Zn_2Cl_6]$ [80]. Whereas previous cross-coupling methodology has necessarily employed a carbonyl compound which can form a very stable ketyl radical or dianion as one of the reacting partners, Pederson's approach utilizes a chelating aldehyde as the fast-reacting component. For example, the chelator 2-methoxybenzaldehyde (**159**) was found to dimerize an order of magnitude faster than 4-methoxybenzaldehyde, which cannot chelate (Scheme 3-65). In a thorough study, Pederson examined aldehydes **162**, which

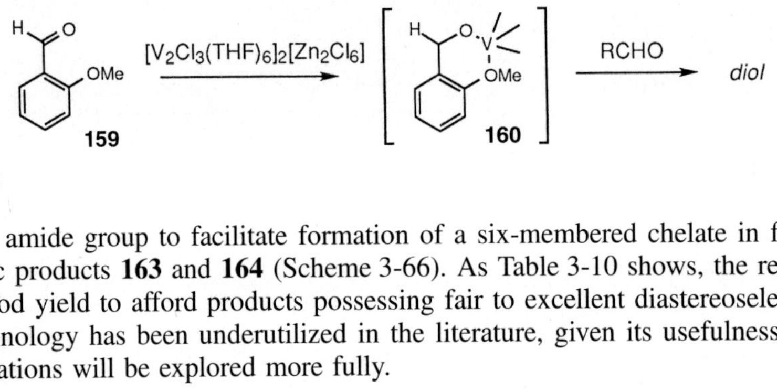

Scheme 3-65

contain an amide group to facilitate formation of a six-membered chelate in forming diastereomeric products **163** and **164** (Scheme 3-66). As Table 3-10 shows, the reactions proceed in good yield to afford products possessing fair to excellent diastereoselectivity. This clever technology has been underutilized in the literature, given its usefulness it is hoped that applications will be explored more fully.

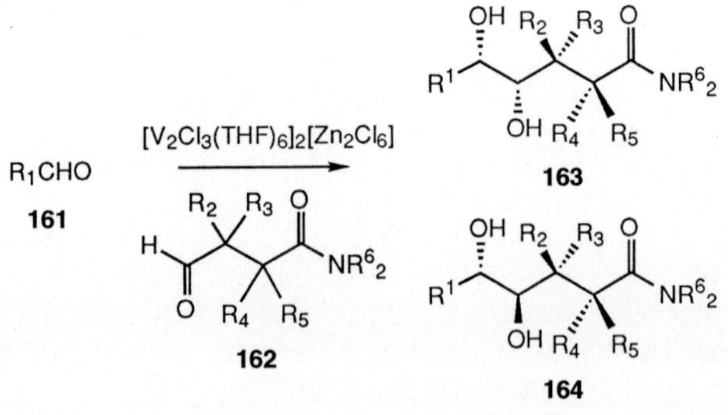

Scheme 3-66

Table 3-10. Cross-coupling of electronically similar aldehydes.

Entry	R^1	R^2	R^3	R^4	R^5	R^6	ds(A:B)	Yield (%)
1	$Ph(CH_2)_2$	H	H	H	H	Bn	4:1	67
2	i-Pr	H	H	H	H	Bn	8:1	73
3	t-Bu	H	H	H	H	Bn	>100:1	42
4	$Ph(CH_2)_2$	H	H	H	H	i-Pr	7:1	56
5	n-Pr	H	Me	H	H	Me	5:1	61
6	i-Pr	H	Me	H	H	Bn	14:1	81
7	t-Bu	H	H	H	Me	Bn	4:1	80
8	$Ph(CH_2)_2$	Me	Me	H	H	Me	>50:1	81
9	$Ph(CH_2)_2$	H	H	Me	Me	Bn	2:1	57

In very recent work Pederson has published a useful, general pinacol cross-coupling reaction of 2-[*N*-(Alkoxycarbonyl)amino] aldehydes (**165**) and aliphatic aldehydes using the $[V_2Cl_3(THF)_6]_2$ $[Zn_2Cl_6]$ reagent to produce pharmacologically interesting *syn,syn*-3-[*N*-(alkoxycarbonyl)amino] 1,2-diols (**166**) with high *syn/anti* diastereoselectivity (Scheme 3-67) [81]. The high selectivity of the reaction can be explained by invoking a reactive complex in which the aldehyde **165** is bound in a multi-point fashion (either two-point **167** or three-point **168**) to octahedral vanadium, along with the aliphatic aldehyde partner. Multi-point bonding is expected to produce an especially rigid reactive assembly, thus accounting for the high observed diastereoselectivities.

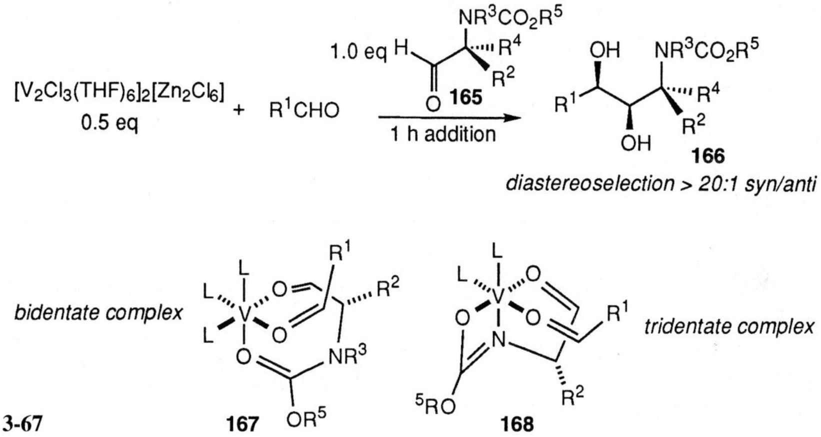

Scheme 3-67

3.7 The Keto Ester Coupling

Although the McMurry reaction has been expanded to include the intramolecular coupling of keto esters, these substrates have proven to be more problematic, requiring individual optimizations for each particular reaction case, especially for big rings (Scheme 3-68) [7]. The difficulties stem in part from the fact that the keto component is much more reactive as a ketyl radical than the ester component, so that quenching of the ketone may occur long before it reacts with the ester. Similarly, aldehydes are not competent substrates for coupling with esters, because of their high reactivity.

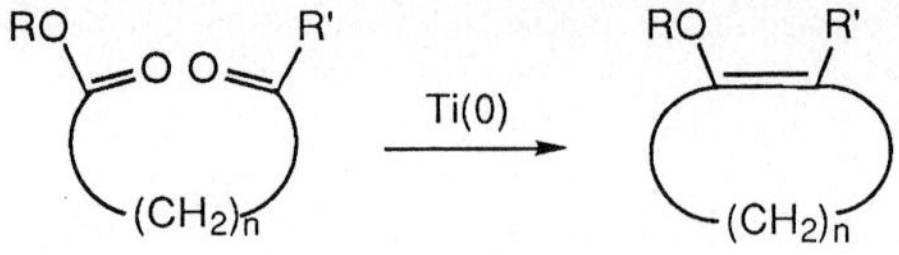

Scheme 3-68

The $TiCl_3$/LAH reagent combination is found to be generally superior to the $TiCl_3$/Zn–Cu derived procedures, and the presence of triethylamine in the reaction pot has been found in some instances to be beneficial, presumably by inhibiting the proteolysis of the product enol ether in the otherwise Lewis acidic reaction medium. Shown in Scheme 3-69 are keto ester couplings of 2-(*n*-carboethoxyalkyl)-cyclohexanones **169** and open-chain keto esters **171**, leading to products **170** and **172** of varying ring size. Not surprisingly, yields are good for smaller rings (n = 6,7, ca 80%), and fall off to 50–60% for larger rings.

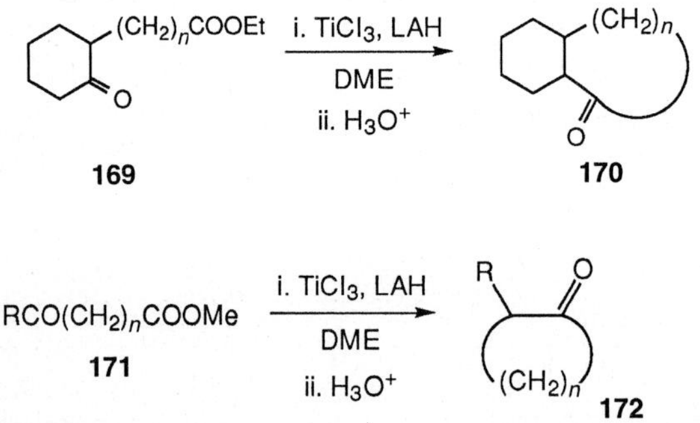

n	*Ring size*	*Yield, %*
2	5	68
3	6	80
4	7	82
5	8	52
6	9	50
11	14	54

R	*n*	*Ring size*	*Yield, %*
Me	8	10	50
Et	9	11	45
Et	10	12	63
Me	11	13	60

Scheme 3-69

Example of a McMurry Keto Ester Coupling: 2-Methylcyclotridecanone (172, *n* = 11, R = Me) [7]. A black slurry of the coupling reagent was prepared by adding LAH (114 mg, 3.0 mmol) to a stirred suspension of $TiCl_3$ (925 mg, 6.0 mmol) in 40 mL dry DME under an argon atmosphere. The mixture was stirred for 10 min at room temperature, triethylamine (0.17 mL, 1.2 mmol) was added, and the mixture was refluxed for 1.5 h. Methyl 13-oxotetradecanoate (**171**, n = 11, R = Me, 115 mg, 0.6 mmol) in 20 mL DME was then added to the refluxing slurry over a 24 h period via syringe pump. After a further 3 h reflux period, the reaction mixture was cooled to room temperature, diluted with 12 mL ether, and quenched by cautious addition of 6 mL methanol and 6 mL water. The mixture was further diluted with a pentane/ether mixture, passed rapidly through Florisil, washed with brine, dried ($MgSO_4$), and concentrated at the rotary evaporator. The crude product was then stirred for 3 h in dilute ethanolic/aqueous HCl, reisolated, and purified by chromatography on silica gel to yield 2-methyl-cyclotridecanone (**172**, n = 11, 62 mg, 60%).

McMurry has applied the keto ester methodology to the synthesis of isocaryophyllene (**176**, Scheme 3-70) [82]. The original synthetic target was caryophyllene itself (**174**), which differs from **176** in the geometry of the trisubstituted double bond. However, upon coupling, the double bond of transiently formed **174** isomerizes to the *Z*-isomer present in **176**. This represents the only known case of double bond isomerization during a coupling step.

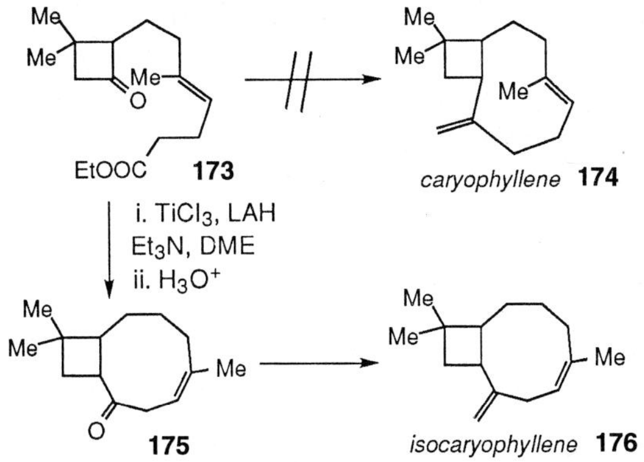

173 → (i. $TiCl_3$, LAH, Et_3N, DME; ii. H_3O^+) → 175 → isocaryophyllene 176

173 ⇸ caryophyllene 174

Scheme 3-70

More recent work by Fürstner, however, has shown that the $TiCl_3/C_8K$ reagent combination is very effective for keto ester couplings, and may be a method of choice for these substrates [22, 23]. The protocol was found to be excellent for the synthesis of both indoles and furans, as shown in Scheme 3-71 [83]. It should be noted that C_8K is a commercially available solid, although it can be easily prepared as described below in the context of preparing the low-valent titanium reagent.

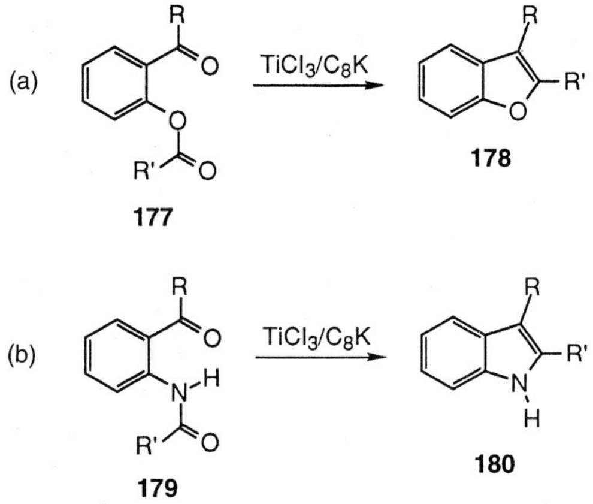

(a) 177 → ($TiCl_3/C_8K$) → 178

Entry	R	R'	Yield, %
A	Ph	Ph	88
B	Ph	Me	85
C	Me	Ph	80
D	H	Ph	89

(b) 179 → ($TiCl_3/C_8K$) → 180

Entry	R	R'	Yield, %
A	Ph	Ph	90
B	Ph	Me	87
C	Me	Ph	75
D	Me	Me	70
E	H	Ph	90
F	Ph	*t*-Bu	84

Scheme 3-71

Preparation of Titanium on Graphite. A 200 mL two-necked flask, equipped with a Teflon-coated magnetic stirring bar and a reflux condenser connected to an argon line, was charged with graphite (3.0 g, 250 mmol) and heated to 150–160°. Freshly cut potassium (1.20 g, 31 mmol) was added in pieces with vigorous stirring at that temperature until the bronze-colored potassium-graphite laminate (C_8K) was formed (5–10 min). After being cooled to room temperature it was suspended in anhydrous THF (40 mL) and $TiCl_3$ (2,31 g, 15 mmol) was added to this mixture causing the solvent to boil. When the exothermic reaction had subsided, the slurry was heated for 1.5 h to ensure complete reduction of the starting material.

McMurry-Type Reactions of Acyloxycarbonyl Compounds: Representative Procedure. A solution of substrate **177**, entry A (1.51 g, 5 mmol) in anhydrous THF (60 mL) was added dropwise over a period of 8 h to a freshly prepared refluxing suspension of titanium-graphite in THF (40 mL) under argon. The mixture was allowed to cool to room temperature and filtered over a short plug of silica gel. The inorganic residue was washed with ethyl acetate (50 mL) in several portions, the filtrate evaporated in vacuo, and the crude product purified by column chromatography using toluene/ethyl acetate (25/1) as eluent, thus affording 2,3-diphenylbenzo[b]furan (**178**, entry A) as colorless crystals (1.23 g, 88%).

McMurry-Type Reactions of Acylamidocarbonyl Compounds: Representative Procedure. A solution of compound **179**, entry B (1.20 g, 5 mmol) in THF (10 mL) was added at once to a freshly prepared boiling suspension of titanium-graphite under argon. Reflux was continued for 3 h and the mixture was filtered over silica after being cooled to room temperature. Washing of the inorganic solids with four portions of ethyl acetate (50 mL) and evaporation of the solvent, followed by column chromatography with toluene as eluent afforded crystalline 2-methyl-3-phenylindole (**180**, entry B) (901 mg, 87%).

3.8 Tandem Couplings

There have been several interesting cases of tandem couplings in which an initial intermolecular reaction is followed by an intramolecular macrocyclization reaction. Although the overall yields are generally modest, the strategy is attractive since four carbon-carbon bonds are formed in the process. Vögtle has recently performed a cyclodimerization of the diketone **181** to form macrocyclic diarylhexatriene **182** (Scheme 3-72) [84]. The yield of the reaction was a modest 20%; however, the result is remarkable nonetheless given the complexity of the process. The overall yield in the reaction is undoubtedly suppressed as a result of both *E*- and *Z*-isomer formation in the initial step. For macrocyclization to occur, *Z*-isomer formation is required.

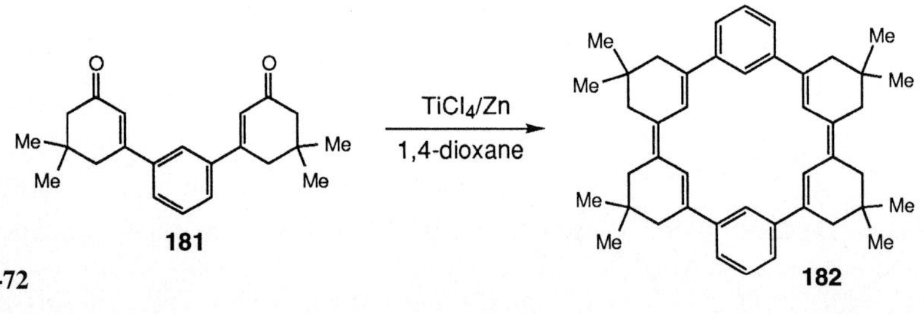

181 → ($TiCl_4$/Zn, 1,4-dioxane) → **182**

Scheme 3-72

The tandem strategy has been more commonly used to assemble aromatic porphycene structures from appropriate dialdehydes. Porphycenes are of current interest as effective photosensitizers for biomedical applications, including photodynamic tumor therapy and virus inactivation. For example, the synthesis of dithiaporphycene **184** proceeded from dialdehyde **183** in 8% yield (Scheme 3-73) [85].

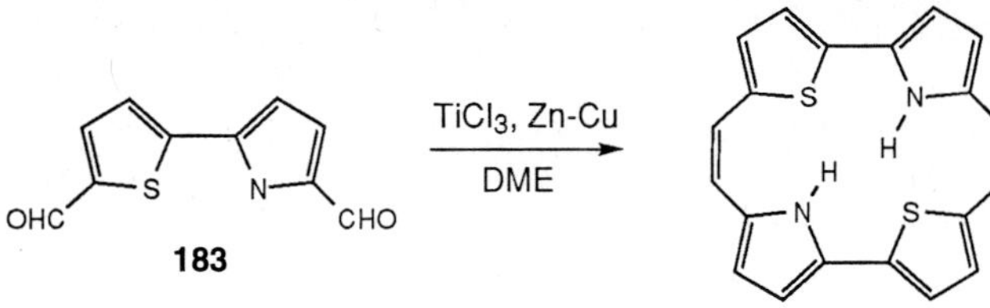

Scheme 3-73

3.9 The Allyl/Benzyl Alcohol McMurry Coupling

Allylic and benzylic alcohols undergo coupling to afford good yields of various bibenzyls and 1,5-dienes as products (Scheme 3-74) [86]. Table 3-11 shows four examples of this type of coupling, for example 1-cycloheptenol reacts to afford the dimerized product in 87% yield. Disubstituted benzyl alcohols perform especially well in the reaction, often affording products in high yields. Unfortunately the reaction has been employed in relatively few instances, and seems to be underutilized. McMurry originally published the procedure utilizing the $TiCl_3$/LAH reagent system employing a slight excess of LAH to

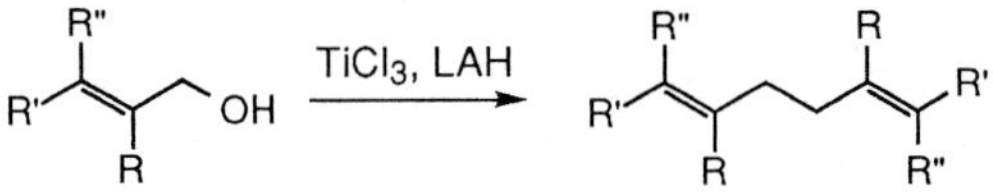

Scheme 3-74

Table 3-11. Coupling of allylic and benzylic alcohols by $TiCl_3$/LAH.

Entry	Substrate	Product	Yield (%)
A	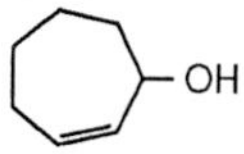	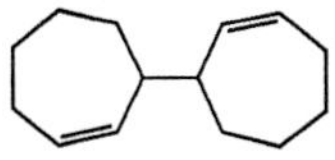	87
B	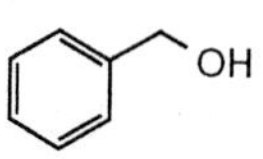	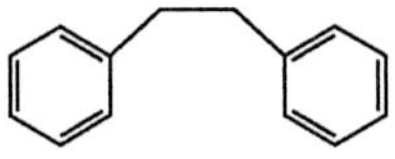	78
C	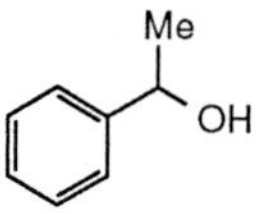	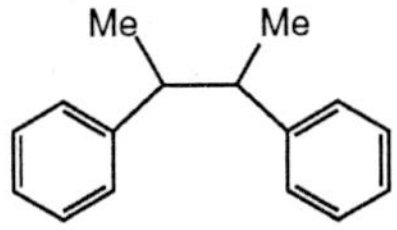	68
D	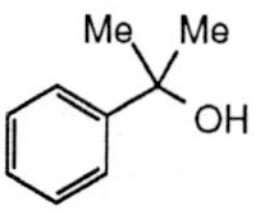	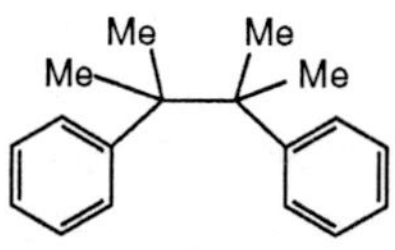	95

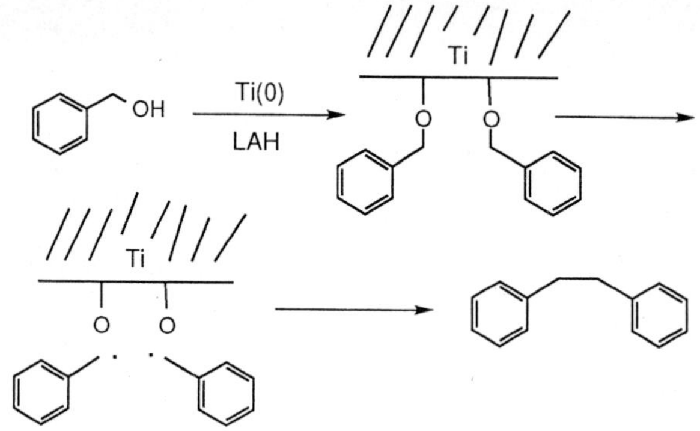

Scheme 3-75

act as a base in deprotonating the alcohol substrate as a necessary prelude to reaction. Presumably Zn–Cu couple would not work as efficiently in this instance, unless the reagent could effect deprotonation through evolution of hydrogen gas. Given the stoichiometry of McMurry's protocol ($TiCl_3$:LAH = 15:20), he has proposed that the active reagent contains Ti^{II}. The reaction may proceed through a free allyl or benzyl radical intermediate, which dimerizes to form the product (Scheme 3-75). If this is the case, then this process should be relatively ineffective in making large rings, since the template effect of the metal surface would be missing. In fact, when the bis(allylic) diol **185** was subjected to the reaction conditions, no cyclized product was observed, but 80% of the deoxygenated diene **186** was isolated (Scheme 3-76) [87].

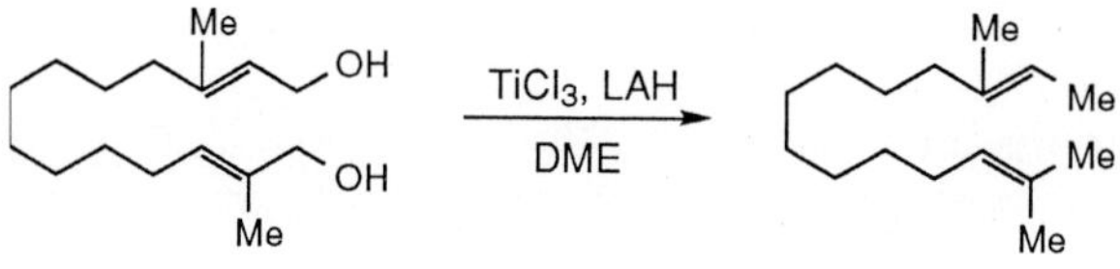

Scheme 3-76

3.10 The Reaction of Other Functional Groups

Although the keto ester coupling has proven to be synthetically useful, the coupling of two esters to yield an acyloin intermediate, followed by deoxygenation to afford an alkyne, has not been successful (Scheme 3-77) [88]. Treatment of esters with the active

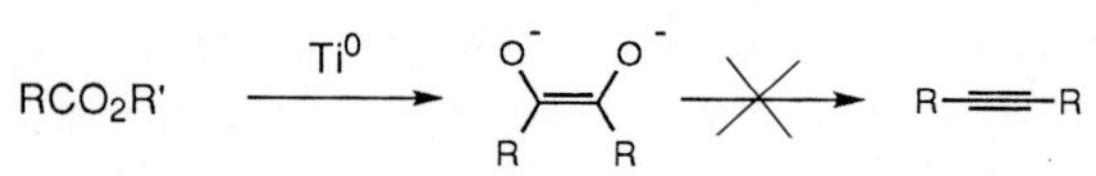

Scheme 3-77

reagent results in a complex mixture of products containing only trace amounts of acetylene. However, other functional groups do react to give useful products. For example, the thioketone **187** reacts smoothly to afford tetraphenylethylene (**40**) in 90% yield (Scheme 3-78) [6].

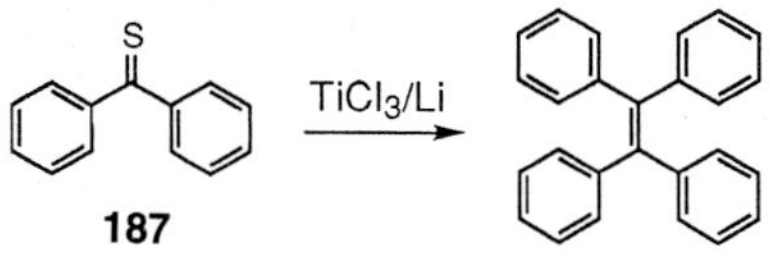

Scheme 3-78

In an interesting variation by McMurry that presaged some of the Pederson work, bis(imine) **188** afforded alkene **189** in 14% yield when treated with the $TiCl_3$/Li reagent combination (Scheme 3-79) [6].

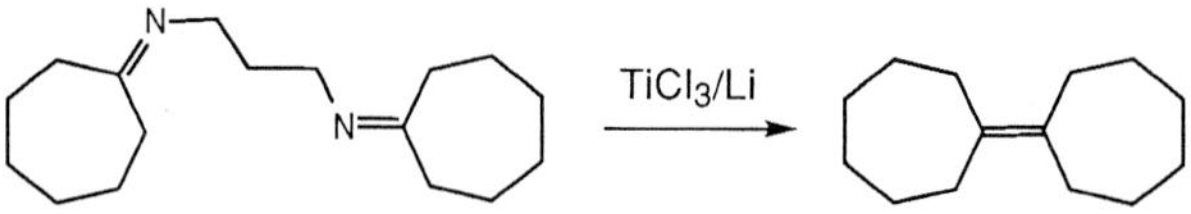

Scheme 3-79

Functional Group Reactivities. Shown in Table 3-12 is a summary of the most common organic functional groups and their reactivity with low-valent titanium. The symbol I means that under most conditions the functional group reacts rapidly with the McMurry

Table 3-12. Reactivities of functional groups with the McMurry reagent.

Functional group		Compat-ibility	Product	Functional group		Compat-ibility	Product
–CHO	aldehyde	I	alkene	–OCOO–	carbonate	S	
(epoxide ring, O)	epoxide	I	alkene	$–CONR_2$	amide	S	
 X, OH	halohydrin	I	alkene	–COR	ketone	S	alkene
				–CN	nitrile	S	
(quinone ring, O, O)	quinone	I	hydro-quinone	–COOH	carboxylic acid	S	
				–C≡C–	alkyne	S	arene, diana
 HO, OH	diol	I	alkene	–COOR	ester	S	enol ether (with –COR)
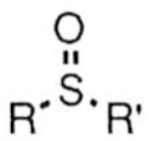 R–S(=O)–R'	sulfoxide	I	sulfide	–TsO	tosylate	C	
				–OH	alcohol	C	
$–NO_2$	nitro	I	amine	–OR	ether	C	
–C=NOH	oxime	I	imine	–OSiPR3	silyl ether	C	
				$–SiR_3$	alkyl silane	C	
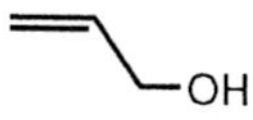 (CH2=CH–CH2–OH)	allylic alcohol	I	diene	$–NR_3$	amine	C	
$–OCR_2C–$	acetal	S		–RC=CR′–	alkene	C	
–X	halide	S		–S–	sulfide	C	

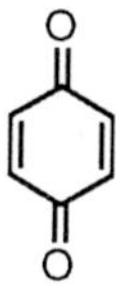

reagent. In the cases where the reactivity leads to a major isolable product, that product is listed. The symbol S means that the functional group is stable in some cases but not others. The symbol C means that the functional group is compatible with the reagent under standard conditions of the reaction.

3.11 The McMurry Reaction in Polymer Synthesis

There are only a very few instances in which the McMurry reaction has been exploited in polymer synthesis. In order for a polymerization to be successful, the spacer unit (Y, Scheme 3-80) must be rigid enough to prevent competing intramolecular cyclization from occurring.

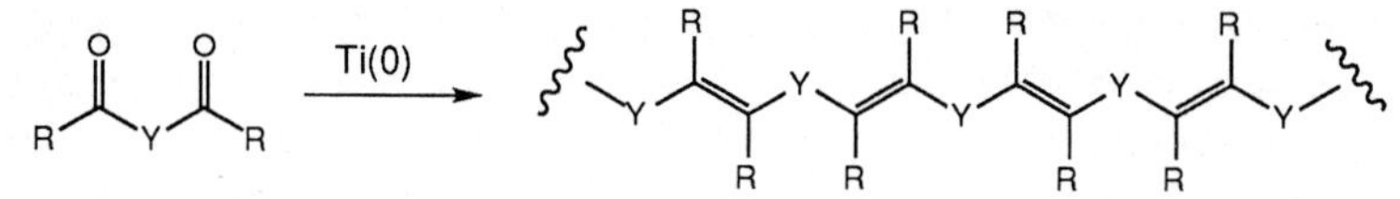

Scheme 3-80

As a good example of this principle, the ferrocene dialdehyde **190** has been polymerized to afford the organometallic polymer **191** (Scheme 3-81) [89]. Cooke et al. have polymerized benzene-1,4-dialdehyde (**192**) to give the pinacol **193**, which is more soluble in organic media than the corresponding polyene **194** (Scheme 3-82) [90].

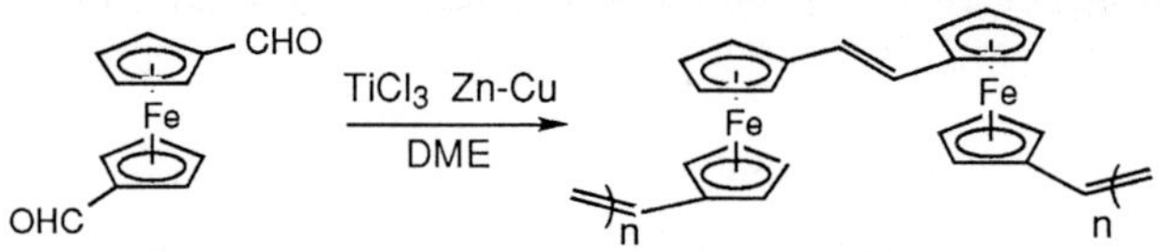

Scheme 3-81

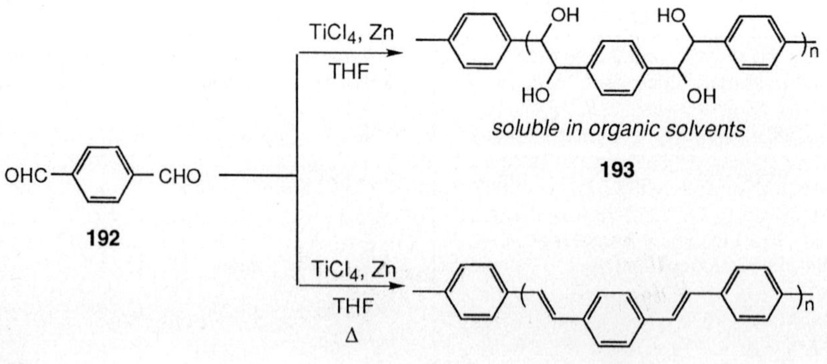

Scheme 3-82

In general, though, this approach to polymer synthesis remains ripe for development; for example, the McMurry coupling could be used to advantage in the synthesis of a wide variety of substituted polyacetylenes, which are of intense interest in both the academic and industrial communities (Scheme 3-83).

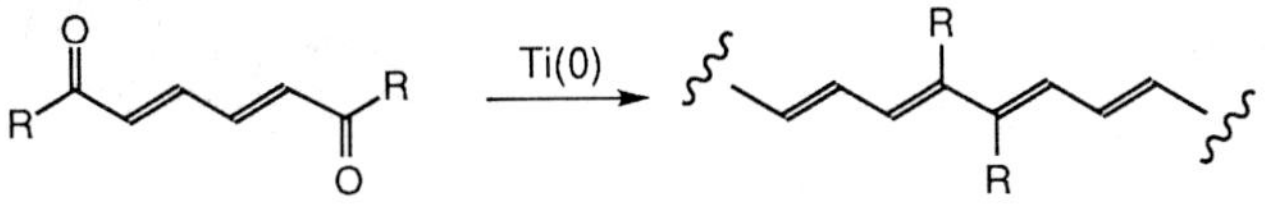

Scheme 3-83

3.12 Conclusion and Future Directions

One of the big challenges yet to be met in the McMurry reaction is the highly predictable, diastereoselective synthesis of large ring pinacols. Future efforts will undoubtedly target a general solution to this problem. For example, it would be interesting to see the Pederson methodology applied to large-ring substrates including natural products. In this instance a single transition metal center or discrete metal complex may serve to act as the effective template for macrocyclization, much as the titanium metal surface does.

The paramount utility of the McMurry reaction is and will remain its astonishing capability in the synthesis of large or strained ring systems. Of current interest is the synthesis of biologically-active cyclic polypeptides. Unfortunately, oftentimes these interesting substances are only available through capricious and costly macrocyclization procedures. Here the McMurry pinacol reaction could work under well-defined low temperature conditions, in which the amide groups are unreactive, to produce cyclic polypeptides.

Acknowledgements. The author would like to gratefully acknowledge Prof. John McMurry (Cornell), Dr. Nathan Siemers (Cornell), and Dr. Scott J. Miller (Harvard) for helpful discussions.

References

[1] J. E. McMurry, M. P. Fleming, *J. Am. Chem. Soc.* **1974**, *96*, 4708–4709.
[2] For reviews of the McMurry reaction, see: (a) J. E. McMurry, *Chem. Rev.* **1989**, *89*, 1513–1524; (b) J. E. McMurry, *Acc. Chem. Res.* **1983**, *16*, 405–411; (c) P. Welzel, *Nachr. Chem. Tech. Lab.* **1983**, *31*, 814–816; (d) Y.-H. Lai, *Org. Prep. Proced. Int.* **1980**, *12*, 361–391.
[3] Rieke has published a comprehensive review on carbonyl coupling reactions using transition metals, lanthanides, and actinides: B. E. Kahn, R. D. Rieke, *Chem. Rev.* **1988**, *88*, 733–745.
[4] S. Tyrlik, I. Wolochowicz, *Bull. Soc. Chim. Fr.* **1973**, 2147–2148.
[5] T. Mukaiyama, T. Sato, J. Hanna, *J. Chem. Lett.* **1973**, 1041–1044.
[6] J. E. McMurry, M. P. Fleming, K. L. Kees, L. R. Krepski, *J. Org. Chem.* **1978**, *43*, 3255–3266.
[7] J. E. McMurry, D. D. Miller, *J. Am. Chem. Soc.* **1983**, *105*, 1660–1661.
[8] J. E. McMurry, J. G. Rico, *Tetrahedron Lett.* **1989**, *30*, 1169–1172.
[9] Personal communication from J. E. McMurry; see also: J. E. McMurry, *Acc. Chem. Res.* **1983**, *16*, 405–411.
[10] J. E. McMurry, M. P. Fleming, K. L. Kees, L. R. Krepski, *J. Org. Chem.* **1978**, *43*, 3255–3266.
[11] J. E. McMurry, M. P. Fleming, unpublished results. The transition metals screened include Al, Si, Sc, V, Cr, Mn, Fe, Co, Ni, Cu, Zr, Nb, Mo, Sn, and W.
[12] R. Dams, M. Malinowski, H. J. Geise, *Bull Soc. Chim. Belg.* **1982**, *91*, 149–152.

[13] Rieke has found that UCl_4 with lithium naphthalenide as reductant is useful for the coupling of aryl ketones: B. E. Kahn, R. D. Rieke, *Organometallics* **1988**, *7*, 463–469.
[14] G. P. Boldrini, D. Savoia, E. Tagliavini, C. Trombini, A. Umani-Ronchi, *J. Organomet. Chem.* **1985**, *280*, 307–312.
[15] J. E. McMurry, M. P. Fleming, *J. Org. Chem.* **1976**, *41*, 896–897.
[16] R. Dams, M. Malinowski, I. Westdrop, H. J. Geise, *J. Org. Chem*, **1982**, *47*, 248–259.
[17] R. Fittig, *Liebigs Ann. Chem.* **1859**, *110*, 23.
[18] Mundy has studied stereochemical aspects of the pinacol reaction; see: (a) B. P. Mundy, R. Srinivasa, Y. Kim, T. Dolph, R. J. Warnet, *J. Org. Chem.* **1982**, *47*, 1657–1661; (b) B. P. Mundy, D. R. Bruss, Y. Kim, R. D. Larsen, R. J. Warnet, *Tetrahedron Lett.* **1985**, *26*, 3927–3931.
[19] K. J. Covert, P. T. Wolczanski, S. A. Hill, P. J. Krusic, *Inorg. Chem.* **1992**, *31*, 66–78.
[20] M. Pasquali, C. Floriani, A. Chiesi-Villa, C. Guastini, *J. Am. Chem. Soc.* **1979**, *101*, 4740–4742.
[21] J. E. McMurry, D. D. Miller, *J. Am. Chem. Soc.* **1983**, *105*, 1660–1661.
[22] A. Fürstner, H. Weidmann, *Synthesis*, **1987**, 1071–1075.
[23] A. Fürstner, R. Csuk, C. Rohrer, H. Weidmann, *J. Chem. Soc. Perkin Trans. I*, **1988**, 1729–1734.
[24] J. E. McMurry, L. R. Krepski, *J. Org. Chem.* **1976**, *41*, 3929–3930.
[25] E. J. Corey, R. L. Danheiser, S. Chandrasekaran, *J. Org. Chem.* **1976**, *41*, 260–265.
[26] D. Lenoir, *Synthesis*, **1977**, 553–554.
[27] J. E. McMurry, T. Lectka, J. G. Rico, *J. Org. Chem.* **1989**, *54*, 3748–3749. The reagent is also often represented with the stoichiometry $TiCl_3(DME)_2$, depending on the amount of DME retained in the solvate. Most importantly, the efficacy of the reagent should not be affected by retained DME.
[28] D. S. Bomse, T. H. Morton, *Tetrahedron Lett.* **1975**, 781–785.
[29] J. E. McMurry, T. Lectka, unpublished results.
[30] (a) G. A. Olah, G. K. Surya Prakash, *J. Org. Chem.* **1977**, *42*, 580–582; (b) D. Lenoir, H. Burghard, *J. Chem. Res. (S)* **1980**, 396–400; (c) D. Lenoir, P. Lemmen, *Chem. Ber.* **1980**, *113*, 3112–3119; (d) G. Böhrer, R. Knorr, *Tetrahedron Lett.* **1984**, *25*, 3675–3678; (e) H. Wenck, A. deMeijere, F. Gerson, R. Gleiter, *Angew Chem., Int. Ed. Engl.* **1986**, *25*, 335–336; (f) J. Janssen, W. Lüttke, *Chem. Ber.* **1982**, *115*, 1234–1243; (g) R. Willem, H. Pepermans, K. Hallenga, M. Gielen, R. Dams, H. J. Geise, *J. Org. Chem.* **1983**, *48*, 1890–1898; (h) G. A. Olah, G. K. Surya Prakash, G. Liang, *Synthesis* **1976**, 318–319; (i) D. Lenoir, D. Malwitz, B. Meyer, *Tetrahedron Lett.* **1984**, *25*, 2965–2968; (j) D. Lenoir, *Chem. Ber.* **1978**, *111*, 411–414; (k) K. Yamamoto, J. Ojima, N. Morita, T. Asao, *Bull. Chem. Soc. Jpn.* **1988**, *61*, 1281–1283.
[31] L. A. Paquette, T.-H. Yan, G. J. Wells, *J. Org. Chem.* **1984**, *49*, 3610–3617.
[32] P. L. Coe, C. E. Scriven, *J. Chem. Soc., Perkin Trans. 1* **1986**, 475–477.
[33] J. Shani, A. Grazit, T. Livshitz, S. Biran, *J. Med. Chem.* **1985**, *28*, 1504.
[34] J. E. McMurry, K. L. Kees, *J. Org. Chem.* **1977**, *42*, 2655–2656.
[35] T. Eguchi, T. Terachi, K. Kakinuma, *Tetrahedron Lett.* **1993**, *34*, 2175–2178. For the synthesis of another 36-membered ring without recourse to high-dilution see: A. Fürstner, G. Seidel, *Synthesis*, **1995**, 63–68; A. Fürstner, A. Hupperts, *J. Am. Chem. Soc.* **1995**, *117*, 4468–4475.
[36] J. E. McMurry, P. Kocovsky, *Tetrahedron Lett.* **1985**, *26*, 2171–2172.
[37] J. E. McMurry, G. K. Bosch, *J. Org. Chem.* **1987**, *52*, 4885–4893.
[38] D. Robinson, C. West, *Biochemistry* **1970**, *9*, 70.
[39] C. B. Jackson, G. Pattenden, *Tetrahedron Lett.* **1985**, *26*, 3393–3396.
[40] J. E. McMurry, G. K. Bosch, *Tetrahedron Lett.* **1985**, *26*, 2167–2170.
[41] J. E. McMurry, J. R. Matz, K. L. Kees, *Tetrahedron* **1987**, *43*, 5489–5498.
[42] J. E. McMurry, J. R. Matz, *Tetrahedron Lett.* **1982**, *23*, 2723–2724.
[43] J. E. McMurry, J. R. Matz, K. L. Kees, P. A. Bock, *Tetrahedron Lett.* **1982**, *23*, 1777–1780.
[44] I. Ben, L. Castedo, J. M. Saa, J. A. Seijas, R. Suau, G. Tojo, *J. Org. Chem.* **1985**, *50*, 2236–2240.
[45] B. W. Disanayaka, A. C. Weedon, *J. Chem. Soc, Chem. Commun.* **1985**, 1282.
[46] B. W. Disanayaka, A. C. Weedon, *J. Org. Chem.* **1987**, *52*, 2905–2910.
[47] Y.-J. Wu, D. J. Burnell, *Tetrahedron Lett.* **1988**, *29*, 4369–4372.
[48] U. Berlage, J. Schmidt, U. Peters, P. Welzel, *Tetrahedron Lett.* **1987**, *28*, 3091–3094.
[49] D. L. J. Clive, C. Zhang, K. S. K. Murthy, W. D. Hayward, S. Daigneault, *J. Org. Chem.* **1991**, *56*, 6447–6458.
[50] D. L. J. Clive, K. S. K. Murthy, A. G. H. Wee, S. Siva Prasad, G. V. J. da Silva, M. Majewski, P. C. Anderson, C. F. Evans, R. D. Haugen, L. D. Heerze, J. R. Barrie, *J. Am. Chem. Soc.* **1990**, *112*, 3018–3028.
[51] J. E. McMurry, C. N. Hodge, *J. Am. Chem. Soc.* **1984**, *106*, 6450.
[52] J. E. McMurry, T. Lectka, C. N. Hodge, *J. Am. Chem. Soc.* **1989,** *112*, 8867–8872.
[53] J. E. McMurry, T. Lectka, *Acc. Chem. Res.* **1992**, *25*, 47–53.

[54] J. E. McMurry, T. Lectka, *J. Am. Chem. Soc.* **1990**, *112,* 869–870.
[55] J. E. McMurry, T. Lectka, *J. Am. Chem. Soc.* **1993**, *115,* 10167–10173.
[56] J. E. McMurry, T. Lectka, unpublished results.
[57] T. S. Sorensen, S. M. Whitworth, *J. Am. Chem. Soc.* **1990**, *112,* 8135–8144.
[58] J. E. McMurry, G. J. Haley, J. R. Matz, J. C. Clardy, G. Van Duyne, R. Gleiter, W. Schäfer, D. H. White, *J. Am. Chem. Soc.* **1984**, *106*, 5018–5019.
[59] J. E. McMurry, G. J. Haley, J. R. Matz, J. C. Clardy, G. Van Duyne, R. Gleiter, W. Schäfer, D. H. White, *J. Am. Chem. Soc.* **1986**, *108*, 2932–2938.
[60] To date, neutral homoaromaticity has not been conclusively demonstrated in a molecule: G. C. Christoph, J. L. Muthard, L. A. Paquette, M. C. Bohm, R. Gleiter, *J. Am. Chem. Soc.* **1978**, *100,* 7782–7784.
[61] J. E. McMurry, G. J. Haley, J. R. Matz, J. C. Clardy, J. Mitchell, *J. Am. Chem. Soc.* **1986**, *108,* 515.
[62] D. H. R. Barton, B. J. Willis, *J. Chem. Soc. Perkin Trans. I* **1972**, 305.
[63] J. E. McMurry, R. Swenson, *Tetrahedron Lett.* **1987**, *28*, 3209–3212.
[64] P. E. Eaton, P. G. Jobe, K. Nyi, *J. Am. Chem. Soc.* **1980**, *102*, 6636–6638.
[65] J. E. McMurry, K. Ramig, unpublished results. For a recent development see: A. Fürstner, G. Seidel, B. Gabor, C. Kopiske, C. Krüger, R. Mynott, *Tetrahedron* **1995**, *51*, 8875–8888.
[66] J. E. McMurry, N. O. Siemers, *Tetrahedron Lett.* **1993**, *34,* 7891–7894.
[67] M. Lipton, W. C. Still, *J. Comput. Chem.* **1988**, *9,* 343–355.
[68] J. E. McMurry, J. G. Rico, Y. N. Shih, *Tetrahedron Lett.* **1989**, *30,* 1173–1176.
[69] J. E. McMurry, R. G. Dushin, *J. Am. Chem. Soc.* **1989**, *111,* 8928–8929.
[70] J. E. McMurry, R. G. Dushin, *J. Am. Chem. Soc.* **1990**, *112,* 6942–6949.
[71] J. E. McMurry, N. O. Siemers, *Tetrahedron Lett.* **1993**, *35,* 4505–4508.
[72] K. C. Nicolaou, Z. Yang, E. J. Sorensen, M. Nakada, *J. Chem. Soc. Chem. Commun.* **1993**, 1024–1026.
[73] K. C. Nicolaou, Z. Yang, J. J. Liu, H. Ueno, P. G. Nantermet, R. K. Guy, C. F. Claiborne, J. Renaud, E. A. Couladouros, K. Paulvannan, E. J. Sorensen, *Nature* **1994**, *367,* 630–634.
[74] D. J. Kempf, T. J. Sowin, E. M. Dohery, S. M. Hannick, L. Codavoci, R. F. Henry, B. E. Green, S. G. Spanton, D. W. Norbeck, *J. Org. Chem.* **1992**, *57,* 5692–5700.
[75] T. Mukaiyama, H. Sugimura, T. Ohno, S. Kobayashi, *Chem. Lett.* **1989**, 1401–1404.
[76] A. Clerici, O. Porta, *J. Org. Chem.* **1982**, *47,* 2852–2856.
[77] E. J. Roskamp, S. F. Pederson, *J. Am. Chem. Soc.* **1987**, *109,* 3152–3154.
[78] L. E. Manzer, *Inorg. Chem.* **1977**, *16,* 525.
[79] E. J. Roskamp, S. F. Pederson, *J. Am. Chem. Soc.* **1987**, *109,* 6551–6553.
[80] J. H. Freudenberger, A. W. Konradi, S. F. Pederson, *J. Am. Chem. Soc.* **1989**, *111,* 8014–8016.
[81] A. W. Konradi, S. J. Kemp, S. F. Pederson, *J. Am. Chem. Soc.* **1994**, *116,* 1316–1323.
[82] J. E. McMurry, D. D. Miller, *Tetrahedron Lett.* **1983**, *24*, 1885–1888.
[83] A. Fürstner, D. N. Jumbam, *Tetrahedron* **1992**, *48,* 5991–6010. For a simplified "instant" procedure see: A. Fürstner, A. Hupperts, A. Ptock, E. Janssen, *J. Org. Chem.* **1994**, *59*, 5215–5229. A. Fürstner, A. Ernst, *Tetrahedron* **1995**, *51*, 773–786; A. Fürstner, A. Ptock, H. Weintritt, R. Goddard, C. Krüger, *Angew. Chem.* **1995**, *107*, 725–728.
[84] F. Vögtle, C. Thilgen, *Angew. Chem. Int. Ed. Engl.* **1990**, *29,* 1162–1164.
[85] G. De Munno, F. Lucchesini, R. Neidlein, *Tetrahedron* **1993**, *49,* 6863–6872.
[86] J. E. McMurry, M. Silvestri, *J. Org. Chem.* **1975**, *40*, 2687–2688.
[87] J. E. McMurry, M. G. Silvestri, M. P. Fleming, T. Hoz, M. W. Grayston, *J. Org. Chem.* **1978**, *43,* 3249–3255.
[88] J. E. McMurry, unpublished results.
[89] R. Bayer, T. Poehlmann, O. Nuyken, *Makromol. Chem. Rapid Commun.* **1985**, *14,* 359–64.
[90] A. W. Cooke, K. B. Wagener, *Synthetic Metals* **1989**, *29,* E525-E530.
[91] (a) L. E. Aleandri, B. Bogdanovic, A. Gaidies, D. J. Jones, S. Liao, A. Michalowicz, J. Rozière, A. Schott, *J. Organomet. Chem.* **1993**, *459*, 87–93. (b) L. E. Aleandri, S. Becke, B. Bogdanovic, D. J. Jones, J. Rozière, *J. Organomet. Chem.* **1994**, *472*, 97–112. See also Chapter 8 of this book.

4 Ultrasound-Induced Activation of Metals: Principles and Applications in Organic Synthesis

Jean-Louis Luche and Pedro Cintas

4.1 Introduction

Organic synthesis without metals cannot be imagined, since reactions using them rank among the more important methods for the formation of carbon–carbon bonds. However, chemists often find at the bench that mixing a metal and an organic partner does not necessarily produce a reaction. "Activation" is frequently required as a first step on the way to the desired reagent or reaction. Several methods have been introduced to give access to new organometallic reagents, either via direct processes, or indirectly via auxiliaries, easily prepared reagents which can then undergo metal exchange. Practical experience has led to a deep insight into the advantages and limitations of such methods. Numerous articles and specific reviews dealing with particular methods have been published, and this book contains discussions on these points. A recent review [1] and a book [2] by one of us give a general overview of this subject.

In this chapter we aim to bring to the attention of synthetic chemists the possibilities offered by sonochemistry for metal activation. Sonochemistry is now a well-established branch of chemistry with relevant theoretical and experimental advances [3, 4]. The effectiveness of sonication in this field was perceived at the beginning of the fifties by Renaud [5], but significant applications in synthesis are much more recent. Ultrasound offers inherent advantages (and of course limitations) compared with other activation agents, and the methodology has proven to be a safe, quick, and easy way to produce active materials under very mild conditions. Reviews and books contain pertinent discussions on this topic [3, 4, 6–12].

An in-depth study of the principles of sonochemistry lies beyond the scope of this survey. However some theoretical aspects will first be considered, to allow a better understanding of the results and to familiarize beginners with the technique and equipment [13]. More importantly, the aim is to demonstrate that the interaction of acoustic waves with a chemical system is far from being merely a rather complicated way of achieving agitation or surface cleaning, but involves complex physical phenomena which are presently a matter of advanced research.

The symbolism adopted here is)))) for sonochemical reactions, and ⤸ for reactions effected in the absence of ultrasound, also called "silent" reactions, in accordance with internationally accepted usage.

4.2 The Physical and Chemical Effects of Ultrasound

The propagation of an acoustic pressure wave in a liquid consists of an alternation of compressions and rarefactions. When the amplitude of the wave reaches a critical value,

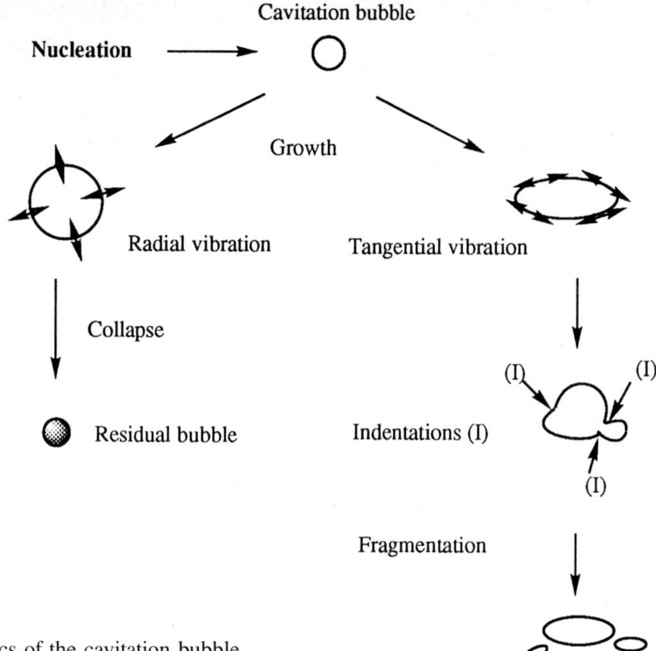

Fig. 4-1. Dynamics of the cavitation bubble.

the Blake threshold, the intermolecular van der Waals forces are not strong enough to maintain the cohesion of the liquid during the rarefaction phase. The structure is broken and a cavity filled with gases and/or vapors of the surrounding liquid is created, preferentially where small solid particles or gas bubbles are present. A description of these complex phenomena accessible to organic chemists is given in [14]. After this nucleation and growth, the subsequent evolution of the bubble forms the basis of sonochemistry. A pictorial representation is given in Fig. 4-1.

4.2.1 Dynamics of the Cavitation Bubble. Transient and Stable Cavitation

After their creation the bubbles grow to 2–10 times the initial diameter of a few microns and undergo radial vibration. Some of them become highly unstable. After a few cycles of rarefaction and compression (transient cavitation), they collapse violently in 10^{-5}–10^{-7} s. Under these adiabatic conditions the content in the residual bubble is subjected to high temperatures (ca. 5000 K), and pressures (several hundreds of bars) generating a "hot spot" [15, 16]. Sonochemistry should therefore be closely related to flash thermolysis, but many experimental results cannot be explained solely by this approach.

Bubbles undergo not only radial vibrations but also surface deformations resulting from tangential forces on the boundary layer, leading to molecular redistribution and reorientation. After several tens or hundreds of cycles (stable cavitation), the deformed bubbles fragment into smaller gaseous inclusions. During these events, considerable electrical fields [17] are generated in the separation zones (indentations) of the "daughter" bubbles from the "mother" large bubble. High energy electrons are emitted in these zones [18] and the bubble bears an overall negative charge [19]. The electrons propagating across the bubble can be captured by the vapors, the solvent, or molecules in solution. This theory,

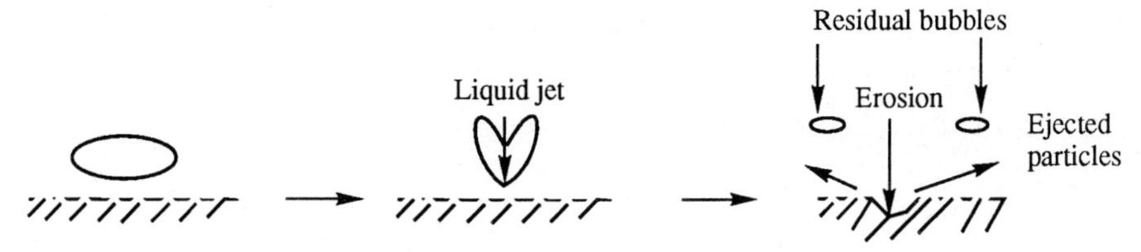

Fig. 4-2. Cavitational erosion.

which predates that of the "hot spot" [20], emphasizes the electrical consequences of cavitation, which, from the chemist's viewpoint, should be ionization and formation of radicals or radical ions from the irradiated compounds, a kind of "mass spectrometry" chemistry.

The probable coexistence of "hot spot" and "electrical discharge" effects reflects the complexity of cavitation and explains the unique character of sonochemistry. The relative influence of these intricate phenomena should be frequency dependent [21], an aspect recently reexamined by Italian sonochemists [22].

4.2.2 Cavitation in Heterogeneous Solid-Liquid Systems

In the recent development of synthetic sonochemistry, heterogeneous reactions were much more intensively studied than homogeneous ones, essentially because cavitation appears to be more easily predictable when produced in two-phase systems. In the strongly deformed cavitation bubble, a liquid jet propagates at a high velocity before violently hitting the interface [8, 23, 24]. On solids, erosion takes place [25], with removal of particles and/or permanent deformation of the surface (Fig. 4-2). In terms of chemistry, what is produced can be described as an "activation" of the surface.

4.2.2.1 The Physical Nature of Activation

Usually, a metal surface is not pure. A solid coating of oxides, hydroxides, carbonates, and sometimes nitrides, constitutes a "passivation" layer which prevents the reagent in the solution from reaching the metal [26]. Activation consists of reestablishing the contact between the partners. In a study of the formation of Grignard reagents, the rate of the subsequent reaction was shown to be proportional to the active surface area [27], a result

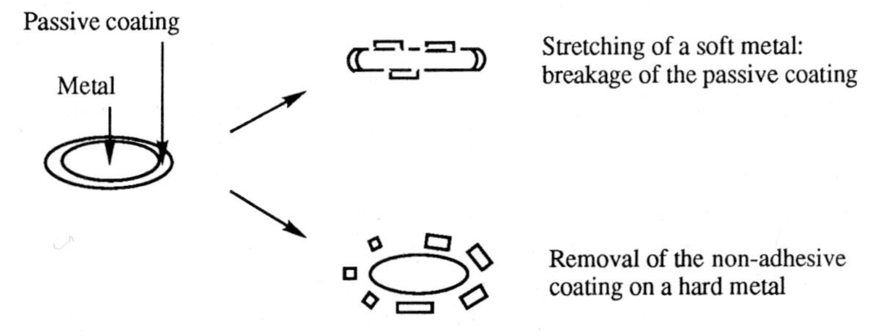

Fig. 4-3. Schematic mechanism for the activation of metals.

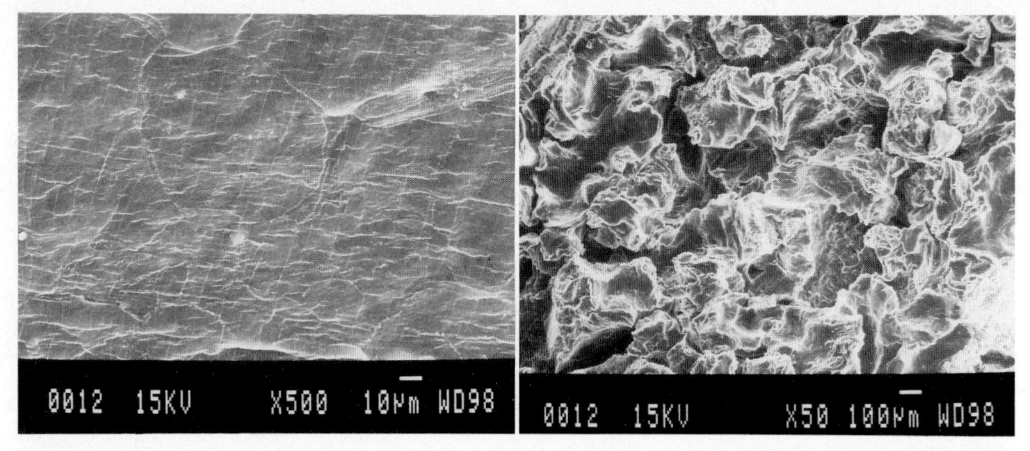

Fig. 4-4. Micrograph of a lithium sample. Left: initial state. Right: after 5 min sonication. (Photograph: INP Grenoble and URA CNRS 332, reproduced by permission of the American Chemical Society).

which can reasonably be extrapolated to most reactions of metals. Two mechanisms, sketched in Fig. 4-3, can be envisaged for surface depassivation, according to the mechanical properties (e.g. the hardness [28]) of the metal itself and its coating.

When sonicated, soft materials such as alkali metals undergo permanent plastic deformation, regardless of the nature of the passivation layer. Electron microscopy of a lithium surface (hardness 0.6 Mohs) reveals very important changes in morphology [29]. The surface area is increased, and numerous irregularities appear (Fig. 4-4). Sodium and potassium, softer than lithium, can even be dispersed to give highly reactive fine powders [30]. Sonication of magnesium, harder than lithium (2.3 Mohs), seems to affect only the superficial layers [31]. Fracturing of the oxide (5 Mohs), not necessarily accompanied by its removal, offers a better explanation for the experimental observations.

A second factor is the adhesion of the passivating layer to the underlying metal. The role of ultrasound should be to overcome this adhesion. Aluminum (hardness 2.7 Mohs) is thought to undergo plastic surface deformations which fracture its hard (9 Mohs) strongly adhering oxide layer, before the metal structure itself is broken. Zinc is probably activated via another mechanism, as its oxide is not firmly attached. Indeed, Auger spec-

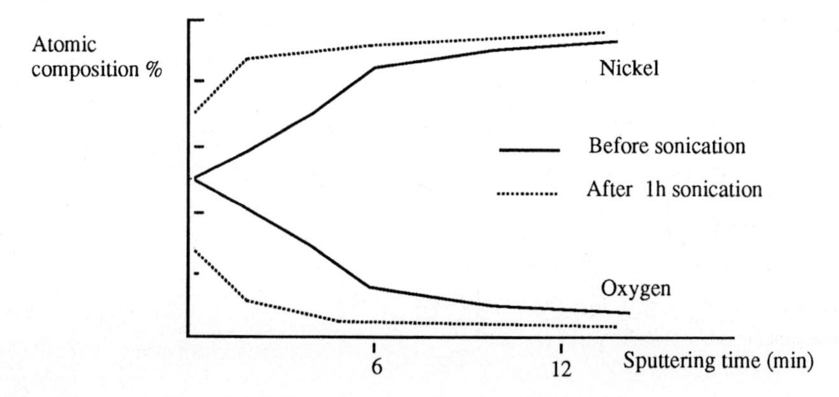

Fig. 4-5. Surface composition of nickel powder determined by Auger spectroscopy.

troscopy [8] reveals that sonication reduced the oxygen content of the surface layer, and the size of the particles, especially the larger ones, is slightly reduced. Copper [32] and nickel [33] (Fig. 4-5) show similar phenomena.

The catalytic activity of a sonicated nickel powder for hydrogenation is increased by a factor of 10^5, but is lost after prolonged sonication. Collision between particles are believed to produce local fusion at the contact point, and reagglomeration. This interpretation was later challenged [34] and an alternative explanation can be proposed based on tribochemistry [35]. Lastly, when hard metals with hard adhesive oxide are involved, such as molybdenum and tungsten, sonochemical activation no longer takes place [8].

4.2.2.2 The Chemical Component of the Sonochemical Activation

Having undergone a first step establishing the physical contact between the reaction partners, the metallic surface is assumed to react. Let us consider as an example a reaction starting from an organic halide. The initial step is an electron transfer from the metal to the carbon-halogen bond, taking place on the surface [36, 37]. In a study of the Barbier reaction in the presence of lithium (see Section 4.10.1.), it was noted that not all the results could be interpreted solely on the basis of mechanical effects [29]. Experiments revealed that the initial electron transfer itself is enhanced by sonication, which then has a true *chemical* role (Fig. 4-6), a conclusion consistent with results obtained later.

The promotion of single electron transfers (SET) is regarded as a general characteristic of sonochemistry. In reactions with metals, the origin of this effect is still unclear. Theoretical calculations showed that, for non-transition metals, the energy gap between the valence and conduction bands (in the language of organic chemists, the HOMO and LUMO) is reduced by sonication [38]. An interpretation of the sonochemically enhanced reactivity based on this result should however be treated with caution, since the frequencies used in this study, up to 10^{11} Hz, are much higher than those employed in sonochemistry. On the other hand, the creation of structural defects by ultrasound is expected to have other consequences besides mechanical ones. Physicists know that in disordered metallic structures the electrons do not circulate freely, but are located at lattice defects where the wave-function reaches maximum values [39]. Correspondingly, it was shown that the more "reactive" metal atoms are those with a minimum number of neighbors [40]. It is intuitively understood that an atom placed on an edge or a defect has unsatisfied valences and is less tightly bound to the lattice. After the initial electron transfer, the question of whether the intermediate radical or radical anion remains at the metal surface or not [36, 37] does not affect the last step, during which an atom or ion must be extracted from the metal (Fig. 4-7).

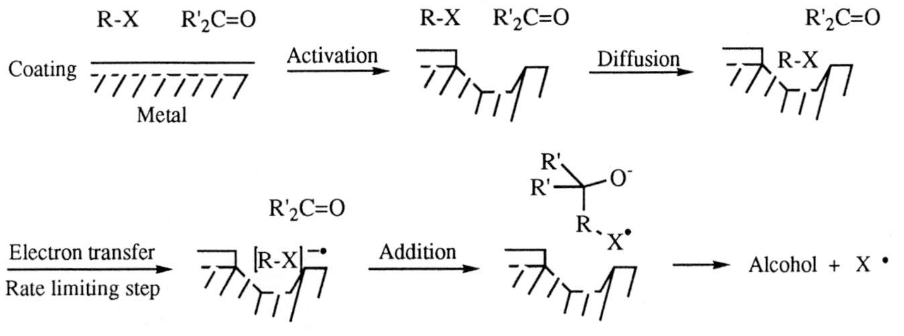

Fig. 4-6. Mechanism of the sonochemical Barbier reaction.

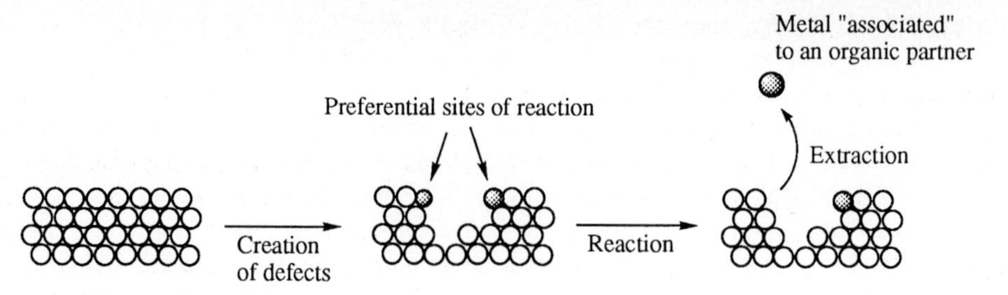

Fig. 4-7. Preferential reaction sites on a metal surface.

It can be suggested that this extraction depends on the sublimation energy [28]; the lower the energy, the easier the reaction. Literature data for reactive metals such as K, Li, Zn, Na, Mg ($\Delta H_{subl.}$ from 21 to 35 kJ mol^{-1}), metals with medium reactivity (Al, Cu; $\Delta H_{subl.}$ ~ 80 kJ mol^{-1}), and less reactive metals (Ti, Mo, W; $\Delta H_{subl.}$ from 110 to 200 kJ mol^{-1}), support this interpretation.

4.2.3 The Relationship between Sonochemical and Mechanochemical Activation

The surface area and reactivity of a metal increase when it is subjected to mechanical stresses. Contraintuitively, the surface effect has only minor consequences, whereas the accumulation of energy in lattice defects is considered to be responsible for up to 90% of the increase in reactivity [35]. At the same time, high energy electrons are ejected by the collision and can induce chemical reactions (Fig. 4-8) [41]. For instance, styrene polymerizes when subjected to ball milling in a metallic reactor [42].

The catalytic properties of nickel powders can be enhanced in the same way. Prolonged mechanical treatment results in the loss of the catalytic property by a relaxation process [43], an effect similar to the observations mentioned above [33].

From the description of the effects of the cavitational collapse on metallic surfaces, an analogy with micro-hammering can be drawn. Shock waves and microstreaming produce lattice defects and surface vacancies, increasing the total energy of the system, which can then relax, among other pathways, by easier chemical reactions. The analogy between mechanochemical and sonochemical reactions thus seems plausible. In this context it is worth considering Rieke's active metal preparation method [44], which should also produce such energy-rich disorganised states. Since the reductions require prolonged heating, some deactivation can result from a thermal relaxation process (annealing), which is minimized in reactions performed under sonication in shorter times at lower temperatures [45].

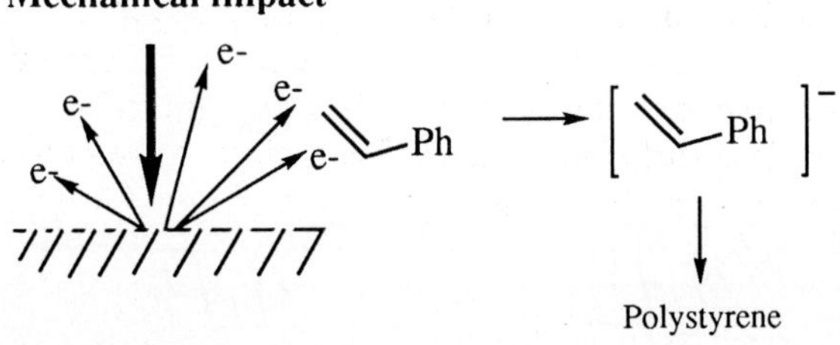

Fig. 4-8. Tribochemical polymerization of styrene.

4.2.4 Sonochemical Reactivity, a General Approach

In order to place sonochemistry in a general context, some fundamental aspects are briefly discussed here. Many sonochemical processes cannot be understood on the basis of cleaning and erosion effects, and a true picture of how the non-quantified mechanical energy is transformed into chemical energy cannot be derived from the "hot spot" theory alone. In the absence of a general understanding, a classification leading to empirical rules was proposed with the purpose of predicting the behavior of a given system [46].

The *first type of sonochemical reactions* are those occurring in solution. Cavitation phenomena generate reactive intermediates such as radicals or radical-ions. Ionic processes remain unaffected. Sonochemical switching, a term coined by Ando [47] for describing the *difference in nature* of the reaction products in the presence and absence of ultrasound, can occur. Sonochemical switching is often an indication of the existence of a SET process in competition with an ionic one.

The *second type* includes heterogeneous ionic reactions. The influence of ultrasound is purely physical and sonochemical switching is not expected. Although the ionic pathway remains unchanged by sonication, rates and yields can be substantially improved by what may be called "ultrasonic agitation". This class of reactions can also be described as "false sonochemistry".

The *third type* of sonochemical reactions are heterogeneous radical reactions, or processes that can follow either an ionic or a SET mechanism. They are influenced by the physical *and* chemical effects of sonication, and the nature and ratio of the products represent the relative importance of the two different mechanisms. Many of these reactions involve metals and the results have led to important conclusions concerning the reaction mechanisms.

4.3 Sonochemical Reactions of Inorganic Compounds

4.3.1 Activation of Metals

As examples of the morphological modifications produced by sonication of metals [8, 25], finely dispersed alkali metals are prepared by ultrasonic irradiation of the solids in a cleaning bath. Thus, a piece of potassium metal in dry toluene or xylene rapidly transforms to a silvery blue "colloidal" suspension [30]. Sodium forms dispersions in xylene but not in toluene nor THF. Metallations and reductions can be achieved with these reagents (see Sections 4.7.1 and 4.7.2). Activation of molten sodium as an emulsion in an inert solvent was described in 1959 [48]. Particles of ca. 1 μm average size are produced as a blue-purple suspension. Margulis described similar results and observed a frequency effect [49]. Smaller particles of a few μm in size are produced at 44 kHz as compared to 22 kHz and a steady state is reached after ca. 20 min. The experimental set-up for the production of highly active zinc powders, less than 100 μm in size, by sonication of the molten metal was also described [50]. The surface state of samarium was not studied, but its reaction with iodine is strongly accelerated by sonication in THF, producing SmI_3 in a few minutes [51].

$$2\ Sm + 3\ I_2 \xrightarrow[\text{)))), r.t., 5 min}]{\text{THF}} 2\ SmI_3(THF)_3 \xrightarrow[\text{)))), 20 min, r.t., 100\%}]{\text{Sm, Hg (cat)}} 3\ SmI_2 \qquad (4.1)$$

Reduction of the latter to the useful diiodide in the presence of mercury is brought to completion in ca. 20 min [Eq. (4.1)]. Palladium, platinum, and rhodium blacks prepared by sonochemical reduction of aqueous salts exhibit a surface area increase by ca. 60% and a higher activity. Numerous papers on this subject have been published, but are not always readily accessible [52].

4.3.2 Activation by Cementation

Formation of a surface alloy (cementation) is one of the activation methods commonly used by organic chemists, typical examples being zinc–copper or –silver couples, and sodium or aluminum amalgams. In recent years the Zn–Cu couple was found to be easily produced by sonication of zinc dust with copper iodide [53]. An organic solvent such as THF, methanol, ethanol, *n*- or *i*-propanol, may be used, but in any case 5–40% of water is essential. The deep black couple forms in a few minutes and can be used directly for conjugate addition reactions (see Section 4.10.3). Similarly the zinc-nickel couple is obtained from nickel salts and zinc [54] (see Section 4.6.1.). Cementation of cobalt on zinc in water suspension was studied under sonication [55]. The temperature should be kept above 50 °C, otherwise the reduction of water by zinc is dominant and zinc hydroxide is formed. Whether or not a couple is formed from a sonicated mixture of zinc and magnesium powders was not determined. Such a metallic system has been used for an intramolecular reductive coupling (see Section 4.6.2) [56].

4.3.3 Reduction and Sonolysis of Metallic Compounds

The sonochemically modified Rieke procedure was applied to a series of metal chlorides (Mg, Zn, Cd, Ni, Fe, Pd, Co, Cu, Pb) [45, 57, 58]. In this method the reduction can be carried out with lithium instead of potassium under milder conditions (less than 40 min at room temperature compared with several hours in refluxing THF or DME). For salts insoluble in THF, addition of an electron carrier (naphthalene) is recommended. Reactive nickel produced from nickel acetylacetonate and sodium reacts in situ with cyclooctadiene to yield the $Ni(COD)_2$ complex [59]. The useful low-valent (or metallic?) titanium is similarly obtained by sonicating $TiCl_3$ and lithium at 30 °C, whereas the silent reaction requires reflux [Eq. (4.2)] [60].

$$TiCl_3 + Li \xrightarrow[))))]{\text{THF, 30 °C}} \text{“Ti”} \tag{4.2}$$

$$WCl_6 \xrightarrow[\text{10 °C,)))), 40\%}]{\text{Na, CO (1 atm.), THF}} W_2(CO)_{10}^{2-} \tag{4.3}$$

$$VCl_3(THF) \xrightarrow[\text{THF, 10 °C,)))), 35\%}]{\text{Na, CO (4.4 atm.)}} V(CO)_6^- \tag{4.4}$$

Silent reaction: 160 °C, 200 atm.

Reduction of transition metal halides by sodium, under a carbon monoxide atmosphere forms the corresponding metal carbonyl [Eqs. (4.3), (4.4)] [61]. This process constitutes an important improvement, since it is efficient even under low pressures of CO (1–4 bar), in contrast to the silent method which requires much higher pressures, even when using Rieke's metals.

Colloidal silver is prepared by sonicating the perchlorate or nitrate at 200 kHz (a rather unexplored frequency domain) in water containing an alcohol or a surfactant [62]. The authors interpret this atypical reaction as a reduction of silver ions by radicals formed by sonolysis of the alcohol or the surfactant (Scheme 4-1).

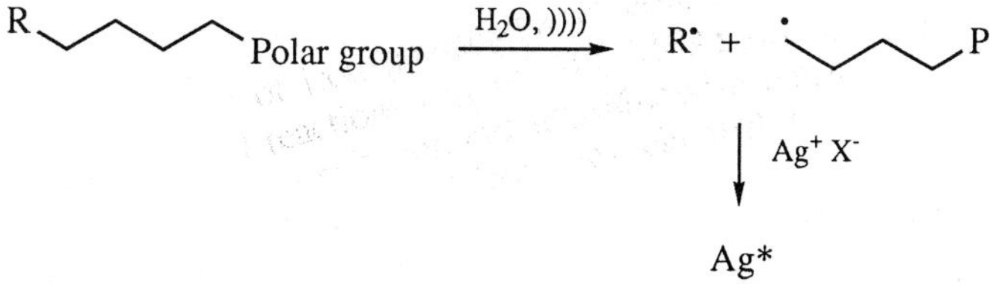

R–(chain)–Polar group $\xrightarrow{H_2O,\ ))))}$ $R^{\bullet}$ + •(chain)–P; $\xrightarrow{Ag^+ X^-}$ Ag*

Scheme 4-1

$Cp^*Fe(CO)_2Br$ $\xrightarrow[)))),\ r.t.]{K,\ THF}$ $Cp^*Fe(CO)_2K$ $\xrightarrow{C_2H_5\text{-}I}$ $Cp^*Fe(CO)_2C_2H_5$

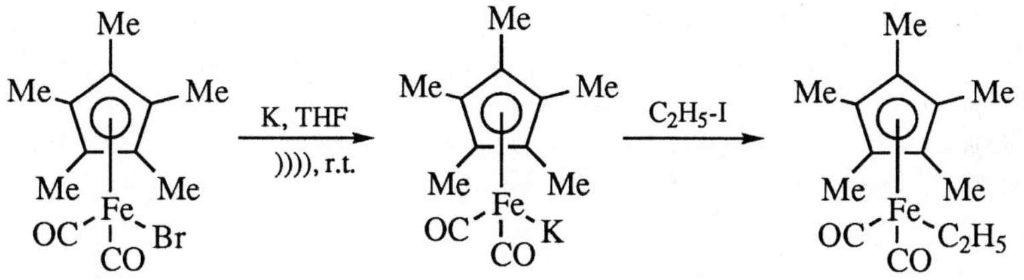

Scheme 4-2

The sonochemical reductions of metal derivatives have been extended to transition metal complexes. For instance, the bromo compound shown in Scheme 4-2 is readily metallated, and the resulting anion alkylated [63]. Similarly, arene ruthenium complexes are prepared by zinc reduction of the dichloride (Scheme 4-3) [64].

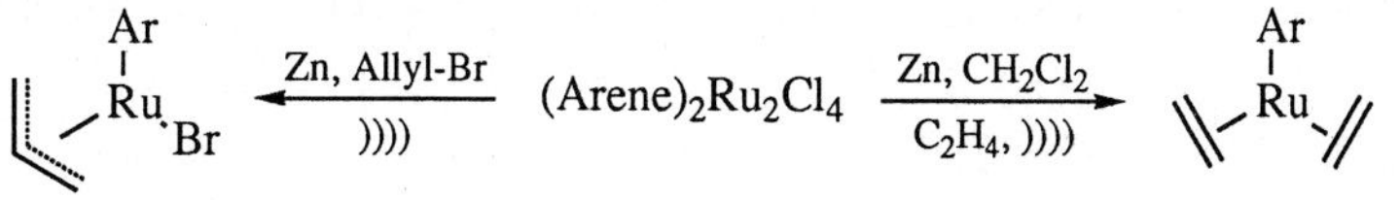

(allyl)Ru(Ar)Br $\xleftarrow[))))]{Zn,\ Allyl\text{-}Br}$ $(Arene)_2Ru_2Cl_4$ $\xrightarrow[C_2H_4,\ ))))]{Zn,\ CH_2Cl_2}$ $(C_2H_4)_2Ru(Ar)$

Scheme 4-3

Sonolysis of iron pentacarbonyl constitutes one of the earliest studied homogeneous sonochemical reactions [Eq. (4.5)] [65]. Amorphous iron is obtained in solvents chosen to provide a high cavitation energy, otherwise clusterification is the preferred pathway. This highly reactive form of iron was claimed to have interesting catalytic properties for Fischer–Tropsch synthesis and hydrogenation, the activities being increased by a factor of 10–30 in comparison with microcrystalline iron.

$$Fe_3(CO)_{12} \xleftarrow[)))),\ r.t.]{Heptane} Fe(CO)_5 \xrightarrow[)))),\ r.t.]{Decaline} Fe^* \quad (4.5)$$

4.4 Metals in Organic Sonochemistry: Preliminary Remarks

According to the empirical rules [46], the initial step of electron transfer from a metal to a substrate should be sensitive to sonication. Indeed, the applications of sonochemistry in organometallic syntheses agree with this assumption. In the following discussions, a classification related to the substrate to which the electron is transferred was chosen, even if in some instances the actual mechanism is more or less obscure or controversial. From reactions in which the transfer occurs without bond breakage, to the cases where a bond is cleaved, the sonochemical methods provide the synthetic chemist with an extremely diversified field of investigation.

4.5 Electron Transfer to Conjugated Hydrocarbons

Most of the results published in this domain deal with the formation of arene–alkali metal compounds, either as an end goal or for a further synthetic use, according to the general scheme shown in Scheme 4-4.

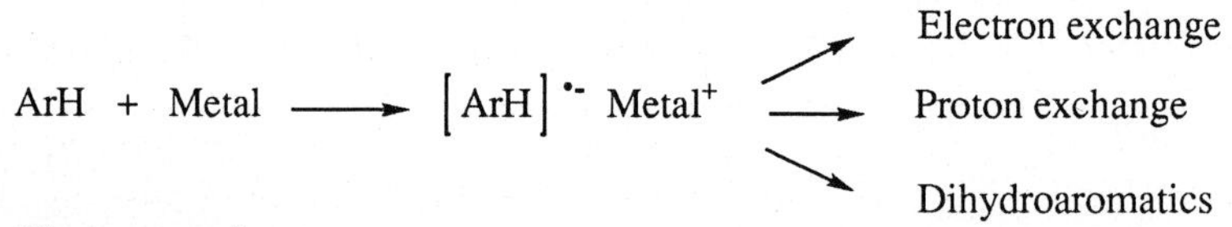

$$\mathrm{ArH + Metal} \longrightarrow [\mathrm{ArH}]^{\bullet -}\ \mathrm{Metal}^{+} \longrightarrow \begin{cases} \text{Electron exchange} \\ \text{Proton exchange} \\ \text{Dihydroaromatics} \end{cases}$$

Scheme 4-4. Electron transfers to arenes.

4.5.1 Aromatic Radical Anions

The role of sonication in the formation of arene–metal compounds was discovered as early as 1957. Benzoquinoline-sodium was prepared in 45 min at room temperature instead of a 2-days reflux in ether. The experimental set-up has been described in the original publication [66]. Extensions using laboratory cleaning baths were also described. Naphthalene-, anthracene-, biphenyl-, or acenaphthylene-lithium, or the analogous sodium compounds, can thus be prepared in non-specialized laboratories [67, 68]. In some cases non-ethereal solvents were used, in the presence of *N,N′*-tetramethylethane- or propanediamine [69]. Naphthalene–sodium thus obtained reduces the carbon–sulfur bond in sulfones much more efficiently under sonication than the usual thermal reaction (Scheme 4-5a) [51].

Since radical anions derived from arenes are intermediates in Birch-type reductions, attempts were made to improve some of these reactions [68, 70]. Phenols sonicated in the presence of lithium and trimethylsilyl chloride (TMSCl) are reduced to the dihydro-bis-trimethylsilyl derivatives, easily reoxidized to the bis-silylated aromatic compound (Scheme 4-5b) [70]. Although the reaction rate is unusually slow for a sonochemical reaction, good yields of the expected product are obtained, while the silent reaction is even slower and gives poorer yields.

Among the arene–metal reagents, the highly useful lithium 4,4′-di(*t*-butyl)biphenyl (LiDBB) [71] has frequently been used in sonochemistry. This radical anion, easily formed by a single-electron transfer from the metal to DBB, has been used as an electron carrier by Fry et al. in a series of reactions [72–75].

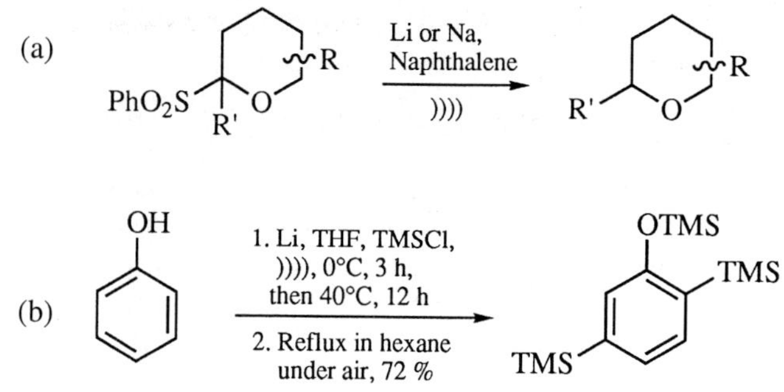

(a) PhO_2S, R', O, R — Li or Na, Naphthalene,)))) → R', O, R

(b) OH — 1. Li, THF, TMSCl,)))), 0°C, 3 h, then 40°C, 12 h; 2. Reflux in hexane under air, 72 % → OTMS, TMS, TMS

Scheme 4-5 TMS = Trimethylsilyl

An "umpolung" Barbier condensation with xanthene-9-one and tropylium bromide was effected via the formation of the ketone dianion from lithium and catalytic amounts of DBB (Scheme 4-6) [72]. The normal approach using an organometallic derived from tropylium bromide is not feasible due to the antiaromatic character of the anion. Similarly, dianions of aromatic ketones (Scheme 4-7) undergo alkylation to alcohols [73]. Long irradiation times lead to the reduction of the alkylated alkoxide to the hydrocarbon. The reaction does not proceed in the absence of DBB or ultrasound.

Extension to aliphatic esters was accomplished (Scheme 4-8). A complex pathway takes place from the initial ketyl radical anion, which ends with the formation of a symmetrical ketone when lithium is used. In contrast, with sodium, an acyloin coupling takes place [74]. From an aromatic ester the ketyl will couple to the α-diketone (Scheme 4-9) [75]. The same reaction also occurs in the absence of LiDBB, but sonication times are

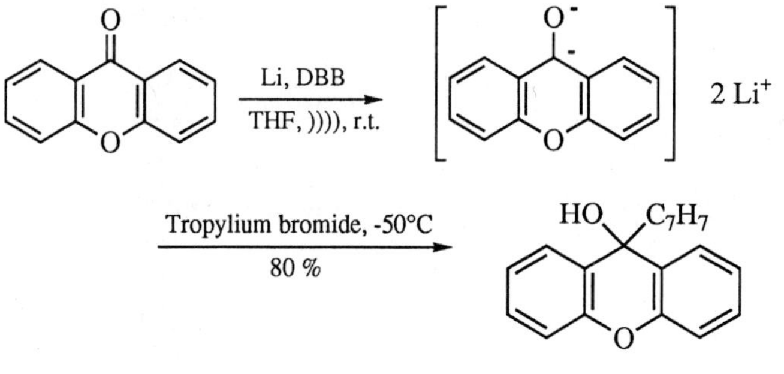

Li, DBB, THF,)))), r.t. → $2\ Li^+$

Tropylium bromide, -50°C, 80 % → HO, C_7H_7

Scheme 4-6

Ar, Ar, O — Li, DBB, THF,)))), r.t. → Ar, Ar, OLi, Li — H_2O → Ar, Ar, OLi, H

RX

Ar, Ar, OLi, R — Li,)))) → Ar, Ar, Li, R — H_2O → Ar, Ar, H, R

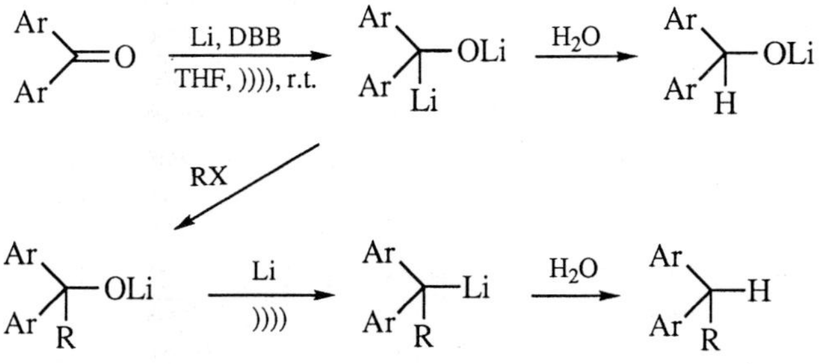

Scheme 4-7. Aromatic ketones reduction and deoxygenation.

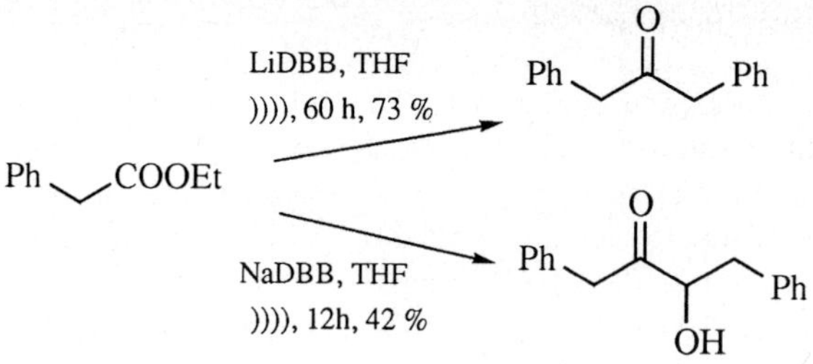

Ph⌄COOEt — LiDBB, THF,)))), 60 h, 73 % → PhCH$_2$COCH$_2$Ph
Ph⌄COOEt — NaDBB, THF,)))), 12h, 42 % → PhCH$_2$COCH(OH)CH$_2$Ph

Scheme 4-8

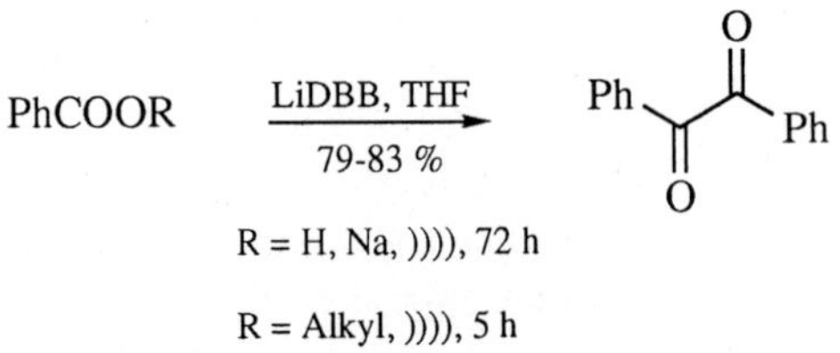

PhCOOR — LiDBB, THF, 79-83 % → PhCOCOPh

R = H, Na,)))), 72 h

R = Alkyl,)))), 5 h

Scheme 4-9

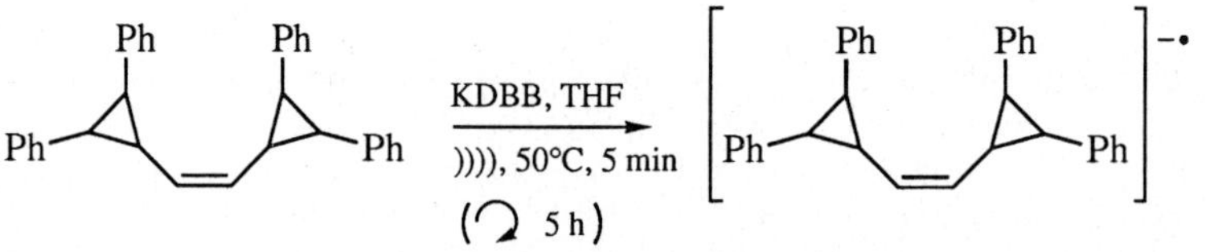

KDBB, THF,)))), 50°C, 5 min (5 h)

Scheme 4-10

very long and the yields much lower. The potassium analog (KDBB) exhibits improved reducing properties for some olefins (Scheme 4-10) [76]. Its preparation consists in adding DBB to excess potassium in THF under sonication at 50 °C for 5 min. The formation rate of KDBB is largely dependent on the temperature and the power of the generator. The deep green color of the radical anion appears immediately at 50 °C, or in 30 min at 0 °C. Under stirring, completion of the reaction occurs in 5 h.

The complex between magnesium and anthracene is a versatile and useful dianion [77], extensively studied by Bogdanovic et al. [78] and having many applications. The formation and subsequent thermolysis of the adduct are both highly favored by sonication [79]. The resulting active dispersion of magnesium is used in the preparation of Grignard reagents [80]. In the example shown (Scheme 4-11) the allylic chloride reacts with the activated metal at –60 °C in THF, then cyclizes on heating at 65 °C for 1 h.

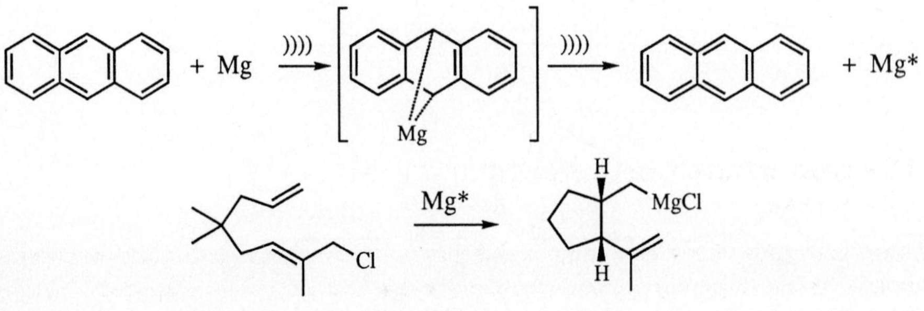

+ Mg —)))) → [Mg adduct] —)))) → + Mg*

Mg* → MgCl

Scheme 4-11. Activation of magnesium by anthracene.

4.5.2 Radical Anions from Dienes

Radical anions from conjugated dienes can also be prepared under sonication and used for the preparation of lithium amides. The direct attack of the N–H bond of amines is generally not possible even under sonication, due to its low acidity (see however Section 4.7.3), and the presence of an electron carrier is necessary. In industrial preparations styrene is used [81], but the by-product ethylbenzene contaminates the reaction product. Isoprene was found to be more convenient, since its reduction product is the easily eliminated 2-methyl-butene. Sonication using a bath allows quantities of base up to 0.1 mole to be obtained readily from the amine and lithium, preferably containing 2% sodium (Scheme 4-12) [82]. Without ultrasound, isoprene undergoes competitive addition reactions.

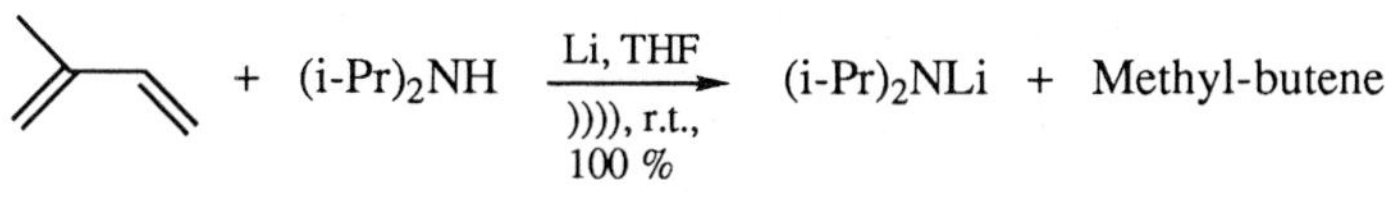

+ $(i\text{-}Pr)_2NH$ —Li, THF;)))), r.t., 100 %→ $(i\text{-}Pr)_2NLi$ + Methyl-butene

Scheme 4-12

This preparation can be effected with the substrate to be deprotonated present in the initial mixture. With carboxylic acids, a rapid reaction takes place leading to carboxylate dianions in a few minutes at room temperature. With a leaving group present in the ω position, cyclization occurs in good yields, higher for the ω-chloro compounds [83]. When applied to a dipeptide precursor (Scheme 4-13), experiments run with LDA, pre-formed by the usual methods or generated sonochemically in situ, gave higher stereoselectivity with the sonochemical method.

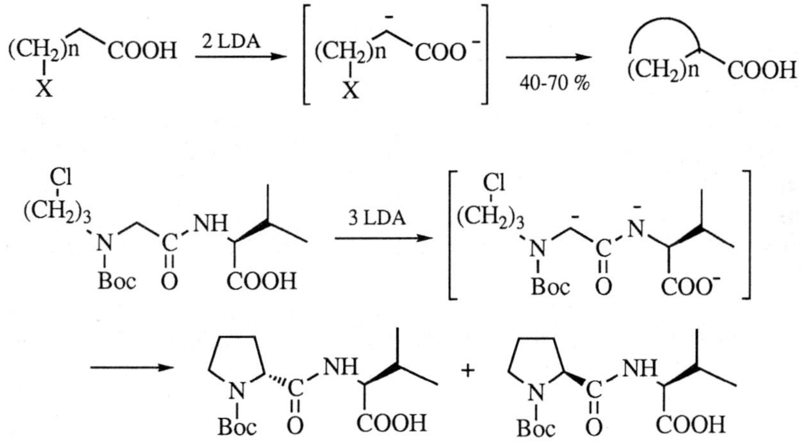

$(CH_2)_n$COOH (X) —2 LDA→ [$(CH_2)_n$COO$^-$ (X)] —40-70 %→ $(CH_2)_n$COOH

Cl$(CH_2)_3$N(Boc)–C(=O)–NH–CH(COOH) —3 LDA→ [Cl$(CH_2)_3$N(Boc)–C(=O)–N$^-$–CH(COO$^-$)] → N(Boc)–C(=O)–NH–CH(COOH) + N(Boc)–C(=O)–NH–CH(COOH)

Scheme 4-13. Cyclization of functional carboxylic acids by in situ generated LDA.

4.6 Electron Transfers to Multiple Bonds

Sonication generally increases the rates and yields of C=C bond reduction. Reductions of C=O bonds by the dissolving metal process are known to follow a complex mechanism [84]. In any case however, the first step should be an electron transfer generating the ketyl radical-anion, and should be accelerated by sonication [46]. Although in most instances the qualitative reaction outcome is not modified by sonication, a sonochemical switching was found in one case.

4.6.1 Reductions of Carbon–Carbon Double Bonds

The enhanced reactivity of the sonicated zinc–acetic acid system is reported in two papers [85, 86]. The examples given in Scheme 4-14 illustrate the advantages of the method, namely very short times, high yields, and excellent selectivity, since non-conjugated olefins remain intact.

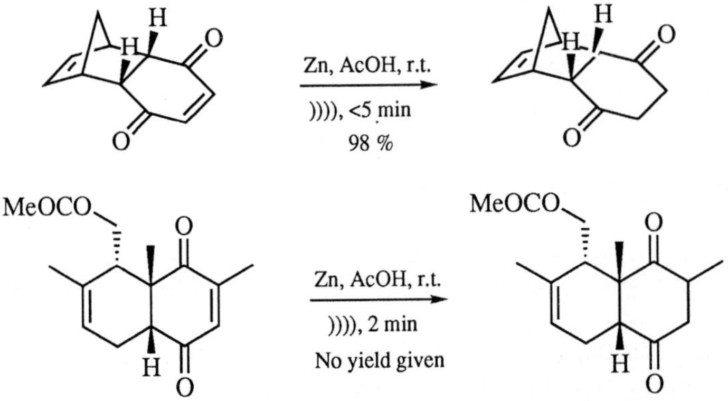

Zn, AcOH, r.t.
)))), <5 min
98 %

MeOCO
Zn, AcOH, r.t.
)))), 2 min
No yield given

Scheme 4-14. Reduction of conjugated C=C bonds.

A new process for the "hydrogenation" of olefins was found with the zinc–nickel chloride system in the form of an aqueous alcohol dispersion [54, 87]. The chemical role of nickel seems to be threefold: activation of zinc by cementation, easier reduction of water to hydrogen, then catalysis of the hydrogenation step [88]. The selectivity of the system can be modified by adjusting the conditions [54]. With the zinc–nickel couple (9:1) in aqueous ethanol, carvone is reduced to the tetrahydro compound, but in a slightly basic medium only the conjugated olefin is reduced (Scheme 4-15). With a 1:1 Zn–Ni couple, the presence of hydrogen gas is necessary, sonication becomes useless and only the isolated olefin is reduced. Another example of this method is shown in Scheme 4-16.

In a total synthesis of the tumor promoter indolactam V from 4-aminoindole, a key step was the ultrasonic reduction of an enamino group with magnesium powder in methanol (Scheme 4-17). This reaction, described as "clean and expedient", accomplished the desired reduction quantitatively [89].

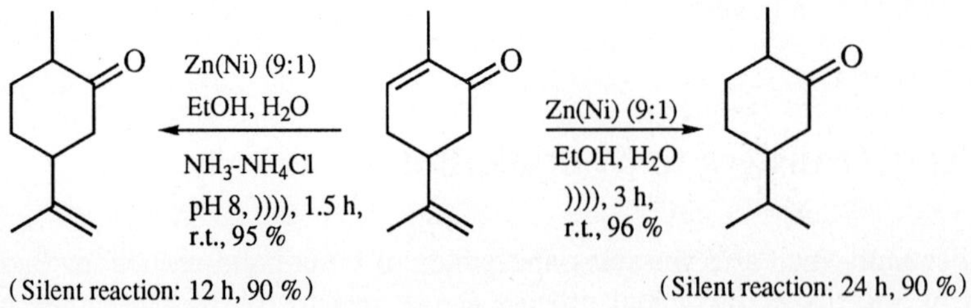

Zn(Ni) (9:1)
EtOH, H_2O
NH_3-NH_4Cl
pH 8,)))), 1.5 h,
r.t., 95 %
(Silent reaction: 12 h, 90 %)

Zn(Ni) (9:1)
EtOH, H_2O
)))), 3 h,
r.t., 96 %
(Silent reaction: 24 h, 90 %)

Scheme 4-15

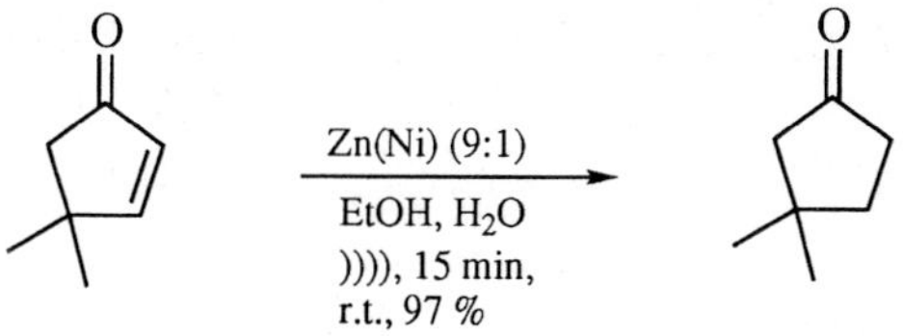

Zn(Ni) (9:1)
EtOH, H_2O
)))), 15 min,
r.t., 97 %

Scheme 4-16 (c. f. conventional hydrogenation with H_2 + Pd/C: 2 days, 60 %)

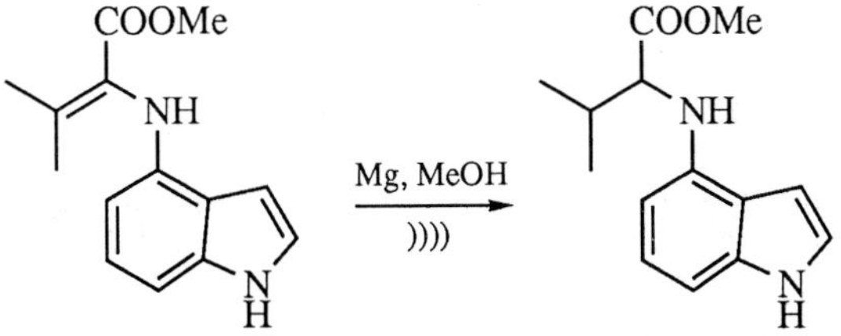

Mg, MeOH
))))

Scheme 4-17

4.6.2 Reductions of Carbonyl Groups

The general picture of this process is shown in Scheme 4-18. In the case of stable ketyls, e.g. from benzophenone and alkali metals, it is easily observed that sonication greatly accelerates their formation [90].

Quinones and α-diketones are easily reduced by zinc and TMSCl [91]. The yields are not always improved, but the rate is increased about ten-fold compared with the silent process (Scheme 4-19). As expected, the process occurs more efficiently in THF than in diethyl ether, due to a higher cavitation energy. In an example using a ketone, Huffman et al. observed that the stereoselectivity of camphor reduction is the same either sonochemically in THF or silently in liquid ammonia with the usual three alkali metals

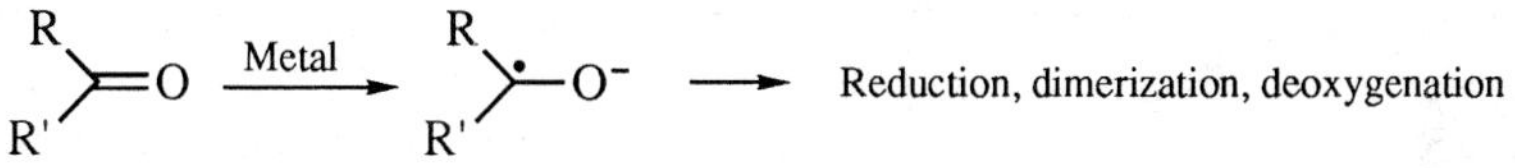

Metal ⟶ Reduction, dimerization, deoxygenation

Scheme 4-18. Electron transfer to carbonyl groups.

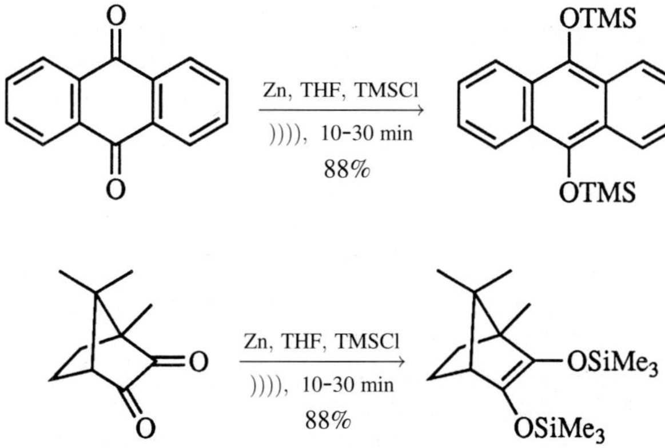

Zn, THF, TMSCl
)))), 10–30 min
88%

Zn, THF, TMSCl
)))), 10–30 min
88%

Scheme 4-19. Reduction of quinones.

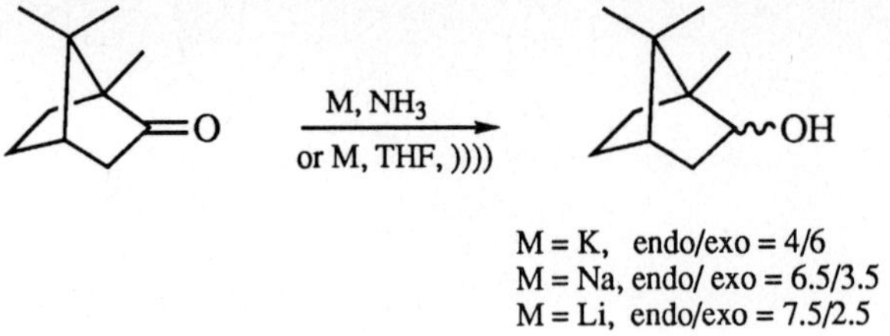

M, NH_3 or M, THF,))))

M = K, endo/exo = 4/6
M = Na, endo/ exo = 6.5/3.5
M = Li, endo/exo = 7.5/2.5

Scheme 4-20

(Scheme 4-20) [92]. The authors analyze the results and conclude that the mechanism does not involve a ketone dianion.

Using aluminum in liquid ammonia, Sato et al. reduced various aromatic ketones to alcohols [93]. The competing pinacolization can be suppressed by addition of ammonium chloride (Scheme 4-21). When an imide is sonicated with aluminum in aqueous THF, one C=O group is reduced to give a hydroxylactam in excellent yield, without disturbing functional chains on the nitrogen atom (Scheme 4-22) [94].

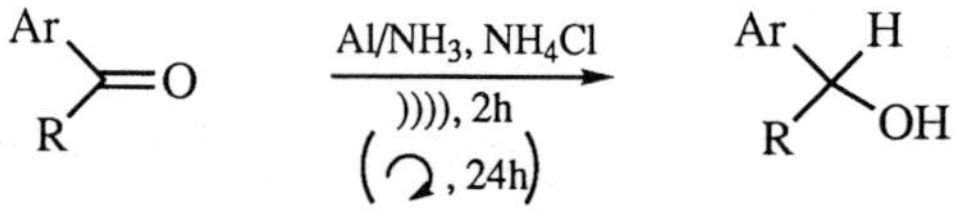

$Al/NH_3, NH_4Cl$,)))), 2h (↻, 24h)

Scheme 4-21

$Al(Hg), THF, H_2O$,)))), 15°C, 15-20 min, 70-95 %

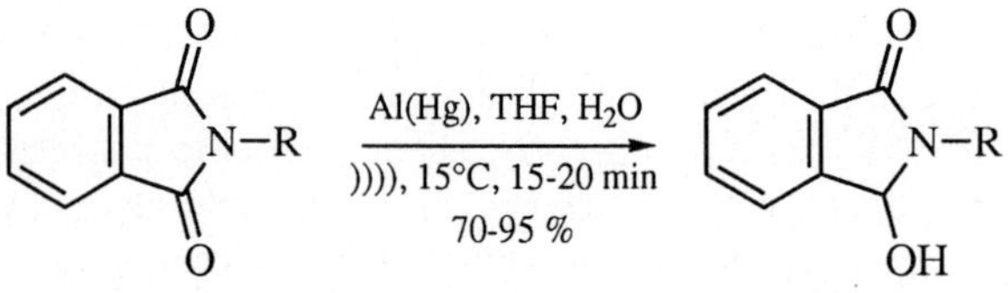

Scheme 4-22 R = $PhCH_2$, (*E*)-CH_2CH_2-CH=CH-COOEt, $(CH_2)_3$-I

Pinacolization is one of the so-called "easy" reactions for which no general satisfactory solution exists in synthesis. Thus, although the attempted sonochemical improvements have met with only limited success, they deserve a short mention (Scheme 4-23). Aromatic aldehydes and ketones give pinacols when sonicated with low-valent titanium in DME [60]. However, the reaction is highly sensitive to the nature of the solvent. In apolar media the reduction of titanium trichloride does not occur, and in THF the McMurry coupling to olefins is the preferred pathway. With aluminum or zinc and aluminum chloride, dimerization of the ketyl is followed by rearrangements to benzopinacolones [95]. Boudjouk et al. reported that aryl- or vinyl-substituted carbonyl compounds readily dimerize in the presence of TMSCl and zinc in ethers to give *O*-silylated pinacols in good yields [96]. Sonochemical yields are generally higher than those of the silent reaction. Hydrolysis in the presence of boron trifluoride rearranges the products to pinacolones, while tetrabutylammonium fluoride (TBAF) deprotects the pinacols almost quantitatively.

If the initial ketyl radical anion is next to a functionality able to undergo addition or coupling, cyclization can occur. One of these reactions is the acyloin condensation of diesters. Classically, this reaction is performed with sodium in refluxing solvents. The condensation can be improved by the presence of TMSCl, the so-called Rühlmann procedure. A considerable improvement was achieved by the use of technical grade TMSCl

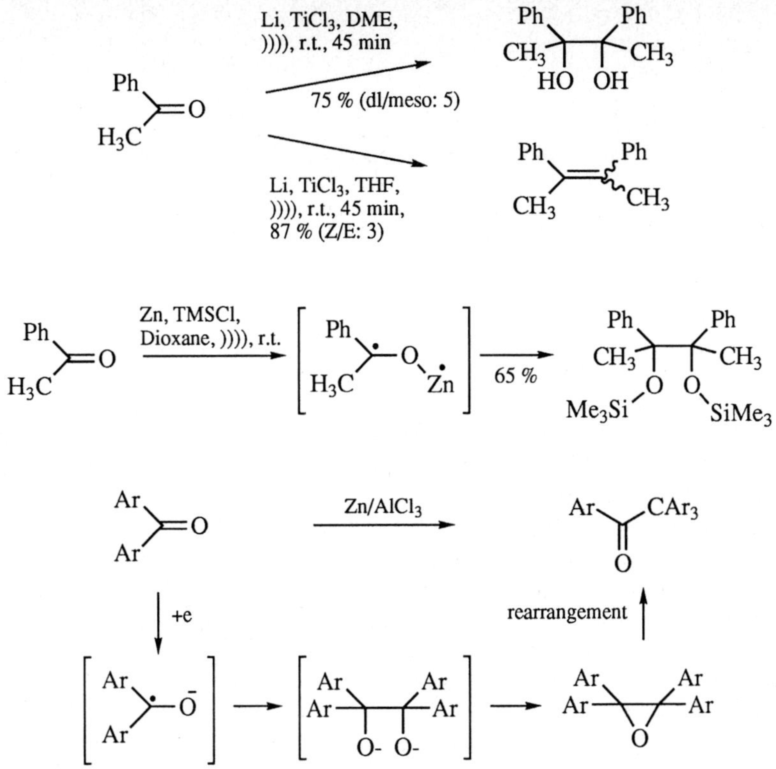

Li, $TiCl_3$, DME,)))), r.t., 45 min

75 % (dl/meso: 5)

Li, $TiCl_3$, THF,)))), r.t., 45 min, 87 % (Z/E: 3)

Zn, TMSCl, Dioxane,)))), r.t.

65 %

Zn/$AlCl_3$

+e

rearrangement

Scheme 4-23. Sonochemical pinacolizations.

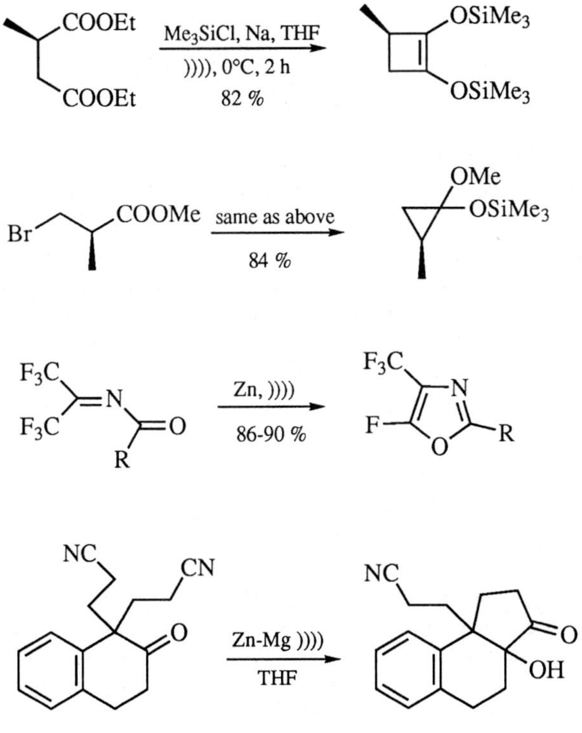

Me_3SiCl, Na, THF)))), 0°C, 2 h 82 %

same as above 84 %

Zn,)))) 86-90 %

Zn-Mg)))) THF

Scheme 4-24

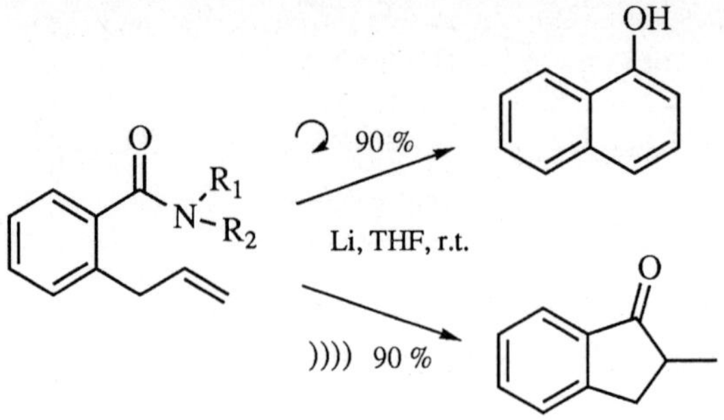

90 %
Li, THF, r.t.
)))) 90 %

Scheme 4-25. Reductive cyclizations of *o*-allylbenzamide.

and ultrasonically dispersed sodium [97]. Interestingly, a chiral center next to the carbonyl is preserved (Scheme 4-24a). Cyclizations are also successful from β-haloesters, which lead to the cyclopropyl derivatives in high yields (Scheme 4-24b). The same mechanism probably applies to the cyclization of aryl oxazabutadienes bearing trifluoromethyl groups, in which a fluorine ion is a leaving group (Scheme 4-24c) [98]. Using a zinc–magnesium mixture with di(cyanoethyl)tetralone, an annelation occurs (Scheme 4-24d) [56]. Another example is provided by the formation of the indanone nucleus from *o*-allyl benzamides (Scheme 4-25), a case of sonochemical switching [99]. The initial ketyl cyclizes to 2-methylindanone and liberates an amide ion, which deprotonates the allyl moiety. The resulting carbanion cyclizes to α-naphthol. If the first step of the process is sufficiently accelerated by sonication, the ketyl is generated much more rapidly and only cyclization to 2-methylindanone is obtained.

A more complex evolution of ketyl intermediates occurs in deoxygenation reactions, among which the Clemmensen procedure is recognized as a powerful tool [100]. Since it necessitates strong protic acids, sonochemical improvements were proposed to obtain milder conditions. A first attempt, using various types of zinc in the presence of hydroiodic acid at room temperature, remained inconclusive [101]. In a more convincing approach, 3-oxosteroids were mildly reduced under high-intensity irradiation [102]. High yields are readily obtained at room temperature in acetic acid or acetic acid–water. Selec-

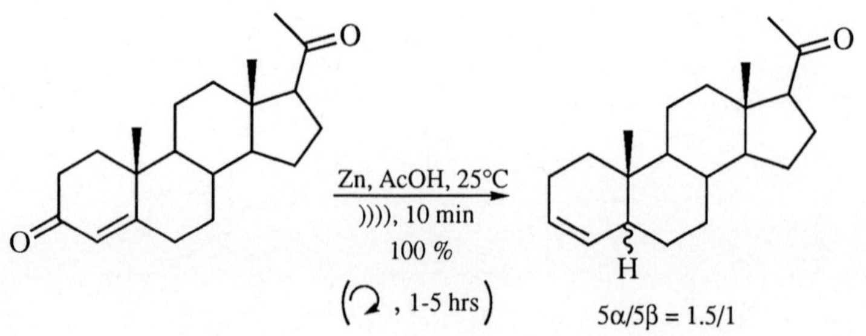

Zn, AcOH, 15°C
)))), 15 min
87 %

Zn, AcOH, 25°C
)))), 10 min
100 %
(, 1-5 hrs)
$5\alpha/5\beta = 1.5/1$

Scheme 4-26

tivity in favor of the 3-oxo group in the presence of 17- and 20-oxo groups is noticeable (Scheme 4-26). The same authors report the use of ultrasonically activated zinc in acetic acid for the reduction of 3-keto-Δ^4 steroids (Scheme 4-26b) [103]. By-products are practically absent, but the reaction lacks selectivity at the C-5 position.

4.7 Electron Transfers to Single Bonds (Excluding Carbon–Halogen Bonds)

The transfer of one electron to a single bond leads to a "three-electron" bond, a radical anion. Its stability is strongly dependent on the nature of the atoms and the structure of the acceptor, but generally breakage occurs. Under this heading are reductive processes of various C–heteroatom and C–C bonds, and also a few examples of heteroatom–heteroatom bonds.

4.7.1 Cleavage of the C–H Bond

Apart from reactions of alcohols with alkali and alkaline-earth metals, direct hydrogen–metal exchange is not easy. However, lithium metallates phenylacetylene under sonication (Scheme 4-27a) [68]. Ultrasonically dispersed potassium (UDP) deprotonates at the α position in relation to an activating group. The Dieckmann and Thorpe–Ziegler cyclizations of hexane- and heptanedioate esters or the corresponding dinitriles proceed readily at room temperature (Scheme 4-27b) [30]. With succinate esters, dimerization is preferred (Scheme 4-27c) [104]. These reactions contrast with the acyloin condensation mentioned

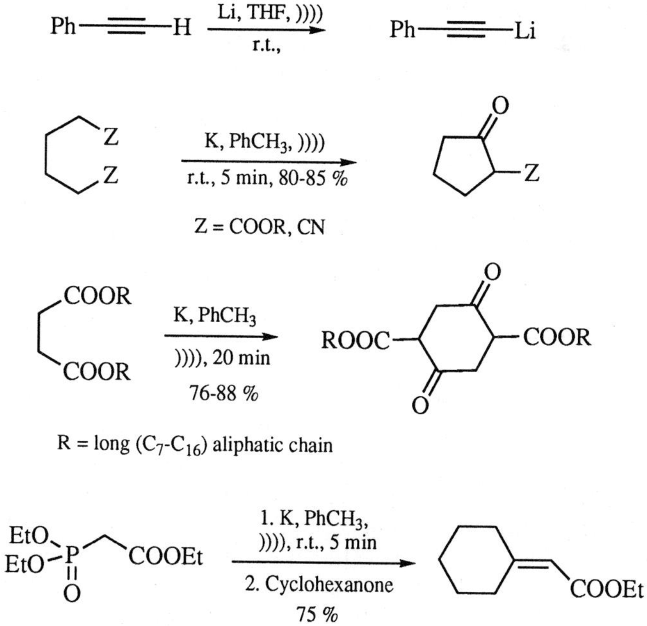

Ph–≡–H → (Li, THF,)))), r.t.,) Ph–≡–Li

Z, Z → (K, $PhCH_3$,)))), r.t., 5 min, 80-85 %) → 2-Z-cyclopentanone

Z = COOR, CN

COOR, COOR → (K, $PhCH_3$,)))), 20 min, 76-88 %) → ROOC–(cyclohexanedione)–COOR

R = long (C_7-C_{16}) aliphatic chain

$(EtO)_2P(O)CH_2COOEt$ → (1. K, $PhCH_3$,)))), r.t., 5 min; 2. Cyclohexanone; 75 %) → cyclohexylidene–COOEt

Scheme 4-27

above [97], since UDP acts as a base, whereas sodium in the presence of TMSCl behaves as a SET agent. The former behavior is also illustrated in the generation of Wittig-Horner reagents (Scheme 4-27d) [30].

4.7.2 Reduction of Carbon–Carbon and Carbon–Heteroatom Single Bonds

Only one case of C–C bond reduction has been reported, namely the decyanation of *vic*-dinitriles to olefins by sodium [105]. In the example given in Scheme 4-28 the authors mention that the yields may seem poor (30%), but alternative methods lead to overreduction to the saturated compound.

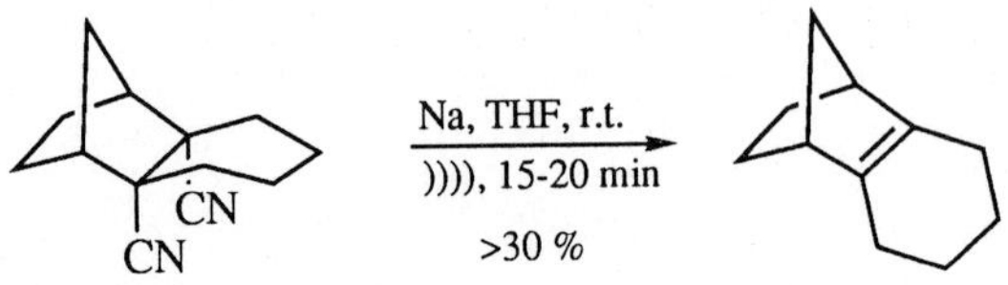

CN CN

Na, THF, r.t.
)))), 15-20 min
>30 %

Scheme 4-28

Sonication helps greatly in the reduction of C–O bonds by metals [106, 107]. For instance, one epoxidic C–O bond in epoxyketones is cleaved by aluminum amalgam without ketone reduction (Scheme 4-29) [107]. Optimization of the acoustic intensity and temperature is also described in the same paper.

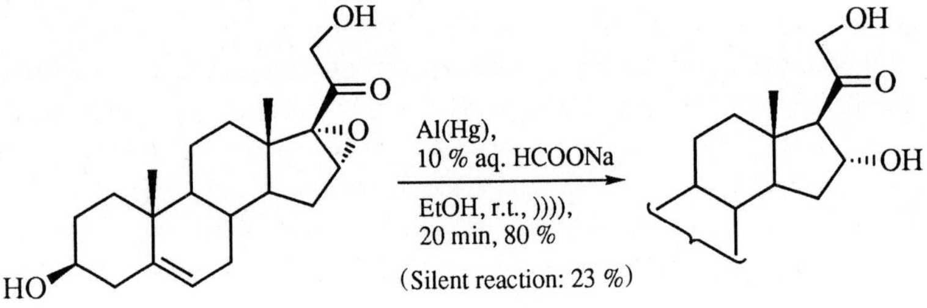

OH O O HO

Al(Hg),
10 % aq. HCOONa
EtOH, r.t.,)))),
20 min, 80 %
(Silent reaction: 23 %)

OH O OH

Scheme 4-29

The reduction of carbon–sulfur bonds by UDP was reported in a series of papers by Chou et al. (Scheme 4-30) [108–111]. Substituted 3-sulfolenes undergo this reaction to give dienes with excellent selectivity and short reaction times [108]. However, bubbling nitrogen into the mixture reduces the rate and, when using the *cis*-2,5-disubstituted 3-sulfolene, isomerization of the kinetic *(E,Z)* diene to the more stable *(E,E)* diene occurs. The mechanism is not clear, but a reductive pathway requiring 2 equivalents of metal must be ruled out, since the reaction is effective using less than 1 equivalent.

Starting from 2-sulfolenes leads to another type of cleavage. No reaction occurs after 20 h stirring, but under sonication the S–C(sp_2) bond is reduced. Alkylation of the intermediate anion gives a γ-unsaturated sulfone. This type of cleavage is also found in sulfolanes, with similar results. EPR spectra of the mixture suggest an electron transfer from the metal to a sulfur–oxygen bond [109]. In addition to these results in aprotic media, Chou et al. found that the presence of water significantly accelerates the C–S bond cleavage in saturated cyclic sulfones that are otherwise unreactive [110]. In a further development, the reactions of 2-methyl-2-sulfolene with UDP followed by methylation were improved by addition of phenol as a proton donor (Scheme 4-31) [110, 111].

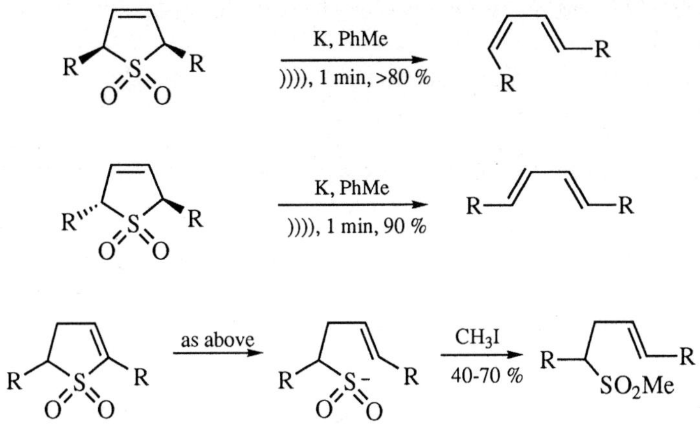

K, PhMe
)))), 1 min, >80 %

K, PhMe
)))), 1 min, 90 %

as above

CH_3I
40-70 %

SO_2Me

Scheme 4-30. Reductive cleavage of sulfolenes.

Similarly, the cleavage of carbon–sulfur or carbon–selenium bonds in cyclic onium salts with magnesium as the electron source was reported (Scheme 4-32) [112]. In Scheme 4-32b an interesting change in selectivity results from the use of a different metal and solvent.

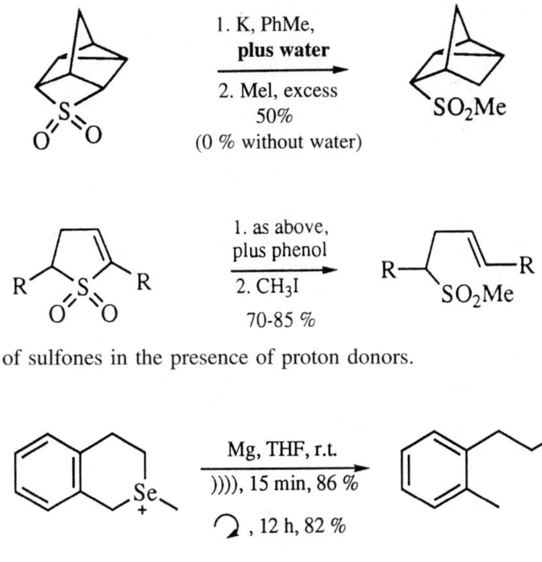

1. K, PhMe, **plus water**
2. MeI, excess
50%
(0 % without water)

SO_2Me

1. as above, plus phenol
2. CH_3I
70-85 %

SO_2Me

Scheme 4-31. Cleavage of sulfones in the presence of proton donors.

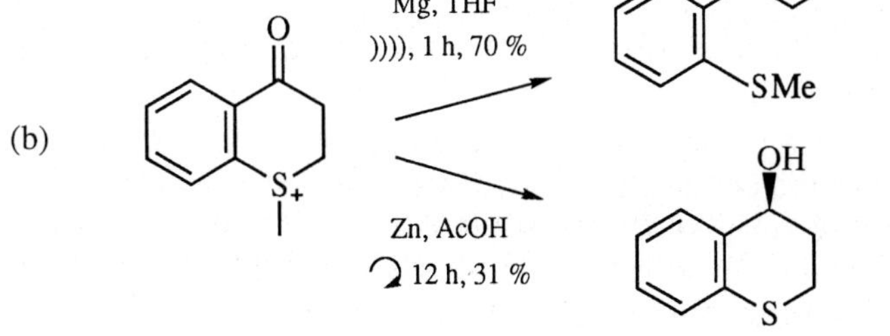

(a)

Mg, THF, r.t.
)))), 15 min, 86 %
, 12 h, 82 %

(b)

Mg, THF
)))), 1 h, 70 %

SMe

Zn, AcOH
12 h, 31 %

OH

Scheme 4-32

4.7.3 Reduction of Various Bonds Involving Heteroatoms

Generation of alkali metal amides (see Section 4.7.2) is generally not feasible by direct reduction of the N–H bond. An exception is the attack of potassium by 1,3-propanediamine which occurs rapidly, especially in the presence of catalytic amounts of ferric nitrate [113]. The resulting reagent, KAPA, was used for isomerization of acetylenes to the terminal position. NAPA, the sodium analog prepared in the same manner, exhibits similar properties (Scheme 4-33).

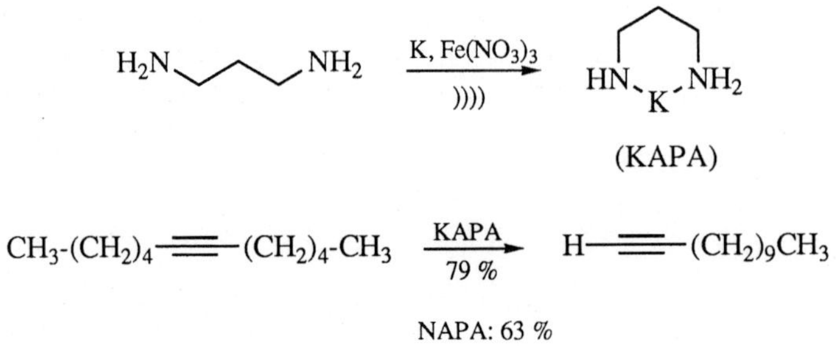

H_2N–$(CH_2)_3$–NH_2 $\xrightarrow[))))]{K, Fe(NO_3)_3}$ HN–K–NH_2 (KAPA)

CH_3-$(CH_2)_4$–≡–$(CH_2)_4$-CH_3 $\xrightarrow[79\ \%]{KAPA}$ H–≡–$(CH_2)_9CH_3$

NAPA: 63 %

Scheme 4-33. Alkyne isomerization with alkali 3-aminopropanamide.

Other heteroatomic bond cleavage reactions also have synthetic applications [114–116]. One of these is the reduction of diphenyldiselenide by sodium under ultrasonic irradiation at 50 kHz to give sodium phenylselenide [51, 114]. Catalytic amounts of benzophenone accelerate the reaction. The authors discuss the advantages of the reagent prepared by this method [51], such as milder conditions, ease in manipulation, and improved nucleophilic properties (Scheme 4-34).

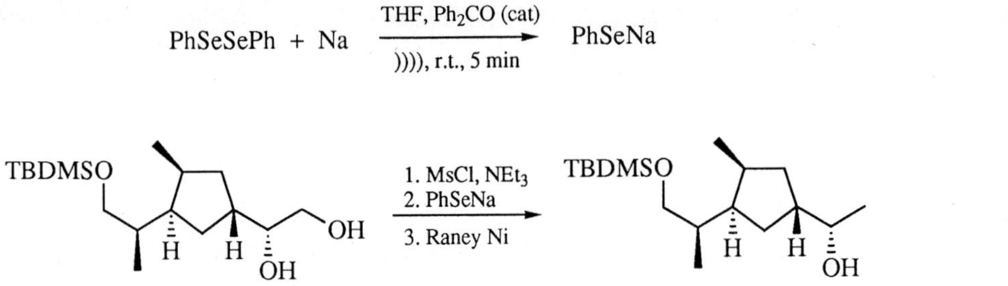

PhSeSePh + Na $\xrightarrow[)))),\ r.t.,\ 5\ min]{THF,\ Ph_2CO\ (cat)}$ PhSeNa

TBDMSO … H H OH OH $\xrightarrow[3.\ Raney\ Ni]{1.\ MsCl,\ NEt_3;\ 2.\ PhSeNa}$ TBDMSO … H H OH

Scheme 4-34

TBDMSO = t-butyldimethylsilyoxy

4.8 The Carbon–Halogen Bond. Formation of Organometallics

The importance of this reaction needs no comment. In his 1950 paper, Renaud [5] showed that sonication is efficient in the preparation of lithium, magnesium, aluminum, and mercury compounds, even in undried diethyl ether. This discovery remained practically unknown and unexploited for many years, one reason for this being that cheap, easily accessible ultrasound generators were not commercially available. A second mention of such reactions appeared in 1959, with the preparation of phenyl-sodium from ultrasonically activated sodium and chlorobenzene [48]. Modern organometallic sonochemistry stems from these initial discoveries that were virtually unnoticed at that time.

4.8.1 Formation of Organoalkali and Grignard Reagents

The preparation of organolithium and magnesium reagents can be considered as a milestone for sonochemistry [68, 117]. Organometallic reagents can be prepared without previous activation of the metal and in commercial undried diethyl ether using cleaning baths. The induction period is usually suppressed [Eq. (4.6)]. Primary organolithiums are obtained in good yields (61–95%) from alkyl bromides with Li wire or Li–2% Na sand. Secondary and tertiary alkyl bromides require longer reaction times (>1 h) [118–120].

$$\text{R–X} + 2\,\text{Li} \xrightarrow[\text{))))}]{\text{THF, r.t.}} \text{R–Li} + \text{LiX} \tag{4.6}$$

R=Et, nBu, *i*Pr, 2-propenyl, Ph, cyclopentyl, ...

Magnesium activation is rather easy by sonication [5, 117], even in wet solvents. For instance, initiation of the reaction of 2-bromopentane in ether half-saturated with water occurs in a few minutes, whereas without sonication no reaction is observed after stirring for several hours [121]. This finding may be of interest for industrial applications. The kinetics of the formation of butyl magnesium bromide in anhydrous media have been studied [122]. The higher rates observed were assigned to the destruction of the passivation layer on the metal and to the absence of redeposition of the reaction products on the active surface. Functionalized reagents and some others, inaccessible by conventional methods, can be prepared smoothly (Scheme 4-35) [123, 124].

Although the formation of Grignard reagents has been studied by several authors, to the best of our knowledge only one paper mentions a sonochemical effect in a carbonyl group addition step in a homogeneous solution (Scheme 4-36) [125].

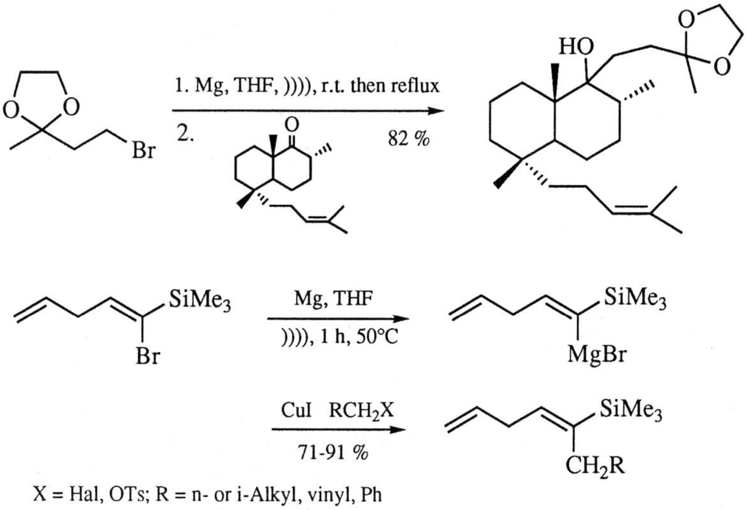

1. Mg, THF,)))), r.t. then reflux
2.
82 %

Mg, THF
)))), 1 h, 50°C

CuI RCH$_2$X
71-91 %

X = Hal, OTs; R = n- or i-Alkyl, vinyl, Ph

Scheme 4-35. Sonochemical Grignard syntheses.

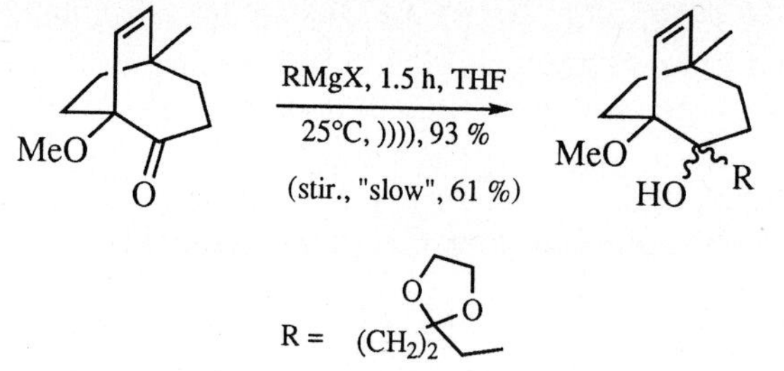

MeO O RMgX, 1.5 h, THF 25°C,)))), 93 % (stir., "slow", 61 %) MeO HO R

R = $(CH_2)_2$ (dioxolane)

Scheme 4-36

4.8.2 Transmetallation with Sonochemically Prepared Organometallic Reagents

Since lithium and magnesium reagents are often used as precursors to other organometallics via metal exchange, the feasibility of a direct, one-pot process for less easily accessible organometallics by a simultaneous sonochemical preparation and metal exchange was studied and proved to be successful. Zinc organometallics illustrate this methodology.

Their one-step preparation from lithium, alkyl halides, and a zinc salt is easily accomplished [(Eq. (4.7)] [126]. Alkyl, vinyl, or aryl reagents are obtained in short times and excellent yields. Sonication in a cleaning bath is sufficient to generate diarylzincs, but the probe technique is necessary in the other cases. Their thermal stability is higher than that of organocopper derivatives, and in the presence of nickel acetylacetonate as a catalyst, they undergo conjugate addition to α-enones, even sterically congested ones in cases where copper reagents are ineffective (Scheme 4-37a) [127].

$$R\text{–}X + ZnBr_2 + Li \underset{))))}{\longrightarrow} R_2Zn \quad (4.7)$$

Solvent: THF, THF-PhMe, ether

Using a different stoichiometry in the preparation, the organozinc halide is obtained, which reacts with trichloroacetyl chloride to give trichloromethyl ketones (Scheme 4-37b) [128]. No Wurtz-type coupling is observed even for benzylic bromides.

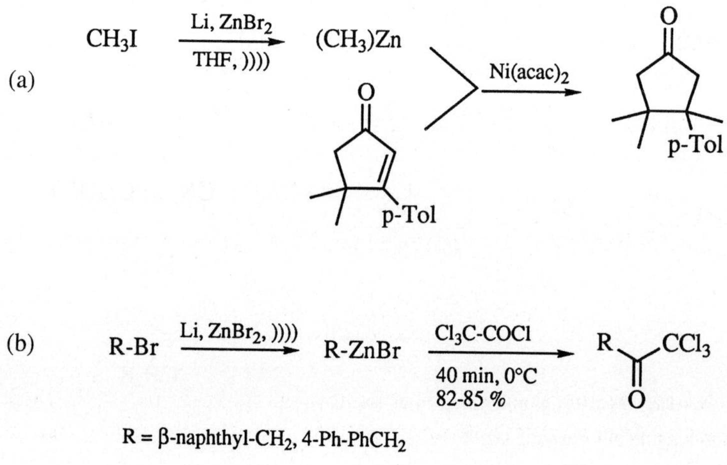

(a) CH_3I $\xrightarrow[\text{THF,))))}]{\text{Li, ZnBr}_2}$ $(CH_3)Zn$ + enone (O, p-Tol) $\xrightarrow{Ni(acac)_2}$ product (O, p-Tol)

(b) R-Br $\xrightarrow{\text{Li, ZnBr}_2\text{,))))}}$ R-ZnBr $\xrightarrow[\text{40 min, 0°C; 82-85 \%}]{Cl_3C\text{-}COCl}$ $R\text{-}CO\text{-}CCl_3$

R = β-naphthyl-CH_2, 4-Ph-$PhCH_2$

Scheme 4-37

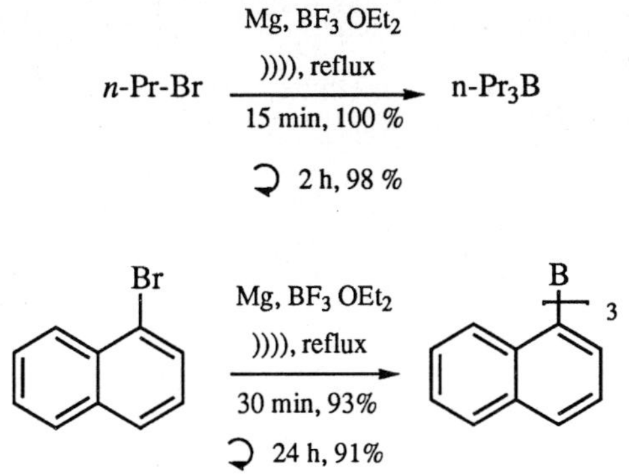

$$n\text{-Pr-Br} \xrightarrow[\text{15 min, 100 \%}]{\text{Mg, BF}_3\text{ OEt}_2\text{,)))), reflux}} \text{n-Pr}_3\text{B}$$

(silent: 2 h, 98 %)

$$\text{1-bromonaphthalene} \xrightarrow[\text{30 min, 93\%}]{\text{Mg, BF}_3\text{ OEt}_2\text{,)))), reflux}} (\text{1-naphthyl})_3\text{B}$$

(silent: 24 h, 91%)

Scheme 4-38. Grignard mediated synthesis of boranes.

Brown et al. improved the preparation of organoboranes by sonication, in terms of ease of manipulation, shorter reaction times, and higher yields and purity (Scheme 4-38). Triorganoboranes are obtained via the in situ formation of Grignard reagents and metal exchange with boron trifluoride etherate [129]. Trinaphthylborane, inaccessible by hydroboration, is obtained in 93% yield in only 15 min, whereas the silent process requires 24 h for an equivalent yield.

Different products are formed in diethyl ether and in THF, with or without sonication. The reaction applied to *n*-butyl bromide in ether gives tri-*n*-butylborane quantitatively, but in THF under identical conditions an ate complex is formed. The reactivities of the halides follow the normal sequence I > Br > Cl. Iodides give side reactions, and therefore bromides are preferred. The metal of choice is magnesium, since lithium gives only Wurtz-type products.

4.8.3 Direct Access to Organozinc Reagents

Recent reviews mention the preparation of organozinc reagents by the direct attack of the metal under sonication [130, 131]. Aryl iodides with an electron-withdrawing substituent in the *ortho* position yield the organozinc reagents when sonicated in polar solvents [132]. Long irradiation times are necessary but the yields are convenient [Eq. (4.8)]. Reduction occurs when this reaction is carried out with a Zn–Ni–NaI system in an aqueous medium [Eq. (4.9)] [133].

$$\text{R-Ph-I} \xrightarrow[\text{)))), 30–50 °C, 5–20 h}]{\text{Zn, solvent}} \text{R-Ph-ZnI} \xrightarrow{\text{R}'\text{-PhX, Pd(0)}} \text{R-Ph-Ph-R}' \quad (4.8)$$

R = *o*-COOMe, *o*-CN, *m*-COOMe

Solvent: *N*-methylpyrrolidone, DMF, tetramethylurea

$$\text{R-X} \xrightarrow[\text{)))), 0.5–4 h, 81–93\%}]{\text{Zn, NiCl}_2\text{, NaI, HMPA, H}_2\text{O, 60 °C}} \text{R-H} \quad (4.9)$$

More interesting is the preparation of Reformatsky reagents [134–136]. The usual one-step process is discussed later (Section 4.10.5), and therefore only the preparation of

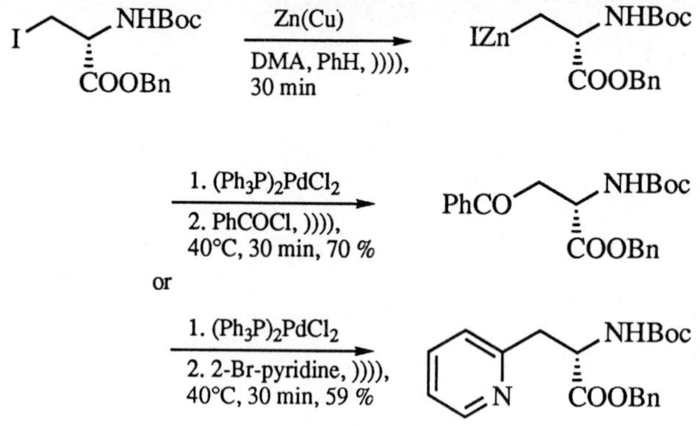

I–CH₂–CH(NHBoc)–COOBn → Zn(Cu), DMA, PhH,)))), 30 min → IZn–CH₂–CH(NHBoc)–COOBn

1. $(Ph_3P)_2PdCl_2$
2. PhCOCl,)))), 40°C, 30 min, 70 % → PhCO–CH₂–CH(NHBoc)–COOBn

or

1. $(Ph_3P)_2PdCl_2$
2. 2-Br-pyridine,)))), 40°C, 30 min, 59 % → 2-pyridyl–CH₂–CH(NHBoc)–COOBn

Scheme 4-39. Synthesis of β-acyl or -arylaminoacids.

the reagents in a separate step is mentioned here, such as those derived from β-iodoaminoacid compounds (Scheme 4-39) [134]. In the presence of a zinc–copper couple the reagent is formed smoothly in benzene–dimethylacetamide solution using a cleaning bath. It reacts with acyl chlorides or aryl iodides in the presence of palladium catalysts to give acyl- or arylaminoacid derivatives. An α,α-difluoro Reformatsky reagent was also prepared by sonication, giving much cleaner reactions than in the one-step process (Scheme 4-40) [135].

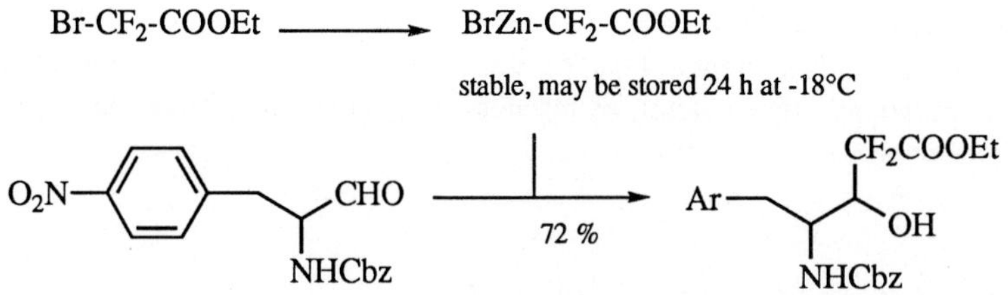

$Br\text{-}CF_2\text{-}COOEt \longrightarrow BrZn\text{-}CF_2\text{-}COOEt$

stable, may be stored 24 h at -18°C

O_2N–C₆H₄–CH₂–CH(NHCbz)–CHO → 72 % → Ar–CH₂–CH(NHCbz)–CH(OH)–CF_2COOEt

Scheme 4-40

4.8.4 Other Reactions

Organoaluminum sonochemistry is not highly developed at the present time despite the synthetic and economic interest. Aluminum, a ductile metal, is easily activated by sonication, as evidenced by the use of foils to determine the efficiency of cleaning baths [6]. Alkyl halides react with aluminum at room temperature in THF to give the sesquihalide (Scheme 4-41), unless magnesium is added. Examples with bromomethane or -ethane have been published [137–139]. Under similar conditions with stirring but without sonication, no reaction occurs. Triethyl aluminum has been used in sonochemical transmetallation reactions yielding zinc and boron compounds.

Commercially available palladium black reacts with allyl bromides and iodides to give π-allyl palladium complexes in good yields (Scheme 4-42) [140]. The procedure is simple, no chemical preactivation is necessary, and the yields are moderate to good. Allyl chloride or acetate, geranyl bromide, and 2,3-dibromopropene fail to give the desired π-allyl complexes under similar conditions.

Some organometallics were recently prepared by sonoelectrochemistry, another exciting and still futuristic research area [141]. Sonication in a cleaning bath facilitates elec-

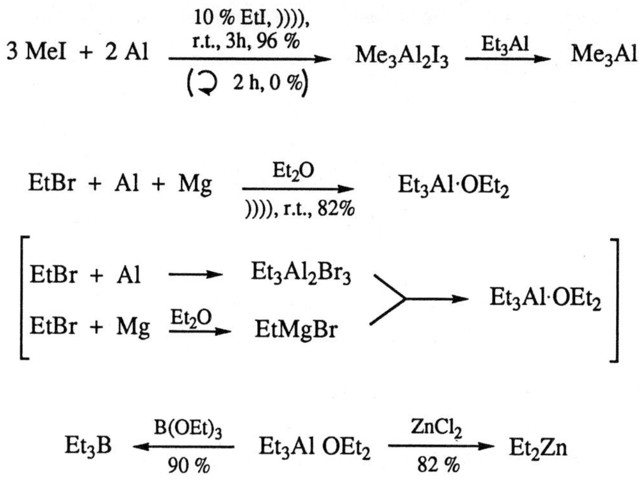

$$3\,MeI + 2\,Al \xrightarrow[(\circlearrowright\ 2\,h,\,0\,\%)]{10\,\%\,EtI,\,)))),\ r.t.,\,3h,\,96\,\%} Me_3Al_2I_3 \xrightarrow{Et_3Al} Me_3Al$$

$$EtBr + Al + Mg \xrightarrow[)))),\ r.t.,\,82\%]{Et_2O} Et_3Al{\cdot}OEt_2$$

$$\left[\begin{array}{l} EtBr + Al \longrightarrow Et_3Al_2Br_3 \\ EtBr + Mg \xrightarrow{Et_2O} EtMgBr \end{array}\right\rangle \longrightarrow Et_3Al{\cdot}OEt_2 \Bigg]$$

$$Et_3B \xleftarrow[90\,\%]{B(OEt)_3} Et_3Al\ OEt_2 \xrightarrow[82\,\%]{ZnCl_2} Et_2Zn$$

Scheme 4-41. Preparation of organoaluminum derivatives.

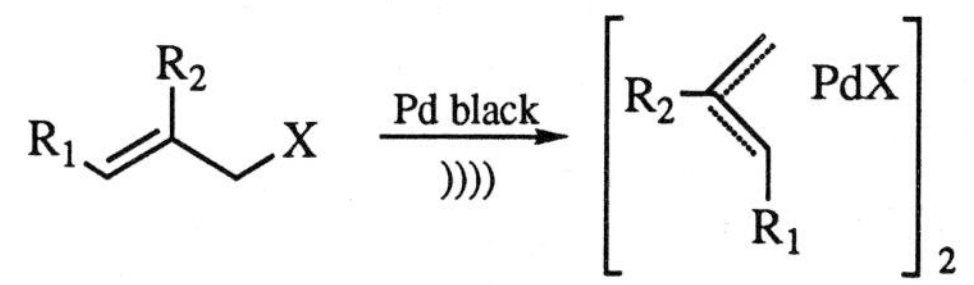

$$R_1CH{=}C(R_2)CH_2X \xrightarrow[))))]{Pd\ black} [(R_1,R_2\text{-allyl})PdX]_2$$

$R_1 = H, CH_3, Ph, COOMe$; $R_2 = H, CH_3$
X = Br: 55-60°C, 80-92%
X = I: 25-30°C, 87%

Scheme 4-42

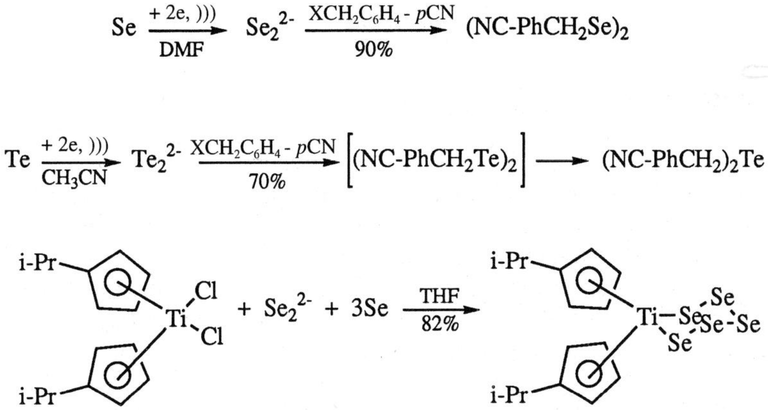

$$Se \xrightarrow[DMF]{+2e,\,)))} Se_2^{2-} \xrightarrow[90\%]{XCH_2C_6H_4\text{-}pCN} (NC\text{-}PhCH_2Se)_2$$

$$Te \xrightarrow[CH_3CN]{+2e,\,)))} Te_2^{2-} \xrightarrow[70\%]{XCH_2C_6H_4\text{-}pCN} [(NC\text{-}PhCH_2Te)_2] \longrightarrow (NC\text{-}PhCH_2)_2Te$$

$$(i\text{-}PrC_5H_4)_2TiCl_2 + Se_2^{2-} + 3Se \xrightarrow[82\%]{THF} (i\text{-}PrC_5H_4)_2TiSe_5$$

Scheme 4-43. Sonoelectrochemistry of elemental chalcogenides.

troreduction of elemental selenium or tellurium to give their anions, which react further with electrophiles to give organoselenium or organotellurium compounds in good yields (60–90%) [142], with considerable advantage over the traditional electrochemical synthesis using electroreduction with sacrificial electrodes (Scheme 4-43).

4.9 Organic Reactions Using Sonochemically in-situ Generated Organometallics

Besides the reactions that make use of organometallics prepared in a separate step, many transformations may be carried out with a reagent prepared in situ, with the advantages of time saving and, more importantly, the ability to use reagents of limited stability, such as carbenoid organometallics. Examples will be examined in this section.

4.9.1 Deprotonations by in-situ Generated Organoalkali-Metal Reagents

Alkyl-lithium reagents, commonly used in deprotonation reactions, can be prepared and simultaneously reacted in the presence of a substrate [143, 144]. For example, sonication of *t*-butyl chloride, lithium, and 1,3-dithiane in THF provides quantitatively in a few minutes the carbanion, which can be trapped by an electrophile. Anisole, acetylenic compounds, and amines can be successfully deprotonated by this method (Scheme 4-44). A surprising reaction "in cascade", the preparation of isobutyric acid dianion, was

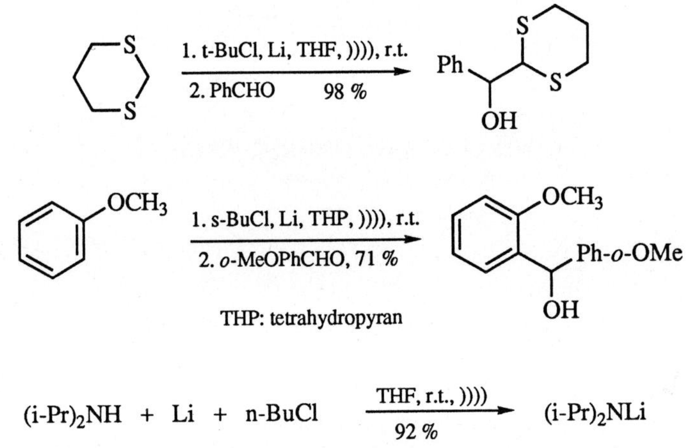

1. t-BuCl, Li, THF,)))), r.t. 2. PhCHO 98 %

OCH_3 — 1. s-BuCl, Li, THP,)))), r.t. 2. *o*-MeOPhCHO, 71 % — OCH_3, Ph-*o*-OMe, OH

THP: tetrahydropyran

$(i\text{-}Pr)_2NH$ + Li + n-BuCl —THF, r.t.,)))); 92 %→ $(i\text{-}Pr)_2NLi$

Scheme 4-44. Deprotonations with in-situ generated butyllithium.

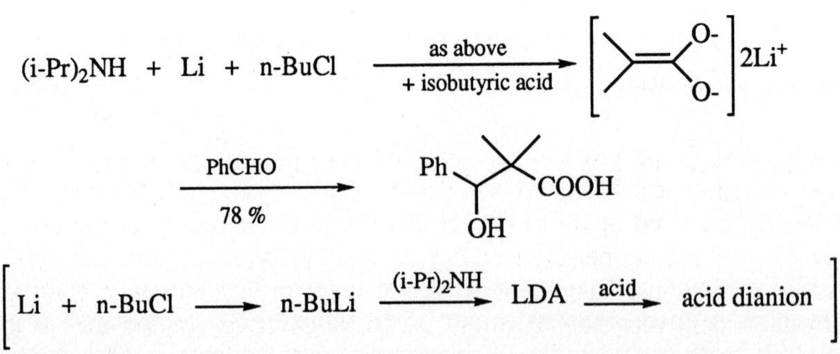

$(i\text{-}Pr)_2NH$ + Li + n-BuCl —as above + isobutyric acid→ [O-, O-] 2Li$^+$

—PhCHO; 78 %→ Ph, COOH, OH

[Li + n-BuCl → n-BuLi —$(i\text{-}Pr)_2NH$→ LDA —acid→ acid dianion]

Scheme 4-45. A "cascade" reaction.

achieved by generating butyllithium from its precursors, followed by deprotonation of diisopropylamine to LDA, then deprotonation of the acid. Quenching with benzaldehyde gave the hydroxyacid in 78% yield (Scheme 4-45) [143].

4.9.2 Wurtz- and Ullmann-Type Coupling Reactions Forming Carbon–Carbon Bonds

The synthetically important Wurtz and Ullmann couplings of organic halides [145] frequently require drastic conditions. Substantial improvements regarding the reaction conditions were obtained by sonochemical irradiation.

Alkyl and aryl halides give homocoupling products with lithium in THF, when sonicated in a bath at room temperature (Scheme 4-46) [146]. Although yields are usually moderate to good, little or no reaction takes place in the absence of ultrasound. The coupling of bromo-toluenes and -pyridines provides a mixture of isomeric products [147], suggesting the intermediacy of radical species, which was confirmed by inhibition of the reaction by radical scavengers.

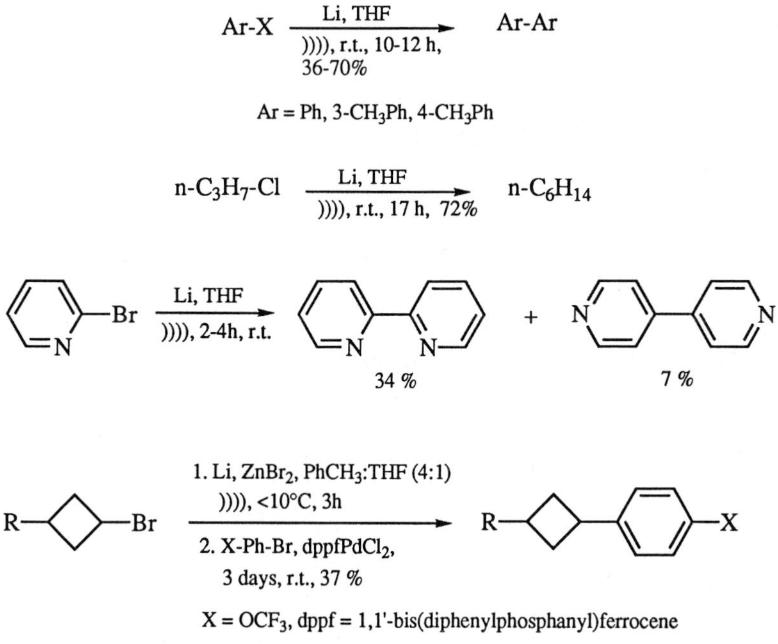

Ar-X —(Li, THF;)))), r.t., 10-12 h, 36-70%)→ Ar-Ar

Ar = Ph, 3-CH_3Ph, 4-CH_3Ph

n-C_3H_7-Cl —(Li, THF;)))), r.t., 17 h, 72%)→ n-C_6H_{14}

2-bromopyridine —(Li, THF;)))), 2-4h, r.t.)→ 2,2'-bipyridine (34 %) + 4,4'-bipyridine (7 %)

R-cyclobutyl-Br —(1. Li, $ZnBr_2$, $PhCH_3$:THF (4:1),)))), <10°C, 3h; 2. X-Ph-Br, dppfPdCl_2, 3 days, r.t., 37 %)→ R-cyclobutyl-C_6H_4-X

X = OCF_3, dppf = 1,1'-bis(diphenylphosphanyl)ferrocene

Scheme 4-46. Examples of Wurtz coupling.

Frequently, Wurtz-type processes are carried out with copper or zinc, sometimes prepared by the sonochemical Rieke's method [45]. One of these, a heterocoupling involving a sonochemically prepared organozinc derivative, was described by Dehmlow [148]. Fundamental aspects of the copper-induced reaction of aryl halides were examined by Lindley et al. [32, 149]. Almost quantitative yields are obtained with copper flakes in four-fold excess in DMF under probe sonication. A ca. 50-fold increase in rate over the silent process is observed at the unusually low temperature of 60 °C (Scheme 4-47a). The decrease of the metal particle size is insufficient to explain this effect, and a breakdown of inter-

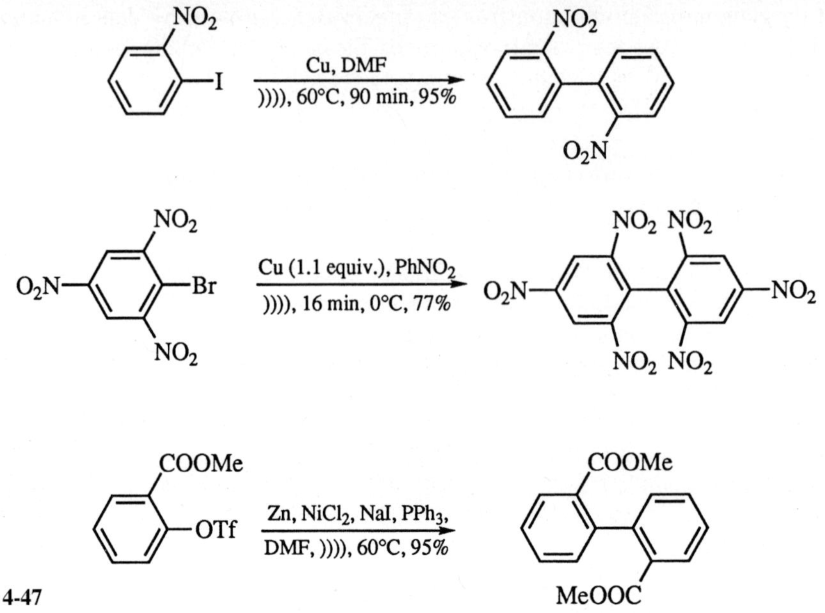

NO2
I
Cu, DMF
)))), 60°C, 90 min, 95%
NO2
O2N
NO2
O2N
Br
NO2
Cu (1.1 equiv.), PhNO2
)))), 16 min, 0°C, 77%
NO2 NO2
O2N
NO2
NO2 NO2
COOMe
OTf
Zn, NiCl2, NaI, PPh3,
DMF,)))), 60°C, 95%
COOMe
MeOOC

Scheme 4-47

mediates and/or the desorption of the products was also invoked. The highly hindered picryl bromide can be coupled at or below room temperature by copper powder used in only 10% excess, to give either the coupling or reduction products depending on the solvent and stoichiometry (Scheme 4-47b) [150]. Sonications were effected with a pulsed wave, a technique not often used. Aryl triflates, and to a lesser extent tosylates, afford biaryls by treatment with sonochemically in-situ generated nickel(0) complexes (Scheme 4-47c) [151].

Vinyl and allyl bromides undergo cross-coupling with perfluoroalkyl halides in the presence of zinc plus palladium(0) or (II) complexes (Scheme 4-48) [152]. In cases involving substituted allyl bromides, the perfluoroalkyl group is introduced regioselectively (>95%) at the γ-position.

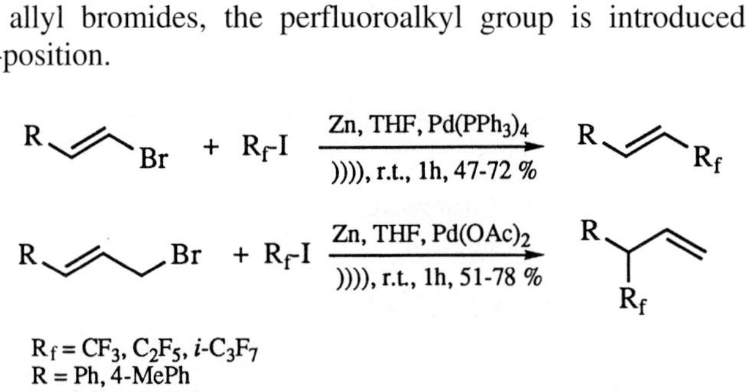

R
Br
+ Rf-I
Zn, THF, Pd(PPh3)4
)))), r.t., 1h, 47-72 %
R
Rf
R
Br
+ Rf-I
Zn, THF, Pd(OAc)2
)))), r.t., 1h, 51-78 %
R
Rf
Rf = CF3, C2F5, i-C3F7
R = Ph, 4-MePh

Scheme 4-48. Perfluoroalkylation of vinylic and allylic bromides.

4.9.3 Coupling Reactions Involving Silicon, Tin, and Germanium

Heterocoupling reactions with organosilicon compounds have been reviewed by Lukevics et al. [11], Boudjouk [153], and Margulis [154]. Examples involving functional halides are shown in Scheme 4-49 (see also Scheme 4-5b, Section 4.5.1) [138, 155, 156].

Alkyltrichlorosilanes have been successfully coupled, as exemplified by n-hexyltrichlorosilane using the liquid sodium–potassium alloy [Eq. (4-11)] [165].

$$nC_6H_{13}SiCl_3 \xrightarrow[))))]{\text{Na, K, pentane}} (nC_6H_{13}Si)_n \tag{4.11}$$

Trialkylchlorostannanes and lithium react more easily than their silicon counterparts [156, 166]. A facile and stereoselective preparation of allylstannanes by irradiation of tributylchlorostannane and allyl chloride in the presence of magnesium occurs with preservation of the original stereochemistry (Scheme 4-52). Without sonication, Wurtz homocoupling becomes noticeable. The method also applies to the synthesis of benzyltributylstannane from benzyl chloride and Bu_3SnCl, and of tetra-allylstannane from allyl chloride and tin tetrachloride [166].

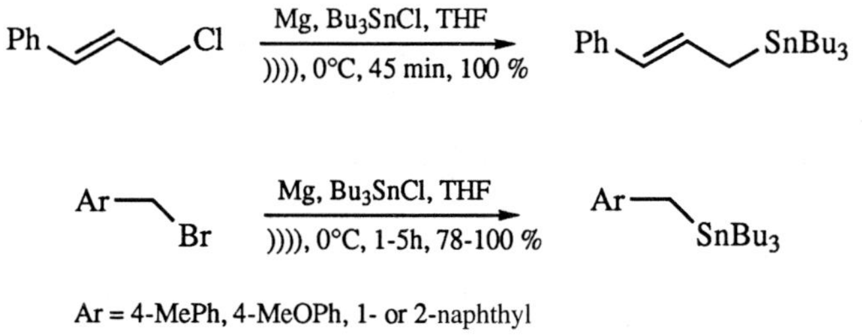

Ph–CH=CH–CH₂Cl → (Mg, Bu_3SnCl, THF;)))), 0°C, 45 min, 100 %) → Ph–CH=CH–CH₂–$SnBu_3$

Ar–CH₂Br → (Mg, Bu_3SnCl, THF;)))), 0°C, 1-5h, 78-100 %) → Ar–CH₂–$SnBu_3$

Ar = 4-MePh, 4-MeOPh, 1- or 2-naphthyl

Scheme 4-52. Synthesis of stannanes.

4.9.4 Addition of Organozinc Reagents to Multiple Bonds

In-situ generated organozinc reagents add to unsaturated compounds in carbometallation reactions. For example, allylic halides add to terminal alkynes in the presence of zinc at 45–50 °C in THF to afford moderate to good yields (48–81%) of 1,4-dienes (Scheme 4-53a) [167]. Similarly, terminal alkynes and isoprene undergo hydroperfluoroalkylation with perfluoroalkyl iodides and zinc in the presence of copper, palladium or

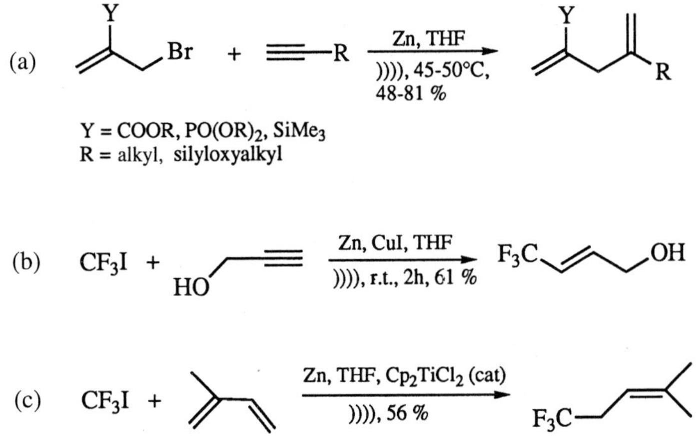

(a) allylic bromide (Y) + ≡–R → (Zn, THF;)))), 45-50°C, 48-81 %) → 1,4-diene (Y, R)

Y = COOR, $PO(OR)_2$, $SiMe_3$
R = alkyl, silyloxyalkyl

(b) CF_3I + HO–CH₂–C≡CH → (Zn, CuI, THF;)))), r.t., 2h, 61 %) → F_3C–CH=CH–CH₂OH

(c) CF_3I + isoprene → (Zn, THF, Cp_2TiCl_2 (cat);)))), 56 %) → F_3C–CH₂–CH=C(CH₃)₂

Scheme 4-53

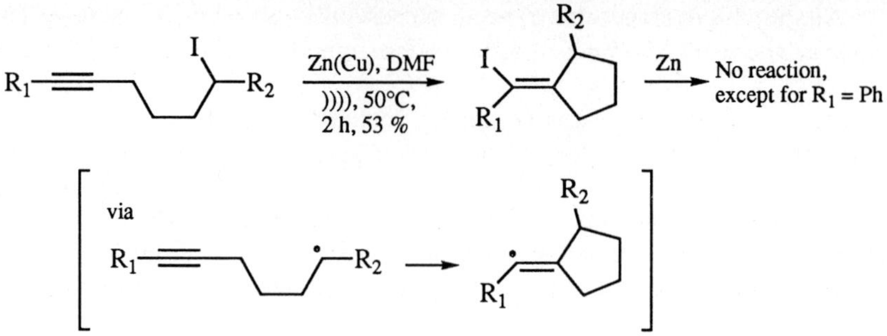

Zn(Cu), DMF
)))), 50°C,
2 h, 53 %
Zn
No reaction,
except for R_1 = Ph
via

Scheme 4-54

titanium catalysts [152, 168]. Yields are moderate to good (52–74%) but stereoselectivity is low (Scheme 4-53b, c). Intramolecular reaction also occurs with δ-iodoalkynes (Scheme 4-54) [169], giving the vinylic iodo compound, an example of radical cyclization analogous to the atom transfer method developed by Curran [170].

Carbenoid reagents add to olefins in the Simmons–Smith cyclopropanation, one of the first organometallic reactions studied under sonication (Scheme 4-55). The unpredictable induction period is avoided and reproducible yields are obtained using various zinc systems [171]. A medium scale reaction using 1.8 kg zinc metal has been described, and various examples can be found in the literature [172–178].

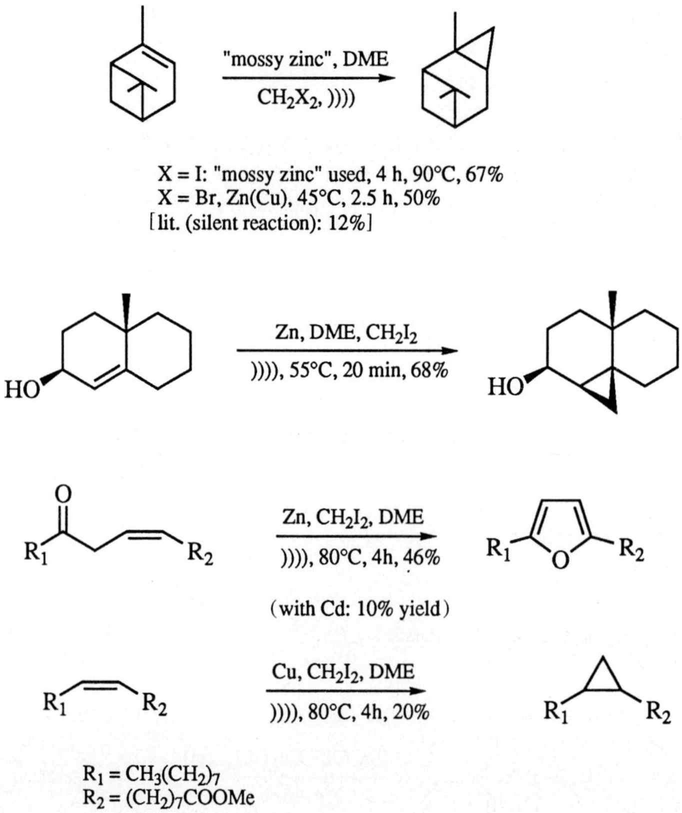

"mossy zinc", DME
CH_2X_2,))))

X = I: "mossy zinc" used, 4 h, 90°C, 67%
X = Br, Zn(Cu), 45°C, 2.5 h, 50%
[lit. (silent reaction): 12%]

Zn, DME, CH_2I_2
)))), 55°C, 20 min, 68%

Zn, CH_2I_2, DME
)))), 80°C, 4h, 46%
(with Cd: 10% yield)

Cu, CH_2I_2, DME
)))), 80°C, 4h, 20%

$R_1 = CH_3(CH_2)_7$
$R_2 = (CH_2)_7COOMe$

Scheme 4-55. Simmons–Smith cyclopropanations.

An allylic alcohol undergoes the reaction with a stereoselectivity resulting from the assistance of the hydroxyl group, unchanged by sonication [175]. In the area of fatty acids chemistry, when applied to a long chain enone ester, the ultrasound-assisted reaction gave a furanoid derivative, with an identical result on replacing zinc by cadmium. Substrates without the keto group undergo clean cyclopropanation, although in low yield [176].

The method of activating the zinc can sometimes be important. For example, the cyclopropanation of dimethyl maleate using a zinc–cobalt couple is stereospecific, but a *cis*-*trans* mixture is obtained when activation is effected with nickel (Scheme 4-56) [177].

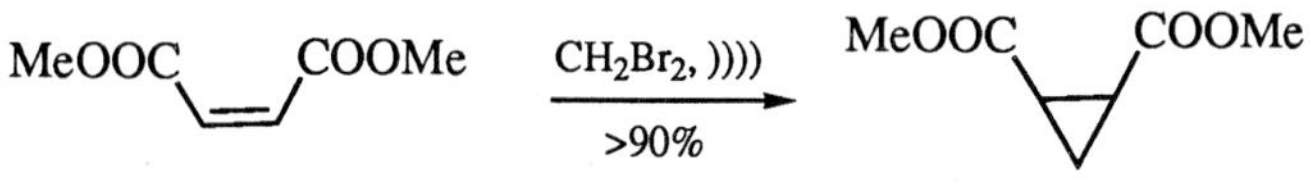

MeOOC COOMe $\xrightarrow[>90\%]{CH_2Br_2,\ ))))}$ MeOOC COOMe

Scheme 4-56 Zn(Co): *cis* isomer. Zn(Ni): isomer mixture

Cyclopropylidenes generated from *gem*-dihalocyclopropanes and metals (Li, Na, or Mg) via a SET mechanism undergo the usual rearrangement to allenes or insertion into a C–H bond [179, 180]; *gem*-dibromocyclopropanes are preferred to the chloro compounds in the reaction with lithium or magnesium. With ultrasonically prepared sodium sand, the reaction time is further reduced, and chloro compounds may be used. Trapping by olefins yields spirohydrocarbons. 1,1-Dihalo-olefins treated in the same manner yield unusual addition products (Scheme 4-57) [181].

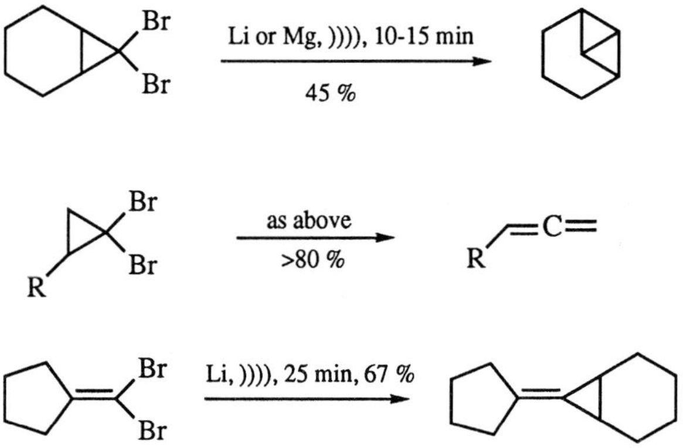

Br, Br — Li or Mg,)))), 10-15 min; 45 %

Br, Br, R — as above; >80 % — R =C=

Br, Br — Li,)))), 25 min, 67 %

Scheme 4-57. Carbenoids from *gem*-dihalides.

4.9.5 In-situ Preparation of Dichloroketene

Because of the synthetic potential of dichloroketene, improvements for its preparation were sought for and are probably used, even though only two papers mention such studies. Formation of dichloroketene and addition to olefins occur easily from trichloroacetyl chloride and ordinary zinc (Scheme 4-58) [182]. With dioxene the reaction is a convenient route to semisquaric acid, although the yield of the first step is modest [183].

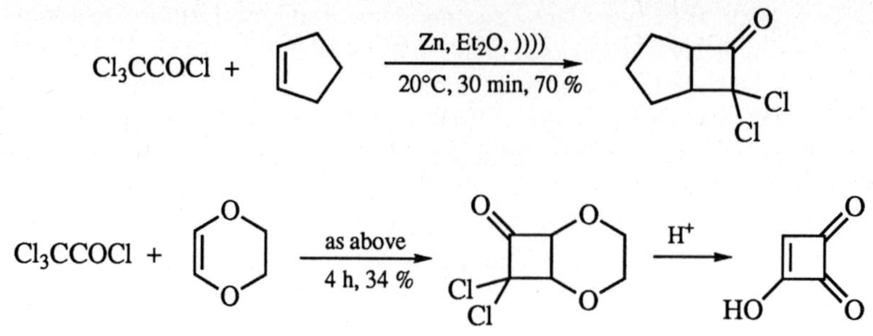

Cl_3CCOCl + cyclopentene $\xrightarrow[\text{20°C, 30 min, 70 \%}]{\text{Zn, Et}_2\text{O,))))}}$ (bicyclic dichlorocyclobutanone: O, Cl, Cl)

Cl_3CCOCl + dioxene $\xrightarrow[\text{4 h, 34 \%}]{\text{as above}}$ (O, Cl, Cl adduct) $\xrightarrow{H^+}$ (HO, O, O product)

Scheme 4-58. Generation of dichloroketene.

4.10 Additions of Organic Halides to Aldehydes and Ketones in the Presence of Metals

This class of reactions is one of the domains of choice for sonochemistry. They encompass not only the Barbier and Reformatsky reactions, but also similar processes, which are reviewed in a recent book [184], where the reader will find a comprehensive treatment.

4.10.1 The Barbier Reaction in the Presence of Lithium or Magnesium

The Barbier reaction of an organic halide and a carbonyl compound in the presence of a metal, which has long been familiar as a one-pot Grignard synthesis, allows one to circumvent the sometimes difficult or tedious preparation of the organometallic reagent. Its applicability was limited and the yields not always satisfactory until lithium replaced magnesium [185]. Sonochemical studies of the reaction further broadened its scope and applications [117].

In a study of its mechanism, it was surprising to note the variation of the reaction rate with temperature. Between –50 and +25 °C, a typical reaction was found to proceed fastest at ca. 0 °C (Fig 4-10) [29]. This result and other experiments (e.g. with preactivated

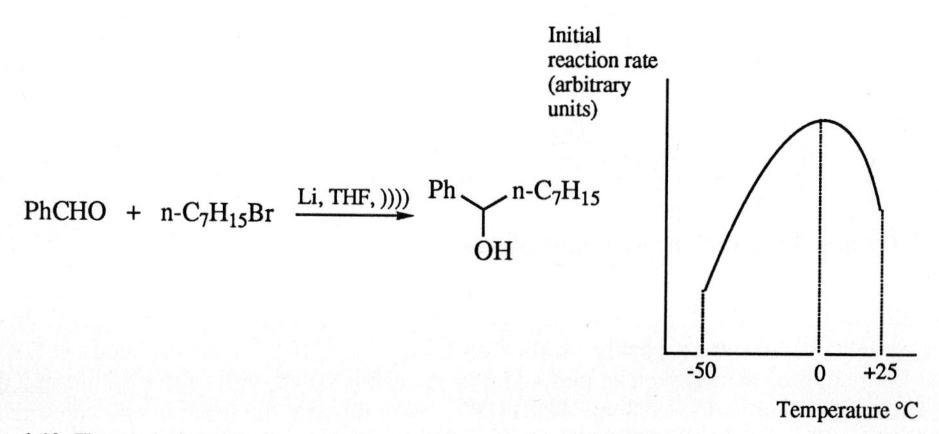

Fig. 4-10. The unusual temperature dependence of the sonochemical Barbier reaction.

metal) helped to confirm that sonication does promote electron transfers from metals (see Section 4.2.2.2). The mechanism was also shown to be slightly different from the admitted in-situ formation of an organolithium species followed by addition to the acceptor. Alkyl chlorides probably give the organolithium compounds, but the actual reactive species from bromo compounds is believed to be a radical anion (Scheme 4-59) [186]. Semi-empirical calculations supported this interpretation [187].

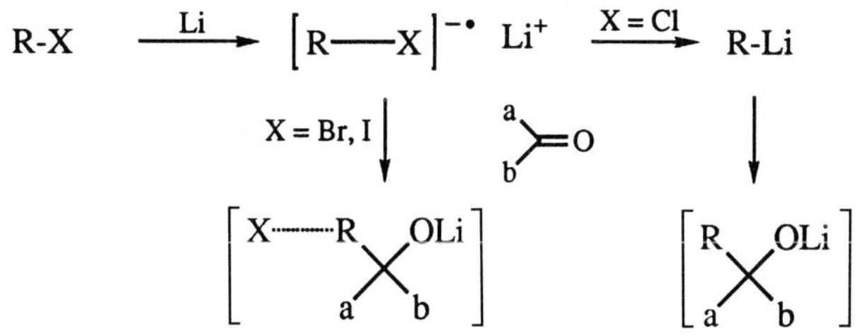

R-X —Li→ [R—X]$^{-\bullet}$ Li$^+$ —X = Cl→ R-Li

X = Br, I

Scheme 4-59

Due to its advantages, synthetic applications have been found and a few examples are shown in Scheme 4-60 [188–197]. One can point to the successful use of benzylic halides [189], the selective 1,2-addition to α-enones [190], and the possibility of using allylic

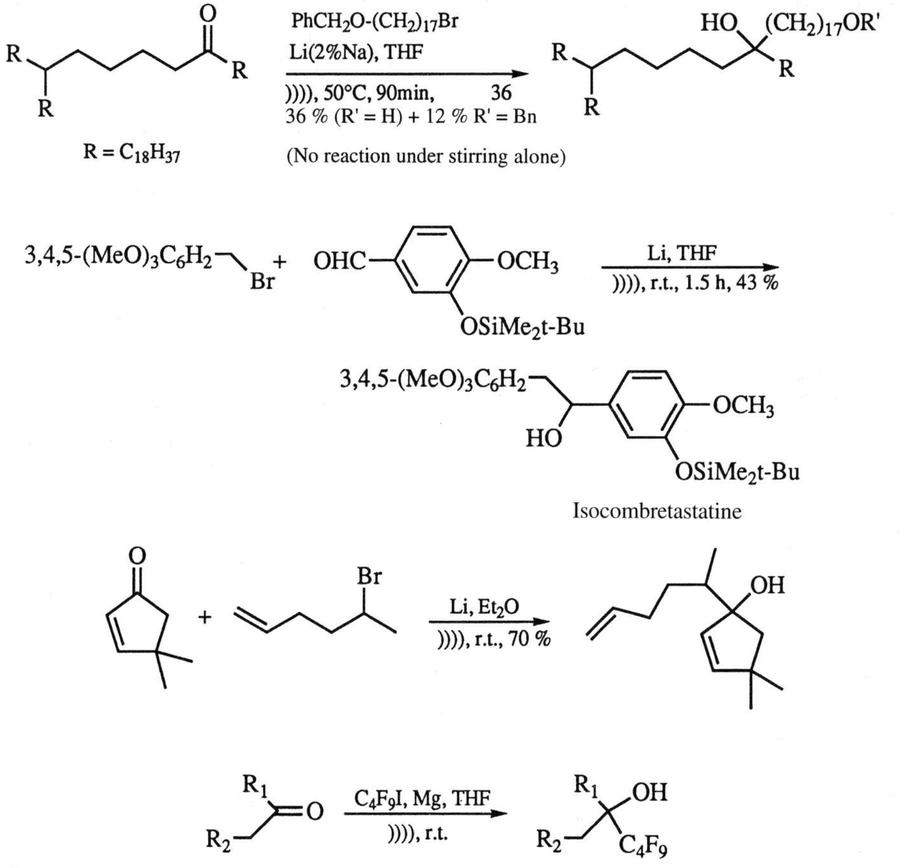

$PhCH_2O$-$(CH_2)_{17}Br$, Li(2%Na), THF,))))), 50°C, 90min, 36 % (R' = H) + 12 % R' = Bn

R = $C_{18}H_{37}$

(No reaction under stirring alone)

3,4,5-$(MeO)_3C_6H_2$ + OHC–C_6H_3(OCH$_3$)(OSiMe$_2$t-Bu); Li, THF,))))), r.t., 1.5 h, 43 %

Isocombretastatine

Li, Et_2O,))))), r.t., 70 %

C_4F_9I, Mg, THF,))))), r.t.

Scheme 4-60. Applications of the sonochemical Barbier synthesis.

phosphates as a source of allyl anions [191]. Magnesium, although not frequently used in these reactions, provides good yields for the preparation of perfluoroalkyl carbinols [192].

Bromochloromethane was used to introduce a chloromethyl group (Scheme 4-61). The actual reagent cannot be the organometallic species, which is highly unstable under the reaction conditions. With ketones, cyclization to the epoxide occurs spontaneously in good yields [193].

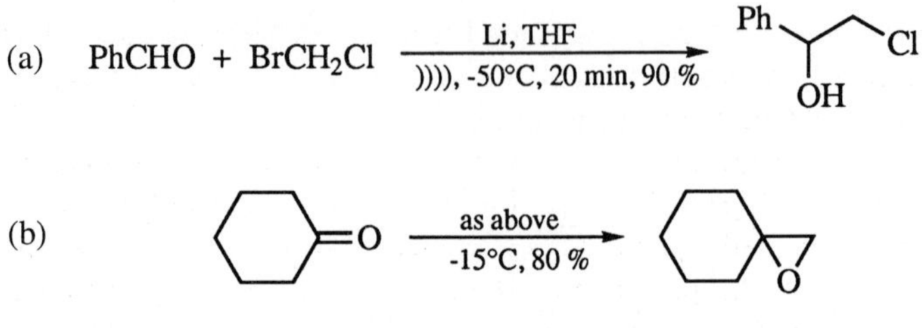

(a) PhCHO + $BrCH_2Cl$ Li, THF /)))), -50°C, 20 min, 90 %

(b) as above / -15°C, 80 %

Scheme 4-61

Intramolecular Barbier processes are generally more difficult to achieve (Scheme 4-62) [194–196]. A vinylic bromide with a keto group in the δ-position was found to cyclize more efficiently under sonication than via the low-temperature halogen-metal exchange with *t*-butyllithium [194], but a similar result was not found in another example [195, 196].

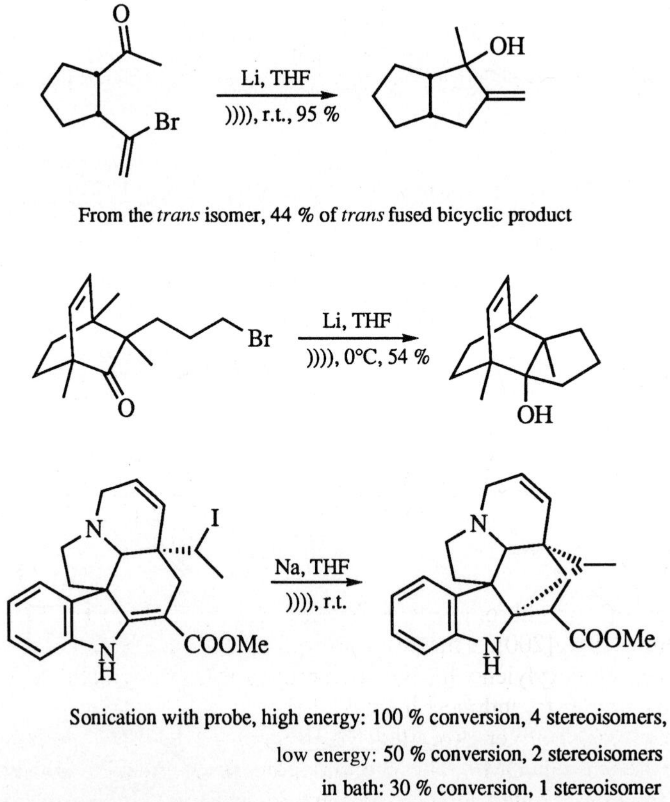

Li, THF /)))), r.t., 95 %

From the *trans* isomer, 44 % of *trans* fused bicyclic product

Li, THF /)))), 0°C, 54 %

Na, THF /)))), r.t.

Sonication with probe, high energy: 100 % conversion, 4 stereoisomers,
low energy: 50 % conversion, 2 stereoisomers
in bath: 30 % conversion, 1 stereoisomer

Scheme 4-62. Intramolecular Barbier reactions.

In the synthesis of vindolinine, a critical annelation step was achieved using sodium. High energy sonication gave a high yield but only moderate stereoselectivity. Irradiation in a cleaning bath led to a lower yield, ca. 30%, but the selectivity was 100% [197].

4.10.2 Barbier Reactions in the Presence of Zinc

In the Barbier reaction it is sometimes necessary to use metals other than lithium or magnesium to obtain different selectivities, or to effect reactions with compounds that give unstable reagents with the more reactive metals.

Addition of various groups to carbonyl compounds in the presence of zinc has been described (Scheme 4-63) [167, 198]. In the example of Scheme 4-63b the stereoselectivity seems to remain unchanged by sonication. The reaction of 4-bromo-2-sulfolene with various carbonyl compounds in the presence of the zinc–silver couple, gives addition at C-4, but with magnesium, the transposed product is obtained in lower yields (Scheme 4-63c) [199].

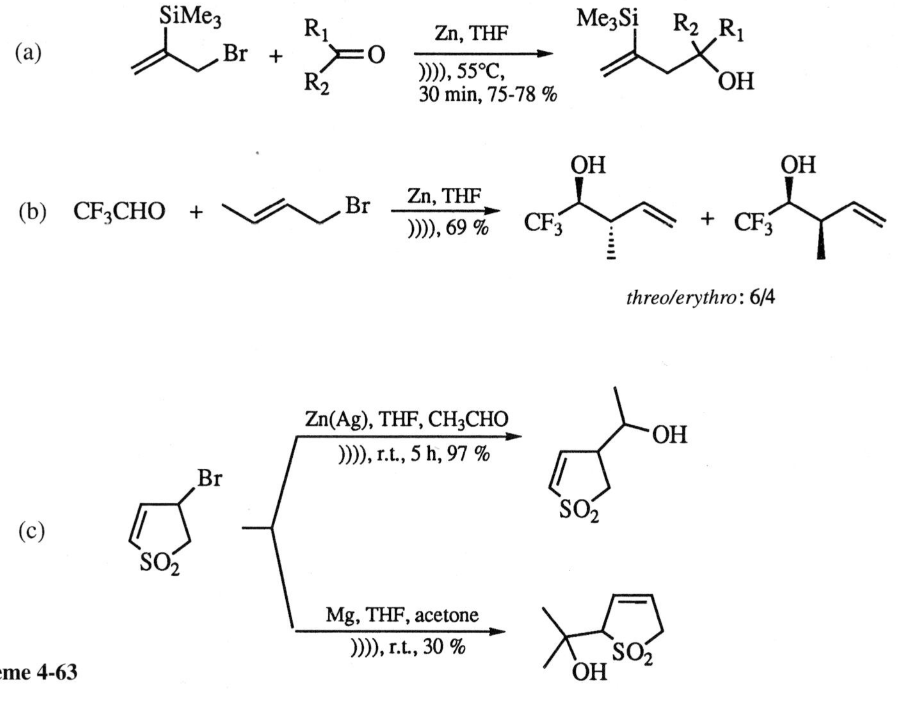

(a) $SiMe_3$ Br + R_1 R_2 =O — Zn, THF;)))), 55°C, 30 min, 75-78 % → Me_3Si R_2 R_1 OH

(b) CF_3CHO + Br — Zn, THF;)))), 69 % → CF_3 OH + CF_3 OH

threo/erythro: 6/4

(c) Br SO_2 — Zn(Ag), THF, CH_3CHO;)))), r.t., 5 h, 97 % → OH SO_2

Mg, THF, acetone;)))), r.t., 30 % → OH SO_2

Scheme 4-63

The reaction of α,α'-dibromo-*o*-xylene with an activated olefin and zinc gives cyclic adducts (Scheme 4-64) [200]. This reaction, described by the authors as a Diels–Alder cycloaddition via the *o*-xylylene, has been employed in carbohydrate chemistry to produce ring compounds similar to anthracyclinones [201].

The sonochemical Barbier procedure has been used to effect perfluoroalkylations of aldehydes and ketones in DMF. The β-elimination observed with lithium or magnesium derivatives is thereby avoided [152]. The method was applied to a glyceraldehyde ketal [202]. Perfluoroalkylation effected with arene tricarbonylchromium aldehydes in the pre-

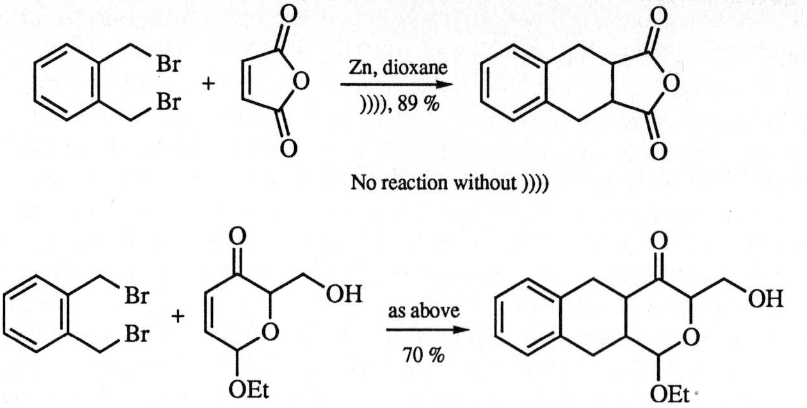

Zn, dioxane

))))), 89 %

No reaction without))))

as above

70 %

Scheme 4-64. "Diels-Alder–like" addition of *o*-xylylene to electron-deficient olefins.

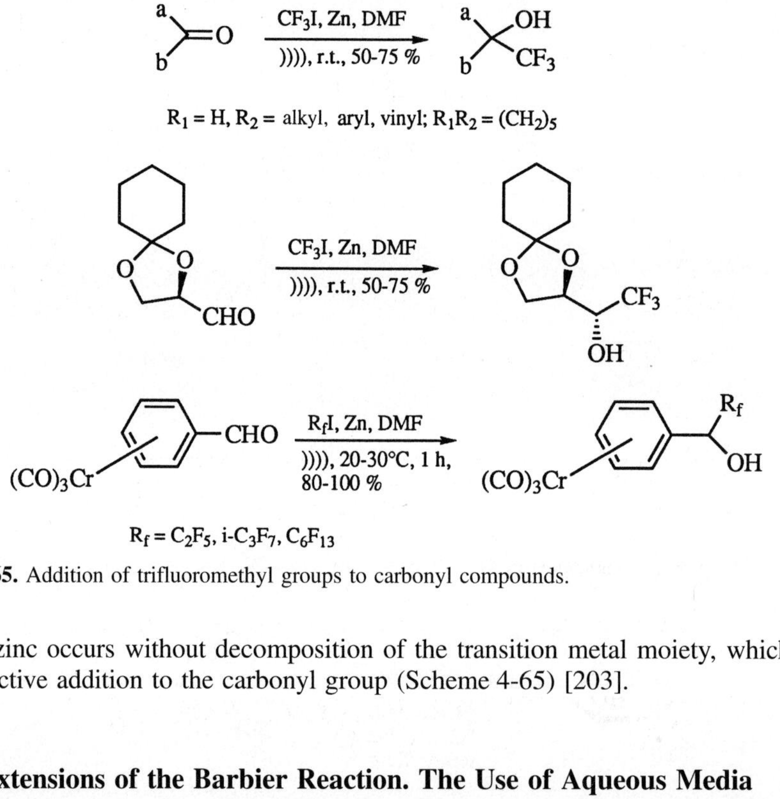

CF_3I, Zn, DMF

))))), r.t., 50-75 %

R_1 = H, R_2 = alkyl, aryl, vinyl; R_1R_2 = $(CH_2)_5$

CF_3I, Zn, DMF

))))), r.t., 50-75 %

R_fI, Zn, DMF

))))), 20-30°C, 1 h, 80-100 %

R_f = C_2F_5, i-C_3F_7, C_6F_{13}

Scheme 4-65. Addition of trifluoromethyl groups to carbonyl compounds.

sence of zinc occurs without decomposition of the transition metal moiety, which induces stereoselective addition to the carbonyl group (Scheme 4-65) [203].

4.10.3 Extensions of the Barbier Reaction. The Use of Aqueous Media

Some organic reactions are now conducted in water and the so-called "hydrophobic" effects are, at least in part, responsible for the reactivities observed [204]. Up to quite recently, however, the idea of performing metal-mediated reactions in an aqueous medium seemed absurd. However, sonochemical research led to the unexpected discovery that metal-induced reactions in aqueous media are not only possible, but also efficient for synthesis. "True" organometallics are not thought to be formed; instead the reaction apparently proceeds through radical species on the metal surface rather than in solution. In general, sonication is of considerable benefit in these reactions. This methodology was successfully applied mainly with zinc, copper, tin, and indium.

The first example found was the allylation of aldehydes and ketones, effected in aqueous THF in the presence of zinc [205]. The yields are not always satisfactory, and replacement of the THF-water solvent by THF with saturated aqueous NH_4Cl greatly improved the method. Sonication becomes unnecessary, and better yields result from stirring at room temperature (Scheme 4-66). Allylic bromides usually react faster than the chlorides. Selectivity in favor of aldehydes is high, which permits additions on this group even in the presence of unprotected keto groups [206]. Saturated alkyl halides remain unreactive under these conditions. Unlike zinc, the reactions with tin require sonication in all cases, and allylic chlorides do not react at all [207]. The use of zinc or tin appears to be complementary, tin providing higher selectivities and yields in some cases. With regard to regioselectivity, the allylic group becomes attached to the substrate at the more highly substituted carbon [205]. The reaction is ideal for the allylation of water-soluble molecules, such as unprotected carbohydrates (Scheme 4-67) [207].

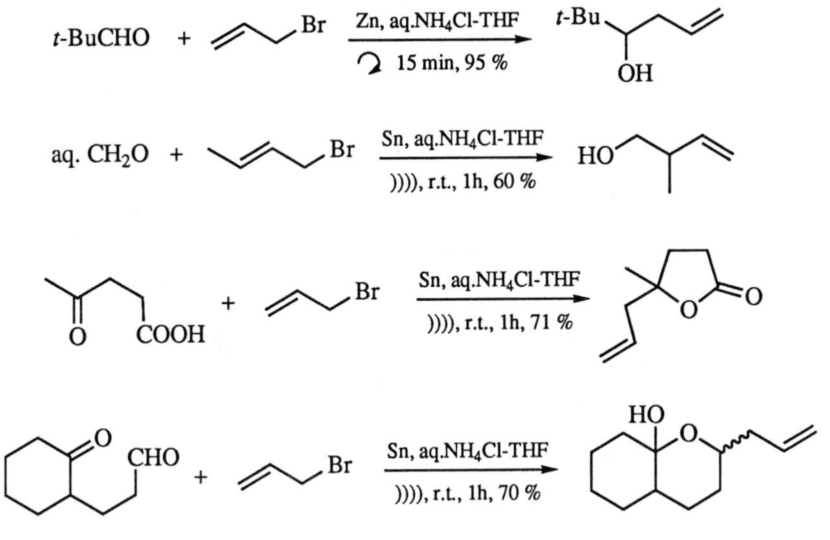

t-BuCHO + Br → (Zn, aq.NH_4Cl-THF, 15 min, 95 %) t-Bu, OH

aq. CH_2O + Br → (Sn, aq.NH_4Cl-THF,)))), r.t., 1h, 60 %) HO

O, COOH + Br → (Sn, aq.NH_4Cl-THF,)))), r.t., 1h, 71 %) O, O

O, CHO + Br → (Sn, aq.NH_4Cl-THF,)))), r.t., 1h, 70 %) HO, O

Scheme 4-66. Allylations in aqueous media.

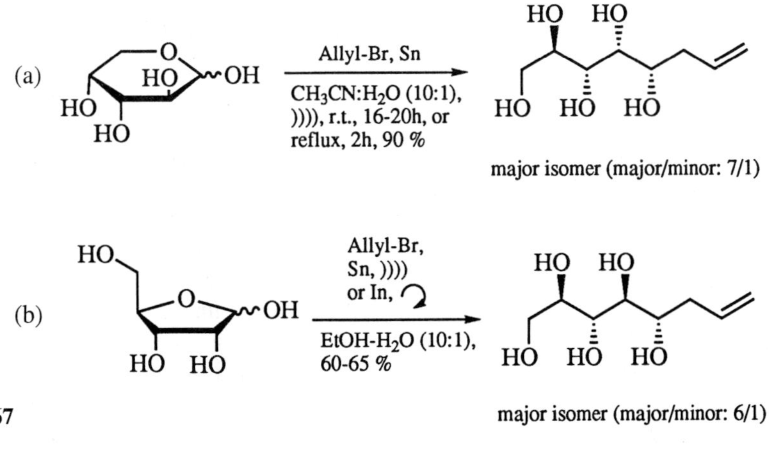

(a) O, HO, OH, HO, HO → (Allyl-Br, Sn; CH_3CN:H_2O (10:1),)))), r.t., 16-20h, or reflux, 2h, 90 %) HO HO, HO HO HO

major isomer (major/minor: 7/1)

(b) HO, O, OH, HO HO → (Allyl-Br, Sn,)))) or In; EtOH-H_2O (10:1), 60-65 %) HO HO, HO HO HO

Scheme 4-67

major isomer (major/minor: 6/1)

Conjugate additions of alkyl groups to electron-deficient olefins occur in aqueous media in the presence of a zinc–copper couple prepared as mentioned in Section 4.3.2 [53]. The solvents that may be used, mentioned in Section 4.3.2, play an important role, not through the properties usually considered in organic synthesis, but through their ability to be structured and to absorb a maximum of acoustic energy [208]. These conjugate

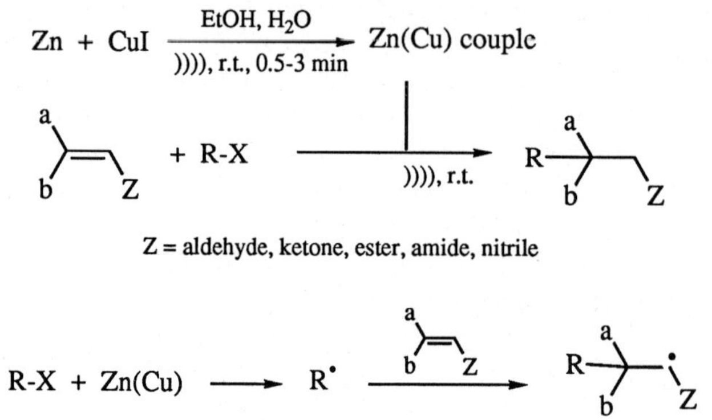

Zn + CuI → (EtOH, H_2O;)))), r.t., 0.5-3 min) Zn(Cu) couple

a, b, Z + R-X → ()))), r.t.) R, a, b, Z

Z = aldehyde, ketone, ester, amide, nitrile

R-X + Zn(Cu) → R$^{\bullet}$ → (a, b, Z) R, a, b, Z

Scheme 4-68. Conjugate additions of alkyl halides to electron-deficient olefins.

(a) Cl-$(CH_2)_6$-I (i) or HO-$(CH_2)_{12}$-I (ii) + COOEt → Zn(Cu), EtOH, H_2O, 9:1,)))), r.t., 2 h, 91 %; (i) 69 %; (ii) 41 % → Cl-$(CH_2)_6$ COOEt; HO-$(CH_2)_{12}$ COOEt

(b) Br (Ad-Br) + O → Zn(Cu), EtOH, H_2O, 9:1,)))), r.t., 2 h, 91 % → O Ad

(Initiation by Bu_3SnH: 33% yield)

(c) t-C_4H_9-I → Zn(Cu), H_2O,)))), 25°C, 75 % → t-$C_4H_9^{\bullet}$ ← $PhCH_3$, AIBN, 110°C, 55 % ← t-C_4H_9, O, O–N, S

t-$C_4H_9^{\bullet}$ + Ph, CN, CN → Ph, t-C_4H_9, CN, CN + Ph, t-C_4H_9, CN, CN

5.6:1

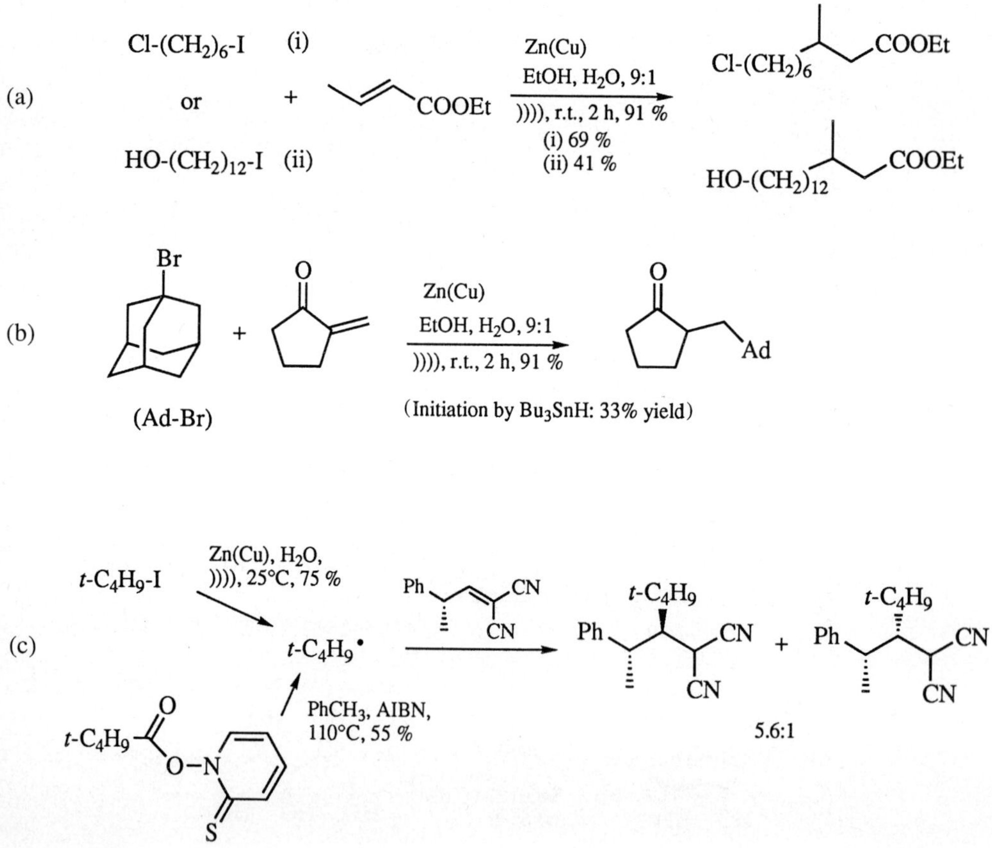

Scheme 4-69

additions, the principle of which is shown in Scheme 4-68, probably follow a free radical pathway [208, 209].

Secondary and tertiary alkyl bromides and iodides give excellent yields, but iodides must be used with primary derivatives. Chlorides always remain unchanged, and a hydroxyl group does not inhibit the reaction, offering possibilities for the addition of functionalized groups (Scheme 4-69a) [210, 211]. The activated olefin can be an aldehyde, ketone, ester, nitrile, or amide. The synthetic potential is illustrated in a few examples. This method was found in some instances superior to the usual Bu_3SnH/AIBN method (Scheme 4-69b) [212]. A comparison between this sonochemical reaction and the Barton method showed that higher yields are obtained in the former but the steric courses are virtually the same [Scheme 4-69(c)] [209].

Spanish authors applied these conjugate additions to the synthesis of vitamin D_3 analogs (Scheme 4-70a) [211]. The reaction succeeds even in cases known to be difficult in

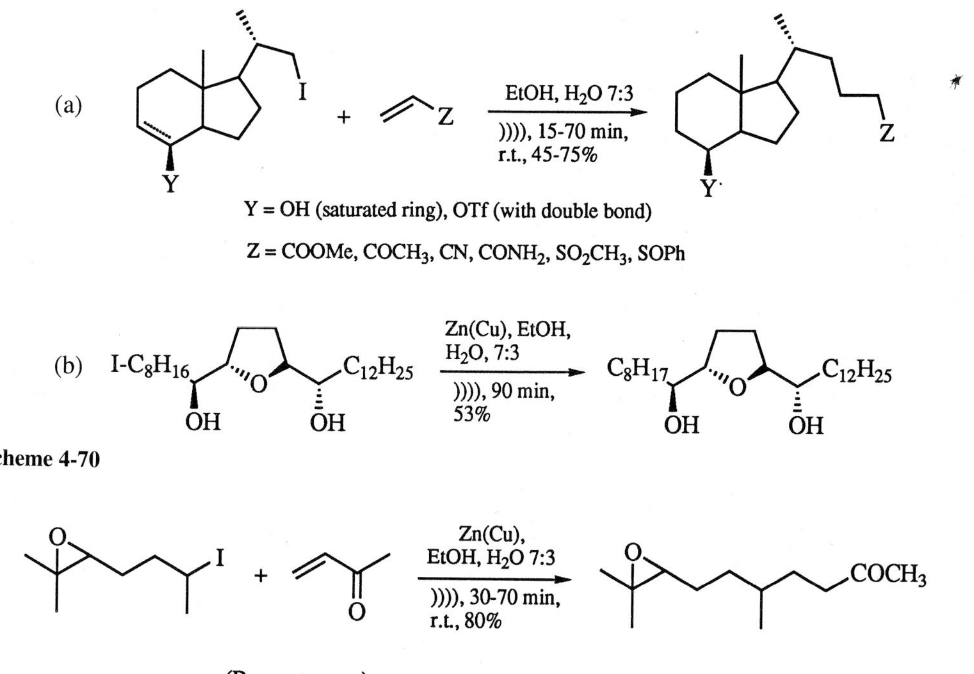

(a) I + Z; EtOH, H_2O 7:3;)))), 15-70 min, r.t., 45-75%; Y; Z; Y

Y = OH (saturated ring), OTf (with double bond)

Z = COOMe, $COCH_3$, CN, $CONH_2$, SO_2CH_3, SOPh

(b) I-C_8H_{16}, O, $C_{12}H_{25}$, OH, OH; Zn(Cu), EtOH, H_2O, 7:3;)))), 90 min, 53%; C_8H_{17}, O, $C_{12}H_{25}$, OH, OH

Scheme 4-70

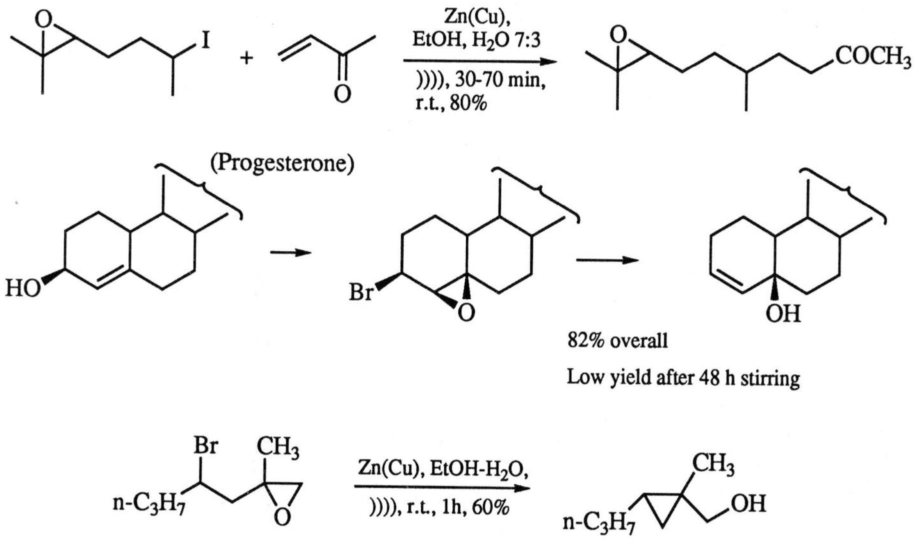

O, I + O; Zn(Cu), EtOH, H_2O 7:3;)))), 30-70 min, r.t., 80%; O, $COCH_3$

(Progesterone)

HO → Br, O → OH

82% overall

Low yield after 48 h stirring

Br, CH_3, n-C_3H_7, O; Zn(Cu), EtOH-H_2O,)))), r.t., 1h, 60%; CH_3, n-C_3H_7, OH

Scheme 4-71. Reactions of epoxyalkyl halides.

conventional organometallic synthesis, such as vinylsulfoxides. Although in most examples the iodide occupies a primary position, the yields are generally excellent, even in the presence of a free hydroxyl group or a vinylic triflate. Another example of an attempted synthetic application is given in Scheme 4-70b. In this case only reduction of the iodide was observed [213]. This undesired pathway is a common minor side-reaction in most cases.

The protocol has been applied to epoxyalkyl halides, which would give rise to severe difficulties in the usual organometallic syntheses (Scheme 4-71) [214–216]. Depending on the number of carbons between the halide and the epoxide ring, the chemical evolution can differ substantially, with either preservation of the ring or its opening to give an allylic alcohol or a cyclopropyl carbinol.

4.10.4 Additions to Trivalent Functionalities

The versatility of the sonochemical Barbier reactions was further demonstrated in condensations with less common electrophiles. The Bouveault aldehyde synthesis by formylation of alkyl or aryl lithium reagents suffers numerous side-reactions. In the ultrasonic modification, aldehydes are obtained in moderate to good yields from alkyl and aryl halides in the presence of lithium (containing 2% Na) and DMF [217]. Examples are given in Scheme 4-72 [218]. An unusual frequency effect was observed in the reaction run in diethyl ether. Irradiations at 50 kHz remain unsuccessful, but at 500 kHz the expected products are ontained in 30–45 min. Lithiation of the intermediate α-amino alkoxide and reaction with electrophiles yields *ortho*-substituted benzaldehydes in a one-pot process. Similar results were obtained in the Barbier reaction of isocyanates (Scheme 4-73) [219].

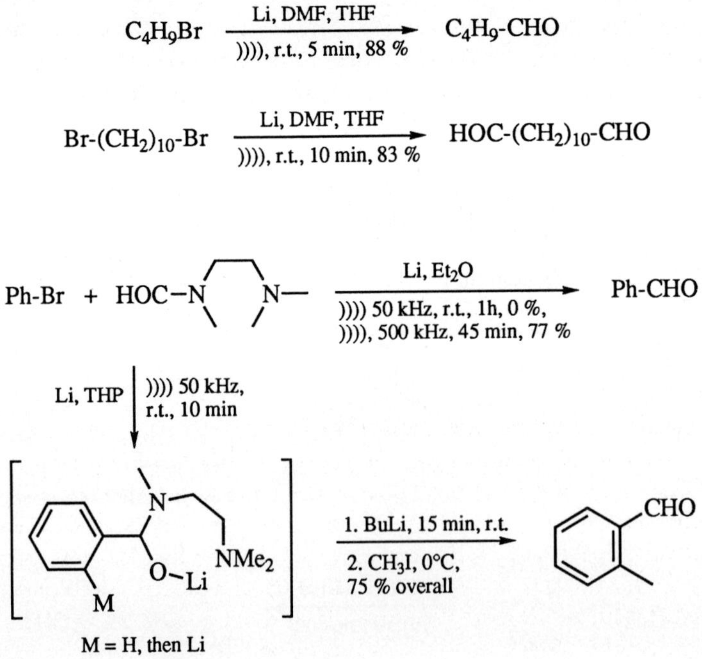

C_4H_9Br —(Li, DMF, THF;)))), r.t., 5 min, 88 %)→ C_4H_9-CHO

Br-$(CH_2)_{10}$-Br —(Li, DMF, THF;)))), r.t., 10 min, 83 %)→ HOC-$(CH_2)_{10}$-CHO

Ph-Br + HOC–N(piperazine)N– —(Li, Et_2O;)))) 50 kHz, r.t., 1h, 0 %,)))), 500 kHz, 45 min, 77 %)→ Ph-CHO

Li, THP |)))) 50 kHz, r.t., 10 min

NMe₂ O–Li M

M = H, then Li

1. BuLi, 15 min, r.t. 2. CH_3I, 0°C, 75 % overall → CHO

Scheme 4-72. The Bouveault reaction.

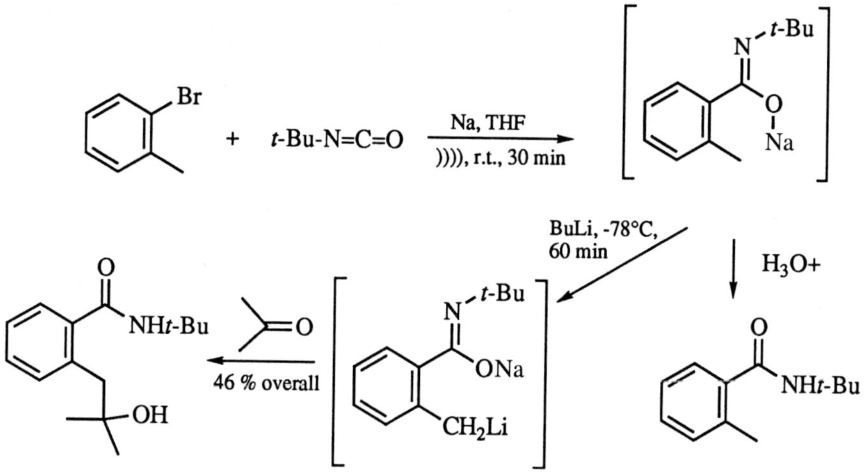

Scheme 4-73

4.10.5 Reactions from α-Halocarbonyl Compounds

The reactivity of α-halocarbonyl compounds with metals (generally zinc but other metals can also be used) has been widely explored and used in synthesis. One of the first sonochemical reactions reported in the literature was the reduction of α,α'-dibromo ketones with ultrasonically dispersed mercury in acids. A probable 2-hydroxyallyl cation intermediate leads to α-acetoxy ketones and the corresponding dehalogenated ketone as the major products (Scheme 4-74) [220]. The process is comparable to an electrochemical reduction at a mercury cathode, but the total yield of products is lower (30–75%) than in the electrochemical reaction (>90%). Reduction of asymmetrical dibromo ketones usually gives two isomeric α-acetoxy ketones, but selectivity can be obtained by replacing acetic acid by trimethyl- or triethylacetic acid.

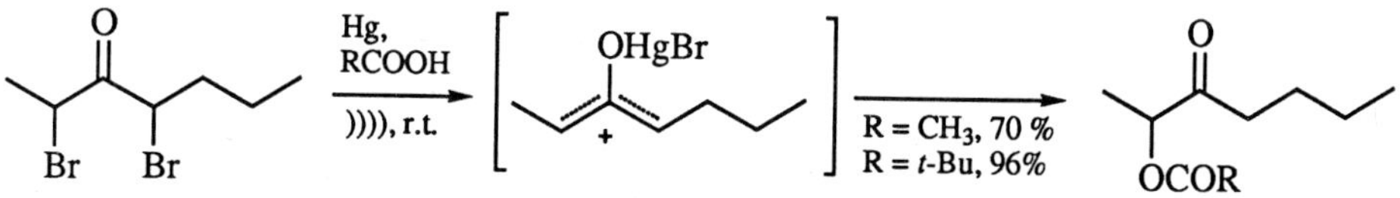

Scheme 4-74

A synthetically interesting transformation of α,α'-dibromo ketones, applied in the synthesis of a variety of natural products, is the Noyori or Hoffmann–Noyori reaction [221, 222]. Classically, polybromo ketones react with iron carbonyl in the presence of olefins or dienes in a [3 + 2] and [3 + 4] fashion, to give five- or seven-membered carbocycles respectively. Joshi and Hoffmann designed a sonochemical variant to prepare the allyl cation from a dibromo ketone and a diene in the presence of copper or the zinc–copper couple. Reactions are completed in 2 h in dioxane solvent at 5–10 °C and yields range from 75 to 80% (Scheme 4-75). The cycloaddition has been recently extended to anthracenes under the same conditions.

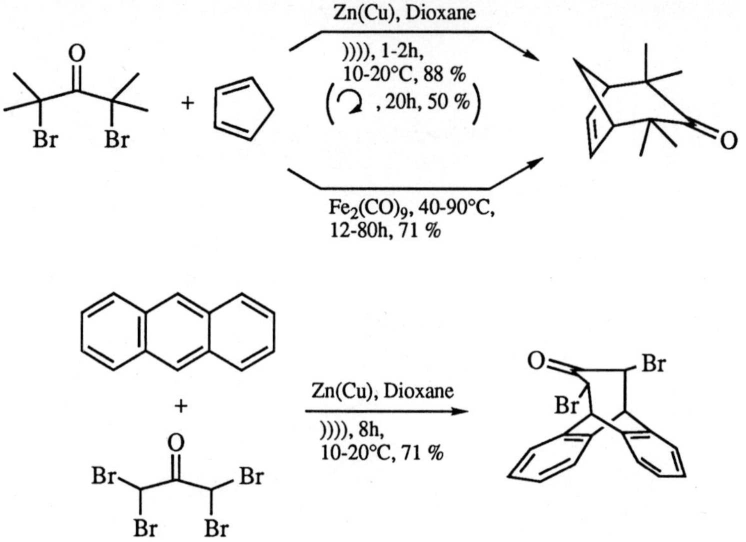

Zn(Cu), Dioxane
)))), 1-2h, 10-20°C, 88 %
(, 20h, 50 %)
Br Br + O
$Fe_2(CO)_9$, 40-90°C, 12-80h, 71 %

+ Br Br Br Br O
Zn(Cu), Dioxane
)))), 8h, 10-20°C, 71 %
O Br Br

Scheme 4-75. Hoffmann [3 + 2]cycloadditions to dienes.

The synthetic utility of the Reformatsky reaction is well established [223]. In any event, metal activation is necessary for success and often plays a crucial role in reaction conditions [1, 2]. Han and Boudjouk observed significant improvements in rate and yield (98% in 5 min!) in typical reactions (Scheme 4-76) [224]. The procedure was further improved by using Rieke's zinc slurries [45] and by probe sonication [225]. The higher reactivity was shown to be due not to an increased surface area, but to morphological and composition changes. Various papers on the synthesis of hydroxyesters [226], ketoesters [227], β-lactams [228, 229], and diverse fluoro compounds (Scheme 4-77) [230, 231] have been published. An interesting application of the Blaise reaction also deserves mentioning [232].

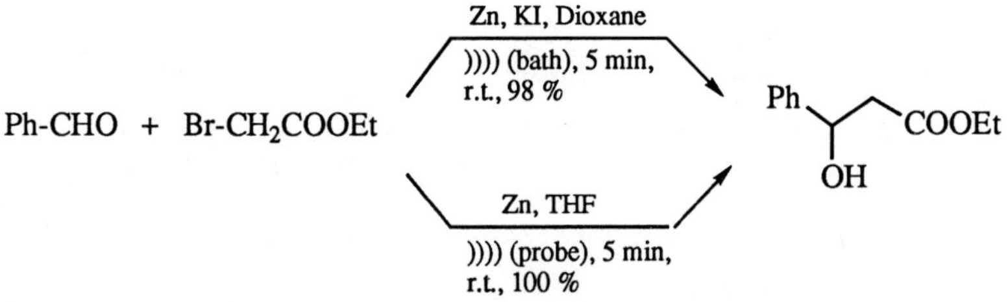

Ph-CHO + Br-CH_2COOEt
Zn, KI, Dioxane
)))) (bath), 5 min, r.t., 98 %
Zn, THF
)))) (probe), 5 min, r.t., 100 %
Ph COOEt OH

Scheme 4-76

Kitazume et al. applied the ultrasound Reformatsky protocol to trifluoroacetaldehyde [198]. Yields of the corresponding β-hydroxy esters are good and the *threo* configuration is favored. Fluorinated β-keto-γ-butyrolactones or tetronic acids have been prepared by a reaction between *O*-trimethylsilyl cyanohydrins and ethyl α-fluoro- (or trifluoromethyl) bromoacetate (Scheme 4-78) [230]. In the absence of ultrasound the reaction was unsuccessful.

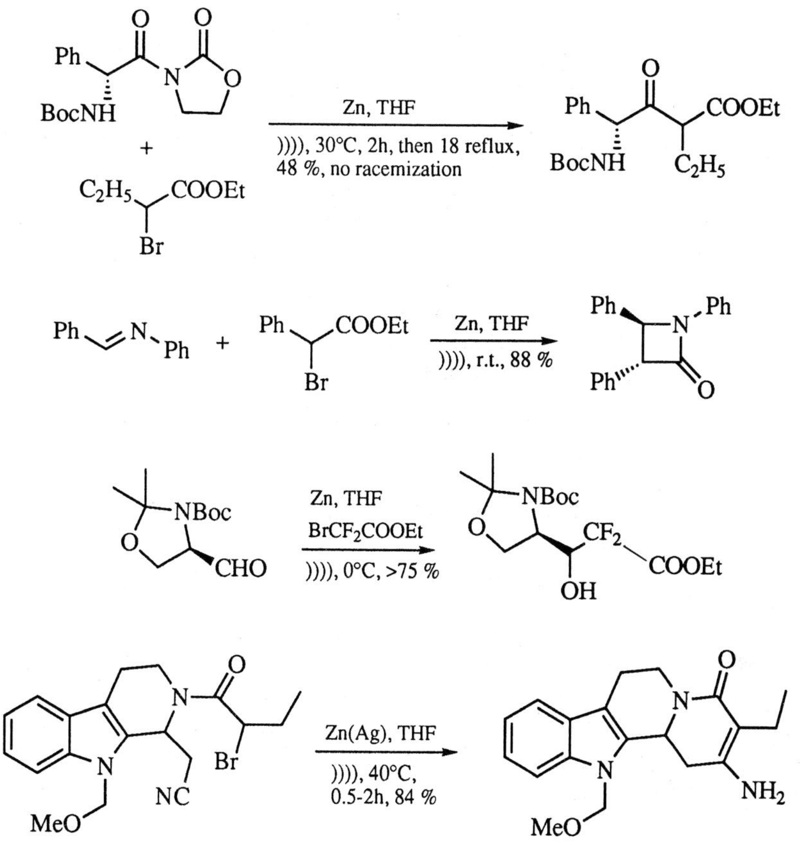

Scheme 4-77. Typical examples of sonochemical Reformatsky reactions.

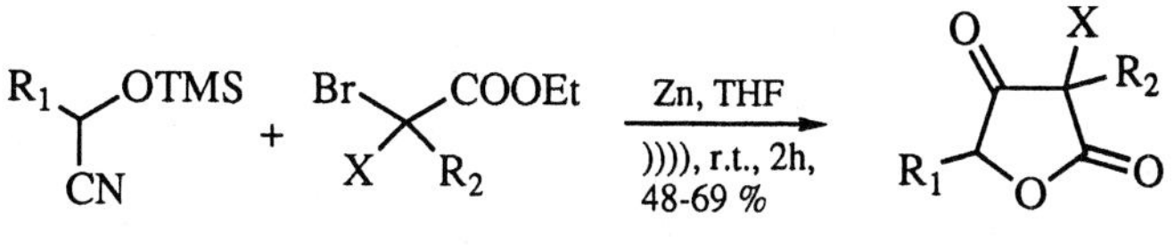

Scheme 4-78 $R_1 = CH_3, C_6H_5$ $R_2 = H, CH_3$ $X = F, CF_3$

4.11 Reactions Using Activated Metal Catalysts

The effect of ultrasonic activation on the properties of palladium, platinum, or rhodium blacks prepared and/or used under sonication has been reviewed, and frequency effects were noted [52]. But no clearcut conclusions can be drawn from this not easily accessible literature, and the area still offers very broad possibilities for new investigations.

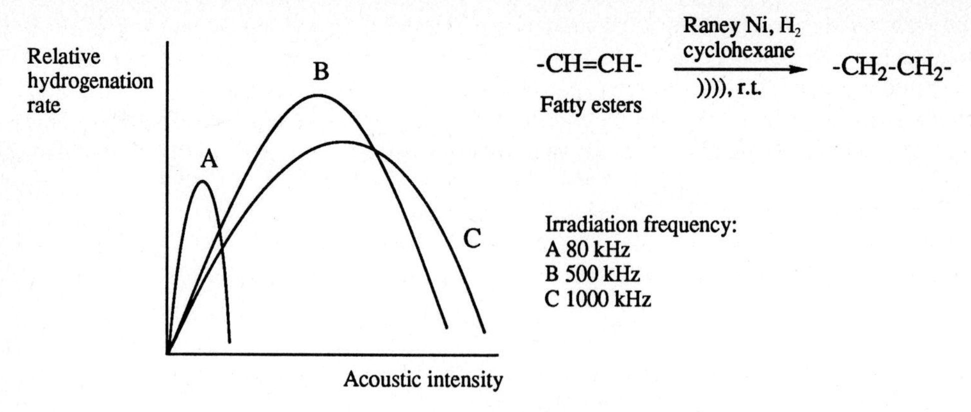

Fig. 4-11. Influence of frequency and intensity on the hydrogenation of fatty esters.

4.11.1 Nickel-Catalyzed Hydrogenations and Hydrogenolyses

Pioneers in this field are Saracco and Arzano, who mentioned in 1968 the role of sonication in the reduction of unsaturated fatty acid esters by Raney nickel in cyclohexane [233]. In a systematic study, important parameters for a rational use of ultrasound were determined, such as the optimal frequency and intensity (Fig. 4-11), and the geometry of the reactors.

By sonication in water, Raney nickel can be activated by removal of aluminum-rich particles [234]. The product is an efficient enantioselective catalyst after reaction with tartaric acid and sodium bromide. The authors state that crystalline domains of the cata-

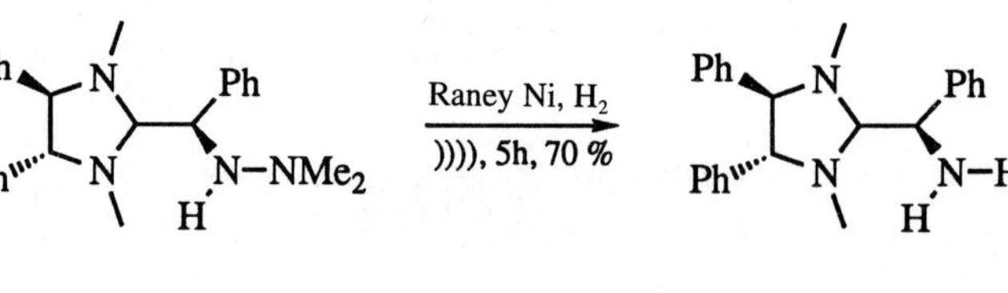

H_2, Ni*, THF, 100°C;)))), 4 h, 86 % (e.e. 91 %) [stirring, 24 h, 70 % (e.e. 90 %)]

Scheme 4-79. Enantioselective hydrogenations on Raney nickel.

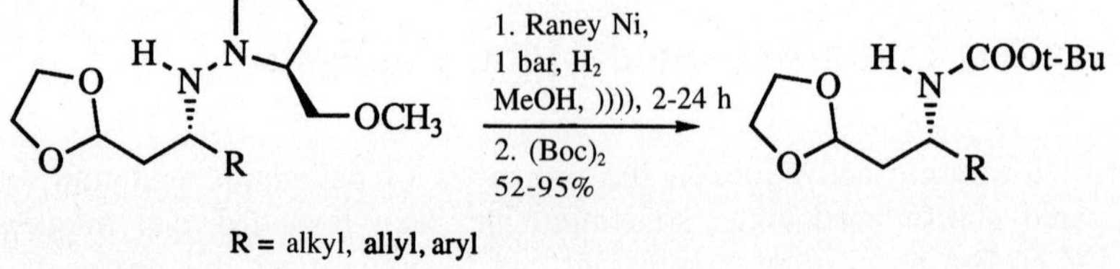

Raney Ni, H_2,)))), 5h, 70 %

1. Raney Ni, 1 bar, H_2, MeOH,)))), 2-24 h; 2. $(Boc)_2$; 52-95%

R = alkyl, allyl, aryl

Scheme 4-80. Hydrogenolysis of hydrazines.

lyst are enantiodifferentiating, whereas the disordered aluminum-rich domains lead to smaller inductions. Scheme 4-79 illustrates the advantages of this modified catalyst.

Sonohydrogenolysis of hydrazines by Raney nickel is performed at room temperature with hydrogen at normal pressure. Steric hindrance seems to have no effect on the process, benzylic C–N bonds resist cleavage, and chiral centers remain intact (Scheme 4-80) [235].

The direct activation of nickel powder has been discussed in Section 4.2.2.1 [33]. The activity of this ultrasonically activated catalyst is comparable to that of Raney nickel, but the former is more selective and does not reduce aldehydes or ketones.

4.11.2 Hydrogenations, Hydrogenolyses, and Hydrosilylations Using Various Transition Metal Catalysts

A variety of catalytic reductive processes has been studied under sonochemical conditions (Scheme 4-81) but a rationale has not yet emerged. Only a descriptive approach can be made at the present time. Palladium under sonication was found to give excellent results for the hydrogenation of difluoro olefins, whereas the silent process was unsuccessful [236]. The hydrogenation of olefins in the presence of Pd/C using formic acid as the source of hydrogen [237] is conducted at room temperature, with quantitative yields in 1 h. Sonication is not necessary for success, but it leads to considerable acceleration.

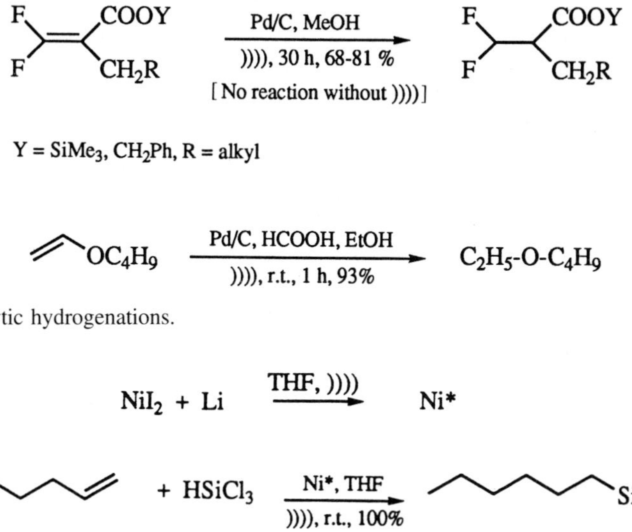

F, F, COOY, CH_2R — Pd/C, MeOH;)))), 30 h, 68-81 % [No reaction without))))] → F, F, COOY, CH_2R

Y = $SiMe_3$, CH_2Ph, R = alkyl

OC_4H_9 — Pd/C, HCOOH, EtOH;)))), r.t., 1 h, 93% → C_2H_5-O-C_4H_9

Scheme 4-81. Catalytic hydrogenations.

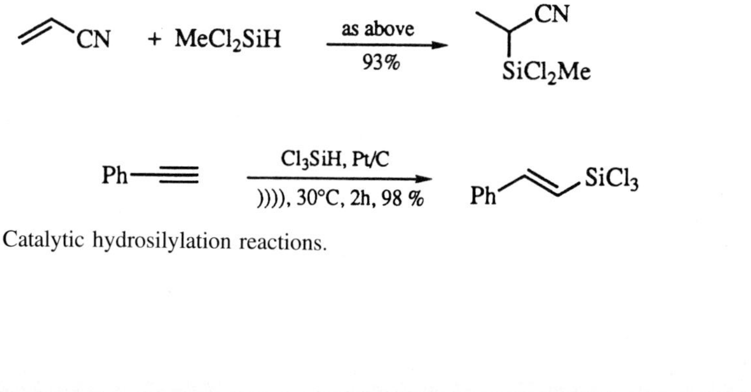

NiI_2 + Li — THF,)))) → Ni*

+ $HSiCl_3$ — Ni*, THF;)))), r.t., 100% → $SiCl_3$

CN + $MeCl_2SiH$ — as above; 93% → CN, $SiCl_2Me$

Ph— — Cl_3SiH, Pt/C;)))), 30°C, 2h, 98 % → Ph, $SiCl_3$

Scheme 4-82. Catalytic hydrosilylation reactions.

The use of activated nickel in hydrosilylation reactions has been investigated [153, 154, 238]. Addition of trichlorosilane to 1-hexene occurs quantitatively in 40 min. The interest of this result lies in the possibility of using such a cheap catalyst instead of platinum compounds. Sonication of hydrosilanes and alkenes with Pt/C catalyst at 30 °C and atmospheric pressure gives hydrosilylation products in high yields (70–96%) in 1–2 h [153, 239]. Phenylacetylene could be hydrosilylated almost quantitatively under these conditions, whereas a very low yield (<5%) was obtained with vigorous stirring after 10–48 h. The catalyst can be easily recycled, although in some cases the presence of polymeric side-products complicates its recovery. Recently, the hydrosilylation of styrene and phenylacetylene with triethylsilane in the presence of rhodium and platinum was mentioned [240]. Scheme 4-82 summarizes some of these hydrosilylation reactions.

4.12 Experimental Sonochemistry

4.12.1 Technical Aspects

In contrast to electro- or photochemistry, the equipment required for sonochemical reactions is rather simple and the know-how readily acquired at the bench. Several types of ultrasound generators, cleaning baths, and probes are available at reasonable prices. Usually the acoustic wave is emitted by a piezoelectric ceramic transducer excited by a high-frequency electric field. The sound wave delivered has a wavelength corresponding to twice the thickness of the ceramic. The consequence is that ultrasonic emitters are necessarily "monochromatic". The frequency can be changed only by changing the emitter. Fortunately, all known frequency effects in sonochemistry are broad-band effects, and no important qualitative differences result from using 20, 40, or 60 kHz waves. Lower frequencies are usually preferred because they permit higher energy emission. On the other hand, frequencies around 500 kHz produce interesting effects which have been less thoroughly explored. New generators are now available (e.g. the Undatim™ high frequency generator), but high-frequency studies of organometallics are still rare.

4.12.1.1 Ultrasonic Cleaning Baths

Using a cleaning bath is rather simple, provided that some practical details are taken into account for optimal results. Most of these baths are equipped with several emitter placed under a stainless steel tank and delivering 30–50 kHz waves. Before operating the generator, it is frequently advisable to check with an oscilloscope the correct matching of the electrical high frequency with the resonance frequency of the emitter. A mismatch gives poorer energy emissions, and can even damage the ceramic. The tank is filled with water, the most appropriate liquid for sound transmission. Experimental know-how for a proper use of these baths is given in detail in [13].

Limitations of the method are easily understood. Experiments below 0 °C cannot be effected. The size of the reaction vessel is limited to small volumes (less than half a liter), not only because of the need for size compatibility with the bath, but also due to its generally low energy output. Nevertheless, many experiments can be conducted with this equipment, and numerous examples can be found in the literature.

4.12.1.2 Horn Generators

Although many simple experiments can be run with baths, quantitative experiments such as kinetic studies require accurate control of energy and temperature. This can be achieved by using horn generators. They are composed of a ceramic coupled to a metal transducer. The emission occurs axially from the tip of the horn. Experimentally, this means that the set-up must ensure that the sonicated substance is accurately placed in the emission zone. In experiments with solids, it will be necessary to fix their position in the emission zone. Items of glassware permitting efficient sonication are described in [13], and an example is shown in Fig. 4-12. Other types of less frequently used generators exist, such as the ultrasonic whistle, commonly used in the food industry, but they need no description here since they seem not to have been used in organometallic syntheses.

Fig. 4-12. Picture of a horn generator and the glassware for reactions with metals. (Photograph: The sonochemistry group, URA CNRS 332).

4.12.2 Typical Sonochemical Syntheses

4.12.2.1 Dieckman Condensation with Ultrasonically Dispersed Potassium

Potassium (78 mg, 2 mmol) is sonicated under an argon atmosphere in 5 mL of dry toluene for 15 min, in an ultrasonic cleaning bath cooled to ca. 15 °C. During this time the metal piece disappears to give a fine silvery blue suspension. Diethyl adipate (202 mg, 1 mmol) in 2 mL of toluene is added via a syringe. The blue color disappears as a moderately exothermic reaction proceeds. After 5 min, the mixture is quenched with ethanol, neutralized with 1 N HCl, and worked up as usual. Silica gel chromatography of the crude oil gives 130 mg (83%) of the pure ketoester, identified by comparison with an authentic sample [30].

4.12.2.2 Lithium Di-isopropylamide

A round bottom flask, with a side arm fitted with an argon balloon, is charged with n-butyl chloride (46.3 g, 0.5 mol), di-*iso*propylamine (freshly distilled on calcium hydride, 50.6 g, 0.5 mol) and dry THF (250 mL). Lithium (wire cut into pieces, 7 g, 1 mol) is added portionwise via the side arm under sonication. After the addition is completed (1 h), sonication is continued for 1 h, then the pale yellow solution is standardized with menthol in the presence of phenanthroline. Calculated yields are in the range 90–92% [143].

4.12.2.3 Barbier Reaction with 1-Bromo-4-chlorobutane

Lithium wire (Aldrich, 0.01% Na, 173 mg, 25 mmol) is placed in the sample holder of the glass vessel shown in Fig. 4-12, which is adapted to the ultrasonic probe with a rubber septum. The system is flushed with argon, then benzaldehyde (530 mg, 5 mmol) and 1-bromo-4-chlorobutane (1714 mg, 10 mmol) in 30 mL of dry THF are added to the flask, which is cooled to –45 °C. Sonication is effected at a medium intensity (ca. 100 W electrical power) for 40 min. The mixture is then hydrolyzed (aq. NH_4Cl) and worked up as usual. The resulting oil is chromatographed (SiO_2) to give 760 mg of pure 5-chloro-1-phenyl-pentan-1-ol [241].

4.12.2.4 Conjugate Addition of 6-Bromo-1-hexene to 1-Buten-3-one

Zinc dust (500 mg, 7.6 mmol) and purified copper iodide (356 mg, 2.4 mmol) in 2 mL distilled water are sonicated under argon in a cleaning bath. The initially grey mixture turns to a deep black, heavy suspension in <5 minutes. To this couple is added with continuous sonication 3 mL of ethanol, then but-1-en-3-one (701 mg, 10 mmol) and 6-iodo-1-hexene (210 mg, 1 mmol) in 2 mL of 2-propanol, over a 1 h period. After a further 2 h sonication, ether is added and the mixture is filtered through a celite pad. The filtrate is washed with brine, then worked up as usual. The crude oil is chromatographed on a short column to give 115 mg (75%) of an oil. VPC analysis reveals the presence of two compounds, 1-decen-9-one and 5-cyclopentyl-pentan-2-one in a 3:1 ratio, identified by the usual analytical methods [242].

4.12.2.5 Hydrogenation of 4,4-Dimethylcyclopent-2-en-1-one

Zinc dust (2.4 g, 36.7 mmol) and nickel chloride hexahydrate (1 g, 4.2 mmol) are sonicated in a 1:1 mixture of ethanol and water (14 mL) in a cleaning bath for a few minutes to generate the deep black couple. The title compound (0.7 g, 6.4 mmol) is then added and sonication is continued for 15 min. The mixture is then filtered through a celite pad and the filtrate worked up as usual. The crude oil is distilled (Kugelrohr) to give pure 3,3-dimethyl-cyclopentanone [243].

4.13 Conclusions

Applications of ultrasound to metal-mediated reactions are now well-established research areas and sonochemical reactions clearly rival traditional processes. The examples given in this account, not exhaustively, constitute evidence for such a statement.

Sonication has profoundly changed certain branches of chemical research, such as the preparation of organometallic reagents and metal-assisted reactions in water, to name a few relevant processes. Prior to this, some metals were unknown to form organometallics by direct metallation, and some reactions were known to be difficult or impossible by conventional methods.

Sonochemistry is not yet a mature discipline, despite extensive research and its varied usefulness. So far, some organic transformations have encountered little or no success under ultrasonic irradiation, even though they are known to be susceptible to thermal and pressure effects, which provides further demonstration of the lack of an entirely satisfactory theoretical understanding. The birth of new interpretations, such as those based on electrical phenomena or other physical effects [35, 244], may be of importance, but to date they are still in an exploratory phase of development. Despite this, sonochemists are proud of the uniqueness of their technique, and will continue to study sonication processes, refining former understanding, and discovering principles leading to new synthetic applications.

Acknowledgements. The authors wish to thank their talented colleagues whose names appear in the references. We are also indebted to Mrs M. J. Luche for help in the bibliographical research. Support from the CNRS, France, and D.G.I.C.Y.T., Spain (PB92-0525-C02-01), is also gratefully acknowledged.

References

[1] A. Fürstner, *Angew. Chem.* **1993**, *105*, 171–197; *Angew. Chem. Int. Ed. Engl.* **1993**, *32*, 164–189.
[2] P. Cintas, *Activated Metals in Organic Synthesis*, CRC Press, Boca Raton, 1993.
[3] *Advances in Sonochemistry* (ed.: T. J. Mason) JAI Press, London, **1990**, Vol. 1; **1991**, Vol. 2; **1993,** Vol. 3.
[4] *Ultrasound, its Chemical, Physical, and Biological Effects* (ed.: K. S. Suslick), VCH, Weinheim, **1988**.
[5] P. Renaud, *Bull. Soc. Chim. Fr.* **1950**, 1044–1045.
[6] B. Pugin, A. T. Turner, in Ref. [3], **1990**, Vol. 1, Ch. 3.
[7] J. L. Luche, *ibid*, Ch. 4.
[8] K. S. Suslick, S. J. Doktycz, *ibid.*, Ch. 6.
[9] P. Boudjouk, in Ref. [4], Ch. 5.
[10] C. Einhorn, J. Einhorn, J. L. Luche, *Synthesis* **1989**, 787–813.
[11] Yu. S. Goldberg, R. Sturkovich, E. Lukevics, *Applied Organomet. Chem.* **1988**, *2*, 215–226.
[12] *Current Trends in Sonochemistry* (ed.: G. J. Price), Royal Society of Chemistry, Cambridge, **1992**.
[13] T. J. Mason, *Practical Sonochemistry*, Ellis Horwood, Chichester, **1991**.
[14] T. J. Mason, J. P. Lorimer, *Sonochemistry. Theory, Applications and Uses of Ultrasound in Chemistry*, Ellis Horwood, Chichester, **1988**.
[15] B. E. Noltingk, E. A. Neppiras, *Proc. Phys. Soc. B (London)* **1950**, *63B*, 674–685.
[16] (a) K. S. Suslick, D. A. Hammerton, R. E. Cline, *J. Am. Chem. Soc.* **1986**, *108*, 5641–5642; (b) E. B. Flint, K. S. Suslick, *Science* **1991**, *253*, 1397–1399.
[17] M. A. Margulis, in Ref. [3], **1990**, Vol. 1, Ch. 2.
[18] T. Lepoint, F. Mullie, *Ultrasonics Sonochemistry* **1994**, *1*, S13–S22.
[19] D. J. Watmough, M. B. Shiran, K. M. Khan, A. P. Sarvazyan, E. P. Khizhnyak, T. N. Pashovkin, *Ultrasonics* **1992**, *30*, 325–331.
[20] Ya. I. Frenkel, *Russ. J. Phys. Chem.* **1940**, *14*, 305–308.
[21] R. M. G. Boucher, *Brit. Chem. Eng.* **1970**, *15*, 363–367.
[22] G. Cum, G. Galli, R. Gallo, A. Spadaro, *Ultrasonics* **1992**, *30*, 267–270.
[23] L. A. Crum, *Proc. 1982 Ultrasonics Symp. 1* **1982**, 1–11.
[24] W. Lauterborn, W. Hentschell, *Ultrasonics* **1985**, *23*, 260–268.
[25] W. J. Tomlinson, in Ref. [3], **1990**, Vol. 1, Ch. 5.

[26] K. Hauffe, *Oxidation of Metals*, Plenum, New York, **1965**.
[27] K. S. Root, J. Deutch, G. M. Whitesides, *J. Am. Chem. Soc.* **1981**, *103*, 5475–5479.
[28] (a) *Handbook of the Physicochemical Properties of the Elements* (ed.: G. V. Samsonov), Plenum, New York, **1968**; (b) C. S. G. Phillips, R. J. P. Williams, *Inorganic Chemistry*, Clarendon, Oxford, **1966**.
[29] J. C. de Souza Barboza, C. Petrier, J. L. Luche, *J. Org. Chem.* **1988**, *53*, 1212–1218.
[30] J. L. Luche, C. Petrier, C. Dupuy, *Tetrahedron Lett.* **1984**, *25*, 753–756.
[31] J. L. Luche, Ref. [3], **1993**, Vol. 3, Ch. 3.
[32] J. Lindley, T. J. Mason, J. P. Lorimer, *Ultrasonics* **1987**, *25*, 45–48.
[33] K. S. Suslick, D. J. Casadonte, *J. Am. Chem. Soc.* **1987**, *109*, 3459–3461.
[34] M. A. Margulis, *Ultrasonics* **1992**, *30*, 152–155.
[35] V. V. Boldyrev, *J. Chim. Phys.* **1986**, *83*, 821- 829.
[36] J. F. Garst, F. Ungvary, R. Batlaw, K. E. Lawrence, *J. Am. Chem. Soc.* **1991**, *113*, 5392- 5397.
[37] C. Hamdouchi, M. Topolski, V. Goedken, H. M. Walborski, *J. Org. Chem.* **1993**, *58*, 3148–3155.
[38] G. E. Grechnev, *Sov. J. Low Temp. Phys.* **1985**, *11*, 55–57.
[39] L. Zupirolli, *La Recherche* **1991**, *22*, 649–654.
[40] C. L. Hill, J. B. Vander Sande, G. M. Whitesides, *J. Org. Chem.* **1980**, *45*, 1020–1028.
[41] H. J. Spangenberg, *Sitzungsberichte der Akad. Wiss. DDR* **1981**, *3N*, 55–64; *Chem. Abstr.* **1982**, *96*, 110823k.
[42] C. V. Oprea, F. Weiner, *Angew. Makromol. Chem.* **1984**, *126*, 89–105.
[43] I. Uhara, S. Kishimoto, T. Hikino, Y. Kageyama, H. Hamada, Y. Numata, *J. Phys. Chem.* **1963**, *67*, 996–1001.
[44] R. D. Rieke et al., this book, Ch. 1.
[45] P. Boudjouk, D. P. Thompson, W. H. Ohrborn, B. H. Han, *Organometallics* **1986**, *5*, 1257–1260.
[46] J. L. Luche, C. Einhorn, J. Einhorn, J. V. Sinisterra-Gago, *Tetrahedron Lett.* **1990**, 31, 4125–4128.
[47] (a) T. Ando, S. Sumi, T. Kawate, J. Ishihara, T. Hanafusa, *J. Chem. Soc. Chem. Commun.* **1984**, 439–440; (b) T. Ando, T Kimura, in Ref. [3], **1991**, Vol. 2, Ch. 6.
[48] M. W. T. Pratt, R. Helsby, *Nature* **1959**, *184*, 1694–1695.
[49] M. A. Margulis, G. P. Los, A. A. Bashkatova, A. G. Beilin, I. I. Skorokhodov, O. I. Zinov'ev, *Russ. J. Phys. Chem.* **1991**, *65*, 1618–1621.
[50] P. Kruus, *Ultrasonics* **1988**, *26*, 216–217.
[51] C. M. R. Low, in Ref. [12], pp. 59–86.
[52] A. N. Maltsev, *Russ. J. Phys. Chem.* **1976**, *50*, 995–1001.
[53] C. Petrier, C. Dupuy, J. L. Luche, *Tetrahedron Lett.* **1986**, *27*, 3149–3152.
[54] C. Petrier, J. L. Luche, *Tetrahedron Lett.* **1987**, *28*, 2347–2350.
[55] P. Kruus, D. A. Robertson, L. A. McMillen, *Ultrasonics* **1991**, *29*, 370–375.
[56] P. Hegarty, J. Mann, *Synlett* **1993**, 553–554.
[57] E. R. Burkhardt, R. D. Rieke, *J. Org. Chem.* **1985**, *50*, 416–417.
[58] W. L. Parker, P. Boudjouk, A. B. Rajkumar, *J. Am. Chem. Soc.* **1991**, *113*, 2785–2786.
[59] D. Walther, C. Pfützenreuter, *Z. Chem.* **1989**, *29*, 146–147.
[60] S. K. Nayak, A. Banerji, *J. Org. Chem.* **1991**, *56*, 1940–1942.
[61] K. S. Suslick, R. E. Johnson, *J. Am. Chem. Soc.* **1984**, *106*, 6856–6858.
[62] Y. Nagata, Y. Watananabe, S. Fujita, T. Dohmaru, S. Taniguchi, *J. Chem. Soc. Chem. Commun.* **1992**, 1620–1622.
[63] (a) C. Roger, M. J. Tudoret, V. Guerchais, C. Lapinte, *J. Organomet. Chem.* **1989**, *365*, 347–350; (b) C. Roger, P. Marseille, C. Salus, J. R. Hamon, C. Lapinte, *J. Organomet. Chem.* **1987**, *336*, C13–C16.
[64] R. S. Bates, A. H. Wright, *J. Chem. Soc. Chem. Commun.* **1990**, 1129–1130.
[65] (a) K. S. Suslick, J. W. Goodale, P. F. Schubert, H. H. Wang, *J. Am. Chem. Soc.* **1983**, *105*, 5781–5784; (b) K. S. Suslick, S. B. Choe, A. A. Cichowlas, M. W. Grinstaff, *Nature* **1991**, *353*, 414–416.
[66] W. Slough, A. R. Ubbelohde, *J. Chem. Soc.* **1957**, 918–919.
[67] T. Azuma, S. Yanagida, H. Sakurai, S. Sasa, K. Yoshino, *Synth. Commun.* **1982**, *12*, 137–140.
[68] P. Boudjouk, R. Sooriyakumaran, B. H. Han, *J. Org. Chem.* **1986**, *51*, 2818–2819.
[69] K. Sugahara, T. Fujita, S. Watanabe, H. Hashimoto, *J. Chem. Technol. Biotechnol.* **1987**, *37*, 95–99; *Chem. Abstr.* **1988**, *108*, 21317x.
[70] (a) A. G. M. Barrett, I. A. O'Neil, *J. Org. Chem.* **1988**, *53*, 1815–1817; (b) A. G. M. Barrett, D. Dauzonne, I. A. O'Neil, A. Renaud, *J. Org. Chem.* **1984**, *49*, 4409–4415.
[71] (a) P. K. Freeman, L. H. Hutchinson, *Tetrahedron Lett.* **1976**, 1849–1852; (b) id., *J. Org. Chem.* **1980**, *45*, 1924–1930.

[72] (a) I. T. Badejo, R. Karaman, N. W. I. Lee, E. C. Lutz, M. T. Mamanta, J. L. Fry, *J. Chem. Soc. Chem. Commun.* **1989**, 566–567; (b) I. T. Badejo, R. Karaman, J. L. Fry, *J. Org. Chem.* **1989**, *54*, 4591–4596.
[73] R. Karaman, D. T. Kohlman, J. L. Fry, *Tetrahedron Lett.* **1990**, *31*, 6155–6158.
[74] R. Karaman, J. L. Fry, *Tetrahedron Lett.* **1989**, *30*, 4935–4938.
[75] (a) R. Karaman, J. L. Fry, *Tetrahedron Lett.* **1989**, *30*, 4931–4934; (b) id., *ibid.*, 6267–6270.
[76] R. Karaman, G. X. He, F. Chu, A. Blasko, T. C. Bruice, *J. Org. Chem.* **1993**, *58*, 438–443.
[77] P. Cintas, in Ref. [2], pp. 52–55 and references cited therein.
[78] L. E. Aleandri, B. Bogdanovic, this book, Ch. 8.
[79] H. Bönnemann, B. Bogdanovic, R. Brinkmann, D. W. He, B. Spliethoff, *Angew. Chem.* **1983**, *95*, 749–750; *Angew. Chem. Int. Ed. Eng.* **1983**, *22*, 728–729.
[80] W. Oppolzer, A. Nakao, *Tetrahedron Lett.* **1986**, *27*, 5471–5474.
[81] M. T. Reetz, W. F. Maier, *Liebigs Ann. Chem.* **1980**, 1471–1473.
[82] A. De Nicola, J. Einhorn, J. L. Luche, *J. Chem. Res. (S)* **1991**, 278; see also Ref. [12], pp. 172–174.
[83] A. De Nicola, C. Einhorn, J. Einhorn, J. L. Luche, *J. Chem. Soc. Chem. Commun.* **1994**, 879–880.
[84] J. W. Huffman in *Comprehensive Organic Synthesis, Vol. 8* (eds.: B. M. Trost, I. Fleming), Pergamon, New York, **1993**, pp. 107–127.
[85] A. P. Marchand, G. M. Reddy, *Synthesis* **1991**, 198–200.
[86] M. B. Le Hocine, D. Do Khac, M. Fétizon, F. Guir, Y Guo, T. Prangé, *Tetrahedron Lett.* **1992**, *33*, 1443–1446.
[87] C. Petrier, J. L. Luche, *Tetrahedron Lett.* **1987**, *28*, 2351–2352.
[88] C. Petrier, J. L. Luche, S. Lavaitte, C. Morat, *J. Org. Chem.* **1989**, *54*, 5313–5317.
[89] S. E. de Laszlo, S. V. Ley, R. A. Porter, *J. Chem. Soc. Chem. Commun.* **1986**, 344–346.
[90] H. S. P. Rao, *J. Chem. Educ.* **1988**, *65*, 931.
[91] P. Boudjouk, J. H. So, *Synth. Commun.* **1986**, *16*, 775–778.
[92] J. W. Huffman, W. P. Liao, R. H. Wallace, *Tetrahedron Lett.* **1987**, *28*, 3315–3318.
[93] R. Sato, T. Nagaoka, T. Goto, M. Saito, *Bull. Chem. Soc. Jpn.* **1990**, *63*, 290–292.
[94] F. A. Luzzio, L. C. O'Hara, *Synth. Commun.* **1990**, *20*, 3223–2234.
[95] R. Sato, T. Nagaoka, M. Saito, *Tetrahedron Lett.* **1990**, *31*, 4165–4168.
[96] J. H. So, M. K. Park, P. Boudjouk, *J. Org. Chem.* **1988**, *53*, 5871–5875.
[97] (a) A. Fadel, J. L. Canet, J. Salaun, *Synlett* **1990**, 89–90; (b) A. Fadel, *Synthesis* **1993**, 503–505.
[98] K. Burger, B. Helmreich, *Chem. Zeitung* **1991**, *115*, 253–255; *Chem. Abstr.* **1992**, *115*, 279878d.
[99] J. Einhorn, C. Einhorn, J. L. Luche, *Tetrahedron Lett.* **1988**, *29*, 2183–2184.
[100] S. Yamamura, S. Nishiyama in *Comprehensive Organic Synthesis, Vol. 8* (eds.: B. M. Trost, I. Fleming), Pergamon, New York, **1993**, pp. 307–313.
[101] W. Preston Reeves, J. A. Murry, D. W. Willoughby, W. J. Friedrich, *Synth. Commun.* **1988**, *18*, 1961–1966.
[102] J. A. R. Salvador, M. L. Sae Melo, A. S. Campos Neves, *Tetrahedron Lett.* **1993**, *34*, 361–362.
[103] J. A. R. Salvador, M. L. Sae Melo, A. S. Campos Neves, *Tetrahedron Lett.* **1993**, *34*, 357–360.
[104] S. L. Vorob'eva, N. N. Korotkova, *J. Chem. Res. (S)* **1993**, 34–35.
[105] O. De Lucchi, N. Piccolrovazzi, G. Modena, *Tetrahedron Lett.* **1986**, *27*, 4347–4350.
[106] O. G. Orazov, O. G. Safiev, D. Kurbanov, V. V. Zorin, Yu. K. Khekimov, D. L. Rakhmankulov, *Izv. Akad. Nauk Turkm. SSR, Ser. Fiz. Tekh. Khim. Goel. Nauk* **1989**, *4*, 108–109; *Chem. Abstr.* **1990**, *113*, 6242d.
[107] M. J. S. Miranda Moreno, M. L. Sae Melo, A. S. Campos Neves, *Tetrahedron Lett.* **1993**, *34*, 353–356.
[108] (a) T. S. Chou, M. L. You, *J. Org. Chem.* **1987**, *52*, 2224–2226; (b) id., *Tetrahedron Lett.* **1985**, *26*, 4495–4498.
[109] T. S. Chou, S. H. Hung, M. L. Peng, S. J. Lee, *J. Chin. Chem. Soc.* **1991**, *38*, 283–287.
[110] T. S. Chou, S. H. Hung, M. L. Peng, S. J. Lee, *Tetrahedron Lett.* **1991**, *32*, 3551–3554.
[111] T. S. Chou, S. Y. Chang, *J. Org. Chem.* **1992**, *57*, 5015–5017.
[112] T. Kataoka, K. Tsutsumi, K. Kano, K. Mori, M. Miyake, M. Yokota, H. Shimizu, M. Hori, *J. Chem. Soc. Perkin Trans. I* **1990**, 3017–3025.
[113] T. Kimmel, D. Becker, *J. Org. Chem.* **1984**, *49*, 2494–2496.
[114] S. V. Ley, Y. A. O'Neil, C. M. R. Low, *Tetrahedron* **1986**, *42*, 5363–5368.
[115] T. S. Chou, J. J. Huang, C. H. Tsao, *J. Chem Res. (S)* **1985**, 18–19.
[116] D. Lei, P. Gaspar, *Polyhedron* **1991**, *10*, 1221–1225.
[117] J. L. Luche, J. C. Damiano, *J. Am. Chem. Soc.* **1980**, *102*, 7927–7928.
[118] D. D. Sternbach, J. W. Hughes, D. F. Burdi, B. A. Banks, *J. Am. Chem. Soc.* **1985**, *107*, 2149–2153.
[119] G. Mehta, N. Krishnamurthi, *J. Chem. Soc. Chem. Commun.* **1986**, 1319–1321.

[120] G. Berube, A. G. Fallis, *Tetrahedron Lett.* **1989**, *30*, 4045–4048.
[121] J. D. Sprich, G. S. Lewandos, *Inorg. Chim. Acta* **1983**, *76*, L241–L242.
[122] A. Tuulmets, K. Heinoja, *Organic Reactivity* **1990**, *27*, 63–73.
[123] R. Yamaguchi, H. Kawasaki, M. Kawanisi, *Synth. Commun.* **1982**, *12*, 1027–1037.
[124] H. Hagiwara, H. Uda, *J. Chem. Soc. Chem. Commun.* **1988**, 815–817.
[125] T. Uyehara, J. Yamada, T. Furata, T. Kato, *Chem. Lett.*. **1986**, 609–612.
[126] (a) C. Petrier, J. C. de Souza Barboza, C. Dupuy, J. L Luche, *J. Org. Chem.* **1985**, *50*, 5761–5767; (b) J. C. De Souza Barboza, C. Petrier, J. L. Luche, *Tetrahedron Lett.* **1985**, *26*, 829–830.
[127] J. L. Luche, C. Petrier, J. P. Lansard, A. E. Greene, *J. Org. Chem.* **1983**, *48*, 3837–3839.
[128] E. J. Corey, J. O. Link, Y. Shao, *Tetrahedron Lett.* **1992**, *33*, 3435–3438.
[129] (a) H. C. Brown, U. S. Racherla, *Tetrahedron Lett.* **1985**, *26*, 4311–4314; (b) H. C. Brown, U. S. Racherla, *J.Org. Chem.* **1986**, *51*, 427–432.
[130] E. Erdik, *Tetrahedron* **1987**, *43*, 2203–2212.
[131] P. Knochel, R. D. Singer, *Chem. Rev.* **1993**, *93*, 2117–2188.
[132] K. Takagi, *Chem. Lett.* **1993**, 469–472.
[133] J. Yamashita, Y. Inoue, T. Kondo, H. Hashimoto, *Bull. Chem. Soc. Jpn.* **1985**, *58*, 2709–2710.
[134] (a) R. F. W. Jackson, N. Wishart, A Wood, K. James, M. J. Wythes, *J. Org. Chem.* **1992**, *57*, 3397 to 3404; (b) R. F. W. Jackson, K. James, M. J. Wythes, A. Wood, *J. Chem. Soc. Chem. Commun.* **1989**, 644–645.
[135] J. M. Altenburger, D. Schirlin, *Tetrahedron Lett.* **1991**, *32*, 7255–7258.
[136] E. Nietschmann, O. Böge, A. Tzschach, *J. Prakt. Chem.* **1991**, *333*, 281–284.
[137] A. V. Kuchin, R. A. Nureshev, G. A. Tolstikov, *Zh. Obshch. Khim.* **1983**, *53*, 2519–2527; *Chem. Abstr.* **1984**, *100*, 103426f.
[138] (a) K. Liou, P. Yang, Y. Lin, *J. Organomet. Chem.* **1985**, *294*, 145–149; (b) P. H. Yang, K. F. Liou, Y. T. Lin, *J. Organomet. Chem.* **1986**, *307*, 273–278.
[139] Y. T. Lin, *J. Organomet. Chem.* **1986**, *317*, 277–283.
[140] Y. Inoue, J. Yamashita, H. Hashimoto, *Synthesis* **1984**, 244.
[141] For typical experiments in sonoelectrochemistry, see Ref. [13], pp. 131–139 and references therein.
[142] (a) B. Gautheron, G. Tainturier, C. Degrand, *J. Am. Chem. Soc.* **1985**, *107*, 5579–5581; (b) G. Tainturier, B. Gautheron, C. Degrand, *Organometallics* **1986**, *5*, 942–946.
[143] J. Einhorn, J. L. Luche, *J. Org. Chem.* **1987**, *52*, 4124–4126.
[144] A. Banerji, S. K. Nayak, *Current Science,* **1989**, *58*, 249–252.
[145] M. Xi, B. E. Bent, *J. Am. Chem. Soc.* **1993**, *115*, 7426–7433.
[146] B. H. Han, P. Boudjouk, *Tetrahedron Lett.* **1981**, *22*, 2757–2758.
[147] (a) G. A. Price, A. A. Clifton, *Tetrahedron Lett.* **1991**, *32*, 7133–7134; (b) A. G. Osborne, K. J. Glass, M. L. Staley, *Tetrahedron Lett.* **1989**, *30*, 3567–3568; (c) A. G. Osborne, A. A. Clifton, *Monatsh. Chem.* **1991**, *122*, 529–532.
[148] E. V. Dehmlow, S. Büker, *Chem. Ber.* **1993**, *126*, 2759–2763.
[149] J. Lindley, J. P. Lorimer, T. J. Mason, *Ultrasonics* **1986**, *24*, 292–293.
[150] K. A. Nelson, H. G. Adolph, *Synth. Commun.* **1991**, *21*, 293–305.
[151] J. Yamashita, Y. Inoue, T. Kondo, H. Hashimoto, *Chem. Lett.* **1986**, 407–408.
[152] (a) T. Kitazume, N. Ishikawa, *Chem. Lett.* **1982**, 137–140; (b) id., *J. Am. Chem. Soc.* **1985**, *107*, 5186–5191.
[153] P. Boudjouk, in Ref. [12], pp. 110–122.
[154] O. I. Zinovev, A. M. Margulis, in Ref. [3], **1993**, Vol. 3, pp. 165–207.
[155] Yu. Goldberg, R. Sturkovich, E. Lukevics, *Synth. Commun.* **1993**, *23*, 1235–1238.
[156] R. Bowser, R. S. Davidson, in Ref. [12], pp. 50–58.
[157] P. Boudjouk, B. H. Han, *Tetrahedron Lett.* **1981**, *22*, 3813–3814.
[158] P. D. Lickiss, R. Lucas, *J. Organomet. Chem.* **1993**, *444*, 25–28.
[159] P. Boudjouk, U. Samaraweera, R. Sooriyakumaran, J. Chrusciel, K. R. Anderson, *Angew. Chem.* **1988**, *100*, 1406–1407; *Angew. Chem. Int. Ed. Engl.* **1988**, *27*, 1355–1356.
[160] P. Boudjouk, B. H. Han, K. R. Anderson, *J. Am. Chem. Soc.* **1982**, *104*, 4992–4993.
[161] P. Boudjouk, E. Black, R. Kumarathasan, *Organometallics* **1991**, *10*, 2095–2096.
[162] M. A. Margulis, G. P. Los, O. I. Zinovev, *Russ. J. Phys. Chem.* **1991**, *65*, 1614–1618.
[163] H. K. Kim, K. Matyjaszewski, *J. Am. Chem. Soc.* **1988**, *110*, 3321–3323.
[164] (a) G. J. Price, *J. Chem. Soc. Chem. Commun.* **1992**, 1209–1210; (b) id., in Ref. [12] pp. 87–109.
[165] P. A. Bianconi, T. W. Weidman, *J. Am. Chem. Soc.* **1988**, *110*, 2342–2344.
[166] Y. Naruta, Y. Nishigaichi, K. Maruyama, *Chem. Lett.* **1986**, 1857–1860.
[167] (a) P. Knochel, J. F. Normant, *Tetrahedron Lett.* **1984**, 25, 1475–1478; (b) id., *ibid.*, 4383–4386.
[168] T. Kitazume, N. Ishikawa, *Chem. Lett.* **1982**, 1453–1454.

[169] J. K. Crandall, T. A. Ayers, *Organometallics* **1992**, *11*, 473–477.
[170] D. P. Curran in *Comprehensive Organic Synthesis, Vol. 4* (eds.: B. M. Trost, I. Fleming, M. F. Semmelhack) Pergamon, Oxford, **1991**, Vol. 4, Ch. 4.2.
[171] O. Repic, S. Vogt, *Tetrahedron Lett.* **1982**, *23*, 2729–2732.
[172] O. Repic, in Ref. [13], pp. 118–121.
[173] E. C. Friedrich, J. M. Domek, R. Y. Pong, *J. Org. Chem.* **1985**, *50*, 4640–4642.
[174] N. S. Zefirov, K. A. Lukin, S. F. Poutanskii, M. A. Margulis, *Zh. Org. Khim.* **1987**, *23*, 1799–1800; *Chem Abstr.* **1988**, *108*, 221299d.
[175] D. L. J. Clive, S. Daigneault, *J. Org. Chem.* **1991**, *56*, 3801–3814.
[176] M. L. K. Jie, W. L. K. Lam, *J. Chem. Soc. Chem.Commun.* **1987**, 1460–1461.
[177] X. Xu, Z. Li, Y. Na, G. Liu, *Yingyong Huaxue* **1987**, *4*, 73–75; *Chem. Abstr.* **1988**, *109*, 54337c.
[178] G. Etemad-Moghadam, M. Rifqui, P. Layrolle, J. Berlan, M. Koenig, *Tetrahedron Lett.* **1991**, *32*, 5965–5968.
[179] L. Xu, F. Tao, T. Yu, *Tetrahedron Lett.* **1985**, *26*, 4231–4234.
[180] L. Xu, T. Yu, F. Tao, S. Wu, *Sci. Sin. Ser. B* **1988**, *31*, 897–908; *Chem. Abstr.* **1989**, *110*, 231026u.
[181] L. Xu, F. Tao, U. H. Brinker, *Acta Chem. Scand.* **1992**, *46*, 650–653.
[182] G. Mehta, H. S. P. Rao, *Synth. Commun.* **1985**, *15*, 991–1000.
[183] M. Fetizon, I. Hanna, *Synthesis* **1990**, 583–584.
[184] C. Blomberg, *The Barbier Reaction and Related One Step Processes*, Springer, Berlin, **1993**.
[185] P. J. Pearce, D. H. Richards, N. F. Scilly, *J. Chem. Soc. Perkin Trans. I* **1972**, 1655–1660.
[186] J. C. de Souza Barboza, J. L. Luche, C. Petrier, *Tetrahedron Lett.* **1987**, *28*, 2013–2016.
[187] A. Moyano, M. A. Pericas, A. Riera, J. L. Luche, *Tetrahedron Lett.* **1990**, *31*, 7619–7622.
[188] R. Cloux, G. Defayes, K. Foti, J. C. Dutoit, E. Kovats, *Synthesis* **1993**, 909–913.
[189] (a) I. C. Burkow, R. K. Sydnes, D. C. N. Ubeda, *Acta Chem. Scand.* **1987**, *B41*, 235–244; (b) S. B. Singh, G. R. Pettit, *Synth. Commun.* **1987**, *17*, 877–892.
[190] (a) T. Uyehara, J. Yamada, K. Ogata, T. Kato, *Bull. Chem. Soc. Jpn.* **1985**, *58*, 211–216; (b) M. Ihara, M. Katogi, K. Fukumoto, T. Kametani, *J. Chem. Soc. Chem. Commun.* **1987**, 721–722.
[191] S. Araki, Y. Butsugan, *Chem. Lett.* **1988**, 457–458.
[192] G. Rong, R. Keese, *Tetrahedron Lett.* **1990**, *31*, 5617–5618.
[193] C. Einhorn, C. Allavena, J. L. Luche, *J. Chem. Soc. Chem. Commun.* **1988**, 333–334.
[194] B. M. Trost, B. P. Coppola, *J. Am. Chem. Soc.* **1982**, *104*, 6879–6881.
[195] R. L. Snowden, P. Sonnay, *J. Org. Chem.* **1984**, *49*, 1464–1465.
[196] W. Zhang, P. Dowd, *Tetrahedron Lett.* **1993**, *34*, 2095–2098.
[197] G. Hugel, D. Cartier, J. Levy, *Tetrahedron Lett.* **1989**, *30*, 4513–4516.
[198] (a) T. Kitazume, *Ultrasonics* **1990**, *28*, 322–325; (b) T. Kitazume, N. Ishikawa, *Chem. Lett.* **1981**, 1679–1680.
[199] H. H. Tso, T. S. Chou, S. C. Hung, *J. Chem. Soc. Chem. Commun.* **1987**, 1552–1553.
[200] (a) B. H. Han, P. Boudjouk, *J. Org. Chem.* **1982**, *47*, 751–752; (b) id., *ibid.* 1453–1454.
[201] S. Chew, R. J. Ferrier, *J. Chem. Soc. Chem. Commun.* **1984**, 911–912.
[202] Y. Hanazawa, J. Uda, Y. Kobayashi, Y. Ishido, T. Taguchi, M. Shiro, *Chem. Pharm. Bull.* **1991**, *39*, 2459–2461.
[203] A. Solladie-Cavallo, D. Farkhani, S. Fritz, T. Lazrak, J. Suffert, *Tetrahedron Lett.* **1984**, *25*, 4117–4120.
[204] C. J. Li, *Chem. Rev.* **1993**, *93*, 2023–2035.
[205] (a) C. Petrier, J. L. Luche, *J. Org. Chem.* **1985**, *50*, 910–912; (b) C. Einhorn, J. L. Luche, *J. Organomet. Chem.* **1987**, *322*, 177–183.
[206] C. Petrier, J. Einhorn, J. L. Luche, *Tetrahedron Lett.* **1985**, *26*, 1449–1452.
[207] E. Kim, D. M. Gordon, W. Schmid, G. M. Whitesides, *J. Org. Chem.* **1993**, *58*, 5500–5507.
[208] (a) J. L. Luche, C. Allavena, *Tetrahedron Lett.* **1988**, *29*, 5369–5372; (b) J. L. Luche, C. Allavena, C. Petrier, C. Dupuy, *Tetrahedron Lett.* **1988**, *29*, 5373–5374.
[209] B. Giese, W. Damm, M. Roth, M. Zehnder, *Synlett* **1992**, 441–443.
[210] C. Dupuy, C. Petrier, L. Sarandeses, J. L Luche, *Synth. Commun.* **1991**, *21*, 643–651.
[211] (a) J. Pérez-Sestelo, J. L. Mascarenas, L. Castedo, A. Mourino, *J. Org. Chem.* **1993**, *58*, 118–123; (b) J. Péres-Sestelo, J. L. Mascarenas, L. Castedo, A. Mourino, *Tetrahedron Lett.* **1994**, *35*, 275–278.
[212] M. Ohno, K. Ishizaki, S. Eguchi, *J. Org. Chem.* **1988**, *53*, 1285–1288.
[213] B. Figadere, J. C. Harmange, L. X. Hai, A. Cavé, *Tetrahedron Lett.* **1992**, *33*, 5189–5192.
[214] L. A. Sarandeses, A. Mourino, J. L. Luche, *J. Chem. Soc. Chem. Commun.* **1991**, 818–820.
[215] L. A. Sarandeses, A. Mourino, J. L. Luche, *J. Chem. Soc. Chem. Commun.* **1992**, 798–799.
[216] L. A. Sarandeses, J. L. Luche, *J. Org. Chem.* **1992**, *57*, 2757–2760.
[217] C. Petrier, A. L. Gemal, J. L. Luche, *Tetrahedron Lett.* **1982**, *23*, 3361–3364.
[218] J. Einhorn, J. L. Luche, *Tetrahedron Lett.* **1986**, *27*, 1791–1792.

[219] J. Einhorn, J. L. Luche, *Tetrahedron Lett.* **1986**, *27*, 501–504.
[220] (a) A. J. Fry, G. S. Ginsburg, *J. Am. Chem. Soc* **1979**, *101*, 3927–3932; (b) A. J. Fry, W. A. Donaldson, G. S. Ginsburg, *J. Org. Chem.* **1979**, *44*, 349–352.
[221] N. N. Joshi, H. M. R. Hoffmann, *Tetrahedron Lett.* **1986**, *27*, 687–690.
[222] H. M. R. Hoffmann, U. Karama, *Chem. Ber.* **1992**, *125*, 2803–2807.
[223] (a) A. Fürstner, *Synthesis*, **1989**, 571–590; (b) M. W. Rathke, P. Weipert in *Comprehensive Organic Synthesis, Vol. 2* (eds.: B. M. Trost, I. Fleming, C. H. Heathcock), Pergamon, Oxford, 1991, Ch. 1.8.
[224] B. H. Han, P. Boudjouk, *J. Org. Chem.* **1982**, *47*, 5030–5032.
[225] K. S. Suslick, S. J. Doktycz, *J. Am. Chem. Soc.* **1989**, *111*, 2342–2344.
[226] C. L. Lim, B. H. Han, *J. Korean Chem. Soc.* **1991**, *35*, 762–764; *Chem. Abstr.* **1992**, *116*, 128122x.
[227] C. Kashima, X. C. Huang, Y. Harada, A. Hosomi, *J. Org. Chem.* **1993**, *58*, 793–794.
[228] A. K. Bose, K. Gupta, M. S. Manhas, *J. Chem. Soc. Chem. Commun.* **1984**, 86–87.
[229] N. Oguni, T. Tomago, N. Nagata, *Chem. Express* **1986**, *1*, 495–497.
[230] T. Kitazume, *Synthesis* **1986**, 855–857.
[231] K. S. Kim, L. Qian, *Tetrahedron Lett.* **1993**, *34*, 7195–7196.
[232] R. L. Beard, A. I. Meyers, *J. Org. Chem.*, **1991**, *56*, 2091–2096.
[233] G. Saracco, F. Arzano, *Chimica e Industria (Milano)* **1968**, *50*, 314–318.
[234] A. Tai, T. Kikukawa, T. Sugimura, Y. Inoue, T. Ozawa, S. Fujii, *J. Chem. Soc. Chem. Commun.* **1991**, 795–796.
[235] (a) A. Alexakis, N. Lensen, P. Mangeney, *Synlett* **1991**, 625–626; (b) D. Enders, M. Klatt, R. Funk, *Synlett* **1993**, 226–228.
[236] T. Kitazume, T. Ohnogi, H. Miyaushi, T. Yamazaki, S. Watanabe, *J. Org. Chem.* **1989**, *54*, 5630–5632.
[237] P. Boudjouk, B. H. Han, *J. Catalysis* **1983**, *79*, 489–492.
[238] P. Boudjouk, B. H. Han, J. R. Jacobsen, B. J. Hauck, *J. Chem. Soc. Chem. Commun.* **1991**, 1424–1425.
[239] (a) D. Jang, D. Shin, B. H. Han, *J. Korean Chem. Soc.* **1991**, *35*, 745–749; *Chem. Abstr.* **1992**, *116*, 129015b; (b) B. H. Han, P. Boudjouk, *Organometallics* **1983**, *2*, 769–771.
[240] B. H. Han, D. H. Shin, S. Y. Cho, *Bull. Korean Chem. Soc.* **1985**, *6*, 320–331; *Chem. Abstr.* **1986**, *104*, 129304x.
[241] J. C. de Souza-Barboza, Ph. D. Thesis, University of Grenoble (France), **1988**.
[242] M. J. Aurell, J. L. Luche, to be published.
[243] M. J. Luche, unpublished results.
[244] M. Vinatoru, personal communication.

5 Preparation and Applications of Functionalized Organozinc Reagents

Paul Knochel

5.1 Preparation of Functionalized Organozinc Reagents

5.1.1 Introduction

The insertion of zinc metal into alkyl iodides was the first method available for preparing organozinc compounds. As early as 1849, Frankland obtained diethylzinc by the reaction of ethyl iodide with zinc [1]. A range of diorganozincs and alkyl zinc halides were subsequently prepared by this method [2]. However, the discovery by Grignard of an efficient and general preparation of organomagnesium halides which have a higher reactivity towards most organic electrophiles led to the virtual abandonment of organozinc reagents in organic chemistry. Only a few classes of organozinc species such as zinc enolates (Reformatsky reaction) [3, 4] or zinc carbenoids [5–9] survived this loss of interest. The low reactivity of zinc organometallics, which was perceived by organic chemists of the first half of this century as a disadvantage, turned out to be a formidable advantage in modern organic chemistry where selectivity is a concept more important then reactivity. It was soon shown that a range of polyfunctional organozinc reagents can be prepared and that they react with electrophiles in the presence of the appropriate catalyst giving excellent yields [10]. In fact, a unique set of polyfunctional reagents, containing most functional groups met in organic chemistry, could be prepared and used for carbon-carbon bond forming reactions. The purpose of this chapter is to present the various preparative methods available for the synthesis of polyfunctional organozinc reagents and to define their scope and their limitations. Applications of copper- and palladium-catalyzed reactions in organic synthesis will then be described. Finally, the use of polyfunctional diorganozincs in asymmetric synthesis using a chiral titanium catalyst will be covered.

5.1.2 The Direct Insertion of Zinc Metal into Organic Substrates

The oxidative addition of organic halides to zinc constitutes a very general method for preparation of organozinc halides. This reaction was performed under a wide range of reaction conditions [2, 10]; however, the use of THF as solvent led to the most convenient and most general preparation. This solvent was first introduced by Gaudemar [11] and allowed preparation under mild conditions (0 to 40 °C) of allylic, benzylic, and alkyl zinc halides starting from allylic and benzylic bromides and primary or secondary alkyl iodides. The preparation of allylzinc bromide was especially interesting (Scheme 5-1), since this constitutes a very convenient allylation reagent [12]. Although Gaudemar did not use functionalized organic substrates, all the experimental conditions for an extension to such substrates had been worked out as early as 1962. In 1988 it was finally shown that Gau-

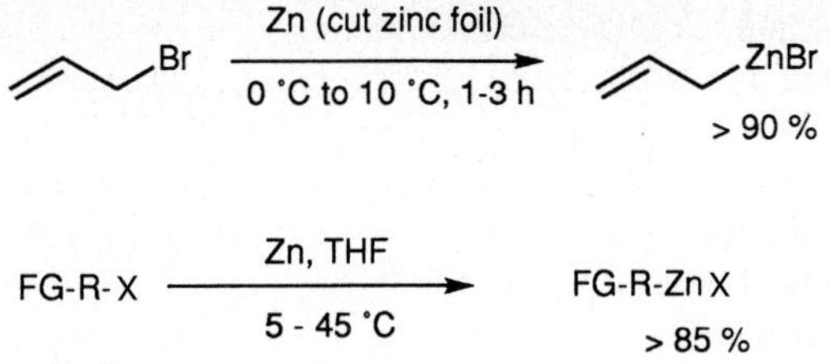

Zn (cut zinc foil)

Br → ZnBr

0 °C to 10 °C, 1-3 h

> 90 %

FG-R-X → FG-R-ZnX (Zn, THF; 5 - 45 °C) > 85 %

X = Br, I;
FG = CO_2R, enoate, CN, enone, halide, $(RCO)_2N$, $(TMS)_2N$, RNH, NH_2, RCONH, $(RO)_3Si$, $(RO)_2P(O)$, RS, RS(O), RSO_2, PhCOS;
R = alkyl, aryl, benzyl, allyl

Scheme 5-1

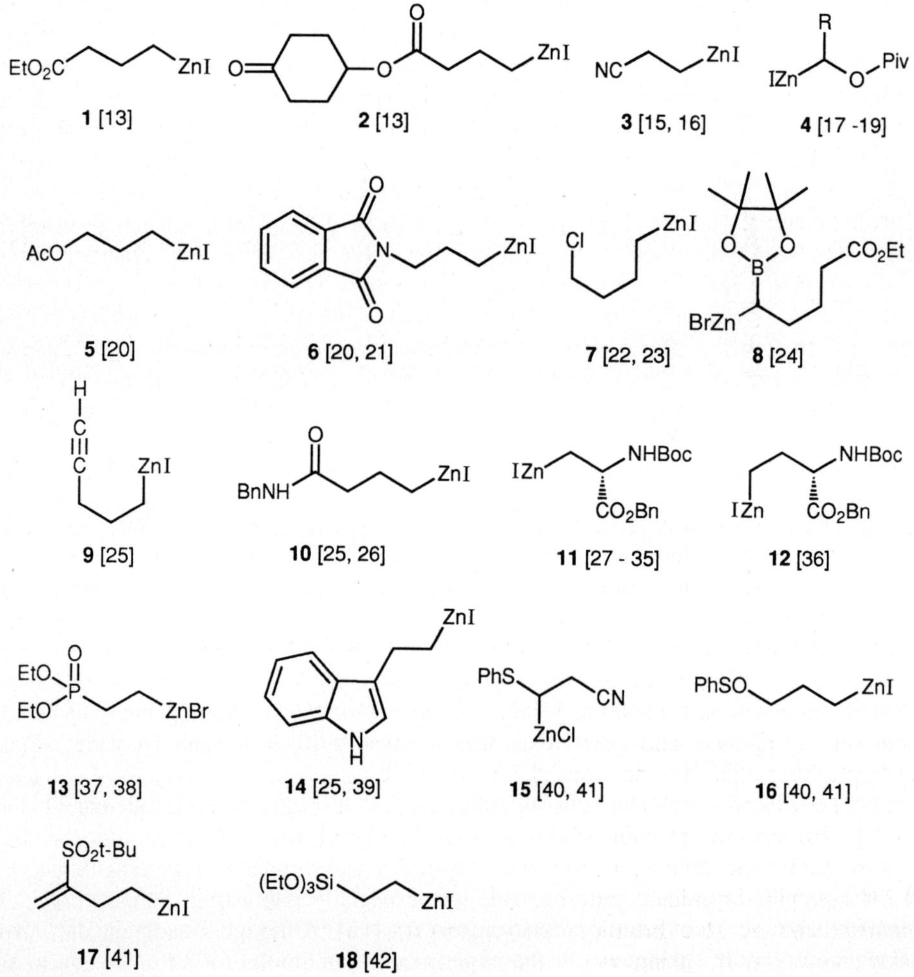

EtO_2C ZnI — **1** [13]

O, O, ZnI — **2** [13]

NC ZnI — **3** [15, 16]

R, IZn, O, Piv — **4** [17 -19]

AcO ZnI — **5** [20]

O, N, ZnI, O — **6** [20, 21]

Cl ZnI — **7** [22, 23]

O, B, O, CO_2Et, BrZn — **8** [24]

H, C, C, ZnI — **9** [25]

O, BnNH, ZnI — **10** [25, 26]

IZn, NHBoc, CO_2Bn — **11** [27 - 35]

NHBoc, IZn, CO_2Bn — **12** [36]

O, EtO, P, EtO, ZnBr — **13** [37, 38]

ZnI, N, H — **14** [25, 39]

PhS, CN, ZnCl — **15** [40, 41]

PhSO ZnI — **16** [40, 41]

SO_2t-Bu, ZnI — **17** [41]

$(EtO)_3Si$ ZnI — **18** [42]

Scheme 5-2

demar's method could be applied effectively to functionalized organic halides [13]. Instead of using cut zinc foil, it was found advantageous to employ zinc dust (325 mesh) which had been successively activated with 1,2-dibromomethane (3–5 mol %) and chlorotrimethylsilane (1 mol %) [13, 14]. Under these conditions a broad variety of alkyl zinc halides containing almost all functional groups encountered in organic chemistry can be prepared; the organic zinc halides **1–18** illustrate the broad scope of the method (Scheme 5-2). Certain experimental conditions such as optimal concentration and temperature must be observed in order to obtain high yields. The reactions must be performed under strictly anhydrous conditions and under an inert atmosphere. The alkyl iodide has to be added as a 2–3 M THF solution in order to obtain fast reactions and high conversions. Whereas secondary iodides react usually at 25 °C, the formation of zinc reagents from primary halides often requires reaction temperatures between 40 and 50 °C. Interestingly, if the functional group present in the organic halide has electron-acceptor character (e.g. phosphonate, **13**) [37, 38], or is a boronic ester (e.g. **8**) [24] or a cyanide (e.g. **3**) [15, 16], faster insertion reactions are observed and the use of the corresponding alkyl *bromide* (instead of the alkyl iodide) is possible. The method can be extended to the preparation of alkenyl zinc iodides. In these cases, since the C(sp_2)–I bond is stronger, the insertion reaction is more sluggish and succeeds only if an electron-withdrawing substituent is attached to the unsaturated system of the molecule. Thus, the direct insertion of zinc to pure (*E*)-1-iodo-1-octene proceeds only at 70 °C in DMF and affords an *E*/*Z* mixture of octenylzinc iodides [43]. On the other hand, the insertion of zinc dust into 3-iodo-2-cyclohexen-1-one (**19**) in THF is complete within 1 h at 25–50 °C, affording the corresponding zinc reagent **20** in over 85% yield (Scheme 5-3) [44, 45]. Similarly, the alkenyl zinc reagents **21–23** bearing respectively an ester, sulfonyl, and keto group have been obtained [44, 45]. The insertion of zinc dust into aromatic iodides is generally not possible in THF, and the use of more polar solvents such as *N*,*N*-dimethylacetamide (DMAC) is usually required [43]. A further improvement can be made by using a reactive zinc powder deposited on graphite, and treatment of a THF slurry of zinc chloride containing 10 mol % of silver acetate with C_8K [46] produces a suspension of activated zinc on graphite which reacts rapidly with aromatic and heteroaromatic iodides in THF (Scheme 5-4) [47]. Even more reactive is the highly activated zinc obtained on reduction of zinc chloride by lithium naphthalenide (Rieke zinc) [48–53]. This highly reactive zinc powder converts even aromatic *bromides* into the corresponding aryl zinc halides (Scheme 5-5) [48]. An alternative method for preparing activated zinc consists of using sodium metal dispersed on an inert support such as titanium oxide (TiO_2) as reducing agent. This cheap reducing agent converts zinc chloride into a zinc powder deposited on TiO_2 which displays good reactivity. Although not as active as Rieke zinc, it allows the preparation of secondary zinc bromides starting from the corresponding secondary alkyl bromides under mild conditions (Scheme 5-6) [54]. The preparation of benzylic zinc halides can be performed very efficiently and proceeds with significantly less formation of Wurtz coupling byproducts than in the preparation of the corresponding magnesium or lithium benzylic species [55–57]. Thus, slow addition of a benzylic bromide or chloride such as **24** to zinc powder provides the corresponding benzylic zinc organometallic **25** in over 90% yield (Scheme 5-7) [56]. Typically, the reaction proceeds within 2–3 h at 5 °C for benzylic bromides, but requires 3 h at 65 °C for benzylic chlorides [55, 56]. *Secondary* benzylic zinc bromides insert zinc dust with significant amounts of Wurtz coupling; however, by using zinc powder deposited on titanium oxide [54], almost no Wurtz coupling is observed with these substrates (Scheme 5-8). Whereas allylic bro-

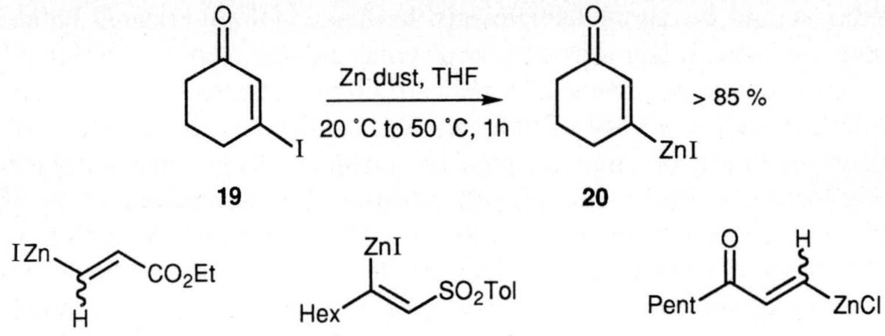

Zn dust, THF

20 °C to 50 °C, 1h

> 85 %

ZnI

19 **20**

IZn CO_2Et H

ZnI Hex SO_2Tol

O H Pent ZnCl

21 **22** **23**

Scheme 5-3

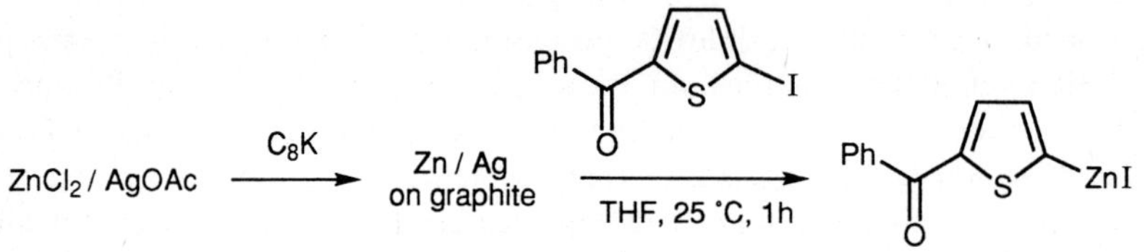

$ZnCl_2$ / AgOAc $\xrightarrow{C_8K}$ Zn / Ag on graphite $\xrightarrow{\text{THF, 25 °C, 1h}}$

Ph O S I

Ph O S ZnI

Scheme 5-4

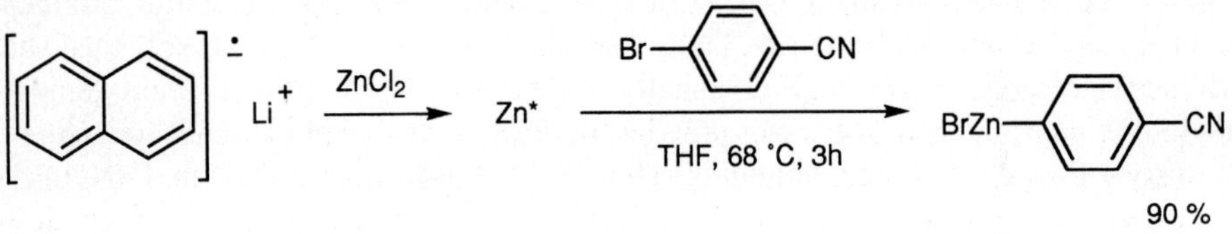

Li^+ $\xrightarrow{ZnCl_2}$ Zn* $\xrightarrow{\text{THF, 68 °C, 3h}}$ BrZn–C₆H₄–CN

Br–C₆H₄–CN

90 %

Scheme 5-5

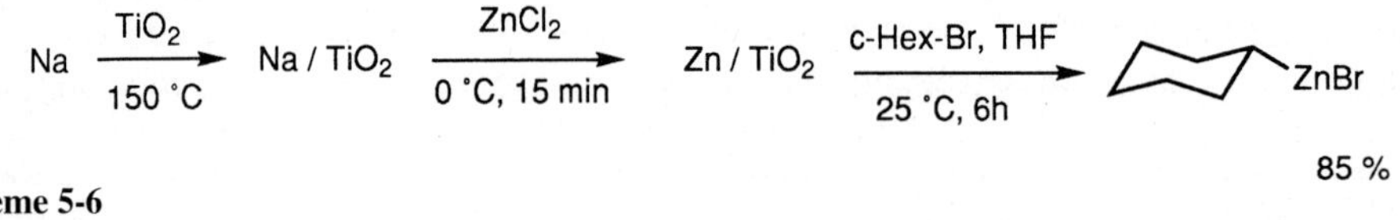

Na $\xrightarrow[150\ °C]{TiO_2}$ Na / TiO_2 $\xrightarrow[0\ °C,\ 15\ min]{ZnCl_2}$ Zn / TiO_2 $\xrightarrow[25\ °C,\ 6h]{\text{c-Hex-Br, THF}}$ ZnBr

85 %

Scheme 5-6

OAc Cl OAc

Zn, THF

65 °C, 3h

OAc ZnCl OAc

> 80 %

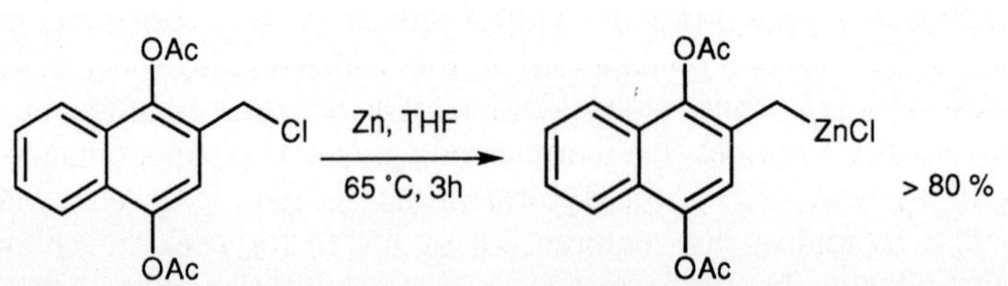

24 **25**

Scheme 5-7

mides bearing electron-donating substituents have a high tendency to give homocoupling by-products, ethyl (2-bromomethyl)acrylate (**26**) can be converted to the corresponding functionalized allylic zinc bromide **27** in excellent yield if granular zinc (30 mesh) is used (Scheme 5-9) [57–60]. Besides THF, other solvents such as ethyl acetate [2, 61] or mixtures of benzene with DMAC or hexamethylphosphoramide (HMPA) [62–72] have been used (Scheme 5-10). The preparation of polyfluorinated zinc iodides has proved to be especially efficient in dioxane (Scheme 5-11) [73–81] or in DMF [82–84] or other polar solvents [85] (Scheme 5-12). The above examples demonstrate that the activation of the zinc metal is very important [14]. Reproducible results are easily obtained by using zinc dust activated by chlorotrimethylsilane and dibromoethane [13]. The nature of the leaving group is also crucial and under standard conditions only alkyl iodides and in some cases alkyl bromides can be used. However, by performing the reactions in a polar solvent such as dimethylpropyleneurea (DMPU) in the presence of a catalytic amount of lithium iodide and a stoichiometric amount of lithium bromide, a smooth zinc insertion is observed and chloro-substituted mesylates can be selectively converted to the chloro-substituted zinc reagents (Scheme 5-13) [86]. Interestingly, whereas the benzylic bromide **28** (X = Br, Scheme 5-14) gives solely the Wurtz-coupling product, the corresponding phosphate **29** ($X = OP(O)(OEt)_2$) leads quantitatively to the benzylic reagent **30** [86]. The method can be extended to the preparation of allylic zinc species, and geranyl phosphate (**31**, Scheme 5-15) undergoes insertion by zinc powder in DMPU at 30 °C without appreciable dimerization. The resulting allylic zinc compound can be allylated to give **32** in good yields without the addition of any copper or palladium catalyst [86]. The insertion of zinc into organic halides with ultrasonic activation [87] constitutes an excellent preparation of aromatic organozinc halides, and has found applications in natural products syntheses (Scheme 5-16) [86–92]. Finally, electrochemical activation [93] may be a valuable approach to polyfunctional aromatic zinc derivatives starting from cheap and readily available aromatic chlorides [94–97]. Thus, the electrolysis of *p*-trifluoromethylchlorobenzene (**33**, Scheme 5-17) in the presence of a catalytic amount of a nickel–bipyridine complex and a sacrificial zinc anode provides the corresponding aryl-zinc reagent **34**, which can be coupled with aryl bromides or iodides in the presence of catalytic amounts of palladium(0) complexes [98], affording polyfunctional biphenyl [95]. Very recently, the dramatic effect of a catalytic amount of lead on the reaction of zinc metal with diiodomethane was studied [99]. A novel catalytic effect of lead on the reduction of (iodomethyl)zinc iodide with zinc metal leading to a geminal dizinc compound was found [100].

5.1.3 The Halide–Zinc Exchange Reaction

The halide–zinc exchange reaction is complementary to the direct insertion of zinc dust into an organic halide. The first example reported in the literature was the preparation of bis-(iodomethyl)zinc using the reaction of diiodomethane with diethylzinc [9, 101, 102]. This reaction is very fast due to the presence of the second iodine atom in diiodomethane, which facilitates the exchange reaction (Scheme 5-18) [101, 103]. In its absence only a sluggish reaction is observed in a solvent such as THF. However, by performing the reaction in the absence of a solvent at 50–55 °C, a smooth iodine–zinc exchange is observed (Scheme 5-19) [104]. It was soon found that the presence of a catalytic amount of a copper(I) salt such as copper iodide or copper cyanide enhances the reaction rate and makes

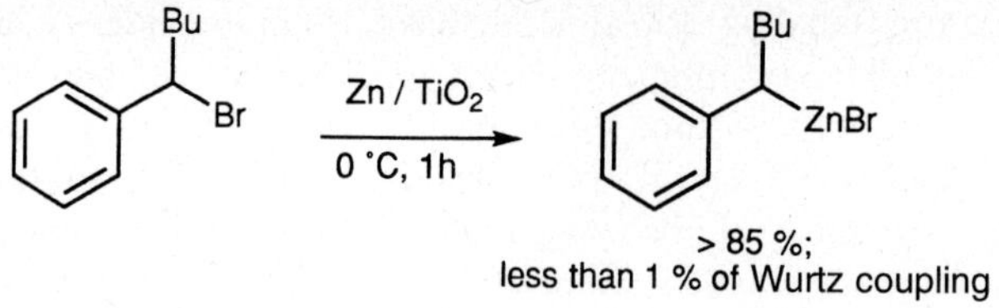

Bu Br → Bu ZnBr

Zn / TiO_2, 0 °C, 1h

> 85 %; less than 1 % of Wurtz coupling

Scheme 5-8

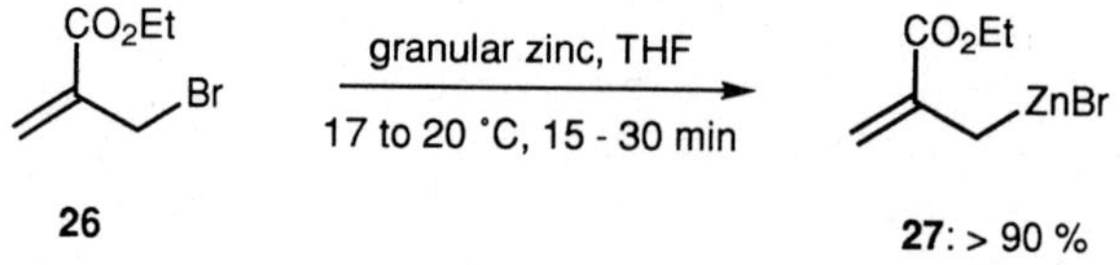

CO_2Et Br → CO_2Et ZnBr

granular zinc, THF, 17 to 20 °C, 15 - 30 min

26 → 27: > 90 %

Scheme 5-9

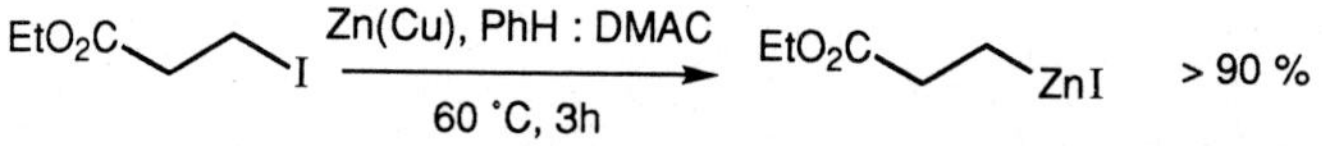

EtO_2C I → EtO_2C ZnI > 90 %

Zn(Cu), PhH : DMAC, 60 °C, 3h

Scheme 5-10

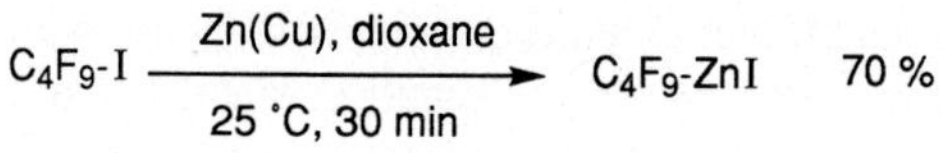

C_4F_9-I → C_4F_9-ZnI 70 %

Zn(Cu), dioxane, 25 °C, 30 min

Scheme 5-11

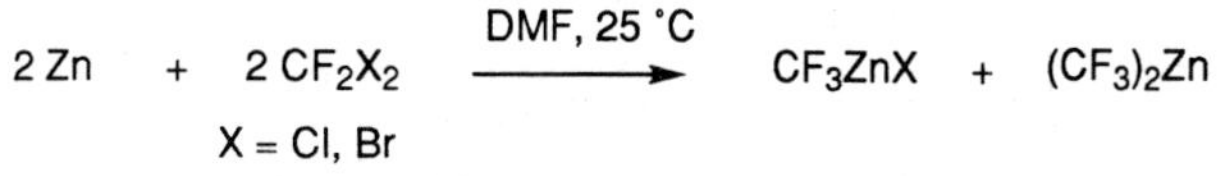

2 Zn + 2 CF_2X_2 (X = Cl, Br) → CF_3ZnX + $(CF_3)_2Zn$

DMF, 25 °C

Scheme 5-12

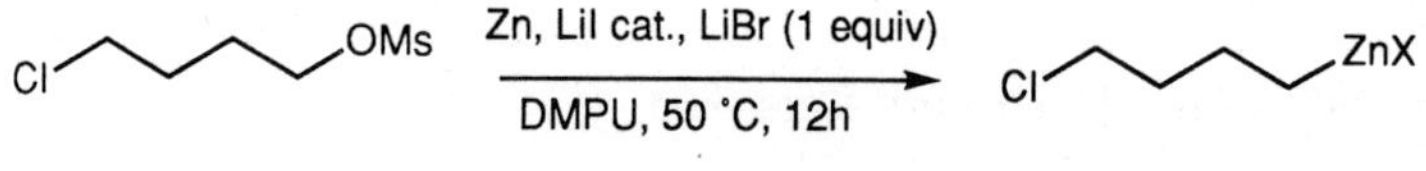

Cl OMs → Cl ZnX

Zn, LiI cat., LiBr (1 equiv), DMPU, 50 °C, 12h

Scheme 5-13

X → ZnX

Zn, LiI cat., DMPU

28 : X = Br
29 : X = OP(O)(OEt)$_2$

30 : 100 %

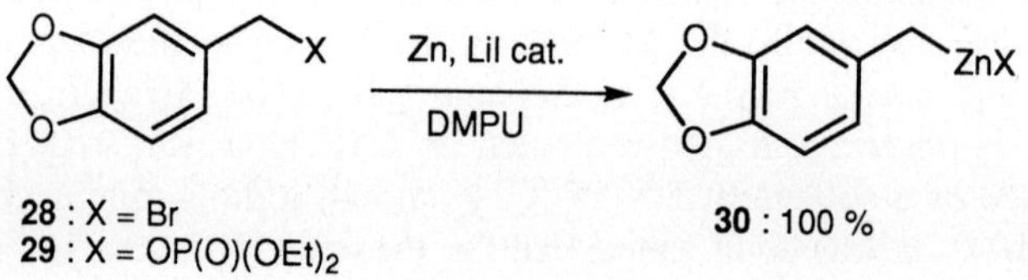

Scheme 5-14

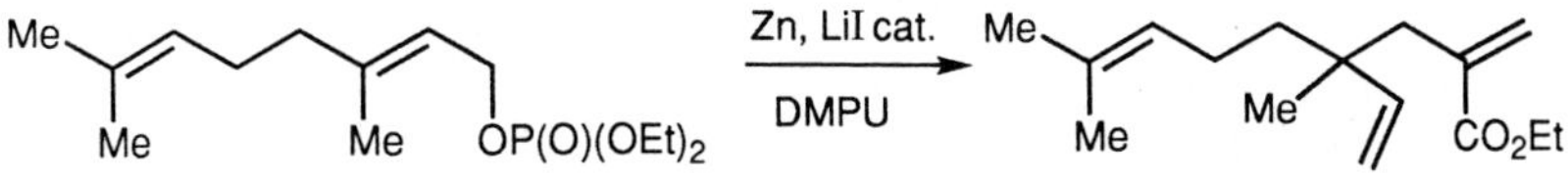

31 → (Zn, LiI cat., DMPU) → 32 : 69 %

Scheme 5-15

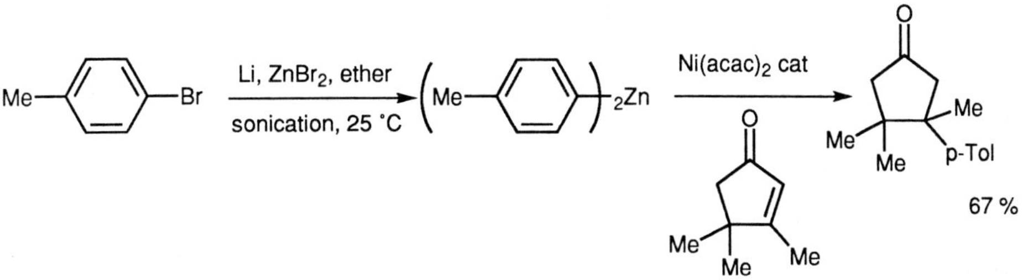

$Me-C_6H_4-Br$ → (Li, $ZnBr_2$, ether; sonication, 25 °C) → $(Me-C_6H_4)_2Zn$ → ($Ni(acac)_2$ cat) → 67 %

Scheme 5-16

33 → (electrolysis (0.15 A); DMF, $ZnBr_2$, Bu_4NBr cat. $Ni(BF_4)_2bipy_3$ cat) → 34 → (Pd(0) cat) → $MeO-C_6H_4-C_6H_4-CF_3$ 83 %

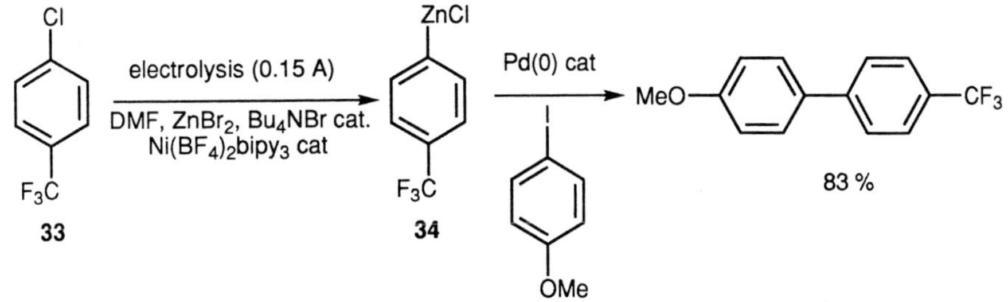

Scheme 5-17

$$Et_2Zn + 2\ ICH_2I \xrightarrow[-20\ ^\circ C]{THF} 2\ EtI + (ICH_2)_2Zn$$

Scheme 5-18

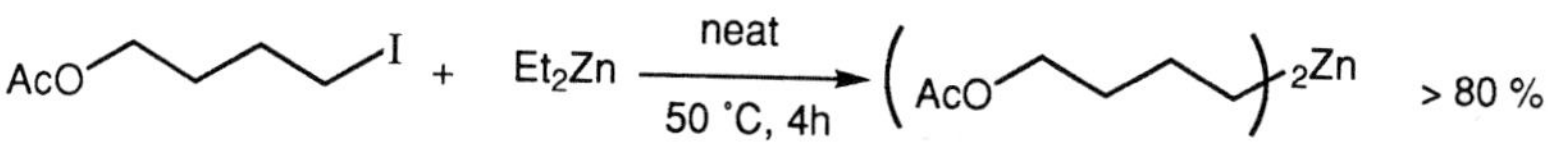

$$AcO(CH_2)_5I + Et_2Zn \xrightarrow[50\ ^\circ C,\ 4h]{neat} (AcO(CH_2)_5)_2Zn \quad > 80\ \%$$

Scheme 5-19

$$Et_2Zn + CuI \longrightarrow EtCu + EtZnI$$
$$EtCu \longrightarrow Et\bullet + Cu(0)$$
$$Et\bullet + R\text{-}I \longrightarrow Et\text{-}I + R\bullet$$
$$R\bullet + Et_2Zn \longrightarrow R\text{-}ZnEt_2\bullet$$
$$R\text{-}ZnEt_2\bullet \longrightarrow R\text{-}ZnEt + Et\bullet$$

Scheme 5-20

it possible to scale up the reaction under excellent conditions [104, 105]. The iodine–zinc exchange reaction constitutes an excellent preparation of functionalized dialkylzincs, and has found many important applications in catalytic asymmetric synthesis [105]. The copper(I) catalysis has been explained as due to a radical chain mechanism (Scheme 5-20) [105]. Other transition metals such as palladium [106, 107], nickel [108], manganese [109], or iron [109] also catalyze the halogen–zinc exchange; however, in all these cases the product is not a dialkylzinc (R_2Zn) but rather an alkyl-zinc halide (RZnX), as shown by gravimetric titrations [106]. Thus, the reaction of octyl iodide with diethylzinc in the presence of CuI (0.2 mol %) produces Oct_2Zn in over 80% yield, whereas performing the reaction in THF in the presence of a palladium(II) salt instead of copper iodide produces, under very mild conditions (25 °C, 2 h), octylzinc iodide in 80% yield (Scheme 5-21) [106]. A tentative mechanism has been proposed for this catalysis. The reduction of the palladium(II) precatalyst with diethylzinc generates the active palladium(0) catalyst, which then undergoes an oxidative addition with octyl iodide giving an (iodo)(octyl)palladium(II) complex (**35**), which exchanges its iodide ligand with diethylzinc providing the mixed octyl(ethyl)palladium(II), **36**. This intermediate does not undergo significant cross-coupling which would lead to decane, but rather exchanges its organic ligands again with diethylzinc producing the mixed dialkyl-zinc **37**, which through a further exchange with ethylzinc iodide gives octylzinc iodide. The newly formed diethylpalladium(II) complex (**38**) decomposes to give ethane and ethylene [106, 107], generating the palladium(0) catalyst (Scheme 5-21). The insertion of the palladium(0) catalyst into to octyl iodide seems to occur via a free radical mechanism, since the palladium (or nickel) catalysis is more efficient with alkyl iodides as substrates than with alkyl sulfonates. The postulated free radical nature of this insertion has been exploited for synthetic purposes. Thus, if 6-iodo-1-hexene (**39**, Scheme 5-22) is treated with diethylzinc in the presence of a palladium(II) or nickel(II) catalyst, an efficient radical ring closure takes place leading to the cyclized organozinc halide **40** which can be further trapped with an electrophile [106]. By using a

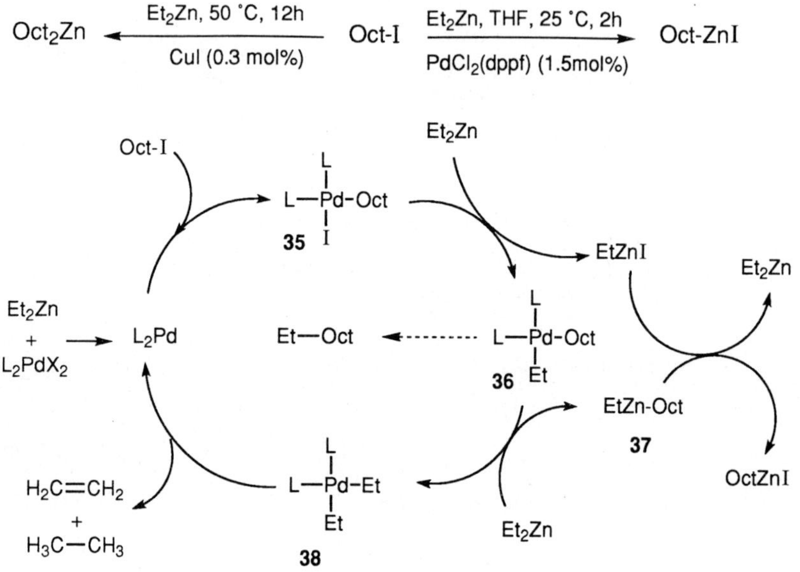

Scheme 5-21

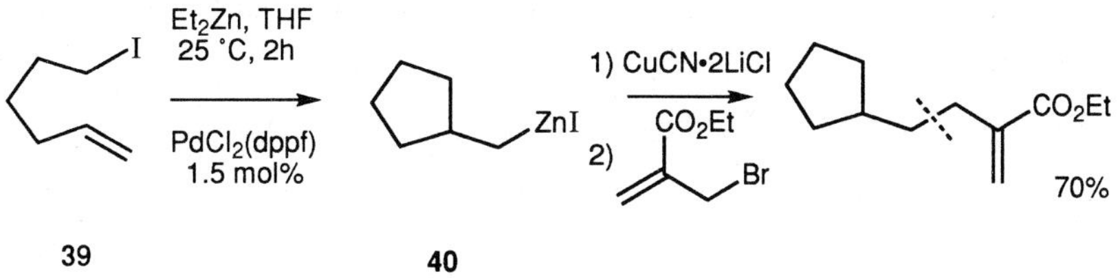

Et2Zn, THF 25 °C, 2h; PdCl2(dppf) 1.5 mol%

1) CuCN•2LiCl 2) CO2Et / Br

70%

39 40

Scheme 5-22

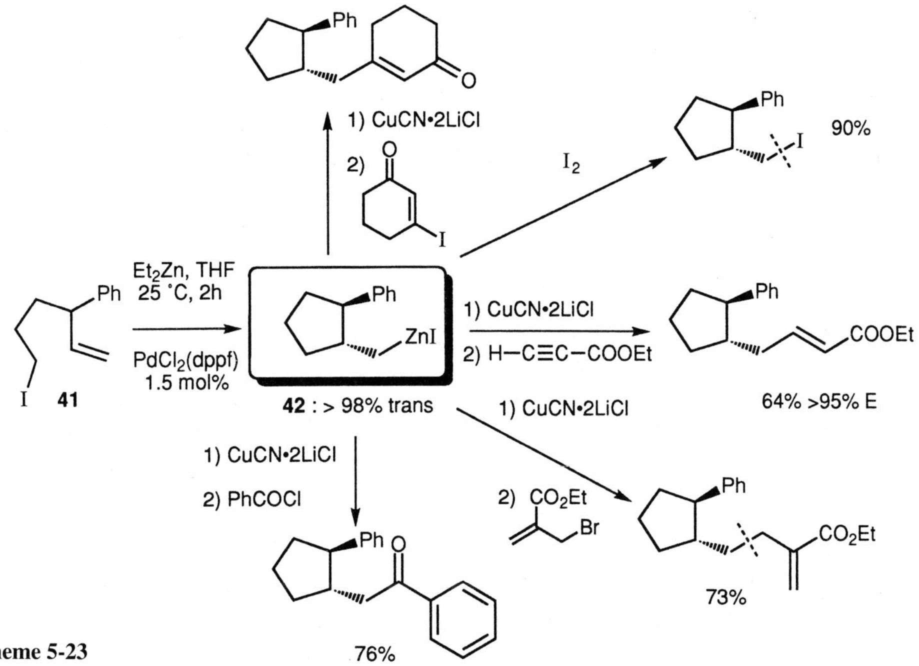

42 : > 98% trans

41

Et2Zn, THF 25 °C, 2h; PdCl2(dppf) 1.5 mol%

1) CuCN•2LiCl 2) 3-iodocyclohexenone

I2 — 90%

1) CuCN•2LiCl 2) H—C≡C—COOEt — 64% >95% E

1) CuCN•2LiCl 2) PhCOCl — 76%

1) CuCN•2LiCl 2) CO2Et / Br — 73%

Scheme 5-23

Et2Zn (2 equiv.) PdCl2(dppf) 1.5 mol% 25 °C, 2 h

1) CuCN•2LiCl 2) COOEt / Br

44 43 — 75 : 25 — 62 %

(1 : 1) mixture

Et2Zn (2 equiv.) PdCl2(dppf) 1.5 mol% 25 °C, 5 h

1) CuCN•2LiCl 2) COOEt / Br

46 45 — 100 : 0 — 94 : 6 — 71 %

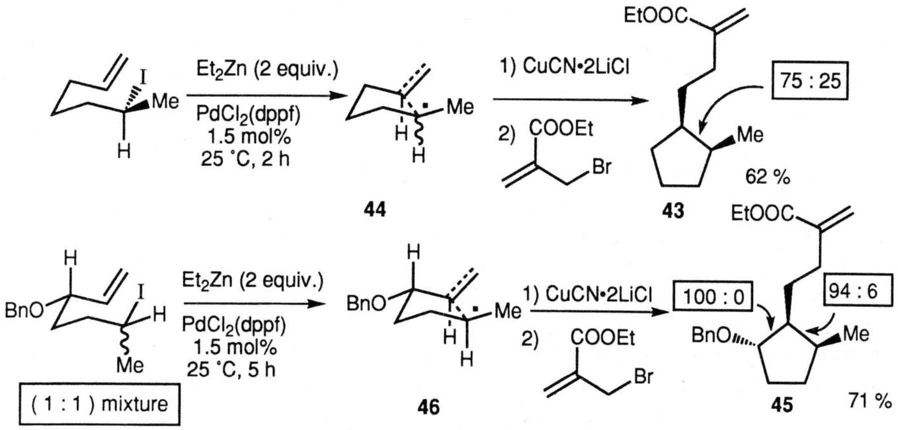

Scheme 5-24

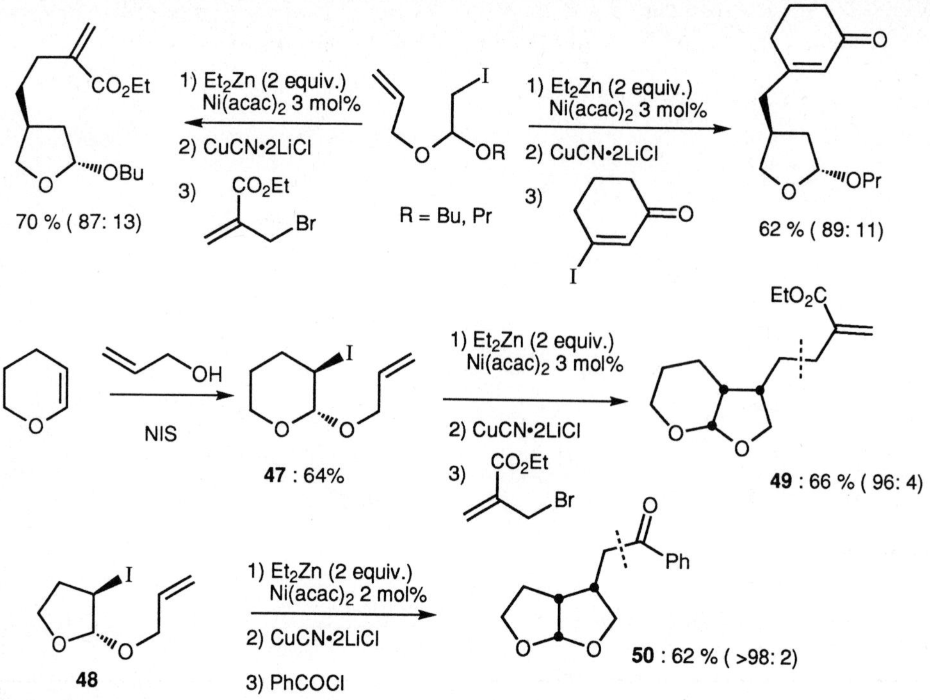

Scheme 5-25

substituted substrate such as **41**, a *trans*-1,2-disubstituted cyclopentylmethylzinc iodide (**42**) is produced with high diastereoselectivity (Scheme 5-23) [106, 107]. The use of a secondary alkyl iodide furnishes preferentially the *cis*-1,2-disubstituted cyclopentane (e.g. **43**) via a transition state such as **44** (Scheme 5-24). An excellent diastereoselectivity is observed if an additional substituent is introduced at position 3 in the cyclization substrate, leading to a cyclopentane derivative such as **45** via the transition state **46**, with an almost complete stereocontrol of three adjacent chiral centers [107]. Interestingly, the nickel-catalyzed iodine–, or bromine–zinc exchange reaction provides an entry to substituted tetrahydrofurans (Scheme 5-25) [112]. The use of unsaturated cyclic iodoacetals such as **47** or **48** provides bicyclic heterocyclic molecules such as **49** or **50** with an excellent *endo* selectivity [112]. This diastereoselectivity can be explained by assuming that the oxygen [O-1]-substituent prefers an axial conformation due to an anomeric stabilization, whereas the carbon chain attached to C-3 will occupy an equatorial position. The radical transition state **51** (Scheme 5-26) will therefore give preferentially the cyclized radical **52** which leads to the *endo* product **53** [112]. This cyclization reaction can be used to prepare *trans*-4,5-disubstituted butyrolactones such as **54** (see **54a–c**) as shown in Scheme 5-27 [112]. As an application, a synthesis of the antitumor antibiotic (–)-methylenolactocin **55** has been performed (Scheme 5-28) [113]. In the key cyclization step involving the ring closure of the bromoacetal **56**, the resulting organometallic intermediate is treated with oxygen in the presence of chlorotrimethylsilane leading directly to the *trans* aldehyde **57** via the zinc hydroperoxide **58**. The aldehyde **57** is readily oxidized to the lactone **59** which has already been converted to (–)-methylenolactocin [114]. Benzylic bromides can be converted by a palladium(0)-catalyzed bromine–zinc exchange to benzylic zinc bromides and can be

Scheme 5-26

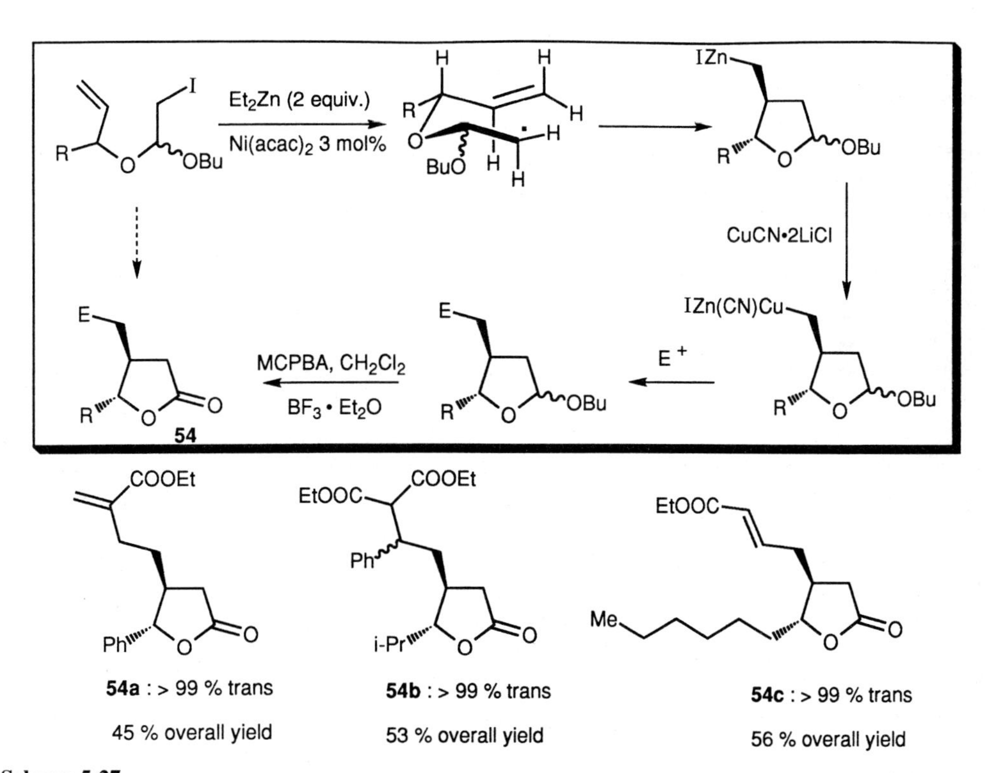

Scheme 5-27

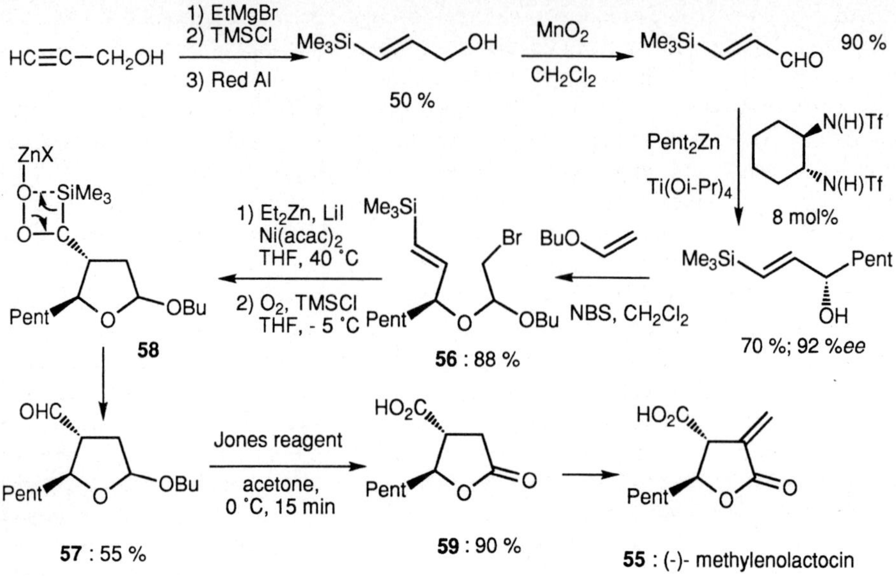

Scheme 5-28

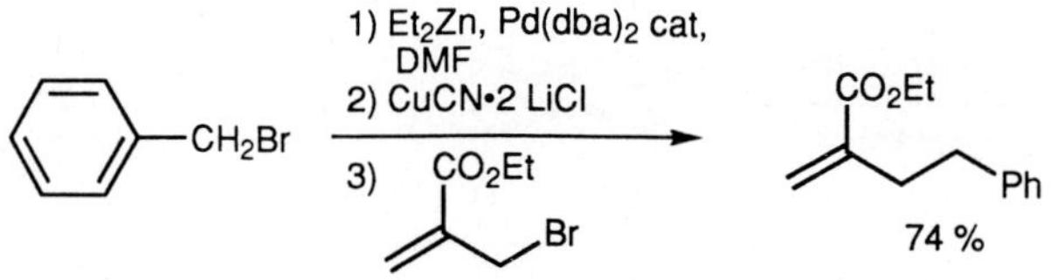

Scheme 5-29

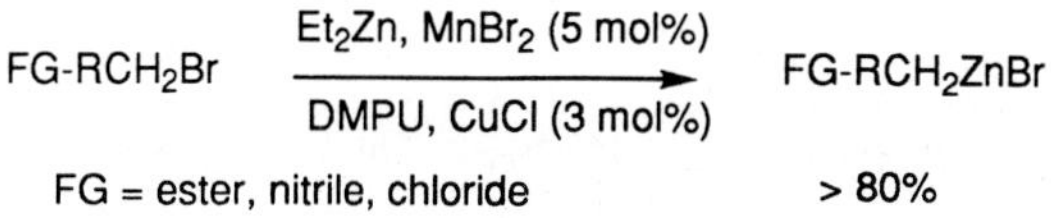

Scheme 5-30

allylated in the presence of $CuCN \cdot 2\,LiCl$ (Scheme 5-29) [106, 115]. Non-activated functionalized alkyl bromides undergo a bromide–zinc exchange using a mixed metal catalysis with manganese(II) bromide and copper(I) chloride in a polar solvent such as DMPU [116–118] under very mild conditions (25 °C, 4–10 h; Scheme 5-30) [109].

5.1.4 The Boron–Zinc Exchange Reaction

The synthesis of a broad range of functionalized organoboranes has been reported over the last fifty years. These compounds are ideal precursors for the formation of dialkyl-zincs because of their excellent thermal stability and high-yield preparations. Some 30 years ago it was reported that the treatment of Bu_3B with Et_2Zn produces Bu_2Zn

and Et_3B in almost quantitative yield. Although the reaction is an equilibrium (Scheme 5-31) [119], the higher reactivity of diethylzinc shifts the equilibrium sufficiently to the right, so that ca. 85% of dibutylzinc is formed. By distilling off the volatile Et_3B (b.p. = 95 °C), a further equilibrium shift occurs, providing isolated Bu_2Zn in excellent yield [119]. The reaction has been applied in the past to the preparation of diallyl- and dibenzyl zinc with great success (Scheme 5-32) [120, 121]. More recently, a synthesis of dialkyl zincs and their addition to aldehydes has been described [122–124]. This method led to a very elegant and efficient catalytic asymmetric synthesis of a (*R*)-(−)-muscone precursor (**60**) using (+)-DAIB (**61**) as a chiral catalyst (Scheme 5-33) [124]. The preparation of polyfunctional dialkyl zincs by the transmetallation of organoboranes has also been extensively investigated [125–129] and it has been shown that the exchange reaction using diethyl(alkyl)boranes as starting materials leads to excellent results. This class of boranes is readily prepared by using diethylborane, **62** (Köster's reagent) [130], as hydroborating agent. Diethylborane is simply prepared by mixing borane–methyl sulfide complex and triethylborane in the molar ratio 1:2 (Scheme 5-34) [126]. As a hydroboration agent it gives the same regioselectivity as more bulky hydroboration agents such as disiamylborane or thexylborane [130], and leads to organoboranes which are convenient to handle and easy to analyze spectroscopically. Thus, hydroboration of *β*-pinene with diethylborane (**62**) affords in almost quantitative yield the corresponding diethyl(alkyl)borane, which on treatment with diethylzinc (2 equiv., 0 °C, 0.5 h) produces the desired dialkyl zinc

$$3\ Et_2Zn + 2\ Bu_3B \rightleftharpoons 3\ Bu_2Zn + 2\ Et_3B$$

Scheme 5-31

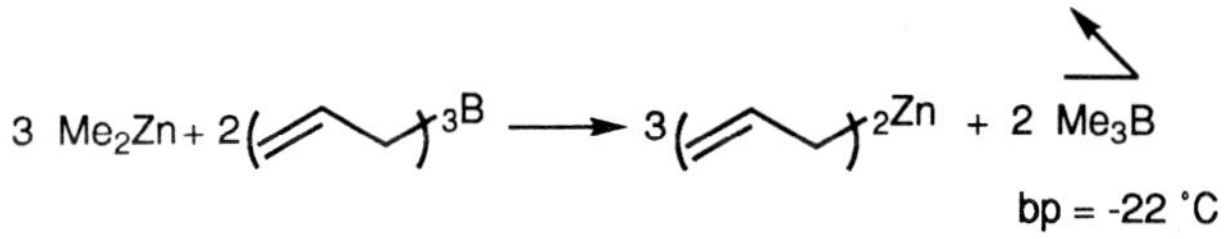

3 Me_2Zn + 2(allyl)$_3$B → 3(allyl)$_2$Zn + 2 Me_3B (bp = -22 °C)

Scheme 5-32

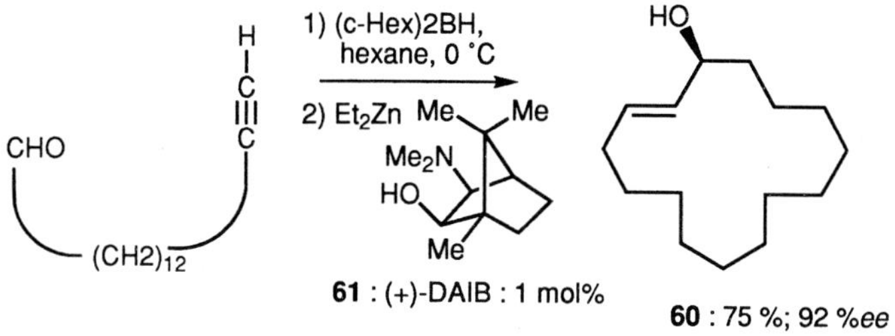

OHC–$(CH_2)_{12}$–C≡C–H → 1) (c-Hex)2BH, hexane, 0 °C; 2) Et_2Zn, **61** : (+)-DAIB : 1 mol% → **60** : 75 %; 92 %*ee*

Scheme 5-33

$$2\ Et_3B + BH_3 \cdot Me_2S \xrightarrow{25\ °C} 3\ Et_2BH$$

62

Scheme 5-34

reagent **63** in over 90% yield (Scheme 5-35). Similarly (–)-cinchonidin (**64**) has been converted into the corresponding zinc reagent **65** (>80% yield) [126]. The reaction can be applied to prepare a variety of polyfunctional dialkyl zincs (**66–70**) directly from the corresponding olefins **71–75** using the hydroboration-transmetallation sequence (Scheme 5-36) [126]. Many of these zinc reagents cannot be prepared by any other known method. It is remarkable that a sensitive function such as an acrylate is tolerated in the organometallic compound **68**. Also the presence of two acidic protons at the alpha position relative to the nitro function seems not to interfere with the formation of dialkyl zincs (**69**, **70**) [126]. The thermal stability of **70** is somewhat limited and the reagent has to be used for further transformations as soon as possible

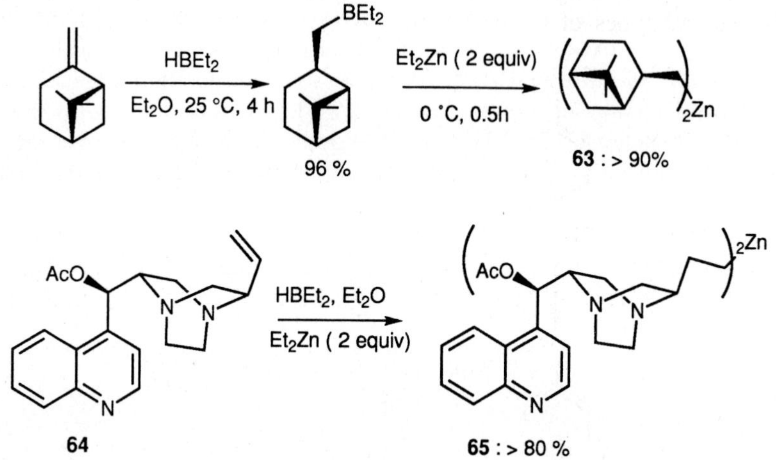

BEt_2 ; $HBEt_2$; Et_2O, 25 °C, 4 h ; 96 % ; Et_2Zn (2 equiv) ; 0 °C, 0.5h ; $_2Zn$; **63** : > 90%

AcO ; N ; $HBEt_2$, Et_2O ; Et_2Zn (2 equiv) ; $_2Zn$; **64** ; **65** : > 80 %

Scheme 5-35

FG-R ; 1) Et_2BH **62** ; 2) Et_2Zn (2 equiv.) ; (FG-R)$_2Zn$

OCOPh ; I ; CO_2Et ; NO_2 ; Me ; Me ; NO_2

71 ; **72** ; **73** ; **74** ; **75** ; i

$_2Zn$; $PhCO_2$; I ; CO_2Et ; NO_2 ; Me ; Me ; NO_2

66 ; **67** ; **68** ; **69** ; **70**

i: Et_2BH (1 equiv), then Et_2Zn (2 equiv)

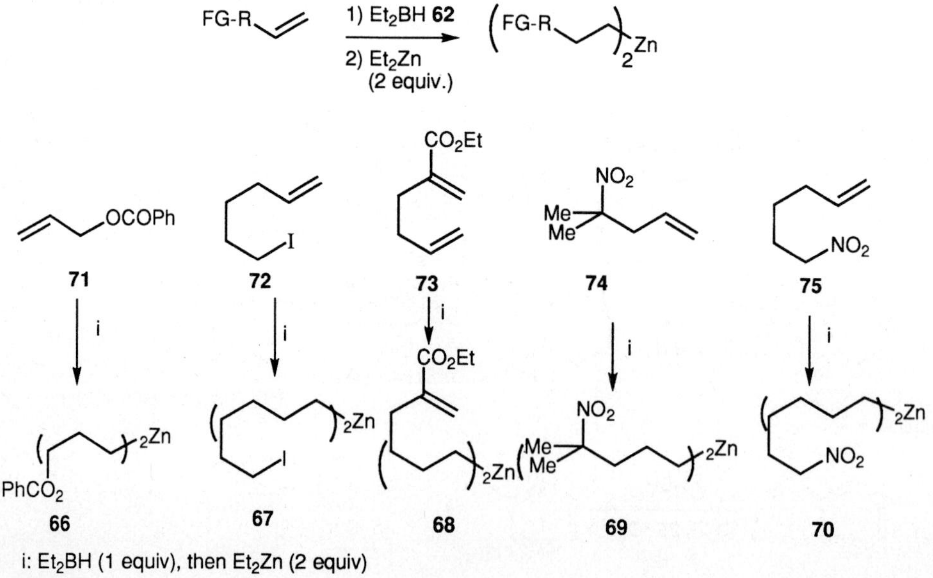

Scheme 5-36

after its generation. Of special interest is the preparation of dialkyl zincs using functionalized dienes as starting materials, since in this case the corresponding alkyl iodides required for a direct insertion of zinc would not even be available for these substrates. The boron–zinc exchange represents a truly unique preparation of zinc organometallics [128]. Thus, the oxygenated dienes **76–78** are regiospecifically hydroborated and transmetalled to yield the new zinc reagents **79–81** (Scheme 5-37) [128]. Interestingly, the triisopropylsilyl enol ether moiety survives the reaction of **81** with electrophiles and subsequent work-up, and can then be used for a new carbon–carbon bond-forming reaction in a further step [128]. Since diastereomerically pure organoboranes can be easily prepared, it was tempting to use these compounds for preparing diastereomerically pure diorganozincs. Unfortunately, a loss of stereochemistry was observed with several types of organoboranes (Scheme 5-38) [131].

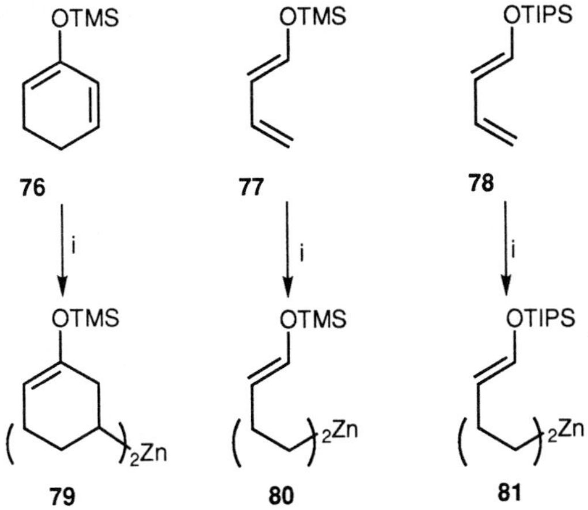

Scheme 5-37 i: Et_2BH (1 equiv), then Et_2Zn (2 equiv)

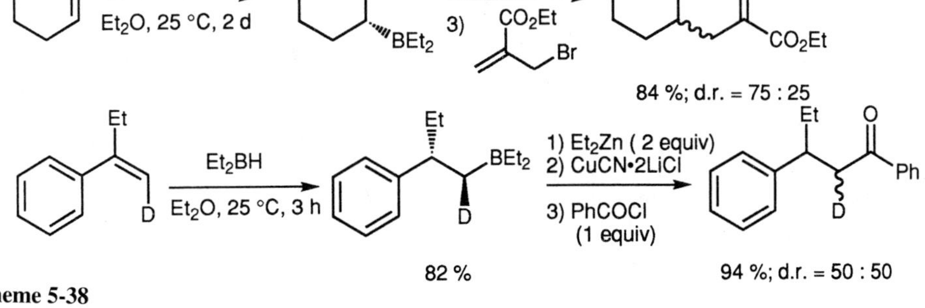

Scheme 5-38

The boron–zinc exchange reaction appears to be a very general and efficient method for preparing polyfunctional diorganozincs, which are especially useful in the asymmetric synthesis of polyfunctional secondary alcohols (see Section 5.2.4). A related mercury–zinc exchange reaction has also been reported in the literature [132, 133]; however, due to the high toxicity of many diorganomercury compounds this reaction may not find broad application.

5.1.5 The Lithium–Zinc Transmetallation

The direct insertion of zinc into alkenyl and aryl iodides is a difficult reaction, as discussed in Section 5.1.2, and alternative syntheses of these classes of zinc reagents would be welcome. The halogen–lithium exchange followed by a lithium–zinc transmetallation offers such an alternative [134]. Thus, whereas alkyl lithiums are too reactive to tolerate most organic functionalities [135–138], alkenyl- and aryl lithiums are far less reactive, and a broad range of functional groups such as halide [139–143], sulfone [144], epoxide [145], ester [146–154], nitro [155–157], or cyanide [153] can be present in alkenyl- and aryl lithiums at low temperatures [158–163]. A subsequent transmetallation of these lithium species with zinc bromide or iodide produces functionalized aryl- or alkenyl zinc halides which have considerably better thermal stability and can in many cases be handled at room temperature without decomposition (Scheme 5-39) [134, 164]. For example, 1,2,2-trifluoroethenyllithium (**82**) is a highly unstable lithium carbenoid which decomposes readily even at –78 °C, but its transmetallation with zinc(II) salt produces a zinc reagent **83** which is stable at 25 °C (Scheme 5-40) [81, 165–174]. It is also possible by this method to prepare aryl- and alkenyl-zinc species not available by the direct insertion of zinc. For example, the presence of an azido group inhibits the direct zinc insertion, but an iodine-lithium exchange followed by a transmetallation produces the desired (*E*)-alkenyl zinc iodide, **84**, which after further transmetallation with CuCN · 2 LiCl adds to ethyl propiolate in high yields (Scheme 5-41) [134, 164, 175].

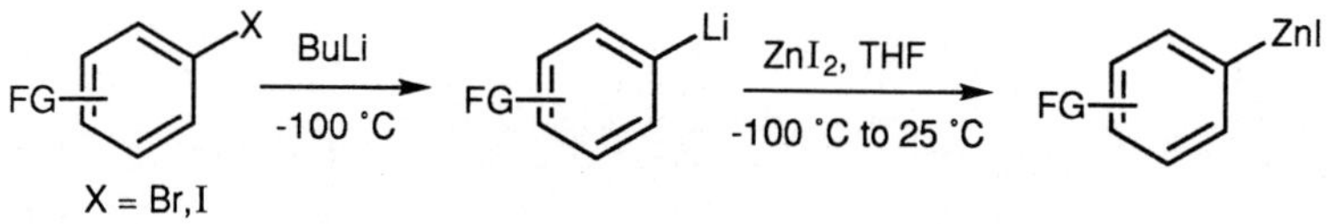

FG–C$_6$H$_4$–X (X = Br, I) → (BuLi, -100 °C) → FG–C$_6$H$_4$–Li → (ZnI_2, THF, -100 °C to 25 °C) → FG–C$_6$H$_4$–ZnI

Scheme 5-39

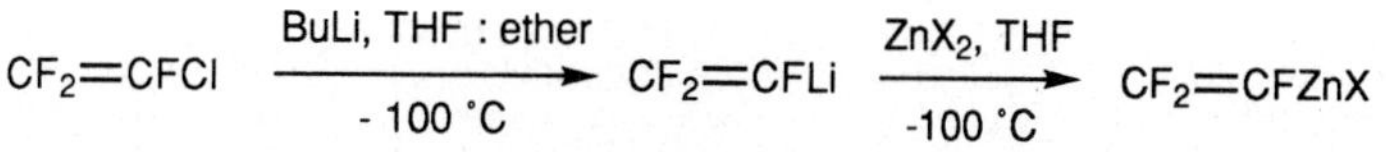

$CF_2{=}CFCl$ → (BuLi, THF : ether, - 100 °C) → $CF_2{=}CFLi$ (**82**) → (ZnX_2, THF, -100 °C) → $CF_2{=}CFZnX$ (**83**)

Scheme 5-40

N_3-alkenyl iodide → (1) BuLi, - 100 °C; 2) ZnI_2, -90 °C) → **84** (ZnI) → (1) CuCN•2LiCl; 2) HC≡C–CO_2Et) → dienyl CO_2Et, 81 % (100 % E, E)

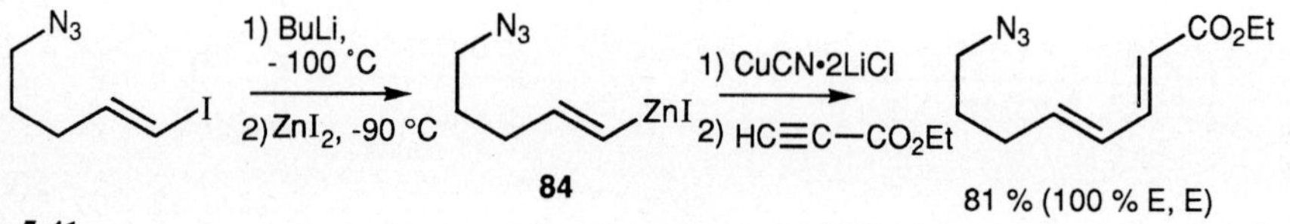

Scheme 5-41

5.1.6 Carbon–Carbon Bond-Forming Reactions for the Preparation of Organozinc Compounds

5.1.6.1 Homologation of Zinc–Copper Compounds with Zinc Carbenoids

Zinc carbenoids such as iodomethylzinc iodide have found extensive applications as cyclopropanation agents [5–8, 176–189]. Recently, they have also be used to homologate copper reagents of the type FG–RCu · MX_2 (M = Mg, Zn) by adding one (or more) methylene units (Scheme 5-42) [103, 190–194], leading to new organocopper–zinc (or- magnesium) reagents of the type FG–RCH_2Cu · MX_2. The nature of the organic moiety attached to copper can be quite diverse. It has been found that various enolates of ketones, nitriles, or aldehydes can be converted to the corresponding homoenolates [195] in high yields [190]. Interestingly, the use of bis(iodomethyl)zinc makes it possible in some cases to produce the double methylene homologation product (Scheme 5-43) [190]. In general, the methylene homologation reaction proceeds cleanly if the reactivity of the starting organocopper reagent is substantially different from the reactivity of the methylene homologated product. If this condition is not fulfilled, a polymerization reaction occurs [190].

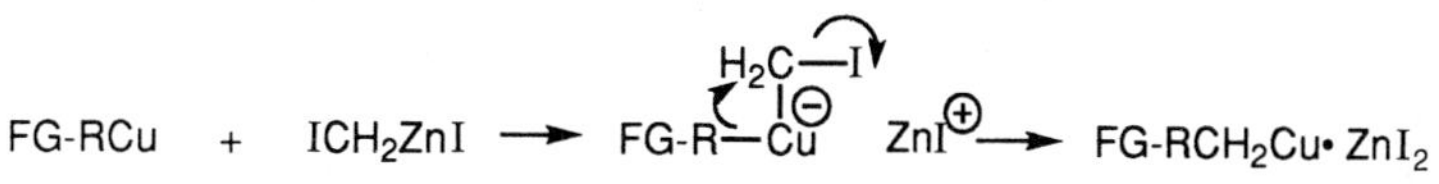

FG-RCu + ICH_2ZnI → FG-R–$Cu^{\ominus}$(H_2C–I) $ZnI^{\oplus}$ → FG-RCH_2Cu• ZnI_2

Scheme 5-42

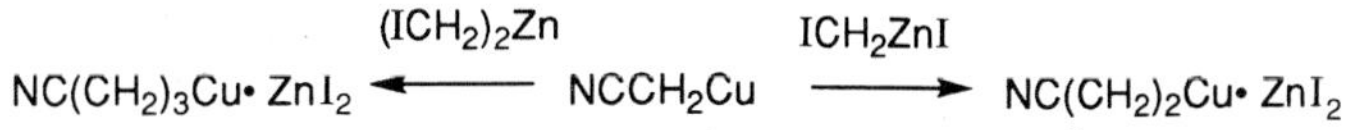

$NC(CH_2)_3Cu$• ZnI_2 ←[$(ICH_2)_2Zn$]— $NCCH_2Cu$ —[ICH_2ZnI]→ $NC(CH_2)_2Cu$• ZnI_2

Scheme 5-43

Of special interest for synthetic applications is the homologation of alkenyl and alkynyl copper species. This method gives a unique approach to functionalized allylic (and propargylic) zinc–copper reagents. Since these organometallics are very reactive, the reaction has to be performed in the presence of an electrophile such as an aldehyde or a ketone so that no further methylene homologation occurs. This reaction makes it possible to produce a range of polyfunctional homoallylic alcohols such as **85** (Scheme 5-44) [103, 191]. The reaction can be used to prepare highly substituted α-methylenebutyrolactones such as **86** with high *cis* diastereoselectivity (Scheme 5-45) [103, 194]. By adding a zinc–copper reagent bearing a keto function to the propiolic ester, a bicyclic butyrolactone (e.g. **87**) can be obtained (Scheme 5-46) [103, 194]. Similarly, alkynyl copper can be homologated. If no electrophile is present to trap the very reactive propargylic zinc–copper intermediate, a multiple homologation occurs and up to four methylene units can be inserted, leading to dienyl-coppers such as **88** (Scheme 5-47) [103, 192]. In the presence of an electrophile, the multiple homologation is not observed and interesting cyclization reactions can take place (Scheme 5-48) [103]. A novel rearrangement of α-silylated lithium carbenoids such as **89** by reaction with a dialkyl zinc provides an entry to a range of α-silylated zinc reagents such as **90** (Scheme 5-49) [196–197]. Lithium zincates such as **91** undergo a halide–zinc exchange reaction with various types of 1,1-dibromides, or induce selective deprotonations leading to zinc carbenoids which undergo a 1,2-migration [198–206] to provide new organozinc reagents (Scheme 5-50) [207–213]. Polyfluorinated zinc–copper reagents have been obtained by a selective double insertion of difluorocarbene to phenylcadmium halides [214].

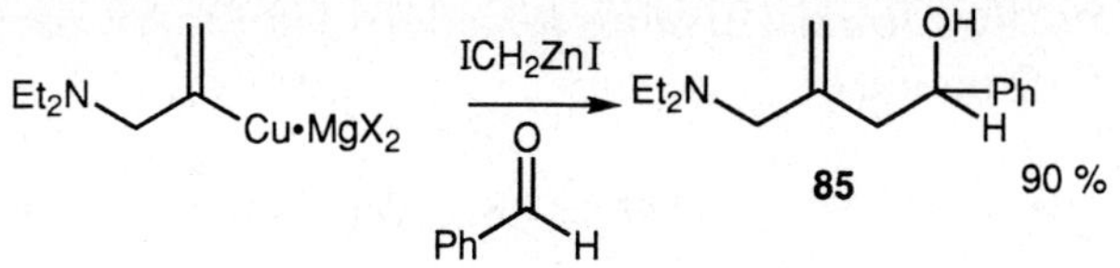

Et_2N Cu•MgX_2 → ICH_2ZnI, O, Ph H → Et_2N OH Ph H **85** 90 %

Scheme 5-44

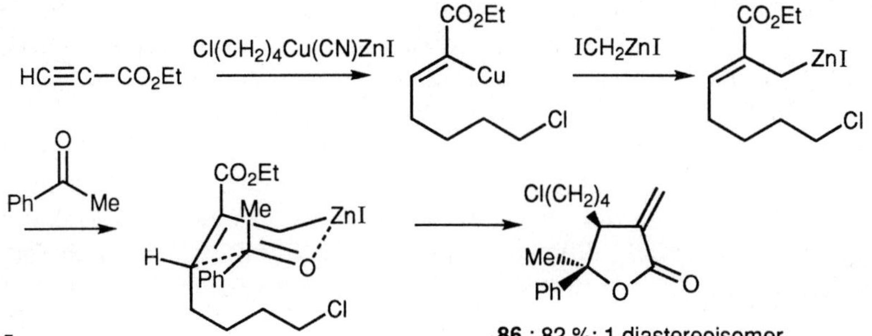

HC≡C–CO_2Et → $Cl(CH_2)_4Cu(CN)ZnI$ → CO_2Et Cu Cl → ICH_2ZnI → CO_2Et ZnI Cl

O Ph Me → CO_2Et Me ZnI H Ph O Cl → $Cl(CH_2)_4$ Me Ph O O

86 : 82 %; 1 diastereoisomer

Scheme 5-45

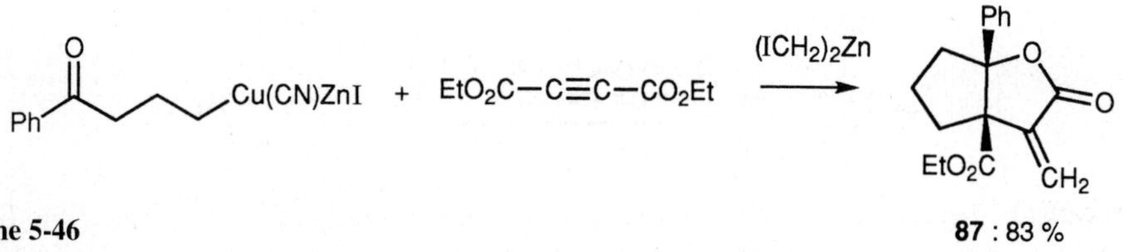

O Ph Cu(CN)ZnI + EtO_2C–C≡C–CO_2Et → $(ICH_2)_2Zn$ → Ph O O EtO_2C CH_2

87 : 83 %

Scheme 5-46

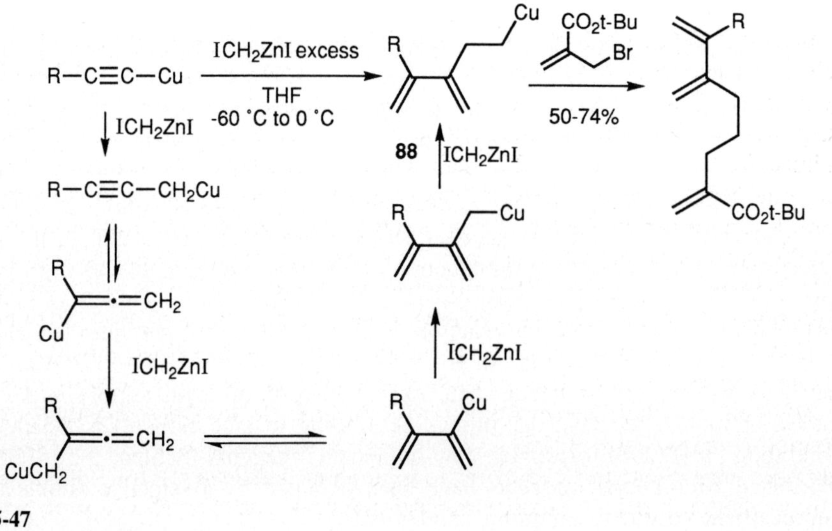

R–C≡C–Cu → ICH_2ZnI excess, THF, -60 °C to 0 °C → R Cu **88** → CO_2t-Bu, Br, 50-74% → R CO_2t-Bu

R–C≡C–Cu ↓ ICH_2ZnI → R–C≡C–CH_2Cu ⇅ R Cu =•=CH_2 ↓ ICH_2ZnI → R $CuCH_2$ =•=CH_2 ⇌ R Cu → ICH_2ZnI → R Cu → ICH_2ZnI → **88**

Scheme 5-47

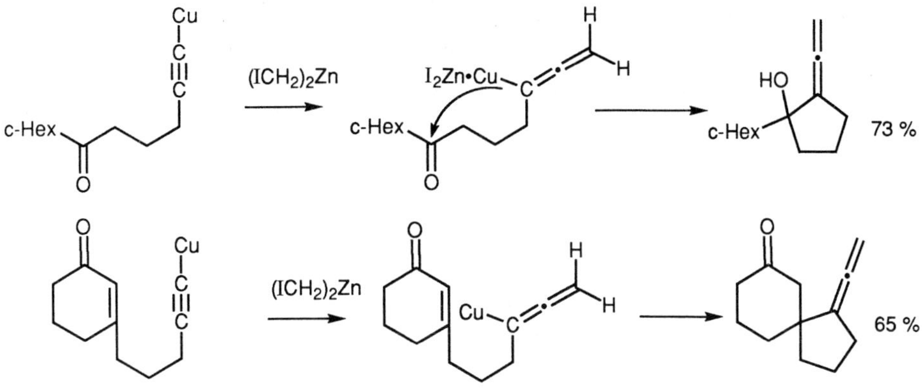

Scheme 5-48

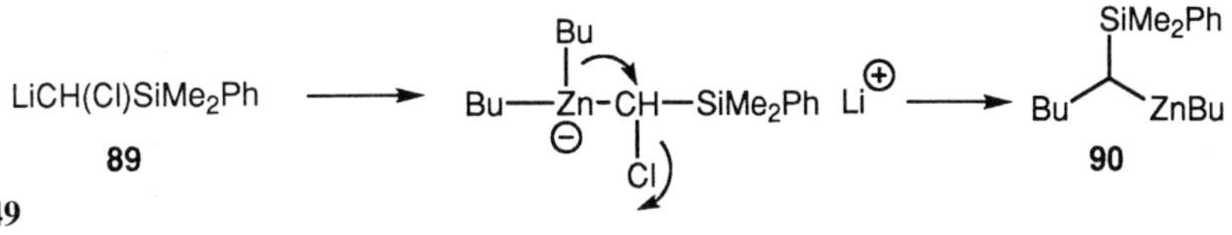

Scheme 5-49

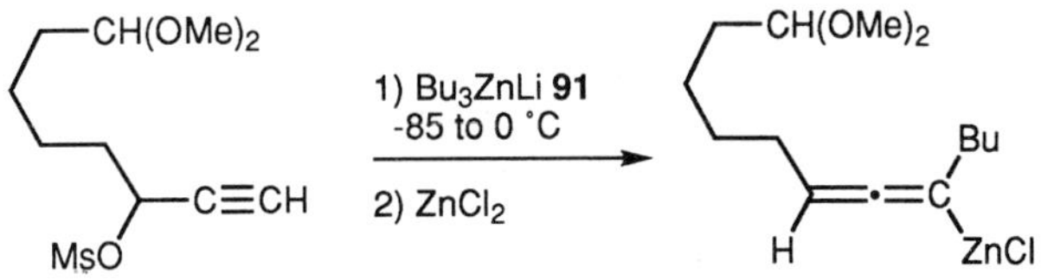

Scheme 5-50

5.1.6.2 The Use of Bimetallic Reagents for the Preparation of Organozinc Compounds

1,n-Bimetallic compounds ($n = 1$, 4–6) have been used to prepare organozinc halides by the selective reaction of the bimetallic reagent with one equivalent of an electrophile E^1 [39, 215–219]. This method provides new functionalized organozincs which can in a second step react with a different electrophile E^2 (Scheme 5-51). Thus, the treatment of the zinc bimetallic **92** with CuCN · 2 LiCl (1 equiv.) and pentane (30 volume %) leads to a copper–zinc heterobimetallic species which reacts selectively with cyclohexenone in the presence of chlorotrimethylsilane [13, 220–230] affording the polyfunctional zinc–copper reagent **93**, this can be further trapped by a second more reactive electrophile such as allyl bromide (Scheme 5-52) [215]. The scope of the reaction is broad, and a range of 1,n-bimetallics ($n = 4$, 5, 6) have been found to react selectively with two different electrophiles [215]. The reactivity of zinc and magnesium 1,1-bimetallics is unique. These reagents are prepared by the allylzincation of alkenylmagnesium or lithium derivatives **94** [231, 232] and can be cleanly reacted with several electrophiles affording organozinc compounds (Scheme 5-53) [216, 217]. Benzylic zinc–magnesium 1,1-bimetallic compounds have also been prepared [219]. The α-stannylated alkyl zinc halides have been used for a diasterometric synthesis of aldol products such as **95** (Scheme 5-54) [218].

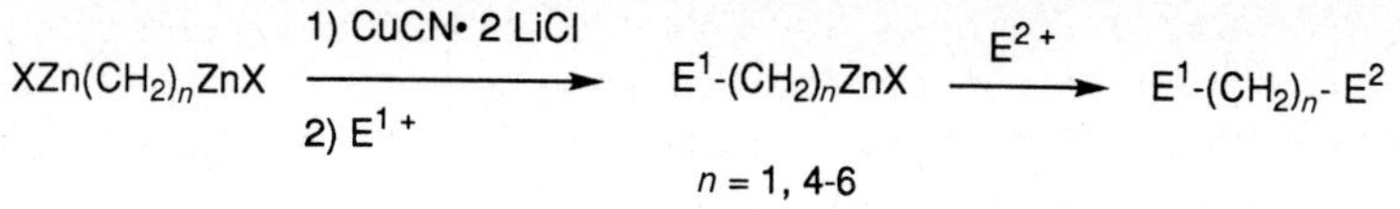

$XZn(CH_2)_nZnX$ —1) CuCN• 2 LiCl; 2) E^{1+}→ E^1-$(CH_2)_nZnX$ —E^{2+}→ E^1-$(CH_2)_n$-E^2

n = 1, 4-6

Scheme 5-51

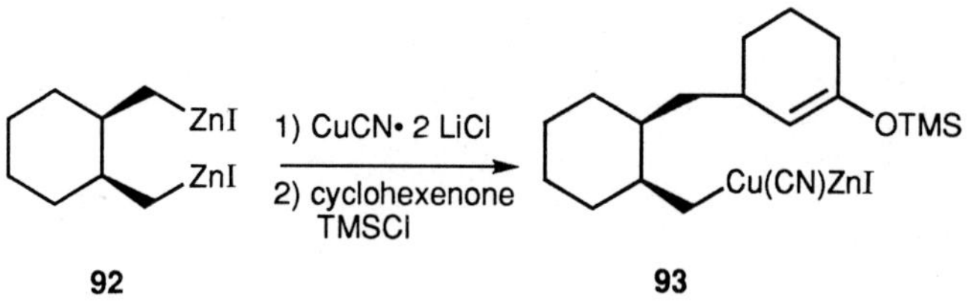

92 —1) CuCN• 2 LiCl; 2) cyclohexenone, TMSCl→ 93

93 —allyl bromide→ 74 %

Scheme 5-52

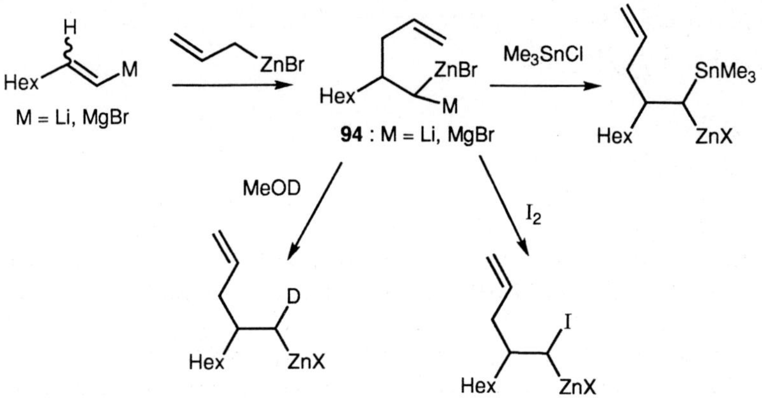

M = Li, MgBr

94 : M = Li, MgBr

Reagents: allylzinc bromide; Me_3SnCl; MeOD; I_2

Scheme 5-53

1) THF, 35 °C; 2) O_2, TMSCl

cis : trans = 88 :12

95 : 56 %; d.r. = 88 : 12

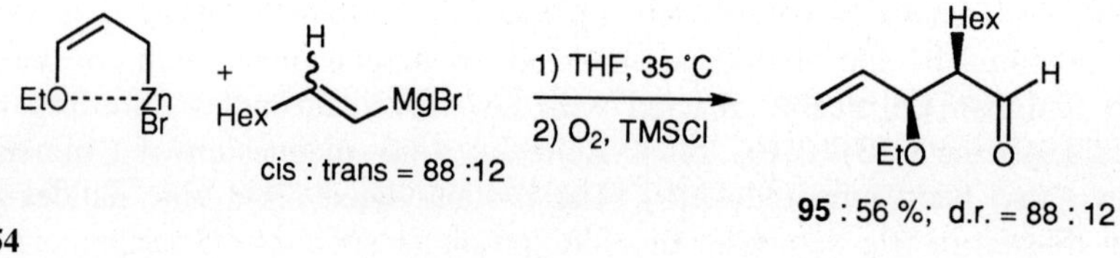

Scheme 5-54

Interestingly, the incorporation of an α-alkoxy substituent into the alkenyl organometallic (as in **96**) to control the stereochemistry in the formation of the corresponding zinc–magnesium 1,1-bimetallic has been elegantly demonstrated by Marek, showing that the control of up to three stereocenters in an acyclic chain is possible, as in compound **97** (Scheme 5-55) [233–238]. Finally, bimetallic reagents can be used to perform cyclization reactions, and the reaction of a 1,4-heterobimetallic with oxalyl chloride produces a cyclopentene derivative (**98**) in satisfactory yield (Scheme 5-56) [39].

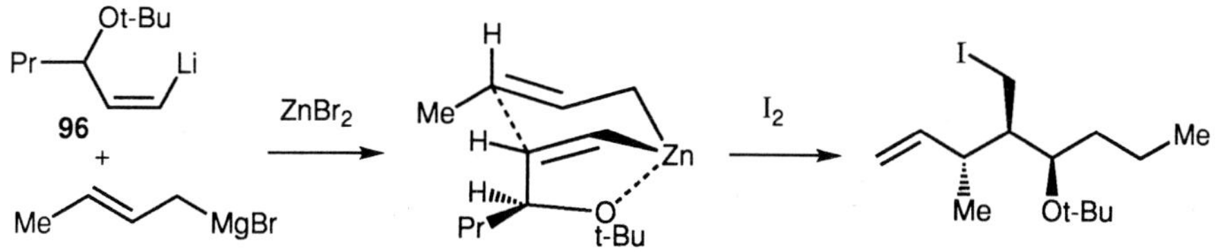

Scheme 5-55 **97** : 56 %; d.r. > 95 : 5

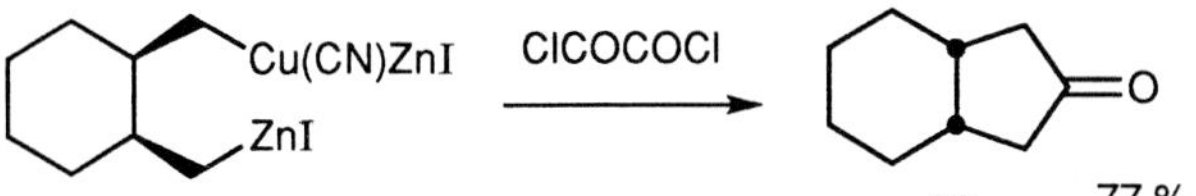

Scheme 5-56 **98** 77 %

5.2 Applications of Organozinc Compounds in Organic Synthesis

5.2.1 The Reactivity of Organozinc Derivatives

As may be expected from their high functional group tolerance, organozinc compounds are very unreactive towards organic electrophiles. The only such reactions that have proved useful in synthesis [10] are those with oxygen, affording hydroperoxides such as **99** (Scheme 5-57) [2, 239], with halides [2], and with pseudo-halides such as tosyl cyanide affording nitriles (Scheme 5-58) [42]. On the other hand, organozinc compounds readily undergo transmetallation reactions with metallic salts [2], giving intermediate transition metal organometallics. These usually have an even stronger carbon–metal bond, but due to the presence of d-orbitals they can react via new mechanisms (which were not available for the initial zinc reagents) with a range of organic electrophiles. Especially useful for synthetic applications are transmetallations to copper or palladium intermediates, and additions of diorganozincs to aldehydes which are catalyzed by chiral titanium complexes. These reactions greatly enhance the initially narrow scope of organozinc chemistry [2], and should be considered as important ways for constructing polyfunctional organic molecules. One of the major advantages of this chemistry is the broad tolerance of a variety of functional groups, which avoids wasteful, expensive, and time-consuming protection–deprotection steps, and allows a wide variety of functional group interconversions [10]. The following sections will discuss in detail these copper-, palladium-, and titanium-catalyzed reactions of organozinc compounds with electrophiles.

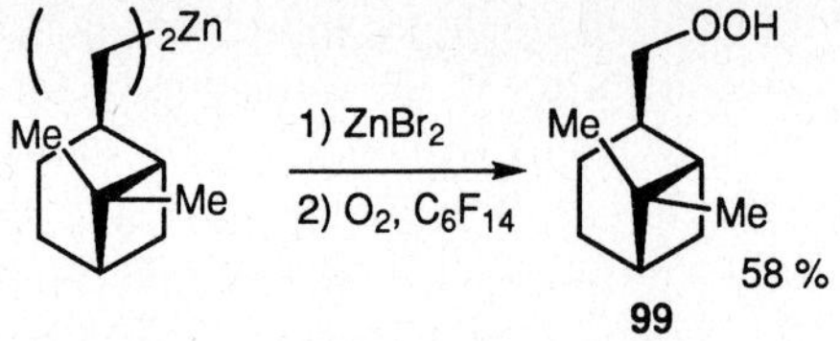

Scheme 5-57

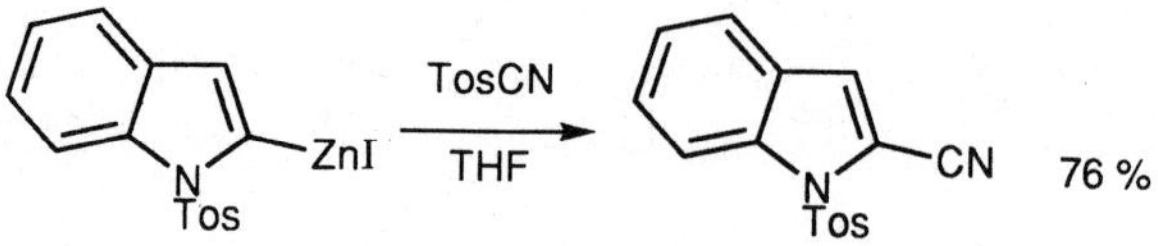

Scheme 5-58

5.2.2 Copper(I)-Mediated Reactions of Organozinc Compounds

5.2.2.1 Substitution Reactions

Direct substitution reactions using organozinc species are difficult, and the presence of a transition metal catalyst is often necessary. Triorganotin halides can in some cases react directly with an organozinc reagent (Scheme 5-59) [240]. A more general reaction is that of copper–zinc compounds, prepared by the transmetallation of organozinc halides, with the THF-soluble salt CuCN · 2 LiCl [13]. Under these conditions high yields of polyfunctional stannanes are obtained (Scheme 5-60) [241]. Reactive organic halides such as allylic chlorides, bromides, or phosphates react with organozinc compounds in the presence of *catalytic* amounts of copper(I) salts. This is a very fast and efficient reaction which proceeds in high yields. The regioselectivity of the allylation is usually excellent, and the S_N2' substitution product is obtained with high selectivity (Scheme 5-61) [13, 48,

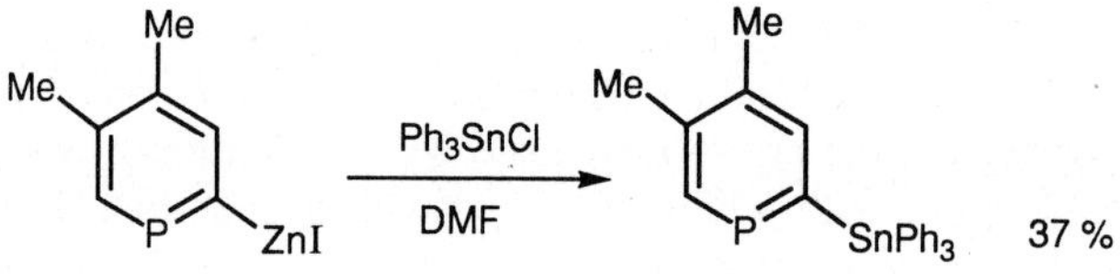

Scheme 5-59

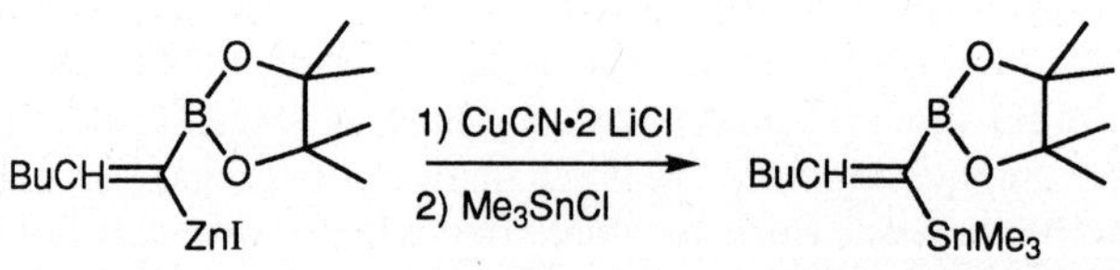

Scheme 5-60

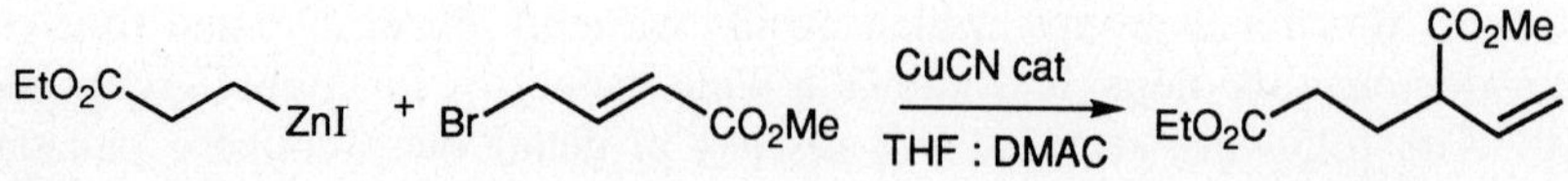

Scheme 5-61

68, 195, 242–245]. In contrast, the copper–zinc carbenoid $ICH_2Cu \cdot ZnI_2$ generated in situ leads to a selective S_N2 substitution of geranyl bromide, **100** (Scheme 5-62) [17]. Also, nickel- or palladium-catalyzed allylation reactions of organozinc compounds tend to produce the substitution product formed from the least substituted end of the allylic system (Scheme 5-63) [15]. Several applications for the preparation of complex molecules such as isocarbocyclins have been reported (Scheme 5-64) [246]. A multiple substitution can be accomplished by using a 1,3-dichloro-substituted propene derivative such as **101**. The reaction allows the introduction of two nucleophiles in the 1,3-positions of a carbon chain via two successive S_N2' substitutions (Scheme 5-65) [247]. Propargylic halides react with zinc–copper compounds leading to polyfunctional allenes (Scheme 5-66) [32, 68, 128]. Cationic transition metal polyene complexes such as pentadienyliron and pentadie-

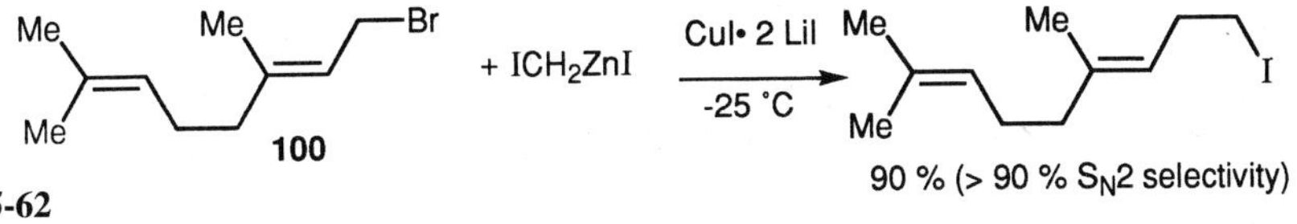

100 + ICH_2ZnI → ($CuI \cdot 2\ LiI$, -25 °C) 90 % (> 90 % S_N2 selectivity)

Scheme 5-62

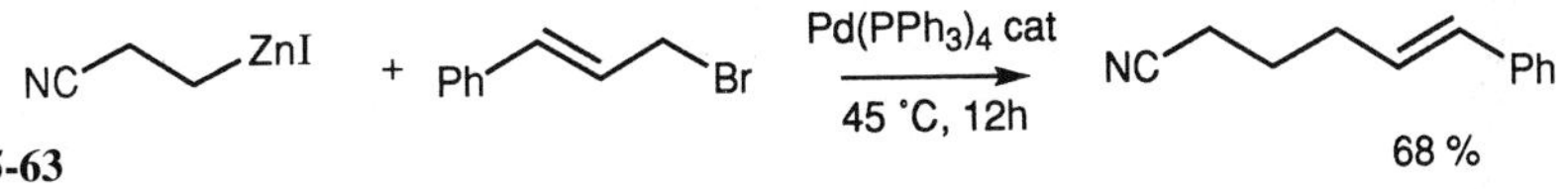

$NC(CH_2)_3ZnI$ + Ph–CH=CH–CH_2Br → ($Pd(PPh_3)_4$ cat, 45 °C, 12h) NC…Ph 68 %

Scheme 5-63

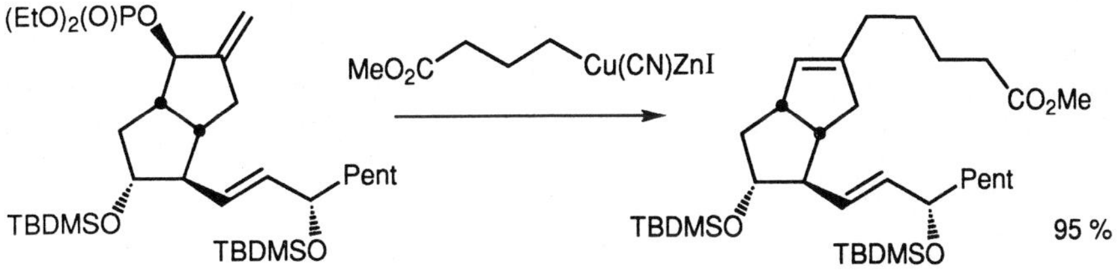

$(EtO)_2(O)PO$ … TBDMSO, Pent, TBDMSO + $MeO_2C(CH_2)_4Cu(CN)ZnI$ → CO_2Me, Pent, TBDMSO, TBDMSO 95 %

Scheme 5-64

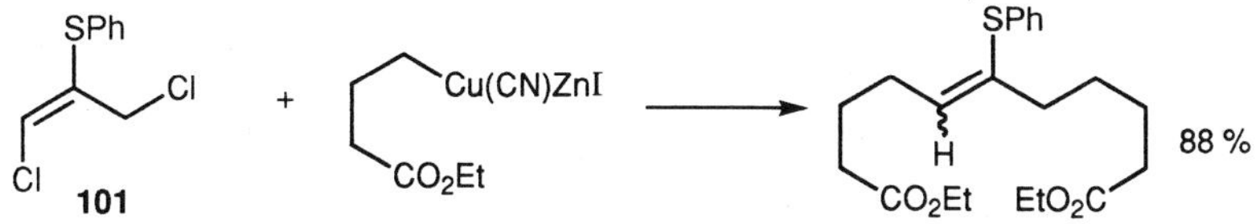

SPh, Cl, Cl **101** + Cu(CN)ZnI … CO_2Et → SPh, H, CO_2Et, EtO_2C 88 %

Scheme 5-65

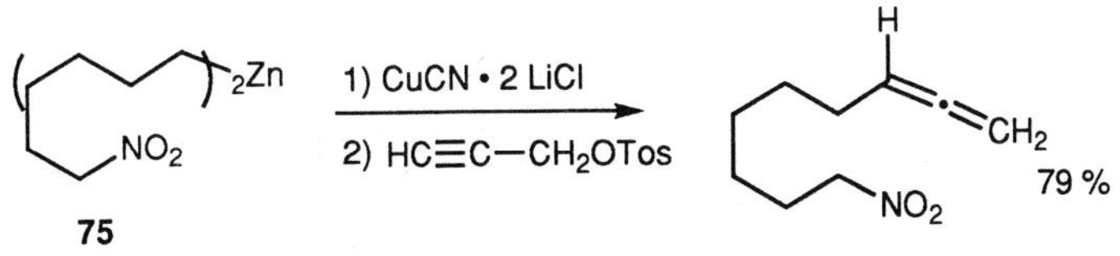

75 (NO_2…)$_2$Zn → 1) $CuCN \cdot 2\ LiCl$ 2) $HC{\equiv}C{-}CH_2OTos$ → H, CH_2, NO_2 79 %

Scheme 5-66

nylmolybdenum complexes (e.g. **102**) react stereoselectively with functionalized zinc–copper reagents furnishing polyfunctional metal complexes (Scheme 5-67) [248–252]. A subsequent anionic cyclization can be initiated in some cases, leading to interesting bicyclic molecules [251, 252]. Alkynyl iodides and bromides are also very good substrates, and cross-coupling reactions with various zinc–copper reagents occur at low temperatures producing polyfunctional acetylenes [253, 254]. It provides a useful entry to substituted ynols such as **103** (Scheme 5-68) [254]. Unlike alkynyl halides, alkenyl iodides or bromides react with zinc–copper compounds only if polar solvents such as DMPU [116–118] or *N*-methylpyrrolidone (NMP) are used, or if the alkenyl halide bears an electron-withdrawing substituent. Thus, the reaction of the zinc–copper reagent **104** with the functionalized alkenyl iodide **105** produces *stereospecifically* the polyfunctional diene **106** in 65% yield (Scheme 5-69) [255]. Alkenyl halides or sulfones bearing an electron-withdrawing group undergo a substitution via an addition-elimination mechanism [10]. The reaction of zinc–copper compounds with diethyl(phenylsulfonylmethylidene)malonate **107** leads to polyfunctional alkylidenemalonates such as **108** (Scheme 5-70) [256, 257]. A similar addition-elimination has been performed with *E*-(ethylsulfonyl)nitroethylene (**109**) and provides a very convenient synthesis of pure (*E*)-

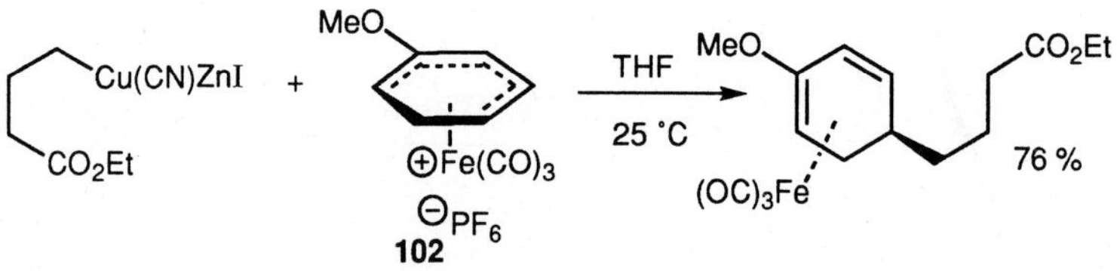

Cu(CN)ZnI + MeO ⊕Fe(CO)3 ⊖PF6 **102** → THF, 25 °C → MeO (OC)3Fe CO2Et 76 %

Scheme 5-67

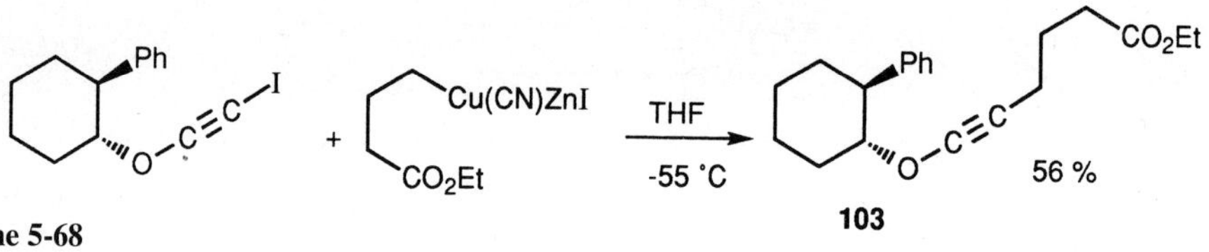

Ph O–C≡C–I + Cu(CN)ZnI CO2Et → THF, -55 °C → Ph O–C≡C CO2Et **103** 56 %

Scheme 5-68

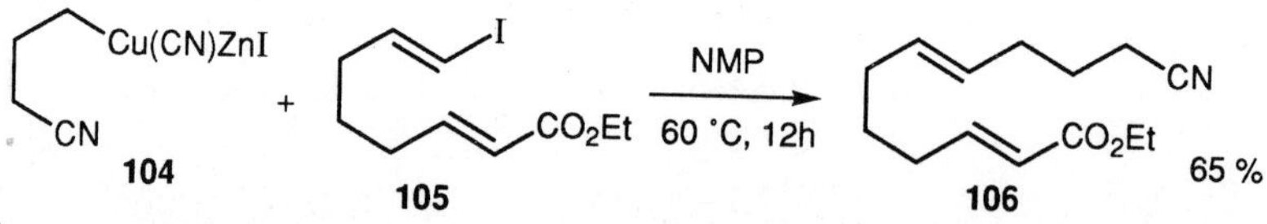

Cu(CN)ZnI CN **104** + I CO2Et **105** → NMP, 60 °C, 12h → CN CO2Et **106** 65 %

Scheme 5-69

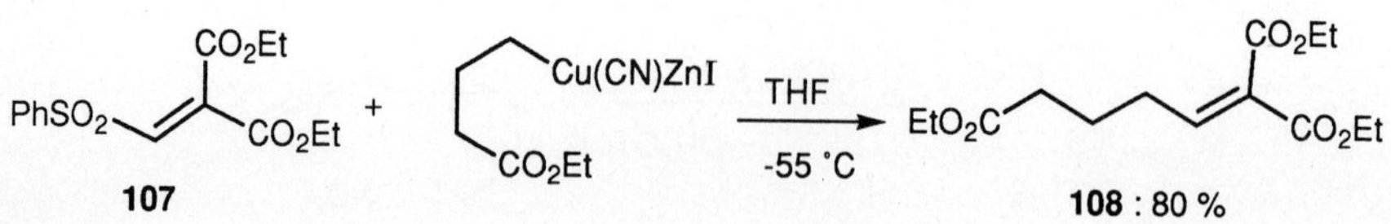

PhSO2 CO2Et CO2Et **107** + Cu(CN)ZnI CO2Et → THF, -55 °C → EtO2C CO2Et CO2Et **108** : 80 %

Scheme 5-70

nitro olefins such as **110** (Scheme 5-71) [19, 38]. The reaction of zinc–copper species with β-iodoenones proceeds especially well and allows the synthesis of a variety of polyfunctional enones [10]. Polyfunctional squaric acid derivatives are of interest for the preparation of new materials. They can be prepared by the reaction of 3,4-dichlorocyclobutene-1,2-dione (**111**) with zinc-copper organometallics (Scheme 5-72) [258]. Benzylic halides are less reactive than allylic halides, and under the previously developed standard conditions, no product is obtained. However, a high yield of the cross-coupling product is obtained by performing the reaction in DMPU using a mixed magnesium–zinc–copper reagent (Scheme 5-73) [259]. The reaction can be extended to cross-coupling reactions between primary or secondary organozinc compounds and functionalized primary halides (Scheme 5-74) [259]. Remarkably, the hydrogen atoms at the α-position relative to the nitro group are not deprotonated under the mild reaction conditions [259]. The copper(I)-catalyzed reaction of organozinc halides with acid chlorides allows the preparation of a range of polyfunctional ketones (Scheme 5-75) [10, 47, 48]. The acylation can be per-

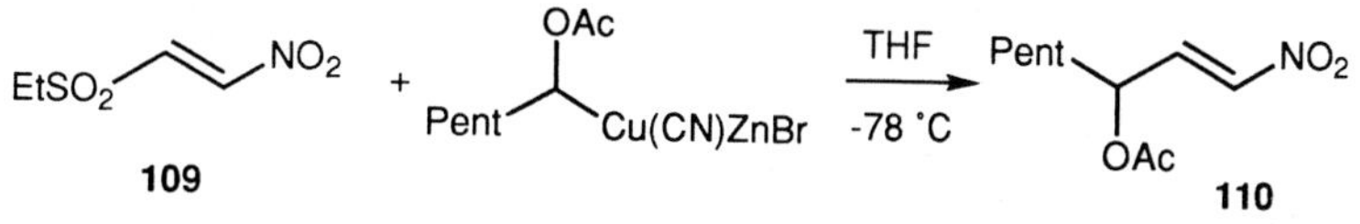

Scheme 5-71

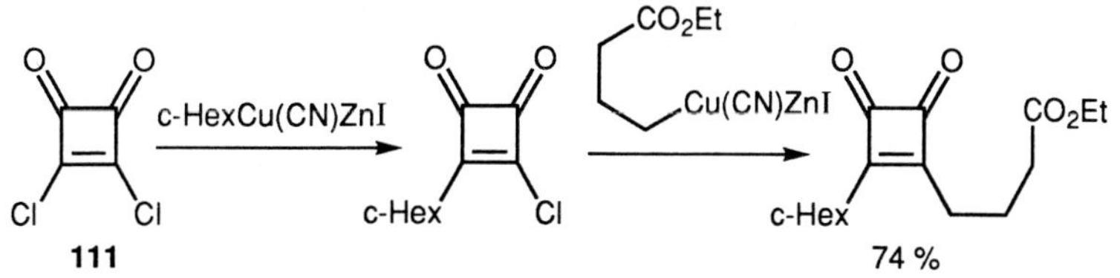

Scheme 5-72

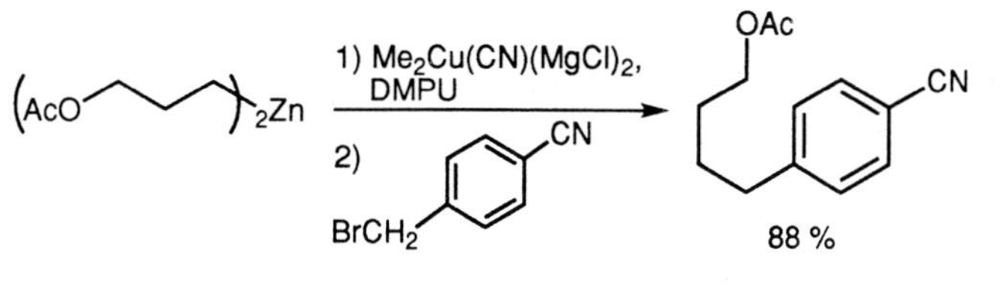

Scheme 5-73

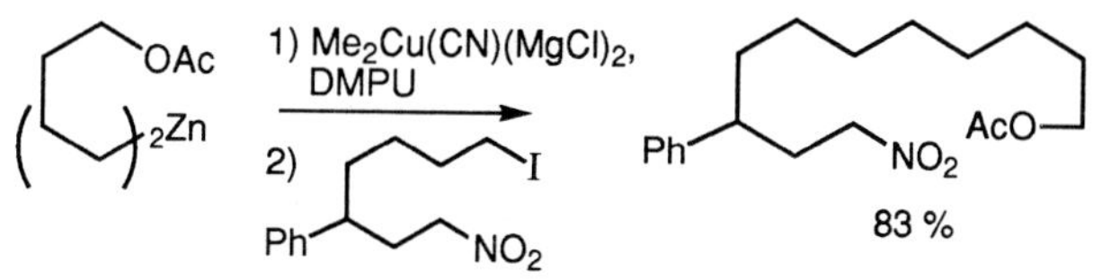

Scheme 5-74

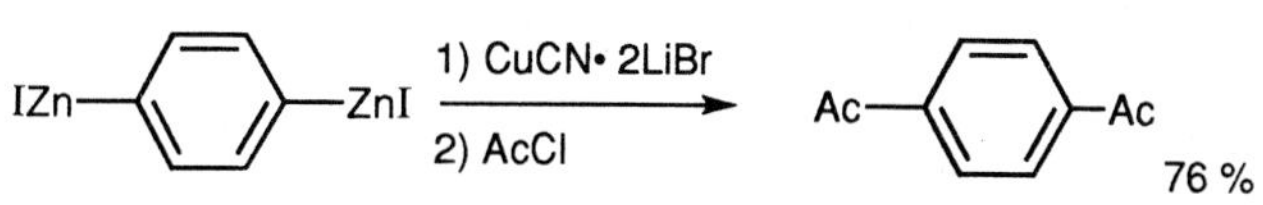

Scheme 5-75

formed either by using stoichiometric amounts of copper salts ($CuCN \cdot 2\,LiCl$) or in the presence of catalytic quantities of copper salts [13]. Whereas allylic zinc reagents add twice to esters or acid chlorides [11, 12, 260], a selective monoaddition is observed with nitriles as substrates under Barbier conditions (Scheme 5-76) [261–264].

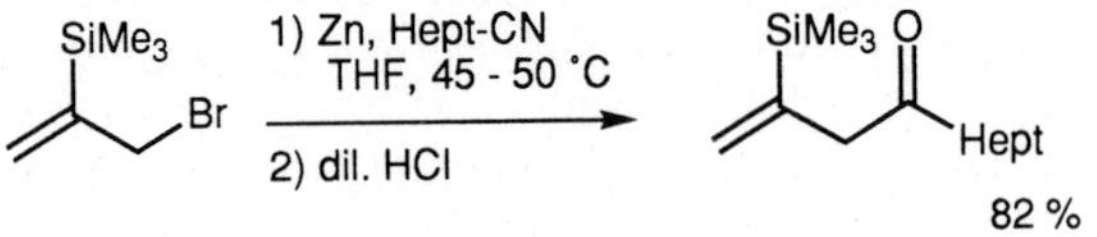

SiMe3 Br — 1) Zn, Hept-CN THF, 45 - 50 °C; 2) dil. HCl → SiMe3 O Hept 82 %

Scheme 5-76

5.2.2.2 Addition Reactions

Organozinc derivatives undergo addition reactions to carbonyl derivatives or to double bonds far less readily than organolithiums or Grignard reagents. Allylic and to some extent propargylic zinc compounds are reactive enough to add to a variety of carbonyl derivatives [260]. Additions of allylic reagents to imines can occur with very high diastereoselectivity (Scheme 5-77) [57–60, 265, 266]. The polyfunctional allylic zinc compound **112** developed by Klumpp has been used in elegant preparations of a range of heterocyclic and carbocyclic compounds (Scheme 5-78) [267–272]. Propargylic zinc derivatives do not react selectively with aldehydes or ketones, and often give a mixture of allenic and homopropargylic alcohols [260, 273, 274]. The regioselectivity depends strongly on the nature of the substituents; 1-trimethylsilylpropargyl zinc reagents add to aldehydes with excellent regioselectivity and diastereoselectivity leading to *anti*-homopropargylic alcohols (Scheme 5-79) [275]. Alkyl zinc derivatives react slowly with aldehydes [2]; however, an alkenyl zinc chloride has been added to a protected α-amino-aldehyde in satisfactory yield

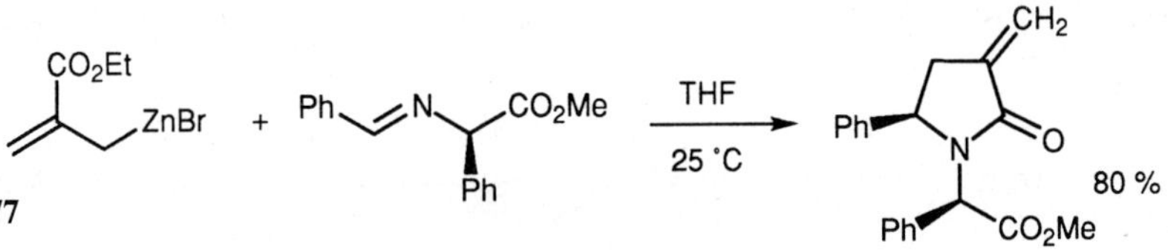

CO2Et ZnBr + Ph N CO2Me Ph — THF 25 °C → CH2 Ph N O Ph CO2Me 80 %

Scheme 5-77

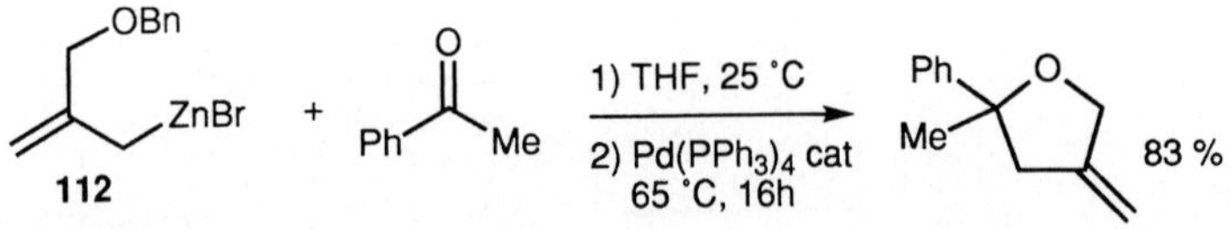

OBn ZnBr 112 + O Ph Me — 1) THF, 25 °C; 2) Pd(PPh3)4 cat 65 °C, 16h → Ph O Me 83 %

Scheme 5-78

SiMe3 Pr ZnCl + O Pr H — THF, 25 °C → OH SiMe3 Pr Pr 90 %

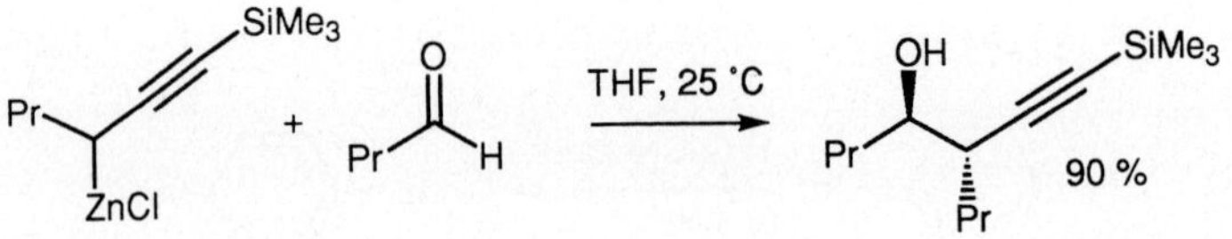

Scheme 5-79

(Scheme 5-80] [276]. Zinc homoenolates have also been added to aldehydes [277]. The rate of addition can be improved by performing the addition in the presence of Lewis acids. Interesting results have been reported using chlorotrimethylsilane [71] (Scheme 5-81) or titanium alkoxides [70, 277, 278]. Boron trifluoride etherate seems to be a very powerful catalyst for this reaction [20]. With a conjugated unsaturated aldehyde such as cinnamaldehyde, the nature of the addition product depends strongly on the nature of the catalyst used. With boron trifluorate etherate only the 1,2-addition product **113** is obtained, whereas in the presence of chlorotrimethylsilane the selective formation of the 1,4-addition product **114** is observed (Scheme 5-82) [20]. The addition of zinc–copper reagents to aldehydes bearing a chiral center at the α-position proceeds with a diastereoselectivity similar to that observed with organotitanium reagents (Scheme 5-83) [20, 279–281]. By using a non-complexing solvent such as dichloromethane, alkyl zinc iodides react directly with activated α-oxygenated aldehydes such as **115** leading to the addition product with excellent diastereoselectivity in the "matched" case (Scheme 5-84) [282]. 2-Bromomethyl-

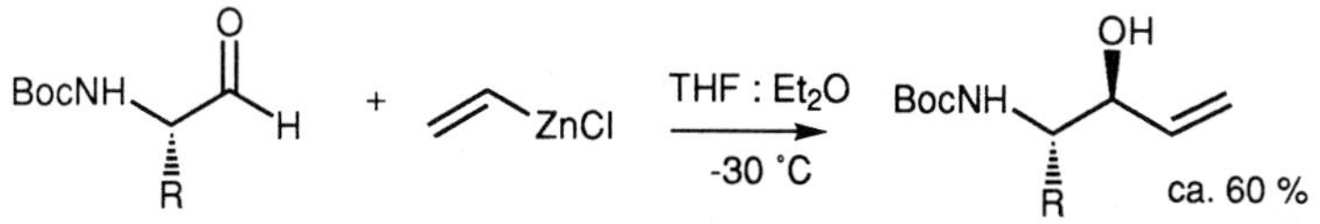

BocNH, R, O, H + ZnCl; THF : Et_2O, -30 °C; BocNH, R, OH; ca. 60 %

Scheme 5-80

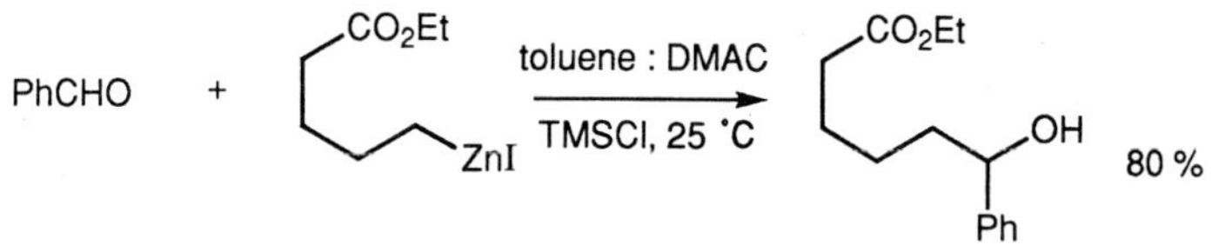

PhCHO + CO_2Et, ZnI; toluene : DMAC, TMSCl, 25 °C; CO_2Et, OH, Ph; 80 %

Scheme 5-81

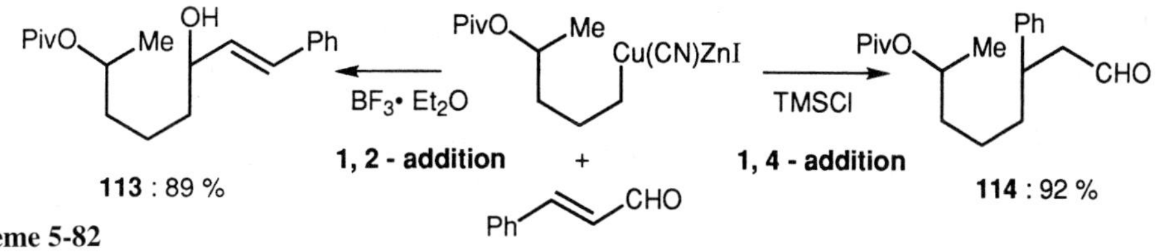

PivO, Me, Cu(CN)ZnI + Ph, CHO; BF_3• Et_2O, **1, 2 - addition** → PivO, Me, OH, Ph, **113** : 89 %; TMSCl, **1, 4 - addition** → PivO, Me, Ph, CHO, **114** : 92 %

Scheme 5-82

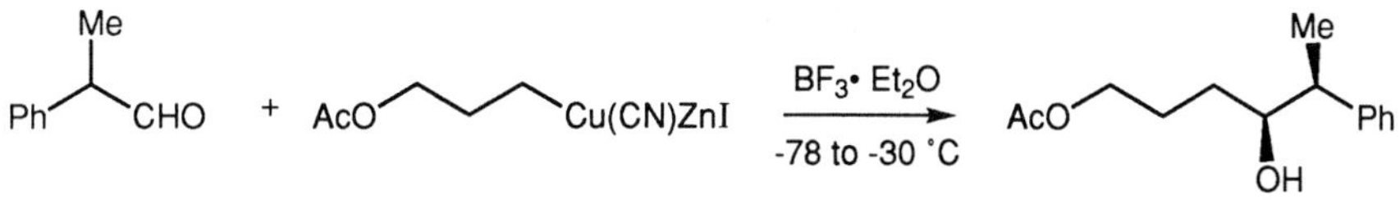

Ph, Me, CHO + AcO, Cu(CN)ZnI; BF_3• Et_2O, -78 to -30 °C; AcO, Me, Ph, OH; 77 %; d.r. = 83 : 17

Scheme 5-83

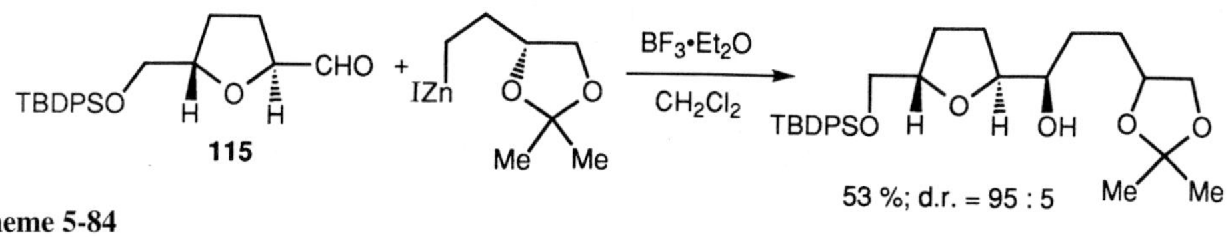

TBDPSO, H, O, H, CHO, **115** + IZn, O, O, Me, Me; BF_3•Et_2O, CH_2Cl_2; TBDPSO, H, O, H, OH, O, O, Me, Me; 53 %; d.r. = 95 : 5

Scheme 5-84

oxazoles such as **116** can be easily converted to the corresponding zinc reagents. The enolate character of these reagents confers on them a high reactivity, and their additions to various aldehydes or ketones proceed with excellent yields (Scheme 5-85) [283]. Synthetic applications reported include the preparation of polyoxygenated metabolites of unsaturated fatty acids (Scheme 5-86) [284] and (perfluoralkyl)arylcarbinols [285]. Ketones do not add most alkyl zinc derivatives, even in the presence of a Lewis acid catalyst [2, 10]. Similarly, the addition to imines is difficult and can be achieved only with particularly reactive substrates [286–291]. The *N*-acylation of imine derivatives leads to immonium salts which readily add zinc or zinc–copper reagents (Scheme 5-87) [292–302]. Whereas the direct addition of zinc–copper reagents to carbon monoxide does not occur, the addition of polyfunctional dialkyl zincs to chromiumpentacarbonyl–tetrahydrofuran complex (Scheme 5-88) produces an alkyl-chromium intermediate **117** [303–305], which undergoes

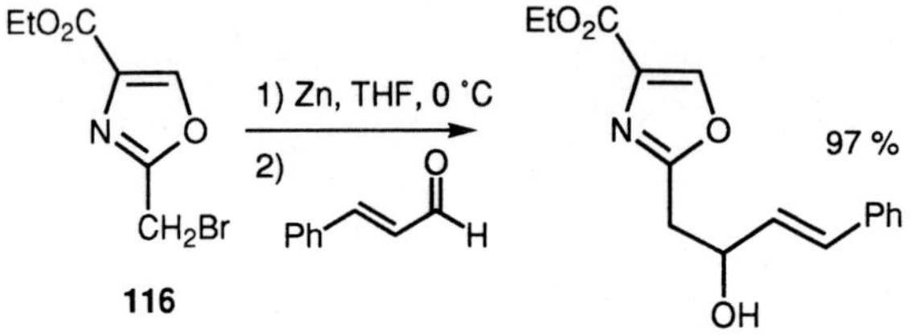

Scheme 5-85

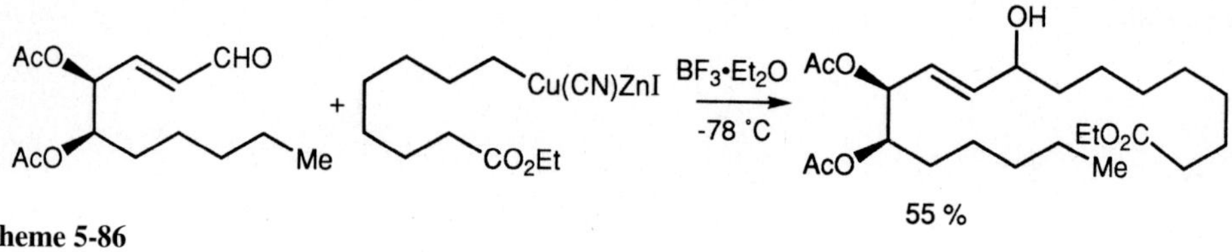

Scheme 5-86

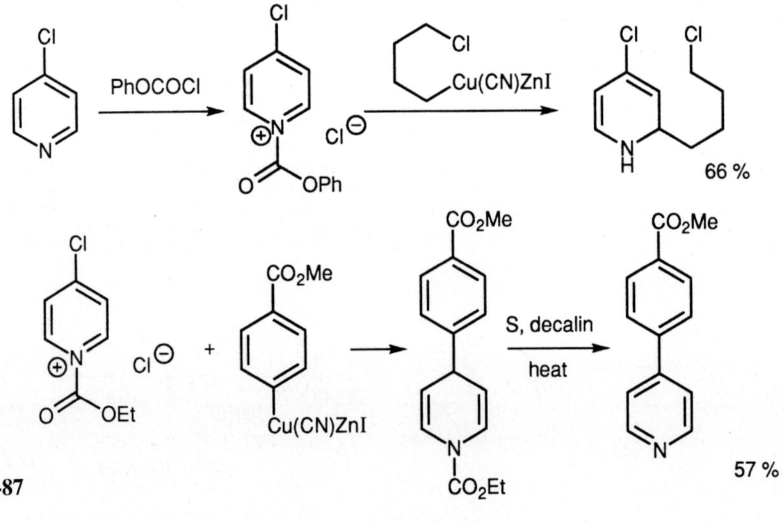

Scheme 5-87

a 1,2-migration leading to an acyl-metalate **118**. Trapping this with a Meerwein salt in dichloromethane furnishes a Fischer carbene complex **119** in 39% overall yield. This reaction sequence allows the preparation of a range of new polyfunctional chromium-carbene complexes [306]. Lithium- and magnesium-derived organocopper reagents add to various activated double bonds (Michael addition) in excellent yield and regioselectivity [307–309]. It has been shown that zinc–copper reagents prepared by the transmetallation of organozinc reagents with $CuCN \cdot 2\,LiCl$ [13, 310] add to β-monosubstituted enones in the presence of chlorotrimethylsilane [220–230], leading selectively to the 1,4-adduct (Scheme 5-89) [13]. The addition to β-disubstituted enones does not proceed under these conditions. However, the use of a polar solvent facilitates the addition to these hindered Michael-acceptors. In such solvents methyl acrylate gives the expected Michael-adduct (Scheme 5-90) [69]. Instead of using a polar solvent, it is also possible to activate the sterically hindered enone through the addition of boron trifluoride etherate. In the presence of this Lewis acid the 1,4-addition can be performed in THF with satisfactory yields (Scheme 5-91) [23]. Al-

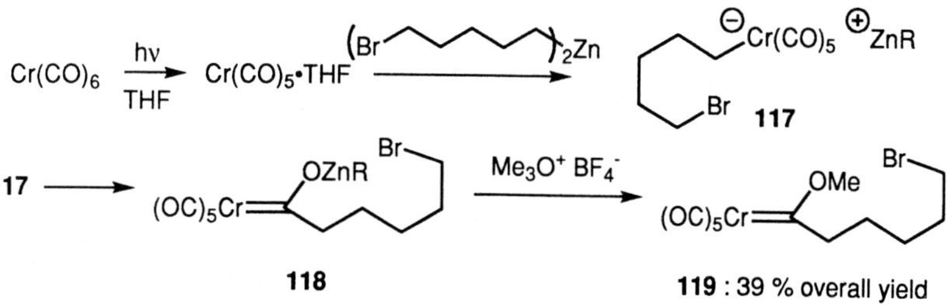
Cr(CO)6
hv
THF
Cr(CO)5•THF
(Br)2Zn
Cr(CO)5
ZnR
Br
117
117
OZnR
Br
(OC)5Cr
Me3O+ BF4-
OMe
Br
(OC)5Cr
118
119 : 39 % overall yield

Scheme 5-88

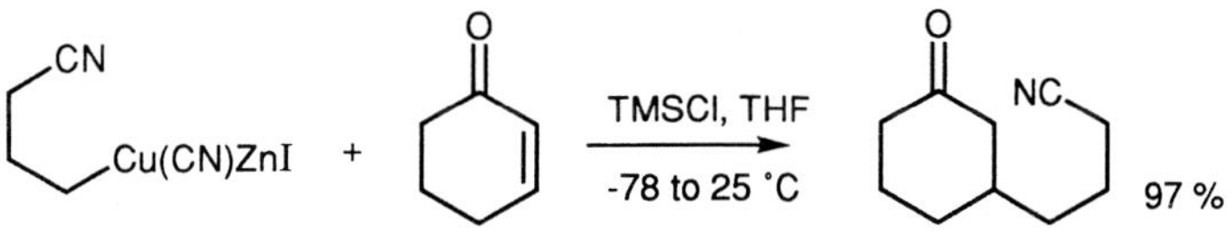
CN
Cu(CN)ZnI
+
O
TMSCl, THF
-78 to 25 °C
O
NC
97 %

Scheme 5-89

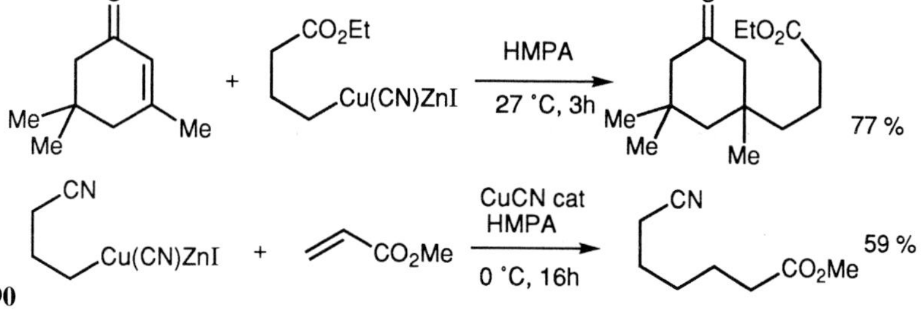
O
Me
Me
Me
+
CO2Et
Cu(CN)ZnI
HMPA
27 °C, 3h
O
EtO2C
Me
Me
Me
77 %
CN
Cu(CN)ZnI
+
CO2Me
CuCN cat
HMPA
0 °C, 16h
CN
CO2Me
59 %

Scheme 5-90

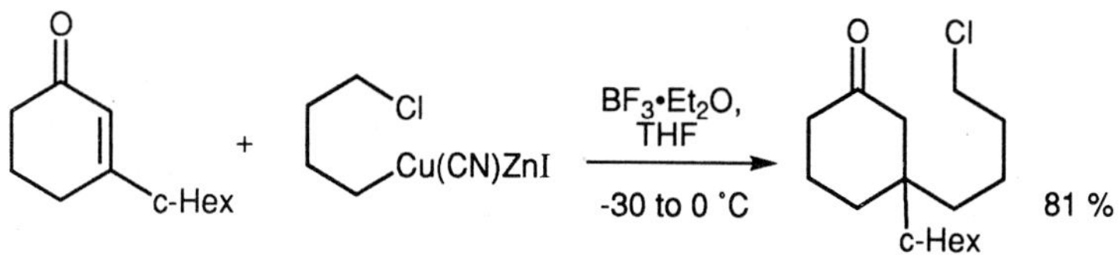
O
c-Hex
+
Cl
Cu(CN)ZnI
BF3•Et2O,
THF
-30 to 0 °C
O
Cl
c-Hex
81 %

Scheme 5-91

kenyl-copper species prepared from alkenyl-zirconium derivatives can be added to enones. The transmetallation from zirconium to zinc and copper is performed by using lithium trimethylzincate, a soft but highly reactive transmetallating agent (Scheme 5-92) [311]. Alkynyl zinc halides undergo 1,4-addition to enones in the presence of *tert*-butyldimethylsilyl triflate [312]. The intramolecular 1,4-addition of an organozinc allows the preparation of bicyclic ketones (Scheme 5-93) [313]. The conjugated addition of functionalized zinc–copper reagents has found several applications in the field of natural products chemistry [314–316], as illustrated by the synthesis of prostaglandin derivatives such as **120** (Scheme 5-94) [314]. Although the Michael addition of zinc–copper organometallics to unsaturated esters occurs only slowly, the addition to alkylidenemalonates proceeds smoothly (Scheme 5-95) [164]. Nitro-olefins are excellent Michael-acceptors and add many types of nucleophilic reagents [317–321]; although alkyl-lithiums can be added to some nitro-olefins with satisfactory yields [322–324], the addition of lithium cuprates has been far less successful [325–326]. This may be due to the high reactivity of the resulting

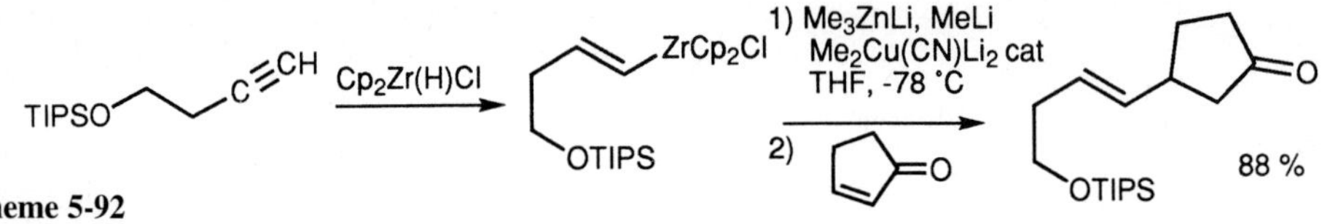

TIPSO, C≡CH, $Cp_2Zr(H)Cl$, $ZrCp_2Cl$, OTIPS, 1) Me_3ZnLi, MeLi, $Me_2Cu(CN)Li_2$ cat, THF, -78 °C, 2), O, OTIPS, 88 %

Scheme 5-92

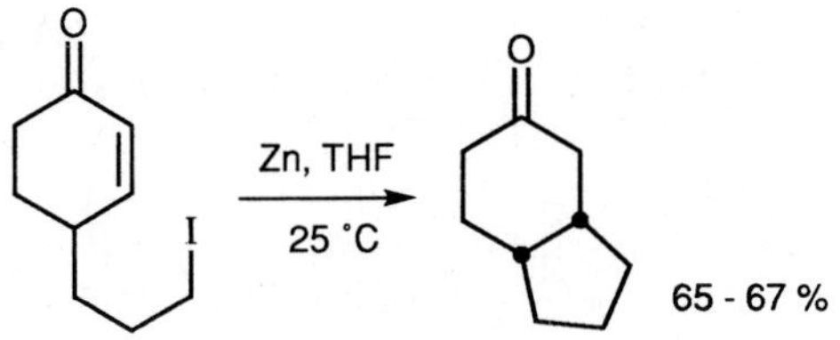

O, I, Zn, THF, 25 °C, O, 65 - 67 %

Scheme 5-93

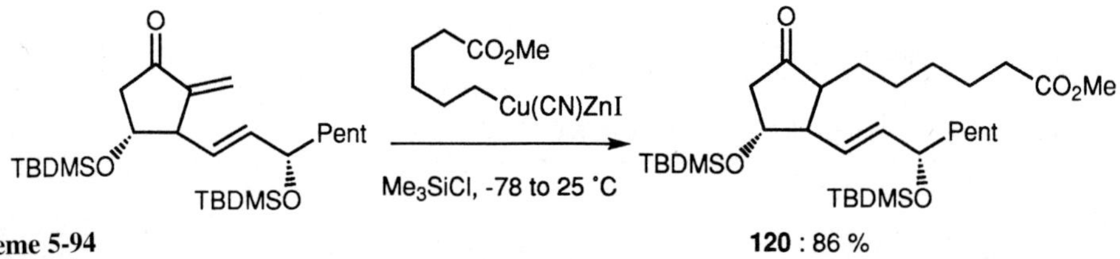

O, TBDMSO, Pent, TBDMSO, CO_2Me, Cu(CN)ZnI, Me_3SiCl, -78 to 25 °C, O, CO_2Me, TBDMSO, Pent, TBDMSO

Scheme 5-94

120 : 86 %

t-BuO, O, Cu(CN)ZnI, +, c-Hex, CO_2Et, CO_2Et, THF, 25 °C, c-Hex, CO_2Et, t-BuO, CO_2Et, O, 79 %

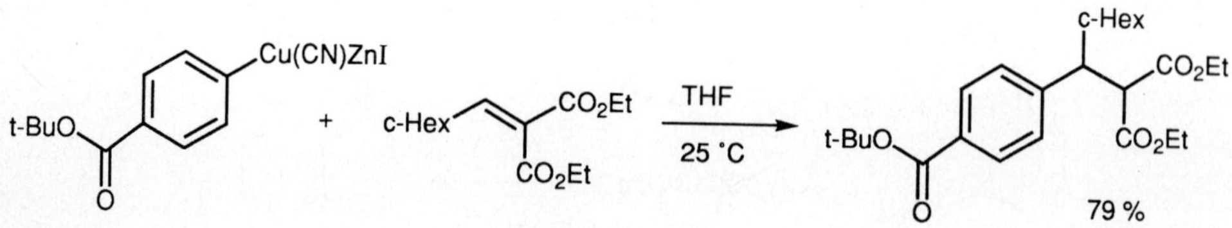

Scheme 5-95

lithium nitronates obtained after addition, which polymerize the starting nitro-olefin. In strong contrast, copper–zinc organometallics add to a range of nitro-olefins affording polyfunctional nitro compounds with excellent yields (Scheme 5-96) [22, 38, 104, 327]. Interestingly, the intermediate zinc–copper nitronates can be readily oxidized with ozone, affording polyfunctional ketones such as **121** (Scheme 5-97) [327]. The presence of a leaving group in the β-position relative to the nitro group allows the preparation of nitro-olefins (Scheme 5-98) [22, 38, 104, 322–324, 327]. A rapid synthesis of the nitrotriene **122** can be achieved by the addition of the dienyl zinc–copper reagent **123** to the β-sulfonylnitroethylene **124**. Its efficient cyclization on silica gel provides stereospecifically the bicyclic nitro derivative **125** (Scheme 5-99) [22, 38]. β-Disubstituted nitro-olefins are difficult to prepare by a nitro-aldol reaction due to the reversibility of this reaction. Denmark has developed an ingenious synthesis of these molecules using the addition of polyfunctional zinc–copper organometallics to a nitro-olefin followed by a phenylselenation and

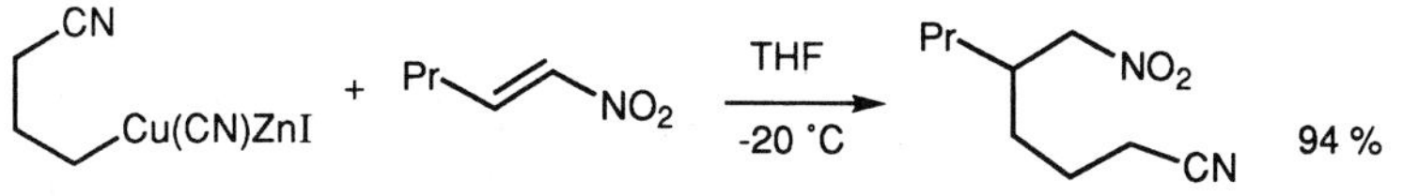

CN
Cu(CN)ZnI + Pr NO2
THF
-20 °C
Pr NO2 CN 94 %

Scheme 5-96

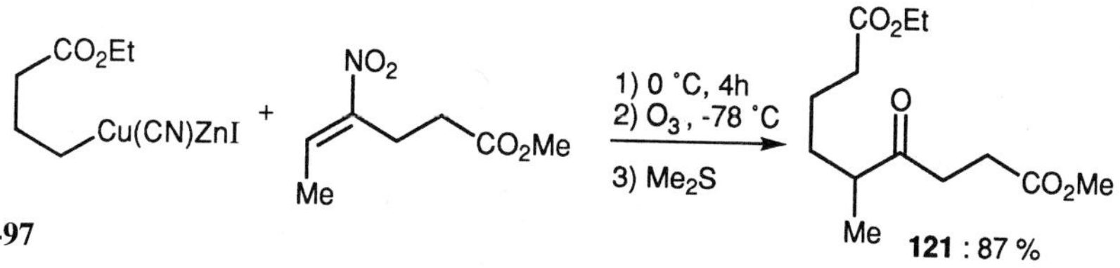

CO2Et
Cu(CN)ZnI + NO2 Me CO2Me
1) 0 °C, 4h
2) O3, -78 °C
3) Me2S
CO2Et O Me CO2Me **121** : 87 %

Scheme 5-97

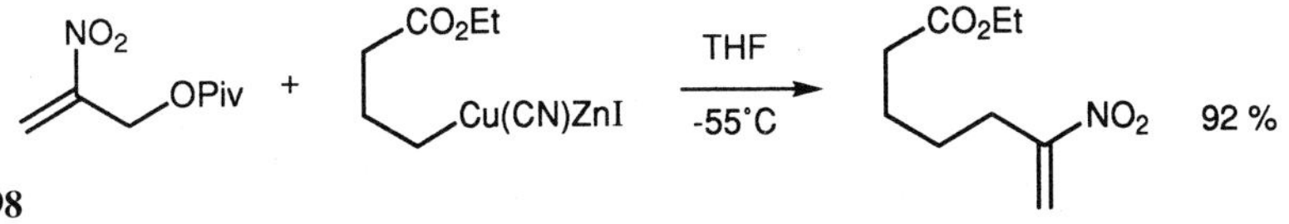

NO2 OPiv + CO2Et Cu(CN)ZnI
THF
-55°C
CO2Et NO2 92 %

Scheme 5-98

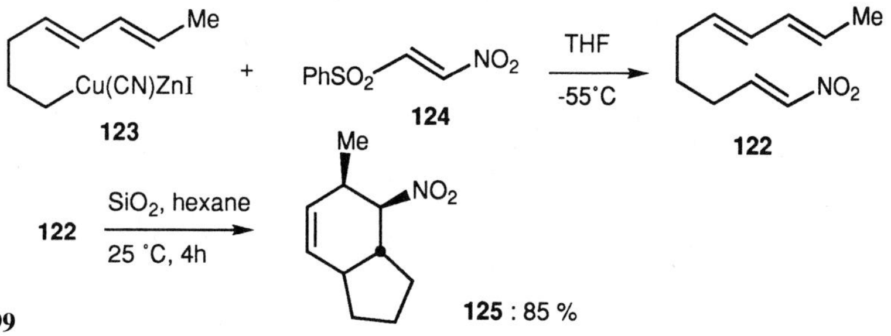

Me Cu(CN)ZnI **123** + PhSO2 NO2 **124**
THF
-55°C
Me NO2 **122**

122
SiO2, hexane
25 °C, 4h
Me NO2 **125** : 85 %

Scheme 5-99

oxidative elimination (Scheme 5-100) [328]. Acetylenic esters are excellent Michael-acceptors and add zinc–copper reagents. By performing the addition to ethyl propiolate at low temperature, one obtains only the (*E*) unsaturated esters, e.g. **126** (Scheme 5-101). On warming up, a loss of the stereochemistry of the intermediate α-carbethoxyalkenyl-copper is observed. By performing the reaction in the presence of chlorotrimethylsilane a stereochemically pure α-silylated unsaturated ester such as **127** is obtained [253]. The addition to higher acetylenic esters does not proceed at low temperature, the reaction mixture must be allowed to warm up. This usually results in isomerization of the alkenyl-copper intermediate and produces an *E*/*Z* mixture of unsaturated esters [253]. Other acetylenic carboxylic derivatives, such as dimethylacetylenedicarboxylate [19, 25, 195] or propiolamide (**128**) similarly add zinc–copper compounds (Scheme 5-102) [25].

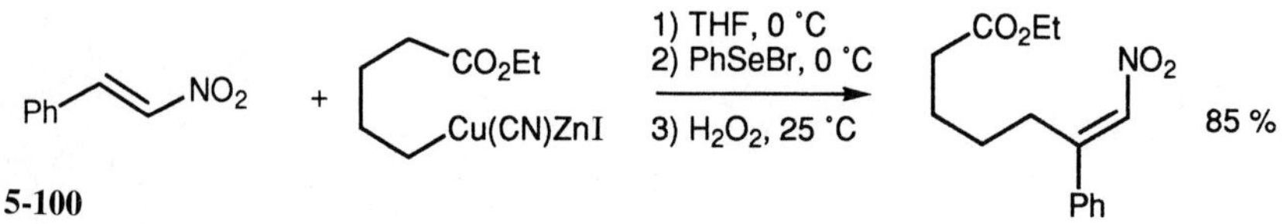

Scheme 5-100

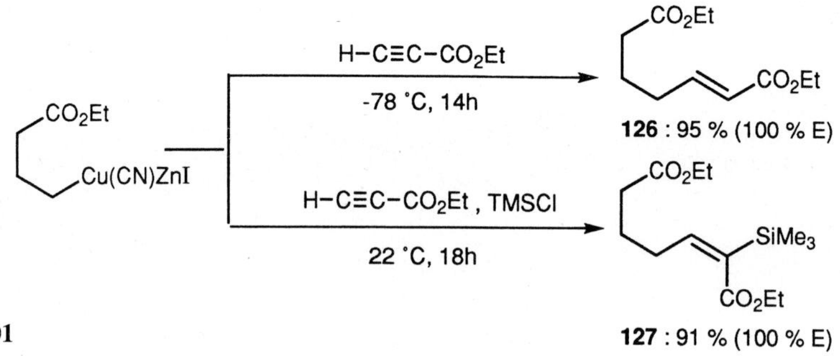

Scheme 5-101

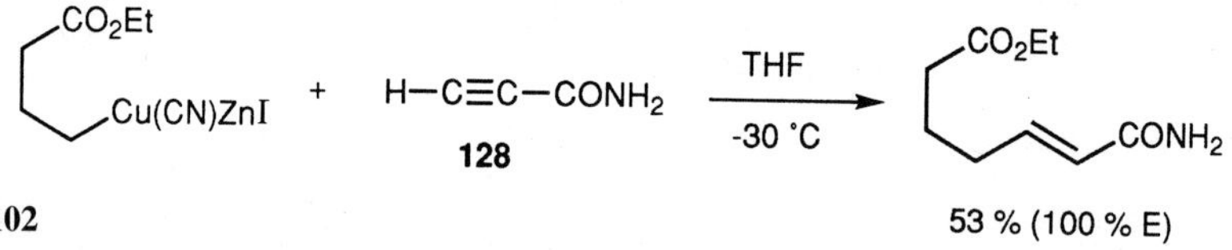

Scheme 5-102

By performing the addition of bis-(2-carbethoxyethyl)zinc to an acetylenic ester in a solvent mixture containing HMPA, a cyclization occurs after the addition to the alkyne, providing a 2-carbethoxycyclopentenone (Scheme 5-103) [329–331]. This new cyclization method has been applied by Crimmins to prepare a precursor of ($\pm$)-bilobalide [331]. The addition of functionalized copper–zinc reagents to less activated alkynes can be realized by treating the zinc–copper species with $Me_2Cu(CN)Li_2$ [332–333]. Thus, the addition to 1-thiomethylalkynes produces polyfunctional trisubstituted alkenylthioethers (Scheme 5-104) [332]. The addition to non-activated alkynes [332] is not possible, but acetylene itself adds *secondary* polyfunctional zinc–copper reagents and after iodolysis affords (*Z*)-disubstituted alkenyl iodides in fair yield (Scheme 5-105) [332]. By performing the reaction intramolecularly, the addition to a disubstituted alkyne can be realized under these reaction conditions. It allows the stereoselective preparation of cyclopentane derivatives such as **129** in a satisfactory yield (Scheme 5-106) [332]. Substituted zinc malonates can be added to non-activated alkynes providing β,γ-unsaturated malonates (Scheme 5-107) [334]. As expected, the addition of zinc and copper organometallics to non-activated olefins and alkynes is difficult [334–337]. The use of allylic zinc reagents leads to the best results [338–342], and Klumpp has reported an efficient preparation of bicyclic cyclopentanes using this reaction (Scheme 5-108) [343]. Recently, Marek has shown that the intramolecular carbometallation of olefins can be realized by carefully controlling the reaction conditions (solvent, salts). The zinc reagent has been generated by using Rieke's method [48–53], and allows the preparation of various cyclopentane derivatives [344–345].

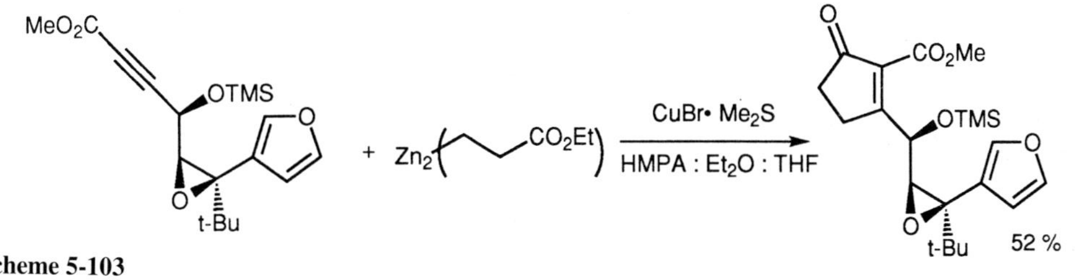

MeO2C, OTMS, O, t-Bu + Zn_2(~CO2Et) CuBr• Me2S / HMPA : Et2O : THF → O, CO2Me, OTMS, O, t-Bu 52 %

Scheme 5-103

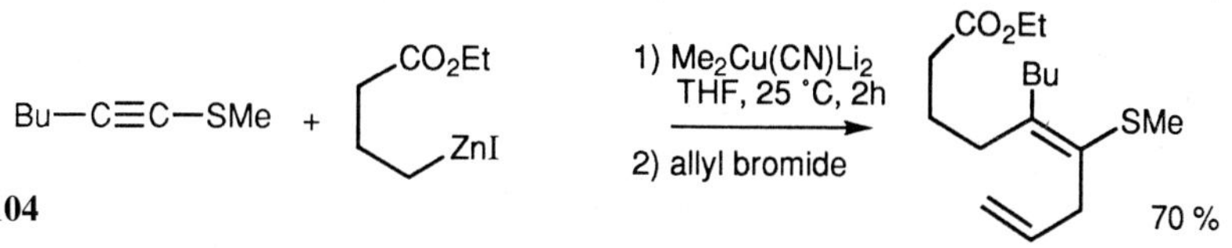

Bu—C≡C—SMe + CO2Et, ZnI 1) Me2Cu(CN)Li2 THF, 25 °C, 2h 2) allyl bromide → CO2Et, Bu, SMe 70 %

Scheme 5-104

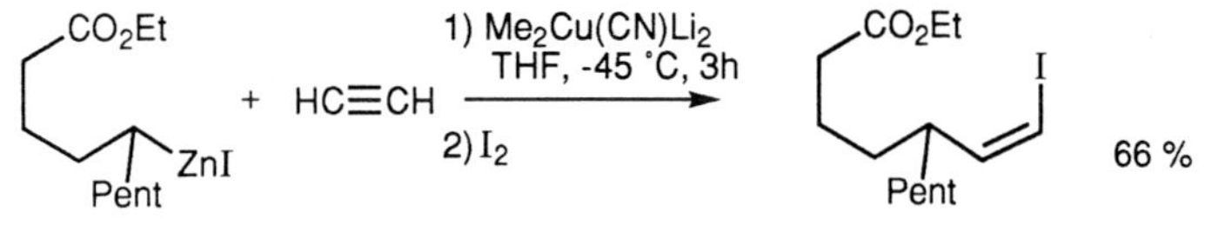

CO2Et, ZnI, Pent + HC≡CH 1) Me2Cu(CN)Li2 THF, -45 °C, 3h 2) I2 → CO2Et, I, Pent 66 %

Scheme 5-105

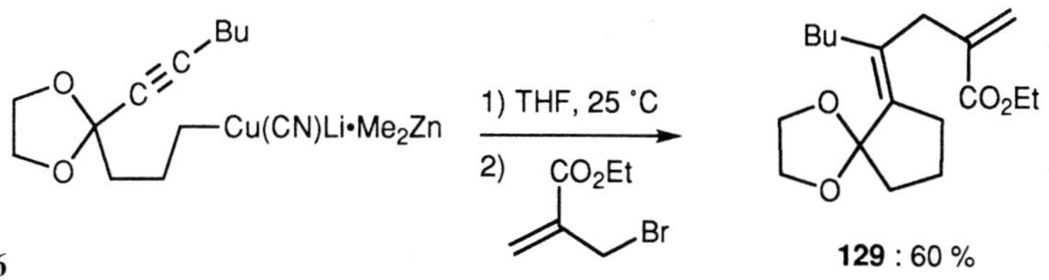

O, O, C≡C–Bu, Cu(CN)Li•Me2Zn 1) THF, 25 °C 2) CO2Et, Br → Bu, O, O, CO2Et **129** : 60 %

Scheme 5-106

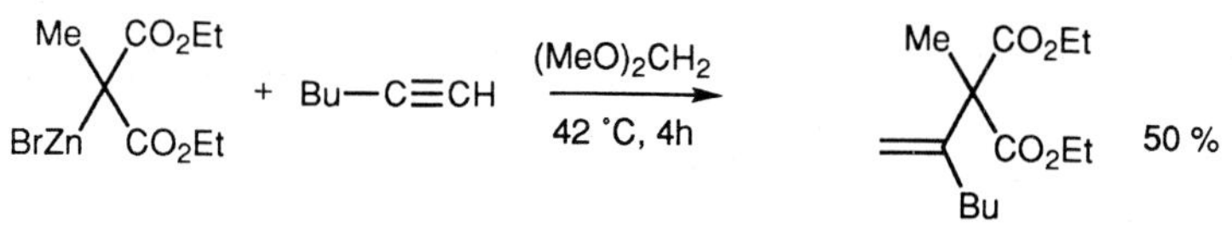

Me, CO2Et, BrZn, CO2Et + Bu—C≡CH (MeO)2CH2 42 °C, 4h → Me, CO2Et, CO2Et, Bu 50 %

Scheme 5-107

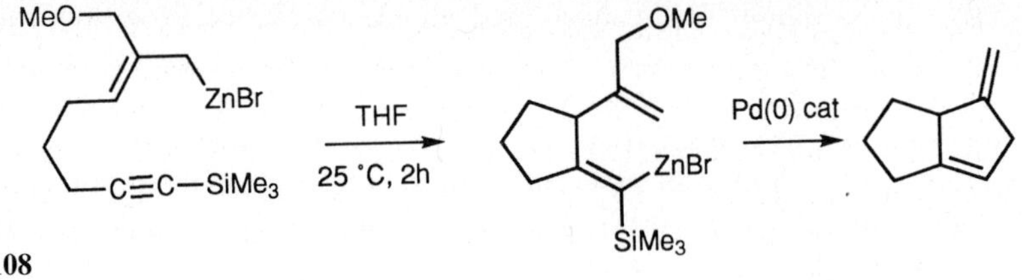

Scheme 5-108

5.2.3 Palladium(0)-Catalyzed Reactions of Organozinc Compounds

Negishi demonstrated 20 years ago that various organozinc derivatives can be coupled with aryl, heteroaryl, alkenyl, and acyl halides in the presence of catalytic amounts of palladium(0) salts [98, 346–351]. This cross-coupling reaction combines a reasonable reactivity of the zinc organometallic with a broad functional group tolerance and has found numerous synthetic applications. Polyfunctional zinc reagents [351–365] bearing groups such as a silylated acetylene [351], alkenyl-silane [353–362], allyl-silane [362], 1-alkoxyacetylene [354], polythiophene [355], various substituted aromatic rings [47, 48, 356, 361] or heterocyclic rings [357–361], ester [48, 95, 134, 364, 365], nitrile [48, 134], ketone [67], protected α-aminoester [27–30, 35, 36], stannane [365], or boronic ester [241, 367–369] have been used. The cross-coupling of zinc homoenolates with aromatic or alkenyl halides proceeds especially well by using bis-(tris-*o*-tolylphosphine)palladium(II) dichloride (**130**) as a catalyst (Scheme 5-109) [366]. The catalyst **130** can be used with other zinc reagents with good success (Scheme 5-110) [66]. Recently, some other catalysts such as $PdCl_2$(dppf) [370] and bis-(tris-*o*-furylphosphine)palladium(II) dichloride [371] have been reported. These new complexes having loosely-bound phosphine ligands give excellent results [370, 371]. Besides unsaturated halides, alkenyl and aryl triflates can be used [67, 363, 372]. This is especially advantageous when the starting triflate is readily available (Scheme 5-111) [67]. New α-amino acids such as **131** have been pre-

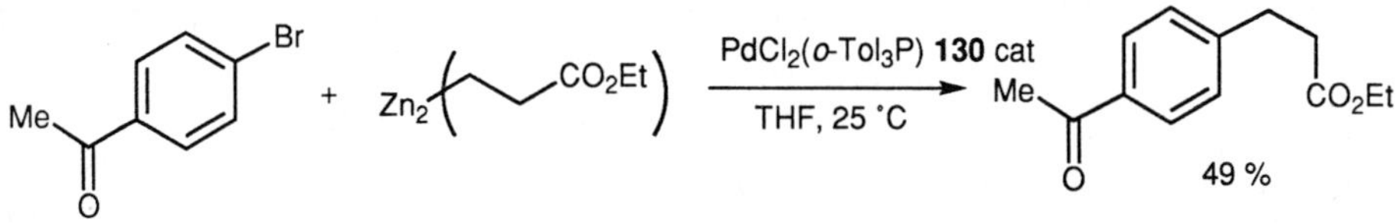

Scheme 5-109

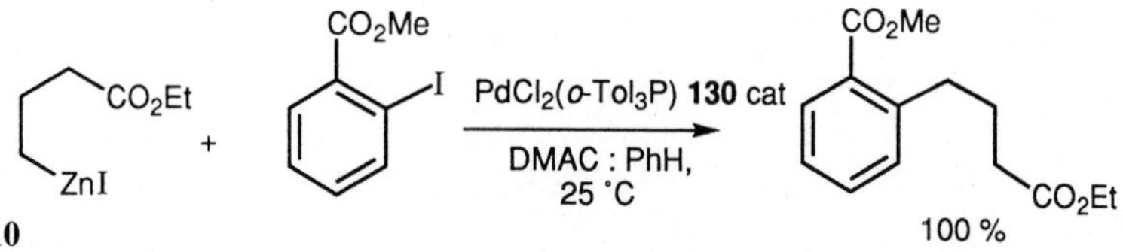

Scheme 5-110

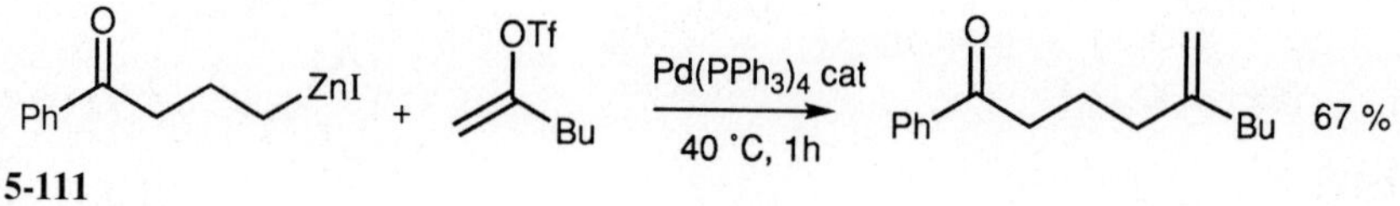

Scheme 5-111

pared using the serine-derived organozinc iodide **132** (Scheme 5-112) [27–30, 34, 35]. Polyfluorinated dienes and enynes can be readily prepared by palladium-catalyzed cross-coupling reactions of polyfluorinated alkenyl halides (Scheme 5-113) [165–171]. Intermediate polyfunctional allylic boronic esters are obtained by the palladium(0)-catalyzed coupling between an alkenyl iodide and the α-boronylzinc reagent **133**, and after intramolecular carbonyl addition these afford cyclized products such as **134** (Scheme 5-114) [367–369]. A palladium(0)-catalyzed acylation allows the synthesis of polyfunctional ketones [347–348]. In many cases functionalized substrates have been used (Scheme 5-115) [48, 64, 72]. The acylation of the zinc reagent **132** with various acid chlorides produces new amino-acids such as **135** (Scheme 5-116) [27–36]. Interestingly, reaction of **135** with phenyl chloroformate affords the double acylation product **136**. Further extensions of the Negishi coupling reaction will certainly be reported in the future [373].

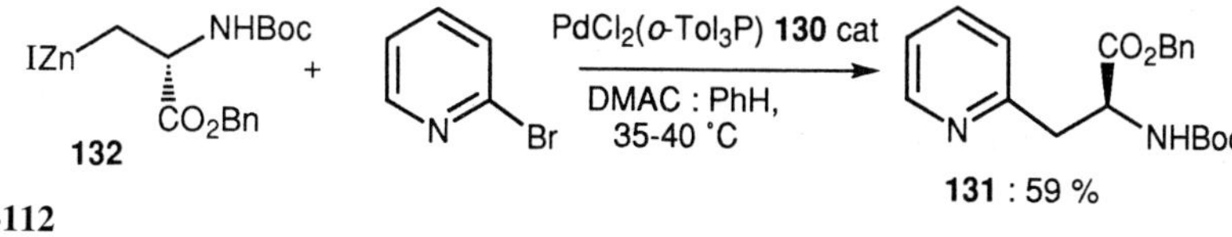

IZn NHBoc CO2Bn **132** + 2-bromopyridine; $PdCl_2(o\text{-}Tol_3P)$ **130** cat; DMAC : PhH, 35-40 °C; **131** : 59 %

Scheme 5-112

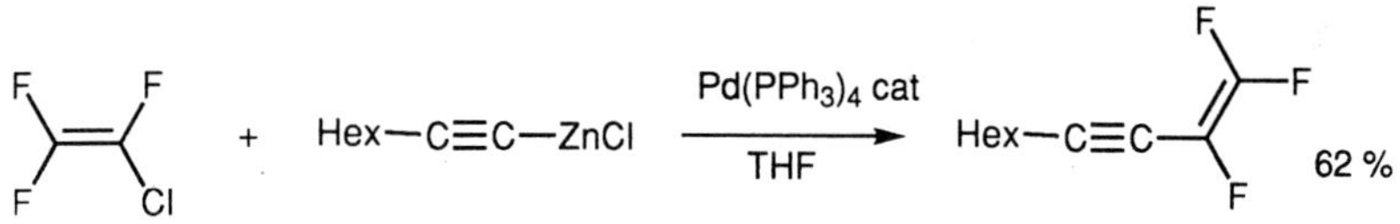

Hex—C≡C—ZnCl; $Pd(PPh_3)_4$ cat; THF; 62 %

Scheme 5-113

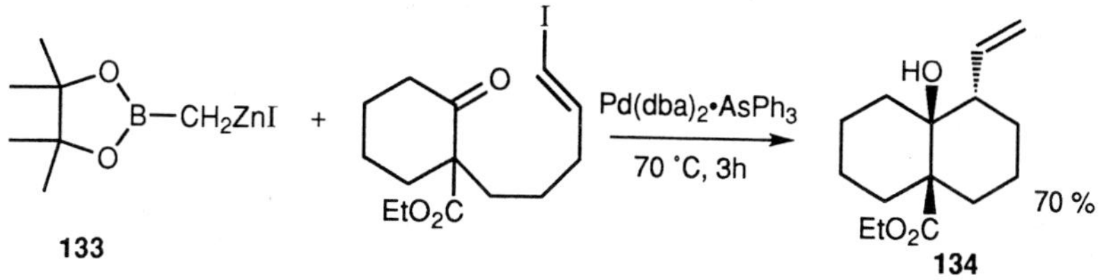

B—CH_2ZnI **133**; EtO_2C; $Pd(dba)_2$•$AsPh_3$; 70 °C, 3h; HO; EtO_2C **134**; 70 %

Scheme 5-114

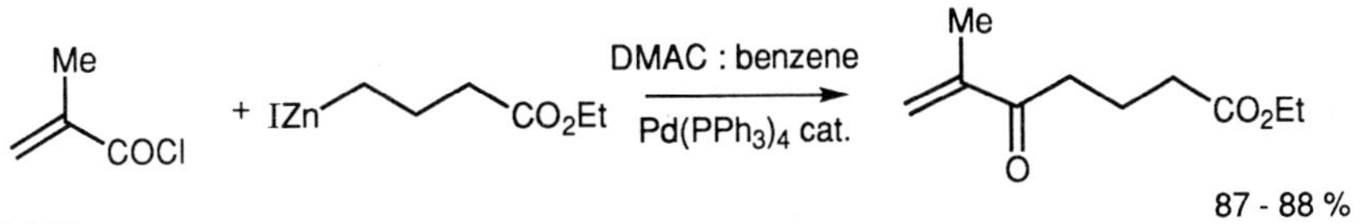

Me COCl + IZn CO2Et; DMAC : benzene; $Pd(PPh_3)_4$ cat.; Me, O, CO_2Et; 87 - 88 %

Scheme 5-115

135 : 90 %; COCl; Pd(0) cat.; IZn NHBoc CO_2Bn **132**; PhOCOCl; Pd(0) cat.; BnO_2C NHBoc; **136** : 45 %

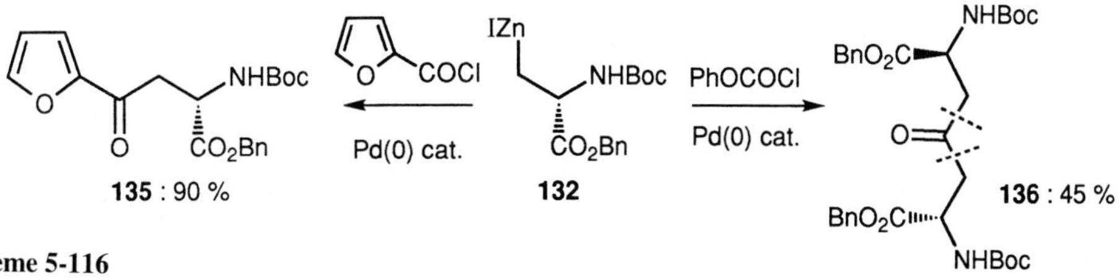

Scheme 5-116

5.2.4 Titanium(IV)-Catalyzed Reactions of Organozinc Compounds and their Applications in Asymmetric Synthesis

It has been shown that diethylzinc can be added to aldehydes in the presence of chiral catalysts such as **61** or **137–139** with high enantioselectivity (Scheme 5-117) [374–397]. The addition of other diorganozincs, especially functionalized organometallics, is of great synthetic interest. Since diethylzinc is among the most reactive organozinc reagents, the use of very active catalysts is essential if higher dialkyl zincs or functionalized zinc reagents are to be added. Yoshioka and Ohno [278] as well as Seebach [376–384] found that the addition of titanium alkoxides produces titanium complexes of very high catalytic activity which allow the addition of diethylzinc to various aldehydes with high enantioselectivity and under very mild conditions. These catalysts also allow the addition of higher dialkyl zincs of polyfunctional zinc reagents. Seebach found that the transmetallation of alkenyl magnesium halides with zinc chloride in ether followed by the addition of 1,3-dioxane constitutes a convenient method for preparing higher salt-free dialkyl zincs [380]. A few functionalized dialkyl zincs can be obtained by this method and have been added to aldehydes in the presence of $\alpha,\alpha,\alpha',\alpha'$-tetraaryl-1,3-dioxolane-4,5-dimethanol (TADDOL, **137**) with high enantioselectivity (Scheme 5-118) [380]. Functionalized propargylic alcohols with high optical purity can be prepared by the addition of bis-(3-butenyl)zinc to acetylenic aldehydes using this catalyst (Scheme 5-119) [382]. Interestingly, lithium and magnesium reagents can also be directly converted to trialkoxy-titanium organometallics. This approach allows a more efficient transfer of the organic moiety to the aldehyde, since by using dialkyl zincs no more than one of the two organic groups is transferred to the aldehyde [383]. The iodine–zinc exchange reaction (Section 5.1.3) and the boron-zinc transmetallation (Section 5.1.4) are very general methods for preparing polyfunctional zinc organometallics, and the reagents obtained in this way have found extensive applications in asymmetric additions to aldehydes (Scheme 5-120) [104, 390–397).

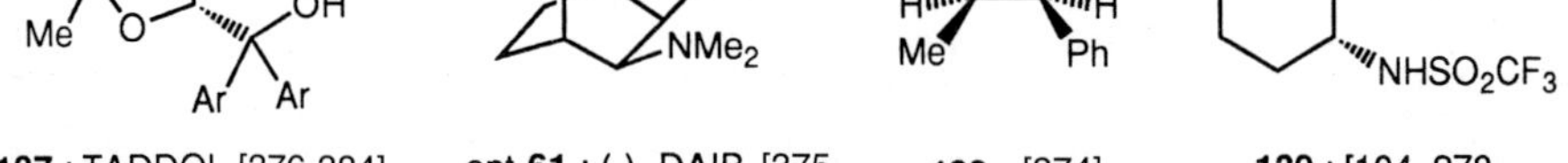

RCHO + Et_2Zn —chiral catalyst→ R–*CH(OH)–Et
high enantioselectivity

137 : TADDOL [376-384]   ent-**61** : (-)- DAIB [375, 385-387]   **138** : [374]   **139** : [104, 278, 388-397]

Scheme 5-117

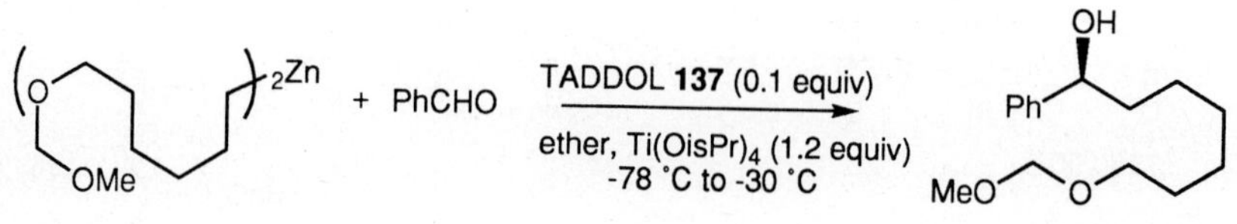

+ PhCHO —TADDOL **137** (0.1 equiv); ether, $Ti(OiPr)_4$ (1.2 equiv), -78 °C to -30 °C→

68 %; 84 %*ee*

Scheme 5-118

The bis-triflamide **139** is an excellent catalyst for these reactions and allows large structural variations of the substrates. The enantioselectivities are good for aromatic or aliphatic aldehydes. α,β-Unsaturated aldehydes react well and allow the preparation of a broad range of polyfunctional allylic alcohols with high enantioselectivity. The presence of an α-substituent in the unsaturated aldehyde enhances the enantioselectivity (Scheme 5-121) [395]. In its absence, however, it is possible to obtain excellent results by replacing the cocatalyst $Ti(O\textit{i}Pr)_4$ by $Ti(O\textit{t}Bu)_4$ (Scheme 5-122) [391]. Enantioselective and diastereoselective addition reactions can be performed and molecules with C_2 or C_3 symmetry such as **140** [393] or **141** [394] have been prepared (Scheme 5-123). Aldol products such as **142** have been obtained by the catalytic asymmetric addition of dialkyl zincs to β-alkoxyaldehydes such as **143** followed by a simple sequence of functional group interconver-

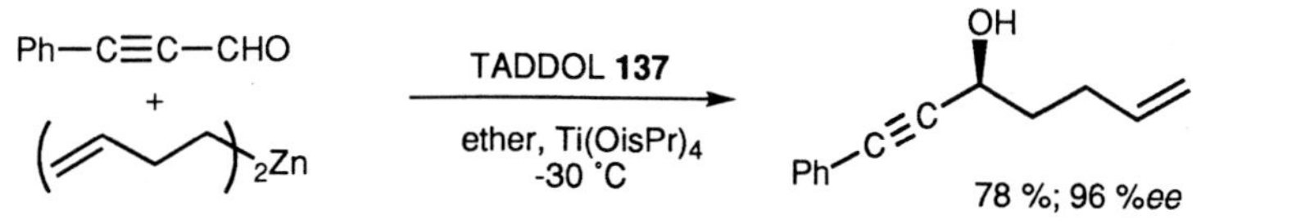

Ph—C≡C—CHO + (allyl…)$_2$Zn; TADDOL **137**; ether, $Ti(OisPr)_4$, -30 °C; 78 %; 96 %ee

Scheme 5-119

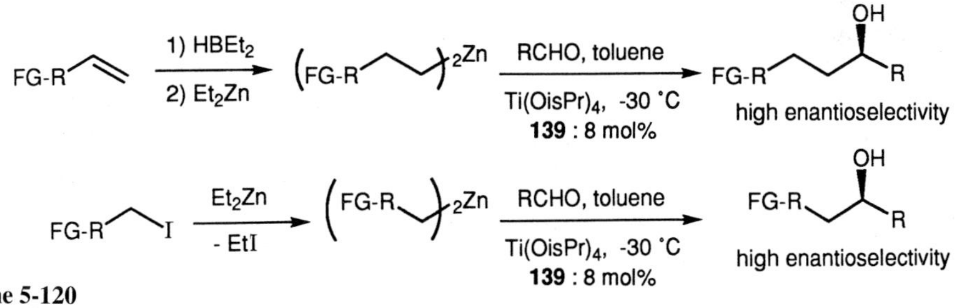

FG-R (alkene); 1) $HBEt_2$; 2) Et_2Zn; $(FG\text{-}R\ldots)_2Zn$; RCHO, toluene; $Ti(OisPr)_4$, -30 °C; **139** : 8 mol%; OH; high enantioselectivity

FG-R…I; Et_2Zn; - EtI; $(FG\text{-}R\ldots)_2Zn$; RCHO, toluene; $Ti(OisPr)_4$, -30 °C; **139** : 8 mol%; OH; high enantioselectivity

Scheme 5-120

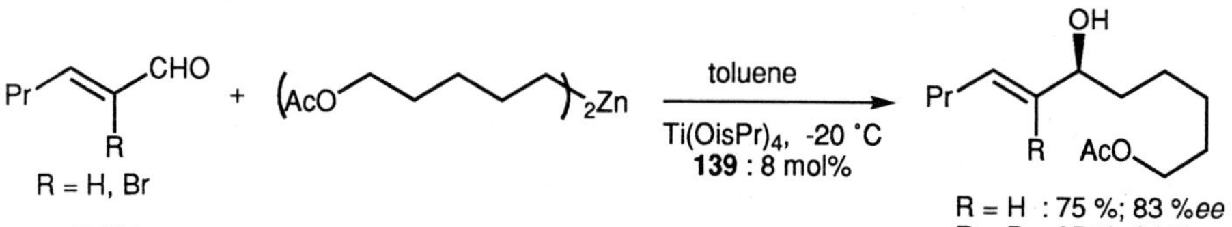

Pr…CHO, R; R = H, Br; + $(AcO\ldots)_2Zn$; toluene; $Ti(OisPr)_4$, -20 °C; **139** : 8 mol%; OH; R AcO

R = H : 75 %; 83 %ee
R = Br : 95 %; 94 %ee

Scheme 5-121

TIPSO…CHO; Me_2Zn, $Ti(Ot\text{-}Bu)_4$; **139** : 8 mol%; - 20 °C; TIPSO…OH, Me; 79 %; 92 %ee (with $Ti(Oi\text{-}Pr)_4$; 94 %; 55 %ee)

Ph…CHO; - 20 °C; Ph…Me, HO; 66%; 89 %ee (with $Ti(Oi\text{-}Pr)_4$; 97 %; 25 %ee)

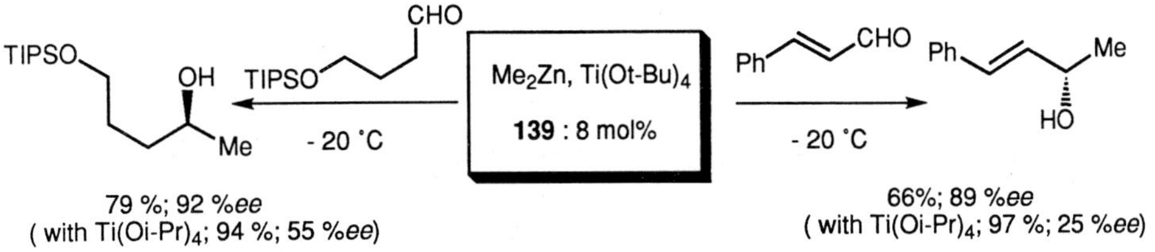

Scheme 5-122

sions (Scheme 5-124) [395]. By using α-alkoxyaldehydes such as **144**, epoxides can be obtained with good enantioselectivities (Scheme 5-125) [396]. Applications to the synthesis of natural products have been reported. Thus, the addition of bis(5-bromopentyl)zinc, prepared via a boron–zinc transmetallation, to the long-chain aldehyde **145** provides the bromo-alcohol **146** with 92% *ee* (Scheme 5-126). Alkylation of this with $Bu_2Cu(CN)Li_2$ produces 10-nonacosanol (ginnol, **147**) in three stages (61% yield and 92% *ee*) [127].

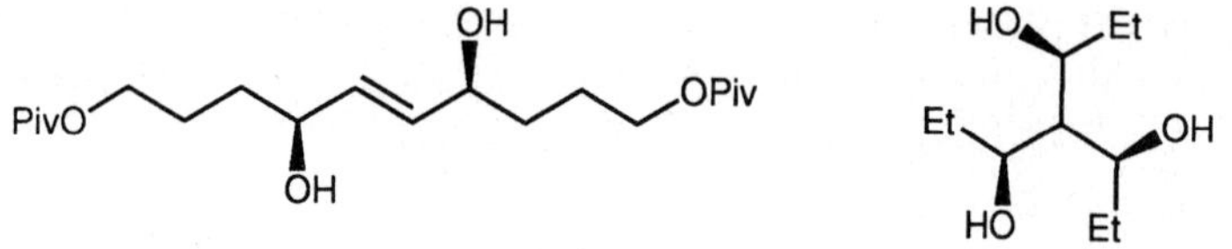

PivO OH OH OPiv **140**; HO Et Et OH HO Et **141**

Scheme 5-123

TIPSO (CHO) **143** + $(AcO(CH_2)_4)_2Zn$; 1) toluene, $Ti(Oi\text{-}Pr)_4$, -20 °C; **139** : 8 mol%; 2) TBDPSCl; 3) CF_3CO_2H; 4) PCC → OHC, OTBDPS, AcO **142** : 50 % overall yield, 91 %*ee*

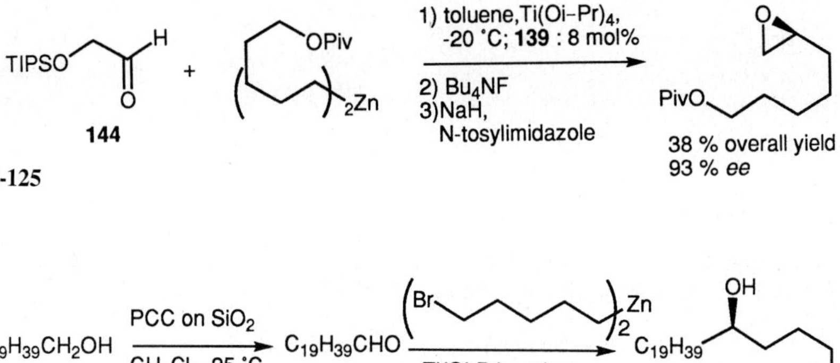

Scheme 5-124

TIPSO (CHO) **144** + $(PivO(CH_2)_4)_2Zn$; 1) toluene, $Ti(Oi\text{-}Pr)_4$, -20 °C; **139** : 8 mol%; 2) Bu_4NF; 3) NaH, N-tosylimidazole → PivO (epoxide) 38 % overall yield, 93 % *ee*

Scheme 5-125

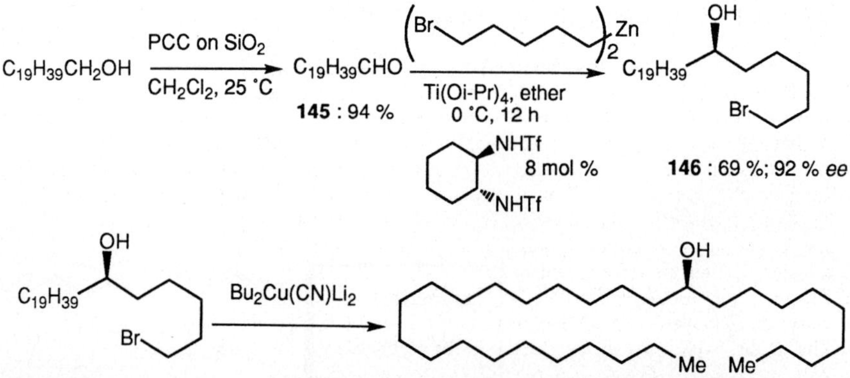

$C_{19}H_{39}CH_2OH$ → (PCC on SiO_2, CH_2Cl_2, 25 °C) $C_{19}H_{39}CHO$ **145** : 94 % → ($(Br(CH_2)_5)_2Zn$, $Ti(Oi\text{-}Pr)_4$, ether, 0 °C, 12 h; NHTf/NHTf cyclohexane 8 mol %) $C_{19}H_{39}$ OH Br **146** : 69 %; 92 % *ee*

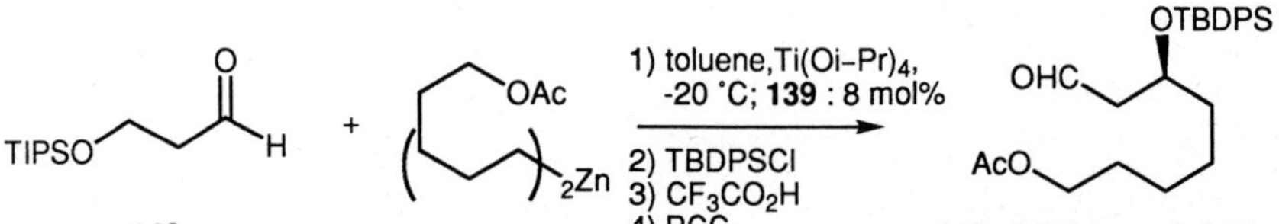

$C_{19}H_{39}$ OH Br → ($Bu_2Cu(CN)Li_2$) OH Me Me **147** : 95 %; 92 % *ee*

61 % overall yield

Scheme 5-126

References

[1] E. Frankland, *A. Chem.* **1849,** *71,* 171, 213.
[2] K. Nützel, *Methoden der Organischen Chemie, Metallorganische Verbindungen Be, Mg, Ca, Sr, Ba, Zn, Cd.* Volume 13/2a. Thieme, Stuttgart, **1973**, pp. 552–858.
[3] A. Fürstner, *Synthesis* **1989**, 571–590.
[4] R. L. Shriner, *Org. React.* **1942,** *1,* 1–37.
[5] H. E. Simmons, R. D. Smith, *J. Am. Chem. Soc.* **1958**, *80*, 5323–5332.
[6] H. E. Simmons, R. D. Smith, *J. Am. Chem. Soc.* **1959**, *81*, 4256–4264.
[7] E. P. Blanchard, H. E. Simmons, *J. Am. Chem. Soc.* **1964**, *86*, 1337, 1347–1356.
[8] H. E. Simmons, T. L. Cairns, A. Vladuchik, C. M. Hoiness, *Org. React.* **1972**, *20*, 1–111.
[9] J. Furukawa, N. Kawabata, *Adv. Organomet. Chem.* **1974**, *12*, 83–134.
[10] P. Knochel, R. D. Singer, *Chem. Rev.* **1993**, *93*, 2117–2188.
[11] M. Gaudemar, *Bull. Soc. Chim. Fr.* **1962**, 974–987.
[12] L. Miginiac in *The Chemistry of the Metal–Carbon Bond, Vol. 3* (eds.: F. R. Harley, S. Patai), Wiley, New York, **1985**, p. 99.
[13] P. Knochel, M. C. P. Yeh, S. C. Berk, J. Talbert, *J. Org. Chem.* **1988**, *53*, 2390–2392.
[14] E. Erdik, *Tetrahedron* **1987**, *43*, 2203–2212.
[15] M. C. P. Yeh, P. Knochel, *Tetrahedron Lett.* **1988**, *29*, 2395–2396.
[16] T. N. Majid, M. C. P. Yeh, P. Knochel, *Tetrahedron Lett.* **1989**, *30*, 5069–5072.
[17] P. Knochel, T.-S. Chou, H. G. Chen, M. C. P. Yeh, M. J. Rozema, *J. Org. Chem.* **1989**, *54*, 5202–5204.
[18] T.-S. Chou, P. Knochel, *J. Org. Chem.* **1990**, *55*, 4791–4793.
[19] P. Knochel, T.-S. Chou, C. Jubert, D. Rajagopal, *J. Org. Chem.* **1993**, *58*, 588–599.
[20] M. C. P. Yeh, P. Knochel, L. E. Santa, *Tetrahedron Lett.* **1988**, *29*, 3887–3890.
[21] M. C. P. Yeh, H. G. Chen, P. Knochel, *Org. Synth.* **1991**, *70*, 195–203.
[22] C. Retherford, M. C. P. Yeh, I. Schipor, H. G. Chen, P. Knochel, *J. Org. Chem.* **1989**, *54*, 5200–5202.
[23] M. C. P. Yeh, P. Knochel, W. M. Butler, S. C. Berk, *Tetrahedron Lett.* **1988**, *29*, 6693–6696.
[24] P. Knochel, *J. Am. Chem. Soc.* **1990**, *112*, 7431–7433.
[25] H. P. Knoess, M. T. Furlong, M. J. Rozema, P. Knochel, *J. Org. Chem.* **1991**, *56*, 5974–5978.
[26] R. Duddu, M. Eckhardt, M. Furlong, H. P. Knoess, S. Berger, P. Knochel, *Tetrahedron* **1994**, *50*, 2415–2432.
[27] M. J. Dunn, R. F. W. Jackson, J. Pietruszka, N. Wishart, D. Ellis, M. J. Wythes, *Synlett* **1993**, 499–500.
[28] R. L. Dow, B. M. Bechle, *Synlett* **1994**, 293–294.
[29] R. F. W. Jackson, K. James, M. J. Wythes, A. Wood, *J. Chem. Soc. Chem. Commun.* **1989**, 644–645.
[30] R. F. W. Jackson, M. J. Wythes, A. Wood, *Tetrahedron Lett.* **1989**, *30*, 5941–5944.
[31] R. F. W. Jackson, A. Wood, M. J. Wythes, *Synlett* **1990**, 735–736.
[32] M. J. Dunn, R. F. W. Jackson, *J. Chem. Soc. Chem. Commun.* **1992**, 319–320.
[33] R. F. W. Jackson, N. Wishart, M. J. Wythes, *J. Chem. Soc. Chem. Commun.* **1992**, 1587–1589.
[34] R. F. W. Jackson, N. Wishart, M. J. Wythes, *Synlett* **1993**, 219–220.
[35] R. F. W. Jackson, N. Wishart, A. Wood, K. James, M. J. Wythes, *J. Org. Chem.* **1992**, *57*, 3397–3404.
[36] J. L. Fraser, R. F. W. Jackson, B. Porter, *Synlett* **1994**, 379–380.
[37] C. Retherford, T.-S. Chou, R. M. Schelkun, P. Knochel, *Tetrahedron Lett.* **1990**, *31*, 1833–1836.
[38] C. Retherford, P. Knochel, *Tetrahedron Lett.* **1991**, *32*, 441–444.
[39] For a review see: P. Knochel, M. J. Rozema, C. E. Tucker, C. Retherford, M. Furlong, S. AchyuthaRao, *Pure Appl. Chem.* **1992**, *64*, 361–369.
[40] S. AchyuthaRao, C. E. Tucker, P. Knochel, *Tetrahedron Lett.* **1990**, *31*, 7575–7578.
[41] S. AchyuthaRao, T.-S. Chou, I. Schipor, P. Knochel, *Tetrahedron* **1992**, *48*, 2025–2043.
[42] I. Klement, K. Lennick, C. E. Tucker, P. Knochel, *Tetrahedron Lett.* **1993**, *34*, 4623–4626.
[43] T. N. Majid, P. Knochel, *Tetrahedron Lett.* **1990**, *31*, 4413–4416.
[44] C. JanakiramRao, P. Knochel, *J. Org. Chem.* **1991**, *56*, 4593–4596.
[45] P. Knochel, C. JanakiramRao, *Tetrahedron* **1993**, *49*, 29–48.
[46] A. Fürstner, *Angew. Chem. Int. Ed. Engl.* **1993**, *32*, 164–189.
[47] A. Fürstner, R. Singer, P. Knochel, *Tetrahedron Lett.* **1994**, *35*, 1047–1050.
[48] L. Zhu, R. M. Wehmeyer, R. D. Rieke, *J. Org. Chem.* **1991**, *56*, 1445–1453.
[49] L. Zhu, R. D. Rieke, *Tetrahedron Lett.* **1991**, *32*, 2865–2866.
[50] R. D. Rieke, S. J. Uhm, P. M. Hudnall, *J. Chem. Soc. Chem. Commun.* **1973**, 269–270.
[51] R. D. Rieke, P. T.-J. Li, T. P. Burns, S. T. Uhm, *J. Org. Chem.* **1981**, *46*, 4323–4324.
[52] R. D. Rieke, S. J. Uhm, *Synthesis* **1975**, 452–453.
[53] R. D. Rieke, *Science* **1989**, *246*, 1260–1264.

[54] H. Stadtmüller, B. Greve K. Lennick, A. Chair, P. Knochel, *Synthesis* **1995**, 69–72.
[55] S. C. Berk, P. Knochel, M. C. P. Yeh, *J. Org. Chem.* **1988**, *53*, 5789–5791.
[56] S. C. Berk, M. C. P. Yeh, N. Jeong, P. Knochel, *Organometallics* **1990**, *9*, 3053–3064.
[57] N. E. Alami, C. Belaud, J. Villiéras, *Tetrahedron Lett.* **1987**, *28, 59–60.*
[58] N. El Alami, C. Belaud, J. Villiéras, *J. Organomet. Chem.* **1988**, *348*, 1–9.
[59] J. Villiéras, M. Rambaud, *Synthesis* **1982**, 924–926.
[60] J. Villiéras, M. Villiéras, *Janssen Chimica Acta* **1993**, *11 (3)*, 3–8.
[61] H. Hunsdieker, H. Erlbach, E. Vogt, German Patent 722467, **1942**; *Chem. Abstr.* **1943**, *37*, P 5080.
[62] Y. Tamaru in *New Trends in Organometallic Chemistry* (ed.: H. Sakurai), Tohoku University, Sendai, Japan, **1990**, 304–307.
[63] Y. Tamaru, H. Ochiai, Z. Yoshida, *Tetrahedron Lett.* **1984**, *25*, 3861–3864.
[64] Y. Tamaru, H. Ochiai, T. Nakamura, K. Tusbaki, Z. Yoshida, *Tetrahedron Lett.* **1985**, *26*, 5559–5562.
[65] Y. Tamaru, H. Ochiai, F. Sanada, Z. Yoshida, *Tetrahedron Lett.* **1985**, *26*, 5529–5532.
[66] Y. Tamaru, H. Ochiai, T. Nakamura, Z. Yoshida, *Tetrahedron Lett.* **1986**, *27*, 955–958.
[67] Y. Tamaru, H. Ochiai, T. Nakamura, Z. Yoshida, *Angew. Chem.* **1987**, *99*, 1193–1195; *Angew. Chem. Int. Ed. Engl.* **1987**, *26*, 1157.
[68] H. Ochiai, Y. Tamura, K. Tsubaki, Z. Yoshida, *J. Org. Chem.* **1987**, *52*, 4418–4420.
[69] Y. Tamuru, H. Tanigawa, T. Yamamoto, Z. Yoshida, *Angew. Chem.* **1989**, *101*, 358–360; *Angew. Chem. Int. Ed. Engl.* **1989**, *28*, 351.
[70] H. Ochiai, T. Nishihara, Y. Tamaru, Z. Yoshida, *J. Org. Chem.* **1988**, *53*, 1343–1344.
[71] Y. Tamaru, T. Nakamura, M. Sakaguchi, H. Ochiai, Z. Ochiai, Z. Yoshida, *J. Chem. Soc. Chem. Commun.* **1988**, 610–611.
[72] Y. Tamaru, H. Ochiai, T. Nakamura, Z. Yoshida, *Org. Synth.* **1988**, *67*, 98–104.
[73] H. Blancou, P. Moreau, A. Commeyras, *Tetrahedron*, **1977**, *33*, 2061–2067.
[74] R. N. Haszeldine, E. G. Walaschewski, *J. Chem. Soc.* **1953**, 3607–3610.
[75] W. T. Miller, E. Bergmann, A. H. Fainberg, *J. Am. Chem. Soc.* **1957**, *79*, 4159–5164.
[76] R. D. Chambers, W. K. R. Musgrave, J. Savory, *J. Chem. Soc.* **1962**, 1993–1999.
[77] H. Blancou, A. Commeyras, *J. Fluorine Chem.* **1982**, *20*, 255–265.
[78] S. Benefice, H. Blancou, A. Commeyras, *J. Fluorine Chem.* **1983**, *23*, 47–55.
[79] S. Benefice-Malouet, H. Blancou, A. Commeyras, *J. Fluorine Chem.* **1985**, *30*, 171–187.
[80] For an excellent review see: D. J. Burton, *Tetrahedron* **1992**, *48*, 189–275.
[81] M. J. Dunn, R. F. W. Jackson, G. R. Stephenson, *Synlett* **1992**, 905–906.
[82] For an excellent review see: D. J. Burton, Z.-Y. Yang, P. A. Morken, *Tetrahedron* **1994**, *50*, 2993–3063.
[83] D. J. Burton, D. M. Wiemers, *J. Am. Chem. Soc.* **1985**, *107*, 5014–5015.
[84] D. M. Wiemers, D. J. Burton, *J. Am. Chem. Soc.* **1986**, *108*, 832–834.
[85] J. Grondin, P. Sebban, P. Vottero, H. Blancou, A. Commeyras, *J. Organomet. Chem.* **1989**, *362*, 237–242.
[86] C. Jubert, P. Knochel, *J. Org. Chem.* **1992**, *57*, 5425–5431.
[87] C. Einhorn, J. Einhorn, J.-L. Luche, *Synthesis* **1989**, 787–813.
[88] C. Petrier, J. C. de Souza Barbosa, C. Dupuy, J.-L. Luche, *J. Org. Chem.* **1985**, *50*, 5761–5765.
[89] J. C. de Souza Barbosa, C. Petrier, J.-L. Luche, *Tetrahedron Lett.* **1985**, *26*, 829–830.
[90] E. J. Corey, J. O. Link, Y. Shao, *Tetrahedron Lett.* **1992**, *33*, 3435–3438.
[91] C. Einhorn, J.-L. Luche, *J. Organomet. Chem.* **1987**, *322*, 177–183.
[92] J.-L. Luche, C. Einhorn, J. Einhorn, J. V. Sinisterra-Gago, *Tetrahedron Lett.* **1990**, *31*, 4125–4128.
[93] J. Chaussard, J.-C. Folest, J.-Y. Nedelec, J. Perichon, S. Sibille, M. Troupel, *Synthesis* **1990**, 369–381.
[94] S. Sibille, V. Ratovelomanana, J. Perichon, *J. Chem. Soc. Chem. Commun.* **1992**, 283–284.
[95] S. Sibille, V. Ratovelomanana, J.-Y. Nédélec, P. Périchon, *Synlett* **1993**, 425–426.
[96] N. Zylber, J. Zylber, Y. Rollin, E. Dunach, J. Perichon, *J. Organomet. Chem.* **1993**, *444*, 1–4.
[97] J. J. Habeeb, A. Osman, D. G. Tuck, *J. Organomet. Chem.* **1980**, *185*, 117–172.
[98] E. Negishi, L. F. Valente, M. Kobayashi, *J. Am. Chem. Soc.* **1980**, *102*, 3298–3299.
[99] K. Takai, T. Kakiuchi, K. Utimoto, *J. Org. Chem.* **1994**, *59*, 2671–2673.
[100] K. Takai, T. Kakiuchi, Y. Kataoka, K. Utimoto, *J. Org. Chem.* **1994**, *59*, 2668–2670.
[101] J. Furukawa, N. Kawabata, J. Nishimura, *Tetrahedron Lett.* **1966**, 3353–3354.
[102] S. Sawada, Y. Inouye, *Bull. Chem. Soc. Jpn.* **1969**, *42*, 2669–2672.
[103] A. Sidduri, M. J. Rozema, P. Knochel, *J. Org. Chem.* **1993**, *58*, 2694–2713.
[104] M. J. Rozema, S. AchyuthaRao, P. Knochel, *J. Org. Chem.* **1992**, *57*, 1956–1958.
[105] M. J. Rozema, C. Eisenberg, H. Lütjens, R. Ostwald, K. Belyk, P. Knochel, *Tetrahedron Lett.* **1993**, *34*, 3115–3118.
[106] H. Stadtmüller, R. Lentz, C. E. Tucker, T. Stüdemann, W. Dörner, P. Knochel, *J. Am. Chem. Soc.* **1993**, 115, 7027–7028.

[107] H. Stadtmüller, C. E. Tucker, A. Vaupel, P. Knochel, *Tetrahedron Lett.* **1993**, *34*, 7911-7914.
[108] P. Knochel, unpublished results, 1992–1993.
[109] I. Klement, K. Chau, G. Cahiez, P. Knochel, *Tetrahedron Lett.* **1994**, *35*, 1177–1180.
[110] R. Sustmann, J. Lau, *Chem. Ber.* **1986**, *119*, 2531–2541.
[111] R. Sustmann, J. Lau, M. Zipp, *Rec. Trav. Chim. Pays. Bas* **1986**, *105*, 356–359.
[112] A. Vaupel, P. Knochel, *Tetrahedron Lett.* **1994**, *35*, 8349–8352.
[113] A. Vaupel, P. Knochel, *Tetrahedron Lett.* **1995**, *36*, 231–232.
[114] M. B. M. de Azevedo, M. M. Munta, A. E. Greene, *J. Org. Chem.* **1992**, *57*, 4567–4569.
[115] K. A. Agrios, M. Srebnik, *J. Org. Chem.* **1993**, *58*, 6908–6910.
[116] T. Mukhopadhyay, D. Seebach, *Helv. Chim. Acta* **1982**, *65*, 385–391.
[117] D. Seebach, A. K. Beck, T. Mukhopadhyay, E. Thomas, *Helv. Chim. Acta* **1982**, *65*, 1101–1133.
[118] M. Bengtsson, T. Liljefors, *Synthesis* **1988**, 250–252.
[119] L. I. Zakhakin, O. Y. Z. Okhlobystin, *Obsc. Chim.* **1960**, *30*, 2134; Engl.: **1960**, *30*, 2109; *Chem. Abstr.* **1961**, *55*, 9319a.
[120] K.-H. Thiele, P. Zdunneck, *J. Organomet. Chem.* **1965**, *4*, 10–17.
[121] K.-H. Thiele, G. Engelhardt, J. Köhler, M. Arnstedt, *J. Organomet. Chem.* **1967**, *9*, 385–393.
[122] M. Srebnik, *Tetrahedron Lett.* **1991**, *32*, 2449–2452.
[123] W. Oppolzer, R. N. Radinov, *Helv. Chim. Acta* **1992**, *75*, 170–172.
[124] W. Oppolzer, R. N. Radinov, *J. Am. Chem. Soc.* **1993**, *115*, 1593–1594.
[125] F. Langer, J. Waas, P. Knochel, *Tetrahedron Lett.* **1993**, *34*, 5261–5264.
[126] F. Langer, A. Devasagayaraj, P.-Y. Chavant, P. Knochel, *Synlett* **1994**, 410–412.
[127] L. Schwink, P. Knochel, *Tetrahedron Lett.* **1994**, *35*, 9007–9010.
[128] A. Devasagayaraj, L. Schwink, P. Knochel, *J. Org. Chem.* **1995**, *60*, 3311–3317.
[129] H. Eick, F. Langer, P. Knochel, to be published.
[130] R. Köster, G. Griasnow, W. Larbig, P. Binger, *Liebigs Ann. Chem.* **1964**, *672*, 1–34.
[131] F. Langer, P. Knochel, unpublished results, 1994.
[132] E. Frankland, D. F. Duppa, *Ann. Chem.* **1864**, *130*, 117–126.
[133] M. J. Rozema, D. Rajagopal, C. E. Tucker, P. Knochel, *J. Organomet. Chem.* **1992**, *438*, 11–27.
[134] C. E. Tucker, T. N. Majid, P. Knochel, *J. Am. Chem. Soc.* **1992**, *114*, 3983–3985.
[135] E. Negishi, *Organometallics in Organic Synthesis, Vol. 1*, Wiley, New York, **1980**.
[136] B. J. Wakefield, *Organolithium Methods,* Academic Press, London, **1988**.
[137] J. C. Stowell, *Carbanions in Organic Synthesis*, Wiley, New York, **1979**.
[138] L. Brandsma, H. Verkruijsse, *Preparative Polar Organometallic Chemistry, Vol. 1*, Springer, Berlin, **1987**.
[139] E. Piers, V. Karunaratne, *Can. J. Chem.* **1984**, *62*, 629–631.
[140] E. Piers, V. Karunaratne, *J. Chem. Soc. Chem. Commun.* **1983**, 935–936.
[141] E. Piers, V. Karunaratne, *J. Chem. Soc. Chem. Commun.* **1984**, 959–960.
[142] E. Piers, V. Karunaratne, *J. Org. Chem.* **1983**, *48*, 1774–1776.
[143] E. Piers, B. W. A. Yeung, *J. Org. Chem.* **1984**, *49*, 4567–4569.
[144] C. Najera, M. Yus, *J. Org. Chem.* **1988**, *53*, 4708–4715.
[145] P. E. Peterson, D. J. Nelson, R. Risener, *J. Org. Chem.* **1986**, *51*, 2381–2382.
[146] J. L. Herrmann, M. H. Berger, R. H. Schlessinger, *J. Am. Chem. Soc.* **1979**, *101*, 1544–1549.
[147] M. M. Midland, A. Tramontano, J. R. Cable, *J. Org. Chem.* **1980**, *45*, 28–29.
[148] W. E. Parham, L. D. Jones, Y. Sayed, *J. Org. Chem.* **1975**, *40*, 2394–2399.
[149] W. E. Parham, L. D. Jones, *J., Org. Chem.* **1976**, *41*, 1187–1191.
[150] W. E. Parham, L. D. Jones, *J. Org. Chem.* **1976**, *41*, 2704–2706.
[151] W. E. Parham, R. M. Piccirilli, *J. Org. Chem.* **1977**, *42*, 257–260.
[152] W. E. Parham, D. W. Boykin, *J. Org. Chem.* **1977**, *42*, 260–263.
[153] W. E. Parham, C. K. Bradsher, *Acc. Chem. Res.* **1982**, *15*, 300.
[154] V. Bolitt, C. Mioskowski, S. P. Reddy, J. R. Falck, *Synthesis* **1988**, 388–389.
[155] G. Köbrich, P. Buck, *Chem. Ber.* **1970**, *103*, 1412–1419.
[156] P. Buck, G. Köbrich, *Chem. Ber.* **1970**, *103*, 1420–1430.
[157] J. F. Cameron, J. M. J. Fréchet, *J. Am. Chem. Soc.* **1991**, *113*, 4303–4313.
[158] D. Caine, A. S. Frobese, *Tetrahedron Lett.* **1978**, 5167–5170.
[159] E. Negishi, L. D. Boardman, J. M. Tour, H. Sawada, C. L. Rand, *J. Am. Chem. Soc.* **1983**, *105*, 6344–6346.
[160] S. Sengupta, V. Snieckus, *J. Org. Chem.* **1990**, *55*, 5680–5683.
[161] L. Duhamel, P. Duhamel, D. Enders, W. Karl, F. Leger, J. M. Poirier, G. Raabe, *Synthesis* **1991**, 649–654.
[162] M. P. Cooke, *J. Org. Chem.* **1984**, *49*, 1144–1146.

[163] M. P. Cooke, R. K. Widener, *J. Org. Chem.* **1987**, *52*, 1381–1397.
[164] G. Cahiez, P. Venegas, C. E. Tucker, T. N. Majid, P. Knochel, *J. Chem. Soc. Chem. Commun.* **1992**, 1406–1408.
[165] J. P. Gillet, R. Sauvêtre, J.-F. Normant, *Synthesis* **1986**, 538–543.
[166] J.-P. Gillet, R. Sauvêtre, J.-F. Normant, *Tetrahedron Lett.* **1985**, *26*, 3999–4002.
[167] F. Tellier, R. Sauvêtre, J.-F. Normant, *J. Organomet. Chem.* **1986**, *303*, 309–315.
[168] F. Tellier, R. Sauvêtre, J.-F. Normant, *J. Organomet. Chem.* **1987**, *328*, 1–13.
[169] F. Tellier, R. Sauvêtre, J.-F. Normant, Y. Dromzee, Y. Jeannin, *J. Organomet. Chem.* **1987**, *331*, 281–298.
[170] P. Martinet, R. Sauvêtre, J.-F. Normant, *J. Organomet. Chem.* **1989**, *367*, 1–10.
[171] B. Jiang, Y. Xu, *J. Org. Chem.* **1991**, *56*, 7336–7340.
[172] P. A. Morken, H. Lu, A. Nakamura, D. J. Burton, *Tetrahedron Lett.* **1991**, *32*, 4271–4274.
[173] S. W. Hansen, T. D. Spawn, D. J. Burton, *J. Fluorine Chem.* **1987**, *35*, 415–420.
[174] P. A. Morken, D. J. Burton, *J. Org. Chem.* **1993**, *58*, 1167–1172.
[175] I. Klement, P. Venegas, G. Cahiez, P. Knochel, to be published.
[176] J. A. Staroscik, B. Rickborn, *J. Org. Chem.* **1972**, *37*, 738–740.
[177] N. Kawabata, T. Nakagawa, T. Nakao, S. Yamashita, *J. Org. Chem.* **1977**, *42*, 3031–3035.
[178] C. R. Johnson, M. R. Barbachyn, *J. Am. Chem. Soc.* **1982**, *104*, 4290–4291.
[179] B. Fabisch, T. N. Mitchell, *J. Organomet. Chem.* **1984**, *269*, 219–221.
[180] R. Bartnik, G. Mloston, *Synthesis* **1983**, 924–929.
[181] I. Arai, A. Mori, H. Yamamoto, *J. Am. Chem. Soc.* **1985**, *107*, 8254–8256.
[182] E. Mash, K. A. Nelson, *J. Am. Chem. Soc.* **1985**, *107*, 8256–8258.
[183] A. Mori, I. Arai, H. Yamamoto, *Tetrahedron* **1986**, *42*, 6447–6458.
[184] T. Sugimura, T. Futagawa, A. Tai, *Tetrahedron Lett.* **1988**, *29* 5775–5778.
[185] T. Sugimura, T. Futagawa, M. Yoshikawa, A. Tai, *Tetrahedron Lett.* **1989**, *30*, 3807–3810.
[186] E. A. Mash, K. A. Nelson, S. Van Deusen, S. B. Hemperly, *Org. Synth.* **1989**, *68*, 92–103.
[187] E. A. Mash, D. S. Torok, *J. Org. Chem.* **1989**, *54*, 250–253.
[188] E. A. Mash, S. B. Hemperly, K. A. Nelson, P. C. Heidt, S. Van Deusen, *J. Org. Chem.* **1990**, *55*, 2045–2055.
[189] R. H. Newman-Evans, R. J. Simon, B. K. Carpenter, *J. Org. Chem.* **1990**, *55*, 695–711.
[190] P. Knochel, N. Jeong, M. J. Rozema, M. C. P. Yeh, *J. Am. Chem. Soc.* **1989**, *111*, 6474–6476.
[191] P. Knochel, S. AchyuthaRao, *J. Am. Chem. Soc.* **1990**, *112,* 6146–6148.
[192] M. J. Rozema, P. Knochel, *Tetrahedron Lett.* **1991**, *32*, 1855–1858.
[193] M. R. Burns, J. K. Coward, *J. Org. Chem.* **1993**, *58*, 528–532.
[194] S. AchyuthaRao, P. Knochel, *J. Am. Chem. Soc.* **1992**, *114*, 7579–7581.
[195] E. Nakamura, S. Aoki, K. Sekiya, H. Oshino, I. Kuwajima, *J. Am. Chem. Soc.* **1987**, *109*, 8056–8066.
[196] E. Negishi, K. Akiyoshi, *J. Am. Chem. Soc.* **1988**, *110*, 646–647.
[197] E. Negishi, K. Akiyoshi, B. O'Connor, K. Takagi, G. Wu, *J. Am. Chem. Soc.* **1989**, *111*, 3089–3091.
[198] J. A. Miller, *J. Org. Chem.* **1989**, *54*, 998–1000.
[199] P. Kocienski, S. Wadmann, K. Cooper, *J. Am. Chem. Soc.* **1989**, *111*, 2363–2365.
[200] M. Mortimore, P. Kocienski, *Tetrahedron Lett.* **1988**, *27*, 3357–3360.
[201] K. Ramig, M. Bhupathy, T. Cohen, *J. Org. Chem.* **1989**, *54*, 4404–4412.
[202] P. Charreau, M. Julia, J.-N. Verpeux, *J. Organomet. Chem.* **1989**, *379*, 201–210.
[203] T. J. Michnick, D. S. Matteson, *Synlett* **1991**, 631–632.
[204] T. Nguyen, E. Negishi, *Tetrahedron Lett.* **1991**, *32*, 5903–5906.
[205] J. Villiéras, A. Reliquet, J.-F. Normant, *J. Organomet. Chem.* **1978**, *144*, 1–12, 263–269.
[206] J. Villiéras, A. Reliquet, J.-F. Normant, *Synthesis* **1978**, 27–29.
[207] T. Harada, K. Hattori, T. Katsuhira, A. Oku, *Tetrahedron Lett.* **1989**, *30*, 6035, 6039–6040.
[208] T. Harada, T. Katszhira, K. Hattori, A. Oki, *J. Org. Chem.* **1993**, *58*, 2958–2965.
[209] T. Harada, D. Hara, K. Hattori, A. Oku, *Tetrahedron Lett.* **1988**, *29*, 3821–3824.
[210] T. Harada, Y. Kotani, T. Katsuhira, A. Oku, *Tetrahedron Lett.* **1991**, *32*, 1573–1576.
[211] T. Harada, T. Katsuhira, A. Oku, *J. Org. Chem.* **1992**, *57*, 5805–5807.
[212] T. Harada, T. Katsuhira, K. Hattori, A. Oku, *J. Org. Chem.* **1993**, *58*, 2958–2965.
[213] T. Katsuhira, T. Harada, K. Maejima, A. Osada, A. Oku, *J. Org. Chem.* **1993**, *58*, 6166–6168.
[214] Z.-Y. Yang, D. M. Wiemers, D. J. Burton, *J. Am. Chem. Soc.* **1992**, *114*, 4402–4403.
[215] S. AchyuthaRao, P. Knochel, *J. Org. Chem.* **1991**, *56*, 4591–4593.
[216] P. Knochel: "Carbometallation of Alkenes and Alkynes", in *Comprehensive Organic Chemistry, Vol. 4*, (ed.: B. M. Trost), Pergamon, Oxford, **1991**, pp. 865–911.
[217] P. Knochel, J.-F. Normant, *Tetrahedron Lett.* **1986**, *27*, 1039–1042, 1043–1046, 4427–4430, 5727–5730.

[218] P. Knochel, C. Xiao, M. C. P. Yeh, *Tetrahedron Lett.* **1988**, *29*, 6697–6700.
[219] P. Knochel, M. C. P. Yeh, C. Xiao, *Organometallics* **1989**, *8*, 2831–2835.
[220] C. Chuit, J. P. Foulon, J.-F. Normant, *Tetrahedron* **1981**, *37*, 1385–1389; *ibid.* **1980**, *36*, 2305–2310.
[221] M. Bourgain-Commercçon, J. P. Foulon, J.-F. Normant, *J. Organomet. Chem.* **1982**, *228*, 321–326.
[222] E. J. Corey, N. W. Boaz, *Tetrahedron Lett.* **1985**, *26*, 6015–6018, 6019–6022.
[223] A. Alexakis, J. Berlan, Y. Besace, *Tetrahedron Lett.* **1986**, *27*, 1047–1050.
[224] Y. Horiguchi, S. Matsuzawa, E. Nakamura, I. Kuwajima, *Tetrahedron Lett.* **1986**, *27*, 4025–4028.
[225] E. Nakamura, S. Matsuzawa, Y. Horiguchi, I. Kuwajima, *Tetrahedron Lett.* **1986**, *27*, 4029–4032.
[226] M. Bergdahl, E.-L. Lindstedt, M. Nilsson, T. Olsson, *Tetrahedron* **1988**, *44*, 2055–2062.
[227] M. Bergdahl, E.-L. Lindstedt, M. Nilsson, T. Olsson, *Tetrahedron* **1989**, *45*, 535–543.
[228] M. Bergdahl, E.-L. Lindstedt, T. Olsson, *J. Organomet. Chem.* **1989**, *365*, C11–C14.
[229] M. Bergdahl, M. Nilsson, T. Olsson, *J. Organomet. Chem.* **1990**, *391*, C19–C22.
[230] M. Bergdahl, M. Eriksson, M. Nilsson, T. Olsson, *J. Org. Chem.* **1993**, *58*, 7238–7244.
[231] M. Gaudemar, *C.R. Hebd. Sceances Acad. Sci., Ser. C* **1971**, *273*, 1669–1673.
[232] M. Bellasoued, Y. Frangin, M. Gaudemar, *Synthesis* **1977**, 205–208.
[233] I. Marek, J.-M. Lefrançois, J.-F. Normant, *Tetrahedron Lett.* **1991**, *32*, 5969–5972.
[234] I. Marek, J.-F. Normant, *Tetrahedron Lett.* **1991**, *32*, 5973–5976.
[235] I. Marek, J.-M. Lefrançois, J.-F. Normant, *Tetrahedron Lett.* **1992**, *33*, 1747–1748.
[236] I. Marek, J.-M. Lefrançois, J.-F. Normant, *Synlett* **1992**, 633–635.
[237] D. Beruben, I. Marek, L. Labaudinière, J.-F. Normant, *Tetrahedron Lett.* **1993**, *34*, 2303–2306.
[238] J.-F. Normant, J.-C. Quirion, *Tetrahedron Lett.* **1989**, *30*, 3959–3962.
[239] I. Klement, P. Knochel, *Synlett* **1995**, in press.
[240] H. T. Teunissen, F. Bickelhaupt, *Tetrahedron Lett.* **1992**, *33*, 3537–3538.
[241] J. R. Waas, S. AchyuthaRao, P. Knochel, *Tetrahedron Lett.* **1992**, *33*, 3717–3720.
[242] E. Nakamura, K. Sekiya, I. Kuwajima, *Tetrahedron Lett.* **1987**, *28*, 337–340.
[243] E. Nakamura, K. Sekiya, M. Arai, S. Aoki, *J. Am. Chem. Soc.* **1989**, *111*, 3091–3093.
[244] M. Arai, T. Kawasuji, E. Nakamura, *J. Org. Chem.* **1993**, *58*, 5121–5129.
[245] K. Sekiya, E. Nakamura, *Tetrahedron Lett.* **1988**, *29*, 5155–5156.
[246] T. Tanaka, K. Bannai, A. Hazato, M. Koga, S. Kurozumi, Y. Kato, *Tetrahedron* **1991**, *47*, 1861–1876.
[247] H. G. Chen, J. L. Gage, S. D. Barrett, P. Knochel, *Tetrahedron Lett.* **1990**, *31*, 1829–1832.
[248] M. C. P. Yeh, M.-L. Sun, S.-K. Lin, *Tetrahedron Lett.* **1991**, *32*, 113–116.
[249] M. C. P. Yeh, S.-J. Tau, *J. Chem. Soc. Chem. Commun.* **1992**, 13–15.
[250] M. C. P. Yeh, C.-J. Tsou, C.-N. Chuang, H.-C. Lin, *J. Chem. Soc. Chem. Commun.* **1992**, 890–891.
[251] M. C. P. Yeh, B.-A. Shen, H.-W. Fu, S. I. Tau, L. W. Chuang, *J. Am. Chem. Soc.* **1993**, *115*, 5941–5952.
[252] M. C. P. Yeh, C.-N. Chuang, *J. Chem. Soc. Chem. Commun.* **1994**, 703–704.
[253] M. C. P. Yeh, P. Knochel, *Tetrahedron Lett.* **1989**, *30*, 4799–4802.
[254] H. Sörensen, A. E. Greene, *Tetrahedron Lett.* **1990**, *31*, 7597–7598.
[255] S. Marquais, G. Cahiez, P. Knochel, *Synlett* **1994**, 849–850.
[256] C. E. Tucker, S. AchyuthaRao, P. Knochel, *J. Org. Chem.* **1990**, *55*, 5446–5448.
[257] C. E. Tucker, P. Knochel, *Synthesis* **1993**, 530–536.
[258] A. Sidduri, N. Budries, R. M. Laine, P. Knochel, *Tetrahedron Lett.* **1992**, *33*, 7515–7518.
[259] C. E. Tucker, P. Knochel, *J. Org. Chem.* **1993**, *58*, 4781–4782.
[260] P. Knochel in *Comprehensive Organic Synthesis, Vol. 1* (eds.: B. M. Trost, I. Fleming, S. L. Schreiber), Pergamon, Oxford, **1991**, pp. 211–229.
[261] P. Knochel, J.-F. Normant, *Tetrahedron Lett.* **1984**, *25*, 4383–4386.
[262] G. Rousseau, J. M. Conia, *Tetrahedron Lett.* **1981**, *22*, 649–652.
[263] G. Rousseau, J. Drouin, *Tetrahedron* **1983**, *39*, 2307–2310.
[264] P. Knochel, J.-F. Normant, *J. Organomet. Chem.* **1986**, *309*, 1–23.
[265] Y. A. Dembélé, C. Belaud, P. Hitchcock, J. Villiéras, *Tetrahedron: Asymmetry* **1992**, *3*, 351–354.
[266] Y. A. Dembélé, C. Belaud, P. Hitchcock, J. Villiéras, *Tetrahedron: Asymmetry* **1992**, *3*, 511–514.
[267] J. van der Louw, J. L. van der Baan, H. Stichter, G. J. J. Out, F. Bickelhaupt, G. W. Klumpp, *Tetrahedron Lett.* **1988**, *29*, 3579–3580.
[268] J. van der Louw, J. L. van der Baan, F. Bickelhaupt, G. W. Klumpp, *Tetrahedron Lett.* **1987**, *28*, 2889–2892.
[269] J. van der Louw, J. l. van der Baan, H. Stieltjes, F. Bickelhaupt, G. W. Klumpp, *Tetrahedron Lett.* **1987**, *28*, 5929–5932.
[270] J. van der Louw, J. L. van der Baan, H. Stichter, G. J. J. Out, F. J. J. de Kanter, F. Bickelhaupt, G. W. Klumpp, *Tetrahedron* **1992**, *48*, 9877–9900.
[271] J. van der Louw, J. L. van der Baan, G. J. J. Out, F. J. J. de Kanter, F. Bickelhaupt, G. W. Klumpp, *Tetrahedron* **1992**, *48*, 9001–9916.

[272] T. A. van der Heide, J. L. van der Baan, F. Bickelhaupt, G. W. Klumpp, *Tetrahedron Lett.* **1992**, *33*, 475–476.
[273] J. L. Moreau in *The Chemistry of Ketenes, Allenes and Related Compounds* (ed.: S. Patai), Wiley, New York, **1980**, Part 1, pp. 363–414.
[274] H. Yamamoto in *Comprehensive Organic Synthesis, Vol. 2* (eds.: B. M. Trost, I. Fleming, S. L. Schreiber), Pergamon, Oxford, **1991**, pp. 81–98.
[275] G. Zweifel, G. Hahn, *J. Org. Chem.* **1984**, *49*, 4565–4567.
[276] W. J. Thompson, T. J. Tucker, J. E. Schwering, J. L. Barnes, *Tetrahedron Lett.* **1990**, *31*, 6819–6822.
[277] A. E. DeCamp, A. T. Kawaguchi, R. P. Volante, I. Shinkai, *Tetrahedron Lett.* **1991**, *32*, 1867–1870.
[278] H. Takahashi, T. Kawakita, M. Ohno, M. Yoshioka, S. Kobayashi, *Tetrahedron* **1992**, *48*, 5691–5700.
[279] M. T. Reetz, *Organotitanium Reagents in Organic Synthesis*, Springer, Berlin, **1986**.
[280] M. T. Reetz, *Topics Curr. Chem.* **1982**, *106*, 1.
[281] B. Weidmann, D. Seebach, *Angew. Chem. Int. Ed. Engl.* **1983**, *22*, 31.
[282] U. Koert, H. Wagner, U. Pidun, *Chem. Ber.* **1994**, *127*, 1447–1457.
[283] A. R. Gangloff, B. Akermark, P. Helquist, *J. Org. Chem.* **1992**, *57*, 4797–4799.
[284] P. Quinton, T. LeGall, *Tetrahedron Lett.* **1991**, *32*, 4909–4912.
[285] A. Solladié-Cavallo, D. Farkhani, S. Fritz, T. Lazrak, J. Suffert, *Tetrahedron Lett.* **1984**, *25*, 4117–4120.
[286] E. Wissing, R. W. A. Havenith, J. Boersma, G. van Koten, *Tetrahedron Lett.* **1992**, *33*, 7933–7936.
[287] F. H. van der Steen, H. Kleijn, J. T. B. H. Jastrebski, G. van Koten, *J. Org. Chem.* **1991**, *56*, 5147–5158.
[288] H. L. van Maanen, J. T. B. H. Jastrzebski, J. Verweij, A. P. G. Kieboom, A. L. Spek, G. van Koten, *Tetrahedron: Asymmetry* **1993**, *4*, 1441–1444.
[289] K. Soai, T. Hatanaka, T. Miyazawa, *J. Chem. Soc. Chem. Commun.* **1992**, 1097–1098.
[290] A. R. Katritzky, P. A. Harris, *Tetrahedron: Asymmetry* **1992**, *3*, 437–442.
[291] K. Soai, T. Suzuki, T. Shono, *J. Chem. Soc. Chem. Commun.* **1994**, 317–318.
[292] D. L. Comins, S. O'Connor, *Tetrahedron Lett.* **1987**, *28*, 1843–1846.
[293] M.-J. Shiao, K.-H. Liu, L.-G. Lin, *Synlett* **1992**, 655–656.
[294] W.-L. Chia, M.-J. Shiao, *Tetrahedron Lett.* **1991**, *32*, 2033–2034.
[295] T.-L. Shing, W.-L. Chia, M.-J. Shiao, T.-Y. Chau, *Synthesis* **1991**, 849–850.
[296] M.-J. Shiao, W.-L. Chia, T.-L. Shing, T. J. Chow, *J. Chem. Res. (S)* **1992**, 247–247.
[297] C. Agami, F. Couty, J.-C. Daran, B. Prince, C. Puchot, *Tetrahedron Lett.* **1990**, *31*, 2889–2892.
[298] C. Agami, F. Couty, M. Poursoulis, J. Vaissermann, *Tetrahedron* **1992**, *48*, 431–442.
[299] C. Andrés, A. Gonzales, R. Pedrosa, A. Perez-Encabo, S. Garcia-Granda, M. A. Salvado, F. Gomez-Beltran, *Tetrahedron Lett.* **1992**, *33*, 4743–4746.
[300] D. L. Comins, M. A. Foley, *Tetrahedron Lett.* **1988**, *29*, 6711–6714.
[301] J.-L. Bettiol, R. J. Sundberg, *J. Org. Chem.* **1993**, *58*, 814–816.
[302] J. Yamada, H. Sato, Y. Yamamoto, *Tetrahedron Lett.* **1989**, *30*, 5611–5614.
[303] C. P. Casey, S. W. Polichnowski, *J. Am. Chem. Soc.* **1978**, *100*, 7565–7568.
[304] J. M. Maher, R. P. Beatty, N. J. Cooper, *Organometallics* **1985**, *4*, 1354–1361.
[305] I. Lee, N. J. Cooper, *J. Am. Chem. Soc.* **1993**, *115*, 4389–4390.
[306] H. Stadtmüller, P. Knochel, *Organometallics* **1995**, *14*, 3163–3166.
[307] G. Posner, *Org. React.* **1972**, *19*, 1–113; *ibid.* **1975**, *22*, 253–400.
[308] B. H. Lipshutz, *Synthesis* **1987**, 325–342.
[309] B. H. Lipshutz, S. Sengupta, *Org. React.* **1992**, *41*, 135–631.
[310] T. Stemmler, J. E. Penner-Hahn, P. Knochel, *J. Am. Chem. Soc.* **1993**, *115*, 348–350.
[311] B. H. Lipshutz, M. R. Wood, *J. Am. Chem. Soc.* **1993**, *115*, 12625–12626.
[312] S. Kiwe, J. M. Lee, *Tetrahedron Lett.* **1990**, *31*, 7627–7630.
[313] B. S. Bronk, S. J. Lippard, R. L. Danheiser, *Organometallics* **1993**, *12*, 3340–3349.
[314] H. Tsujiyama, N. Ono, T. Yoshino, S. Okamoto, F. Sato, *Tetrahedron Lett.* **1990**, *31*, 4481–4484.
[315] T. Yoshino, S. Okamoto, F. Sato, *J. Org. Chem.* **1991**, *56*, 3205–3207.
[316] K. Miyaji, Y. Ohara, Y. Miyauchi, T. Tsuruda, K. Arai, *Tetrahedron Lett.* **1993**, *34*, 5597–5600.
[317] D. Seebach, E. W. Colvin, F. Lehr, T. Weller, *Chimia* **1979**, *33*, 1–18.
[318] A. Yoshikoshi, M. Miyashita, *Acc. Chem. Res.* **1985**, *18*, 284–290.
[319] A. G. M. Barrett, G. G. Graboski, *Chem. Rev.* **1986**, *86*, 751–762.
[320] G. Rosini, R. Ballini, *Synthesis* **1988**, 833–847.
[321] R. Tamura, A. Kamimura, N. Ono, *Synthesis* **1991**, 423–434.
[322] P. Knochel, D. Seebach, *Tetrahedron Lett.* **1981**, *2*, 3223–3226.
[323] P. Knochel, D. Seebach, *Tetrahedron Lett.* **1982**, *23*, 3897–3900.
[324] D. Seebach, P. Knochel, *Helv. Chim. Acta* **1984**, *67*, 261–283.
[325] S. B. Bowlus, *Tetrahedron Lett.* **1975**, 3591–3592.

[326] A.-T. Hansson, M. Nilsson, *Tetrahedron* **1982**, *38*, 389–391.
[327] C. Jubert, P. Knochel, *J. Org. Chem.* **1992**, *57*, 5431–5438.
[328] S. E. Denmark, L. R. Marcin, *J. Org. Chem.* **1993**, *58*, 3850–3856.
[329] M. T. Crimmins, P. G. Nantermet, *J. Org. Chem.* **1990**, *55*, 4235–4237.
[330] M. T. Crimmins, P. G. Nantermet, B. W. Trotter, I. M. Vallin, P. S. Watson, L. A. McKerlie, T. L. Reinhold, A. W.-H. Cheung, K. A. Steton, D. Dedopoulou, J. L. Gray, *J. Org. Chem.* **1993**, *58*, 1038–1047.
[331] M. T. Crimmins, D. K. Jung, J. L. Gray, *J. Am. Chem. Soc.* **1993**, *115*, 3146–3155.
[332] S. A. Rao, P. Knochel, *J. Am. Chem. Soc.* **1991**, *113*, 5735–5741.
[333] J.-F. Normant, A. Alexakis, *Synthesis* **1981**, 841–870.
[334] M. T. Bertrand, G. Courtois, L. Miginiac, *Tetrahedron Lett.* **1974**, 1945–1948.
[335] H. Lehmkuhl, O. Olbrysch, *Liebigs Ann. Chem.* **1975**, 1162–1175.
[336] H. Lehmkuhl, I. Döring, H. Nehl, *J. Organomet. Chem.* **1981**, *221*, 123–130.
[337] H. Lehmkuhl, H. Nehl, *J. Organomet. Chem.* **1981**, *221*, 131–136.
[338] E. Negishi, J. A. Miller, *J. Am. Chem. Soc.* **1983**, *105*, 6761–6763.
[339] G. A. Molander, *J. Org. Chem.* **1983**, *48*, 5409–5411.
[340] A. Yanagisawa, S. Habaue, H. Yamamoto, *J. Am. Chem. Soc.* **1989**, *111*, 366–368.
[341] K. Kubota, M. Nakamura, M. Isaka, E. Nakamura, *J. Am. Chem. Soc.* **1993**, *115*, 5867–5868.
[342] E. Negishi, D. E. Van Horn, T. Yoshida, C. L. Rand, *Organometallics* **1983**, *2*, 563.
[343] J. van der Louw, C. M. D. Komen, A. Knol, F. J. J. de Kanter, J. L. van der Baan, F. Bickelhaupt, G. W. Klumpp, *Tetrahedron Lett.* **1989**, *30*, 4453–4456.
[344] C. Meyer, I. Marek, G. Courtemanche, J.-F. Normant, *Synlett.* **1993**, 266–268.
[345] C. Meyer, I. Marek, G. Courtemanche, J.-F. Normant, *Tetrahedron Lett.* **1993**, *34*, 6053–6056.
[346] M. Kobayashi, E. Negishi, *J. Org. Chem.* **1980**, *45*, 5223–5225.
[347] E. Negishi, V. Bagheri, S. Chatterjee, F.-T. Luo, J. A. Miller, A. T. Stoll, *Tetrahedron Lett.* **1983**, *24*, 5181–5184.
[348] R. A. Grey, *J. Org. Chem.* **1984**, *49*, 2288–2289.
[349] E. Negishi, *Acc. Chem. Res.* **1982**, *15*, 340–348.
[350] E. Negishi, A. O. King, N. Okukado, *J. Org. Chem.* **1977**, *42*, 1821–1823.
[351] E. Negishi, H. Matsushita, M. Kobayashi, C. L. Rand, *Tetrahedron Lett.* **1983**, *24*, 3823–3824.
[352] J. G. Millar, *Tetrahedron Lett.* **1989**, *30*, 4913–4914.
[353] S. Hyuga, N. Yamashina, S. Hara, A. Suzuki, *Chem. Lett.* **1988**, 809–812.
[354] A. Löffler, G. Himbert, *Synthesis* **1992**, 495–498.
[355] G. Mignani, F. Leising, R. Meyrueix, H. Samson, *Tetrahedron Lett.* **1990**, *31*, 4732–4746.
[356] R. C. Borner, R. F. W. Jackson, *J. Chem. Soc. Chem. Commun.* **1994**, 845–846.
[357] T. Sakamoto, Y. Kondo, N. Takazawa, H. Yamanaka, *Tetrahedron Lett.* **1993**, *34*, 5955–5956.
[358] T. Sakamoto, Y. Kondo, N. Murata, H. Yamanaka, *Tetrahedron Lett.* **1992**, *33*, 5373–5374.
[359] T. Sakamoto, Y. Kondo, N. Takazawa, H. Yamanaka, *Heterocycles* **1993**, *36*, 941.
[360] T. Sakamoto, Y. Kondo, N. Takazawa, H. Yamanaka, *Tetrahedron Lett.* **1993**, *34*, 5955–5956.
[361] A. S. Bell, D. A. Roberts, K. S. Ruddock, *Synthesis* **1987**, 843–844.
[362] A. Minato, K. Suzuki, K. Tamao, M. Kumada, *Tetrahedron Lett.* **1984**, *25*, 83–86.
[363] C. A. Quesnelle, O. B. Familoni, V. Snieckus, *Synlett* **1994**, 349–350.
[364] T. Sakamoto, S. Nishimura, Y. Kondo, H. Yamanaka, *Synthesis* **1988**, 485–486.
[365] T. Sato, A. Kawase, T. Hirose, *Synlett* **1992**, 891–892.
[366] E. Nakamura, I. Kuwajima, *Tetrahedron Lett.* **1986**, *27*, 83–86.
[367] G. Kanai, N. Miyaura, A. Suzuki, *Chem. Lett.* **1993**, 845.
[368] T. Watanabe, N. Miyaura, A. Suzuki, *J. Organomet. Chem.* **1993**, *444*, C1–C3.
[369] T. Watanabe, M. Sakai, N. Miyaura, A. Suzuki, *J. Chem. Soc. Chem. Commun.* **1994**, 467–468.
[370] T. Hayashi, M. Konishi, Y. Kobori, M. Kumada, T. Higuchi, K. Hirotsu, *J. Am. Chem. Soc.* **1984**, *106*, 158–163.
[371] V. Farina, B. Krishnan, *J. Am. Chem. Soc.* **1991**, *113*, 9585–9595.
[372] K. Ritter, *Synthesis* **1993**, 734–762.
[373] E. Erdik, *Tetrahedron* **1992**, *48*, 9577–9648.
[374] K. Soai, S. Niwa, *Chem. Rev.* **1992**, *92*, 833–856.
[375] R. Noyori, M. Kitamura, *Angew. Chem. Int. Ed. Engl.* **1991**, *30*, 49–69.
[376] A. K. Beck, B. Bastani, D. A. Plattner, W. Petter, D. Seebach, H. Braunschweiger, P. Gysi, L. LaVecchia, *Chimia* **1991**, *45*, 238–244.
[377] B. Schmidt, D. Seebach, *Angew. Chem.* **1991**, *103*, 100–101; *Angew. Chem. Int. Ed. Engl.* **1991**, *30*, 99.
[378] D. Seebach, L. Behrendt, D. Felix, *Angew. Chem. Int. Ed. Engl.* **1991**, *30*, 1008.

[379] B. Schmidt, D. Seebach, *Angew. Chem. Int. Ed. Engl.* **1991**, *30*, 1321.
[380] J. L. v. d. Bussche-Hünnefeld, D. Seebach, *Tetrahedron* **1992**, *48*, 5719–5730.
[381] D. Seebach, D. A. Plattner, A. K. Beck, Y. M. Wang, D. Hunziker, W. Petter, *Helv. Chim. Acta* **1992**, *75*, 2171.
[382] D. Seebach, A. K. Beck, B. Schmidt, Y. M. Wang, *Tetrahedron* **1994**, *50*, 4364–4384.
[383] B. Weber, D. Seebach, *Tetrahedron* **1994**, *50*, 7473–7484.
[384] B. Weber, D. Seebach, *Tetrahedron* **1994**, *50*, 6117–6128.
[385] M. Suzuki, A. Yanagisawa, R. Noyori, *J. Am. Chem. Soc.* **1985**, *107*, 3348–3349.
[386] Y. Morita, M. Suzuki, R. Noyori, *J. Org. Chem.* **1989**, *54*, 1785–1787.
[387] R. Noyori, S. Suga, K. Kawai, S. Okada, M. Kitamura, *Pure Appl. Chem.* **1988**, *60*, 1597.
[388] M. Yoshioka, T. Kawakita, M. Ohno, *Tetrahedron Lett.* **1989**, *30*, 1657–1660.
[389] H. Takahashi, T. Kawakita, M. Yoshioka, S. Kobayashi, M. Ohno, *Tetrahedron Lett.* **1989**, *30*, 7095–7098.
[390] W. Brieden, R. Ostwald, P. Knochel, *Angew. Chem. Int. Ed. Engl.* **1993**, *32*, 582–584.
[391] S. Nowotny, S. Vettel, P. Knochel, *Tetrahedron Lett.* **1994**, *35*, 4539–4540.
[392] R. Ostwald, P.-Y. Chavant, H. Stadtmüller, P. Knochel, *J. Org. Chem.* **1994**, *59*, 4143–4153.
[393] S. Vettel, P. Knochel, *Tetrahedron Lett.* **1994**, *35*, 5849–5852.
[394] H. Lütjens, P. Knochel, *Tetrahedron: Asymmetry* **1994**, *5*, 1161–1162.
[395] P. Knochel, W. Brieden, M. J. Rozema, C. Eisenberg, *Tetrahedron Lett.* **1993**, *34*, 5881–5884.
[396] C. Eisenberg, P. Knochel, *J. Org. Chem.* **1994**, *59*, 3760–3761.
[397] C. Eisenberg, W. Brieden, M. J. Rozema, P. Knochel, to be published.

6 Metal Atom/Vapor Approaches to Active Metal Clusters/Particles

Kenneth J. Klabunde and Galo Cardenas-Trivino

6.1 Atoms, Clusters, and Nanoscale Particles

6.1.1 Introduction

Most metallic elements exist as solid substances, and since they are solids their chemical reactivities depend on intrinsic surface reactivity and surface area. Of course the most stable and least reactive form of the metals is that of perfect, single crystals. Since we are interested in producing high chemical reactivities, it is necessary to wholly or partially defeat crystallization. The ultimate way to do this is to generate free atoms of the metals. These would possess the highest ultimate surface area per unit mass (theoretically about 2000 $m^2\ g^{-1}$) and the atoms would have a coordination number of zero, thereby having no steric restrictions to reaction, and orbitals would be poised and available for reaction. Indeed, free metal atoms as in-situ reagents have proven to be quite useful in chemical synthesis [1, 2].

Unfortunately free metal atoms cannot be bottled and sold as chemical synthons. Neither can dimers, trimers, or small clusters be bottled. So how big do metal clusters have to be before they are stable enough to isolate? The answer to this question depends on the metal and the preparative method. Before discussing this in more detail, let us consider some intriguing recent results concerning preparation and reactivity of free clusters as they grow one atom at a time.

6.1.2 Gas-Phase Metal Clusters

As we consider what shape metal clusters take on as they grow to 10, 50, or 100 atoms, we might imagine irregular defective shapes, or symmetrical, close-packed shapes [3]. Since irregular shapes would possess surface atoms with lower coordination numbers, one would predict higher chemical reactivities. Is there any evidence to support this idea?

In recent years elegant gas phase metal cluster growth processes have been developed [4, 5]. Basically the methods depend on pulsed laser vaporization of metallic elements. The atoms produced are injected into a relatively cold helium stream where metal atom recombination can take place, the bonding energy being drained off by collisions with the He gas. In this way a vast number of metal clusters can be formed and detected by laser ionization and mass spectrometry. Some cluster stoichiometries are more favorable than others and the number of atoms contained in these are called "magic numbers". These often represent closed shell clusters and occasionally MacKay icosahedra (those clusters where one, 13, 55, 147, 309, 561, 932 . . . metal atoms can be completely encapsulated in close-packed structures, e.g. as in Fig. 6-1). However, Riley and co-workers [6] have shown that both open and closed-shell clusters do form, and that gas phase annealing can change the reactivity of certain iron clusters towards ammonia or water. Such results sug-

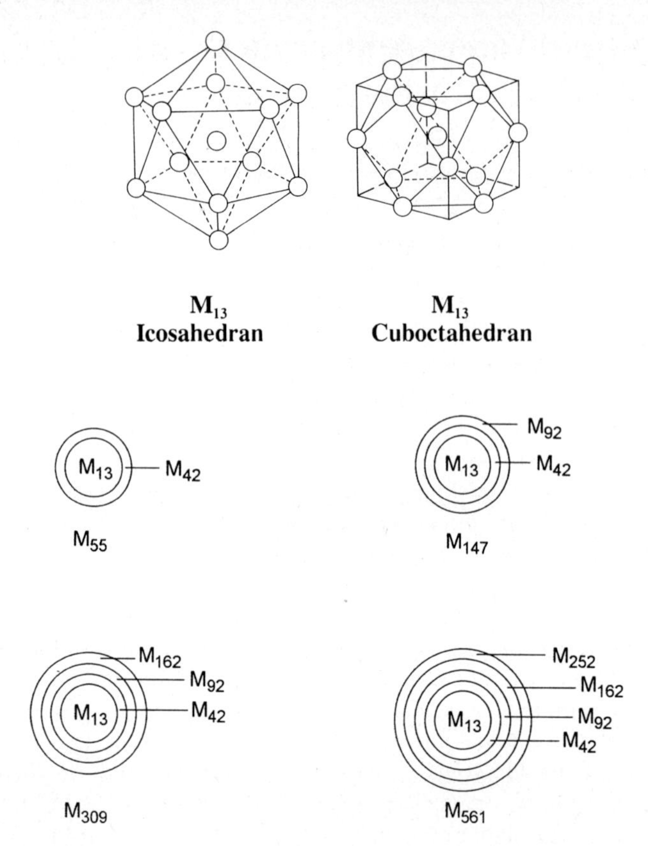

Fig. 6-1. MacKay icosahedra for metal clusters M_{13}, M_{55}, M_{147}, M_{309}, and M_{561}. M_{13} is the smallest cluster with an interior atom. Each successive icosahedron has an additional monatomic layer of atoms (with no vacancies). (In parts reproduced with permission of Academic Press) [3].

gest that initially clusters of irregular shape were formed and that upon annealing their shape changed so that a lower energy, more close-packed form was produced. Similar behavior has been reported by Jarrold and co-workers [7] in their studies of aluminium clusters.

Smalley and co-workers [8] reported somewhat related results where it was shown that copper clusters that are open-shell have a high tendency to react so that they can become closed-shell species. For example, consider the Cu_6, Cu_7^+, and Cu_{17}^+ clusters where shell closing (8 valence electrons) could occur by chemisorption of one CO molecule. Indeed, these species do form quite stable CO adducts with considerable exothermicity [8].

6.1.3 Metal Clusters/Particles Formed in Low-Temperature Matrices

Although studies of gas phase clusters are teaching us a great deal about the formation, stability, and reactivities of small metal clusters, this approach does not allow isolation of multi-gram amounts of materials. Growth of clusters in matrices has a broader appeal since, depending on experimental conditions, the smallest clusters (dimers, trimers, tetramers) can be trapped and spectroscopically analyzed [9, 10], or much larger clusters can be prepared and isolated in multi-gram quantities as ultrafine powders [11].

Table 6-1. Properties of ultrafine particles of nickel prepared by metal vapor–solvent codepositions (SMAD method).

Sample[a]	Surface area ($m^2\ g^{-1}$)	Crystallite size (nm)[b]	Appearance[c]	Sintering temp. (°C)[d]
Ni–pentane from 500:1 ratio (non-ferromagnetic)	80–100	2–3	Rough pieces	260–280
Ni–pentane from 100:1 ratio (ferromagnetic)	30–60	3–8	Rough pieces	220–240
Ni–hexane from 100:1 ratio (ferromagnetic)	45	2–8	5×20 μm rough pieces	
Ni–toluene from 400:1 ratio (non-ferromagnetic)		1-2		340–360
Ni–toluene from 100:1 ratio (ferromagnetic)	100	2	0.5–0.8 μm spheres	250–300
Ni–THF from 100:1 ratio (non-ferromagnetic)	400	2–3	0.5–1.5 μm spheres	200–300
Ni–C_4F_8	90			
Ni–Xe	20			
Ni–SF_6		3–5		

a) The ratio refers to the molar ratio of solvent to Ni vapor codeposited at 77 K. After warming and vacuum removal of solvent the dry powder was isolated. b) By X-ray diffraction (XRD) and application of the Scherrer equation using peak width at half-height. c) By scanning electron microscopy (SEM). d) The temperature at which rapid crystal growth occurred.

In the current context the isolation of ultrafine powders is of most interest. The formation of these nanoscale particles from single atoms is a fascinating process. Attempts to understand it have been made and these will be discussed later. At this point it is useful to simply point out that the host matrix is crucial in mediating the metal atom growth process. Thus, if the host is a good ligating material, the metal atoms form mononuclear coordination compounds. If the host has intermediate ligating properties, occasionally small clusters are formed [12], and if the host has poor ligating properties, larger clusters form [13], and finally, if the host has no ligating properties large metal crystallites are formed (e.g. Table 6-1) [14].

Of course the formation of *cluster compounds* (strongly ligated, stable polynuclear metal clusters that are soluble) [15] is a subject in its own right that will not be considered here. Primarily we will restrict discussion to the formation, chemistry, and physics of the ultrafine powders produced by the metal atom aggregation method. Therefore, it is now appropriate to consider the experimental methods used in their preparation.

6.2 Experimental Metal Atom/Vapor Chemistry. The Solvated Metal Atom Dispersion (SMAD) Method to Ultrafine Powders*

6.2.1 Metal Vaporization Methods

A wide variety of methods has been developed for vaporization of the metallic elements [1, 2, 16, 17]. Temperatures ranging from a few hundred degrees to 2000 °C are necessary, depending on the metal in question. Electrical resistive heating of crucibles is the most

---

* Parts of this section are reprinted with permission of the American Chemical Society and Elsevier Science Publishing Co. Inc.

Fig. 6-2. A typical metal vapor synthesis reactor made of glass and utilizing resistive heating. (A version of this model is available through Kontes Glass Company).

common method. The crucibles are usually made of tungsten wire covered with alumina cement, and these work well for common metals of interest including Mg, Ca, Ba, Sr, Mn, Fe, Co, Ni, Pd, Cu, Ag, Au, Zn, Cd, Hg, Ga, In, Tl, Ge, Sn, Pb, and the lanthanides. For more difficult elements, electron beam methods have been developed, for example for Ti, Pt, B, Al, Mo, W, Nb, and others [18, 19].

Figs. 6-2 to 6-4 show photographs of the most common types of metal vapor reactors and typical resistive heating vaporization sources. Detailed descriptions have been given elsewhere [17, 18].

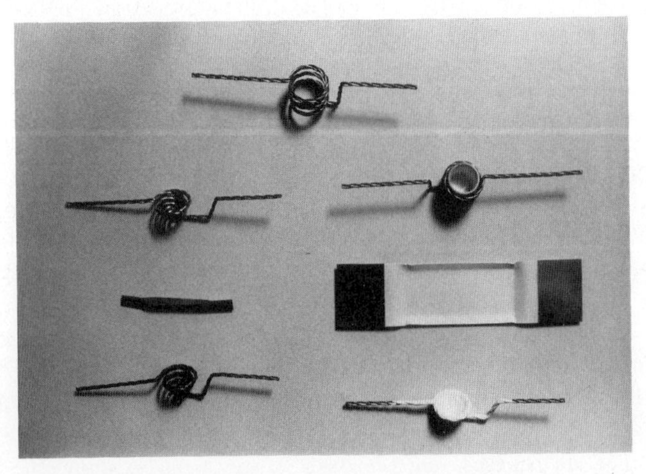

Fig. 6-3. Several types of commercially available metal vaporization sources.

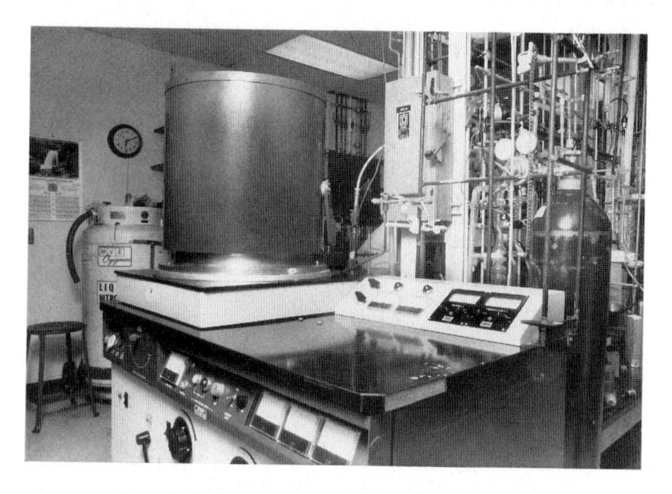

Fig. 6-4. An electron beam metal vapor synthesis codeposition reactor. (A version of this model is available from Planer Instruments).

Typical preparations of ultrafine metal powders using a resistive heating metal vapor reactor are described below, along with their use in synthetic schemes.

6.2.2 Preparation of Ultrafine Nickel in Pentane, Ni–FOP (Nickel–Fragments of Pentane)*

Small nickel pieces in the form of shot (about 2 g, 99.5% purity) were loaded into a crucible made of 1.5 mm Mo wire coated with alumina cement. The crucible was insulated with alumina wool (Saffil, ICI). The reactor was assembled with the crucible in place and pumped down to $<10^{-3}$ torr. The reactor vessel was cooled to 77 K using a dewar of liquid nitrogen. Pentane (freshly distilled from Na–benzophenone under Ar) was allowed to evaporate and enter the reactor as a vapor where it condensed on the inner walls of the reactor chamber. After 20–30 mL of liquid pentane had been introduced in this way, Ni vaporization was begun by heating the crucible resistively using a current of about 100 A at 3–6 V. The white-hot crucible vaporized molten Ni for 2 hours while 150 mL of pentane was cocondensed, and 1.5–2.0 g of Ni was deposited. A black frozen matrix was formed. The liquid nitrogen dewar was removed and the reaction flask isolated from the vacuum system. The matrix was allowed to melt yielding a black Ni–pentane slurry. This was stirred magnetically as it warmed, then transferred via a Teflon cannula into a glass filter frit. Filtration yielded a black ultrafine powder (1.5 g) which was dried under vacuum for several hours. Elemental analysis typically indicated an M:C:H ratio of approximately 5:1:1.3, the carbon being in the form of fragments of the pentane, especially C_1 fragments, and not carbidic. Powder X-ray diffraction showed broad absorptions due to Ni metal with crystallite sizes of less than 2 nm (20 Å) as calculated by the Scherrer equation and surface areas ranging from 30 to 60 $m^2\ g^{-1}$.

* Adapted from Hooker et al. [20].

6.2.3 Preparation of Nickel Carbide (Ni_3C) from Ultrafine Nickel Particles (Ni–FOP) [20]

A 100 mg sample of Ni–FOP was placed in a quartz tube (length = 15 cm, diameter = 0.8 cm) sealed at the bottom. About 10 drops of deoxygenated 1-octene were added. The tube was connected to an inert atmosphere line so that the quartz tube was under a head of pure argon. Connection of the gas supply to a bubbler prevented excess pressure build-up in the apparatus. A thermocouple was inserted directly into the Ni–FOP sample. The tube was placed in a hot silicone oil bath and heated to 190 °C for 30 minutes. Then the sample was cooled to room temperature, and evacuated to remove any residual volatiles. The resulting black powder was pure Ni_3C according to XRD analysis, with average crystallite size of 7 nm and surface area of 13 $m^2 g^{-1}$.

6.2.4 Preparation of Non-Solvated Ethylcadmium Iodide (EtCdI) from Ultrafine Cadmium Particles [11a]

Cadmium slurries were prepared by codeposition of cadmium vapor (~7 g) with about 50 mL of solvent in about 1 h (diglyme, *p*-dioxane, THF, toluene, or hexane). Brown to black matrices, which turned black on melt-down, were initially formed. The slurry was again cooled to –196 °C and 7.8 g of C_2H_5I (0.05 mol) distilled in. The upper portion of the reactor was wrapped with 3–4 turns of Tygon tubing with water flowing through it, which served as a means of condensing the solvent and EtI vapors. The slurry was refluxed under reduced pressure overnight. With toluene as solvent, a colorless solution and white precipitate were formed (Cd completely consumed). Removal of toluene and other volatiles under vacuum left yellow-white crude EtCdI. The volatile fractions contained about 3% $(C_2H_5)_2Cd$, as determined by hydrolysis with 10% HCl. The nonvolatile EtCdI was removed by washing with dry degassed THF (several 10 mL portions). The combined fractions were filtered under nitrogen, and solvent was removed under vacuum until some solid separated. At this point the solution was cooled at –10°C overnight. A 50% yield of EtCdI (THF solvated) was realized in this way, as determined by isolating it, drying under vacuum, and weighing in a Vacuum Atmosphere inert atmosphere box. Hydrolysis of the solid with 10% HCl-H_2O yielded 1 mol of ethane per mol of EtCdI, as identified by GLC, MS, and IR.

6.2.5 Preparation of Methyltin Iodides [CH_3SnI_3, $(CH_3)_2SnI_2$, $(CH_3)_3SnI$] from Ultrafine Tin Particles [11a]

Tin slurries were prepared by the codeposition of tin vapor (~0.5 g) with about 50 mL of solvent in about 1 h. The colors of the matrices varied depending on whether the solvent was toluene (yellow), dioxane (brown-black), THF (black), or hexane (black). On warming the solution, finely divided black slurries were formed and were transferred under fast N_2 flush to a 100 mL one-necked round-bottom flask equipped with a condenser, magnetic spin bar, and inert gas inlet. Methyl iodide (50 mmol) was added via syringe and the reaction mixture heated to reflux for 21 h. All of the metal was consumed, and a yellow precipitate was formed (SnI_4). The reaction mixture was filtered to remove SnI_4, the toluene was evaporated under vacuum, the resultant viscous liquid ($(CH_3)_xSnI_{4-x}$) was dissolved in 25 mL of CCl_4, and a weighed amount of CH_3CN was added to the solution as an NMR standard. An NMR spectrum of the solution revealed the presence of singlet resonances for CH_3SnI_3 ($\delta = 2.30$, 72%), $(CH_3)_2SnI_2$ ($\delta = 1.56$, 22%), and $(CH_3)_3SnI$ ($\delta = 0.82$, 6%) with an overall yield of 51% of organotin compounds based on the amount of tin vaporized. A 14% yield of SnI_4 was also isolated.

6.2.6 Coupling Reactions of Organohalides Using Ultrafine Nickel Particles [11a]

Nickel slurries were prepared as described for tin slurries, except that ~1 g of Ni was codeposited with about 60 mL of solvent in about 1 h. The colors of the matrices were: toluene (red-brown), THF (yellow), and hexane (black). Generally, THF was employed as the slurry solvent, the resulting very fine black slurry was transferred by syringe to a three-necked round-bottom flask equipped with a reflux condenser, addition funnel, stirring bar, and N_2 bubbler, and about 40 mmol of alkyl halide was slowly added (15 min) followed by reflux for about 20 h. The volatile products formed (no stable organometallics were isolable) were fractionated by trap-to-trap distillation through cold traps at −77, −116, and −196 °C. The gases were analyzed by PV measurements and GLC techniques.

When less volatile products were being analyzed, no trap-to-trap distillation was carried out. The reaction mixture was simply filtered and GLC analyses carried out. For example, biphenyl and bibenzyl were analyzed on a 5 ft × 0.25 in SE-30 column at 170 °C (bibenzyl was used as an internal standard when biphenyl was being analyzed and vice versa).

6.2.7 Grignard Reagent Preparation Using Solvated Metal Atom Dispersed Magnesium (SMAD-Mg)

6.2.7.1 Active Magnesium Preparation

In a typical active Mg preparation, 3.5 g Mg was vaporized in one hour from an aluminium oxide crucible in a tungsten wire heater (Mathis Co., C5 crucible, B 10) at 5.5 V, 100 A, connected to water-cooled copper electrodes. This apparatus was contained in a vacuum vessel (Kontes, 3000 mL, Kontes Glass Cat. No. K-612000 bottom, K-613200 top). During the Mg evaporation ca. 20 mL of THF (or hexane) was codeposited (vessel immersed in liquid nitrogen). Upon completion of the deposition the vessel was isolated from the vacuum pump and allowed to warm to room temperature. The Mg was isolated by surrounding the vessel with a polyethylene glove bag, flushing with argon, filling the vessel with argon, and transferring the Mg–THF to a weighed tube with a stopcock. The Mg–THF slurry could be used directly, or the THF could be pumped off leaving dry active Mg.

6.2.7.2 Manipulation of the Active Magnesium under Inert Atmosphere

After preparation of the Mg–THF slurry, the THF could be pumped off. Then a desired portion of the dry active Mg could be transferred to a weighed reaction flask in an argon-filled glove bag, followed by reweighing of the Grignard reaction flask to determine the charge of Mg.

In the cases where the Mg–THF slurry was used directly for Grignard preparations, it was only possible to make a rough estimate of the Mg transferred to the Grignard reaction flask. This was done by shaking the Mg–THF slurry vigorously and quickly pouring off a known volume, then the estimate could be made (with the assumption that the Mg was evenly dispersed after violent agitation).

The following section gives descriptions of typical Grignard reactions carried out.

6.2.7.3 Grignard Reactions with Active Magnesium

Bromopentafluorobenzene: in a typical Grignard reaction, 0.31 g (13 mmol) active Mg was transferred under argon to a one-necked 50 mL flask equipped with a serum cap for removal of liquid aliquots by syringe. After weighing, the flask was charged with 8 mL THF under argon. Then the flask, while under a head of argon, was cooled to –30 °C and 1.78 g (7.2 mmol) bromopentafluorobenzene was syringed in. Aliquots (0.20 mL) were taken intermittently with a cold syringe and transferred into 0.5 mL THF samples at –78 °C containing a few crystals of iodine (with methanol quenching in some cases). After warming and agitation, GLPC analyses were carried out on a 5 ft × 1/4 in column (20% Carbowax 20 M).

The preceding general technique was used for chlorobenzene, 2-chloroallylbenzene, allyl bromide, bromopentafluorobenzene, and trifluoromethyl bromide with indicated reaction temperatures of 25 °C, 25 °C, –40 °C, –30 °C, and –30 °C respectively.

6.3 The Matrix Clustering Process. Ultrafine Powders from Solvated Metal Atoms

6.3.1 Controlling Features

Now that the experimental method has been described it is appropriate that we consider the cluster/particle formation process in more detail.

When metal atoms are co-deposited with a very large excess of organic solvent the atoms are effectively frozen in place. However, if lesser amounts of solvent are employed some small clusters are also formed during the co-deposition process. This is because during freeze-down the atoms possess some mobility since they have kinetic energy and are very small in size. At this point the ligating ability of the solvent is particularly important. For example, if Ni is co-deposited with THF, mononuclear solvated metal atoms are formed even with low THF/Ni ratios. However, with pentane in similar pentane/Ni ratios significant clustering takes place during the deposition process. Thus, it has been shown that the polarity of the solvent plays a role in the efficiency of atom solvation; qualitatively an ordering for effective solvation is diglyme > dioxane > THF > toluene > pentane > xenon.

After the deposition is complete no further cluster growth occurs until the matrix is warmed. This process is also critical in determining the final particle size. It has been found that *slow* warming (several hours) leads to smaller particles. And during this process the ligating power of the solvent also has an effect.

We can summarize by listing three controlling criteria in particle growth:
1. solvent choice (more polar, more strongly ligating solvents favor smaller particles);
2. solvent/metal ratio (higher ratios favor smaller particles);
3. matrix warm-up rate (slow warming favors smaller particles).

In order to better understand these observations let us consider data concerning some mono- and bimetallic co-deposition and warm-up experiments.

6.3.2 Nickel with Various Solvents (Example of a Monometallic System)

In the case of Ni several solvents have been compared, specifically pentane, hexane, toluene, and THF [13, 20, 21]. Using a pentane/Ni molar ratio above 500, followed by warming to room temperature over 1–2 hours and vacuum removal of pentane, a black non-ferromagnetic Ni–FOP powder was formed. On the other hand, when a ratio of about 100/1 was used the resulting powder was ferromagnetic with larger crystallites (see Table 6-1) [13, 21–24]. Depositions of hexane with Ni gave similar results. Scanning electron microscopy revealed that the Ni–hexane particles were fairly uniform in size but had a very rough appearance [22]. In fact, the appearance of the particles was greatly dependent on the solvent employed (e.g. Fig. 6-5).

Nickel–toluene and Ni–THF depositions yielded *spherically* shaped particles of a nearly noncrystalline nature. More organic residue (fragments of the solvent) remained

(a)

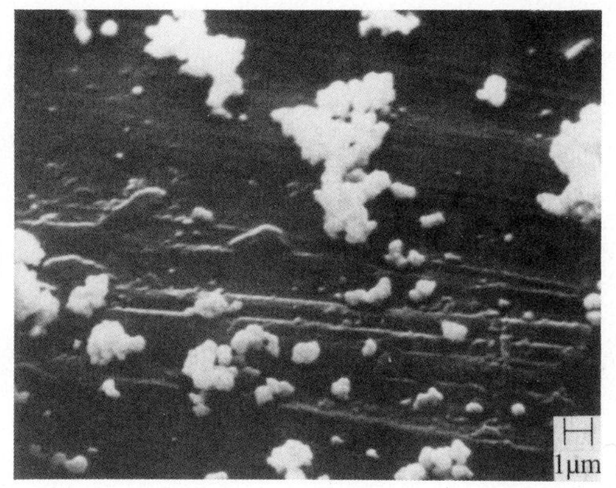

(b)

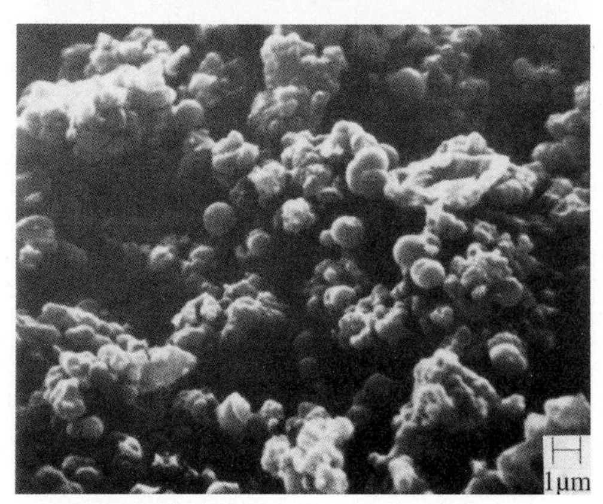

(c)

Fig. 6-5. Scanning electron micrographs of fresh Ni–solvent particles prepared by the solvated metal atom dispersion (SMAD) method. (a) Ni–hexane, (b) Ni–toluene, (c) Ni–THF. (Reproduced with permission of the American Chemical Society) [22].

bound to the Ni particles than in the Ni–alkane cases. With toluene if a 400/1 toluene/Ni ratio was used, non-ferromagnetic particles were obtained. In the case of Ni–THF the ultrafine powders were invariably non-ferromagnetic and the amount of organic material retained was the highest.

Since particle sizes were large compared with crystallite sizes it is clear that each particle is made up of a collection of nanocrystals. Thus, size and shape are clearly different depending on the solvent used and on the solvent/metal ratio. And the chemistry and catalytic properties are also different, as will be discussed later in this chapter.

6.3.3 Gold–Tin with Various Solvents (Example of a Bimetallic System)

In an attempt to learn more about the cluster growth process for solvated metal atoms, bimetallic systems that can form a series of intermetallic compounds have been considered [25]. Our reasoning was as follows: during the process of nanoparticle growth, solvated metal atoms must decompose thermally with the formation of metal dimers, trimers, etc., and in this way particle nucleation and growth occurs. Does this occur with kinetic control and irreversibility of each step, or is some selectivity involved, implying reversibility of some steps with inherent thermodynamic control? We reasoned that if strictly kinetic growth was dominant, a statistical mixture of M and M′ would form, whereas if thermodynamic control was dominant, thermodynamically favorable intermetallic compounds should form [Eqs. (6.1) and (6.2)].

$$n\mathrm{M(solv)} + m\mathrm{M'(solv)} \xrightarrow[\text{control}]{\text{kinetic}} \mathrm{M}_n\mathrm{M}'_m\mathrm{(solv)} \tag{6.1}$$

$$n\mathrm{M(solv)} + m\mathrm{M'(solv)} \xrightleftharpoons[\text{control}]{\text{thermodynamic}} \mathrm{M}_x\mathrm{M}'_y\mathrm{(solv)} + \mathrm{M}_{n-x}\mathrm{(solv)} + \mathrm{M}'_{m-y}\mathrm{(solv)} \tag{6.2}$$

For this study we chose the Au–Sn combination for several reasons: (1) the relatively low reactivity of Au and Sn with solvents; (2) ease of vaporization; (3) the fact that gold and tin form several well characterized intermetallics such as Au_5Sn, AuSn, and $AuSn_2$; (4) the ability to use Mössbauer spectroscopy to monitor the state of tin. Of course, we were also able to vary solvents and warm-up rates [25].

In this study gold and tin were usually vaporized simultaneously from two separate crucibles and co-condensed with an excess (about 100:1) of the vapor of the solvent of choice. Upon warm-up the atom clustering took place, mainly in the cold liquid solvent as it melted and warmed. Two warm-up procedures were used: rapid, 0.5 h from 77 K to room temperature; slow, 3 h from 77 K to room temperature.

X-ray powder diffraction (XRD) studies showed that when a 1:1 ratio of Au:Sn was used, the major product was particles of AuSn, with smaller amounts of Sn and Au_5Sn particles. Such results suggest that a small but significant amount of selectivity occurs during particle growth. Other experiments also support this idea. Layering experiments, where gold and tin were evaporated at different times, were carried out. A layer of frozen Sn atoms/acetone was covered by a layer of frozen Au/acetone with about the same molar ratio of Au:Sn overall. Upon warming and clustering, exactly the same product distribution was formed. These results show that the clustering process takes place after solvent melt-down and mixing. If this were not true, larger amounts of Sn and Au particles should have formed due to the proximity of these species with each other in the initial frozen matrix.

These results also allowed our experimental design to be simplified. It is difficult to heat two vaporization crucibles simultaneously so that both metals vaporize at the desired rate. It is much easier to load metals into one crucible and just vaporize all the metal. Even though one metal usually vaporizes faster than the other, our results with Au and Sn show that this did not really affect the product outcome since the clustering process clearly takes place after solvent melting and mixing. Therefore, many of our experiments were carried out by the "one crucible–complete evaporation" method.

The next set of experiments was done to assess the results of solvent variation. We compared pentane, acetone, toluene, ethanol, cyclohexane, hexane, and diethylether. Results with fast warm-up (0.5 h) and slow warm-up (3 h) were compared. From all these experiments several generalizations were possible [25]:

- The product mix of AuSn, Sn, and Au_5Sn did not change with solvent variation (AuSn > Sn ~ Au_5Sn).
- The surface area and crystallite sizes did change with solvent variation but the range was not large (13–26 nm).
- No correlation with the melting temperature of the solvent.
- No correlation with solvent dielectric constant.
- A very rough correlation with viscosity; the more viscous the solvent, the smaller the crystallite size.
- A strong variation with warm-up procedure; slow warm-up gave small crystallites with larger surface areas. For example, with pentane the crystallite size was 16 nm for fast warm-up and 10 nm for slow warm-up, and the specific surface area was 17 $m^2\,g^{-1}$ for fast warm-up and 24 $m^2\,g^{-1}$ for slow warm-up. The effect of warm-up rate was most pronounced for the solvents of lowest viscosity.

These results suggest that the cluster growth process is only slightly selective for formation of thermodynamically stable intermetallic phases, but this small selectivity is not affected by solvent change. Warm-up procedure has the most pronounced effect particularly for the lowest viscosity solvents. Perhaps this can be rationalized if the growth process occurs within a rather narrow temperature range just when the melting solvent becomes less viscous. For slow warm-up the viscosity gradient would be less steep, and so in the critical viscosity range the cluster growth would be more selective in the sense that the formation could be partly reversible and the competing reaction with the host solvent, which tends to stop cluster growth, would also be more favored as time allowed.

Warming rate also had an effect on the mix of products. Fig. 6-6 shows differential scanning calorimertry (DSC) curves for Au–Sn–acetone powders produced by fast warm-up and slow warm-up. (DSC provides information about exothermic and endothermic reactions and phase changes upon controlled heating, and can be quite helpful when dealing with intermetallic compounds.) Endothermic processes show upward peaks while exothermic are downward. The sharp endothermic peaks represent phase changes in the individual ultrafine particles and correspond to melting points of the various compositions. In the fast warm-up experiment a mix of Sn, Au_5Sn, AuSn, and other intermetallics is present. However, in the slow warm-up only Au_5Sn and AuSn are present. This result strongly suggests that under the correct (and strict) cluster growth conditions product selectivity is possible to a degree.

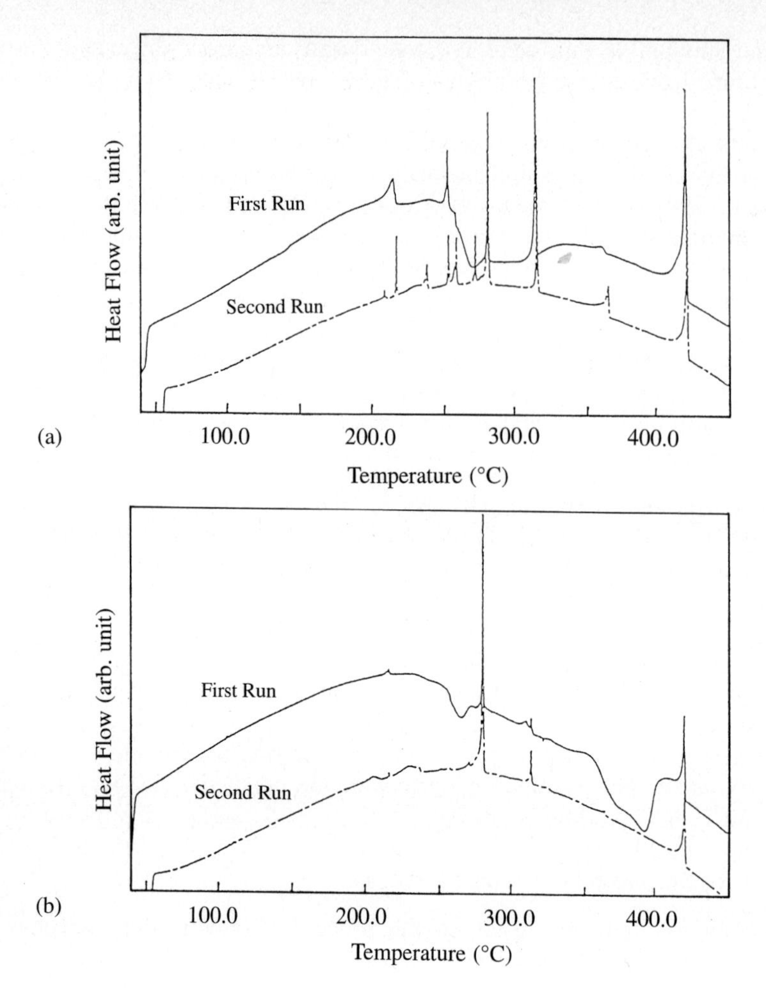

Fig. 6-6. Differential scanning calorimetry (DSC) curves of Au–Sn particles prepared in acetone: (a) fast warm-up, (b) slow warm-up.

6.3.4 Summary of the Clustering Process

Scheme 6-1 summarizes all of the observations in a cluster growth scheme. The following points can be made:

- Upon slight warming and matrix softening and melt-down, solvated atoms become labile and begin to oligomerize.
- In a certain narrow range of viscosity and temperature this oligomerization process is at least partially reversible and a competition is established between further cluster growth, declustering, and solvent reaction/ligation.
- If the temperature is raised quickly through this critical narrow range of temperature and viscosity, the cluster growth process becomes dominant and a rapid growth process occurs that is controlled by statistics of collisions.
- As the particles grow they become less mobile with size increase, and eventually the solvent reaction/ligation process catches up and stops further growth.

– The observations could be summed up by saying that slow warm-up allows a "milder" clustering process to occur, while fast warm-up causes a "wilder" process to occur.

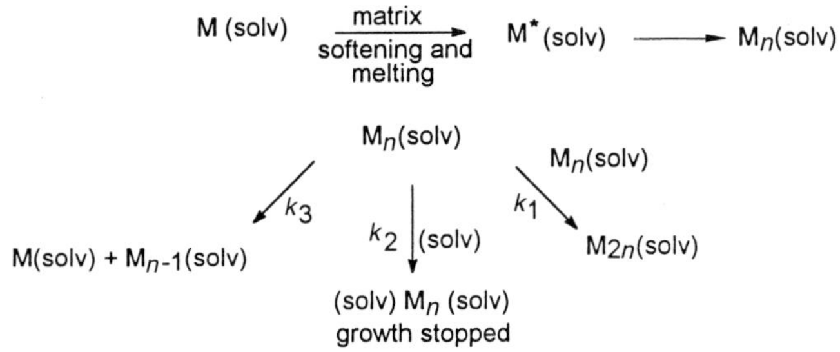

M (solv) —matrix softening and melting→ M* (solv) ——→ M_n(solv)

M_n(solv)

k_3 ↙ M(solv) + M_{n-1}(solv)

k_2 ↓ (solv) → (solv) M_n (solv) growth stopped

k_1 ↘ M_n(solv) → M_{2n}(solv)

Scheme 6-1

6.4 Reaction Chemistry of Solvated Metal Atom Dispersed (SMAD) Powders

6.4.1 Introduction

In this section the chemistry of ultrafine powders prepared by metal atom/vapor techniques is considered. Metal *atom* chemistry will not be covered since it has been reviewed before [1, 2] and does not fit the general subject matter of this book.

Since the active metal powders prepared by metal vapor methods are formed from solvated metal atoms, we have named them "solvated metal atom dispersed" or SMAD powders.

6.4.2 Reactions with Organic Halides

6.4.2.1 Mechanistic Considerations

Since we are interested in understanding the reaction chemistry of metal clusters, it is appropriate that the reactivity of *atoms* vs. clusters first be addressed. Although there are very few reports that deal directly with comparisons of atom vs. cluster reactivity, there are numerous qualitative observations suggesting that under similar conditions, small clusters are often more reactive than metal atoms [26, 27]. For example Ni atoms do not react with alkanes at 77 K, but clusters that form from the atoms on slight warming do react. Similar results were found in many transition metal atom/cluster–alkane matrix studies. In the gas phase, atoms and clusters of V, Fe, Co, Ni, Cu, Nb, Mo, Ru, Pd, W, Ir, Pt, and Al for M_n ($n = 1-14$) have been studied in CO adsorption experiments [26a]. Under the conditions of the experiment metal atoms did not usually react, whereas reactivity toward clusters when $n \geq 5$ was usually increased by a factor of 2. Similarly, reaction cross-section for $Al_n + O_2$ and $Al_n + D_2$ increased with cluster size [26b].

Explanations for increased reactivity of clusters compared with atoms or smaller clusters usually rely on several points:

a) Larger clusters have lower ionization energies (I.E.), and if electron transfer is necessary to initiate a reaction this can aid the situation.

b) Favorable thermodynamics of forming several bonds and being able to get cluster-type molecules (for example, $Al_n + O_2 \rightarrow Al_{n-4} + 2\ Al_2O$).
c) Clusters are more easily polarizable.

Perhaps the most relevant evidence on atoms vs. clusters, at least in the sense of understanding alkyl halides reactions, is that from the study of the low temperature chemistry of Mg, Mg_2, Mg_3, Mg_4 and of Ca, Ca_2, Ca_3, and Ca_x. In this work it was shown that small clusters of magnesium and calcium reacted in low temperature argon matrices containing RX, whereas of Mg and Ca did not [Eq. (6.3)] [27].

$$\underset{\text{argon at 9 K}}{\underset{\text{isolated in}}{Mg/Mg_2/Mg_3}} \xrightarrow{CH_3Br} \underset{(Mg_2/Mg_3\text{ consumed})}{\underset{\text{isolated in argon}}{\text{Mg atoms} + CH_3Mg_2Br + CH_3Mg_3Br}} \tag{6.3}$$

This interesting finding was rationalized as due to several effects. First, the reaction of the clusters is more energetically favorable as pointed out by Jaisien and Dykstra [28]. This is the result of the Mg–Mg bond being very weak in Mg_2, but relatively strong in R–Mg–Mg–X; thereby, *three* bonds are made while only one is broken in the reaction

$$Mg_2 + CH_3Br \rightarrow CH_3MgMgBr$$

Secondly, ionization energy probably also plays a role. Clusters generally have a lower I.E. than atoms, and electron transfer is a likely first step in the reaction [Eq. (6.4)]. Thirdly, cleavage of the C–Br bond may be more favorable in a four-centered transition state, requiring a cluster rather than an atom, as shown in **1**.

$$Mg_2 + CH_3Br \rightarrow Mg_2^+CH_3Br^- \rightarrow CH_3Mg_2Br \tag{6.4}$$

$$\begin{array}{c}(Mg---Mg)^+\\ \backslash \quad\quad / \\ (C--Br)^-\end{array}$$

Further work with a series of CH_3X species and magnesium and calcium showed the following trends in reactivities [27]:

- clusters were more reactive than atoms;
- larger clusters Mg_4, Ca_3, Ca_x were more reactive than smaller clusters;
- calcium species were more reactive than magnesium species;
- the alkyl halide reactivity trend was $CH_3I > CH_3F > CH_3Br > CH_3Cl$.

Considerations of the overall thermodynamics of bond breaking/making, coupled with I.E.s, seemed to explain these trends.

6.4.2.2 Syntheses

Magnesium. Magnesium clusters prepared by the SMAD method in THF are extremely tiny particles and can be handled by syringe as black slurries at room temperature [29, 30].

If the THF was evaporated, the Mg powder left behind was black, pyrophoric in air, and extremely reactive with water. This material could be stored under Ar in the dry state or the THF slurry state for weeks at a time with no noticeable change in appearance or reactivity.

A series of Grignard reagents were prepared at various temperatures using this active form of magnesium. Two significant advantages were that initiation of the reactions took place immediately, and low temperatures could be employed [Eq. (6.5)].

$$RX + Mg^* \xrightarrow{\text{solvent}} RMgX \tag{6.5}$$

The systems studied and the results obtained are described below: Bromobenzene: At 25 °C in THF or Et_2O the reaction was very rapid and yields were high. Chlorobenzene: At 25 °C for 3 hours in THF an 85% yield was obtained. 2-Chloroallylbenzene: At 25 °C for 23 hours in THF an 80% yield was obtained. Bromopentafluorobenzene: At –30 °C for 45 minutes in THF a 77% yield was obtained. Allylbromide: at –40°C for 30 minutes in THF a 61% yield was obtained. Trifluoromethyl bromide: At –30 °C in THF the reaction proceeded smoothly; however, polymeric material $(-CF_2-CF_2-)_n$ formed continuously, and only a 1% yield of CF_3MgBr could be trapped as CF_3I by addition of iodine.

A number of comparisons of reactivity with that of other forms of Mg were carried out. Chlorobenzene and 2-chloroallylbenzene were allowed to stir under Ar at 25 °C with Mg chips, Mg powder, and Mg filings freshly prepared by filing a Mg rod in an argon atmosphere. No reactions took place in any of these control experiments, demonstrating that the SMAD-Mg was clearly more reactive and had the added advantage of immediate reaction initiation, even at low temperature.

The high reactivity and storability of SMAD-Mg is presumably due to the protection of the clean rough Mg surface by adsorbed solvent. The solvent also contributes to limiting particle growth during formation, as well as discouraging particle growth (sintering) during storage.

In order to assess the sensitivity of the SMAD-Mg to solvent, a series of experiments on Mg preparations with different solvents was carried out:

1. Magnesium obtained by the SMAD method using THF;
2. SMAD-Mg–THF followed by THF evaporation and readdition of THF to the dry powder;
3. SMAD-Mg–hexane followed by hexane evaporation and addition of THF.

With all three of these samples the rate of Grignard formation with chlorobenzene was about the same. It can be concluded that either alkanes or ethers will allow formation of the active clean Mg surfaces needed to initiate Grignard formation and enable the reaction to continue.

The importance of being able to initiate and carry out Grignard preparations at low temperatures has been further demonstrated by studies of strained ring systems [31, 32]. For example, cyclopropylmethylmagnesium bromide was prepared at –75 °C using SMAD-Mg–THF, and this very low temperature greatly discouraged rearrangement to 3-butenylmagnesium bromide (Scheme 6-2). Similarly, benzocyclobutenylmethylmagne-

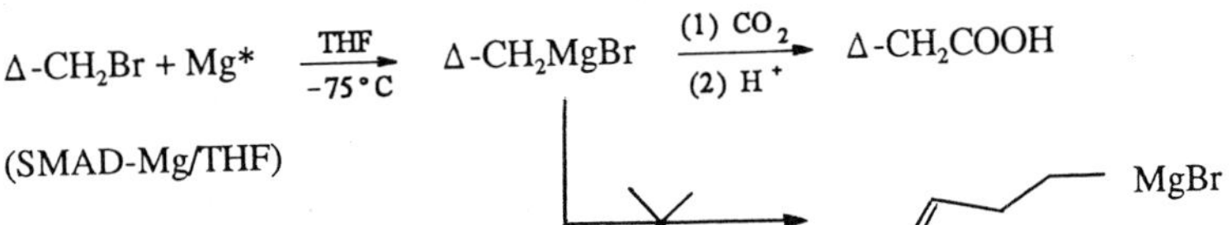

$$\Delta\text{-}CH_2Br + Mg^* \xrightarrow[-75\,°C]{THF} \Delta\text{-}CH_2MgBr \xrightarrow[(2)\ H^+]{(1)\ CO_2} \Delta\text{-}CH_2COOH$$

(SMAD-Mg/THF)

MgBr

Scheme 6-2

sium bromide has been prepared with SMAD-Mg–Et_2O, and the low temperature used discouraged rearrangement to the ring-opened isomer (Scheme 6-3) [31, 32].

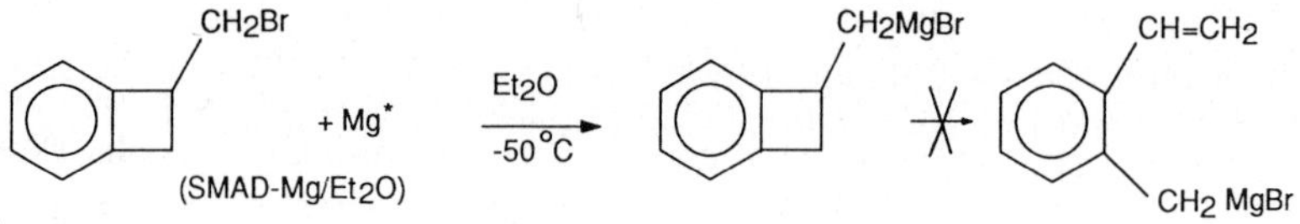

$$\text{benzocyclobutenyl-}CH_2Br + Mg^* \text{ (SMAD-Mg/}Et_2O) \xrightarrow[-50\,^\circ C]{Et_2O} \text{benzocyclobutenyl-}CH_2MgBr \not\rightarrow \text{o-}(CH{=}CH_2)C_6H_4CH_2MgBr$$

Scheme 6-3

Zinc and Cadmium. Other organohalide reactions with SMAD metals have dealt with zinc and cadmium powders. Again, the ultrafine slurries could be handled by syringe. Cadmium slurries in diglyme, dioxane, THF, hexane, and toluene were prepared by co-condensing about 9 g of Cd vapor with ~60 mL of solvent. Colored matrices were formed that turned black upon melt-down. Alkyl iodides were added to the slurries followed by warming and then refluxing at reduced pressure [Eq. (6.6)] [11, 30].

$$\text{Cd vapor} + \text{hexane} \xrightarrow{77\text{ K}} \xrightarrow{\text{warm}} Cd_n\text{–hexane} \xrightarrow{RI} RCdI + R_2Cd + CdI_2 \qquad (6.6)$$

Yields of organocadmium compounds ranged from 55 to 83%, with diglyme as solvent giving the highest yield. A mixture of RCdX and R_2Cd was produced with RCdI usually the major product.

In the case of SMAD-Zn slurries, reaction with alkylbromides yielded R_2Zn species (R = Et, *n*-Pr, and *n*-Bu) that could be isolated by vacuum distillation. Yields ranged from 20 to 100%, with toluene solvent giving the lowest and diglyme giving the highest yields [11]. Methyl iodide also reacted very efficiently. An example of particular relevance is the preparation of pure $(CH_3)_2Zn$ in hexane solvent; this product can be freed of zinc salts simply by distillation [Eq. (6.7)].

$$\text{Zn–hexane} + CH_3I \rightarrow \underset{\text{(distilled off with hexane)}}{(CH_3)_2Zn} + ZnI_2 \qquad (6.7)$$

Similarly, Zn–diglyme, –dioxane, –THF, and –hexane slurries could be used in Simmons–Smith chemistry for cyclopropanation.

Yields of norcarane were not very high and were strongly dependent on solvent, polar solvents giving higher yields (~25%) after short reflux times. The advantages are that CH_2Br_2 rather than CH_2I_2 can be used, the procedure can be carried out in the absence of hydroxylic solvents or acids, and no prior metal cleansing is necessary (e.g. Scheme 6-4).

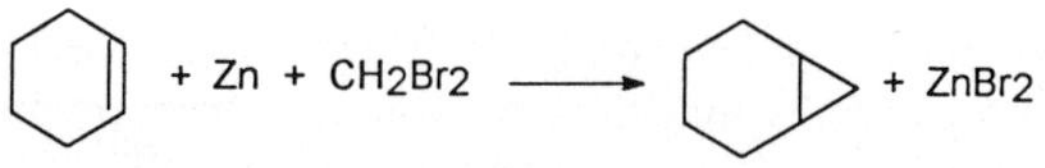

$$\text{cyclohexene} + Zn + CH_2Br_2 \longrightarrow \text{norcarane} + ZnBr_2$$

Scheme 6-4

Thus, the SMAD method provides a convenient way to prepare pure R_2Zn and R_2Cd/RCdX. Solvent does play a critical role, with the more polar solvents giving the highest yields. However, even nonpolar solvents can be used with some sacrifice in yield. This may be acceptable when polar solvents are to be avoided. These metal powders can be stored anaerobically in dry or slurry form for many months without adverse effect on their reactivity. In addition, the SMAD metals are reactive enough in the ether solvents to allow avoidance of the highly polar solvents such as DMSO, DMF, or HMPA.

It would be further advantageous if the yields of reactions in the ether type solvents could be improved. Rieke and coworkers [33] have shown that addition of salts such as KI has a beneficial effect on reactivities with alkyl halides. It could be speculated that this effect was due to better dispersion of the metal on high surface area crystals of KCl or KI. In order to test this, a series of experiments was carried out in which the solvated metal atom solutions were allowed to deposit on high surface area supports of KCl, KI, and Al_2O_3, in much the same way as has been used to prepare highly dispersed SMAD catalysts [34]. In particular, Cd–THF and Cd–diglyme matrices were allowed to melt down and disperse on the supports, and then reactions with n-propyl bromide were carried out [11]. It was found that large amounts of KI dramatically increased the yield of *n*-PrCdX. If this were simply due to the higher dispersion of the Cd on the KI surface, then other supports should work as well. However, KCl and Al_2O_3 did not enhance the Cd-slurry activity. Neither did catalytic amounts of KI. Therefore, it was concluded that the higher yields were simply due to an S_N2 process where *n*-PrBr was converted to *n*-PrI by excess KI before reaction with Cd. Indeed, a control experiment with *n*-PrI showed that under the same reaction conditions in the absence of cadmium, *n*-PrBr was converted to *n*-PrI in one hour [Eq. (6.8)].

$$n\text{-PrBr} \xrightarrow{\text{KI}} n\text{-PrI} \xrightarrow{\text{Cd}} n\text{-PrCdI} \qquad (6.8)$$

Aluminum and Indium. Other metal powders prepared by the SMAD procedure have shown high reactivities with organohalides [11]. For example, aluminum slurries in toluene, xylene, or hexane showed high reactivities towards C_6H_5Br and C_6H_5I [Eq. (6.9)].

$$\text{Al}_n\text{–toluene} + C_6H_5Br \xrightarrow[\text{5 h}]{\text{reflux under reduced pressure}} \underset{75\%}{(C_6H_5)_3Al_2Br_3} \qquad (6.9)$$

Likewise, rapid reactions with $(C_6H_5)_2Hg$ were observed [Eq. (6.10)].

$$\text{Al}_n\text{–toluene} + (C_6H_5)_2Hg \rightarrow (C_6H_5)_3Al + Hg \qquad (6.10)$$

In the case of aluminum, ether type solvents could not be used due to extensive deoxygenation of the solvent by the Al atoms/clusters.

Activated indium could also be prepared in a variety of solvents, and reaction with EtI under reflux at reduced pressure yielded a mixture of compounds [Eq. (6.11)].

$$\text{In}_n\text{–diglyme} + \text{EtI} \xrightarrow[\text{21 h}]{\text{reflux under reduced pressure}} \underbrace{Et_2InI + EtInI_2}_{64\%} + InI_3 \qquad (6.11)$$

Tin and Lead. Tin can also be activated in the SMAD method [11]. In diglyme a fine slurry could be handled by syringe. The toluene slurry was less fine, but still highly reactive. Reaction with CH_3I followed by reflux overnight yielded a mixture [Eq. (6.12)].

$$\text{Sn}_n\text{–toluene} + \text{MeI} \xrightarrow[\text{16 h}]{\text{reflux}} \underset{72\%}{MeSnI_3} + \underset{22\%}{Me_2SnI_2} + \underset{6\%}{Me_3SnI} + SnI_4 \qquad (6.12)$$

overall yield of organotin compounds = 51%

In dioxane or THF the formation of Me_2SnI_2 was much more favored, up to 50% of the total. Lead was also prepared in an active form. In diglyme Me_3PbI was formed [Eq. (6.13)].

$$Pb_n\text{–diglyme} + MeI \xrightarrow[24\ h]{reflux} \underset{10\%}{Me_3PbI} + PbI_2 \tag{6.13}$$

Although the yield was low, this was the first example of the direct reaction of lead metal with MeI in the absence of an activator such as sodium metal.

Nickel. Nickel-SMAD powders have also been treated with alkyl and aryl halides [11]. The Ni_n–THF slurry was particularly reactive and could be manipulated by syringe. When this black slurry was treated with organohalides at 40–50 °C, extensive reactions took place. Products were invariably due to the decomposition of intermediate "RNiX" species [Eqs. (6.14), (6.15)].

$$Ni_n\text{–THF} + CH_2I_2 \xrightarrow[7\ h]{40\text{–}50\ °C} C_2H_4 \quad (55\%\ \text{yield with } 100\%\ \text{selectivity}) \tag{6.14}$$

$$Ni_n\text{–THF} + \text{n-PrBr} \xrightarrow[20\ h]{40\text{–}50\ °C} \underset{31\%}{CH_3CH{=}CH_2} + \underset{69\%}{CH_3CH_2CH_3} \quad (\text{overall yield } 22\%) \tag{6.15}$$

The exclusive formation of C_2H_4 from CH_2I_2 suggests that the expected intermediate CH_2INiI is stable enough that two such molecules can react with each other to yield C_2H_4 and 2 NiI_2.

In the cases of benzyl chloride and allyl bromide rapid reactions took place producing mainly coupling products but also reduced product [Eq. (6.16)].

$$Ni_n\text{–THF} + C_6H_5CH_2Cl \rightarrow \underset{88\%}{C_6H_5CH_2CH_2C_6H_5} + \underset{12\%}{C_6H_5CH_3} \quad (\text{overall yield } 42\%) \tag{6.16}$$

In general the highest yields of coupling products resulted from "stabilized" RNiX species (benzyl, allyl, CH_2I). Presumably then, reactive intermediates live long enough so that intermolecular coupling can occur. In contrast, the analogous presumed intermediates CH_3NiX, EtNiX, and PrNiX immediately decomposed yielding disproportionation and radical abstraction products, and no coupling products were formed.

6.4.3 Reactions with Hydrogen

6.4.3.1 Magnesium

Imamura and coworkers have reported that Mg_n–THF SMAD powders react efficiently with H_2, and proposed this form of magnesium as a good H_2 storage medium [35]. In this work only a relatively small amount of THF was employed: for example, 1 g of Mg vapor was solvated with 5 mL of THF. The THF was evaporated leaving a fine powder of Mg. This material was outgassed at 200 °C to about 10^{-5} torr for 2 hours, and then exposed to H_2 pressures of up to 460 torr at 200 or 250 °C for several hours. No additional activation steps were needed.

These dry SMAD Mg–THF samples were extremely reactive with H_2, with about 20% of the Mg converted to MgH_2 over 20 hours. Furthermore, the Mg–THF sample was extremely reactive even at 0 °C for $H_2 + D_2 \rightarrow HD$ with a rate of 1.35×10^{19} molecules s^{-1} g^{-1}. It was noted that activated forms of normal Mg powder showed essentially no activity for H_2 adsorption or H/D exchange under the same conditions. Indeed, preliminary kinetic studies showed that the apparent activation energy for the H_2 absorption process was about 42 kJ mol^{-1} for the Mg–THF sample, whereas pure samples of activated normal Mg have been known to yield much higher values, for example, 96–117 kJ mol^{-1}.

The high reactivity of these SMAD Mg–THF samples is apparently due to a morphology of layer-type flakes, and to the preservation of active surface sites by the strong adsorption of THF and fragments of THF. After heating at 200 °C the elemental composition of the starting Mg–THF sample was $Mg(C_xH_yO_z)_{0.027}$. Higher temperature pyrolysis caused the THF and small amounts of alkanes/alkenes/aldehydes to be released. Apparently the presence of small amounts of these carbonaceous species also aids the hydridation process.

6.4.4 Catalysis by SMAD Metals (without Catalyst Supports)

6.4.4.1 Introduction

Although there is a great deal of published literature on SMAD catalysts [34], most of that work has dealt with supported catalysts. Here we will only consider the unsupported metal particles and situations where only ultrafine metal particles and solvents are present (no Al_2O_3, SiO_2, C, MgO, etc. as catalyst supports).

As discussed earlier the morphology of metal particles prepared by the SMAD method is highly dependent on the solvent employed. Shape, roughness, crystallinity, and particle size are changed with solvent variation. We have also seen that the reactivity of the SMAD metals is solvent dependent. Therefore, it should not come as a surprise that catalytic properties are also dependent on solvent.

6.4.4.2 Nickel

In hydrogenation studies of 1-heptene, norbornene, and benzene it was found that SMAD Ni–pentane, Ni–hexane, and Ni–toluene were very active, whereas Ni–THF SMAD powder was completely inactive under the same conditions [21]. The most active catalyst was Ni–hexane which slightly exceeded the rates for a commercial sample of Raney Ni. Interestingly, if THF was added to the Ni–hexane powder, followed by pump-off of excess THF, the catalytic activity fell to 20% of the original value but did not fall to zero. This results shows that THF strongly adsorbs to the Ni surface and inhibits chemisorption of alkene or benzene and probably H_2 as well. Indeed, in separate experiments it was shown that none of the adsorbed THF could be displaced by 1-hexene, toluene, diethylamine, or diethylether. Stronger ligands were able to displace some of the THF (up to 15%) in the order triethylphosphite > acetophenone > ethanol.

It should be noted, however, that THF did not completely poison the Ni_n–hexane catalyst. This result is reasonable since the Ni_n–THF SMAD system would have THF as well as fragments of THF adsorbed, whereas Ni–hexane (THF added later) probably only had THF bound.

An even more revealing study was carried out in a recirculation reactor where 1-butene or 1,3-butadiene with H_2 or D_2 were circulated over small amounts of Ni–pentane, Ni–toluene, and Ni–THF dry powders [21]. It was observed that relative *activities* for hydrogenation and isomerization varied greatly. But even more importantly, catalytic *selectivities* varied as well.

Several catalytic processes were examined: (1) hydrogenation vs. isomerization of 1-butene; (2) complete *vs.* partial hydrogenation of 1,3-butadiene; and (3) H_2–D_2 exchange.

Figs. 6-7 and 6-8 compare the selectivities of Ni–pentane and Ni–toluene for hydrogenation of 1,3-butadiene. In the case of Ni–pentane this process was rapid with both double bonds reacting immediately. Note that with Ni–toluene the rate dropped by a factor of 40, and butane was a product only after the diene had been converted to butenes. Also a nonequilibrium mixture of butenes was formed with 1-butene favored.

This was followed by equilibration and finally selective hydrogenation of 1-butene to butane. So in the Ni–toluene case, 1,3-butadiene was chemisorbed and only one double bond reacted with chemisorbed hydrogen at which point 1-butene was released.

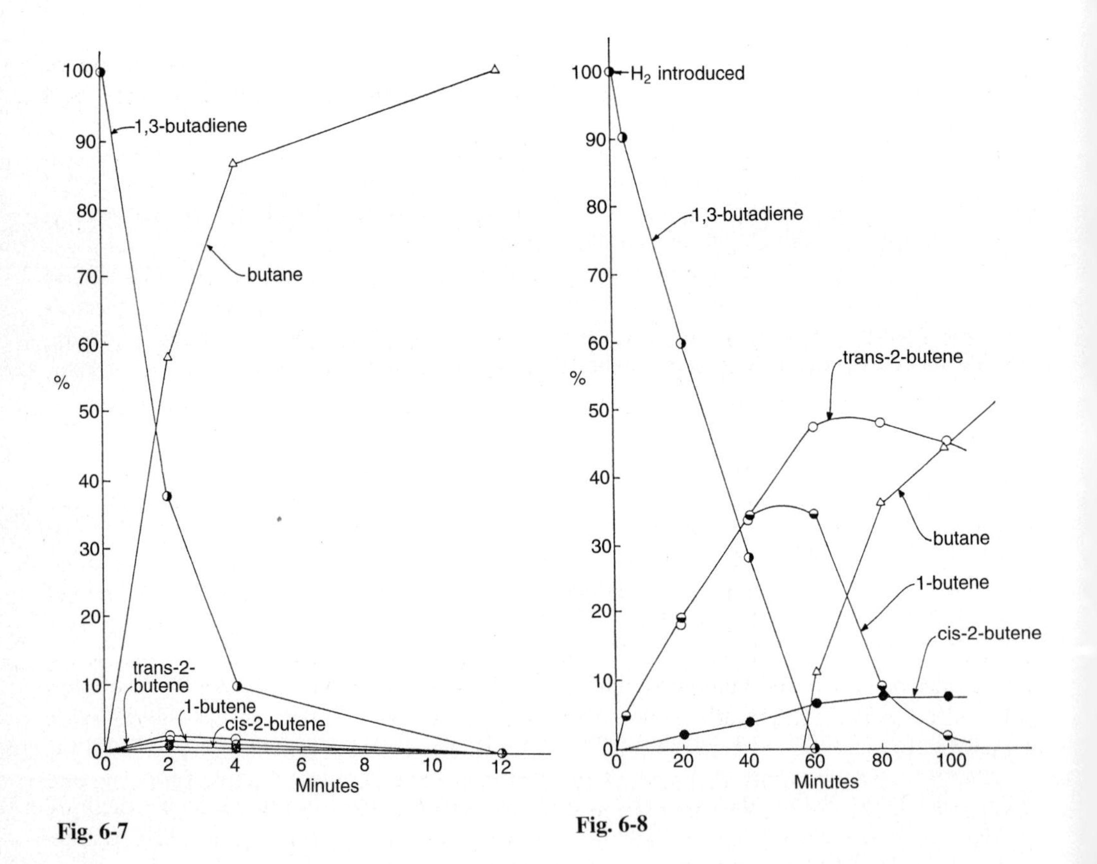

Fig. 6-7. Hydrogenation of 1,3-butadiene over a Ni–pentane SMAD powder (also referred to as Ni–FOP). (Reproduced with permission of Academic Press) [21].

Fig. 6-8. Hydrogenation of 1,3-butadiene over a Ni–toluene SMAD powder. (Reproduced with permission of Academic Press) [21].

Table 6-2. Rates of hydrogenation and isomerization of 1-butene and 1,3-butadiene under recirculation reactor conditions. (20 torr alkene or diene, 40–50 torr H_2, 0 °C). (Reprinted with permission of Academic Press) [21].

Catalyst	Isomerization of 1-butene		Hydrogenation of 1-butene		Hydrogenation of 1,3-butadiene		Selectivity[a]
	(mmol min^{-1} g^{-1})	Relative rate	(mmol min^{-1} g^{-1})	Relative rate	(mmol min^{-1} g^{-1})	Relative rate	
Ni–pentane[b]	0.039	0.14	4.38	2.1	3.64	1.8	×
Ni–toluene[c]	0.21	0.75 (red 0.091)[d]	0.11	0.052 (red 33)[d]	0.088	0.043	○
Ni–THF	0.0077	0.028	0.0070	0.0033	0.028	0.014	△
Raney Ni[e]	0.28	(1.0)	2.13	(1.0)	2.02	(1.0)	×

a) (○) Butane was formed after 1,3-butadiene was completely consumed; (×) butane was formed when 1,3-butadiene was still present; (△) intermediate between ○ and ×. b) Ferromagnetic sample. c) Ferromagnetic sample. d) Prior reduction of catalyst under 40 torr of H_2 at 100 °C for 2 h. e) Grace Chemical Co., No. 28.

In the case of Ni–THF, hydrogenation of 1,3-butadiene went by a complex pathway. 1-Butene was selectively formed followed by isomerization to mainly *t*-2-butene, and finally butane was slowly formed.

Similar patterns were observed for comparative studies of 1-butene hydrogenation and isomerization. Table 6-2 summarizes the rate data. Basically, Ni–pentane was an especially active hydrogenation catalyst and a relatively poor isomerization catalyst. Ni–toluene was considerably less active in hydrogenation, but was a superior isomerization catalyst. And Ni–THF was much less active than either of the others, but was quite selective.

An interesting finding with the Ni–toluene system was that if it was treated with H_2 at 100 °C for 2 h prior to being used as a catalyst, its isomerization activity went down while its hydrogenation activity went way up. This would suggest that fragments of toluene bound to the surface are responsible for the selectivity. High temperature H_2 probably removes some of these, opening up sites that are more active in hydrogenation.

With the Ni–THF system another set of experiments was carried out where 1,3-butadiene was hydrogenated in the presence of a mixture of H_2 and D_2. Some of the 1,3-butadiene picked up d_1 and d_2 before reduction, indicating an H–D exchange in competition with reduction. Other products showed an approximately statistical uptake of deuterium, which shows that H–D exchange competes very strongly with uptake of H_2 or D_2.

These data combined with many others indicate that catalytic activities and selectivities can be varied and tailored by choice of SMAD solvent. The catalysts formed upon solvated atom aggregation are stable at relatively high temperatures and have good aging properties.

The degrees of activity and selectivity must depend on: (1) the final particle size and surface area; (2) the nature and strength of adsorption of organic species and fragments by the metal crystallites; such organic species form during cluster growth by solvent fragmentation on the growing, highly reactive cluster; (3) the influence of the organic medium on the growth of specific (and perhaps characteristic) microcrystalline metal faces which are catalytically active [21].

Although the cluster formation process has been discussed earlier in this chapter, it is important to reemphasize one aspect of it, that of reaction with the host organic medium (in this case pentane, toluene, or THF). Experimentally it is known that pentane, the least reactive of the solvents employed, reacts chemically with small Ni clusters. This conclusion is based on two observations: (1) By varying the pentane/Ni ratio of the codeposition from 500/1 to 100/1 the crystallite size of the resulting Ni powder changed drastically. In either case plenty of pentane was available for reaction. However, very large dilution impedes cluster growth. This indicates that competitive pathways must exist where Ni_n growth competes with the reaction of Ni_n with pentane which yields strongly bound organic fragments (Scheme 6-5). Thus, high dilution serves to encourage the k'_{1-4} processes in Scheme 6-5 at the temperatures at which k_{1-4} are effective. (2) The dry Ni–pentane powders possess a large number of bound organic fragments that are not in the form of pentane, but are fragments of pentane. Furthermore, for the powders prepared in the larger excess of pentane the ratio of these fragments to Ni is much greater.

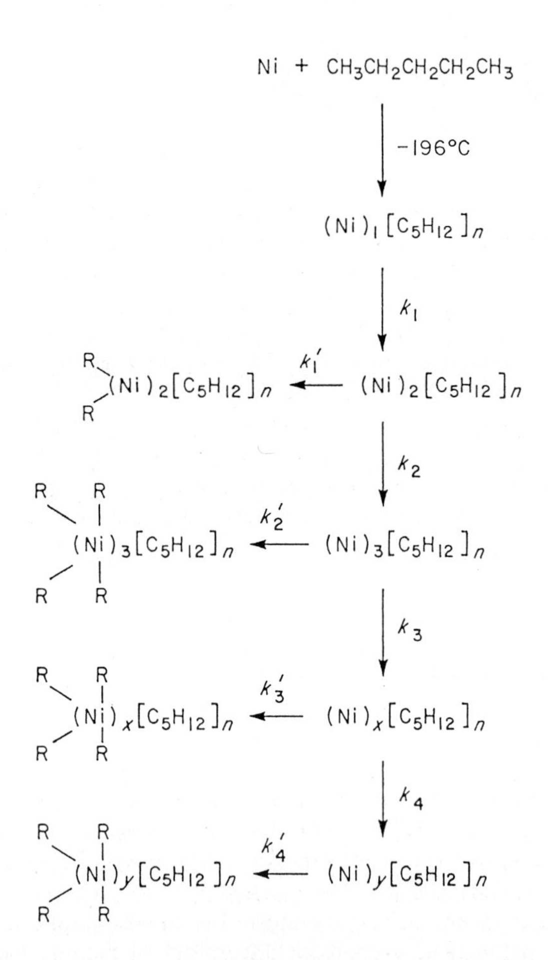

Scheme 6-5

Similar growth–dilution behavior has been found with many other solvents and metals [14, 24, 25, 36, 37]. There can be no doubt that the forming clusters are reacting rather extensively with the host solvents, even at quite low temperatures. What makes this especially interesting from a catalyst synthesis point of view is that it is the clusters with the most bound organics that are often the most active or selective. And in general the more polar solvents yield the highest surface concentration of carbonaceous materials, and produce the most selective catalysts. For example, Ni–toluene is a very good isomerization catalyst whereas Ni–pentane is not (but is an active hydrogenation catalyst).

Finally, a few words about catalyst structure are in order.

- *Ni–pentane or –hexane*: After formation various organic groups (C_1, C_2, C_3, C_4, etc.) are bound to the surface of the clusters. Upon treatment with olefin, arene, and/or H_2 some of these groups are displaced making available good hydrogenation sites. These sites can be quenched by addition of a polar solvent such as THF.
- *Ni–toluene*: Benzene, toluene, phenyl groups, and possibly C_1 groups are bound to the surface of the starting catalyst. There is a strong competition for sites with incoming olefin, arene, or H_2.
- *Ni–THF*: Bound organic groups are numerous and include chemisorbed THF and its fragments. A very strong competition is set up when olefin, arene, or H_2 are added.

6.4.4.3 Other Metals

The SMAD method has also been used to prepare a series of rare-earth metallic catalysts. For example, Sm and Y fine particles were prepared by codeposition with organic solvents [38]. Very unusual catalytic selectivities were observed. Alkenes, dienes, and arenes were readily hydrogenated over these SMAD metal particles, but alkynes were unreactive under the same conditions. Kinetic studies showed that the H_2 chemisorption step seemed to be rate-determining. Another unusual finding was that the reaction $H_2 + D_2 \rightarrow 2HD$ did *not* take place over these catalytic particles. This result and the selectivity data show that these rare-earth SMAD catalyst particles function by different mechanisms than analogous transition metal systems.

Combinations of La and Ni as bimetallic SMAD catalysts have also been studied [39]. After codeposition of the two metal vapors with organic solvents, warming, and depositing on a SiO_2 catalyst support, the average diameter of the bimetallic (La–Ni) particles was about 4 nm. X-ray photoelectron spectroscopy showed that the nickel was metallic whereas the La existed as both metal and La_2O_3. Catalytic activities were improved by the presence of La.

6.4.5 Formation of Carbides

6.4.5.1 Introduction

In the previous section we have seen that during cluster growth from solvated metal atoms some reaction of the growing clusters with the solvent host takes place. The final product is an ultrafine "pseudoorganometallic" powder, since some solvent and solvent fragments remain bound to the metal particles. The metal surface of this material is in a chemically and catalytically active form. It turns out that the carbonaceous material is also in an active form, and usually makes up about 2–8% of the total mass. The carbonaceous fragments are mainly C_1 species distributed throughout the nanoscale metallic particles, but are not in the form of interstitial carbidic carbon [13].

So the picture that has evolved for these extremely reactive materials is that of nearly noncrystalline metal particles interspersed with CH_3, CH_2, CH, plus other reactive carbon species, leading to the conclusion that both activated metal and activated carbon exist simultaneously.

The question arises as to how easily these materials could be converted to metal carbides while maintaining their nanoscale particle size. Such materials could be quite useful technologically.

6.4.5.2 Nickel and Palladium

Nickel carbide, Ni_3C, is a rather unusual carbide in that it is not stable at high temperatures, and so it cannot be made by high temperature processes. It is usually prepared by the reaction of activated Ni with CO. Since there is a great body of literature on catalytic applications and coking with Ni, and the SMAD process has also been investigated extensively with Ni, it was decided to study carbide formation with SMAD-nickel particles. Palladium, having similar chemistry and a relatively stable PdC_x phase, was also chosen for study [20].

X-Ray diffraction (XRD) analysis of fresh SMAD-Ni–pentane (referred to as Ni–FOP, or Ni with fragments of pentane) showed the presence of very small Ni crystallites and a small amount of NiO crystallites (due to adventitious oxidation during manipulation and analysis). After heat treatment at 200 °C under Ar, XRD indicated the presence of larger Ni crystallites plus Ni_3C with the amount of NiO diminished almost to zero (e.g. Fig. 6-9 and Table 6-3). There was no evidence of a metastable Ni–C solid solution, as has been observed in molten Ni/C rapidly quenched [40]. After higher temperature treatments of the Ni–FOP only Ni crystallites were detected, with larger crystallites and smaller surface areas.

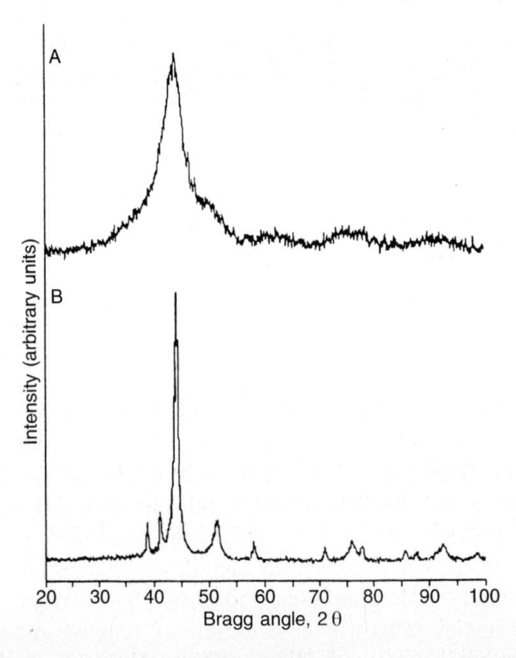

Fig. 6-9. Powder X-ray diffraction patterns of Ni–FOP powder; A. untreated, B. after heating to 200 °C, showing Ni crystallite growth and formation of a small amount of Ni_3C. (Reproduced with permission of the American Chemical Society) [20].

Table 6-3. Particle/crystallite data on Ni–FOP and Pd–FOP samples. (Reprinted with permission of the American Chemical Society) [20].

Sample	Identification	Unheated			200 °C		
		Species detected by XRD	Metal crystallite size (nm)	Surface area ($m^2 g^{-1}$)	Species detected by XRD	Metal crystallite size (nm)	Surface area, ($m^2 g^{-1}$)
1	Ni–FOP	Ni/NiO	1.7	29	Ni/Ni_3C	7.4	22
2	Ni–FOP	Ni/NiO	1.8	62	Ni/Ni_3C	6.2	28
3	Ni/PMMA[a]	Ni/NiO	1.9	21	Ni/Ni_3C	6.9	13
4	Pd–FOP	Pd	3.9	59	Pd	6.3	17

Sample	300 °C			400 °C		
	Species detected by XRD	Metal crystallite size (nm)	Surface area ($m^2 g^{-1}$)	Species detected by XRD	Metal crystallite size (nm)	Surface area, ($m^2 g^{-1}$)
1	Ni	12	19	Ni	15	21
2	Ni	12	22	Ni	16	22
3	Ni	12	19	Ni	19	16
4	PdC_x	7.5	9.0	Pd	8.3	6.6

a) PMMA = Poly(methyl methacylate)

The amount of Ni_3C formed from fresh non-oxygen-pacified Ni–FOP samples ranged from 28 to 40%, the remainder being Ni crystallites. From pacified samples the amount of Ni_3C formed dropped to 18%, but no NiO remained.

Palladium behaved somewhat differently [20]. No PdO was detected, and upon heating to 200 °C crystallite sizes increased. Upon heating to 300 °C a PdC_x phase was formed. Further heating to 400 °C caused this phase to decompose to Pd and graphite.

These results are consistent with the idea that heating the Ni–FOP to 200 °C converted some nickel to Ni_3C. Carbon is a limiting reagent, so only partial conversion could occur:

$$\text{Ni–FOP} \xrightarrow{200\,°\text{C}} Ni_3C + Ni + H_2$$

However, if some NiO was present, a thermodynamically favored reduction took place:

$$\text{Ni–FOP/NiO} \xrightarrow{200\,°\text{C}} Ni + Ni_3C + CO_2 + H_2O$$

The next part of this work was undertaken in an attempt to protect the active Ni–FOP and Pd–FOP powders by adding a polymer coating of poly(methyl methacrylate). Such protection was possible as shown by Auger surface and interior analysis (e.g. Table 6-4), and surface oxidation of the Ni particles was greatly reduced (e.g. Fig. 6-10). However, a complete surprise came when, upon heating the Ni–FOP/PMMA powder to 200 °C, *complete conversion to pure Ni_3C took place* [20].

This finding prompted a study of other organic additives. It was soon learned that many organic additives behaved similarly, including 1-octene, methyl methacrylate, and even *n*-octane. In the cases of 1-octene and methyl methacrylate the carbide formation

Table 6-4. Relative atomic concentrations from Auger electron spectroscopy. (Reprinted with permission of the American Chemical Society) [20].

Sample	Treatment	Species assigned by XRD	Surface %			Interior %		
			O	Ni	C	O	Ni	C
Ni–FOP	Ni–FOP pacified slowly in air	Ni/NiO	54	41	4.9	9.2	80	10
Ni–FOP	Ni–FOP/PMMA pacified slowly in air	Ni/NiO	11	21	67	2.6	62	35
Ni–FOP	Ni–FOP/PMMA heated to 200 °C	Ni/Ni_3C	6.2	37	57	3.2	67	30
Ni–FOP	Ni–FOP/MMA heated to 200 °C	Ni_3C	6.2	32	61	2.8	61	36

temperature was lowered to 174 °C. When no organic additive was present or *n*-octane was the additive, Ni_3C formation began at 181 °C. It was also observed that the onset of Ni_3C formation coincided directly with the reduction of any NiO that was present. Furthermore, the physical properties of the Ni_3C powder were *not* dependent on the organic reagent used.

According to XRD analysis the carbide samples were pure Ni_3C or PdC_x. However, elemental analysis showed the presence of a slight excess of non-carbide carbon: C(found) was 7.7–9.0%; C(calc) is 6.4%. Thus, a typical sample of Ni_3C analyzed as %Ni = 90.7, %C = 7.7, %H $\leq$ 0.5. The XRD results show crystallites in the size range of 19–24 nm, larger than those found without an organic additive. Surface areas ranged from 18 to 26 $m^2\ g^{-1}$.

In attempts to understand this process better, Auger analysis was undertaken (Table 6-4). Such studies showed that the fresh Ni–FOP samples had carbon distributed throughout, whereas oxygen was concentrated near the surface, supporting the idea that oxidation was due to adventitious sources. The Auger analysis along with total elemental analysis showed that the initial Ni–FOP samples did not possess enough carbon to enable all the

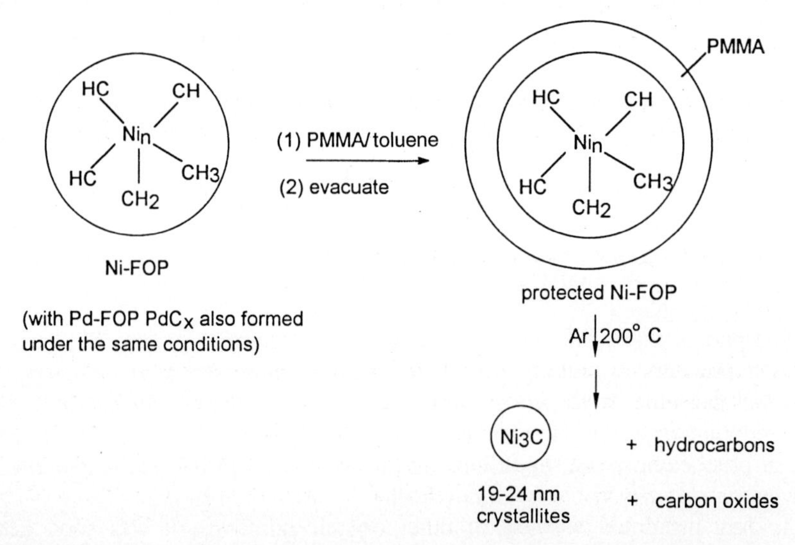

Fig. 6-10. Protection of SMAD-Ni–FOP with poly(methyl methacrylate) (PMMA), followed by conversion to nickel carbide.

metal to be converted to Ni_3C. We might expect the following stoichiometry based on elemental analysis of typical Ni–FOP samples [Eqs. (6.17), (6.18)].

$$\underset{\text{Ni–FOP after pacification}}{Ni_{15}C_3H_4O_6} \xrightarrow{181\ ^\circ C} Ni_3C + 12\ Ni + 2\ H_2O + 2\ CO_2 \quad \text{(balanced equation)} \tag{6.17}$$

$$Ni_{15}C_3H_4O_6 + \underset{\text{excess}}{\text{1-octene}} \xrightarrow{181\ ^\circ C} 5\ Ni_3C + \text{trace } C_{graphite} + 2\ H_2O + 2\ CO_2 + \text{hydrocarbons} \tag{6.18}$$

The Auger analysis also showed that a small amount of surface graphite was present on the Ni_3C particles, and this is logically the source of the slight excess carbon found in elemental analysis.

It is apparent from these results that the Ni particles in Ni–FOP are essentially capable of activating any organic compound so that carbon atoms become available for carbide formation. Cleavage of C–C and C–H bonds on the Ni surface allows carbonaceous groups to form, followed by carbon atom migration into the particle interior. When Ni_3C stoichiometry has been reached, migration stops, and probably a thin layer of carbonaceous material remains on the surface and is converted to graphite.

The ability to form carbidic phases under these relatively mild conditions appears to be unique to metal samples with very small crystallite sizes; indeed Ni–FOP seems to be unique. Off-the-shelf Ni powder or NiO powder did not react at all with excess methyl methacrylate to form Ni_3C. Also, Raney Ni of surface area 80–100 $m^2\ g^{-1}$ did not react. In the case of Rieke Ni, a partial conversion was observed.

Thus, SMAD-Ni powders possess organic carbonaceous fragments throughout. The metal particles also possess many imperfections and dislocations in the metal lattice. This implies that many reactive sites are present for dissociative chemisorption processes to take place, and carbon does not need to migrate far for carbides to form.

The chemical state of the carbon is also important. No enhancement of carbide phase formation was observed when carbon black (50–200 mesh) was employed as the carbon additive. Thus, the carbonaceous species formed by dissociative chemisorption of organic additives are in a special, reactive state.

It can be concluded that carbide formation under such mild conditions depends on having an active form of Ni as well as an active form of carbon. The SMAD process provides both.

6.4.6 Colloidal Metals in Solvents and Polymers

6.4.6.1 Initial Studies

Using the same SMAD technique, stable colloidal metal solutions can be prepared in many cases, and these have been found to be quite useful as precursors to metal films. Thus, by depositing metal atoms with excess solvents at low temperature followed by warming to room temperature, colloidal sols can be produced when the correct metals and solvents are employed. In general the most stable colloids are produced when polar organic solvents such as acetone, ethanol, isopropanol, dimethylformamide (DMF), dimethylsulfoxide (DMSO), or tetrahydrofuran (THF) are employed [41]. The metals that form such colloids are usually the noble ones such as Pd, Pt, and Au.

Actually, nonaqueous colloidal metal can be prepared by metal atom clustering in the gas phase followed by exposing the clusters to an organic solvent [42], or by the SMAD method where the clustering takes place from the solvated metal atom state (Scheme 6-6) [41].

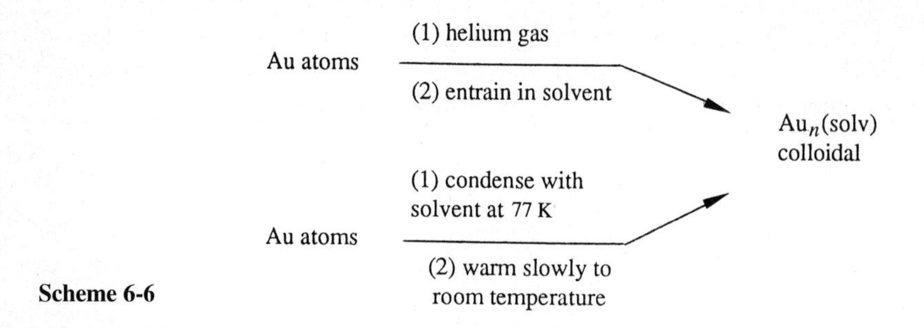

Scheme 6-6

The advantage of these methods as opposed to historical methods of reducing gold salts in aqueous solution is that no reduction byproducts (e.g. halide ions, metal ions, or extraneous organic/inorganic reagents) are present in the final colloidal solution; only metal clusters and pure organic solvent are present.

The majority of metal particles prepared in this way have spherical shapes and crystallite sizes of 4–9 nm, usually about 6 nm. However, in acetone these particles weakly agglomerate to 10–30 nm spherical flocculates.

Palladium colloids prepared by the SMAD method have received extensive study [43]. Several polar organic solvents were successfully employed, including acetone, isopropanol, and acetone–isopropanol mixtures. In these solvents the black colloids were stable indefinitely, even in the presence of air, and particle size ranged from 6 to 8 nm. Electrophoretic mobility studies showed that at least part of the stabilization was the result of the particles acquiring a negative charge, thus creating a potential difference (zeta potential) between the Pd core and the surrounding solvent-bearing counter ions. As expected [44], this zeta potential was destroyed by addition of electrolytes, and particle flocculation and precipitation occurred.

The negative charge that Au and Pd particles acquire in pure polar organic solvents is a very interesting phenomenon, and rather unexpected. It is true that small metal particles, especially gold, would have a rather high electron affinity. However, to strip electrons off the solvent molecules, which apparently occurs, would only be energetically feasible if a collection of solvent molecules serves to stabilize the resultant cation (e.g. Fig. 6-11). As with aqueous colloids, these charged particles would tend to repel each other, thus discouraging aggregation. However, electrolytes would serve to destroy this mode of stabilization.

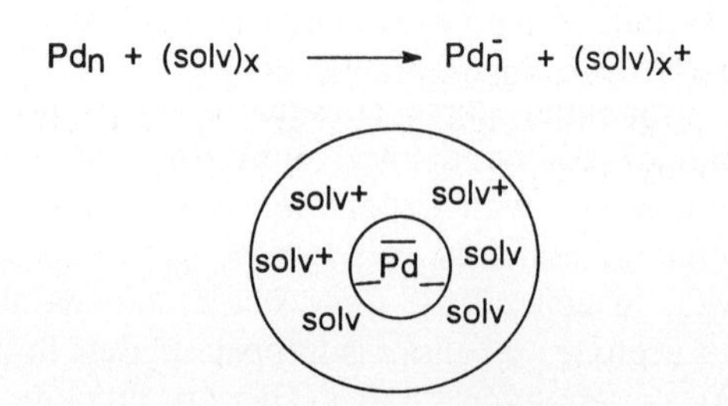

Fig. 6-11. The formation of charged colloidal particles in the SMAD method.

The SMAD method of producing non-aqueous metal colloids often leads to "living colloids". This means that the particle will continue to grow if solvent is slowly removed (Scheme 6-7). After pumping on such a film for three hours at 10^{-3} torr there still remained substantial portions of carbon and hydrogen, according to elemental analysis. Upon heating under vacuum at 100, 200, and 350 °C, first acetone was evolved followed by a mixture of hydrocarbons. The metal-like film initially appeared under the scanning electron microscope like an intertwined mass of strands of spherical particles. Upon heating, this converted to uniform metal films and the conductivity increased [43].

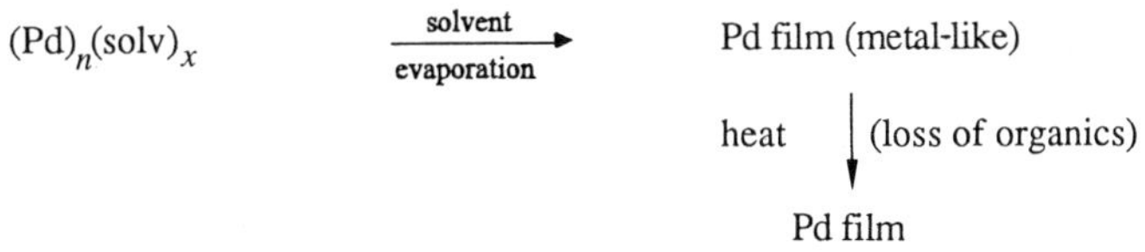

$(Pd)_n(solv)_x \xrightarrow[\text{evaporation}]{\text{solvent}}$ Pd film (metal-like)

heat ↓ (loss of organics)

Pd film

Scheme 6-7

In view of the importance of indium metal and indium oxide thin films as semiconductors [45–47], attempts were also made to prepare colloidal indium in non-aqueous solvents (Scheme 6-8) [48].

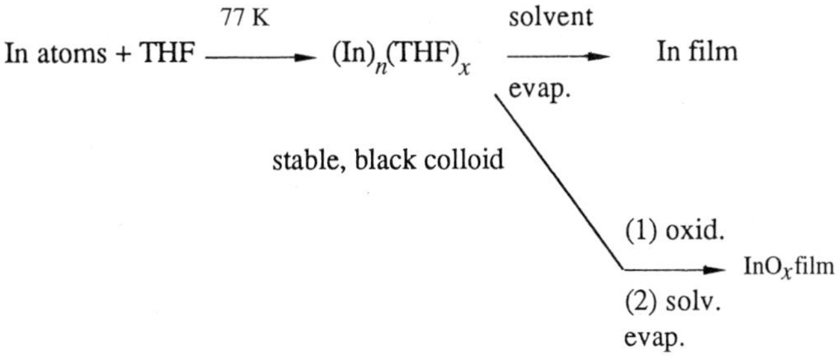

In atoms + THF $\xrightarrow{\text{77 K}}$ $(In)_n(THF)_x$ $\xrightarrow[\text{evap.}]{\text{solvent}}$ In film

stable, black colloid

$\xrightarrow[\text{(2) solv. evap.}]{\text{(1) oxid.}}$ InO_x film

Scheme 6-8

The most stable colloids were obtained with THF, DMF, isopropanol, and DMSO. THF and DMF showed larger particle sizes (20 and 25 nm, respectively).

Since In metal is very oxophilic, it would be expected that those sols would be sensitive to oxygen. Indeed, they were, for example In–isopropanol solution reacted vigorously with air. In fact, rapid exposure can lead to explosive oxidation and care must be taken in their handling. Although this oxidative instability causes a serious handling problem, the In particles can be slowly and controllably oxidized to indium oxide particles and used to prepare films. It is clear that substantial amounts of organic solvent remain coordinated and trapped within the In and In oxide films and the M:C ratios are generally around 2:1.

Another interesting feature is that when ketones were used as solvents, even greater amounts or organic material were incorporated into the films. This suggests that a chemical reaction took place, probably a ketone coupling, a reaction known to take place with reducing metals and carbonyl compounds [Eq. (6.19)].

$$M + R_2C{=}O \rightarrow M{-}O{-}\underset{R}{\overset{R}{C}}{-}\underset{R}{\overset{R}{C}}{-}O{-}M \qquad (6.19)$$

No pure coordination compounds were formed, but colloidal particles with higher contents of organics and smaller particle size resulted. In–isopropanol films showed infrared bands at 3400, 2980, and 1020 cm^{-1} which can be assigned to OH, CH, and C–O. The films exhibited high electrical resistance, probably due to the organic contaminants.

6.4.6.2 Particle Growth and Solvent Studies

In an attempt to understand that growth process and perhaps control colloidal particle sizes, a series of Au–acetone codepositions were carried out [49]. It was found that lower concentrations of Au yielded smaller particles (1–3 nm) while higher concentrations yielded larger ones (4–9 nm). It would seem that particle growth must be controlled by kinetic phenomena, as discussed earlier in this chapter. At low temperatures, as Au atoms begin migration in the softening matrix Au–Au bonds begin to form. This is an irreversible process. And as the Au particles grow, their mobility goes down. But if higher metal concentrations are present, the frequency of encounters is higher and the particles get bigger before the decreased mobility and solution steps are completed (including electron transfer and generation of charged particles).

Similar studies of the Ag–acetone system showed that some control of particle size was possible. However, overall the particles were larger (~30 nm) than with Au (1–9 nm) under similar experimental conditions. In addition, the $(Ag)_n$ colloids were light sensitive, with the light causing enhanced growth and precipitation. The Ag colloids were black while the Au systems were purple.

Films formed from other colloids incorporated substantial amounts of organic residues, part of which could be driven off by heating. Metal conductivity increased after heat treatments, but did not reach that of the pure metals [49].

Other attempts to control particle size and colloid stability have involved the use of fluorocarbon solvents [50]. It seemed reasonable to expect that Au–fluorocarbon colloids would yield cleaner films. Also, perhaps the colloids would possess unique properties due to the low chemical reactivity, polarity, and low surface tension of perfluorocarbon solvents.

Initial work with perfluoroalkanes showed that stable colloids were not formed, and therefore a heteroatom-containing perfluorocarbon was employed, namely perfluoro-*n*-tributylamine (PFTA, called "fluoro-inert", by 3M). Gold vapor–PFTA deposition yielded a brown-black colloid after warming to room temperature. Interestingly, upon extraction with acetone all the gold particles were taken up by the acetone layer. Similar results were found with light alcohols, ketones, aldehydes, and THF, although the particles were not taken up by saturated hydrocarbons, diethyl ether, or water [50].

Remarkably, the PFTA solvent adhered to the gold particles even after extraction into another solvent. These particles tended to strongly repel each other, suggesting that the residual fluorocarbon was still present on the particle surfaces. Indeed, studies of the colloids and films by IR, XPS, pyrolysis-GC-MS, and ^{19}F NMR revealed that this fluorocarbon residue was strongly adhering and in the form of fragments of PFTA. Actually, elemental analyses of room temperature derived acetone-extracted powder yielded an empirical formula $Au_{17}C_{20}H_{30}F_{26}NO_7$. This indicates that both acetone and PFTA fragments were present.

The conclusion from these studies is rather remarkable. Even though gold is an "inert" metal, atoms and small particles of gold are not. The forming particles react extensively with PFTA, yielding stable structutres (e.g. Fig. 6-12).

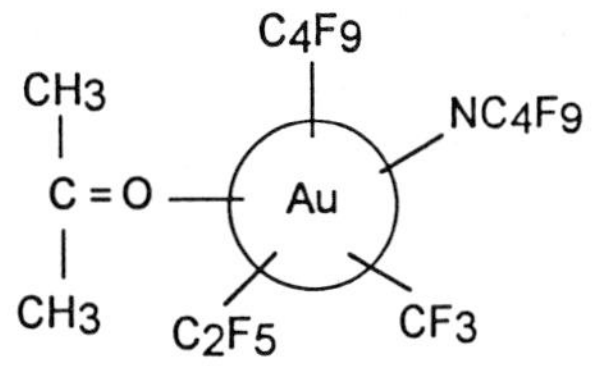

C4F9
CH3
NC4F9
C = O
Au
CH3
C2F5
CF3

Fig. 6-12. Formation of gold colloids in perfluoro-tri-n-butylamine (PFTA); fragments of the PFTA solvent remain bound to the gold particles.

6.4.6.3 Films from Non-Aqueous Metal Colloids

Since the films produced by solvent evaporation are of considerable interest to the electronics and jewelry industries, it is important that the film forming step be understood. In order to gain this understanding, two in-depth studies were carried out.

In the first study Au–acetone and Au–Pd–acetone colloids were produced and films deposited on silver disks by solvent evaporation, then studied in detail by XPS [51]. The Au–acetone films showed evidence of unusual surface species which exhibited a positive shift in the core level binding energy. By applying a biasing potential the surface species were separated from the bulk-like gold particles. These results were consistent with a model of the surface metal clusters as being charged and coated/protected by solvent/solvent fragments.

Fig. 6-13 shows XPS data from the biasing experiments, and Fig. 6-14 summarizes the particle models consistent with the biasing (both + and –).

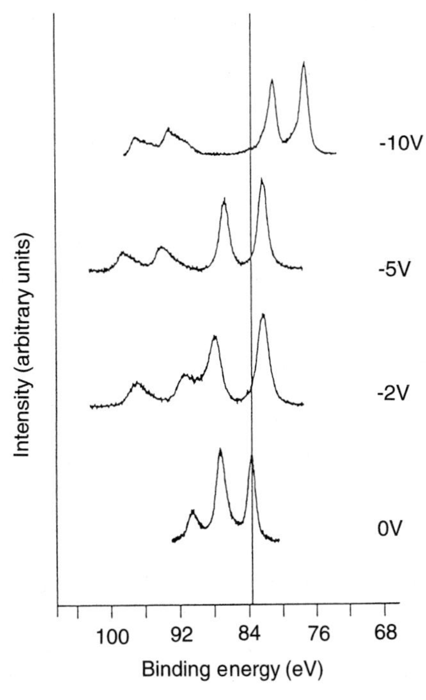

Fig. 6-13. XPS gold 4f electron spectra of a SMAD Au–acetone film under the influence of an applied negative bias potential (shown on right of diagram). (Reproduced with permission of the American Chemical Society) [51].

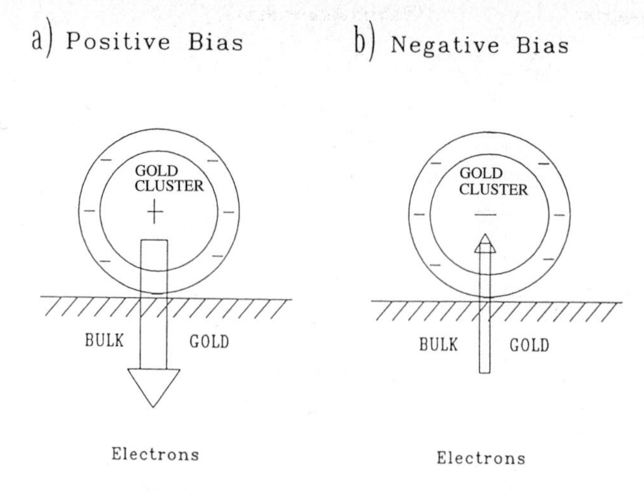

Fig. 6-14. Model to help explain XPS shifts in a SMAD-Au film upon applying positive or negative potentials. The carbonaceous layer around the gold clusters acts as a selective barrier to the direction of electron flow, providing a greater resistance to electron flow into the cluster. (Reproduced with permission of the American Chemical Society) [51].

Perhaps the important points that come out of this work are that:

1. Metal atoms and growing clusters interact strongly with solvents (even fluorocarbons) including breaking bonds and abstracting electrons.
2. The resultant particles, when deposited as films, retain some of the solvent/solvent fragments as well as negative charge (and appropriate counterions).
3. According to electrical potential biasing experiments, the solvent-like coating on a typical particle allows the flow of electrons out of the cluster but does not favor electron flow into the cluster.
4. Only upon heating is the solvent material removed and bulk-like metal film formed.

Although the films formed from these colloidal solutions contain organic material, the fact remains that this is an attractive technique with advantages over direct metal vapor deposition or CVD methods. Therefore, a more detailed study of the films themselves was carried out [52]. The term chemical liquid deposition (CLD) was used to describe this technique, where the liquid colloid is sprayed onto a substrate, the solvent is evaporated, and a metal-like film is produced.

Gold films were formed on copper, aluminum, and silver metals as well as polyethylene, polymethylmethacrylate, and polyphenylenesulfide polymers. In addition, two solvents were employed: acetone and PFTA. The films were initially prepared at room temperature and then treated with ultraviolet light, high energy irradiation, or heat.

Heat treatment was an effective method for increasing the strength of adhesion; usually 200 °C was optimum. Good adhesion and gold-like appearance was possible on copper, aluminum, and silver. In fact, this CLD method appears to be a very nice way to gild silver jewelry with a very thin gold film [53].

The best results, however, were with the polyphenylenesulfide polymer. In this case, 200 °C heat treatment yielded beautiful gold films with excellent adhesion. Apparently this was possible because of the excellent ligating properties of the surface sulfur for gold (Fig. 6-15).

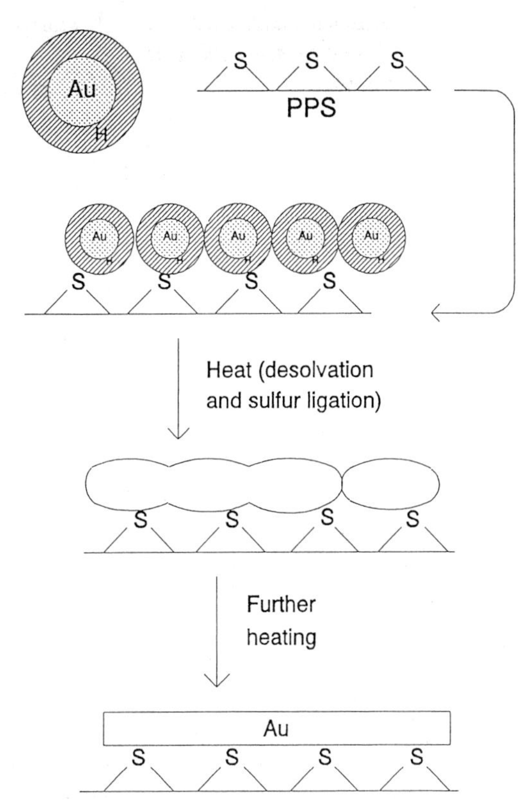

Fig. 6-15. Gold–acetone SMAD film on polyphenylene sulfide polymer; strong adhesion due to sulfur–gold interaction. (H refers to fragments of solvent, in this case acetone). (Reproduced with permission of Gauthier-Villars) [52].

6.4.6.4 Metal Colloids in Polymers

Several research groups have studied the agglomeration of metal atoms in polymeric media. For example, Adrews and Ozin have deposited metal atoms into liquid siloxane oils and have observed arrested cluster growth and have isolated colored oils with increased viscosities [54].

Another approach has been to codeposit metal atoms with appropriate organic monomers in which non-aqueous colloids were formed. These monomers could then be polymerized by conventional means [55, 56]. For example, deposition of gold atoms with styrene at 77 K followed by warming to room temperature yielded a purple colloidal liquid that was stable towards aggregation for several months [56]. Transmission electron microscope (TEM) measurements indicated gold particle sizes ranging from 7 to 15 nm. A polymer initiator was added (AIBN) and the purple liquid was heated at 60 °C. Styrene polymerized, yielding a purple solid of MW 20000–70000, depending on the AIBN concentration. The metal particle size remained the same throughout the polymerization process. However, the concentration of gold did have an effect on the MW of the polymer. The initial stabilization of the colloidal metal was probably due to ligation only, since electrophoresis studies did not indicate charging. Upon generation of a radical initiator, polymerization took place. However, some of the radicals appear to be scavenged by the metal particles, thus affecting the final MW (Fig. 6-16).

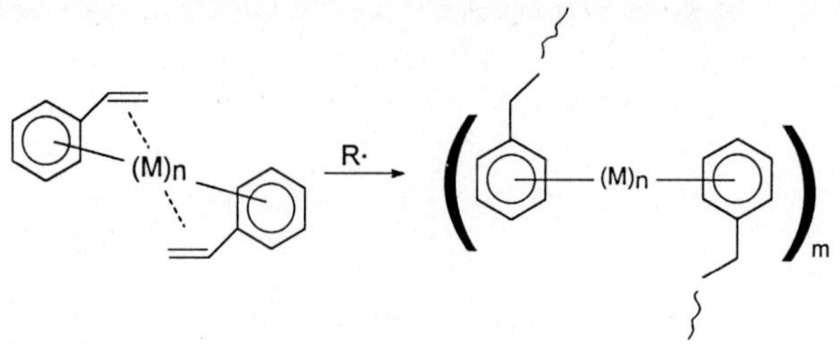

Fig. 6-16. Polymerization of styrene-stabilized metal colloid.

A series of metals have been polymer-encapsulated in this way [57]. Other monomers have also been utilized, such as methyl methacrylate [58], acrylonitrile [59], styrene-methyl methacrylate [60], ethyl methacrylate [61], butyl methacrylate [62], and vinyl acetate [63]. Generally, the metal content ranged from 0.5 to 5% by weight, and MWs were on the order of $10^4 < M_w < 10^6$.

The incorporation of these metal particles into polymers causes significant changes in optical properties as well as changes in thermal decomposition temperatures [64–67], and studies are continuing. Of particular interest is the work in which optically nonlinear polymers gave been doped with gold clusters by the methods discussed above; gold vapor was codeposited with a large excess of a diacetylene monomer such as diphenylbutadiyne, or into a liquid solvent containing the polymerized diyne [68]. Polymers doped with gold clusters were obtained after warm-up and removal of remaining monomer (e.g. Fig. 6-17).

In similar experiments with another diyne, namely 5,7-dodecadiyne-1,12-diol-*co*-bis[((*n*-butoxycarbonyl)methyl)urethane] nonlinear optical measurements indicated a 200-fold enhancement in the third order optical coefficient (relative to free polymer vs gold laden at a 7% metal volume fraction) [68].

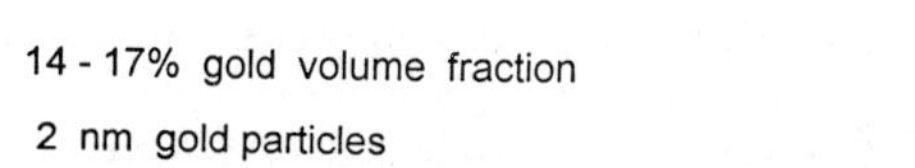

Au atoms + Ph - C≡C-C≡C - Ph $\xrightarrow{77K}$

(Au)n

14 - 17% gold volume fraction

2 nm gold particles

Fig. 6-17. Gold atom/cluster polymerization of a diacetylene derivative.

6.4.7 Magnetic Properties of SMAD Metal Particles (Mono- and Bimetallic)

As we consider the various applications of nanoscale metal particles ("activated metals"), it is important to point out useful and unusual physical properties as well as chemical properties. An important case in point is magnetism. Since ferromagnetism is dependent on long-range order in ferromagnetic metals (Fe, Co, Ni) it would be expected that nanoscale particles would possess magnetic properties different from bulk samples, and may be unique.

Normally studies of magnetic properties of metals fall into the realm of solid state physics. However, when nanoparticles of reactive metals such as Fe, Co, and Ni are needed, chemists have a large role to play. This is particularly true since surface contamination by metal oxides, which are themselves magnetic (either ferromagnetic or antiferromagnetic), can have drastic effects on the magnetic properties of the metal particles themselves. Thus, preparing uniform, protected nanoparticles of Fe, Co, and Ni becomes the realm of the chemist.

Some recent studies using metal vapor methods to prepare ultrafine metallic magnetic particles will now be briefly reviewed.

There are two ways to proceed in such studies. First, the well established gas evaporation method (GEM) could be employed [69–71]. This involves vaporizing the metal into He or Ar gas at a pressure of several torr. Atom agglomeration occurs, and the particle size depends on the pressure of inert gas and the vaporization rate. The resulting particles can be collected on the inner walls of the vaporization chamber, and can be removed as dry powders. However, it is essentially impossible to avoid some surface oxidation during collection and subsequent investigation. Indeed, a recent study reported the synthesis of particles using 0.5 to 8 torr He pressures, and median diameters ranged from 5 to 20 nm depending on the exact pressure used [72]. It was possible to make a plot of saturation magnetization [73], which showed $\sigma = 40$ emu g^{-1} for 5 nm particles and a gradual rise to 160 emu g^{-1} for 20 nm particles (the σ_s value for bulk iron is 220 emu g^{-1}). The coercivity H_c varied greatly with particle diameter and rose sharply with decrease in particle diameter [74].

Electron microscope studies coupled with XRD indicated that the particles were made up of metallic crystalline cores with metal oxide shells. The oxide was believed to be granular and not monocrystalline. It was proposed that the high coercivities were due to a pinning effect resulting from the oxide shell. Thus, the interaction between the superparamagnetic shell and the magnetic iron core was responsible for the high coercivity values at low temperatures and its drastic temperature dependence. Also, the thickness of the oxide shell seemed to have a bearing on σ_s and H_c.

This study and many others indicate that in order to understand these magnetic effects, nanoscale iron particles free of oxide shell must be prepared and studied. How can this be done? In the following paragraphs we describe how the SMAD method has been used, which constitutes the second synthetic approach to be described.

An initial first attempt involved the use of a surfactant for protection of nanoparticles of iron produced by the SMAD process in pentane. Such a process is based on a technology somewhat similar to that used for producing ferrofluids [75, 76], in which magnetic particles are suspended in a carrier solvent with an added surfactant to make the suspension stable. The surfactants are long-chain hydrocarbons with polar end groups, or polymers with polar groups (often carboxylic acids). The surfactant serves three functions: (1) the molecules become adsorbed to the surface of the particle through a polar end group; (2) the long-chain hydrocarbon group sticks out and prevents particle agglomeration; and (3) the organic group enhances solubility in the organic fluid.

Although several research teams have prepared nanoscale metallic particles in polymer- or surfactant-containing oils [54, 75, 77], attempts in this laboratory were aimed at isolation, purification, and characterization of the surfactant protected fine powders. Thus, an Fe–pentane matrix was prepared at 77 K followed by addition of a pentane-oleic acid ($HOOC(CH_2)_7CH{=}CH(CH_2)_7CH_3$] solution [78]. The cold mixture was allowed to warm. Iron clusters formed during the warm-up and melt-down period, and the oleic acid immediately ligated the surface of the particles (the Fe:pentane:oleic acid molar ratio was

1:40:2). The resulting black colloidal solution was syphoned out under Ar and filtered on a fine glass frit under Ar, and the filter cake washed with pentane. Two fractions were collected; the filter cake material consisted of ferromagnetic particles while the pentane solution contained a smaller amount of superparamagnetic iron particles that were "soluble" in pentane, and could be isolated by vacuum removal of pentane (Scheme 6-9).

$$\mathrm{Fe_{vapor}} + \text{pentane} \longrightarrow \text{Fe-pentane} \xrightarrow[\text{(2) warm+melt}]{\text{(1) oleic acid}} \mathrm{Fe_x}\text{-oleic acid/pentane}$$

$$\downarrow$$

$$\mathrm{Fe_y}\text{-oleic} + \mathrm{Fe_z}\text{-oleic}$$

$\mathrm{Fe_y}$ = 8-12 nm

Scheme 6-9 $\mathrm{Fe_z}$ = 2-7 nm

According to TEM studies the diameters of the ferromagnetic particles (Fe_y) were 8–12 nm while those of the superparamagnetic particles (Fe_z) were 2–7 nm. According to XRD the samples were non-crystalline. Exposure to air did not change the nature of the ferromagnetic fraction over a 3-month period.

Magnetic studies revealed that Fe_y had a coercivity of about 60 Oe and $\sigma_s = 55$ emu g^{-1} (bulk iron has $H_c = 0–50$ Oe and $\sigma_s = 220$ emu g^{-1}). Heat treatment of these particles at 360 °C under Ar caused the adsorbed oleic acid to react and oxidize the particles so that Fe_3O_4 was formed. However, at higher temperatures the hydrocarbon portion of the oleic acid reduced the Fe_3O_4 to large crystals of α-Fe [Eq. (6.20)].

$$\mathrm{Fe_y}\text{–oleic} \xrightarrow[\text{Ar}]{360\ ^\circ\text{C}} \mathrm{Fe_3O_4}\text{–organic residue} \xrightarrow[\text{Ar}]{520\ ^\circ\text{C}} \alpha\text{-Fe} \qquad (6.20)$$

Although these studies were moderately successful in that protected nanoparticles of iron metal were obtained, the presence of the oleic acid made interpretation of magnetic data difficult. Therefore, other ways of surface coating SMAD iron particles were needed.

There are several metals that do not under normal conditions form stable alloys with iron [79]. For example Li, Mg, and Ag are immiscible with Fe and no phase diagrams exist for these bimetallic combinations. Of course, alloy formation of two metals is normally carried out by mixing the solid or liquid metals at relatively high temperatures, say in the range 300–1800 °C.

In considering the SMAD method, possibilities present themselves for producing metastable alloys at low temperatures, say from –190 °C to room temperature. Thus, if atoms of Fe and Li were deposited in frozen pentane followed by warming, Fe–Li metastable alloy-like particles would be forced to form since it would not be expected that phase segregations could occur very efficiently at low temperatures (recall earlier discussion of the Au–Sn system). That is, lower temperatures provide kinetic accessibility to metastable phases.

In fact the SMAD method can yield alloy-like particles of immiscible metals, but the degree of success depends on the M–Fe combination under study. With Ag–Fe codeposition a mixture of Ag_n, Fe_n, and $(Fe–Ag)_n$ formed with rather poor selectivity to the bimetallic species. In the case of Li–Fe, particles were formed that were Fe–Li combinations but with partial phase segregation occurring by the time the temperature reached 25 °C [80, 81]. In the case of Mg–Fe, the best alloy-like particles formed, but still with partial phase segregation so that isolated 2–3 nm α-Fe particles exist in a Mg metal matrix (Fig. 6-18) [82].

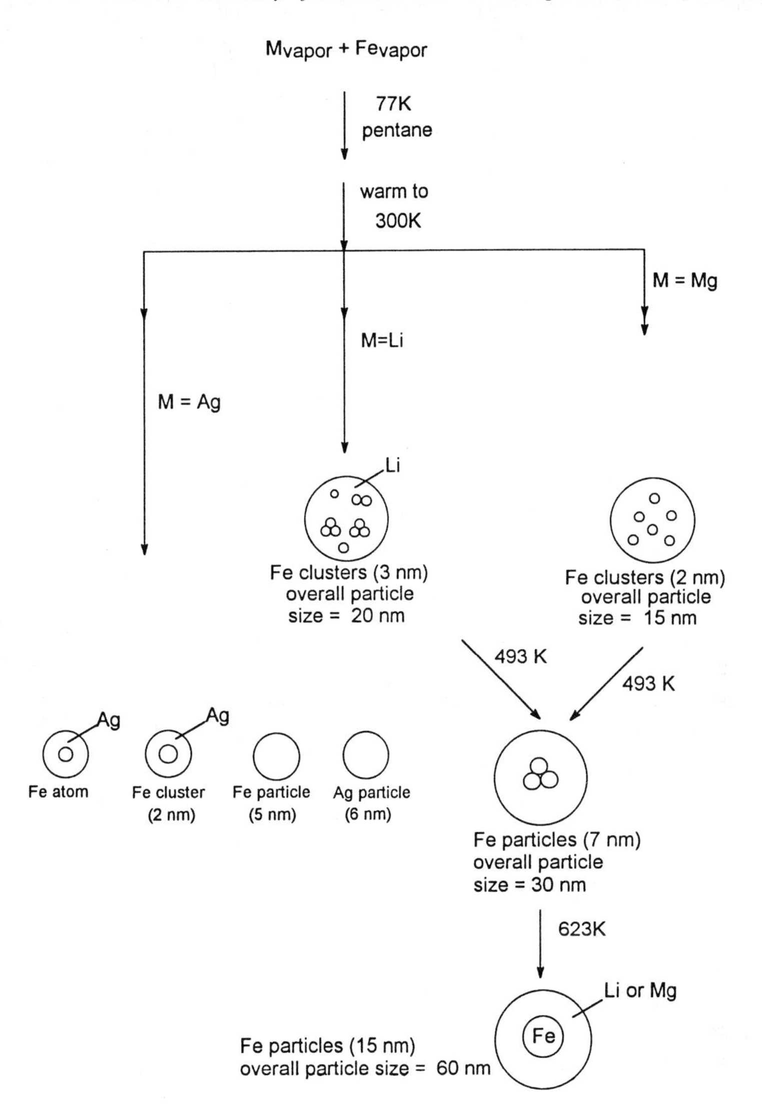

Fig. 6-18. Low temperature metal vapor/matrix method to prepare alloy-like particles of immiscible metals, and their phase separation upon heating.

Thus, with the view of preparing encapsulated, protected nanoparticles of iron metal, the Ag–Fe system was not very successful. With Li–Fe and Mg–Fe, encapsulated iron particles were formed, and the two systems showed subtle but important differences. Figs. 6-19 and 6-20 show Mössbauer spectra of fresh Li–Fe and Mg–Fe samples along with heated samples. In the Li–Fe system the six-line spectrum that is characteristic of ferromagnetic α-Fe was found in both unheated and heated samples. With the Mg–Fe system, however, the fresh unheated sample did not exhibit a six-line spectrum, but instead a doublet that is due to small superparamagnetic particles plus isolated atoms. The heat-treated sample, as expected after α-Fe particle growth, exhibited the six-line spectrum for ferromagnetism.

On the other hand, XRD showed that the fresh Li–Fe and Mg–Fe samples both had average α-Fe crystallite sizes of about 2–3 nm, which should be superparamagnetic and

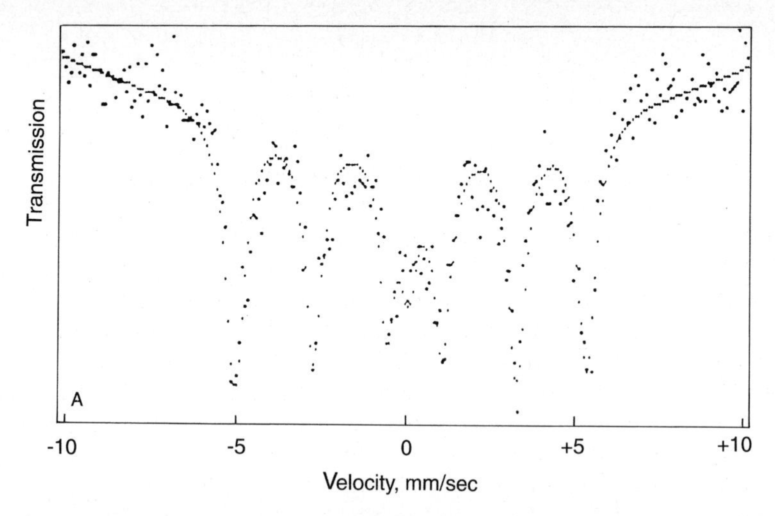

Fig. 6-19. Room temperature Mössbauer spectrum of Fe–Li–pentane SMAD particles (ratio of Fe:Li = 2:5). The broad sextet suggests that the 3 nm α-Fe crystallites in the Li matrix are magnetically interacting, even though they are too small to exhibit ferromagnetism as individual Fe crystallites. (Reproduced with permission of the American Chemical Society) [80].

not ferromagnetic. The only explanation for this seemingly contradictory result is that in the Li–Fe case the 3 nm iron crystallites must be in close proximity as shown in Fig. 6-18. Perhaps there is magnetic interaction between these intimately positioned particles that is not possible in the Mg–Fe system.

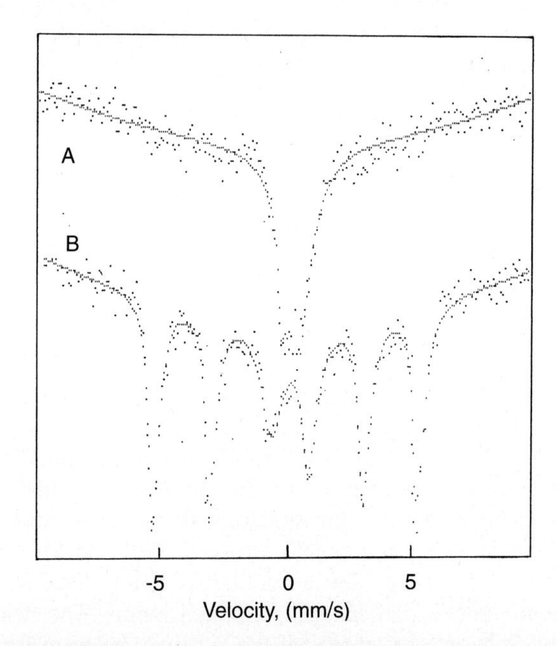

Fig. 6-20. Room temperature Mössbauer spectra of Fe–Mg–pentane SMAD particles; A. fresh, B. after heating to 340 °C under Ar (ratio of Fe:Mg = 4:7 in both cases). The change to a sextet upon heating indicates Fe crystallite growth and the onset of ferromagnetism [82].

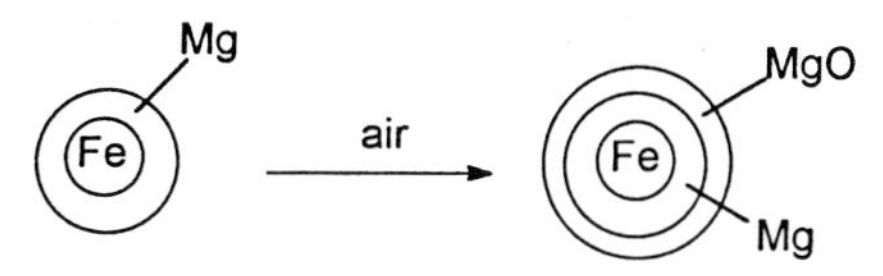

Fig. 6-21. Partial oxidation of a [Mg]Fe nanoparticle to form a protective MgO skin.

So in the Li–Fe and Mg–Fe systems, experiments were successful in protectively encapsulating nanoparticles of metallic iron. Upon careful heat treatment the smaller α-Fe crystallites formed, and excellent environmental protection existed, especially in the Mg–Fe case. Exposure to air only caused a thin "skin" of MgO to form (Fig. 6-21) and the iron was completely protected from oxidation even for months in air.

Now having a series of samples of protected α-Fe nanocrystals of known, controllable size, this encouraged further experiments involving the study of magnetic properties. What was found was most remarkable. When samples with similar α-Fe crystallite sizes were compared, H_c was clearly controlled by the type of surface coating material present (Table 6-5).

It is clear from these results that magnetic properties, in particular H_c, can be controlled by surface chemistry. If magnetic materials such as Fe_3O_4 (ferromagnetic) or FeS (antiferromagnetic) are present, very large H_c values are possible. However, if good protection is present, but with a non-magnetic material such as Mg forming the shell, H_c is very small and near the value for bulk Fe. In the case of Li–Fe, the environmental protection was not as good as with Mg, and the Li shell was converted to Li_2O/Li_2CO_3 and a small amount of FeO_x also eventually formed. The presence of this complex shell material containing some FeO_x probably explains the intermediate H_c value.

A more in-depth discussion of these physical properties is not appropriate here. Suffice it to say that the use of the SMAD method to prepare bimetallic nanoparticles has many possibilities, which are particularly intriguing when "immiscible" bimetallics are used.

Table 6-5. Comparison of coercivities for core-shell particles of Li–Fe, Mg–Fe, Fe_3O_4–Fe, FeS–Fe for α-Fe crystallite sizes of about 3 nm (H_c measured at 10 K)

Shell	Core	H_c (Oe)
Li_2O/Li_2CO_3	Fe	400
Mg/MgO	Fe	70
Fe_3O_4	Fe	3000
FeS	Fe	1000
bulk Fe	Fe	≤ 50

6.5 Conclusions

The solvated metal atom dispersion (SMAD) method of preparing activated metals in the form of nanoscale particles has several interesting and useful features:

1. nearly noncrystalline, highly reactive ultrafine powders can be prepared;
2. metal particles with some solvent fragments adsorbed to the metal particles are obtained;

3. extremely high catalytic activities and unusual selectivities are realized;
4. colloidal metal solutions and polymer matrices can be prepared;
5. bimetallic particles of miscible and immiscible metals can be prepared, and heat treatments can often give controlled desirable phase segregation forming core-shell particles.

Acknowledgement. The support of the National Science Foundation is acknowledged with gratitude.

References

[1] K. J. Klabunde, *Chemistry of Free Atoms and Particles*, Academic Press, New York, **1980**.

[2] K. J. Klabunde, *Free Atoms, Clusters, and Nanoscale Particles*, Academic Press, San Diego, **1994**.

[3] A. W. Olsen, K. J. Klabunde in *Encyclopedia of Physical Science and Technology, Vol. 9*, Academic Press, San Diego, **1992**, pp. 757–793.

[4] (a) T. G. Dietz, M. A. Duncan, D. E. Powers, R. E. Smalley, *J. Chem. Phys.* **1981**, *74*, 6511; (b) M. D. Morse, M. E. Geusic, J. R. Heath, R. E. Smalley, *J. Chem. Phys.* **1985**, *83*, 2293; (c) S. Maruyama, L. R. Anderson, R. E. Smalley, *Rev. Sci. Instrum.* **1990**, *61*, 3686.

[5] S. C. Richtsmeier, E. K. Parks, K. Liu, L. G. Pobo, S. J. Riley, *J. Chem. Phys.* **1985**, *82*, 3659.

[6] (a) E. K. Parks, G. C. Nieman, L. G. Pobo, S. J. Riley, *J. Chem. Phys.* **1988**, *88*, 6260; (b) E. K. Parks, B. H. Weiller, P. S. Bechthold, W. F. Hoffmann, G. C. Nieman, L. P. Pobo, S. J. Riley, *J. Chem. Phys.* **1988**, *88*, 1622; (c) S. J. Riley, Z. *Phys. D: At. Mol. Clusters* **1989**, *12*, 537.

[7] (a) M. F. Jarrold, J. E. Bower, *J. Chem. Phys.* **1987**, *87*, 5728; (b) M. F. Jarrold, J. E. Bower, *J. Am. Chem. Soc.* **1988**, *110*, 70.

[8] (a) M. A. Nygren, P. E. M. Siegbahn, C. Jin, T. Guo, R. E. Smalley, *J. Chem. Phys.* **1991**, *95*, 6181; (b) M. A. Nygren, P. E. M. Siegbahn, *J. Phys. Chem.* **1992**, *96*, 7579; (c) D. W. Liao, K. Balasubramanian, *J. Chem. Phys.* **1992**, *97*, 2548.

[9] (a) D. M. Lindsay, G. A. Thompson, Y. Wang, *J. Phys. Chem.* **1987**, *91*, 2630, and references therein; (b) R. J. VanZee, W. Weltner, Jr., *J. Chem. Phys.* **1990**, *92*, 6976.

[10] M. P. Andrews, G. A. Ozin, *J. Phys. Chem.* **1986**, *90*, 3353, and references therein.

[11] (a) K. J. Klabunde, T. O. Murdock, *J. Org. Chem.* **1979**, *44*, 3901; (b) T. O. Murdock, K. J. Klabunde, *J. Org. Chem.* **1976**, *41*, 1076.

[12] K. J. Klabunde, T. Groshens, M. Brezinski, W. Kennelly, *J. Am. Chem. Soc.* **1978**, *100*, 4437.

[13] S. C. Davis, S. C. Severson, K. J. Klabunde, *J. Am. Chem. Soc.* **1981**, *103*, 3024.

[14] K. J. Klabunde, Y. Tanaka, *J. Mol. Catal.* **1983**, *21*, 57.

[15] D. M. P. Mingos, D. J. Wales, *Introduction to Cluster Chemistry*, Prentice Hall, Englewood Cliffs, **1990**.

[16] (a) P. L. Timms, *Adv. Inorg. Radiochem.* **1972**, *14*, 121; (b) K. J. Klabunde, *Chem. Tech.* **1975**, 624.

[17] K. J. Klabunde, P. L. Timms. P. S. Skell, S. Ittel, *Inorg. Synth.* **1979**, *19*, 57.

[18] P. L. Timms in *Cryochemistry* (eds.: M. Moskovits, G. Ozin), Wiley, New York **1976**.

[19] M. T. Anthony, M. L. H. Green, D. Young, *J. Chem. Soc. Dalton Trans.* **1975**, 1419.

[20] P. Hooker, B. J. Tan, K. J. Klabunde, S. Suib, *Chem. Mater.* **1991**, *3*, 947.

[21] K. J. Klabunde, S. C. Davis, H. Hattori, Y. Tanaka, *J. Catal*ysis **1978**, *54*, 254.

[22] K. J. Klabunde, H. F. Efner, T. O. Murdock, R. Ropple, *J. Am. Chem. Soc.* **1976**, *98*, 1021.

[23] B. A. Scott, R. M. Plecenik, G. S. Gargill, III, T. R. McGuire, S. R. Herd, *Inorg. Chem.* **1980**, *19*, 1252.

[24] S. C. Davis, K. J. Klabunde, *J. Am. Chem. Soc.* **1978**, *100*, 5973.

[25] Y. Wang, Y. X. Li, K. J. Klabunde, "Platinum–Tin and Gold–Tin Bimetallic Particles from Solvated Metal Atoms" in *ACS Sym. Ser. 517* (eds.: M. E. Davis, S. Suib), American Chemical Society, Washington, D.C., **1993**, Ch. 10, pp. 136.

[26] (a) D. M. Cox, K. C. Reichmann, D. J. Trevor, A. Kaldor, *J. Chem. Phys.* **1988**, *88*, 111; (b) M. F. Jarrold, J. E. Bower, *J. Am. Chem. Soc.* **1988**, *110*, 70; (c) K. Kaya, K. Fuke, S. Nonose, N. Kikuchi, *Z. Phys. D: At. Mol. Clusters* **1989**, *12*, 571; (d) K. Fuke, S. Nonose, N. Kikuchi, K. Kaya, *Chem. Phys. Lett.* **1988**, *147*, 479; (e) P. A. Hintz, S. A. Ruatta, S. L. Anderson, *J. Chem. Phys.* **1990**, *92*, 292; (f) K. J. Klabunde, G. H. Jeong, A. W. Olsen, *Selective Hydrocarbon Activation Principles and Progress* (eds: J. A. Davies, P. L. Watson, A. Greenberg, J. F. Liebman), VCH, New York, **1987**, Ch. 13, pp. 433.

[27] (a) Y. Imizu, K. J. Klabunde, *Inorg. Chem*, **1984**, *23*, 3602–3605; (b) K. J. Klabunde, A. Whetten, *J. Am. Chem. Soc.* **1986**, *108*, 6529.
[28] P. G. Jaisien, C. F. Dykstra, *J. Am. Chem. Soc.* **1983**, *105*, 2089.
[29] K. J. Klabunde, H. F. Efner, L. Satek, W. Donley, *J. Organomet. Chem.* **1974**, *71*, 309.
[30] P. Cintas, *Activated Metals in Organic Synthesis*, CRC Press, Boca Raton, **1993**, p. 33.
[31] E. P. Kündig, C. Perret, *Helv. Chim. Acta.* **1981**, *64*, 2606.
[32] W. Oppolzer, E. P. Kündig, P. M. Bishop, C. Perret, *Tetrahedron Lett.* **1982**, *23*, 3901.
[33] R. D. Rieke, *Acc. Chem. Res.* **1977**, 10, 301.
[34] K. J. Klabunde, Y. X. Li, B. J. Tan, *Chem. Mater.* **1991**, *3*, 30.
[35] H. Imamura, T. Nobunaga, M. Kawahigashi, S. Tsuchiya, *Inorg. Chem.* **1984**, *23*, 2509.
[36] (a) K. J. Klabunde, D. Ralston, R. Zoellner, H. Hattori, Y. Tanaka, *J. Catalysis* **1978**, *55*, 213; (b) D. H. Ralston, K. J. Klabunde, *J. Appl. Catal.* **1982**, *3*, 13.
[37] A. W. Olsen, Ph. D. Thesis, Kansas State University, **1989**.
[38] H. Imamura, A. Ohmura, E. Haku, S. Tsuchiya, *J. Catalysis* **1985**, *96*, 139.
[39] S. Wu, W. Huang, W. Zhao, X. Wang, S. Zhang, X. Wang, *Zhongguo Xitu Xuebao* **1991**, *9*, 224 (in Chinese).
[40] S. R. Nishitani, K. N. Ishihara, R. O. Suzuki, P. H. Shingu, *J. Mater. Sci. Lett.* **1985**, *4*, 872.
[41] S. T. Lin, M. T. Franklin, K. J. Klabunde, *Langmuir* **1986**, *2*, 259.
[42] K. Kimura, S. Bandow, *J. Chem. Soc. Jpn.* **1983**, *56*, 3578.
[43] G. Cardenas-Trivino, K. J. Klabunde, E. B. Dale, *Langmuir* **1987**, *3*, 986.
[44] (a) F. Booth, *Prog. Biophys. Chem.* **1953**, *3*, 131; (b) H. B. Bull, *Physical Biochemistry*, 2nd ed., Wiley, New York **1964**.
[45] D. J. Fischer, G. Grant, *Belg. Patent 632,021,* Nov. 18 (1963).
[46] T. Stefanowics, Gadja Szent-Kirrallgine, *Pol. J. Chem.* **1983**, *57*, 707.
[47] R. L. Weither, R. P. Ley, *J. Appl. Phys.* **1976**, *15*, 1.
[48] G. Cardenas-Trivino, K. J. Klabunde, E. B. Dale, "Modelling of Optical Thin Films" in *SPIE, Vol. 821*, **1987**, Society of Photo-Optical Instrumentation Engineers, p. 206.
[49] M. T. Franklin, K. J. Klabunde, *"High Energy Processes in Organometallic Chemistry"*; in *ACS Symp. Ser. 333* (ed: K. S. Suslick), American Chemical Society, Washington, D.C., **1986**, p. 246.
[50] E. B. Zuckerman, K. J. Klabunde, B. J. Oliver, C. M. Sorensen, *Chem. Mater.* **1989**, *1*, 12.
[51] B. J. Tan, P. M. A. Sherwood, K. J. Klabunde, *Langmuir* **1990**, *6*, 105.
[52] K. J. Klabunde, G. Youngers, E. J. Zuckerman, B. J. Tan, S. Antrim, P. M. A. Sherwood, *Eur. J. Solid State Inorg. Chem.* **1992**, *29*, 227.
[53] B. S. Rabinovitch, R. H. Horner, K. J. Klabunde, P. Hooker, *Metalsmith*, **1994**, *14*, 41.
[54] M. P. Andrews, G. A. Ozin, *Chem. Mater.* **1989**, *1*, 174.
[55] R. Wright, 196th National Meeting of the American Chemical Society, Los Angeles, Sept. 1988, Paper 1 and EC, 32.
[56] K. J. Klabunde, J. Habdas, G. Cardenas-Trivino, *Chem. Mater.* **1989**, *1*, 481.
[57] G. Cardenas-Trivino, C. C. Retamal, K. J. Klabunde, *Bol. Soc. Chil. Quim.* **1990**, *35*, 223.
[58] G. Cardenas-Trivino, C. C. Retamal, K. J. Klabunde, *Polymer Bull*, **1991**, *25*, 315.
[59] G. Cardenas-Trivino, C. C. Retamal, *Polymer Bull.* **1991**, *26*, 611.
[60] G. Cardenas-Trivino, C. C. Retamal, K. J. Klabunde, *Polymer Bull.* **1992**, *27*, 383.
[61] G. Cardenas-Trivino, E. J. Acuna, *Polymer Bull.* **1992**, *19*, 1.
[62] G. Cardenas-Trivino, G. E. Salgado, *Polymer Bull.* **1993**, *31*, 23.
[63] G. Cardenas-Trivino, D. C. Munoz, *Makromol. Chem.* **1993**, *194*, 3377.
[64] G. Cardenas-Trivino, C. C. Retamal, D. L. H. Tagle, *Thermochim. Acta.* **1992**, *198*, 123.
[65] G. Cardenas-Trivino, C. C. Retamal, D. L. H. Tagle, *Thermochim. Acta.* **1991**, *176*, 233.
[66] G. Cardenas-Trivino, E. J. Acuna, D. L. H. Tagle, *Thermochim. Acta.* **1994**, in press.
[67] G. Cardenas-Trivino, D. L. H. Tagle, *Thermochim. Acta.* **1992**, *200*, 361.
[68] A. W. Olsen, Z. H. Kafafi, *J. Am. Chem. Soc.*, **1991**, *113*, 7758–7760.
[69] C. Hayashi, *Phys. Today*, December, **1987**, 15; *J. Vac. Sci. Technol.* **1987**, *A5*, 1375.
[70] G. A. Niklasson, *J. Appl. Phys.* **1987**, *62*, 258, and reference therein.
[71] (a) K. Kimura, S. Bandow, *J. Chem. Soc., Jpn.* **1987**, *56*, 3578; (b) S. Yatsuya, S. Kasukabe, R. Uyeda, *Jpn. J. Appl. Phys.* **1973**, *12*, 1675.
[72] S. Gangopadhyay, G. C. Hadjipanayis, B. Dale, C. M. Sorensen, K. J. Klabunde, V. Papaefthymiou, A. Kostikas, *Phys. Rev. B.* **1992**, *45*, 9778.
[73] The maximum strength of the magnetic field induced within a particle by the application of an external magnetic field, σ_s (emu g^{-1}).
[74] The coercivity H_c is a magnetic memory property and represents the negative field strength necessary to bring the particle magnetization back to zero, also called the "coercive force".

[75] M. Kilner, N. Mason, D. M. Lambrick, P. D. Hooker, P. L. Timms, *J. Chem. Soc. Chem. Commun.* **1987**, 356.
[76] S. W. Charles, J. Popplewell, "The Preparation and Properties of Stable Metallic Ferrofluids" *in Thermodynamics of Magnetic Fluids* (ed: B. Berkovsky), Hemisphere, New York, **1978**, p. 1.
[77] T. Furubayashi, I. Nakatani in *Proc. Int. Sym. Phys. Magn. Mat.*, Sendai, Japan, **1987**, p. 182.
[78] C. F. Kernizan, K. J. Klabunde, C. M. Sorensen, G. C. Hadjipanayis, *Chem. Mater.* **1990**, *2*, 70.
[79] O. Kubaschewski, *Iron-Binary Phase Diagrams*, Springer, Berlin, **1982**, pp. 3, 4, 59.
[80] G. N. Glavee, C. F. Kernizan, K. J. Klabunde, C. M. Sorensen, G. C. Hadjipanayis, *Chem. Mater.* **1991**, *3*, 967.
[81] G. N. Glavee, K. Easom, K. J. Klabunde, C. M. Sorensen, G. C. Hadjipanayis, *Chem. Mater.* **1992**, *4*, 1360.
[82] (a) K. J. Klabunde, D. Zhang, G. N. Glavee, C. M. Sorensen, G. C. Hadjipanayis, *Chem. Mater.* **1994**, *6*, 784; (b) D. Zhang, M. S. Thesis, Kansas State University, 1993.

7 Electrochemical Methods in the Synthesis of Nanostructured Transition Metal Clusters

Manfred T. Reetz, Wolfgang Helbig and Stefan A. Quaiser

7.1 Introduction

The electrochemical reduction of transition metal salts in aqueous solution with formation of insoluble powders constitutes an industrial process dating back to the 19th century [1]. Acidic aqueous media are generally used, which means that the powders may contain metal hydrides and/or metal salts as contaminants [1, 2]. The size of the particles, although in the early days not always determined due to the lack of reliable analytical procedures, is rather large, specifically in the μm range [1]. In a number of cases certain properties of the metal powders, such as morphology, bulk density, grain size, particle size, and flowability were shown to depend on a number of factors, including temperature, concentration of the electrolyte, metal ion additives, and current density [1]. For example, deposits of powdery tin from aqueous acidic solutions afford acicular particles at high current densities, while spongy deposits with spherical shapes are obtained at low current densities [1]. By using two different metal salts, codeposition with formation of alloys is possible [3]. It has been claimed that the properties of such electrochemically produced materials may be superior to those of chemically manufactured metals (finer grained, denser and harder structures, often better magnetic properties) [1, 3].

A wide variety of industrial applications have been reported for electrochemically fabricated metal powders and alloys [1, 3]. These include permanent magnets, (e.g. ferromagnetic Fe and Co powders), porous bronze self-lubricating bearings, brass powders, Cu–Pb powders in the manufacture of corrosion resistant and/or anti-abrasive lubricants, and Ti–Al alloy powders in cermet production. In some cases catalytic properties have been studied. A case in point is the electrochemical deposition of palladium in aqueous medium, which leads to Pd powders consisting of α- and β-H–Pd phases; such materials were tested as heterogeneous catalysts in the hydrogenation of cyclohexene [2]. Recent developments include the electrochemical formation of multilayered magnetic materials [4], of magnetic arrays of nickel and cobalt nanowires by deposition into templates with nanometer-sized pores [5], and of gold arrays in nanoporous alumina template membranes [6].

Under different conditions conventional electrochemical metal deposition can be used to coat and protect metal surfaces [1]. Electrochemical processes have also been applied in the recovery of metal powders [1]. Finally, the electrochemical reduction of transition metal salts in the presence of suitable ligands constitutes an alternative method in the synthesis of well-defined metal complexes, the first known example being the electrochemical reduction of $Ni(acac)_2$ in the presence of 1,5-cyclooctadiene with formation of $Ni[COD]_2$ [7].

This chapter does not deal with any of these processes. Rather, emphasis is put on a new development in the area of electrochemical reduction of transition metal salts as initiated in the authors' own laboratories. It concerns the electrochemical synthesis of nanostructured transition metal clusters stabilized by tetraalkylammonium salts [8, 9] and includes information concerning the properties of these materials as investigated by such techniques as high resolution transmission electron microscopy (HRTEM), scanning tunneling microscopy (STM), cyclic voltammetry (CV), and UV-vis-spectroscopy. The magnetic properties of some of the clusters will also be touched on. Since this field is very new, the presentation should be viewed as a preliminary assessment.

7.2 Conventional Syntheses of Transition Metal Clusters Stabilized by Ammonium Salts

Recent interest in transition metal clusters and colloids* in the nanometer size range has arisen from fundamental and practical reasons [10]. Particles in the range 1–10 nm are expected to have unusual electronic properties [11]. Applications in catalysis, photocatalysis, and electrocatalysis have already been reported [10]. The preparation of these materials is generally based on metal vaporization or reduction of transition metal salts by a wide variety of reducing agents, including hydrogen, hydrazine, formaldehyde, ethanol, and trialkylborohydrides. In order to prevent undesired agglomeration, these processes are generally carried out in the presence of stabilizers such as ligands (e.g., phenanthroline), polymers [e.g. poly(vinylpyrrolidone)], or surfactants such as tetraalkylammonium salts [10].

Although some progress in the control of particle size of nanostructured metal clusters has been made by varying such parameters as type of reducing agent, temperature, concentration, and solvent [10, 12], "true control of particle size remains the most attractive goal for the synthetic chemist in this field" [13].

The first report of the stabilization of a metal colloid by a tetraalkylammonium salt, specifically Pt/cetyltrimethylammonium chloride (CTAC), is due to Grätzel in 1979 [14]. An aqueous solution of $PtCl_4$ was reduced with H_2 in the presence of CTAC, affording a clear solution. At about the same time Boutonnet performed similar reactions, characterized the ammonium-stabilized Pt colloids by TEM (3.5–3.7 nm), and used the materials as hydrogenation and isomerization catalysts [15]. Thereafter research in the area of ammonium-protected transition metal clusters was intensified by Esumi [16], Toshima [17], and others [18]. For example, it was shown that H_2PtCl_6 can be reduced by formaldehyde in the presence of dioctadecyldimethylammonium or trioctylmethylammonium chloride with formation of homogeneous dark brown solutions of metal colloids which were characterized by TEM (1.5–2.5 nm) [16]. Ammonium salts were also used in the stabilization of rhodium and gold colloids [18]. A related approach to the synthesis of $N^+R_4X^-$-stabilized transition metal clusters was recently reported by Bönnemann, who reduced various metal salts by trialkylboron hydrides $HBR_3^-N^+R_4$ (see Chapter 9) [19]. The nature of the stabilizing effect of ammonium salts is not well understood. Wiesner was the first to postulate an electrostatic interaction (Fig. 7-1) [20].

In summary, a wide variety of reducing agents can be used to prepare many different $R_4N^+X^-$-stabilized transition metal colloids. However, these methods do not allow for the control of particle size. Furthermore, it should be noted that characterization based solely on TEM does not detect possible side products such as metal hydride species, oxidized forms of the metal, or boron impurities.

* In the literature the distinction between metal cluster and metal colloid is not clear [10], and in this chapter the terms are used interchangeably.

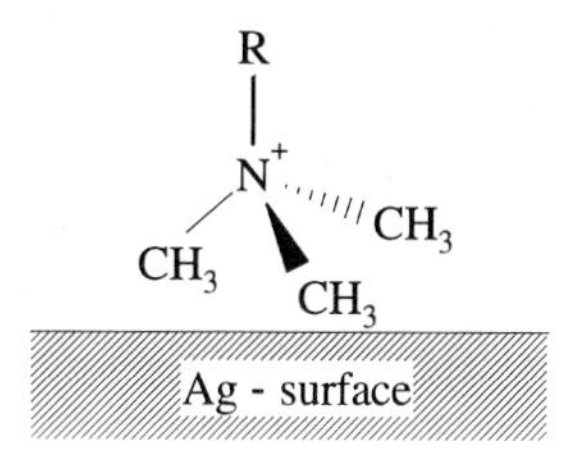

Fig. 7-1. Interaction of ammonium ions with a silver surface [20].

7.3 Electrochemical Synthesis of Metal Clusters Stabilized by Tetraalkylammonium Salts

7.3.1 Syntheses Based on Sacrificial Metal Anodes

7.3.1.1 Palladium Clusters

Our own efforts in the area of ammonium-stabilized clusters were driven by the idea that such materials might be easily accessible in pure form and perhaps in a size-selective manner by electrochemical methods. The initial goal was to transform bulk metal, e.g. a palladium sheet, into a metal cluster stabilized by tetraalkylammonium salts. Accordingly, an inexpensive two-electrode apparatus was devised in which the sacrificial anode is composed of the bulk metal to be transformed into the corresponding metal clusters (Fig. 7-2) [8, 9]. The supporting electrolyte consists of tetraalkylammonium salts, which also serve as stabilizers for the metal clusters. Among other reasons organic solvents were chosen in order to prevent possible metal hydride formation.

Thus, in the overall process the bulk metal anode is oxidized with formation of metal cations which are then reduced at the cathode [Eqs. (7.1)–(7.3)].

$$\text{Anode:} \quad \text{Met}_{\text{bulk}} \rightarrow \text{Met}^{n+} + ne^- \tag{7.1}$$

$$\text{Cathode:} \quad \text{Met}^{n+} + ne^- + \text{stabilizer} \rightarrow \text{Met}_{\text{coll}}/\text{stabilizer} \tag{7.2}$$

$$\text{Sum:} \quad \text{Met}_{\text{bulk}} + \text{stabilizer} \rightarrow \text{Met}_{\text{coll}}/\text{stabilizer} \tag{7.3}$$

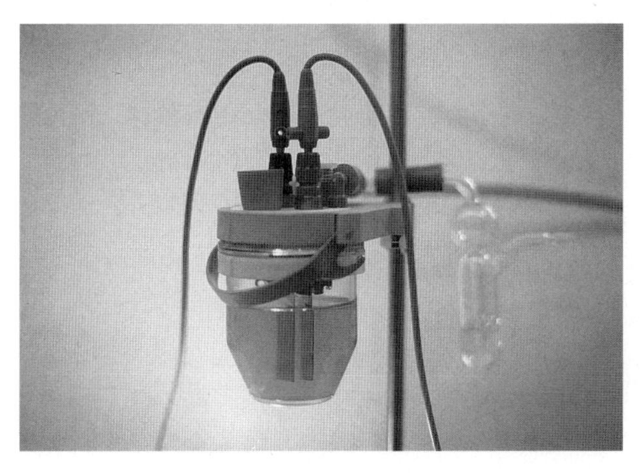

Fig. 7-2. Electrolysis cell used in cluster synthesis [8, 9].

In an initial experiment a commercially available palladium sheet (2×5 cm) as the anode and a platinum sheet (2×5 cm) as the cathode were used, tetraoctylammonium bromide (0.1 M) in acetonitrile/THF (4:1) serving as the supporting electrolyte [8]. Upon applying a current density of 0.1 mA cm^{-2} and an applied voltage of 1 V (vs. counter electrode), the quantitative dissolution of the Pd anode occurred (current yield > 95%). At the same time the formation of a black precipitate was observed. Although it was not certain whether this was undesired palladium powder, in fact it turned out to be a $(nC_8H_{17})_4N^+Br^-$-stabilized palladium cluster which happens to be insoluble in the particular medium used (acetonitrile/THF). Indeed, the black air-stable material could be fully redispersed in THF or toluene in concentrations of up to 1 M (dark brown solutions). These solubility properties are useful in a practical way, because work-up by simple decantation is convenient. The material consists of 72.8% Pd, 19.1% C, 3.27% H, 0.6% N, and 3.98% Br. There is little, if any, oxidized form of Pd present. NMR and MS spectra were shown to be in accord with the presence of $(nC_8H_{17})_4N^+Br^-$. The mean particle size was found to be 4.8 nm, as shown by a TEM analysis (Fig. 7-3c) [8]. In order to see if the current density has any effect upon particle size, the process was repeated using current densities of 0.8 and 5.0 mA cm^{-2}, respectively, under otherwise identical conditions. In an analogous manner materials were obtained having a mean particle size of 3.1 and 1.4 nm, respectively (Fig. 7-3a, b) [8].

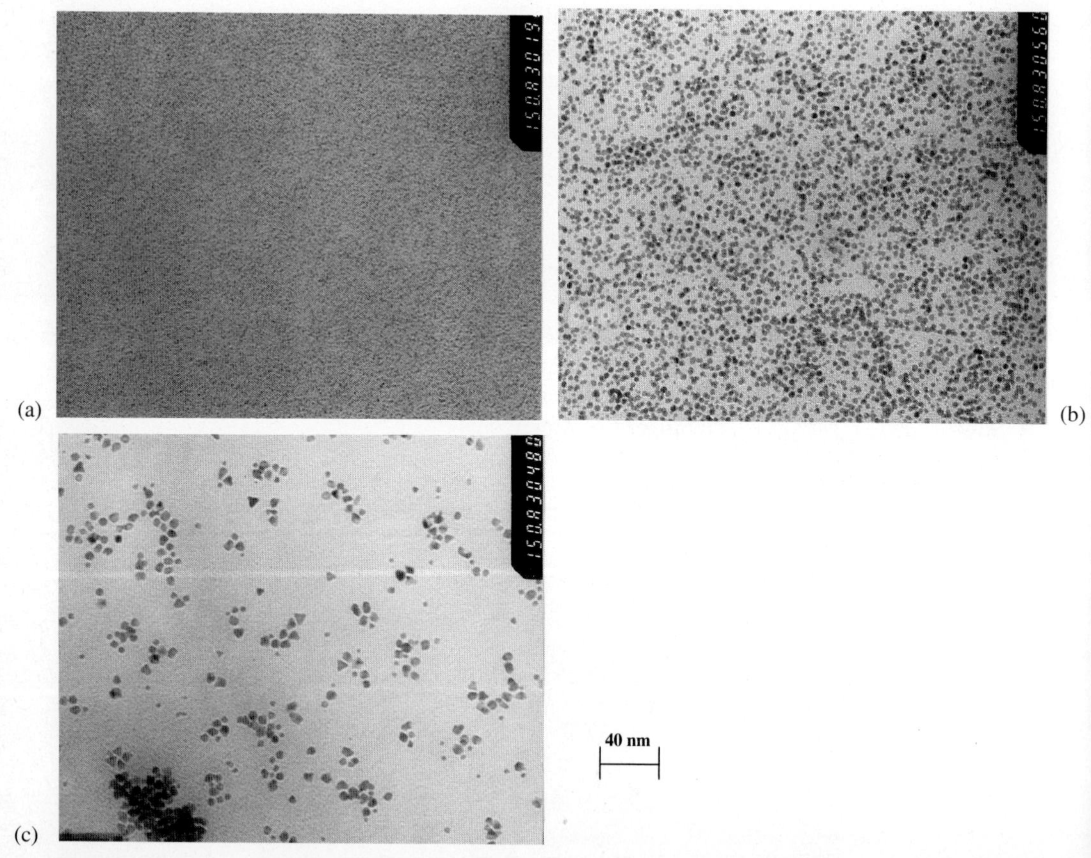

Fig. 7-3. Transmission electron micrographs of Pd clusters prepared at current densities of (a) 5.0, (b) 0.8, and (c) 0.1 mA cm^{-2}; in all cases magnification is 250000. (Reproduced with permission of the American Chemical Society).

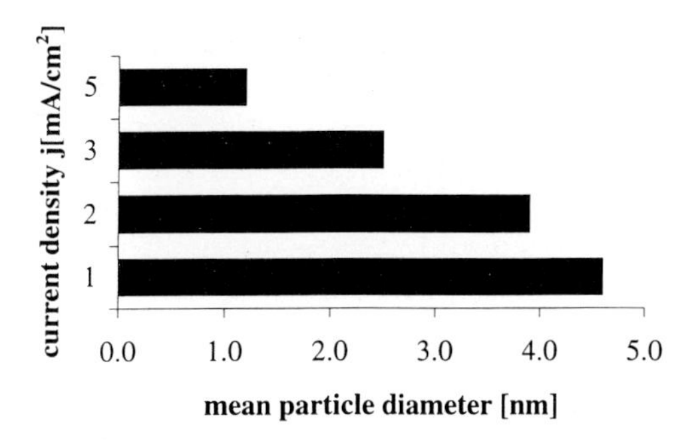

Fig. 7-4. Dependence of average particle size of Pd clusters on the current density during electrolysis; averaged over 200 particles.

The results clearly show that the particle size of the Pd clusters can be controlled by variation of the current density, which is known to be directly proportional to the overpotential at the cathode [21]. With increasing current density and therefore increasing overpotential, the particle size of the colloids decreases (Fig. 7-4). The precipitation of the clusters during the electrochemical process not only simplifies their isolation, it also means that the electrolyte solution does not contain appreciable amounts of particles which could influence the electrochemical properties of the system during electrolysis. The results concerning particle size are reproducible in the ammonium salt concentration range of 0.025 to 1 M, whereas at appreciably lower concentrations mixtures of large colloids (up to 80 nm) are obtained [8]. The size distribution of the clusters, as determined by a statistical analysis of the TEM images, is rather narrow. A typical size distribution diagram is shown in Fig. 7-5.

A wide variety of tetraalkylammonium and phosphonium halides can be used in the preparation of Pd clusters, including chiral salts [9]. Two parameters relating to the geometrical properties of the clusters can be controlled in a defined manner: the size of the metal core by adjustment of the current density, and the thickness of the stabilizing mantle by variation of the size of the alkyl groups in the ammonium salt. Apart from synthetic aspects, these two parameters make systematic combined TEM/STM studies possible (Section 7.3.3.2). By proper choice of the ammonium salt, the solubility in solvents ranging from pentane to water can also be varied at will (Section 7.3.3.1). In-situ immobilization of the colloids is possible by using a suspension of a solid support (e.g. charcoal). Importantly, under otherwise identical conditions the cluster size is the same as in the absence of a support [9].

Although the precise mechanism of cluster formation is difficult to determine, the following gross features are likely [8]. Following oxidation of the Pd sheet, Pd^{II} ions migrate

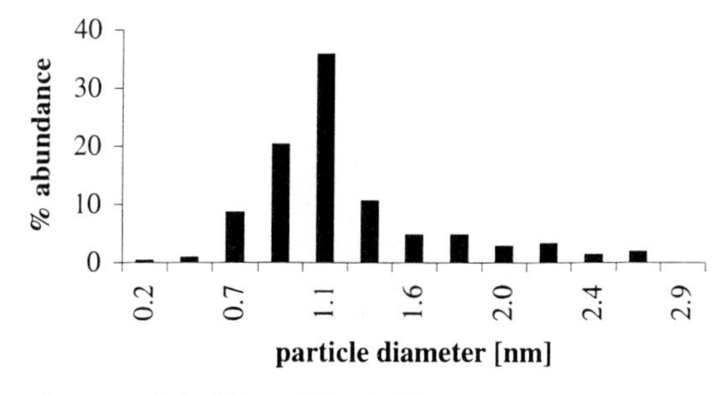

Fig. 7-5. Size distribution diagram of the Pd particles in Fig. 7-3a.

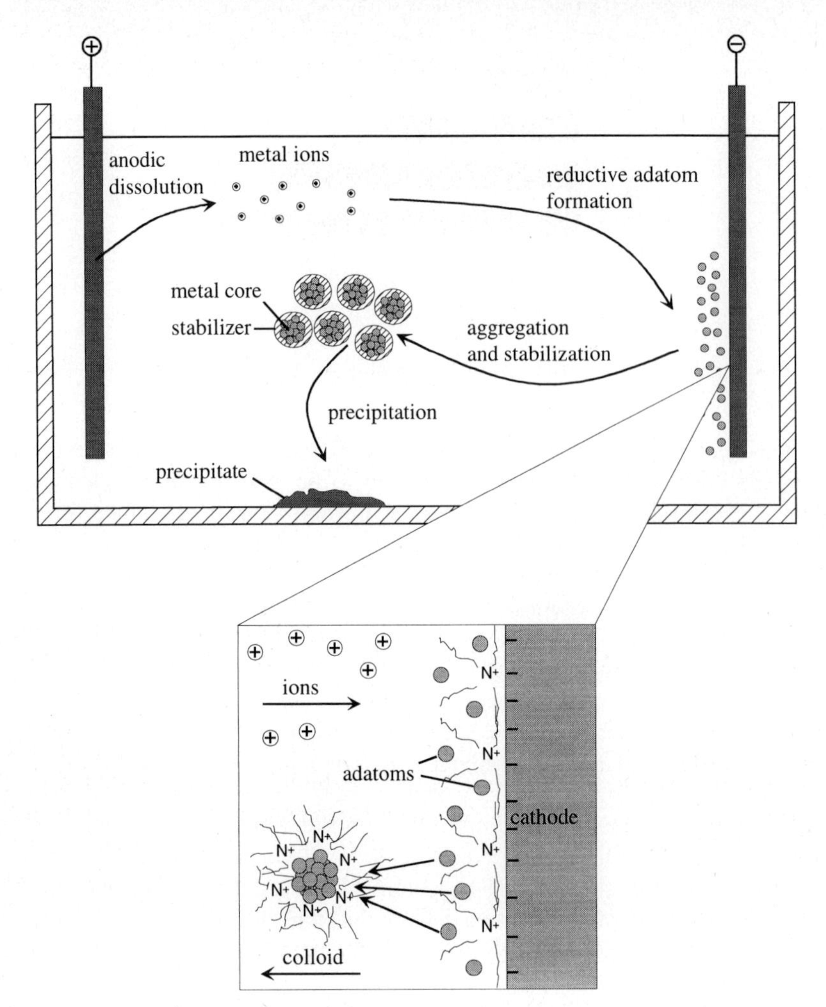

Fig. 7-6. Formation of $R_4N^+X^-$-stabilized metal clusters.

to the cathode where reduction takes place, forming so-called adatoms [21] in the vicinity of ammonium ions near the surface of the electrode. The Pd adatoms form clusters which are protected by the ammonium ions (Fig. 7-6). If solvents are chosen in which the colloids are insoluble, they precipitate out as solid materials.

Why does the cluster size decrease with increasing current density? From the literature on electrochemistry, the critical radius r_{crit} of clusters of adatoms (prior to powder formation) at the cathode should be given by Equation (7.4) [21].

$$r_{\text{crit}} = \frac{2M\gamma}{nF\eta\varrho}, \tag{7.4}$$

where M = molecular weight, γ = surface tension, n = valency, F = Faraday constant, η = overpotential, ϱ = density of the cluster. Thus, r_{crit} is inversely related to the overpotential η, which in turn is directly proportional to the current density [21]. Indeed, upon applying the above equation to palladium using approximate values for some of the parameters, the size range was predicted to be 0.5 to 20 nm, which is in excellent agreement with the experimental results. This means that the ammonium salts are highly effective stabilizers which arrest the growth of particles at the level expected by thermodynamics. In conventional metal powder deposition, stabilizers are absent [1, 3]. Nevertheless, in view of the present results it is likely that a similar mechanism pertains, and that initial clusters of different sizes are actually formed, depending upon the current density. This would mean that the final particle size of such powders, known to be in the μm range [1, 3], depends upon the agglomerization of differently sized clusters.

7.3.1.2 Other Metals

Although the process of screening other metals is still in progress, it can already be seen that the method is applicable to some, but not to all transition metals. For example, it works very well for the production of $R_4N^+X^-$-protected nickel clusters [8]. Materials were readily obtained having average particle sizes of 2.2 and 4.5 nm, respectively, depending upon the applied current density. The nickel content turned out to be in the range 42–46%.

Other metals that were successfully tested include cobalt, iron, copper, silver, and gold [9, 22]. Some of the clusters (e.g., Fe) are very air-sensitive and thus need to be handled under an inert gas atmosphere. In the cases of Cu, Ag, and Au, the ammonium-protected colloids can be dissolved in solvents such as THF, but stability in solution is not as high as in the case of Pd or Ni clusters. Within a few days or weeks, part of the metal precipitates out in the form of nanocrystalline Cu, Ag, or Au powders, and the originally deeply colored solutions gradually become lighter. For certain applications this might actually be convenient. The use of phosphonium salts results in higher stability in solution.

Problems were encountered with certain noble metals such as rhodium, mainly because oxidative dissolution is difficult. On the other hand, the main group element aluminum is consumed quantitatively, but re-reduction at the cathode appears to be incomplete, i.e., the Al clusters are contaminated by oxidized forms of the metal and/or organoaluminum complexes. Thus, these preliminary studies indicate that the scope and limitations of the method depend upon the redox properties of the bulk metal and the corresponding metal ions [22]. It should also be noted that the redox properties of the supporting electrolyte set natural limitations. To some (although limited) extent this can be influenced by the nature of the solvent and by additives (e.g. H_2O).

In yet other cases anodic dissolution and cathodic re-reduction are nearly complete, but the ammonium salts appear to be incapable of stabilizing the clusters [22]. A case in point is zinc. The usual procedure results in the formation of highly dispersed zinc powder. Although the geometrical aspects of ammonium-stabilization have been unraveled by combined TEM/STM investigations (Section 7.3.3.2), it is currently unclear why certain metals are not sufficiently stabilized by ammonium salts.

7.3.1.3 Bimetallic Clusters

Colloids composed of two different metals are of interest in the catalysis of certain organic reactions and in electrocatalysis (fuel cell technology) [10]. In order to apply the electrochemical colloid synthesis to bimetallic clusters, an electrolysis cell composed of two anodes and one cathode was devised. A nickel and a palladium anode were chosen as a first example, with independent control of the current density at each anode [23]. Simultaneous dissolution of the two anodes with formation of $R_4N^+X^-$-stabilized clusters was in fact observed. Although not all parameters have been optimized in these experiments, it is clear that the relative metal content (Ni/Pd) can be controlled to some extent by adjusting the two anodic current densities j(Ni) and j(Pd). For example, at a j(Ni)/j(Pd) ratio of 1.0, a 24:76 relation (by weight) of Ni/Pd was registered, whereas an increase in the current density ratio to 5 led to a higher nickel content (Ni/Pd = 84:16). The sample obtained at a j(Ni)/j(Pd) ratio of 1.0 was characterized by energy dispersive X-ray analysis (EDX). According to this, the material is a true bimetallic colloid, and not a mixture of two different metal clusters. Colloids in the size range 1–4 nm were obtained, depending upon the conditions. More work is necessary in this interesting area, including variation of metals and full characterization by extended X-ray absorption fine structure (EXAFS).

7.3.2 Synthesis Based on the Reduction of Added Transition Metal Salts

As mentioned in Section 7.3.1.2, problems may arise at the anode or at the cathode, depending upon the particular metal. For those metals which do not readily dissolve anodically, a second electrochemical method was recently devised [24, 25]. In this method transition metal salts added to the electrolyte serve as the metal source. Using an electrolysis cell composed of two platinum sheets as anode and cathode and tetrabutylammonium acetate $(nC_4H_9)_4N^+AcO^-$ in THF as the electrolyte, the following salts were successfully reduced: $PtCl_2$, $OsCl_3 \cdot xH_2O$, $RhCl_3 \cdot xH_2O$, and $RuCl_3 \cdot xH_2O$. In all cases the corresponding ammonium-stabilized colloids were formed. TEM studies of these initial samples show that 2–3 nm-sized particles are involved.

The results suggest that this second electrochemical method is complementary to the first process based on anodic dissolution (Section 7.3.1). The colloids described here correspond to those metals which resist smooth anodic dissolution. Interestingly, the salt method can also be applied to metals amenable to anodic dissolution [25]. For example, the use of $Pd(OAc)_2$ as the metal source and $(nC_8H_{17})_4N^+Br^-$ as the electrolyte results in ammonium-stabilized clusters of the type synthesized previously. The salt method can also be employed in the preparation of bimetallic colloids [25]. Exploration of the full scope and limitations is in progress.

7.3.3 Properties of Electrochemically Prepared Metal Clusters Stabilized by Tetraalkylammonium Salts

7.3.3.1 Solubility

Since the type of ammonium salt can be varied at will, it is clear that the solubility of the colloids can easily be controlled. This simple principle is illustrated here using Pd clusters as a representative example [8, 9]. Those clusters having ammonium salts with long lipophilic substituents are readily soluble in hydrocarbons such as pentane, whereas those having short alkyl groups in the ammonium salt (e.g. $(nC_4H_9)_4N^+Br^-$) require polar solvents for complete redispersement. In order to achieve water-solubility, which may be important in certain industrial applications, ammonium salts containing sulfonate moities in the side-chain were chosen. For example, the use of the commercially available betaine

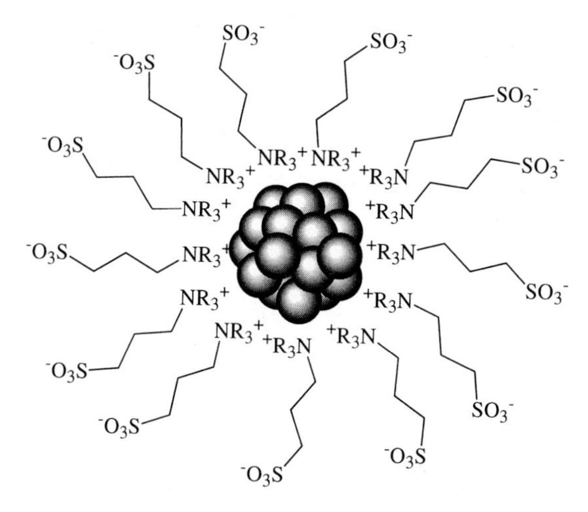

Fig. 7-7. Schematic representation of a water-soluble metal colloid (cross-section).

3-(dimethyldodecylammonio)propane-sulfonate (in combination with additives such as LiCl to ensure conductivity) leads to Pd clusters which are completely soluble in water [8, 9]. The colloid can be pictured as in Fig. 7-7.

7.3.3.2 Geometrical Parameters as Determined by Combined TEM/STM Studies

As mentioned in Section 7.2, Wiesner has postulated electrostatic interactions between a silver surface and tetraalkylammonium ions [20]. However, details of the nature of this type of interaction have not been explored, nor have any geometrical parameters been studied experimentally. For example, it is not clear how close the ammonium ions approach the silver atoms. Although transmission electron microscopy (TEM) constitutes the standard instrumental method in the study of metal colloids, it provides only information regarding the size of the metal cores. Since scanning tunneling microscopy (STM) is a powerful tool in the characterization of surfaces and heterogeneous catalysts [26], it might be expected to be useful in the determination of the outer dimensions of the actual colloidal materials. As shown schematically in Fig. 7-8, the difference between the mean diameters d_{STM} and d_{TEM} should give $2S$, where S represents the thickness of the protective layer [27]. Previous STM studies of gold and palladium clusters stabilized by phenanthro-

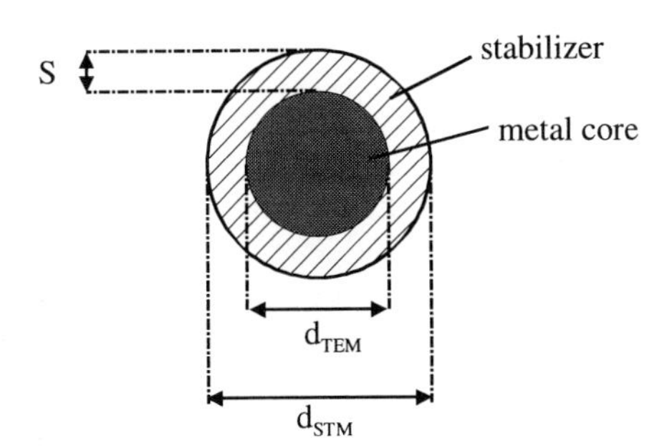

Fig. 7-8. Geometrical parameters of a stabilized metal colloid.

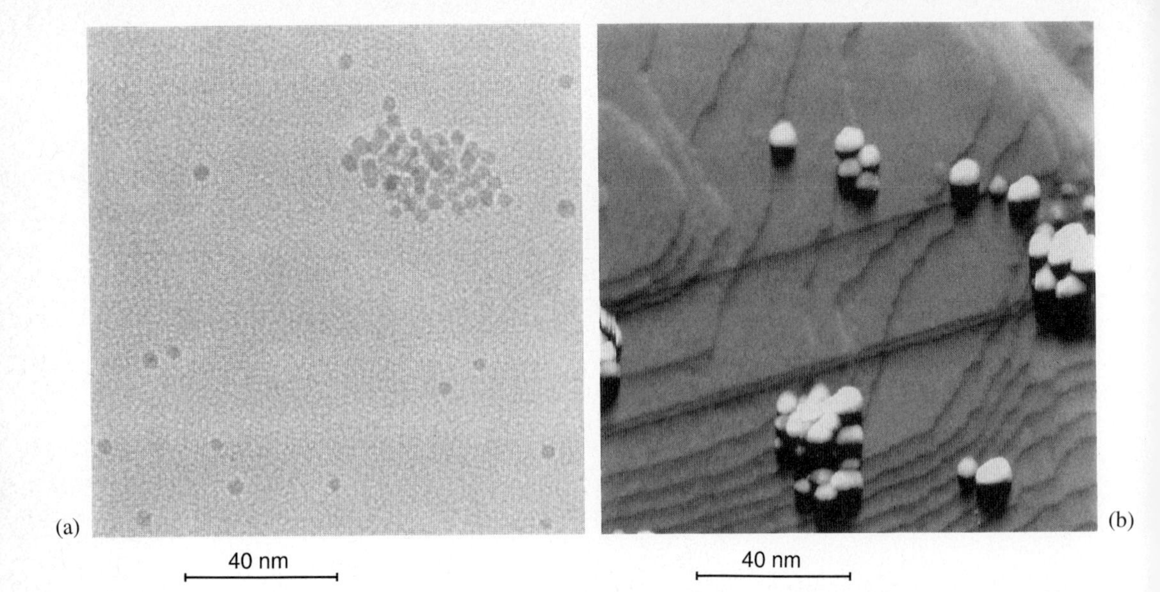

Fig. 7-9. (a) HRTEM image of a 4.1 nm-sized Pd cluster stabilized by $N(nC_8H_{17})_4Br$. (b) STM image of the same colloid. (Reproduced with permission of the American Association for the Advancement of Science).

line derivatives have resulted in the gross characterization of the surface topography, but conclusions regarding d_{STM} have not been possible thus far [28].

In collaboration with the group of U. Stimming at Jülich, palladium clusters were chosen for the first successful combined TEM/STM study of nanostructured metal colloids [27]. Since the size of the metal core and that of the alkyl groups of the ammonium ions can be controlled by adjustment of the current density and by proper choice of the ammonium salts, respectively, the systematic variation of the geometrical parameters shown in Fig. 7-8 is a simple matter. The major problem was to find the best method of sample preparation. After testing several substrates, it was found that quartz slides with 200 nm vapor-deposited gold films are optimal, provided that a flame annealing process, which leads to a smooth gold surface with atomic steps and large atomically flat terraces, is carried out [27]. The ammonium-stabilized clusters were then attached to the surface by dip-coating with the colloidal solutions. Fig. 7-9 shows typical TEM and STM images at the same scale of magnification, specifically of $(nC_4H_9)_4N^+Br^-$-protected Pd clusters with an average diameter d_{TEM} of 4.1 nm. The STM image reveals gold terraces upon which particles with an average diameter d_{STM} of 6.9 nm are clearly visible [27]. Other typical STM images are shown in Fig. 7-10. Each image was analyzed by using about 100 particles for the determination of the average particle diameters d_{TEM} and d_{STM} [27].

In all cases d_{STM} turned out to be larger than d_{TEM}. The results summarized in Fig. 7-11 show that the difference $d_{STM} - d_{TEM}$ in the series of $(nC_8H_{17})_4N^+Br^-$-stabilized clusters is essentially independent of the metal core diameter d_{TEM}. STM analysis shows particles that appear 2.0–2.7 nm larger than the same ones imaged by TEM. Thus, it is the nature of the ammonium salt which contributes to the constant difference. Fig. 7-12 refers to samples having an approximately constant metal core size, but different ammonium salts. The difference $d_{STM} - d_{TEM}$ is now dependent upon the length of the alkyl groups in the ammonium ions. It is also of interest to note that the size distribution diagrams of a given sample, as

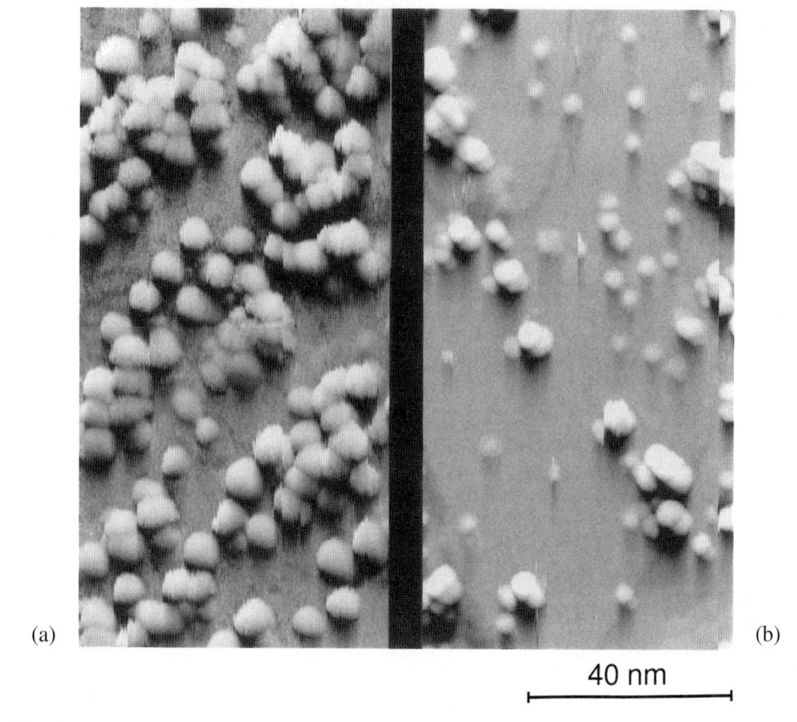

Fig. 7-10. STM images of $N(nC_8H_{17})_4Br$-stabilized Pd clusters. (a) $d_{STM} = 6.9$ nm; $d_{TEM} = 4.1$ nm. (b) $d_{STM} = 4.0$ nm; $d_{TEM} = 2.0$ nm. (Reproduced with permission of the American Association for the Advancement of Science).

derived from STM and TEM, correspond very well to one another. Only small differences occur because STM also detects very small particles (< 1 nm) which are not observable by TEM.

The results of the combined STM/TEM study support the general postulate of Wiesner concerning the interaction of ammonium ions with metal clusters [20]. Assuming that the H-atoms of the α-methylene groups of the ammonium ions are at the outer surface of

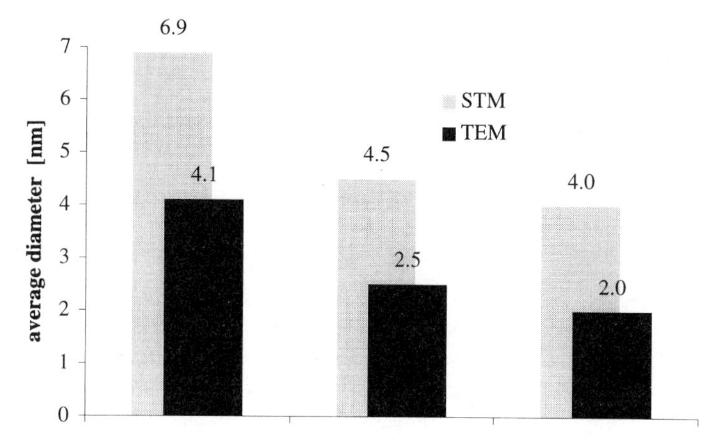

Fig. 7-11. Influence of the metal core size on the STM image [stabilizer: $N(nC_8H_{17})_4Br$].

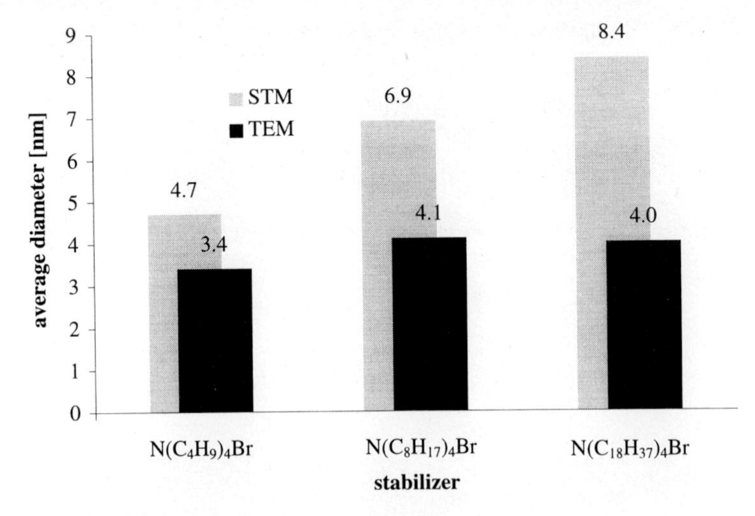

Fig. 7-12. Influence of the stabilizer size on the STM image.

the metal core as shown in Fig. 7-13, it is possible to obtain approximate *S*-values using standard MM2 force field calculations. Table 7-1 shows that the agreement between the calculated and experimental values is excellent [27].

It is clear that STM, if properly applied, is a suitable analytical tool for the visualization of the outer dimensions of a surfactant-protected metal colloid. Furthermore, the combination of TEM and STM provides valuable information regarding geometrical parameters. It is likely that this method will be used in the future to characterize other metal clusters.

One of the questions that also had to be resolved concerns the physics of the STM imaging process [27]. It is clear that the stabilizer becomes visible because of the change in the specific conductivity of the tunneling resistance. However, the change in the specific conductivity can be explained in different ways: For example, it is possible that the stabilizers lower the height of the tunneling barrier. Another possibility concerns an electronic conduction mechanism in the stabilizer. Such a conduction mechanism is reasonable only if the stabilizer contains states within the energy range between

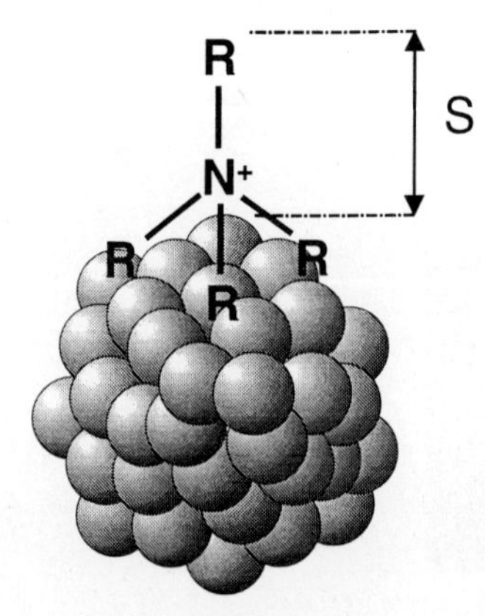

Fig. 7-13. Models for the interaction of tetraalkylammonium ions with Pd clusters.

Table 7-1. Calculated and experimental values of the stabilizer thickness S

R	S (nm) Calculated	Observed[a]
C_4H_8	0.70	0.65
C_8H_{17}	1.10	1.20[b]
$C_{18}H_{37}$	2.40	2.20

a) Average over 100 particles. b) Average over all measured samples.

the Fermi level of the tip and the sample. Currently it is difficult to pin-point such electronic states [27]. Relevant to this discussion is the expectation that conductivity should be sensitive to the tunneling current, which was not observed in the tunneling experiments.

7.3.3.3 UV-Vis Spectra

UV-Vis spectra of several of the $R_4N^+X^-$-protected metal colloids have been recorded [29]. For example, in the case of Pd clusters stabilized by $(nC_8H_{17})_4N^+Br^-$, a broad absorption to short wavelength extending from 1000 to 250 nm is observed (Fig. 7-14). Such behavior has also been noted for other metal colloids [30]. Plasma resonance with defined absorption peaks seems to apply for other metals (Cu, Ag, Au). In the case of Pd clusters it was possible to study the effect of particle size. Fig. 7-14 shows that differences in UV-vis absorption are indeed observed. Upon plotting log A against log λ, the negative slope $k = -\Delta(\log A)/\Delta(\log \lambda)$ of the approximate lines (Fig. 7-15) can be calculated for $\lambda < 450$ nm. For 1.4 nm, 2.3 nm, and 4.1 nm Pd clusters this gives k-values of 1.6, 2.6, and 3.6 respectively. Thus, the larger the particles, the larger the k-values [29], a

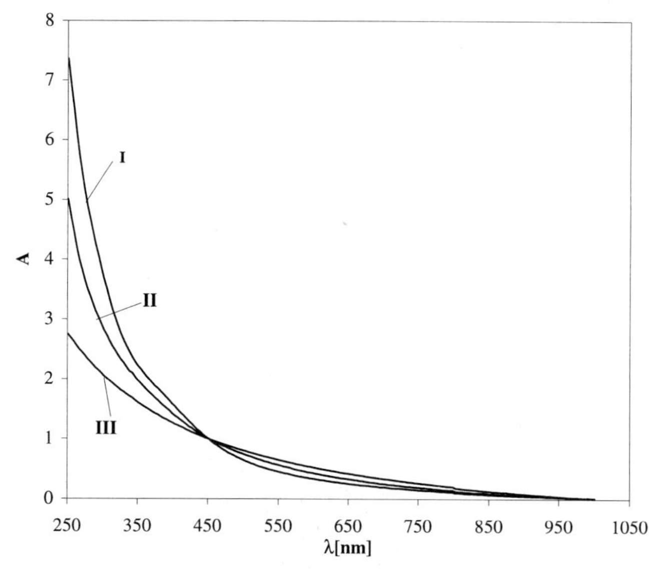

Fig. 7-14. Typical UV-vis spectra of $(nC_8H_{17})_4N^+Br^-$-stabilized Pd colloids in THF (normalized at $\lambda = 450$ nm with $A = 1$). I: 1.4 nm particles; II: 2.3 nm particles; III: 4.1 nm particles.

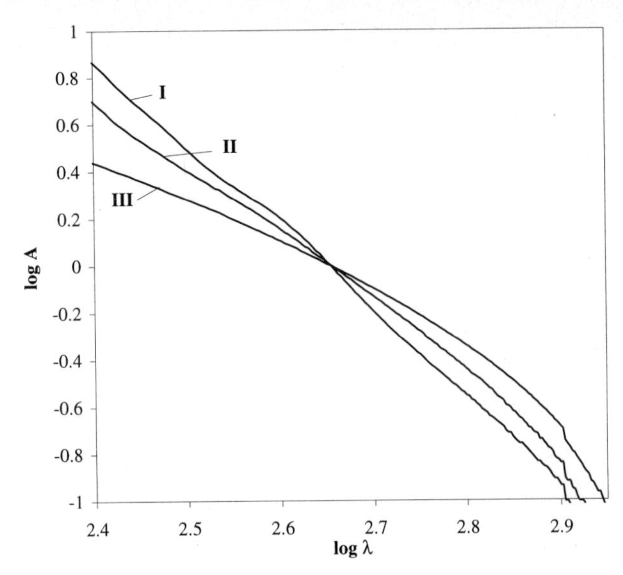

Fig. 7-15. Double logarithmic plots of the spectra shown in Fig. 7-14.

trend that had previously been noted for Pt sols in aqueous solutions [30]. The method can therefore be used to rapidly obtain qualitative information regarding the particle size. A note of caution concerns solutions in which the concentration of the colloids is very high, because under such conditions larger aggregates may form which affect the UV-vis-spectra.

7.3.3.4 Electrochemical Properties as Determined by Cyclic Voltammetry

In connection with the electrochemical metal colloid preparation [8, 9, 24, 25], two points were of special interest: (1) the possible influence of cluster enrichment upon the electrochemical behavior of the electrolyte system during the synthetic process, and (2) the electrochemical behavior of the isolated metal colloids. For these purposes cyclic voltammograms (CVs) of ammonium-stabilized Pd clusters and of the pure ammonium salts themselves in the same solvent were recorded [31]. Curve A in Fig. 7-16 represents the CV of $(nC_4H_9)_4N^+Br^-$ in acetonitrile. As can be seen, the electrolyte is inert within a wide potential range. The flow of current sets in only at very high reduction and oxidation potentials, which is due to decomposition of the electrolyte and/or solvent. Upon adding a 1.4 nm-sized $(nC_4H_9)_4N^+Br^-$-stabilized Pd colloid, drastic changes in the CV occur (Fig. 7-16, curve B). Anodic and cathodic current waves appear.

The similarity of this CV to that of a Pd sheet leads to the plausible assumption that curve B is caused by finely divided palladium at the platinum electrode. However, the positions of the electrochemical oxidation of the Pd electrode and of the Pd colloid differ by about 200 mV. The same effect is observed upon studying the dissolution potential of a Pd electrode with and without added Pd colloids: addition of the colloid results in a cathodic shift of the metal dissolution by about 200 mV. This interesting behavior can be understood on the basis of the previously applied theory used in the explanation of the observed size-selectivity in the formation of Pd clusters [31]: the free energy of a metal colloid, ΔG_{coll}, differs from the free energy ΔG_{bulk} of an infinitely extended metal sheet

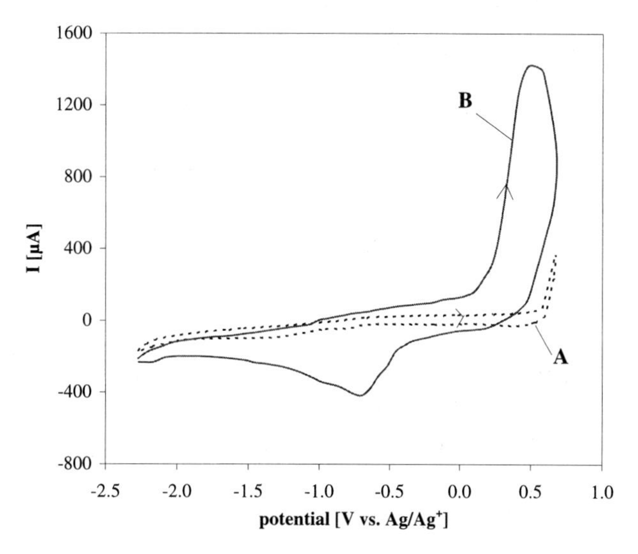

Fig. 7-16. Cyclic voltammograms of a 0.1 M $(nC_4H_9)_4N^+Br^-$ solution in acetonitrile without Pd colloid (curve A) and with Pd colloid (curve B).

(electrode) by the amount ΔG_{surf}, which is the (enhanced) energy of the particles at the surface of the colloid [Eq. (7.5)] [32].

$$\Delta G_{coll} = \Delta G_{bulk} + \Delta G_{surf} \tag{7.5}$$

The electrode potential of the colloid (E_{coll}) and that corresponding to the dissolution of the metal (E_{bulk}) differ by a contribution which is due to the surface effect [Eq. (7.6)].

$$E_{coll} = E_{bulk} - \eta \tag{7.6}$$

The surface contribution η is simply the overpotential, which has already been shown to be linked to the critical radius r_{crit} of the clusters (Section 7.3.1.1) and is given by Equation (7.7).

$$\eta = \frac{-2M\gamma}{nF\varrho r_{crit}} \tag{7.7}$$

where η = overpotential, M = molecular weight, γ = surface tension, n = valency, F = Faraday constant, r_{crit} = critical radius. Simple insertion affords Equation (7.8) for the difference in potentials ΔE:

$$\Delta E = E_{coll} - E_{bulk} = \frac{2M\gamma}{nF\varrho r_{crit}} \tag{7.8}$$

Using an approximate γ-value of 2.5 N m^{-1}, the shifts in the potentials as a function of the particle size can be roughly calculated (Table 7-2) [31]. In the case of the Pd sample used in the above CV study (1.4 nm particles), the observed shift in potential of about 200 mV is in good agreement with the calculated value of 260 mV [31].

Table 7-2. Calculated cathodic potential shift ΔE as a function of particle diameter *d*.

d (nm) ($2r_{crit}$)	$-\Delta E$ (mV)
5	90
2	220
1	450
0.5	910

7.3.3.5 Superparamagnetism

Whereas the bulk ferromagnetic behavior of metals is fairly well understood, including effects such as coupling between neighboring magnetic moments and electronic band filling, the study of the magnetic properties of nanostructured materials is an ongoing endeavor [10, 11]. As systems become very small (typically less than 10 nm), electronic properties change and thermal fluctuations become important. Materials become single domains with atomic magnetic moments sharing a common symmetry axis. This common magnetic orientation may fluctuate thermally or in an external field, so that the entire particle behaves like a single paramagnetic molecule, albeit with a much larger magnetic moment. Bean and Livingston have therefore called this behavior "collective paramagnetism" or "superparamagnetism" [33]. The effect of superparamagnetism has actually been used to determine the sizes of small metal particles [34], including supported metal catalysts [35].

The magnetic properties of electrochemically prepared tetraoctylammonium bromide stabilized cobalt colloid solutions in THF were studied by SQUID (Superconducting Quantum Interference Device) [36] and Gouy balance measurements. The magnetization curves were measured with special emphasis on changes at the freezing point of the organic cluster solutions [37]. Fig. 7-17 shows the magnetization as a function of the applied magnetic field parameter $b\alpha\mu_0 H/T$ for a cobalt colloid having a mean diameter d = 1.9 nm and an average spin moment of 600 $\hbar$. The curves of the liquid phase (T < 200 K) can be reasonably described by the Langevin function [38]. However, upon cooling the colloid below its freezing point, deviations from this behavior occur. These deviations can be understood in terms of magnetic anisotropy effects which occur if the magnetic spin moments of the clusters couple with the crystal lattice [37]. For systems with uniaxial anisotropy, Chantrell recently developed an extension of the Langevin function [39]. Using this extension, anisotropy coupling energies were calculated [37]. Furthermore, the results of magnetic measurements on naked isolated ferromagnetic clusters, as prepared by Becker and deHeer in the gas phase [40], correlate with the magnetic behavior of the electrochemically prepared ammonium-stabilized cobalt colloids.

Concerning applications of micro- and nanostructured magnetic particles, two areas are of special interest, namely magnetic recording techniques and "ferrofluids" [41]. Furthermore, materials composed of nanometer-sized magnetic particles dispersed in a nonmagnetic matrix have magneto-caloric properties that differ from those of the corresponding bulk magnetic materials [42]. The so-called magnetocaloric refrigeration technique seems to have future application as an effective cooling process without the need for ozone-depleting chlorofluorocarbons [43]. The question whether electrochemically prepared mono- or bimetallic colloids having superparamagnetic properties can compete with existing materials is the subject of current studies in our laboratories.

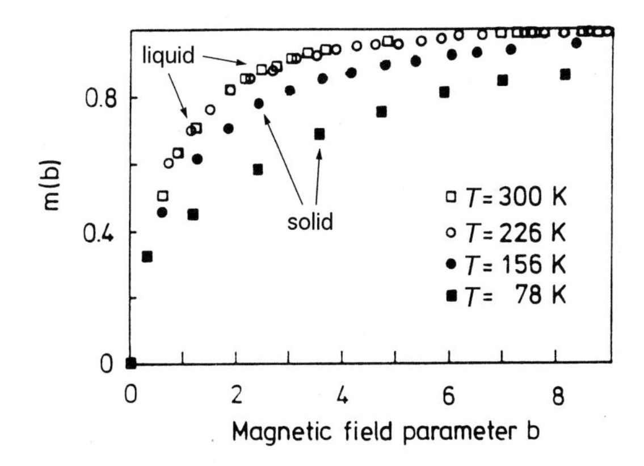

Fig. 7-17. Magnetization as a function of the applied magnetic field parameter b for 1.9 nm-sized Co clusters stabilized by $(nC_8H_{17})_4N^+Br^-$.

7.3.4 Conclusions

This chapter provides an overview of the initial phase of our efforts directed towards the development of electrochemical syntheses of transition metal clusters stabilized by tetraalkylammonium or phosphonium salts. The basic working principle is simple, namely the electrochemical reduction of metal salts in the presence of surfactants which serve as the electrolyte and as the stabilizer. Two approaches have been shown to be successful, depending upon the particular metal used. The first method makes use of a sacrificial anode as the source of the metal. In those cases in which anodic dissolution causes problems, the second approach based on the reduction of added metal salts is the method of choice. Although the full scope and limitations of the two methods still need to be defined, it can safely be said that the salient features of these types of metal colloid syntheses include high yield, absence of undesired side-products such as metal hydrides or boron impurities, easy isolation, variation of solubility (from pentane to water) by the proper choice of ammonium salt, and in some cases control of particle size by adjustment of the current density. Thus, the electrochemical reactions described here appear to be viable alternatives to classical chemical reductions.

The combined use of TEM and STM is a particularly powerful method of characterizing ammonium-stabilized colloids, because details of the geometrical relationship of the inner metal core and the outer stabilizing mantle become visible. It is likely that in the future other types of stabilized clusters will also be studied by this approach. The measurement of magnetic properties, e.g. superparamagnetism, is not only of academic interest, such studies may also be important in the development of advanced materials in electronics.

Another broad area of application concerns catalysis and electrocatalysis (fuel cell technology). Whereas the ammonium-stabilized metal colloids produced conventionally by chemical reductions (Section 7.2) have already been tested in various catalytic hydrogenations, the colloids described in this chapter still need to be screened carefully in similar reactions. Although the materials are formally similar, differences may in fact exist (type and extent of impurities, particle size, immobilization, etc.). Preliminary results in our laboratories show that the electrochemically produced Pd clusters can be used not only in selective olefin hydrogenations but also in C–C bond-forming processes such as Heck reactions. The observation that Pd clusters can be fixed on charcoal during electrochemical synthesis, giving materials that have essentially the same particle size as the non-immobilized samples, may be an important step in the development of future heterogeneous catalysts. The same applies to the convenient preparation of water-soluble betaine-stabilized transition metal clusters, especially in view of the fact that they can be easily immobilized on conventional supports such as charcoal or metal oxides or on electrically conducting materials as SnO_2-layers. The major impact of this method is the observation that the metal clusters do not penetrate deeply into the solid support as in classical methods, where some of the metal is inaccessible for catalysis [25].

References

[1] R. Walker, A. R. B. Sanford, *Chem. Ind.* **1979**, 642; N. Ibl, *Chem.-Ing. Tech.* **1964**, *36*, 601.
[2] N. Ibl, G. Gut, M. Weber, *Electrochim. Acta* **1973**, *18*, 307.
[3] R. Walker, *Chem. Ind.* **1980**, 260.
[4] P. J. Grundy, D. Greig, E. W. Hill, *Endeavor* **1993**, *17*, 154.
[5] T. M. Whitney, J. S. Jiang, P. C. Searson, C. L. Chien, *Science* **1993**, *261*, 1316.
[6] C. A. Foss, G. L. Hornyak, J. A. Stockert, C. R. Martin, *J. Phys. Chem.* **1992**, *96*, 7497.
[7] H. Lehmkuhl, W. Leuchte, *J. Organomet. Chem.* **1970**, *23*, C30.
[8] M. T. Reetz, W. Helbig, *J. Am. Chem. Soc.* **1994**, *116*, 7401.
[9] M. T. Reetz, W. Helbig, patent applied.
[10] G. Schmid, *Clusters and Colloids*, VCH, Weinheim. **1994**; A. Henglein, *J. Phys. Chem.* **1993**, *97*, 5457; S. C. Davis, K. Klabunde, *Chem. Rev.* **1982**, *82*, 153; L. N. Lewis, *Chem. Rev.* **1993**, *93*, 2693; G. Schmid, *Chem. Rev.* **1992**, *92*, 1709; B. C. Gates, L. Guczi, H. Knözinger, *Metal Clusters in Catalysis*, Elsevier, Amsterdam, **1986**.
[11] H. Weller, *Angew. Chem.* **1993**, *105*, 43; *Angew. Chem. Int. Ed. Engl.* **1993**, *32*, 41; A. Henglein, P. Mulvaney, A. Holzwarth, T. E. Sosebee, A. Fojtik, *Ber. Bunsenges. Phys. Chem.* **1992**, *96*, 754; W. P. Halperin, *Rev. Mod. Phys.* **1987**, *58*, 533; K. Ploog, *Angew. Chem.* **1988**, *100*, 611; *Angew. Chem. Int. Ed. Engl.* **1988**, *27*, 593; I. P. Herman, *Chem. Rev.* **1989**, *89*, 1323; G. Nimtz, P. Marquard, H. Gleiter, *J. Cryst. Growth* **1988**, *86*, 66.
[12] Selective synthesis of Pd nanoparticles in the bicontinuous cubic phase of glycerol monooleate: S. Purvada, S. Baral, G. M. Chow, S. B. Qadri, B. R. Ratna, *J. Am. Chem. Soc.* **1994**, *116*, 2135.
[13] J. S. Bradley in *Clusters and Colloids* (ed.: G. Schmid), VCH, Weinheim, **1994**, Ch. 6.
[14] J. Kiwi, M. Grätzel, *J. Am. Chem. Soc.* **1979**, *101*, 7214.
[15] M. Boutonnet, J. Kizling, P. Stenius, G. Maire, *Coll. Surf.* **1982**, *5*, 209; M. Boutonnet, J. Kizling, R. Touroude, G. Maire, P. Sternius, *Appl. Catal.* **1986**, *20*, 163; M. Boutonnet, Thesis, Univ. of Strasbourg, **1977**.
[16] K. Meguro, M. Toriyuka, K. Esumi, *Bull. Chem. Soc. Jpn.* **1988**, *61*, 341.
[17] N. Toshima, T. Takahashi, H. Hirai, *Chem. Lett.* **1985**, 1245.
[18] Y. Sasson, A. Zoran, J. Blum, *J. Mol. Catal.* **1981**, *11*, 293; N. Satoh, K. Kimura, *Bull. Chem. Soc. Jpn.* **1989**, *62*, 1758.
[19] H. Bönnemann, W. Brijoux, R. Brinkmann, E. Dinjus, I. Joussen, B. Korall, *Angew. Chem.* **1991**, *103*, 1344; *Angew. Chem. Int. Ed. Engl.* **1991**, *30*, 1312.
[20] J. Wiesner, A. Wokaun, H. Hoffmann, *Prog. Coll. Polym. Sci.* **1988**, *76*, 271.
[21] Southampton Electrochemistry Group, Instrumental Methods in Electrochemistry, Ellis Horwood, Chichester, **1990**.
[22] M. T. Reetz, W. Helbig, S. A. Quaiser, unpublished results.
[23] M. T. Reetz, W. Helbig, to be published.
[24] M. T. Reetz, S. A. Quaiser, W. Helbig, submitted for publication.
[25] M. T. Reetz, S. A. Quaiser, W. Helbig, patent applied.
[26] G. Binnig, H. Rohrer, C. Gerberg, E. Weibel, *Appl. Phys. Lett.* **1982**, *40*, 178; J. Frommer, *Angew. Chem.* **1992**, *104*, 1325; *Angew. Chem. Int. Ed. Engl.* **1992**, *31*, 1298.
[27] M. T. Reetz, W. Helbig, S. A. Quaiser, U. Stimming, N. Breuer, R. Vogel, *Science* **1995**, *267*, 367.

[28] H. A. Wierenga, L. Soethout, J. W. Gerritsen, B. E. C. van de Leemput, H. van Kempen, G. Schmid, *Adv. Mater.* **1990**, *2*, 482; L. E. C. van de Leemput, J. W. Gerritsen, P. H. H. Rongen, R. T. M. Smokers, H. A. Wierenga, H. van Kempen, G. Schmid, *J. Vac. Sci. Technol.* **1991**, *B9*, 814.
[29] M. T. Reetz, W. Helbig, unpublished results.
[30] D. N. Furlong, A. Launikonis, W. H. F. Sasse, *J. Chem. Soc., Faraday Trans. I* **1984**, *80*, 571.
[31] M. T. Reetz, W. Helbig, to be published.
[32] W. J. Plieth, *J. Phys. Chem.* **1982**, *86*, 3166; W. J. Plieth, *J. Electroanal. Chem.* **1986**, *204*, 343.
[33] C. P. Bean, J. D. Livingston, *J. Appl. Phys.* **1959**, *30*, 1205.
[34] P. W. Selwood, *J. Am. Chem. Soc.* **1956**, *78*, 3893.
[35] P. H. Christensen, S. Mørup, J. W. Niemantsverdriet, *J. Phys. Chem.* **1985**, *89*, 4898.
[36] O. Kahn, *Molecular Magnetism*, VCH, Weinheim, **1993**.
[37] J. A. Becker, S. A. Quaiser, W. Helbig, M. T. Reetz, submitted for publication.
[38] P. W. Selwood, *Chemisorption and Magnetization*, Academic Press, New York, **1975**.
[39] R. W. Chantrell, private communication to J. A. Becker, 1994.
[40] I. M. L. Billas, J. A. Becker, A. Châtelain, W. A. de Heer, *Phys. Rev. Lett.*, **1993**, *71*, 4067.
[41] B. M. Berkowsky, V. F. Medvedev, M. S. Krakov, *Magnetic Fluids – Engineering Applications*, Oxford University Press, Oxford, **1993**.
[42] R. D. Shull in *Superconductivity and its Application* (eds.: H. S. Kwok, D. T. Shaw, M. J. Noughton), *AIP Conf. Proc.* **1993**, *273*, 628.
[43] R. D. Shull, L. H. Bennett, *J. Nanostruct. Mat.* **1992**, *1*, 83.

8 The Magnesium Route to Active Metals and Intermetallics

L. E. Aleandri and B. Bogdanović

8.1 Introduction[1]

Well-known *chemical* routes[2] for the preparation of highly reactive metal powders typically involve the reduction of metal salts with either alkali metals (Rieke method, Chapter 1) [1], potassium–graphite (Chapters 10 and 11) [2], or complex hydrides (Chapter 9) [3], or by electrolysis in a suitable inert solvent (Chapter 7) [4]. The simultaneous *coreduction* of two metal salts by complex hydrides (e.g. $NaBEt_3H$) [3, 5], potassium–graphite [2], or electrolysis (Chapter 7) can be utilized for the production of highly active bimetallic systems such as bimetallic colloids, alloys, and intimate conglomerates of very fine metal particles.

In this chapter, "The Magnesium Route to Active Metals and Intermetallics" — the title of which should also include "active magnesium" itself — novel synthesis routes for organic and inorganic materials will be described, all of which involve as a common characteristic magnesium metal or magnesium compounds in the initial preparative process. Since the early eighties we have investigated the chemical activation of magnesium by exploiting the *reversibility* of the reaction between magnesium, anthracene, and tetrahydrofuran (THF) yielding magnesium–anthracene · 3 THF (MgA, Scheme 8-1) [6] and of the catalytic hydrogenation reaction of magnesium [Eq. (8.1)] [6b, 7].

$$Mg + H_2 \xrightarrow{\text{cat. [7] THF}} MgH_2^* \underset{240\,^\circ C/1\,\text{bar}+H_2}{\overset{>250\,^\circ C/\text{vac.}\ -H_2}{\rightleftarrows}} Mg^* \tag{8.1}$$

The application of active magnesium (Mg*), originating from MgA and catalytically prepared magnesium hydride (MgH_2^*), in synthetic chemistry is discussed in Sections 8.2 and 8.3. MgH_2^*, as well as diethylmagnesium (Et_2Mg), can also be used through the reaction

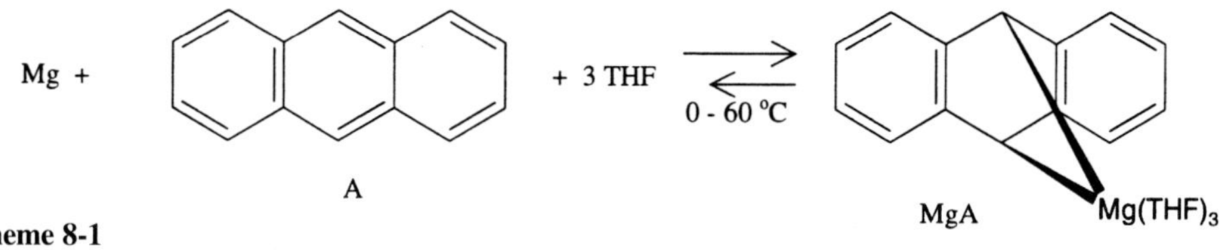

Mg + [A] + 3 THF ⇄ (0 - 60 °C) MgA Mg(THF)$_3$

Scheme 8-1

1 Abbreviations used in this chapter: MgA = magnesium–anthracene·3 THF; MgH_2^* = catalytically prepared magnesium hydride; MgH_2' = catalytically prepared solubilized magnesium hydride; Mg^* = active magnesium powder prepared from MgH_2^* or MgA.

2 In contrast to *physical* methods used for activation of metals such as metal vaporization (Chapter 6), activation by ultrasound (Chapter 4), grinding, etc.

with bis(allyl)metal complexes to prepare both known and new reactive Mg intermetallics and/or their hydrides and carbides. The "wet chemical" synthesis of intermetallics can also be successfully effected through the reaction of metal halides with *excess* amounts of magnesium hydrides, organomagnesium compounds (such as MgA), or magnesium metal. These novel preparative routes will be surveyed in Section 8.4. Lastly, in Section 8.5, the recent (re)discovery of the so-called "Grignard reagents of transition metals" and their applications in the synthesis of highly active nanocrystalline materials including metals, intermetallics, and alloys will be presented.

8.2 Magnesium–Anthracene Systems in Organic, Organometallic, and Inorganic Synthesis

8.2.1 MgA as a Source of *Soluble* Zerovalent Magnesium

MgA possesses a low but measurable solubility in THF (7 and 14 mmol/L at RT and 60 °C, respectively), and suspensions of MgA in THF are stable at RT [6a]. Under such conditions MgA is extremely reactive towards electrophilic substrates. MgA suspended in THF can be described as an ambivalent species which can react with electrophiles as a single-electron donor, as a diorganomagnesium compound (heterolytic Mg–C bond cleavage), or as a two-electron donor undergoing homolytic Mg–C bond cleavage, and hence as a source of "soluble zerovalent magnesium" [8]. The ability of MgA to act as a single-electron donor can be explained by the transformation of MgA into radical anion species, such as the well-characterized radical anion complex $[Mg_2Cl_3(THF)_6]^+ A^{\cdot-}$. In contrast, MgA reveals a high reactivity and tendency to undergo homolytic rather than heterolytic Mg–C bond cleavage due to the weak character of its Mg–C bonds [8, 9]. Stoichiometric reactions with electrophiles, employing MgA as a source of soluble zerovalent magnesium, are generalized in Equations (8.2), (8.3) and (8.4).

$$\text{E–X} + \text{MgA} \rightarrow \text{E–Mg–X} + \text{A} \tag{8.2}$$

$$\text{E–Mg–X} + \text{E–X} \rightarrow \text{E–E} + \text{MgX}_2 \tag{8.3}$$

$$\text{E}'\text{X}_2 + \text{MgA} \rightarrow \text{E:}' + \text{MgX}_2 + \text{A} \tag{8.4}$$

E, E′ = the electrophilic component of the substrate; X = the anionic component of the substrate, e.g. halide ion.

The advantages of using MgA suspensions in THF for reactions with electrophiles compared with any sort of *solid* active magnesium include very low reaction temperatures, even with insoluble (e.g. polymeric) substrates. However, the application of MgA for synthetic chemistry is not limited to those electrophilic substrates which react with MgA according to Equations (8.2) to (8.4). Reactions in which MgA acts as a diorganomagnesium compound (electrophilic attack on the anthracene dianion moiety, like alkyl halides) or as a single-electron donor have been described elsewhere [8, 10].

Typical electrophiles which react according to Equation (8.2) are allyl, propargyl, and benzyl halides [8, 10]. The reaction described in Equation (8.5) allows the preparation of

$$\text{RX} + \text{MgA} \xrightarrow{\text{THF}} \text{RMgX} + \text{A} \qquad \text{R = allyl, propargyl, benzyl} \tag{8.5}$$

the corresponding Grignard reagents at temperatures as low as –78 °C, and in high yields with only minor amounts of side products (e.g. substituted dihydroanthracenes). This method represents an important route for the preparation of allenyl-magnesium chloride from propargyl chloride, since the preparation of this Grignard compound from propargyl chloride and magnesium metal is not possible, even with the most active forms of Mg* (cf. Section 8.3.2). A detailed account of the synthesis of Grignard compounds using MgA is given in [10].

Raston et al. [11] have exploited the reaction between MgA and RX (R = benzyl) to generate, among others, several otherwise nearly inaccessible bi- and polyfunctional benzylic types of Grignard compounds, such as **1**–**4**. Remarkably, Fréchet et al. were able to perform the first direct (i.e. not via lithiated polymer) synthesis of the polymeric Grignard compound **5** by reacting *p*-(chloromethyl)polystyrene with MgA [12]. Starting from **5**, Raston et al. [13] have prepared the first polymer-bound magnesium anthracene.

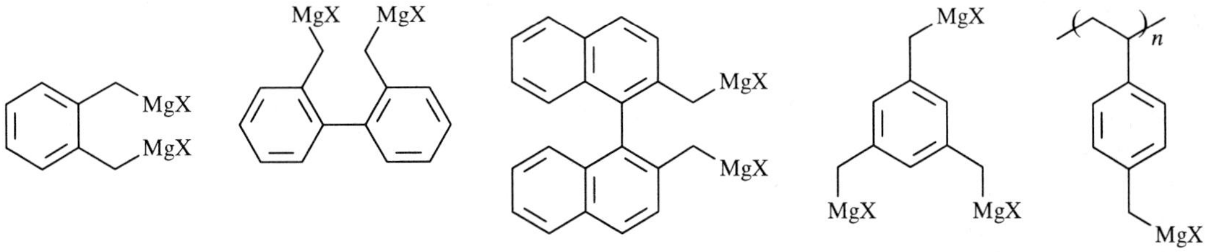

1 **2** **3** **4** **5**

X = Cl, Br

On the basis of the generally accepted mechanism for the formation of Grignard reagents [14], it is probable that the reaction between MgA and RX [Eq. (8.5)] proceeds via a direct (consecutive) electron transfer from MgA to the substrate rather than through the decomposition of MgA to Mg. This is particularly true for the preparation of Grignard compounds from propargyl chloride [10] and the polymer-bound 9-silylated anthracene [13], which do not react with solid active Mg*. The relatively high stability and easy formation of allyl, propargyl, and benzyl chloride radical anions favors the generation of the corresponding Grignards during the reaction with MgA rather than an electrophilic attack on the anthracene dianion moiety [10].

A recent and highly interesting example of the application of MgA as a source of soluble zerovalent magnesium is the preparation of polymer-supported distannane from a polymer-bound tin halide resin as described by Neumann et al. [15] (Scheme 8-2). Polymer-supported distannane can be used as a photochemical regenerative source of stannyl radicals for organic synthesis. MgA can also be used as a reducing agent for the preparation of organo-transition metal complexes [16].

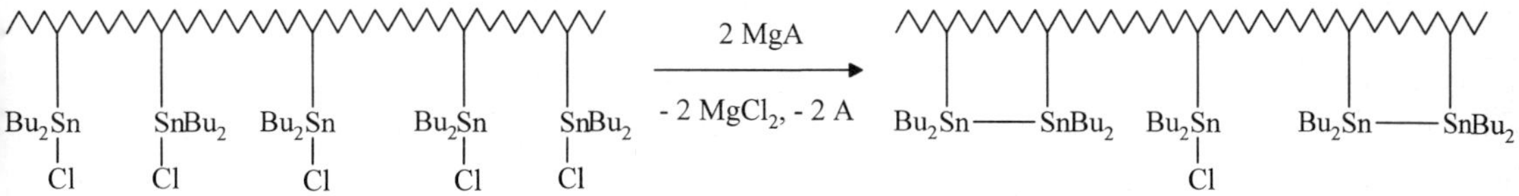

Scheme 8-2

8.2.2 Magnesium–Anthracene–THF and Magnesium–Anthracene–$MgCl_2$–THF Systems as Phase-Transfer Catalysts for Metallic Magnesium

The reaction of MgA to liberate anthracene and highly reactive magnesium [Eqs. (8.2) to (8.4)] can be coupled to the reaction of anthracene in the presence of magnesium in THF to regenerate MgA (Scheme 8-1) which can then be exploited in the catalysis of reactions involving metallic magnesium (Scheme 8-3). In other words, anthracene or MgA in THF can be described as phase transfer catalysts for metallic magnesium [17]. Thus, the addition of a catalytic amount of anthracene (1–5 mol-%) to a suspension of *ordinary* magnesium powder in THF to generate magnesium-anthracene (Scheme 8-1) represents the simplest and certainly the most economical *chemical* means for activating magnesium (which is also suitable for large scale applications). The activating effect of anthracene on magnesium (in THF) persists, unless anthracene is somehow consumed in a side reaction(s). The efficiency of phase transfer reactions can be enhanced considerably through the application of ultrasound treatment [8].

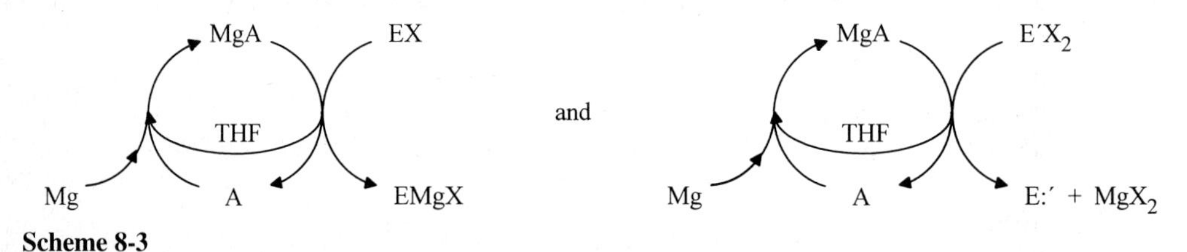

Scheme 8-3

8.2.2.1 Catalytic Synthesis of Magnesium Hydrides

The reaction of magnesium with anthracene in THF to produce MgA was discovered by Ramsden in 1965 [18]. But it took another 14 years to recognize, as a result of the catalytic synthesis of magnesium hydride [7a, 8], that this reaction induces a chemical activation of magnesium. In exploiting MgA for the activation of magnesium, the homogeneous catalytic hydrogenation of magnesium can be achieved via four different routes as shown in Eqs. (8.1) and (8.6) to (8.8). The original catalyst system used to hydrogenate magnesium powder to MgH_2^* [Eq. (8.1)] involved $TiCl_4$ or $CrCl_3$ together with MgA and magnesium in THF [7]. The reduction of $TiCl_4$ or $CrCl_3$ with MgA and magnesium, to give low-valent transition metal species active in hydrogenation, proceeds with the formation of anthracene and $MgCl_2$, both of which are essential for the transfer of magnesium into the liquid (THF) phase. The mechanism of the homogeneous catalytic hydrogenation of magnesium is discussed in Refs. [7, 8].

The activity of $TiCl_4$– or $CrCl_3$–MgA catalysts can be strongly enhanced by using $MgCl_2$ (7–8 mol-% relative to Mg) as a cocatalyst [7c, 8, 19]. The time needed to hydrogenate Mg powder quantitatively [Eq. (8-6)] is then reduced by a factor of ten.

$$Mg + H_2 \xrightarrow{MX_n\text{-MgA-MgCl}_2\text{ cat./THF}} MgH_2^* \tag{8.6}$$

$$MX_n = TiCl_4,\ CrCl_3 \text{ etc.}$$

The enhanced rate of magnesium hydrogenation in the presence of $MgCl_2$ [Eq. (8.6)] can be explained as due to the involvement of the radical anion complex $[Mg_2Cl_3(THF)_6]^+ A^{\bullet-}$

[9] in the magnesium activation/dissolution process [8]. Therefore, for the activation of magnesium metal in THF, catalytic amounts of both anthracene *and* $MgCl_2$ should be used routinely.

When magnesium is hydrogenated in THF using transition metal halide–MgA–$MgCl_2$ catalysts in the presence of a solubilizing agent (typically 5–7 mol-%), a magnesium hydride is produced which is partially or almost completely present in a dissolved form [designated here as MgH'_2; Eq. (8.7)]. The preferred solubilizing agents are either quinuclidine or organomagnesium compounds [20]. Solutions of MgH'_2 in THF can be prepared with concentrations up to 4 molar. MgH'_2 has proven to be an especially useful reagent in the preparation of active Mg intermetallics and their hydrides (see Section 8.4.2).

$$Mg + H_2 \xrightarrow[\text{THF}]{\text{cat. MgCl}_2\text{, sol. agent}} MgH'_2 \qquad (8.7)$$

If the catalytic hydrogenation of magnesium is carried out in the presence of *stoichiometric* amounts of $MgCl_2$ at 0 °C, a THF-soluble hydridomagnesium chloride, **6**, is obtained in nearly quantitative yield [Eq. (8.8)] [21].

$$Mg + MgCl_2 + H_2 \xrightarrow[\text{THF/0°C}]{\text{CrCl}_3\text{-MgA-cat.}} \underset{\mathbf{6}}{[HMgCl \cdot THF]_2} \qquad (8.8)$$

8.2.2.2 Anthracene-Activated Magnesium for the Preparation of Finely-Divided Metal Powders and Transition Metal Complexes

Finely-divided metal powders, as well as many organo-transition metal complexes, form the foundation for a number of useful stoichiometric and catalytic reactions in synthetic chemistry. These reagents are typically accessible through chemical reduction, and magnesium metal due to its low toxicity and general availability, would seem to be the most suitable reducing agent. However, the generally low reactivity of commercial magnesium in the form of powder or turnings precludes its application as a reductant in organic solvents, especially at low temperatures for the reduction of insoluble metal salts. This barrier can be circumvented by activating magnesium with catalytic amounts of anthracene, thus providing a convenient route for the preparation of highly-dispersed metal powders and organotransition metal complexes from their respective metal salts in THF [22].

The reduction of metal salts MX_n by anthracene-activated magnesium to give active metals M* can be represented by Scheme 8-4 (which can be considered an extension of Scheme 8-3). Salts of metals belonging to groups 8–12 can be reduced at room tempera-

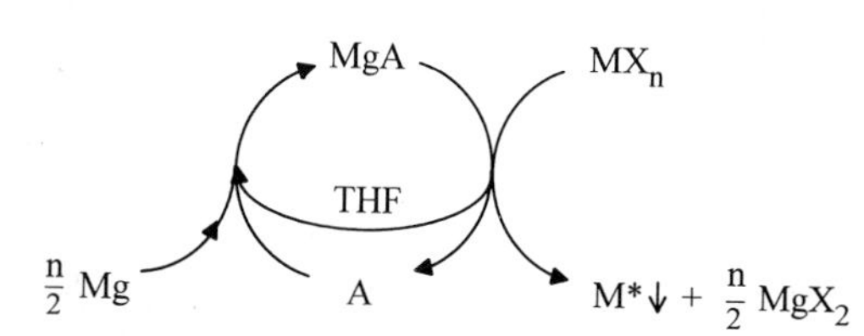

MgA MX$_n$

THF

$\frac{n}{2}$ Mg A M*↓ + $\frac{n}{2}$ MgX$_2$

M = Fe, Cr, Ni, Pd, Pt, Cu, Zn
X = Cl, Br, acac

Scheme 8-4

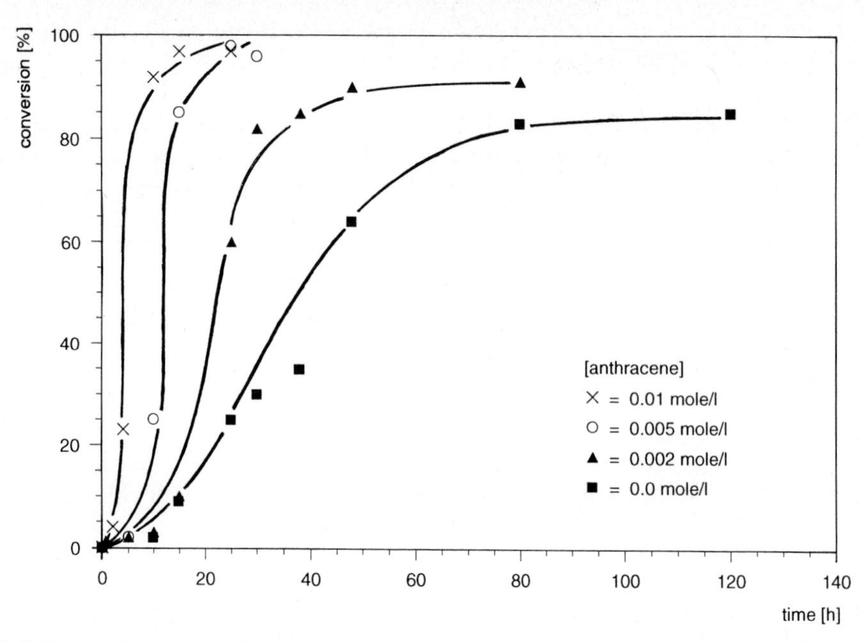

Fig. 8-1. Influence of anthracene on the rate of reduction of $FeCl_2 \cdot 2$ THF to metallic iron (Fe*) with magnesium powder in THF at room temperature (1:1 molar ratio; 0.1 mol of $FeCl_2 \cdot 2$ THF/1 THF). (Reproduced by permission of Elsevier Sequoia [22]).

ture to give slurries of metal powders, M*, along with the corresponding magnesium salts. The strong activating effect of anthracene on the rate and extent of reduction can be exemplified by the RT reduction of $FeCl_2 \cdot 2$ THF to Fe* as shown in Fig. 8-1.

Since $MgCl_2$ and $Mg(acac)_2$ remain dissolved in THF, the resulting metal powders M* can be isolated simply by filtering (Table 8-1). The specific surface areas of active powders were found to be comparable to those obtained by alkali metal reduction [1]. The dispersion of M* can be increased by performing the reduction (Scheme 8-4) in a 35 kHz ultrasound bath. The Ni* obtained from an experiment using ultrasound shows a surface area of 54.7 $m^2 g^{-1}$ as determined by the BET; that of Pd* is 103.4 $m^2 g^{-1}$ and of Cu* 10.5 $m^2 g^{-1}$ [22].

The metal powders listed in Table 8-1 are highly reactive towards oxidative addition reagents, which can be exploited for the synthesis of organometallic compounds. For example, Ni* present as a slurry reacts readily at RT to give a bis(aryl)nickel complex according to Eq. (8.9).

$$2\,Ni^* + 2\,C_6F_5I + 2\,PPh_3 \xrightarrow{41\%} (PPh_3)_2Ni(C_6F_5)_2 + NiI_2 \qquad (8.9)$$

Such active metal powders can also be used as heterogeneous catalysts, for example for hydrogenation reactions. A specific application involves the production of air-stable Ni-doped magnesium powders which show improved properties for the uptake of hydrogen for reversible hydrogen and heat storage in MgH_2/Mg systems [23].

The application of elemental magnesium for the reductive synthesis of organo-transition metal complexes [24, 25] has been reported, but so far a general method has not been developed. However, by reducing transition metal salts with anthracene-activated magne-

Table 8-1. Preparation of finely divided metal powders (M*)[a].

No.	MX_n	g/mmol	Mg (g/mmol)	Anthracene (g/mmol)	THF (ml)	Reaction time (h)	M* (g)	Composition in %					Specific surface area ($m^2 g^{-1}$)
								M	Mg	C	H	X	
1	$NiCl_2$	2.60/20	0.49/20	0.36/2	200	28	1.46	78.9	2.7	12.5	1.3	1.1	7.2
2	$CoCl_2$	5.19/40	0.98/40	0.18/1	200	10	2.62	86.6	3.7	7.0	0.8	3.5	57.3
3	$FeCl_2 \cdot 2$ THF	8.52/40	0.98/40	0.18/1	200	26	2.90	73.9	3.7	7.7	1.1	0.4	96.2
4	$PdCl_2$	8.85/50	1.2/50	0.55/3	400	6	5.33	90.6	6.8	–	–	1.7	57.6
5	CuCl	9.9/100	1.2/50	1.1/6	400	8	5.90	97.7	0.3	–	–	1.6	6.0
6	$ZnCl_2$	13.6/100	2.4/100	1.1/6	500	5	6.15	97.1	0.7	–	–	2.1	8.1
7	CuBr	14.4/100	1.2/50	0.55/3	300	5	6.10	94.6	0.4	1.3	0.4	4.2	6.6
8	$Fe(acac)_3$	11.8/33.3	1.2/50	0.55/3	300	0.5	1.0	74.7	10.9	9.4	1.2	–	96.2

a) Ref. [22], reproduced by permission of Elsevier Sequoia S.A.

Table 8-2. Comparison of the activities of various magnesium samples in the formation of compound **7** according to Scheme 8-6[a].

n	Electrophile (E)	Vapor-activated Mg yield (%)	Anthracene–Mg yield (%)	Rieke Mg yield (%)
1	OH	55	56	55
1	CONHPh	71	67	64
2	OH	57	51	48
2	CONHPh	72	69	64

a) Ref. [26c], reproduced with permission of Pergamon Press Ltd.

sium *in the presence of suitable organic ligands*, a simple preparative route to various transition metal complexes of practical importance, such as metallocenes (Scheme 8-5), olefin–metal(0) compounds, η^3-allyl– and phosphine–metal complexes, can be achieved [22]. Typically the efficiency of this synthetic route can be further enhanced by ultrasound treatment.

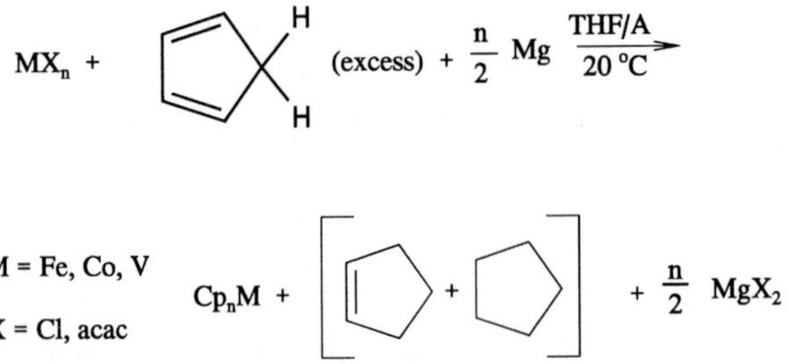

$$MX_n + C_5H_6\ (excess) + \frac{n}{2}\,Mg \xrightarrow[20\,^\circ C]{THF/A} Cp_nM + [C_5H_8 + C_5H_{10}] + \frac{n}{2}\,MgX_2$$

M = Fe, Co, V

X = Cl, acac

Scheme 8-5

8.2.2.3 Catalytic Synthesis of Grignard Compounds

Since allyl, propargyl, and benzyl halides react with MgA in THF to give the corresponding Grignard compounds and anthracene [Eq. (8.5)], it is expected that the presence of anthracene should *catalyze* the Grignard reaction of these organic halides with magnesium.

In a systematic study of the intramolecular magnesium-ene reaction (Scheme 8-6) and its application to the synthesis of natural products [26], Oppolzer et al. [26c] compared different magnesium activation methods in the preparation of substituted allyl Grignard compounds required for a subsequent cyclization step. Under conventional conditions the allyl halides tend to undergo Wurtz coupling in the presence of magnesium turnings. Thus, the activation of magnesium is necessary to achieve high yields of allylmagnesium halides in the first step, a prerequisite for the success of the complete synthesis process. As seen in Table 8-2 [26c], the application of vapor-activated, anthracene-activated, and Rieke magnesium deliver comparable yields of the final product **7** (Scheme 8-6) obtained by reacting the cyclized Grignard compounds with electrophiles.

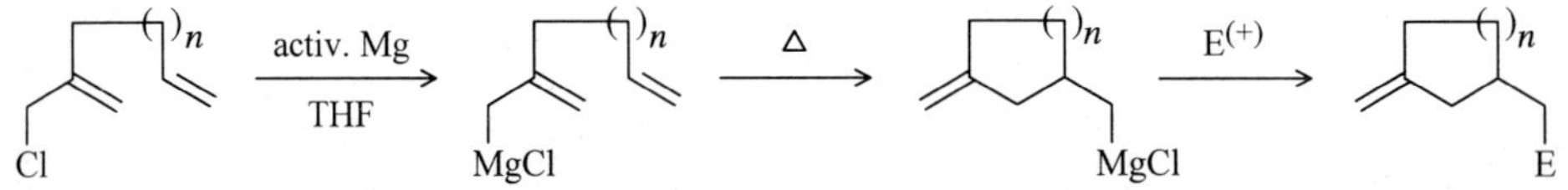

Cl —(activ. Mg, THF)→ MgCl —(Δ)→ MgCl —($E^{(+)}$)→ E **7**

Scheme 8-6

Considering the difficulties in preparing the Grignard reagent from propargyl chloride (cf. Section 8.3.2), it is of interest to be able to catalyze the reaction of propargyl chloride with magnesium by applying the A/MgA system (Scheme 8-7). The reaction is complicated because the primary product of the reaction, allenyl-magnesium chloride (**8**), is unstable and rearranges to **9** and/or **10** [27]. In the presence of 8 mol-% of anthracene, propargyl chloride reacts with commerical Mg powder in THF at 0 °C to give **8** mixed with only small amounts of **9** and **10** [10]. Since propargyl chloride does not react with

$$HC{\equiv}CCH_2Cl + Mg \xrightarrow[0\,°C]{THF\,/\,A} H_2C{=}C{=}CHMgCl \; (\mathbf{8})$$

$$\mathbf{8} \longrightarrow ClMgCH_2C{\equiv}CMgCl \; (\mathbf{9})$$

$$ClMgCH_2C{\equiv}CMgCl \; (\mathbf{9}) \xrightarrow{H_3CC{\equiv}CH} H_3CC{\equiv}CMgCl \; (\mathbf{10})$$

Scheme 8-7

either commercial or active solid magnesium (Mg^*), but does react with MgA to give **8** and anthracene, it is very likely that the generation of **8** takes place via a MgA intermediate.

8.3 Active Magnesium (Mg^*) Powder from MgH_2^* and MgA in Synthetic Chemistry

8.3.1 Preparation and Characterization of Active Magnesium (Mg^*) Powder

A highly reactive, pyrophoric form of magnesium (Mg^*) can be obtained by the dehydrogenation of magnesium hydride, previously prepared via the catalytic hydrogenation of magnesium using homogeneous $TiCl_4$– or $CrCl_3$–MgA catalysts (MgH_2^*; Section 8.2.2.1) [7]. The dehydrogenation of MgH_2^* can be carried out at temperatures above 300 °C under normal pressure, or preferably in vacuo at temperatures above 250 °C [Eq. (8.1)] [6b]. The experimental set-up employed for dehydrogenation of MgH_2^* on a 400 g scale is shown in Fig. 8-2. The experimental procedure is described in detail in [6b]. Mg^* isolated during the dehydrogenation of MgH_2^* in vacuo contains small amounts of C, Ti or Cr, MgO (from the decomposition of THF), and $MgCl_2$. The specific surface area of the Mg^* pro-

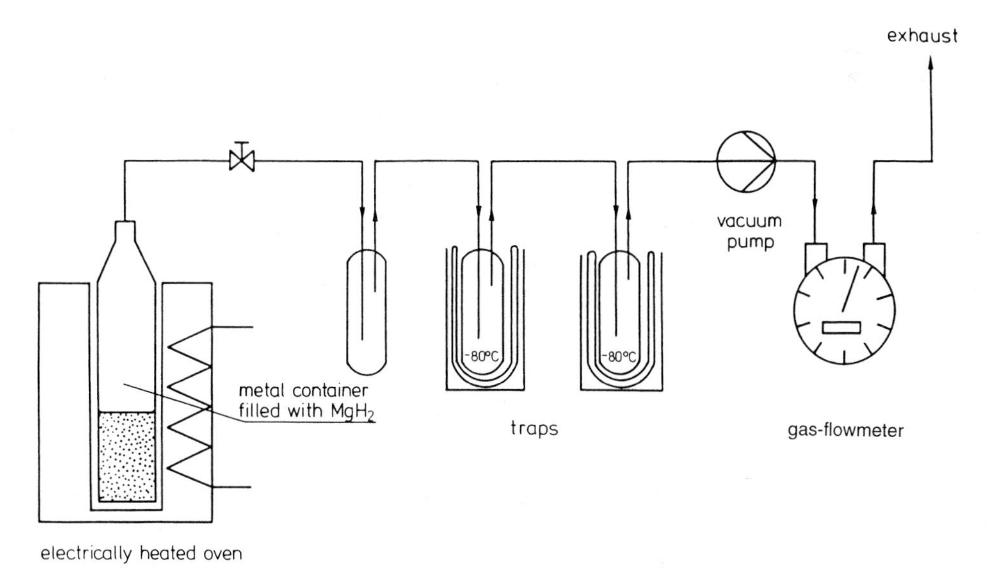

Fig. 8-2. Experimental set-up for dehydrogenation of MgH_2 powder in vacuo [6b].

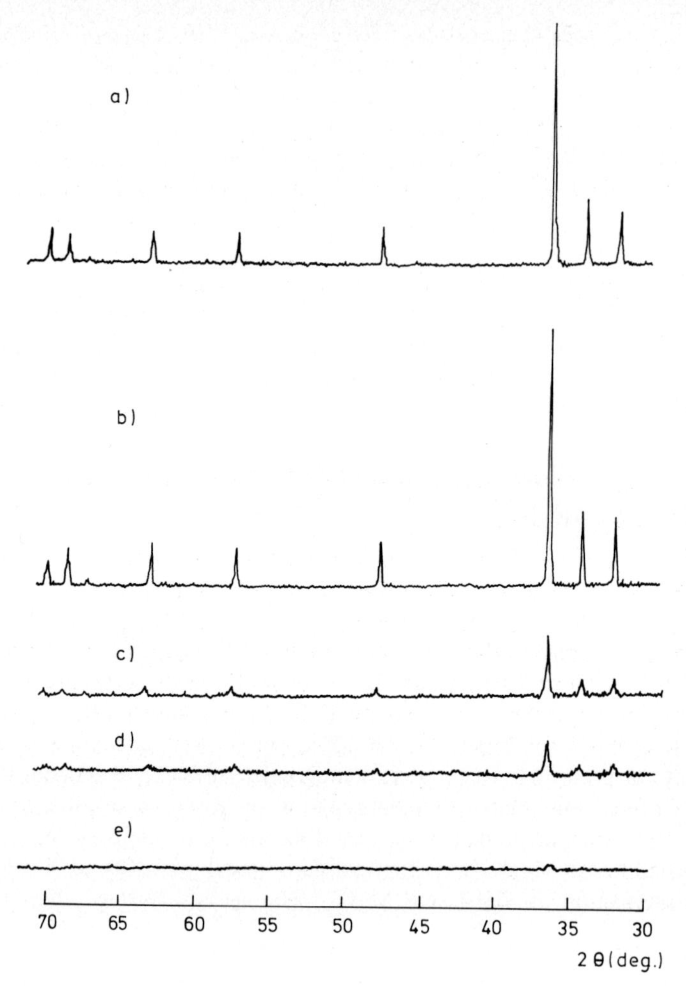

Fig. 8-3. X-ray powder diagrams of various Mg samples (measured in glass capillaries at 0.03°/20 s over 14.8 h). (a) commercial 270 mesh Mg powder; (b) Mg prepared by dehydrogenation of catalytic MgH_2^* in vacuo (Ti catalyst, 60 °C); (c) Mg^* prepared by thermal decomposition of MgA in toluene; (d) Mg^* prepared by thermal decomposition of MgA in vacuo, using MgA prepared at 60 °C; (e) Mg^* prepared by thermal decomposition of MgA in vacuo, using MgA prepared at 20 °C [6b].

duct is dependent on the catalyst used in the initial synthesis of MgH_2^* and ranges from 20 to 60 $m^2 g^{-1}$ [6b].

As a consequence of the finding that the reaction of magnesium with anthracene in THF producing MgA is reversible (Scheme 8-1) [6a], the decomposition of MgA as a method for generating active magnesium (Mg^*) has been investigated in detail [6b]. This process may be accomplished either in solution or by heating solid MgA under vacuum. Stirring a suspension of MgA at room temperature (RT) or with heating in a low-coordinating solvent such as toluene, n-heptane, or ether results in the separation of a bulky, easily filterable, gray precipitate of Mg^* [Eq. 8.10)]. The decomposition of MgA in such solvents can also be effected by ultrasound treatment (US) [6b].

$$\text{MgA} \xrightarrow[\text{heat or US}]{\text{solvent/RT,}} \text{Mg}^* \downarrow + \text{A} + 3\,\text{THF/solvent} \qquad (8.10)$$

After separating the precipitate from anthracene and THF, the Mg^* product can be dried in vacuo and isolated as a gray powder with a specific surface area in the range 13–36 $m^2\,g^{-1}$. The magnesium prepared according to Equation (8.10) does not need to be isolated; it can be used immediately as a Mg^* suspension in the presence of anthracene and THF for chemical reactions, such as a Grignard reaction. In cases where the presence of anthracene and THF is undesirable, Mg^* can be separated from anthracene and THF by filtering and washing, and the wet Mg^* subsequently resuspended in either the same or a different solvent.

A procedure which appears to be particularly suitable for the generation of highly active Mg^* from MgA on a small scale (0.5–1 g) involves the thermal decompositon of solid MgA at 200 °C under high vacuum, whereby THF and anthracene sublime and Mg^* remains as a black pyrophoric powder with a specific surface area between 65 and 109 $m^2\,g^{-1}$ [Eq. (8.11)] [6b].

$$\text{MgA} \xrightarrow[\text{high vacuum}]{150-200\,^\circ\text{C}} \text{Mg}^* + 3\ \text{THF}\uparrow + \text{A}\uparrow \qquad (8.11)$$

The Mg^* samples obtained from the dehydrogenation of MgH_2^* and the decomposition of MgA have been characterized by X-ray powder diffraction (Fig. 8-3). The X-ray powder patterns of Mg^* samples prepared by dehydrogenation of MgH_2^* in vacuo (Fig. 8-3b) look similar to that of commercially available 270 mesh Mg powder (Fig. 8.3a). However, a drastic reduction of diffraction intensity is observed for Mg^* isolated from the decomposition of MgA in toluene (Fig. 8.3c) and in particular for Mg^* samples prepared from the decomposition of MgA in vacuo (Fig. 8.3d, e) [6b]. This suggests that the magnesium particle size in these samples is becoming progressively smaller ($\leq$ 50 Å) [28].

As a measure of the relative chemical reactivity of the various Mg^* samples, especially in the context of their application in Mg/MgH_2 systems as hydrogen- and heat-storage media [23], the rate and extent of their hydrogenation at 240 °C and ambient pressure was studied (Fig. 8-4) [6b]. The highest reactivity with respect to hydrogenation

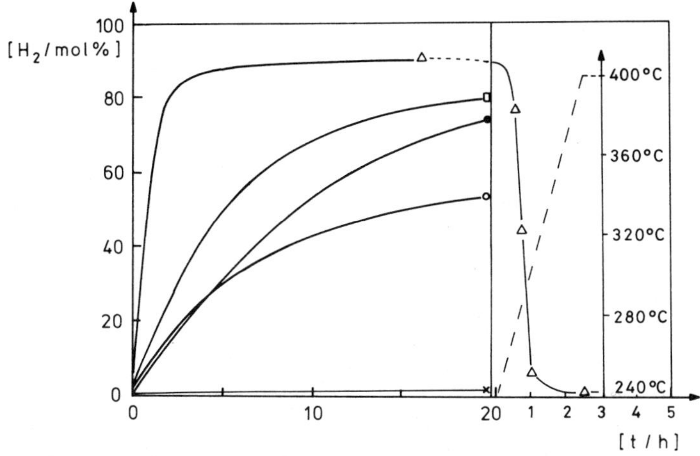

Fig. 8-4. Absorption of hydrogen by various samples of active magnesium (Mg^*) at 240 °C/1 bar H_2. —△— Mg^* prepared by thermal decomposition of MgA in vacuo, using MgA prepared at 20 °C; —□— Mg^* prepared by dehydrogenation of catalytic MgH_2^* in vacuo (Ti catalyst, 60 °C); —●— Mg^* prepared by thermal decomposition of magnesium butadiene in toluene; —○— Mg^* prepared by thermal decomposition of MgA in toluene; —×— Mg^* prepared by dehydrogenation of non-catalytic MgH_2 in vacuo [6b].

was exhibited by the Mg* sample prepared by the thermal decomposition of solid MgA in vacuo [Eq. (8-11)] followed by that of Mg* obtained by dehydrogenating MgH_2^* [Eq. (8.1)]. A Mg sample produced via the dehydrogenation (in vacuo) of non-catalytically prepared MgH_2 revealed virtually no uptake of hydrogen (Fig. 8-4).

These results demonstrate that even among "active Mg" samples great differences in physical properties as well as chemical reactivity exist which are dependent on the initial synthetic route and reaction conditions. Thus, with respect to the application of such active Mg* materials as reagents for synthesis chemistry (see Section 8.3.2), their "preparative history" must be strictly defined in order to achieve reproducible results.

8.3.2 Applications of Mg* Powder in Organic and Organometallic Synthesis

The known applications of Mg* powder from MgH_2^* and MgA in syntheses can be classified according to three substrate types and their respective reactions:

a) organic substrates, insertion of Mg* into carbon–halogen bonds (preparation of Grignard reagents) or carbon–oxygen bonds (ether cleavage reaction);
b) halides of main group elements (B, Si, P), dehalogenation and Mg* insertion;
c) transition metal halides, reduction and the generation of inorganic Grignard reagents through a formal Mg* insertion into the transition metal–halogen bond. (The generation of inorganic Grignard reagents will be presented in Section 8.5).

The preparation of Grignard compounds using Mg* has been studied intensively [6b]. By applying Mg*, isolated from the dehydrogenation of MgH_2^* in vacuo [Eq. (8.1)], the production of alkyl, allyl and aryl Grignard compounds under mild conditions (e.g. RT) can be accomplished in higher yields than by reactions using commercially available magnesium. Furthermore the synthesis can be carried out in various solvents, including toluene to which a small amount of THF (Mg:THF = 1:3) has been added.

Active magnesium, Mg*, generated in situ from MgA in toluene or *n*-heptane [Eq. (8.10); presence of A and ~ 3 THF/RMgX] has been found to be particularly useful for the preparation of Grignard compounds under mild conditions (RT or below) with high yields. This synthetic route is especially suitable for small-scale experiments (for details see [6b]). Reactions of this form of Mg* with organic halides (all those studied except propargyl chloride) lead to high yields of the respective Grignard compound, some of which are known to be difficult to prepare via other methods. Although the reaction between Mg* and propargyl chloride is unsuccessful, the appropriate Grignard reagent can be prepared by reacting the chloride with MgA or commercial Mg powder in the presence of catalytic amounts of anthracene in THF (Sections 8.2.1 and 8.2.2.3).

Isobutyl-, *sec*-butyl-, and *tert*-butyl chlorides do not react with non-activated magnesium in hydrocarbons [29]. In contrast, isobutyl chloride does react with the active magnesium prepared by thermal decomposition of solid MgA in vacuum [Eq. (8.11)] in toluene or n-heptane at 70 °C to produce isobutylmagnesium chloride, but only in low yields. Treatment of *p*-dichlorobenzene with the same type of Mg* in the molar ratio 1:2 in THF affords a mixture of the mono- and di-Grignard compounds. Comparable results have also been obtained by employing Mg* prepared from the reaction between $MgCl_2$ and K in the presence of KI (Rieke procedure) [30] when the molar ratio p-$C_6H_4Cl_2$:Mg* is 1:4.

The "Rieke magnesium" is also active in carbon–oxygen cleavage. Especially interesting is the cleavage of THF, first described by Bickelhaupt et al. [31], which results in the formation of 1-magnesia-2-oxacyclohexane (Scheme 8-8). The same ether cleavage reaction can also be achieved using Mg^* prepared according to Equations (8.1) and (8.11), [6b] and even higher yields can be achieved with Ti catalysts [32, 33].

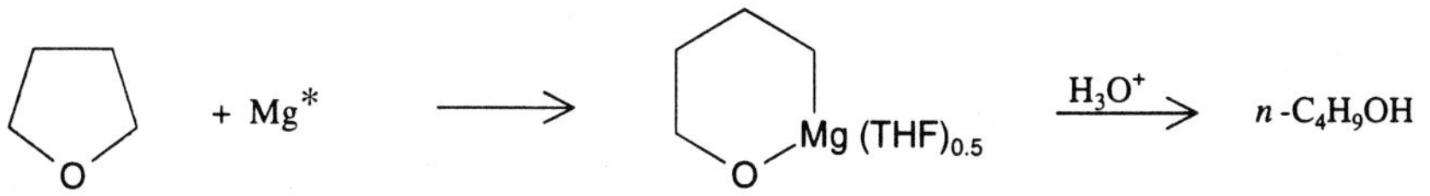

Scheme 8-8

The application of Mg^* is of particular interest in organoboron, organosilicon, and organophosphorous chemistry. By dehalogenating the organoboron compounds **11a, b** with Mg^* [from MgA, Eq. 8.10)] Berndt et al. [34] were able to synthesize the crystalline compounds **12a, b** (Scheme 8-9) which have non-classical structures. Temperature dependent ^{11}B and ^{13}C NMR spectra of the compounds show that in solution they exist in rapid equilibrium with the classical forms **13a, b**. These represent the first evidence of equilibria existing between non-classical and classical boron compounds (**12a, b** ⇌ **13a, b**). Using the same type of Mg^*, open-chain organoboron compounds (**14a–c**) have been transformed into cyclic, non-classical boron compounds **15a–c** (Scheme 8-10) [35].

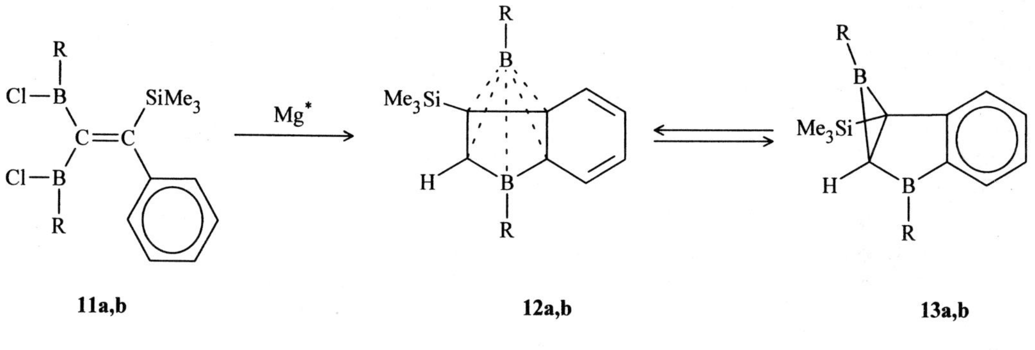

a: R = *t*Bu ; **b**: R = mesityl

Scheme 8-9

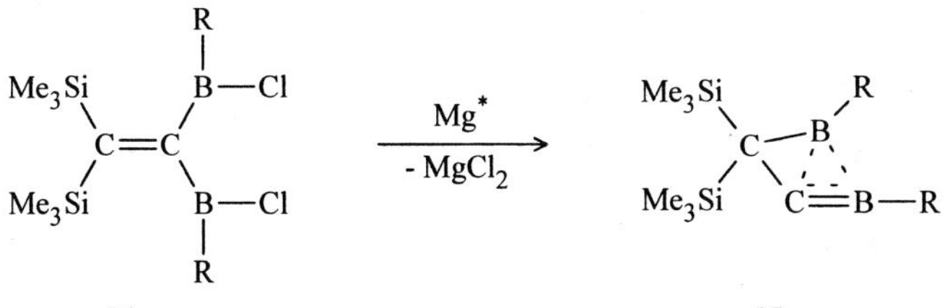

a: R = mesityl **b,c**: R = duryl **c**: Me_3Ge in place of Me_3Si

Scheme 8-10

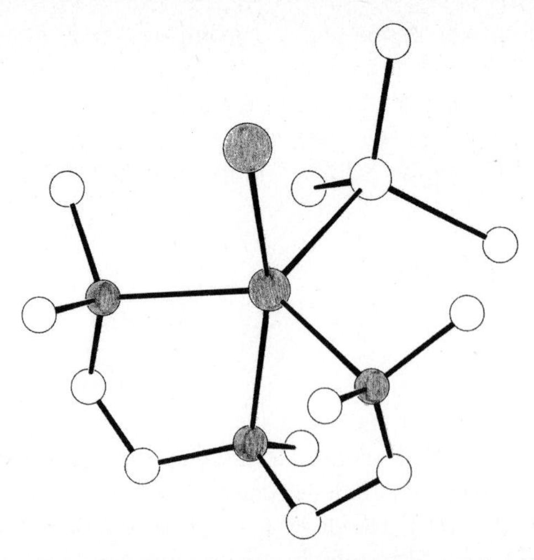

Fig. 8-5. Crystal structure of $Me_3SiMgBr[Me_2NC_2H_4N(Me)C_2H_4NMe_2]$ (**16a**) [36].

Up to now, efforts to obtain an "organosilyl Grignard compound" through the reaction of a silicon halide with magnesium have been unsuccessful. However, by reacting trimethylsilyl bromide with Mg^* (from MgH_2^*) in toluene in the presence of tetramethylethylenediamine (TMEDA) it has proved possible to prepare a trimethylsilylmagnesium bromide-TMEDA adduct (**16**) in about 40% yield together with a minor amount of bis(trimethylsilyl)magnesium–TMEDA [**17**, Eq. (8.12)]. The structure of the analogously prepared complex $Me_3SiMgBr \cdot [Me_2NC_2H_4N(Me)C_2H_4NMe_2]$ (**16a**, Fig. 8-5) has been confirmed by X-ray crystallography [36].

$$Me_3SiBr + Mg^* + \underset{\text{TMEDA}}{Me_2NC_2H_4NMe_2} \xrightarrow{\text{tol.}} \underset{\mathbf{16}}{Me_3SiMgBr \cdot TMEDA} + \underset{\mathbf{17}}{(Me_3Si)_2Mg \cdot TMEDA} \quad (8.12)$$

An initial discouraging result in the field of organophosphorous chemistry eventually led Wilke et al. to the discovery of an extremely efficient chiral ligand for catalytic asymmetric syntheses. Previous experiments had shown that Mg^* (from MgH_2^*) could be employed advantageously for the reduction of halogenophospholenium halides (**18a, b**) to the corresponding phospholenes **19a, b** (Scheme 8-11) [37]. With the goal of constructing the most rigid chiral ligand conceivable for asymmetric catalysis, the synthesis of the chiral azaphospholene **21** according to Scheme 8-12, was proposed. However, when the bromoazaphospholenium bromide **20** was reduced with Mg^*, instead of

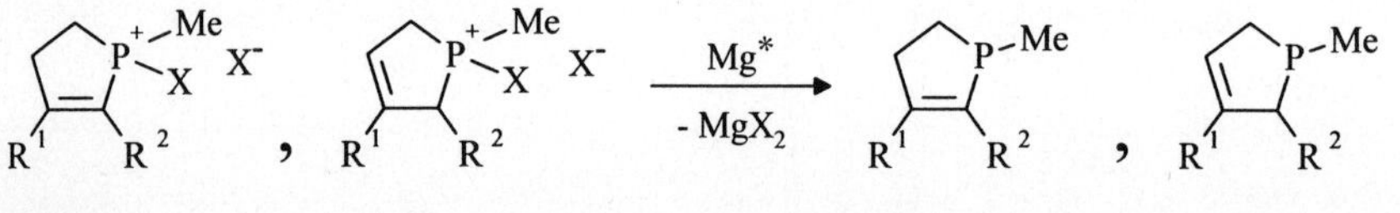

18a, **18b** $\xrightarrow[-\,MgX_2]{Mg^*}$ **19a**, **19b**

Scheme 8-11

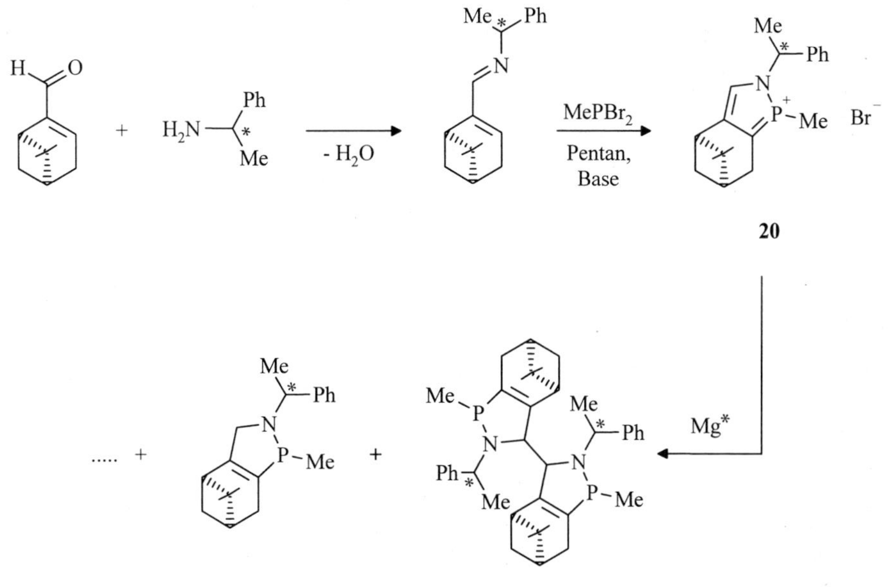

* *R*-Configuration 21 22

Scheme 8-12

the expected **21**, a complicated mixture of a number of unknown phosphorous compounds (as verified by ^{31}P NMR) was isolated [38]. Even though the reaction was a "failure", the product mixture was applied as a chiral modifier for the homogeneous nickel catalyst in a subsequent catalytic hydrovinylation experiment at –70 °C. An enormous catalytic activity, with an almost 100% codimerization selectivity and ~50% *ee*, was observed! In further investigations it was found that the mixture contained a small amount of the previously unknown dimer **22** which was responsible for the extraordinary catalytic activity observed. **22** acts as a chiral ligand which imparts high activity, codimerization selectivity and enantioselectivity to homogeneous nickel catalysts used in hydrovinylation of cyclopentadiene, norbornene, and especially of styrene and its derivatives (*ee*-values of 85–95%) [39].

Mg^* (from MgH_2^*) can also be used as a reducing agent in the synthesis of organotransition metal complexes. According to Jolly et al., the reduction of chromium(+2) [40] or iron(+2) halide phosphine complexes [41] with Mg^* in the presence of conjugated dienes or alkynes provides an easy access to their respective zerovalent complexes. Examples are given in Schemes 8-13 and 8-14. With Mg^* serving as the reducing agent, η^3-allyl metal complexes can be prepared via two routes: an oxidative coupling process in-

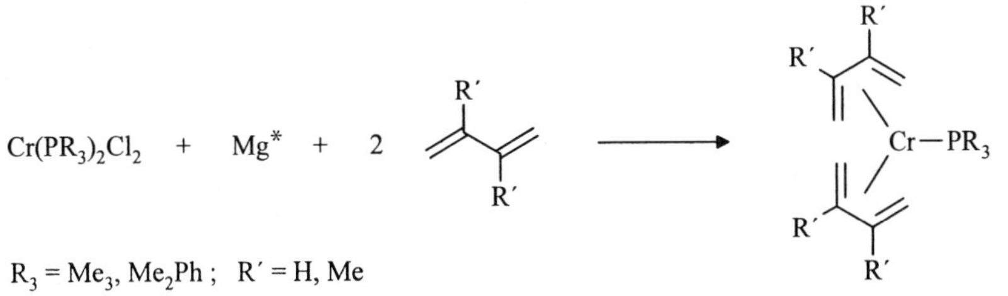

$Cr(PR_3)_2Cl_2$ + Mg^* + 2 [diene] ⟶ [Cr–PR_3 complex]

R_3 = Me_3, Me_2Ph ; R′ = H, Me

Scheme 8-13

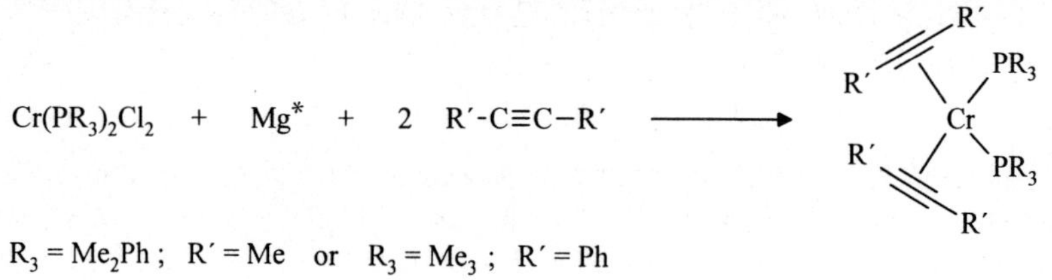

$Cr(PR_3)_2Cl_2$ + Mg^* + 2 R′-C≡C−R′ ⟶ (R′C≡CR′)$_2$Cr(PR_3)$_2$

$R_3 = Me_2Ph$; R′ = Me or $R_3 = Me_3$; R′ = Ph

Scheme 8-14

volving η^5-cyclopentadienyl metal halide complexes and a conjugated diene [42], or the reaction between metal halide–phosphine complexes and allylic halides [41].

While investigating the intermediates generated during the cyclotrimerization of 2-butyne using low-valent chromium catalysts [43], Wilke et al. recently isolated from the reaction of Cp^*CrCl_2 with 2-butyne and Mg^* the highly interesting bimetallic chromacyclopentadiene (**23**) and chromacycloheptatriene (**24**) complexes (Scheme 8-15). These complexes are characterized by fly-over ligands [44]. The reaction gives either **23** or a mixture of **23** and **24**, depending on the solvent used (THF or Et_2O). Complex **23** is prone to add an additional equivalent of Mg^* (or Li), so that after hydrolysis the bimetallic η^3-allylic complex **25** is generated. The monomeric η^5-cyclohexadienylchromium complex **26** is produced when MgH_2^* is employed as the reducing agent instead of Mg^*. Under drastic conditions, both **23** and **24** react with 2-butyne to give the hexamethylbenzene complex **27**. **23** and **24** can therefore be regarded as intermediates in the cyclo-trimerization of 2-butyne to form hexamethylbenzene.

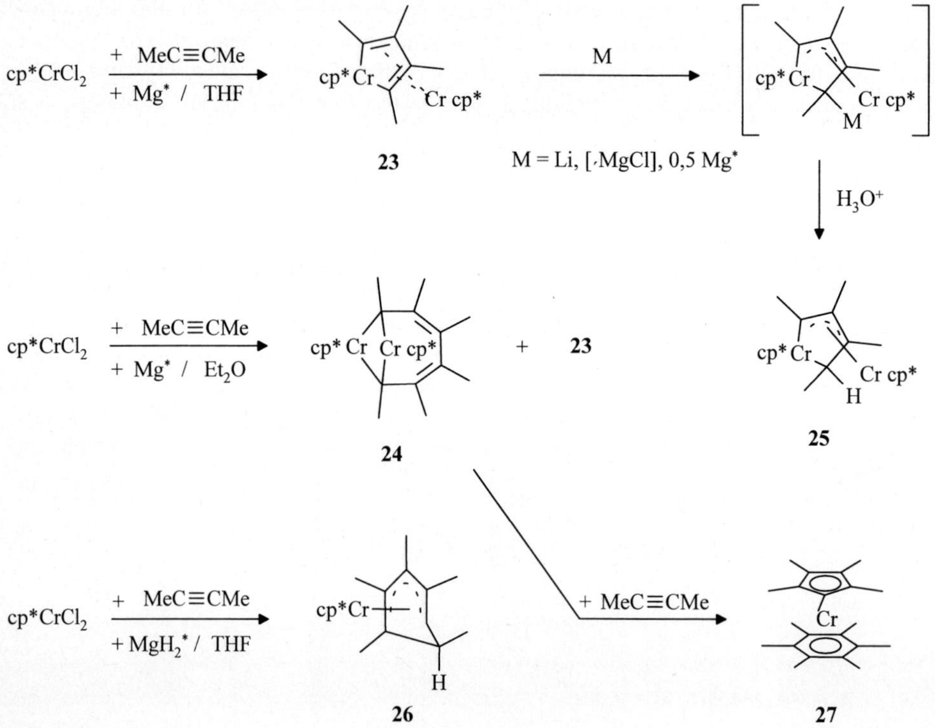

cp*$CrCl_2$ —(+ MeC≡CMe, + Mg^* / THF)→ **23** —(M)→ [intermediate with M] —(H_3O^+)→ **25**

M = Li, [·MgCl], 0,5 Mg^*

cp*$CrCl_2$ —(+ MeC≡CMe, + Mg^* / Et_2O)→ **24** + **23**

cp*$CrCl_2$ —(+ MeC≡CMe, + MgH_2^* / THF)→ **26**

24 —(+ MeC≡CMe)→ **27**

Scheme 8-15

8.4 Wet-Chemical Routes to Magnesium Intermetallics and/or their Hydrides and Carbides

Our research embarked on a new era of development after we realized that Mg* or active Mg compounds such as MgH_2^* or MgA could be utilized to generate *bimetallic* systems, namely Mg intermetallics and/or their hydrides and carbides, through the reaction with an appropriate metal compound in an organic solvent at low temperature. The resulting intermetallic systems are isolated in a highly active, usually X-ray amorphous form [45].

8.4.1 The First Route: Reaction of MgH_2^* or Et_2Mg with Bis(allyl)-Metal Complexes

The production of Mg intermetallics and their hydrides and carbides was initially discovered during experiments aimed at doping MgH_2^* samples with various transition metal compounds in order to improve the dehydrogenation/hydrogenation properties for hydrogen- or heat storage materials [Eq. 8.1)] [45, 46]. When MgH_2^* was doped with Ni or Pd via the reaction with small amounts of bis(η^3-allyl)nickel or bis(η^3-allyl)palladium in THF or toluene, evolution of propene gas was noted at or even below room temperature [46], which indicated that some type of unexpected reaction was occurring. Investigations involving stoichiometric reactions of MgH_2^* with bis(allyl)-metal compounds of Ni, Pd, Pt and Zn in organic solvents revealed, along with propene evolution, the formation of the corresponding Mg intermetallics, ternary Mg hydrides, or carbides in a highly active form [45, 47]. In such stoichiometric reactions with bis(allyl)metal complexes, MgH_2 prepared by non-catalytic hydrogenation of magnesium shows very little reactivity, which underlines the high activity of MgH_2^* [7].

Pathways leading to Mg–Ni intermetallics, hydrides, and carbides from MgH_2^* or HMgCl (**6**) and bis(η^3-allyl)nickel (**28**) (or its homologs) are summarized in Scheme 8-16 [47a]. Depending on the molar ratios of the reactants, finely divided solids that are X-ray amorphous and pyrophoric, with Mg/Ni ratios of 2:1 or 1:1, could be obtained. After

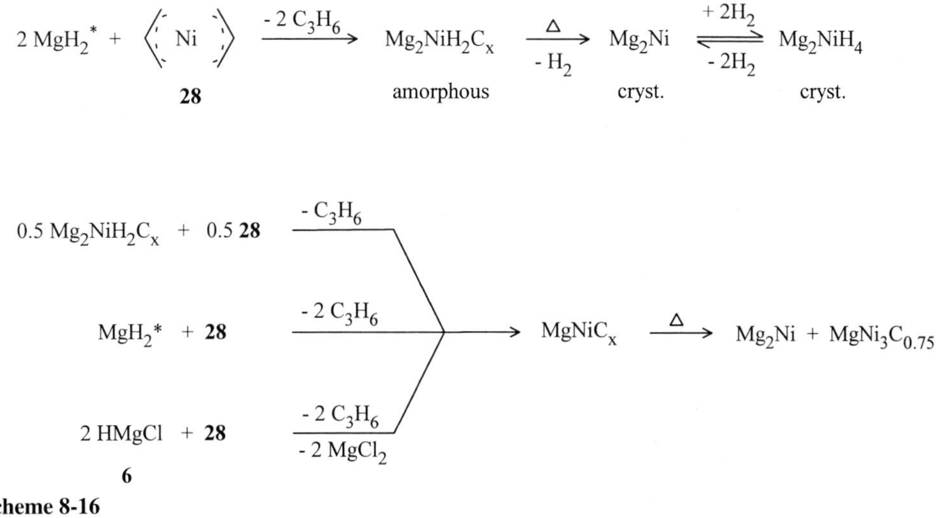

$$2\,MgH_2^* + \underset{\mathbf{28}}{(\eta^3\text{-}C_3H_5)_2Ni} \xrightarrow{-2\,C_3H_6} \underset{\text{amorphous}}{Mg_2NiH_2C_x} \xrightarrow[-H_2]{\Delta} \underset{\text{cryst.}}{Mg_2Ni} \underset{-2H_2}{\overset{+2H_2}{\rightleftharpoons}} \underset{\text{cryst.}}{Mg_2NiH_4}$$

$$0.5\,Mg_2NiH_2C_x + 0.5\,\mathbf{28} \xrightarrow{-C_3H_6} MgNiC_x$$

$$MgH_2^* + \mathbf{28} \xrightarrow{-2\,C_3H_6} MgNiC_x \xrightarrow{\Delta} Mg_2Ni + MgNi_3C_{0.75}$$

$$\underset{\mathbf{6}}{2\,HMgCl} + \mathbf{28} \xrightarrow[-2\,MgCl_2]{-2\,C_3H_6} MgNiC_x$$

Scheme 8-16

thermal treatment and a subsequent hydrogenation/dehydrogenation cycle, the amorphous product $Mg_2NiH_2C_x$ is transformed into a crystalline material corresponding to the well-known Mg_2Ni intermetallic, which reversibly absorbs hydrogen and thus can be used as a hydrogen storage material [48]. Previously, Mg_2Ni has been only prepared by metallurgical methods. The advantages of the material prepared via such a wet-chemical route include high reactivity and dispersion. The X-ray amorphous 1:1 solid $MgNiC_x$ (whose structure is not known) disproportionates to Mg_2Ni and the ternary carbide $MgNi_3C_{0.75}$ upon annealing to 690 °C [49].

Reactions of bis(η^3-allyl)palladium (**29**) with MgH_2^* or Et_2Mg lead to three new intermetallics: Mg_2PdC_y, its hydride $Mg_2PdC_yH_{1-1.4}$, and an amorphous material $MgPdC_wH_x$ (**30**, Scheme 8-17) [47b, c]. The novel crystalline carbide Mg_2PdC_y was initially isolated during the dehydrogenation of the amorphous material $Mg_2PdH_2C_y$, prepared from **29** and MgH_2* (molar ratio 1:2) in THF. The structure of Mg_2PdC_y was elucidated by X-ray powder diffraction. It crystallizes in a face-centered cubic cell and possesses the Ti_2Ni type structure with interstitial carbon. To determine whether the inclusion of carbon stabilizes the formation of this new Mg–Pd alloy, elemental Mg and Pd were heated together in the absence and presence of graphite. The production of the Mg_2Pd intermetallic is clearly favored upon the addition of graphite [47d]. In the presence of hydrogen, Mg_2PdC_y absorbs hydrogen reversibly(!) to give an intermetallic hydride whose structure is not yet determined. The reversible metal hydride/metal system is described in Scheme 8-17.

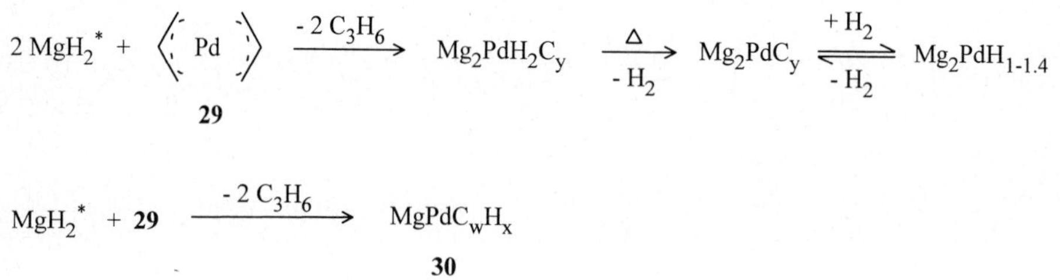

$$2\,MgH_2^* + \mathbf{29} \xrightarrow{-2\,C_3H_6} Mg_2PdH_2C_y \xrightarrow[-H_2]{\Delta} Mg_2PdC_y \underset{-H_2}{\overset{+H_2}{\rightleftharpoons}} Mg_2PdH_{1-1.4}$$

$$MgH_2^* + \mathbf{29} \xrightarrow{-2\,C_3H_6} \underset{\mathbf{30}}{MgPdC_wH_x}$$

Scheme 8-17

The reaction between **29** and MgH_2* in a 1:1 molar ratio in THF at RT and the similar reaction between **29** and Et_2Mg in a 1:1.5 molar ratio produce pyrophoric, X-ray amorphous materials with the general compositon $MgPdC_wH_x$ (**30**, Schemes 8-17 and 8-18). Surprisingly, the products remain amorphous even after annealing at 650 °C. By performing the reaction of **29** with Et_2Mg (molar ratio 1:1.5) at low temperatures in d_8-THF, it was demonstrated by ^1H and ^{13}C NMR spectroscopy that the formation of **30** proceeds through a series of organo-bimetallic intermediates (Scheme 8-18) [47b]. At –60 °C the η^1-η^3-bis(allyl)palladium complex **31** is formed as one ethyl group from Et_2Mg is transferred to **29**. After warming to –10 °C, an exchange of ethyl and allyl groups between Mg and Pd follows to give **32**. The decomposition (at RT) of the third proposed intermediate, **33**, via β-hydrogen and reductive elimination, produces 2 moles of ethene, 1 mole each of ethane and propene and the final solid product **30**. Although **31** and **32** are shown as being ionic, on the basis of the NMR data the presence of bridging alkyl groups or covalent Mg–Pd bonds cannot be ruled out. These results represent the first observation of discrete organo-bimetallic complexes which undergo β-hydrogen and/or reductive elimination to produce intermetallics or intermetallic carbides. It is reasonable to assume that the generation of other intermetallics, hydrides, and carbides from MgH_2^*,

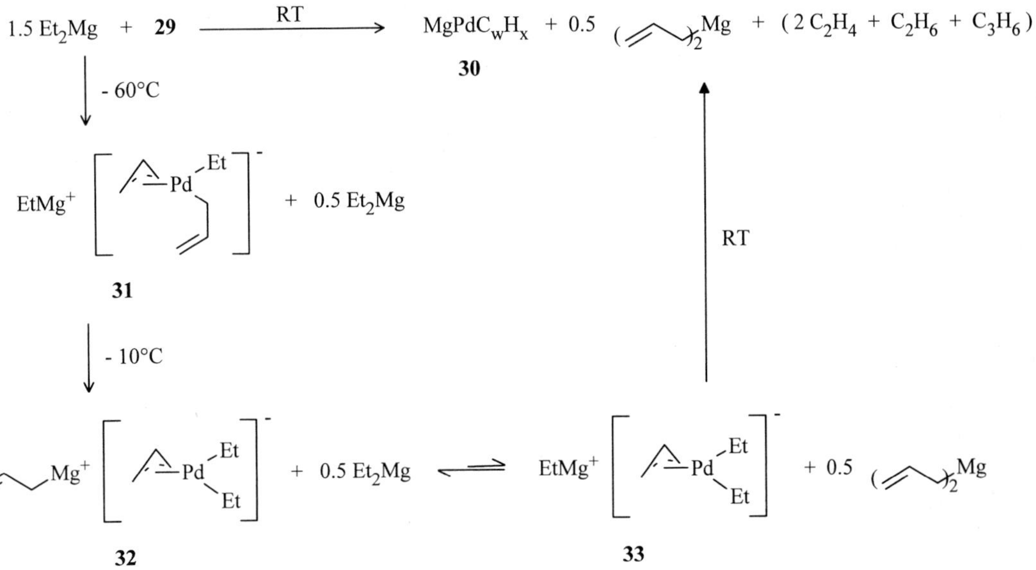

$1.5\ Et_2Mg + \mathbf{29} \xrightarrow{RT} MgPdC_wH_x\ (\mathbf{30}) + 0.5\ (C_3H_5)_2Mg + (2\ C_2H_4 + C_2H_6 + C_3H_6)$

- 60°C: $EtMg^+ [(\eta^3\text{-}C_3H_5)Pd(Et)(\eta^1\text{-}C_3H_5)]^-$ (**31**) + $0.5\ Et_2Mg$

- 10°C: $(C_3H_5)Mg^+ [(\eta^3\text{-}C_3H_5)Pd(Et)_2]^-$ (**32**) + $0.5\ Et_2Mg$ ⇌ $EtMg^+ [(\eta^3\text{-}C_3H_5)Pd(Et)_2]^-$ (**33**) + $0.5\ (C_3H_5)_2Mg$

RT

Scheme 8-18

Mg-alkyls or other metal hydrides and metal-allyl complexes also involve the initial formation of organo-bimetallic complexes analogous to **31**, **32**, and **33**.

The fact that **30** does not react with hydrogen at either normal or elevated pressures indicates that significant amounts of free, finely-divided Mg or Pd (both of which react readily with hydrogen) are not present in the material. Interestingly, **30** could not be crystallized by annealing or hydrogenation. (The known alloys with compositions close to 1:1 are MgPd and $Mg_{0.9}Pd_{1.1}$). It was shown by X-ray absorption spectroscopy that **30** is a Mg–Pd intermetallic and in fact a carbide [47e]. An EXAFS study comparing the crystalline MgPd and $Mg_{0.9}Pd_{1.1}$ alloys as well as $PdC_{0.15}$ and Pd metal showed that the local environment of Pd in **30** closely resembles that of Pd in the alloy $Mg_{0.9}Pd_{1.1}$ (HgMn structural type). The presence of interstitial carbon was definitely confirmed. The location of the carbon in the Mg–Pd framework, based on the EXAFS results, would seem to block any long-range ordering of the metallic sublattice with close contacts between the carbon and the metals. Since carbon (as well as hydrogen) is retained during annealing, the unfavorable conditions for formation of short-range carbon–metal interactions may explain the failure of this system to crystallize even after prolonged heating at 650 °C. (For further discussion concerning the influence of carbon and hydrogen on the formation and structures of Mg–Pd, Mg–Pt, and Mg–Rh systems, see Section 8.4.2.2 and [47e, f].)

Through this wet-chemical synthetic route, a novel Mg–Pt intermetallic not previously obtained via metallurgical techniques was isolated and structurally characterized by X-ray powder diffraction [45, 47f]. (For other new Mg–Pt intermetallics see also Section 8.4.2.2.) The reaction of MgH_2^* with bis(η^3-allyl)platinum (**34**) in a molar ratio of 2:1 in THF proceeds analogously to the reactions with bis(η^3-allyl)nickel or -palladium (Schemes 8-16 and 8-17), i.e. with the evolution of propene and the precipitation of an amorphous Mg–Pt hydride, $Mg_2PtH_2C_x$ (Scheme 8-19). Dehydrogenation of the latter and subsequent annealing affords, as a main product, a tetragonal intermetallic $Mg_2PtC_xH_y$ possessing the Al_2Cu structural type, here designated as t-$Mg_2PtC_xH_y$. The X-ray diffraction pattern reveals that

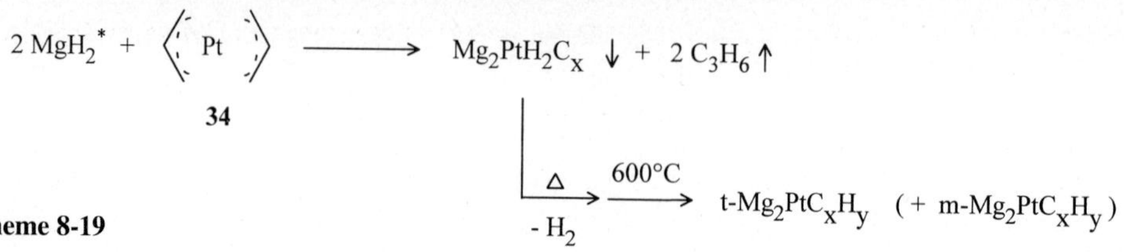

$$2\,MgH_2^* + \text{Pt}(\eta^3\text{-allyl})_2\ (\mathbf{34}) \longrightarrow Mg_2PtH_2C_x \downarrow + 2\,C_3H_6 \uparrow$$

$$Mg_2PtH_2C_x \xrightarrow[-H_2]{\Delta,\ 600°C} t\text{-}Mg_2PtC_xH_y\ (+\ m\text{-}Mg_2PtC_xH_y)$$

Scheme 8-19

the product contains a small amount of yet another unknown Mg–Pt phase, which was later identified as a monoclinic compound, m-$Mg_2PtC_xH_y$. Better synthetic approaches to this second 2:1 phase involve the reaction of $PtCl_2$ with either MgH_2^* or MgH_2' (Section 8.4.2.2). In contrast to Mg_2Ni and Mg_2PdC_y systems (Schemes 8-16 and 8-17), t- and m-Mg_2PtC_x do not react with hydrogen [47f].

The reaction between MgH_2^* and bis(η^1-allyl)zinc (Scheme 8-20) represents a reaction type different from that of MgH_2^* with the bis(η^3-allyl)metal complexes of Ni, Pd, and Pt. The reaction involving zinc is reminescent of the reaction of Et_2Mg with bis(η^3-allyl)palladium (Scheme 8-18) [45]. Although the reactants were employed in a 1:1 molar ratio, the precipitated Mg–Zn hydride was found to contain only one half of the total magnesium present. The remaining half was transformed into the THF-soluble bis(η^1-allyl)magnesium complex via exchange of the allyl and hydride ligands between Mg and Zn. The amorphous hydride $MgZn_2H_2C_y$ crystallizes upon heating at 300 °C and corresponds to the known intermetallic compound $MgZn_2$ [50].

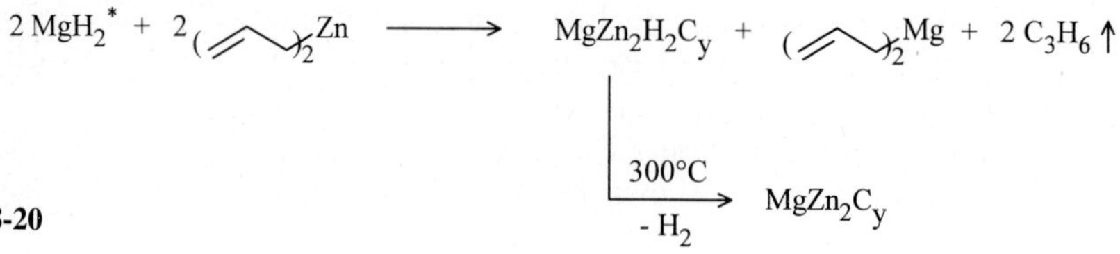

$$2\,MgH_2^* + 2\,(C_3H_5)_2Zn \longrightarrow MgZn_2H_2C_y + (C_3H_5)_2Mg + 2\,C_3H_6 \uparrow$$

$$MgZn_2H_2C_y \xrightarrow[-H_2]{300°C} MgZn_2C_y$$

Scheme 8-20

At this point it should be stressed that this wet-chemical route is not limited to the preparation of *magnesium* intermetallic systems. We have found that hydrides of Li, Ca, Ba, B and Si react with the bis(η^3-allyl)metal complexes of Ni, Pd, and Pt in organic solvents yielding precipitates of the corresponding active intermetallics or their hydrides together with propene gas [45]. A further interesting example is the preparation of a Pt-doped UH_3/U system as described by Mimoun et al. [51]. The reaction of the solid hydride UH_3 with a small amount (3 mol-%) of bis(η^3-methallyl)platinum in boiling toluene proceeds with the evolution of isobutene and the incorporation of Pt into the solid (Scheme 8-21). After thermal dehydrogenation, the resulting Pt-doped active uranium (U^*) can be applied as a reversible hydrogen sponge for dehydrogenation of several cycloalkanes to the corresponding aromatic hydrocarbons, with active Pt–U sites on the surface of the sponge acting as the catalyst. The aromatization of cycloalkanes can thus be ef-

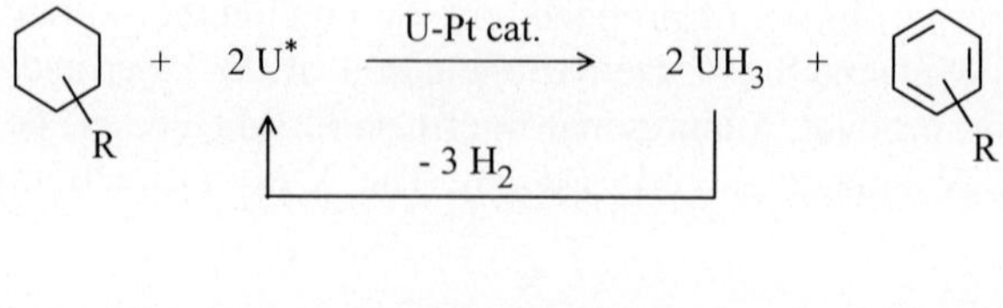

$$\text{R-cyclohexane} + 2\,U^* \xrightarrow{\text{U-Pt cat.}} 2\,UH_3 + \text{R-benzene} \qquad (2\,UH_3 \xrightarrow{-3\,H_2} 2\,U^*)$$

Scheme 8-21

fected under mild conditions which leads to an increased selectivity of these reactions and allows the recovery of pure hydrogen.

Although up to now little investigated, the active intermetallics and their hydrides or carbides, prepared via the wet-chemical routes described here and in the following sections, have promising potential as heterogeneous catalysts and chemical reagents. For instance, the amorphous materials $Mg_2NiC_xH_2$ and $MgNiC_xH_y$ (Scheme 8-16) [47a] as well as the similarly prepared amorphous materials $BaNiC_xH_y$, $BNiC_xH_y$, $B_2NiC_xH_y$, and $SiNiC_xH_y$, were tested as catalysts for the hydrogenation of a mixture of 1-heptene and cyclohexene (Ni:1-heptene = 1:40; 1 bar H_2, RT) and compared with the conventional Raney nickel catalyst. $Mg_2NiC_xH_2$ proved to be the best hydrogenation catalyst, having an activity about one order of magnitude higher than that of Raney nickel. $BNiC_xH_y$ and $B_2NiC_xH_y$ (specific surface areas of 150–160 m^2 g^{-1}) as well as $SiNiC_xH_y$ were about six times as active as Raney nickel [45]. In a different screening test, MgH_2^* was reacted with a minute amount of bis(η^3-allyl)palladium in toluene (Mg:Pd = 1:0.0004) to deposit "Mg_2PdH_2" onto the surface of MgH_2^*. The resulting heterogeneous catalyst promoted rapid stereoselective *cis*-hydrogenation of 3-hexyne (Pd:3-hexyne = 1:12000, 1 bar H_2, RT), yielding quantitatively *cis*-3-hexene of 96% purity [45].

8.4.2 The Second Route: Reaction of Metal Halides with Excess Amounts of Magnesium Hydrides, Organomagnesium Compounds, or Magnesium

A second wet-chemical route to magnesium intermetallic systems is based on the use of readily available metal halides (M^1X_n) instead of bis(allyl)-metal complexes, and can thus be viewed as a significant development. As the reducing agent a series of magnesium-containing compounds can be employed. Particularly effective reagents have proven to be MgH_2^*, MgH_2', Et_2Mg, MgA, anthracene-activated Mg, Mg*, and even in some cases ordinary Mg powder (270 mesh). The reactions were typically carried out in THF [47f, 52].

The essential feature of this novel synthetic route is that the reaction of M^1X_n is performed in an *excess* of the "Mg reagent" in relation to the amount necessary to reduce the metal in M^1X_n to its metallic state or to transform it into a metal hydride. Under such conditions this synthetic route yields two classes of products. One type consists of solid-state Mg intermetallics and their hydrides and carbides, which are the subject of the present section. Products of the second type are observed during the reaction of certain transition metal halides, whereby dark-colored THF solutions are generated containing significant concentrations of the corresponding transition metal, Mg, and halogen [52]. Subsequent investigations have shown that these solutions contain THF-soluble intermetallic species which may be regarded as inorganic Grignard analogues [33, 53]. The latter products and their applications as inorganic Grignard *reagents* will be the topic of Section 8.5.

The production of solid Mg intermetallic systems via this synthetic route can be envisaged as occuring in two stages. In the first stages M^1X_n is transformed either by a Mg hydride reagent to M^1 hydride or by MgR_2 or Mg to the active metal M* (Schemes 8-22 to 8-24, 1st step). Subsequently, during the second step, the M^1 hydride or M^{1*} *in its nascent form* reacts further with the remaining magnesium reagent present in the system (*q* mol) to give a Mg intermetallic or its hydride (Schemes 8-22 to 8-24, 2nd step). In some cases (see below) the synthesis of Mg intermetallics according to Schemes 8-22 to 8-24 is actually carried out in two independent experimental stages. The bimetallic hydride products (Scheme 8-22) can subsequently be thermally dehydrogenated to give the

$$M^1X_n + \frac{n}{2} MgH_2 \xrightarrow{1^{st}\ step} M^1H_{n-p} + \frac{n}{2} MgX_2 + \frac{p}{2} H_2\uparrow$$

$$0 \leq p \leq n$$

$$\xrightarrow[+\,q\,MgH_2]{2^{nd}\ step} \underbrace{Mg_qM^1H_{n-p+2q}} + \frac{n}{2} MgX_2$$

$$\xrightarrow{\Delta} Mg_qM^1 + \left(\frac{n-p}{2} + q\right) H_2\uparrow$$

Scheme 8-22

$$M^1X_n + \frac{n}{2} MgR_2 \xrightarrow{1^{st}\ step} M^{1*} + MgX_2 + \frac{n}{2}\ [(R + H) + (R - H)]\ \left(\text{or } \frac{n}{2}\ R\text{-}R\right)$$

$$\xrightarrow[+\,q\,MgR_2]{2^{nd}\ step} Mg_qM^1$$

Scheme 8-23

$$M^1X_n + \frac{n}{2} Mg \xrightarrow{1^{st}\ step} M^{1*} + \frac{n}{2} MgX_2$$

$$\xrightarrow[+\,q\,Mg]{2^{nd}\ step} Mg_qM^1 + \frac{n}{2} MgX_2$$

Scheme 8-24

corresponding Mg–M^1 intermetallic. The decomposition of organic residues (stemming from the solvent and/or organic groups of the organomagnesium reagents) in the Mg–M^1 hydride products during the dehydrogenation process leads to the ubiquitous presence of carbon (carbide) and hydrogen in the resulting Mg–M^1 intermetallic species (see the discussion at the end of Section 8.4.2.2).

8.4.2.1 $[HTiCl(THF)_{\sim 0.5}]_x$ — a Highly Reactive Titanium Hydride and an Active Species in a McMurry Reaction [54]

Before discussing the various Mg intermetallic systems obtained by this synthetic route, it is of interest to describe an important metal hydride system isolated after the *first* step of the reaction between MgH_2 and M^1X_n, where M^1 is Ti and the "Mg-reagent" is present in a "substoichiometric" amount. A highly reactive, X-ray amorphous titanium hydride, $[HTiCl(THF)_{\sim 0.5}]_x$ (**35**), is produced (together with H_2 gas) during the course of a 1:1 molar reaction between $[TiCl_3(THF)_3]$ and MgH_2^* or MgH_2' in THF [Eq. (8.13)].

$$TiCl_3(THF)_3 + MgH_2^*(MgH_2') \rightarrow \frac{1}{x}\ \underset{\mathbf{35}}{[HTiCl(THF)_{\sim 0.5}]_x} \downarrow + MgCl_2 + 0.5\ H_2\uparrow \qquad (8.13)$$

Through subsequent work we were able to show that the well-known low-valent titanium species obtained by the reduction of $TiCl_3$ with $LiAlH_4$ [55], which is utilized in the McMurry and other reactions [2, 55–57], is in fact **35** [54]. A reinvestigation of the reaction used to generate the solid McMurry reagent has shown that the reaction can be represented by Equation (8.14).

$$[TiCl_3(THF)_3] + 0.5\ LiAlH_4 \xrightarrow[\text{2. pentane}]{\text{1. THF}} \mathbf{35}\downarrow + 0.5\ H_2\uparrow + 0.5\ LiCl + 0.5\ AlCl_3 \qquad (8.14)$$

The precipitate was originally assumed to consist of *metallic* titanium particles with HCl and THF chemisorbed onto the surface [58]. However, according to our X-ray absorption spectroscopy study (EXAFS) at the Ti K-edge, the presence of metallic titanium in **35**, even as nanoparticles, can be excluded. The EXAFS analysis of **35** reveals that the local environment around the Ti absorber consists of oxygen (from THF, 0.4 atom at 2.13 Å), chlorine (1.5 atom at 2.44 Å), and two types of Ti neighbors (1.3 atom at 3.10 Å and 1.9 atom at 4.04 Å). Possible structural models proposed for **35** which are compatible with the EXAFS results are represented in Scheme 8-25 [54].

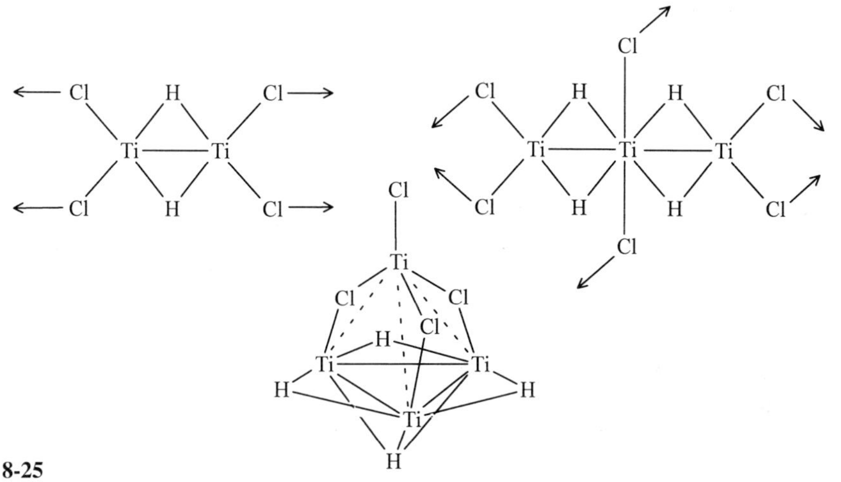

Scheme 8-25

The complex **35** is an active reagent for the coupling of benzophenone to give tetraphenylethene. During the reaction hydrogen is liberated and the inorganic product has been shown to be titanium(+3) oxychloride. Thus the complete McMurry reaction employing **35** as the reagent can be described as in Equation (8.15).

$$Ph_2CO + \mathbf{35} \rightarrow 0.5\ Ph_2C{=}CPh_2 + 0.5\ H_2\uparrow + [TiOCl(THF)_x] \qquad (8.15)$$

The course of the coupling reaction was studied by using $[DTiCl(THF)_{\sim 0.5}]_x$ (D-**35**) as the reagent in a 1:1 molar ratio with Ph_2CO and monitoring the evolution of D_2. At defined intervals, samples were taken from the reaction mixture, hydrolyzed and the products analyzed by GC. The results (Fig. 8-6) show that at the onset of the reaction D_2 is liberated and a greater proportion of benzophenone is transformed into the titanium pinacolate **36** (Scheme 8-26). The formation of **36** during the reaction is confirmed through

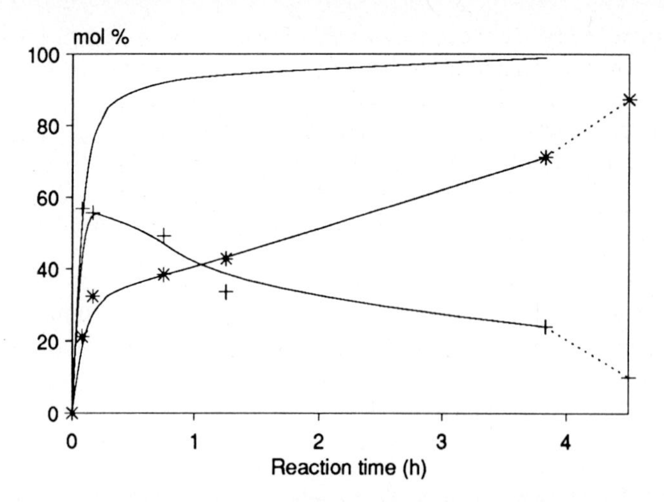

Fig. 8-6. The progress of the coupling reaction of benzophenone in THF at RT using $[DTiCl(THF)_{\sim 0.5}]_x$ (D-**35**) as the reagent (molar ratio 1:1). ∗——∗ mol % of tetraphenylethene; +——+ mol % of pinacol (hydrolysis product of Ti pinacolate); ——— evolution of D_2. After 240 min the reaction mixture was heated at reflux for 40 min whereupon the yield of tetraphenylethene (∗···∗) increases and that of the pinacolate falls (+···+) [54]. Reproduced by permission of Elsevier Sequoia.

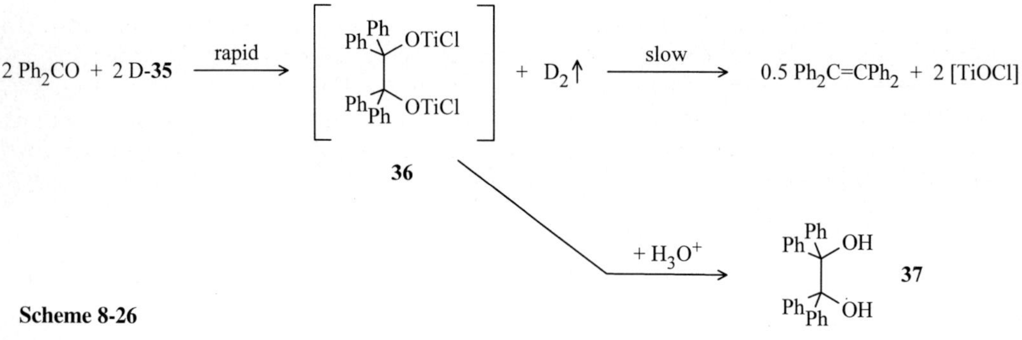

$$2\,Ph_2CO + 2\,\text{D-}\mathbf{35} \xrightarrow{\text{rapid}} \left[Ph_2C(OTiCl){-}C(OTiCl)Ph_2 \right] \; (\mathbf{36}) + D_2\uparrow \xrightarrow{\text{slow}} 0.5\,Ph_2C{=}CPh_2 + 2\,[TiOCl]$$

$$\mathbf{36} \xrightarrow{+H_3O^+} Ph_2C(OH){-}C(OH)Ph_2 \; (\mathbf{37})$$

Scheme 8-26

the detection of its hydrolysis product benzopinacol (**37**) in the protolyzed samples [59]. As the reaction proceeds, the concentration of **36** decreases accompanied by an increase in tetraphenylethene. The concentration of the final product reaches 87% after heating the mixture to reflux.

Titanium is oxidized during the reaction from a formal oxidation state of +2 (in the reagent and intermediate **36**) to +3 (in TiOCl). However, as a result of the homolytic dissociation of the deuterium (or hydrogen) ligand, an additional electron is left over for the reaction. Hence, D-**35** (or **35**) acts like a (strong) *two*-electron reducing agent. Since the reductive coupling of benzophenone with **35** or D-**35** affords tetraphenylethene in nearly the same yield under comparable conditions as that when using the genuine McMurry reagent [55, 58], and further since **35** can be isolated from the reaction of $TiCl_3$ and $LiAlH_4$ (molar ratio 1:0.5) in THF [Eq. (8.14)], it is evident that **35,** not finely-divided metallic titanium, is the active species in the well-known McMurry reagent system $\{TiCl_3 + 0.5\,LiAlH_4\}$.

The recognition that **35** is an active species of the McMurry reagent [54] and the results concerning the nature of the low-valent Ti species derived from Ti halides and magnesium (see Section 8.5.1) [33], confirm the recently expressed view [2] that various structurally different McMurry reagents may exist. Their activity as well as their structure may differ according to the method used for their preparation. For example, the reaction between $[TiCl_3(THF)_3]$ and MgH_2^* (or MgH_2') in a 1:1 molar ratio leads to **35**, whereas the same reaction performed in a 1:1.5 molar ratio gives a highly reactive "TiH_2" [54]. In contrast, the formation of either $[TiMgCl_2 \cdot x\ THF]$ or $[Ti(MgCl)_2 \cdot x\ THF]$ during the reduction of $TiCl_3$ with Mg in THF (Section 8.5.1) [33] is dependent on the reaction time. It is expected that each of these low-valent Ti species should display a unique chemical reactivity. Therefore, their identification may be fruitful for improving selectivity and reproducibility in their application as reagents for syntheses, or may even lead to the discovery of new synthetic reactions.

8.4.2.2 Mg–Pt Intermetallics and their Hydrides from $PtCl_2$ and MgH_2' or MgH_2^*

As an example of the wet-chemical synthesis of intermetallics and their hydrides according to the reaction sequence presented in Scheme 8-22, the preparations of two additional new Mg–Pt intermetallics and their corresponding hydrides will be described [47f, 52]. The preparation of the new t-$Mg_2PtC_xH_y$ compound by the reaction of between MgH_2^* with bis(η^3-allyl)platinum has already been discussed in Section 8.4.1.

A new type of 1:1 Mg–Pt phase can be prepared by reducing $PtCl_2$ with an excess of MgH_2' in THF to yield $MgCl_2$, H_2 and an X-ray amorphous hydride, $MgPtC_xH_y$ (Scheme 8-27). After dehydrogenation and annealing (600 °C), a crystalline product with the composition $Mg_{1.1}PtC_{0.06}H_{0.32}Cl_{0.20}$ was obtained. Based on its X-ray powder pattern, the bimetallic product, denoted as t-$MgPtC_xH_y$, crystallizes in a tetragonal cell having the HgMn structural type [47f].

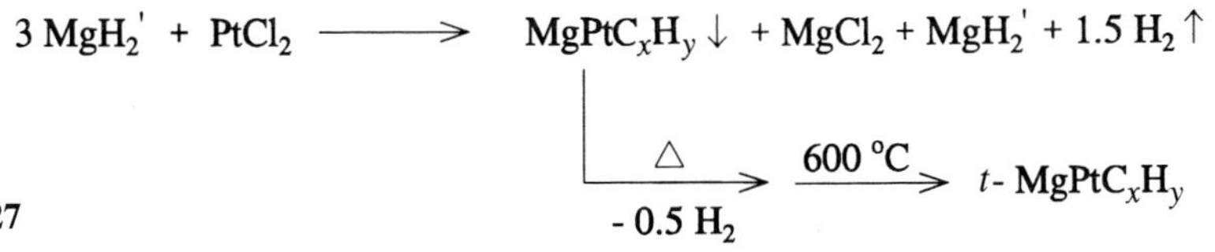

$$3\ MgH_2' + PtCl_2 \longrightarrow MgPtC_xH_y\downarrow + MgCl_2 + MgH_2' + 1.5\ H_2\uparrow$$

$$MgPtC_xH_y \xrightarrow[-0.5\ H_2]{\Delta} \xrightarrow{600\ °C} t\text{-}MgPtC_xH_y$$

Scheme 8-27

A crystalline 2:1 Mg–Pt phase (cf. Scheme 8-19, Section 8.4.1) can be isolated by dehydrogenating and annealing the amorphous products obtained from the reactions of $PtCl_2$ with excess amounts of either solid or solubilized magnesium hydride (Scheme 8-28). The crystalline phase is obtained together with t-$MgPtC_xH_y$ (Scheme 8-27). Its powder diffraction pattern could be indexed to a monoclinic cell and the pattern was refined by the Rietveld technique. The structure of m-$Mg_2PtC_xH_y$ represents a new distorted variant of the high pressure form of titanium, ω-Ti.

The two synthetic routes leading to m-$Mg_2PtC_xH_y$ (together with t-$MgPtC_xH_y$) given in Scheme 8-28 represent two distinctly different ways to the same compound. When both of the reactants are insoluble in THF, i.e. $PtCl_2$ and MgH_2^*, the amorphous product obtained from the RT reaction appears to be a mixture of MgH_2^* and Pt metal. The dehydrogenation of the amorphous material proceeds analogously to that for MgH_2^* with a sharp

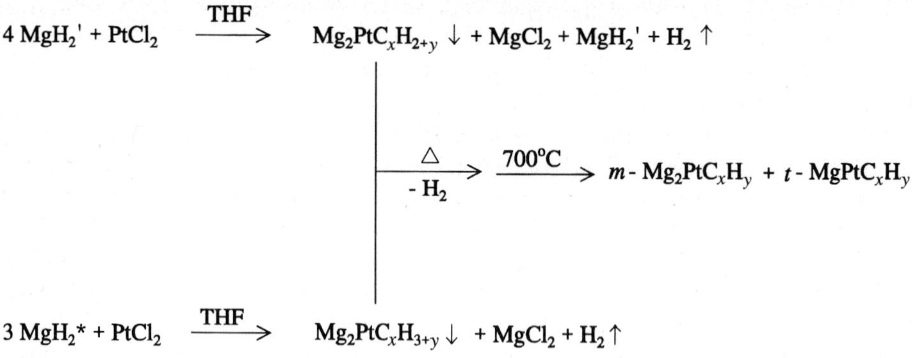

$$4\,MgH_2' + PtCl_2 \xrightarrow{THF} Mg_2PtC_xH_{2+y}\downarrow + MgCl_2 + MgH_2' + H_2\uparrow$$

$$\xrightarrow[-H_2]{\Delta}\ \xrightarrow{700^{\circ}C}\ m\text{-}Mg_2PtC_xH_y + t\text{-}MgPtC_xH_y$$

$$3\,MgH_2{}^* + PtCl_2 \xrightarrow{THF} Mg_2PtC_xH_{3+y}\downarrow + MgCl_2 + H_2\uparrow$$

Scheme 8-28

endothermic reaction at about 340 °C. However, in the situation in which one of the starting compounds is soluble in THF, i.e. MgH_2', the amorphous product isolated from the RT reaction dehydrogenates at a much lower temperature than MgH_2^*, namely at about 225 °C. This observation indicates that MgH_2^* is not present and suggests that a novel hydride system has already been formed during the course of the reaction in solution. The dehydrogenation of the amorphous precursor for t-$Mg_2PtC_xH_y$ (Scheme 8-19, Section 8.4.1) also occurs well below 300 °C, again suggesting that the organometallic reaction in solution is completed with the formation of a bimetallic hydride.

In the magnesium–platinum system, a large number of known and well-characterized phases exist, such as: Mg_6Pt [60], Mg_3Pt [60], MgPt [61], $MgPt_3$ [62], $MgPt_7$ [62] and $MgPt_3C_{0.166}$ [61]. Therefore, it is quite surprising to be able to isolate three new phases in this system, which attests to the versatility of the "wet-chemical" synthetic routes. The structure of the 1:1 Mg–Pt phase obtained by our low-temperature route (Scheme 8-27; tetragonal HgMn structural type) is different from that of the 1:1 alloy MgPt prepared via metallurgical techniques (cubic FeSi-type structure) [61]. Furthermore, by simply varying the ratio of the reactants (MgH_2' and $PtCl_2$, Schemes 8-27 and 8-28) magnesium intermetallics with differing and previously unknown stoichiometries can be prepared in a controlled manner [47f].

The capability of producing new intermetallics using the wet-chemical routes discussed here may stem from several factors. The application of relatively low temperatures may allow for the isolation of thermodynamically unstable compounds not accessible via traditional metallurgical methods. The expression "relatively low temperatures" here refers to both the initial RT reaction in solution and the subsequent thermal treatment. Temperatures employed for annealing did not exceed 700 °C, which is significantly lower than the reaction temperatures of 1100–1200 °C used in the metallurgical preparation of Mg_6Pt, Mg_3Pt, and MgPt [63], and 1200–1400 °C for Mg_3Pt and $MgPt_7$ [62].

Carbon is an additional factor affecting the reaction and the subsequent products, because the initial reactions are performed in organic solvents and in some cases with reactants having organic ligands, e.g. $Pt(\eta^3\text{-}C_3H_5)_2$. In some instances the presence of carbon may be regarded as beneficial, especially when preparing previously unknown compounds. For example, the novel magnesium palladium compound $Mg_2PdC_{0.2-0.3}$ discussed in Section 8.4.1 [47c, d] is stabilized by interstitial carbon. Conversely, the incorporation of carbon into the metal atom lattice can also lead to adverse effects with respect to crystallization, as seen in the amorphous material $MgPdC_xH_y$ (see Section 8.4.1) [47e].

The role played by hydrogen in the synthesis of structurally new intermetallics may be twofold: interstitial hydrogen may allow the production of new bimetallic hydrides, and the dehydrogenation of such hydrides may lead to unique binary intermetallics. Results for the magnesium–rhodium system reported by Yvon et al. [64] provide an illustrative example. By reacting magnesium and rhodium powder at 450 °C under hydrogen pressure, they were able to prepare a new compound stabilized by hydrogen, $Mg_2RhH_{1.1}$, which shows the Ti_2Ni-type structure. Furthermore, upon dehydrogenation of $Mg_2RhH_{1.1}$ another previously unknown binary magnesium rhodium phase, Mg_2Rh, ($MoSi_2$ structural type) was isolated [64]. By reacting $RhCl_3$ with an excess amount of MgH_2^* in THF and subsequently dehydrogenating the intermediate amorphous Mg–Rh hydride, we have also been able to prepare the binary magnesium-rhodium phase Mg_2Rh [52, 65]. (The preparation of Mg–Pd intermetallics has also been effected using this synthetic route) [52].

8.4.2.3 Magnesium Intermetallics from Metal Halides and MgA or Mg*

There is currently a renewed interest in the synthesis of Zintl compounds in solution or via electrochemical routes [66]. We have been able to prepare a series of Zintl-type Mg intermetallics in a highly active state by reacting halides of the metals of main groups 13–15 (M^1X_n) with excess amounts of MgA or Mg* in THF (Table 8-3) [52, 67]. The generalized reactions for the formation of Mg intermetallics from M^1X_n and MgA or Mg* [Eqs. (8.16) and (8.17)] are derived from Schemes 8-23 and 8-24. When MgA is used as the reducing agent [Eq. (8.16)], the progress of the reaction can be monitored by following the increase in anthracene concentration in the reacting solution.

$$M^1X_n + \left(\frac{n}{2} + q\right) MgA \rightarrow Mg_qM^1 \downarrow + \frac{n}{2} MgX_2 + \left(\frac{n}{2} + q\right) A \tag{8.16}$$

$$M^1X_n + \left(\frac{n}{2} + q\right) Mg^* \rightarrow Mg_qM^1 \downarrow + \frac{n}{2} MgX_2 \tag{8.17}$$

The preparation of the Mg–Ga intermetallic compound Mg_2Ga_5 can be achieved through a two-stage reaction. In the first stage [Eq. (8.18)] $GaCl_3$ is reduced by MgA at low temperature to give a highly active Ga (Ga*) suspension in THF. After separating the active metal from $MgCl_2$ and anthracene, Ga* can be converted with excess MgA at RT to Mg_2Ga_5 [Eq. (8.19)].

$$2\ GaCl_3 + 3\ MgA \rightarrow 2\ Ga^* \downarrow + 3\ MgCl_2 + 3\ A \tag{8.18}$$

$$5\ Ga^* + 2.6\ MgA \rightarrow Mg_2Ga_5 \downarrow + 2\ A + 0.6\ MgA \tag{8.19}$$

In contrast, Mg_2Sn can be produced from $SnCl_2$ and MgA in either a single-stage [Eq. (8.16) and Table 8-3] or a two-stage process [Eq. (8.20) and (8.21)].

$$SnCl_2 + MgA \rightarrow Sn^* \downarrow + MgCl_2 + A \tag{8.20}$$

$$Sn^* + 2\ MgA \rightarrow Mg_2Sn \downarrow + 2\ A \tag{8.21}$$

Table 8-3. Preparation of Zintl phases of active magnesium with metals of groups 13–15 by the reaction of metal chlorides (M^1Cl_n) or active metals (M^{1*}) with MgA or Mg* in THF [67].

No.	M^1Cl_n or M^{1*}	Mg reagent	M^1Cl_n: Mg-reagent	Reaction temperature (°C)	Reaction product[a]	Yield[b] (%)	Characterization of reaction product	Specific surface area ($m^2\ g^{-1}$)[c]
1	$GaCl_3$	MgA	2:3	−78 → RT	Ga*			
1a	Ga*	MgA	5:2.6	RT	Mg_2Ga_5	85	X-ray	
2	TlCl	MgA	1:3	RT	MgTl, Tl	81	X-ray; DSC	
3	$GeCl_4$	MgA	1:4	RT	Mg_2Ge, Ge	75	X-ray[d]	135
4	$SnCl_2$	MgA	1:4	RT	Mg_2Sn	95	X-ray; DSC	72.8
5	$PbCl_2$	Mg*[e,f]	1:3	RT	Mg_2Pb	89	X-ray; DSC	24.3
6	$AsCl_3$	Mg*[e]	1:3	RT	Mg_3As_2	86	X-ray; DSC	167.7
7	$SbCl_3$	MgA	1:3	0 → RT	Mg_3Sb_3, Sb	100	X-ray; DSC	153.9
8	$BiCl_3$	Mg*[e]	1:3	RT	Mg_3Bi_2, Bi	96	X-ray	67.9

a) Based on X-ray powder diffraction. b) Based on % of M^1 in the product. c) BET method. d) After heating at 400 °C. e) Prepared according to Eq. (8.10) in toluene. f) Addition of 5 mol % MgA.

The smooth alloying of Sn (or Ga) with Mg in THF *at room temperature* under the influence of MgA is remarkable. In the absence of anthracene or $MgCl_2$, even the active forms of Sn and Mg do not react with one another under such conditions.

Mg intermetallics, as listed in Table 8-3, have previously only been obtained in the bulk form by fusing together the respective metals [68]. The present method allows their preparation as pyrophoric microcrystalline powders with high specific surface areas. The compounds were characterized by X-ray powder diffraction and DSC (Table 8-3).

8.5 Inorganic Grignard Reagents and their Applications in Inorganic Synthesis

Crystalline organometallic complexes containing covalent transition metal–magnesium bonds, such as [CpFe(diphos)MgBr(THF)$_2$] (Cp = η^5-C_5H_5; diphos = 1,2-diphenylphosphinoethane), were first synthesized and structurally characterized by Felkin et al. [69] and Green et al. [70] between 1974 and 1978. Structurally, the Mg complexes resemble their organic counterparts and are thus termed inorganic Grignard analogues.

As already mentioned in Section 8.4.2, we have found that certain transition metal halides react with excess amounts of various magnesium-containing reagents in THF to generate *soluble* bimetallic species, $[M^1(MgCl)_x(MgCl_2)_y(THF)_q]$ where M^1 = Ti or a transition metal of groups 8–10. The presence of direct transition metal–magnesium interactions in these species are assumed. Recently, we have been able to show by X-ray absorption spectroscopy at the Ti K-edge of one such bimetallic species, $[Ti(MgCl)_2(THF)_x]_q$ (**39**, Scheme 8-29) that short titanium to magnesium interatomic distances (2.72 Å) exist. Based on the complete coordination sphere observed for the titanium absorber, the complex represents a new inorganic Grignard analogue [33].

In the THF-soluble M^1–Mg systems, the transition metal is found to be in a low-valent state, actually negatively charged, and thus represents a chemically strong nucleophile. Hence the $[M^1(MgCl)_x(MgCl_2)_y(THF)_q]$ complexes can be utilized as inor-

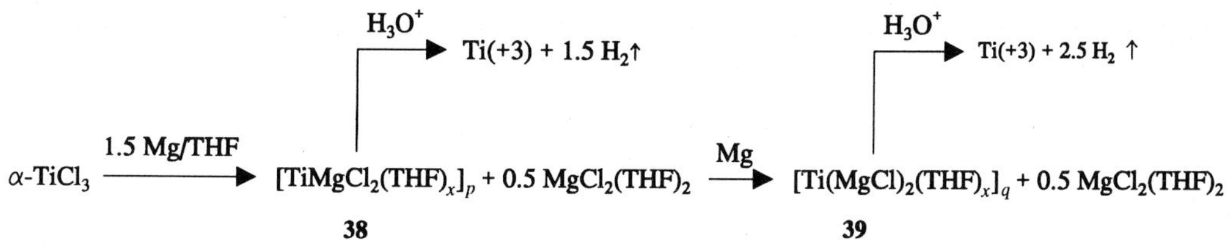

$$\alpha\text{-TiCl}_3 \xrightarrow{1.5\ \text{Mg/THF}} \underset{\mathbf{38}}{[\text{TiMgCl}_2(\text{THF})_x]_p} + 0.5\ \text{MgCl}_2(\text{THF})_2 \xrightarrow{\text{Mg}} \underset{\mathbf{39}}{[\text{Ti(MgCl)}_2(\text{THF})_x]_q} + 0.5\ \text{MgCl}_2(\text{THF})_2$$

$$\mathbf{38} \xrightarrow{H_3O^+} \text{Ti(+3)} + 1.5\ H_2\uparrow \qquad \mathbf{39} \xrightarrow{H_3O^+} \text{Ti(+3)} + 2.5\ H_2\uparrow$$

Scheme 8-29

ganic Grignard *reagents* for the *wet-chemical synthesis* [71] of highly active nanocrystalline metallic and intermetallic materials [52, 53]. Apart from the presence of THF solvent molecules, **39** and the other new transition metal–magnesium systems are characterized by the absence of any stabilizing ligands such as Cp or phosphines. This is particularly advantageous for applying these systems as reagents in inorganic synthetic chemistry.

8.5.1 Preparation and Characterization of Novel Inorganic Grignard Reagents

The reaction between magnesium and the titanium chlorides has been extensively studied, especially with regard to the preparation of catalysts and reagents for various reactions such as the McMurry reaction, or nitrogen or hydrogen fixation. Surprisingly, a number of different stoichiometries have been reported for the resulting bimetallic system, even when similar reaction conditions were employed. We have found that the reduction of α-$TiCl_3$ (or $TiCl_4$) with excess magnesium in THF at RT proceeds in two steps (Scheme 8-29). Previous to our work [33], the occurrence of a two-stage reaction was not recognized, which may be the source of discrepancies in the literature with respect to the stoichiometry of the Ti–Mg system(s).

During the course of the reaction the molar ratio of Mg to Ti in the solution initially levels off at 1.5:1, giving an overall compositon $\{TiMg_{1.5}Cl_3\}$. Protolysis of the black solution delivers 1.5 mol H_2 per mol Ti and Ti^{3+}, indicating that titanium is present in a formal oxidation state of zero. By performing the reaction in a saturated solution of $MgCl_2$ in THF, it could be shown that during the reduction 0.5 mol of *free* $MgCl_2$ per mol of Ti is formed. Thus the solution contains 0.5 equivalent $MgCl_2$ per one equivalent of a bimetallic system having the stoichiometry $\{TiMgCl_2\}$. The complex $[TiMgCl_2(THF)_x]_p$ (**38**, Scheme 8-29) is probably the reagent formerly utilized by Tyrlik et al. [72] and Geise et al. [58] in the McMurry reaction [33]. They reported for the preparation of a "McMurry reagent" from magnesium and α-$TiCl_3$ (or $TiCl_4$) that the optimal reactant molar ratio was 1.7 to 1.

If the system is allowed to react over an extended period of time, the slow uptake of an additional mole of Mg per mole of Ti is observed (Scheme 8-29) leading to the complex $[Ti(MgCl)_2(THF)_x]_q$ (**39**). The existence of **39** as a highly reduced complex is confirmed by its protolysis which delivers 2.5 mol H_2 per mol Ti and indicates that formally titanium has a –2 oxidation state. The second reaction step is most likely reversible [33]. The formation of a complex with the Ti:Mg:Cl ratio 1:2:2 from $TiCl_4$ and Mg in THF was previously reported by Sobota et al. [73].

Compound **39** was investigated by using X-ray absorption spectroscopy at the Ti K-edge. The complex displays the shortest Ti–Mg interatomic distance yet observed, 2.72 Å, which is close to the sum of the Pauling single bond metallic radii (2.69 Å), and thus corresponds to a covalent Ti–Mg bond. The complete environment around the Ti absorber as determined by the EXAFS study includes 1.4 oxygen atom at 2.03 Å and 5 carbon atoms at 3.02 Å (from THF) as well as 2.2 magnesium neighbors and 3.6 distant non-bonding Cl neighbors at 4.05 Å. Hence the coordination sphere of titanium consists of two THF ligands and two magnesium atoms which are bonded to Cl atoms. To maintain the stoichiometry of the complex and to account for the number of nearest neighbors, an oligomeric structure is dictated for **39**. The simplest structural model which would fulfill both requirements is a dimer in which two Ti-centered entities are bridged over the chlorine atoms as shown here [33]. Such an arrangement based on bridging halogen atoms has previously been observed in another inorganic Grignard complex $[Cp_2Mo(H)MgBr(OEt_2)MgBr(C_6H_{11})]_2$ (**40**) [70a, b].

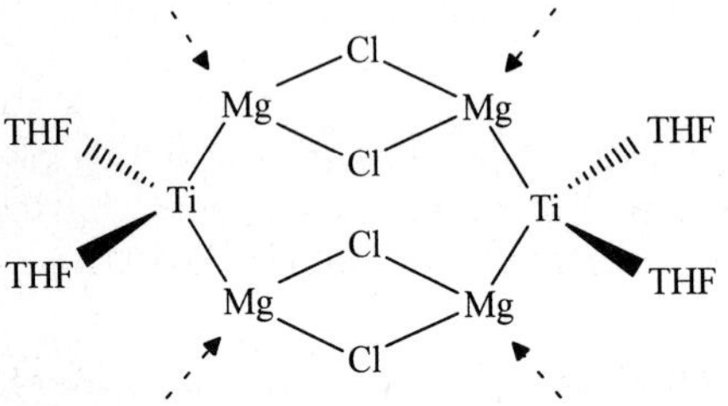

39 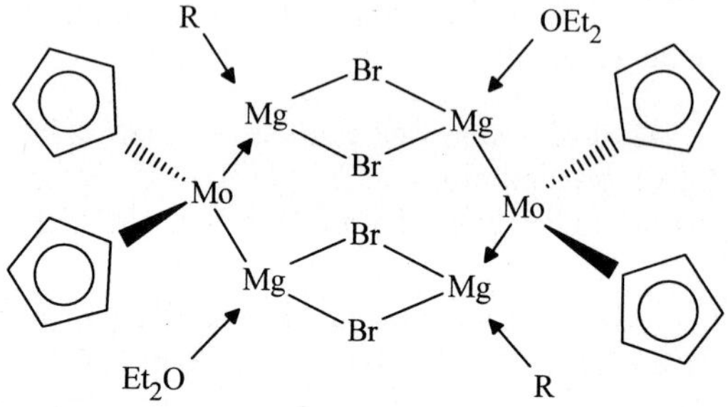 **40**

A nickel–magnesium bimetallic system with the composition $[(NiMgCl)_2\ MgCl_2]$ (**41**) can be isolated in yields of 80–90% (the side-product is elemental nickel) by reducing $NiCl_2$ in THF with excess amounts of Mg* or commercial Mg powder in the presence of an activator, such as anthracene, Mg halides, or treatment with ultrasound (Scheme 8-30). Hydrolysis of **41** delivers about 1.5 mol H_2 per mol Ni in accordance with a formal oxidation state of –1 for Ni. Alcoholysis of **41** in solution results in precipitation of extremely fine Ni*. A THF solution containing $[(FeMgCl)_2 \cdot xMgCl_2]$ can be prepared in an analogous manner, although in much lower yields (13–26%) [52, 53].

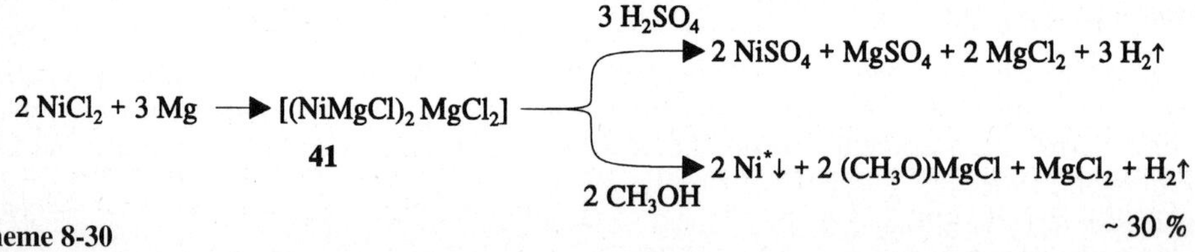

$$2\,NiCl_2 + 3\,Mg \longrightarrow \underset{\mathbf{41}}{[(NiMgCl)_2 MgCl_2]} \begin{cases} \xrightarrow{3\,H_2SO_4} 2\,NiSO_4 + MgSO_4 + 2\,MgCl_2 + 3\,H_2\uparrow \\ \xrightarrow[2\,CH_3OH]{} 2\,Ni^*\downarrow + 2\,(CH_3O)MgCl + MgCl_2 + H_2\uparrow \quad \sim 30\,\% \end{cases}$$

Scheme 8-30

Of particular interest is the preparation of palladium and platinum inorganic Grignard reagents. $PdCl_2$ and $PtCl_2$ react with diethyl magnesium (Et_2Mg) in 1:2 molar ratio, with the elimination of ethyl as an approximately 1:1 mixture of C_2H_4 and C_2H_6, to give almost quantitatively the bimetallic systems **42** and **43** with the composition $[M(MgCl)_2]$ (M = Pd, Pt) in the form of dark-colored THF solutions [Eq. (8.22)]. Most likely, **42** and **43** have oligomeric structures, $[M(MgCl)_2]_x$. However, for simplicity they, along with the other metal Grignard systems described here, will be designated by their most fundamental formula without the coordinating solvent (THF) molecules.

$$MCl_2 + 2\,Et_2Mg \rightarrow [M(MgCl)_2] + 2\,(C_2H_4 + C_2H_6)\uparrow \quad (8.22)$$
$$M = Pd,\ Pt \qquad \mathbf{42}: M = Pd \quad \mathbf{43}: M = Pt$$

Unlike **42**, **43** can also be prepared by a two-stage process. Here $PtCl_2$ reacts with *one* equivalent of Et_2Mg in THF to give a suspension of finely-divided platinum metal (Pt^*) in a $MgCl_2$/THF solution with the liberation of a C_2H_4/C_2H_6 gas mixture (Scheme 8-31). On addition of a second equivalent of Et_2Mg, *dissolution* of Pt^* occurs together with the evolution of a C_2H_4/C_2H_6 gas mixture and formation of **43** [53].

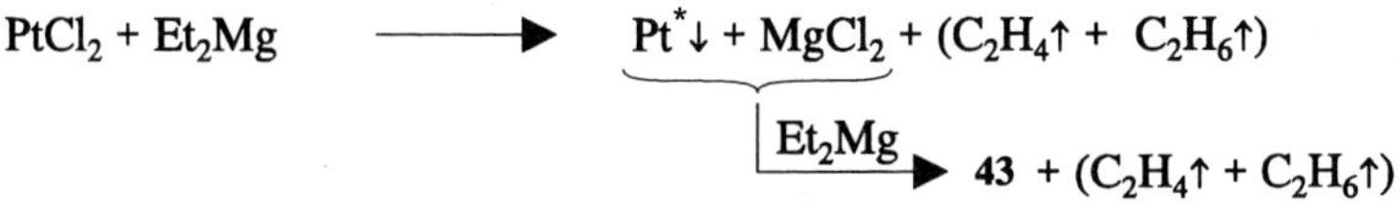

$$PtCl_2 + Et_2Mg \longrightarrow \underbrace{Pt^*\downarrow + MgCl_2}_{} + (C_2H_4\uparrow + C_2H_6\uparrow)$$
$$\xrightarrow{Et_2Mg} \mathbf{43} + (C_2H_4\uparrow + C_2H_6\uparrow)$$

Scheme 8-31

Hydrolysis and alcoholysis of THF solutions containing **42** or **43** results in the quantitative precipitation of finely divided, X-ray amorphous Pd^* or Pt^* accompanied by the evolution of hydrogen and formation of the corresponding Mg^{2+} salts [Eq. [8.23)] [53].

$$\mathbf{42},\ \mathbf{43} + 2\,HX \rightarrow M^*\downarrow + 2\,Mg(X)Cl + H_2\uparrow \quad (8.23)$$
$$30\text{–}50\%$$

The intermetallic phase MgPd can be synthesized from **42**. $MgCl_2$ present in **42** can be removed as the THF-insoluble adduct $MgCl_2 \cdot 2\,C_4H_8O_2$ by adding 1,4-dioxane to a solution containing the bimetallic complex. Addition of ether to the filtrate yields a black precipitate which after annealing exhibits diffuse reflections in its X-ray powder diagram corresponding to the phase MgPd (Scheme 8-32).

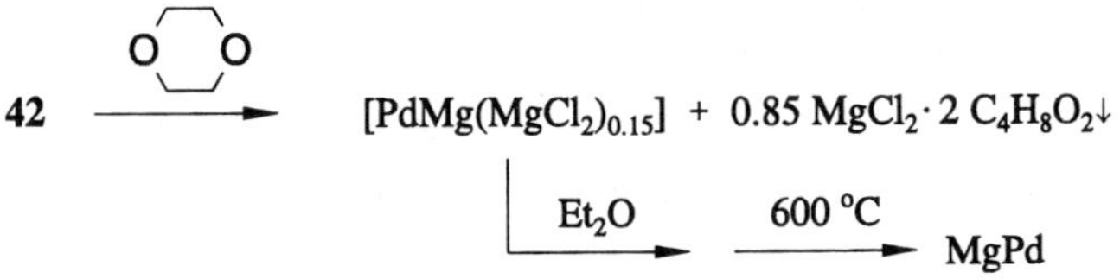

$$\mathbf{42} \xrightarrow{\text{1,4-dioxane}} [PdMg(MgCl_2)_{0.15}] + 0.85\,MgCl_2 \cdot 2\,C_4H_8O_2\downarrow$$
$$\xrightarrow{Et_2O} \xrightarrow{600\,°C} MgPd$$

Scheme 8-32

According to preliminary investigations, the synthesis of THF-soluble inorganic Grignards with the noble metals Ru, Rh, and Ir is possible. The reaction of $RhCl_3$ with Et_2Mg in a molar ratio 1:2 proceeds with the complete elimination of the ethyl groups in the form of a C_2H_4/C_2H_6 gas mixture (4 mol per mol Rh) and the formation of a dark-colored THF solution containing the Rh–Mg Grignard complex **44** [Eq. (8.24)]. When the reaction is performed with an excess of Et_2Mg, i.e. a 1:3 molar ratio of $RhCl_3$ to Et_2Mg, again only 4 mol of a C_2H_4/C_2H_6 gas mixture per mol Rh are liberated. One equivalent of unreacted Et_2Mg remains in solution, as verified by hydrolysis, thus confirming the stoichiometry of the Rh Grignard shown in Equation (8.24).

$$RhCl_3 + 2\,Et_2Mg \rightarrow \underset{\mathbf{44}}{[RhMgCl \cdot MgCl_2]} + 2\,(C_2H_4 + C_2H_6)\uparrow \quad (8.24)$$

By using one or other of the two different reducing agents Mg^* and Et_2Mg, two different metal Grignard systems can be prepared in the Ru– and Ir–Mg–Cl systems. Use of Mg^* gives $[M'(MgCl)_3]$ (M′ = Ru, **45b**; M′ = Ir, **46b**), whereas Et_2Mg gives $[Ru(MgCl)_2 \cdot 0.5\,MgCl_2]$ (**45a**) or $[IrMgCl \cdot MgCl_2]$ (**46a**), as shown in Equations (8.25) to (8.27) and Scheme 8-33.

The generation of the two Ru Grignard analogues **45a** and **45b** is represented in Equations (8.25) and (8.26). Hydrolysis of **45b** results in the precipitation of finely divided Ru metal with the concomitant evolution of 0.9–1.2 mol H_2 per mol Ru.

$$\underset{}{RuCl_3 + 2.5\,Et_2Mg} \rightarrow \underset{\mathbf{45a}}{[Ru(MgCl)_2 \cdot 0.5\,MgCl_2]} + 2.5\,(C_2H_6 + C_2H_4)\uparrow \tag{8.25}$$

$$RuCl_3 + 3\,Mg^* \rightarrow \underset{\mathbf{45b}}{[Ru(MgCl)_3]} \xrightarrow{3HCl} \underset{60-80\%}{Ru^*\downarrow} + 1.5\,H_2\uparrow + 3\,MgCl_2 \tag{8.26}$$

The reaction of $IrCl_3$ with Et_2Mg in the molar ratio 1:3 proceeds similarly to that for the Rh system described above with the evolution of only about 4 mol of a C_2H_4/C_2H_6 gas mixture per mol Ir. Two mols of C_2H_6 per mol Ir are released during the hydrolysis of the reaction mixture indicating the presence of one mol of unreacted Et_2Mg. Thus, the formation of an Ir Grignard compound corresponding to $[IrMgCl \cdot MgCl_2]$ [**46a**, (Scheme 8-33)] is assumed. On the other hand, reactions of $IrCl_3$ with Mg^* (molar ratio

$$IrCl_3 + 3\,Et_2Mg \longrightarrow \underset{\mathbf{46a}}{[IrMgCl \cdot MgCl_2]} + 2\,(C_2H_6 + C_2H_4)\uparrow + Et_2Mg \xrightarrow{H_3O^+} \underset{95\,\%}{2\,C_2H_6}$$

Scheme 8-33

1:3 or higher proportions of Mg^*), are analogous to the corresponding reactions for $RuCl_3$. In every case three mols of Mg react with one mol $IrCl_3$. Hydrolysis of THF solutions of the products generated from the reaction between $IrCl_3$ and 3 or 5 equivalents of Mg^* results in the evolution of 0.8–1.3 mol of H_2 per mol Ir and the precipitation of finely divided metallic Ir. The composition of this second Ir Grignard reagent, **46b**, is therefore given as $[Ir(MgCl)_3]$ [Eq. (8.27)].

$$IrCl_3 + 3\,Mg^* \rightarrow \underset{\mathbf{46b}}{[Ir(MgCl)_3]} \xrightarrow{3HCl} \underset{50-85\%}{Ir^*\downarrow} + 1.5\,H_2\uparrow + 3\,MgCl_2 \tag{8.27}$$

8.5.2 Reactive Alloys, Intermetallics, and Metals via Inorganic Grignard Reagents [52, 53]

The preparation of highly reactive nanocrystalline metals and alloys is possible in a controlled manner by applying the novel THF-soluble inorganic Grignard analogues as Grignard *reagents* for inorganic syntheses. These transition metal Grignards with the generalized formula $[M^1(MgCl)_m]$ react with metal chlorides M^2Cl_n in the molar ratio of $n{:}m$ in THF at RT, generating alloys and intermetallics of the composition $M^1_nM^2_m$ as shown in Equation (8.28). In a similar fashion the reaction of $[M^1(MgCl)_m]$ with its corresponding chloride M^1Cl_n can be employed as a method for the preparation of finely divided metal powders M^1 [Eq. (8.29)].

$$n[M^1(MgCl)_m] + mM^2Cl_n \xrightarrow{\text{THF}} M^1_nM^2_m \downarrow + mnMgCl_2 \qquad (8.28)$$
$$m,\ n = 1, 2, 3$$

$$n[M^1(MgCl)_m] + mM^1Cl_n \xrightarrow{\text{THF}} (m+n)M^1 \downarrow + mnMgCl_2 \qquad (8.29)$$

The reaction of an inorganic Grignard reagent with a metal halide can, however, be more complex than that represented by Equations (8.28) and (8.29), as clearly demonstrated by the preparation of Ti intermetallics detailed below. The coprecipitation of $MgCl_2$ (solubility in THF: ~ 0.5 mol/L at RT) during the reaction, as well as the inclusion of THF solvent molecules in the solid product are unavoidable. However the concentrations of $MgCl_2$ and organics in the solid can be effectively reduced by washing the precipitate with air-free water, and by hydrogenating under pressure, respectively.

The titanium Grignard reagent **39** in THF reacts with a range of metal chlorides M^2Cl_2 (M^2 = Fe, Cr, Sn, Mn, Co, or Ni) in a 1:1 molar ratio to give black air-sensitive, X-ray amorphous precipitates having the general composition $Ti_{1-y}M^2$, $0 \leq y \leq \sim 0.9$ (together with minor amounts of $MgCl_2$ and THF) as well as the zerovalent THF-soluble Ti complex **38** [Eq. (8.30)]. The ratio of Ti to M^2 in the precipitate $Ti_{1-y}M^2$ varies from 1:1 for the Ti–Sn system to < 0.1:1 in the Ti–Ni product.

$$\underset{\mathbf{39}}{[Ti(MgCl)_2(THF)_x]} + M^2Cl_2 \rightarrow \underset{0 \leq y \leq \sim 0.9}{Ti_{1-y}M^2} \downarrow + y\underset{\mathbf{38}}{[TiMgCl_2(THF)_x]} + (2-y)MgCl_2 \qquad (8.30)$$

For the Ti–Fe product, $Ti_{0.7}FeMg_{0.04}Cl_{0.14}(THF)_{0.11}$, it is possible to use Mössbauer (MB) spectroscopy to determine whether the X-ray amorphous solid product is a bimetallic alloy or simply a mixture of Ti and Fe particles. No evidence for elemental iron was found in the MB spectrum (measured at 4.2 K) of the bimetallic Ti–Fe product. By comparing the MB spectrum with those measured for well-known Ti–Fe alloys and with other reference spectra, the features of the spectrum could be definitely assigned to the phases TiFe and $TiFe_2$. According to the areas of the two curves, the ratio of the two phases in the sample is approximately 2:1 (63%:37%).

On the molecular level, the formation of the TiFe intermetallic phase in solution may proceed according to Equation (8.31). The more Fe-rich alloy is produced by a one to one reaction between the Ti Grignard and $FeCl_2$, whereby besides the intermetallic phase $TiFe_2$ and $MgCl_2$, $[TiMgCl_2]$ (**38**) is formed [Eq. (8.32)]. The results of the MB study indicate that both reaction pathways take place to give TiFe and $TiFe_2$. Thus, the reaction between **39** and $FeCl_2$ can be described as the sum of Equations (8.31) and (8.32) to give Equation (8.33). If each pathway [Eqs. (8.31) and (8.32)] proceeds with the same probability, the product ratio of TiFe to $TiFe_2$ should be near 2:1, as shown in Equation (8.33) and found experimentally by MB spectroscopy. The ratio of Ti to Fe

(0.7:1) in the precipitate is also nearly in accordance with that expected from Equation (8.33), i.e. 0.75:1.

$$\underset{\mathbf{39}}{[Ti(MgCl)_2]} + FeCl_2 \rightarrow TiFe\downarrow + 2\,MgCl_2 \qquad (8.31)$$

$$\underset{\mathbf{39}}{[Ti(MgCl)_2]} + FeCl_2 \rightarrow 0.5\,TiFe_2\downarrow + 0.5\underset{\mathbf{38}}{[TiMgCl_2]} + 1.5\,MgCl_2 \qquad (8.32)$$

$$2\,\underset{\mathbf{39}}{[Ti(MgCl)_2]} + 2\,FeCl_2 \rightarrow TiFe\downarrow + 0.5\,TiFe_2\downarrow + 0.5\,\underset{\mathbf{38}}{[TiMgCl_2]} + 3.5\,MgCl_2 \qquad (8.33)$$

Bimetallic X-ray amorphous precipitates with the approximate composition $Ni_2M^2(MgCl_2)_{0.3-0.5}(THF)_{0.1-0.3}$, where M^2 = Fe or Cu, were obtained from reactions of the Ni Grignard reagent **41** (Scheme 8-30) with $FeCl_2$ and $CuCl_2$ in the molar ratio 1:1 in THF. Following annealing at 600 °C, the products were identified by X-ray powder diffraction as Ni–Fe and Ni–Cu alloys.

The Pd Grignard reagent **42** [Eq. (8.22)] reacts with $FeCl_2$ and $TeCl_4$ in the molar ratios 1:1 and 1:0.5 respectively to give X-ray amorphous precipitates with the compositions $PdFeMg_{0.3}Cl_{0.4}C_{0.1}H_{1.7}$ and $Pd_2TeMg_{0.2}Cl_{0.2}C_{0.4}H_{1.7}$. In the case of the Pd–Fe solid-product, it again proved possible by MB spectroscopy to confirm the formation of an Fe alloy. In the Pd–Te solid product, evidence for the presence of the known Pd_2Te phase was obtained from DSC analyses.

Reactions of the Pt Grignard reagent **43** with $FeCl_2$, $CuCl_2$ or $SnCl_2$ in the molar ratio 1:1, and with $RhCl_3$ or $BiCl_3$ in the molar ratio 3:2 in THF give X-ray amorphous precipitates in the corresponding ratios [Eqs. (8.34) and (8.35)].

$$\underset{\mathbf{43}}{[Pt(MgCl)_2]} + M^2Cl_2 \rightarrow PtM^2\downarrow + 2\,MgCl_2 \qquad M^2 = \text{Fe, Cu, Sn} \qquad (8.34)$$

$$3\,\mathbf{43} + 2\,M^2Cl_3 \rightarrow Pt_3M^2_2\downarrow + 6\,MgCl_2 \qquad M^2 = \text{Rh, Bi} \qquad (8.35)$$

The bimetallic products prepared according to Equations (8.34) and (8.35) can be easily crystallized by thermal treatment at 600 or 700 °C under argon. The X-ray powder diagrams of the annealed samples reveal sharp reflections which can be assigned to known intermetallic phases such as PtFe and PtSn. In the case of the Pt–Fe solid, it was also determined by MB spectroscopy that an Fe alloy, not elemental iron, is present in the X-ray amorphous samples *prior* to annealing.

The Ru Grignard reagents **45a** and **45b** react at RT with $PtCl_2$ in either 1:1 or 2:3 molar ratio in THF to yield black precipitates. The X-ray amorphous solids, after washing with air-free water and acetone, reveal the compositions $RuPt(MgCl_2)_{0.07}C_{4.4}H_{5.4}$ and $Ru_2Pt_{2.8}(MgCl_2)_{0.06}C_{2.6}H_{3.0}$, respectively ([Eqs. (8.36) and (8.37)]. According to high resolution TEM photographs, both solids are nanocrystalline and consist of uniform globular particles of 2.5–4.5 nm diameter. After annealing (500–650 °C) the resulting microcrystalline alloys were identified by X-ray powder diffraction as cubic Ru–Pt solid solutions.

$$\underset{\mathbf{45a}}{[Ru(MgCl)_2 \cdot 0.5\,MgCl_2]} + PtCl_2 \rightarrow RuPt \downarrow + 2.5\,MgCl_2 \tag{8.36}$$

$$2\,\underset{\mathbf{45b}}{[Ru(MgCl)_3]} + 3\,PtCl_2 \rightarrow Ru_2Pt_3 \downarrow + 6\,MgCl_2 \tag{8.37}$$

The X-ray amorphous solids $Rh_2Sn_{0.8}(MgCl_2)_{0.5}(THF)_{1.5}$ and $Rh_2Pt(MgCl_2)_{1.5}$ were prepared by reacting the Rh Grignard reagent **44** [Eq. (8.24)] with $SnCl_2$ and $PtCl_2$, respectively. Annealing the amorphous Rh–Pt solid at 700 °C yielded a crystalline material which, based on its X-ray powder pattern, is a cubic alloy $Rh_{0.67}Pt_{0.33}$.

The reaction of the Ir Grignard reagent **46b** [Eq. (8.27)] with $SnCl_2$ in 2:3 molar ratio in THF proceeds with the precipitation of an X-ray amorphous solid with the composition $Ir_2Sn_3(MgCl_2)_{0.4}(THF)_{1.2}$. After washing with water and annealing (600 °C) the solid exhibits in its X-ray powder diagram diffraction lines characteristic of the intermetallic compound Ir_5Sn_7. The DSC analysis of the solid *isolated from the solution* suggests the absence of metallic Sn in the X-ray amorphous product.

8.5.3 Conclusions and Outlook

The previously known methods for the wet-chemical synthesis of nanoparticulate intermetallics and alloys mainly involve the simultaneous coreduction of two (or more) metal salts using various reducing agents in suitable solvents (cf. Section 8.1). According to the method presented in this section, nanoparticulate bimetallic systems can be prepared in a controlled, two-stage process. In the first stage (Section 8.5.1), the appropriate transition metal chloride is converted into a well-defined inorganic Grignard reagent. In the second stage [Section 8.5.2; Eq. (8.28)], the solution of the latter is reacted with a second component, M^2Cl_n, resulting in the precipitation of an X-ray amorphous alloy or intermetallic species. This low-temperature, kinetically controlled process is particularly suitable for the synthesis of metastable systems. By varying the inorganic Grignard reagent used, the M^2Cl_n component, or the synthesis conditions, bimetallic systems with different metal ratios (e.g., RuPt and Ru_2Pt_3) as well as different physico-chemical properties can be prepared. Physical characterization of these bimetallic systems by, for example, Mössbauer spectroscopy and DSC analysis, has shown up to now that the alloy products obtained directly from solution do not contain any metal in elemental form. Removal of residual magnesium salts and solvent from the raw solid products can be successfully achieved. Thermal annealing of the purified solid products has in most cases yielded micro- or nanocrystalline alloys or intermetallics which have subsequently been identified from X-ray powder diffraction.

The generation of homometallic nanoparticles for the preparation of heterogeneous catalysts is also possible using the inorganic Grignard reagents as well as active magnesium compounds (MgA, Et_2Mg) [Scheme 8-31 and Eqs. (8.23), (8.26), (8.27), (8.29)]. Besides their applications in the preparation of highly dispersed mono- or bimetallic particles, inorganic Grignard reagents are highly promising for the synthesis of other inorganic materials. Of particular interest is the question of whether metal–element structures can be built up from inorganic Grignard reagents and various electrophiles other than metal halides.

8.6 Experimental Procedures for Selected Examples

8.6.1 Magnesium Anthracene · 3 THF [74] and the Synthesis of Allenylmagnesium Chloride [10][2]

All reactions and operations are carried out under argon. A mixture of 24.3 g (1.0 mol) of magnesium powder (99%, Eckart Werke PK 31, 270 mesh) and 200 g (1.12 mol) of anthracene (99%, Rütgerswerke AG) is introduced into a three-necked 2 l flask and heated for 10 min at 110 °C in an oven. Thereafter the flask is provided with a magnetic stirring bar and equipped with a glass reflux condenser and a three-way valve. After the reaction flask is evacuated and flushed with argon several times, 1 l of anhydrous THF (refluxed over MgA or Et_2Mg and distilled under argon) and a few drops (~0.05 ml) of ethyl bromide are added. The suspension is then stirred for 24 h at 60 °C to achieve the conversion of Mg to the bulky orange precipitate of MgA · 3 THF. After additional stirring for 12 h at room temperature the suspension is filtered through a glass frit (∅ = 10 cm). MgA · 3 THF is then washed twice with 300 ml of THF and dried for 20 h in vacuo (10^{-3} mbar). Yields of MgA · 3 THF are typically around 90%. It is difficult to prepare MgA completely free of trace amounts of anthracene and elemental magnesium, due to the equilibrium existing between MgA and Mg + A.

The amount of free magnesium present can be determined by measuring the quantity of hydrogen evolved on hydrolysis of a sample of the product. Upon hydrolysis MgA · 3 THF liberates Mg^{2+}, THF and dihydroanthracene (DHA). The ratio MgA:THF:free A in the product can be determined by GC analysis of a sample hydrolyzed by a toluene–CH_3OH mixture (25/1-vol./vol.) containing *n*-octane and *n*-hexadecane as internal standards.

To a stirred cooled (ice bath) suspension of 6.82 g (16.3 mmol) of MgA in 35 ml THF a solution of 1.21 g (1.20 ml, 16.3 mmol) of propargyl chloride in 5 ml of THF was added over a period of 20 min. The mixture was stirred for an additional 10 min at 0 °C. Hydrolysis of a 15.0 ml aliquot of the total of 40 ml of the suspension afforded 120 ml of gas composed of 93.6% C_3H_4 (mixture of allene and propyne) and 6.4% THF (MS analysis). Yield of the Grignard compound: 76.6%.

8.6.2 The Novel Magnesium–Platinum Intermetallic *t*-$MgPtC_xH_y$ (Scheme 8-27) [47f][3]

10 ml (30.3 mmol) MgH'_2 [20] was diluted with 20 ml THF and cooled to –78 °C. 2.6839 g (10.1 mmol) of $PtCl_2$ was added in small aliquots to the stirred MgH'_2–THF solution. The reaction was allowed to warm up to RT over a period of 11 h, during which the onset of gas evolution was noted. After stirring for 24 h at RT, the production of gas ended, signaling the completion of the reaction. The gaseous product, which was collected (325 ml) and analyzed by mass spectrometry, corresponded to 13.5 mmol of H_2. The solid product was filtered, washed twice with THF, and dried under high vacuum. 2.3994 g of a black X-ray amorphous pyrophoric solid was isolated having the net composition $Mg_{1.10}PtC_{0.65}H_{2.98}Cl_{0.21}$. 1.9948 g of the above material was dehydrogenated employing standard conditions (heated from RT to 400 °C at 1 °C/min) yielding 1.1 H/Pt. To deter-

2 Material from [74] is reproduced with permission of Elsevier Publishing Co.
3 Reproduced with permission of R. Oldenbourg Verlag.

mine whether the sample could reabsorb hydrogen, it was heated in an autoclave at 200 °C under 50 bar H_2 pressure for 48 h. No hydrogen uptake was observed, and no further hydrogen release after a second thermolysis step. The sample, still X-ray amorphous, was then annealed at 600 °C for 24 h under argon to give a crystalline product. The composition of the endproduct was found to be $Mg_{1.1}PtC_{0.06}H_{0.32}Cl_{0.20}$.

The X-ray diffraction pattern of the annealed sample, $MgPtC_{0.06}H_{0.32}$, reveals a single set of lines which could be indexed with a primitive tetragonal cell, $a = 2.9566(5)$ and $c = 3.509(1)$ Å. This result suggests that the compound crystallizes with a HgMn-type structure (*t*-MgPt: *P*4/mmm). This structural model was confirmed by the Rietveld refinement: $R_F = 0.054$.

8.6.3 The Titanium Grignard Reagent, $[Ti(MgCl)_2(THF)_x]_2$ (39) (Scheme 8-29) [33]

7.90 g (51.2 mmol) of α-$TiCl_3$ in 500 ml THF was stirred together with an excess amount, 9.72 g (400 mmol), of Mg powder (PK 31) at RT. At defined time intervals the stirring was interupted for 30 min and a sample of the supernatant removed, hydrolyzed, and analyzed for Mg^{2+} (complexometric), Ti^{4+} (photometric), and Cl^- (Volhardt method). After ca. 24 h the titanium concentration in solution reached 97% of the expected concentration based on the initial mass of α-$TiCl_3$ and the Mg/Ti ratio in solution reached a value of 1.5. After an additional 16 h of stirring the concentrations of Ti, Mg, and Cl in solution were 0.102, 0.158, and 0.310 mol L^{-1} respectively and the hydrolysis (4 N H_2SO_4) of an aliquot of the reaction solution delivered 1.52 H_2/Ti. The reaction was stirred for an additional 5 days to allow further reaction with excess magnesium in the system. The Mg/Ti ratio in solution reached a value of 2.5, and hydrolysis of an aliquot of the solution yielded 2.48 H_2/Ti.

It is well known that magnesium cleaves THF under reflux (66 °C) to give oxamagnesium cyclohexane [31], and this reaction is catalyzed by **39**. However, at RT ether cleavage remains insignificant. The amount of n-butanol (formed through the protolysis of the cyclic magnesium hydrocarbon) found in hydrolyzed samples after 7 h of reaction was very low, corresponding to 4–8% of the initial mass of α-$TiCl_3$.

The reaction can be accelerated by ultrasonic treatment. For example, the reaction of 2.433 g (15.77 mmol) α-$TiCl_3$ with 3.5 g (144 mmol) Mg powder in 150 ml THF in an ultrasonic bath maintained at 10 °C is completed within 24 h.

8.6.4 Application of the Platinum Grignard Reagent 43 to the Synthesis of PtFe [Eq. (8.34)] [53]

To a stirred solution of 1.70 g of Et_2Mg (90.6% purity according to the amount of C_2H_6 produced during protolysis; 20.8 mmol) in 100 ml of THF was added at –70 °C over a period of 10 min and in small portions 2.71 g (10.2 mmol) of $PtCl_2$. With continuous stirring the reaction mixture was first kept for 12 h at –70 °C and then allowed over the course of 12 h to warm up gradually to RT. The stirring was continued at RT for 24 h. The solvent was removed in vacuo and the black residue redissolved in 72 ml of fresh THF. Hydrolysis of an aliquot of the solution (10 ml) with dil. HCl gave the result: 0.62 H_2 and 0.25 C_2H_6/Pt. 31.0 ml of the thus prepared THF solution of **43** containing

4.37 mmol of Pt, 8.8 mmol of Mg, and 9.7 mmol of Cl was added over a period of 25 min to a stirred suspension of 0.95 g (4.39 mmol) of $FeCl_2 \cdot 1.3$ THF in 20 ml of THF (the reaction is slightly exothermic!). After stirring 2 days at RT the reaction mixture was heated to reflux for 4 h. The hot suspension was filtered and the isolated solid washed with THF and pentane. After drying under high vacuum at RT, 0.82 g of a black X-ray amorphous solid with the composition $PtFe_{1.3}Mg_{0.6}Cl_{1.4}C_{3.7}H_{8.4}$ was obtained. After annealing for 72 h at 600 °C a crystalline PtFe alloy (the solid shows an overall composition $PtFe_{1.1}Mg_{0.7}Cl_{0.6}C_{0.6}H_{1.3}$) was verified by X-ray diffraction.

References

[1] R. D. Rieke, *Science* **1989**, *246*, 1260; R. D. Rieke, T. P. Burns, R. M. Wehmeyer, B. E. Kahn in *High Energy Processes in Organometallic Chemistry, ACS Symposium Series* **1987**, *333*, 323; A.V. Kavaliunas, A. Taylor, R. D. Rieke, *Organometallics* **1983**, *2*, 377.
[2] A. Fürstner, *Angew. Chem.* **1993**, *105*, 171; *Angew. Chem. Int. Ed. Engl.* **1993**, *32*, 164.
[3] H. Bönnemann, W. Brijoux, R. Brinkmann, R. Fretzen, Th. Joussen, R. Köppler, B. Korall, P. Neiteler, J. Richter, *J. Mol. Catalysis* **1994**, *86*, 129.
[4] M. T. Reetz, W. Helbig, *J. Am. Chem. Soc.* **1994**, *116,* 7401.
[5] J. van Wonterghem, S. Mørup, C. J. W. Koch, S. W. Charles, S. Wells, *Nature* **1986**, *322*, 622.
[6] (a) B. Bogdanović S. Liao, R. Mynott, K. Schlichte, U. Westeppe, *Chem. Ber.* **1984**, *117*, 1378; (b) E. Bartmann, B. Bogdanović, N. Janke, S. Liao, K. Schlichte, B. Spliethoff, J. Treber, U. Westeppe, U. Wilczok, *Chem. Ber.* **1990**, *123*, 1517.
[7] (a) B. Bogdanović, S. Liao, M. Schwickardi, P. Sikorsky, B. Spliethoff, *Angew. Chem.* **1980**, *92*, 845; *Angew. Chem. Int. Ed. Engl.* **1980**, *19*, 818; (b) B. Bogdanović, G. Koppetsch, M. Schwickardi in *Organometallic Syntheses, Vol. 4* (eds.: R. B. King, J. J. Eisch), Elsevier, Amsterdam **1988**, p. 404; (c) B. Bogdanović, P. Bons, S. Konstantinović, M. Schwickardi, U. Westeppe, *Chem. Ber.* **1993**, *126*, 1371.
[8] B. Bogdanović, *Acc. Chem. Res.* **1988**, *21*, 261.
[9] B. Bogdanović, N. Janke, C. Krüger, R. Mynott, K. Schlichte, U. Westeppe, *Angew. Chem.* **1985**, *97*, 972; *Angew. Chem. Int. Ed. Engl.* **1985**, *24*, 960.
[10] B. Bogdanović, N. Janke, H.-G. Kinzelmann, *Chem. Ber.* **1990**, *123*, 1507.
[11] (a) C. L. Raston, G. Salem, *J. Chem. Soc. Chem. Commun.* **1984**, 1702; (b) L. M. Engelhardt, R. I. Papasergio, C. L. Raston, G. Salem, A. H. White, *J. Chem. Soc., Dalton Trans.* **1986**, 789; (c) S. Harvey, P. C. Junk, C. L. Raston, G. Salem, *J. Org. Chem.* **1988**, *53*, 3134; (d) M. J. Gallagher, S. Harvey, C. L. Raston, R. E. Sue, *J. Chem. Soc. Chem. Commun.* **1988**, 289.
[12] S. Itsuno, G. D. Darling, H. D. H. Stöver, J. M. J. Fréchet, *J. Org. Chem.* **1987**, *52*, 4644.
[13] S. Harvey, C. L. Raston, *J. Chem. Soc. Chem. Commun.* **1988**, 652.
[14] C. Hamdouchi, M. Topolski, V. Goedken, H. M. Walborsky, *J. Org. Chem.* **1993**, *58*, 3148.
[15] M. Harendza, K. Leßmann, W. P. Neumann, *Synlett* **1993**, 283.
[16] (a) R. Benn, P. W. Jolly, T. Joswig, R. Mynott, K. P. Schick, *Z. Naturforsch., Part B*, **1986**, *41*, 680; (b) R. Benn, P. W. Jolly, R. Mynott, B. Raspel, G. Schenker, K. P. Schick, G. Schroth, *Organometallics* **1985**, *4*, 1985.
[17] According to Starks: "The general concept of phase-transfer catalysis applies to the transfer of any species from one phase to another (not just anions . . .), provided a suitable catalyst can be chosen and suitable phase compositions and reaction conditions are used". C. M. Starks in *Phase Transfer Catalysis* (ed.: C. M. Starks), ACS Symp. Ser. 326, **1987**, p. 2.
[18] H. E. Ramsden, US Patent 3 354 190, **1967**; *Chem. Abstr.* **1968**, *68*, 114744.
[19] B. Bogdanović, M. Schwickardi, U. Westeppe in *Organometallic Syntheses, Vol. 4* (ed.: R. B. King, J. J. Eisch) Elsevier, Amsterdam **1988**, p. 399.
[20] B. Bogdanović, P. Bons, M. Schwickardi, K. Seevogel, *Chem. Ber.* **1991**, *124*, 1041.
[21] B. Bogdanović, M. Schwickardi, *Z. Naturforsch. Part B* **1984**, *39*, 1001.
[22] H. Bönnemann, B. Bogdanović, R. Brinkmann, D. He, B. Spliethoff, *J. Organomet. Chem.* **1993**, *451*, 23 and references therein.
[23] B. Bogdanović, Th. Hartwig, B. Spliethoff, *Int. J. Hydrogen Energy* **1993**, *18*, 575.

[24] H.-F. Klein, H. H. Karsch, *Chem. Ber.* **1975**, *108*, 944.

[25] W. Gausing, G. Wilke, *Angew. Chem.* **1981**, *93*, 201; *Angew. Chem. Int. Ed. Engl.* **1981**, *20*, 186.

[26] (a) W. Oppolzer, *Angew. Chem.* **1989**, *101*, 39; *Angew. Chem. Int. Ed. Engl.* **1989**, *28*, 38; (b) W. Oppolzer in *Comprehensive Organic Syntheses, Vol. 5* (eds.: B. M. Trost, I. Fleming), Pergamon, Oxford, **1991**, p. 29; (c) W. Oppolzer, P. Schneider, *Tetrahedron Lett.* **1984**, *25*, 3305; (d) W. Oppolzer, A.F. Cunningham, *Tetrahedron Lett.* **1986**, *27*, 5467.

[27] J.-L. Moreau in *The Chemistry of Ketenes, Allenes and Related Compounds* (ed.: S. Patai), Wiley, New York, **1980**, p. 363 and references cited on pp. 365–370.

[28] S. R. Elliott, C. N. R. Rao, J. M. Thomas, *Angew. Chem.* **1986**, *98*, 31; *Angew. Chem. Int. Ed. Engl.* **1986**, *25*, 31.

[29] D. B. Malpass, L. W. Fannin, J. J. Ligi in *Kirk-Othmer Encycl. Chem. Techn., 3rd ed., Vol. 16*, **1981**, pp. 559–561.

[30] R. D. Rieke, S. E. Bales, *J. Am. Chem. Soc.* **1974**, *96*, 1775.

[31] F. Freijee, G. Schat, R. Mierop, C. Blomberg, F. Bickelhaupt, *Heterocycles* **1977**, *7*, 237.

[32] E. Bartmann, *J. Organomet. Chem.* **1985**, *284*, 149; E. Bartmann, *ibid* **1987**, *332*, 19.

[33] L. E. Aleandri, B. Bogdanović, A. Gaidies, D. J. Jones S. Liao, A. Michalowicz, J. Rozière, A. Schott, *J. Organomet. Chem.* **1993**, *459*, 87.

[34] H. Michel, D. Steiner, S. Wocadlo, J. Allwohn, N. Stamatis, W. Massa, A. Berndt, *Angew. Chem.* **1992**, *104*, 629; *Angew. Chem. Int. Ed. Engl.* **1992**, *31*, 607.

[35] A. Berndt, *Angew. Chem.* **1993**, *105*, 1034; *Angew. Chem. Int. Ed. Engl.* **1993**, *32*; 985. C. Wieczorek, J. Allwohn, G. Schmidt-Lukasch, R. Hunold, W. Massa, A. Berndt, *ibid*, **1990**, *102*, 435; *Int. Ed. Engl.* **1990**, *29*, 398.

[36] R. Goddard, C. Krüger, N. Ramadan, A. Ritter, *Angew. Chem.* **1995**, *107*, 1107; *Angew. Chem. Int. Ed. Engl.*, **1995**, *34*, 1030.

[37] H. Kuhn, Dissertation, Bochum University, 1983.

[38] G. Wilke, *Angew. Chem.* **1988**, *100*, 189; *Angew. Chem. Int. Ed. Engl.* **1988**, *27*, 185.

[39] G. Wilke, K. Angermund, G. Fink, C. Krüger, Th. Leven, A. Mollbach, J. Monkiewicz, S. Rink, H. Schwager, K.-H. Walter, *New Aspects of Organic, Chemistry II*, Kodansha, Tokyo **1992**, pp. 1–18.

[40] P. W. Jolly, U. Zakrzewski, *Polyhedron* **1991**, *10, 1427.*

[41] B. Gabor, S. Holle, P. W. Jolly, R. Mynott, *J. Organomet. Chem.* **1994**, *466*, 201.

[42] A. Döhring, R. Emrich, R. Goddard, P. W. Jolly, C. Krüger, *Polyhedron* **1993**, *12*, 2671.

[43] W. Geibel, G. Wilke, R. Goddard, C. Krüger, R. Mynott, *J. Organomet. Chem.* **1978**, *160*, 139; G. Wilke, *Pure Appl. Chem.* **1978**, *50*, 677.

[44] G. Wilke, H. Benn, R. Goddard, C. Krüger, B. Pfeil, *Inorg. Chim. Acta* **1992**, *198–200*, 741–748.

[45] B. Bogdanović, U. Wilczok, US Patents 5 133 929 (1992) and 5 215 710 (1993); priority date: DE 3613532, Apr. 22, 1986.

[46] B. Bogdanović, B. Spliethoff, *Int. J. Hydrogen Energy* **1987**, *12*, 863.

[47] (a) B. Bogdanović, K. H. Claus, S. Gürtzgen, B. Spliethoff, W. Wilczok, *J. Less-Common Met.* **1987**, *131*, 163; (b) B. Bogdanović, S. C. Huckett, U. Wilczok, A. Rufinska, *Angew. Chem.* **1988**, *100*, 1569; *Angew. Chem. Int. Ed. Engl.* **1988**, *27*, 1513; (c) B. Bogdanović, S. C. Huckett, B. Spliethoff, U. Wilczok, *Z. Phys. Chem. (Munich)* **1989**, *162*, 191; (d) D. Noreus, B. Bogdanović, U. Wilczok, *J. Less.-Common Met.* **1991**, *169*, 369; (e) D. J. Jones, J. Rozière, L. E. Aleandri, B. Bogdanović, S. C. Huckett, *Chem. Mater.* **1992**, *4*, 620.; (f) L. E. Aleandri, B. Bogdanović, U. Wilczok, D. Noreus, G. Block, *Z. Phys. Chem. (Munich)*, **1994**, *185*, 131.

[48] (a) J. J. Reilly, R. H. Wiswall, *Inorg. Chem.* **1968**, *7*, 2254; (b) H. Buchner, *Energiespeicherung in Metallhydriden,* Springer, Vienna, **1982**; (c) L. Schlapbach, (ed.), *Hydrogen in Intermetallic Compounds I and II*, (Topics in Applied Physics, Vols. 63 and 67), Springer, Vienna, Berlin, **1988** and **1992**.

[49] (a) L. Hütter, H. H. Stadelmaier, *Acta Metall.* **1958**, *6*, 367; (b) JCPDS File, Swarthmore, PA, Card 28–624, **1978**.

[50] T. Ohba et al., *Acta Crystallogr., Sect. C*; *Cryst. Struct. Commun.* **1984**, *1*, C40.

[51] H. Mimoun, E. Brazi, C. J. Cameron, E. Benazzi, P. Meunier, *New. J. Chem.* **1989**, *13*, 713; cf. also R. L. Burwell, Jr., *Chemtracts-Inorganic Chemistry* **1991**, *3*, 76.

[52] B. Bogdanović, U. Wilczok, Europ. Patent Appl. 0469463 (1992); priority date: July 31, 1990.

[53] L. E. Aleandri, B. Bogdanović, P. Bons, R. A. Brand, Ch. Dürr, A. Gaidies, Th. Hartwig, S. C. Huckett, M. Lagarden, U. Wilczok, *Chem. Mater.* **1995**, *7*, 1153.

[54] L. E. Aleandri, S. Becke, B. Bogdanović, D. J. Jones, J. Rozière, *J. Organomet. Chem.* **1994**, *472*, 97.

[55] J. E. McMurry, M. P. Fleming, *J. Am. Chem. Soc.* **1974**, *96*, 4708; J. E. McMurry, *Acc. Chem. Res.* **1974**, *7*, 281.

[56] J.-M. Pons, M. Santelli, *Tetrahedron* **1988**, *44*, 4295.
[57] (a) B. E. Kahn, R. D. Rieke, *Chem. Rev.* **1988**, *88*, 733; (b) D. Lenoir, *Synthesis* **1989**, 883; c) C. Betschart, D. Seebach, *Chimia* **1989**, *43*, 39; (d) J. E. McMurry, *Chem. Rev.* **1989**, *89*, 1513; (e) M. Garcia, C. del Campo, J. V. Sinisterra, E. F. Llama, *Tetrahedron Lett.* **1993**, *34*, 7973.
[58] R. Dams, M. Malinowski, I. Westdorp, H. J. Geise, *J. Org. Chem.* **1982**, *47*, 248.
[59] It is known in the literature that titanium pinacolates are intermediates in carbonyl coupling reactions yielding alkenes [57(b)–(d)]; see also E. J. Corey, R. L. Danheiser, S. Chandrasekaran, *J. Org. Chem.* **1976**, *41*, 260.
[60] K.-J. Range, P. Hafner, *J. Alloys & Compounds* **1992**, *183*, 430.
[61] H. H. Stadelmeier, W. K. Hardy, *Z. Metallk.* **1961**, *52*, 391.
[62] W. Bronger, W. Klemm, *Z. Anorg. Allg. Chem.* **1962**, *319*, 58.
[63] L. Westin, *Chem. Commun. Univ. Stockholm* **1972**, *5*, 1.
[64] F. Bonhomme, P. Selvam, M. Yoshida, K. Yvon, P. Fischer, *J. Alloys Compounds* **1992**, *178*, 167.
[65] L. E. Aleandri, B. Bogdanović, U. Wilczok, unpublished results.
[66] Review: B. Eisenmann, *Angew. Chem.* **1993**, *105*, 1764; *Angew. Chem. Int. Ed. Engl.* **1993**, *32*, 1693.
[67] L. E. Aleandri, B. Bogdanović, M. Lagarden, K. Schlichte, 24. GDCh-Hauptversammlung, Hamburg, 5–11 Sept. 1993, VCH, Hamburg, **1993**, p. 203.
[68] L. Rösch in *Inorganic Reactions and Methods, Vol. 10* 1st edition (eds.: J. J. Zuckerman, A. P. Hagen), **1987**, p. 327.
[69] (a) H. Felkin, P. J. Knowles, B. Meunier, *J. Chem. Soc. Chem. Commun.* **1974**, 44; (b) H. Felkin, P. J. Knowles, B. Meunier, *J. Organomet. Chem.* **1978**, *146*, 151.
[70] (a) M. L. H. Green, G. A. Moser, I. Packer, F. Petit, R. A. Forder, K. Prout, *J. Chem. Soc. Chem. Commun.* **1974**, 839; (b) K. Prout, R. A. Forder, *Acta Crystallogr.* **1975**, *B 31*, 852; (c) S. G. Davies, M. L. H. Green, K. Prout, A. Coda, V. Tazzoli, *J. Chem. Soc. Chem. Commun.* **1977**, 135.
[71] A. Kornowski, M. Giersig, R. Vogel, A. Chemseddine, H. Weller, *Adv. Mater.* **1993**, *5*, 634.
[72] S. Tyrlik, I. Wolochowicz, *Bull. Soc. Chim. Fr.* **1973**, 2147.
[73] P. Sobota, B. Jezowska-Trzebiatowska, *Coord. Chem. Rev.* **1978**, *26*, 71.
[74] B. Bogdanović, S. Liao, K. Schlichte, U. Westeppe in *Organometallic Syntheses, Vol. 4* (eds.: R. B. King, J. J. Eisch) Elsevier, Amsterdam, **1988**, p. 410.

9 Catalytically Active Metal Powders and Colloids

Helmut Bönnemann and Werner Brijoux

9.1 Introduction

Nanosize metals and alloys, in the form of colloids, are of considerable interest in industrial powder technology [1] and catalysis [2]. Fine metallic powders can be produced by metal evaporation [3], electrochemical processes [4], or chemical reduction of metal salts using naphthalene-activated alkali metals [5], anthracene-activated magnesium [6, 7], or alkali metal tetrahydroborates [8]. The latter reductive route has been widely used for the preparation of transition metal powders; however, the resulting metal powders are contaminated with borides [9]. Recently we have been able to show that by exploiting hydrotriorganoborates, the reduction of transition metal salts can be achieved in organic solvents under mild conditions to give X-ray amorphous nanopowders of metals in groups 6–12 and 14 of the periodic table, free of borides [10]. Furthermore, by co-reducing two metal salts simultaneously the corresponding binary intermetallics can be obtained [10]. The application of hydrotriorganoborates can be extended to the reductive deoxygenation of transition metal oxides and, for example, Fe/Ni ternary spinels to produce metal powders or intermetallics respectively. Specifically we have developed a method to transform needle-shaped iron oxides into acicular iron magnet pigments [11].

Surprisingly, the reduction of $TiCl_4$ by $K[BEt_3H]$ in THF yields ether-soluble Ti^0, stabilized by complexed THF [12]. This discovery led us to extend our preparative concept to include the synthesis of colloidal transition metals in organic media. By treating suspended metal salts in THF with tetraalkylammonium or -phosphonium hydrotriorganoborates, a large variety of colloidal transition metals of groups 6–11 became accessible [13]. The NR_4^+ or PR_4^+ salts, which are formed in a high local concentration at the reduction center, act as efficient ligands and prevent aggregation of the metal. Subsequent chemical studies [14] have revealed that the stabilizing tetraalkylammonium or -phosphonium halide can be pre-reacted with the metal salt, and the resulting intermediate reduced by conventional inorganic or organic reagents to give isolatable metal colloids having a narrow particle size distribution.

The object of this chapter is to review the preparative routes towards the production of nanoscale colloidal metals and alloys utilizing NR_4X and related surfactant molecules as stabilizing agents. In addition, the application of acicular iron powders for magnetic signal recording, and the use of the ether-soluble complex $[Ti^0 \cdot 0.5\ THF]_x$ as a metal-hydrogenation catalyst and a doping agent for heterogeneous catalysts will be discussed, together with the properties of NR_4^+- or PR_4^+-stabilized colloidal noble metals as liquid-phase hydrogenation catalysts.

9.2 Metal and Alloy Powders

9.2.1 Finely Divided Metals and Alloy Powders from Metals Salts

As mentioned in the introduction, the reduction of metal salts by alkali metal tetrahydroborates $M[BH_4]$ (M = Li, Na) in aqueous solution represents a general approach towards the production of transition metal powders [8] which are generally X-ray amorphous [15]. A disadvantage of this chemical procedure lies in the fact that the resulting metal powders are contaminated with borides. This is a result of the reducing potential of all four hydrogen atoms in the BH_4^- anion. For example, the reduction of $NiCl_2$ by $Na[BH_4]$ [Eq. (9.1)], applied on an industrial scale for the deposition of Ni on various supports, gives a nickel powder containing ca. 5 wt % boron in the form of Ni_3B [9].

$$10\,NiCl_2 + 8\,NaBH_4 + 17\,NaOH + 3\,H_2O \rightarrow (3\,Ni_3B + Ni) + 5\,NaB(OH)_4 + 20\,NaCl + 17.5\,H_2\uparrow \quad (9.1)$$

Alkali or alkaline earth metal hydrotriorganoborates, $M[BR_3H]$ [16], may formally be regarded as adducts of MH to BR_3. In contrast to $M[BH_4]$, the organoboron group BR_3 has no reducing properties, but functions solely as a complexing agent to stabilize the metal hydride in organic media. Exploiting this property we were able to develop a process for preparing boride-free metal nanopowders by reducing metal salts with such alkali or alkaline earth metal hydrides in organic solvents [10, 17]. The metal hydrides are solubilized in organic media using BR_3 or $BR_n(OR')_{3-n}$ (R, R′ = alkyl or aryl; $n = 0, 1, 2$) as complexing agents to form hydrotriorganoborates [16] of the general formula $M'H_u \cdot (BR_3)_u$ or $M'H_u \cdot [BR_n(OR')_{3-n}]_u$ (M′ = alkali or alkaline earth metal, $u = 1, 2$). Similarly, hydrotriorganogallates can be prepared using organogallium compounds $GaR_n(OR')_{3-n}$ ($n = 0, 1, 2, 3$) as complexing agents. Metal salts including elements of groups 6–12 and 14 of the periodic table can be successfully reduced with hydrotriorganoborates or gallates in organic solvents between –20 °C and reflux temperatures to give the corresponding metal powders in pure form. With the restriction that the solvent should not react with the hydrides, a great number of organic solvents were found to be suitable for the reduction process [Eq. (9.2)].

$$u\mathrm{MX}_v + v\mathrm{M'(BR_3H)}_u \xrightarrow{\mathrm{THF}} u\mathrm{M}\downarrow + v\mathrm{M'(BR_3X)}_u + \frac{uv}{2}\,\mathrm{H_2}\uparrow \quad (9.2)$$

M = metal powder
M′ = alkali or alkaline earth metal
R = alkyl, C_1–C_6
X = OH, OR, CN, OCN, SCN

Since the borate and gallate complexes formed during the reaction of hydroxides, alkoxides, cyanides, and thiocyanates with triorganoboranes and -gallanes are very soluble in organic solvents, the pure metal powder products can be isolated simply by filtering (Table 9-1). The triorganoboron complexing agent can be regenerated by treating the filtrate with acid. By treatment with HCl/THF, 98% of the triethylborane from $Na[(BEt_3)OH]$ can be recovered.

Sodium hydrotriethylborate in THF possesses a redox potential of –0.77 V, as determined electrochemically (supporting electrolyte: Bu_4NPF_6) [10]. Depending on the parti-

Table 9-1. Preparation of nanocrystalline metal powders in THF with formation of soluble borates [Eq. (9.2)].

No.	Metal salt	Reducing agent	Conditions		Product	
			t (h)	T (°C)	Metal content (%)	Boron content (%)
1	$Fe(OEt)_2$	$NaBEt_3H$	16	65	96.8	0.16
2	$Co(OH)_2$	$NaBEt_3H$	2	23	94.5	0.40
3	$Co(CN)_2$	$NaBEt_3H$	16	65	96.5	0.20
4	$Ni(OH)_2$	$NaBEt_3H$	2	23	94.7	0.13
5	$Ni(OEt)_2$	$NaBEt_3H$	16	65	91.4	0.58
6	CuCN	$LiBEt_3H$	2	23	97.3	0.0
7	CuSCN	$NaBEt_3H$	16	65	95.0	0.23
8	$Pd(CN)_2$	$NaBEt_3H$	16	65	95.5	1.38
9	AgCN	$Ca(BEt_3H)_2$ [a]	2	23	89.6	0.20
10	$Cd(OH)_2$	$NaBEt_3H$	2	23	97.9	0.22
11	$Pt(CN)_2$	$NaBEt_3H$	16	65	87.5	0.93
12	AuCN	$NaBEt_3H$	2	23	97.5	0.0

a) Solvent: diglyme.

cular metal involved, the reaction between the metal salt and the hydride [Eq. (9.2)] proceeds either as a reduction or a metathesis reaction. This can be rationalized on the basis of the HSAB concept [18]. The metal ion M^+ acts as a soft acid and thus competes with the moderately hard acid BR_3 [19] to form the soft basic hydride **1** [Eq. (9.3)]. If M^+ is a softer acid than BR_3, the hydride is transferred to M^+; the resulting [MH] intermediate decomposes spontaneously to give the metal M and hydrogen gas.

$$MX + M'(BR_3H) \xrightarrow{-M'X} \underset{\mathbf{1}}{(M\cdots H\cdots BR_3)} \rightarrow$$

$$[MH] + BR_3 \xrightarrow{-BR_3} M\downarrow + \frac{1}{2} H_2\uparrow \qquad (9.3)$$

Typically the borate or gallate complexes $M'(BR_3X)_u$ generated according to Equation (9.2) are not stable and dissociate to give $M'X_u$ and BR_3, as in Equation (9.4). In most cases the metal halide remains dissolved in the organic phase along with the liberated BR_3. This is especially true in the case of LiCl, LiBr, LiI, and NaI when using THF as the solvent. To obtain pure metal powders according to Equation (9.4) in one step, the cation in the hydride should be selected so that the resulting M′X remains in solution. If the M′ halide precipitates (e.g. NaCl), it must be separated from the metal powder by washing with water.

$$uMX_v + vM'(BR_3H)_u \xrightarrow{THF} uM\downarrow + vM'X_u + uvBR_3 + \frac{uv}{2} H_2\uparrow \qquad (9.4)$$

X = halogen

The reduction of transition or noble metal halides can also be achieved by adding only a “catalytic amount” of BR_3 to $M'H_u$ suspended in an organic solvent. BR_3 reacts with $M'H_u$ to form the hydrotriorganoborate in situ and is liberated following the reduction of

Table 9-2. Preparation of nanocrystalline metal powders in THF [Eq. (9.4)].

No.	Metal salt	Reducing agent	Conditions		Product	
			t (h)	*T* (°C)	Metal content (%)	Boron content (%)
1	$CrCl_3$	$NaBEt_3H$	2	23	93.3	0.3
2	$MnCl_2$	$LiBEt_3H$	1	23	94.0	0.42
3	$FeCl_3$	$LiBEt_3H$	2	23	97.1	0.36
4	CoF_2	$NaBEt_3H$	16	65	96.9	0.0
5	$CoCl_2$	$NaBEt_3H$	16	65	95.1	0.0
6	$CoCl_2$	LiH + 10% BEt_3	16	65	95.8	0.0
7	$CoBr_2$	$LiBEt_3H$	2	23	86.6	0.0
8	$NiCl_2$	$NaBEt_3H$	16	65	96.9	0.0
9	$CuBr_2$	$LiBEt_3H$	2	23	94.9	0.0
10	$CuCl_2$	$Na(Et_2BOMe)H$	2	23	94.7	0.1
11	$ZnCl_2$	$LiBEt_3H$	12	65	97.8	0.0
12	$RuCl_3$	$NaBEt_3H$	16	65	95.2	0.52
13	$RhCl_3 \times 3\,H_2O$	$NaBEt_3H$	2	23	98.1	0.1
14	$RhCl_3$	$LiBEt_3H$	2	23	96.1	0.66
15	$PdCl_2$	$NaBEt_3H$	16	65	98.0	0.29
16	AgF	$NaB(OMe)_3H$	2	23	94.1	0.05
17	AgJ	$NaBEt_3H$	2	23	95.3	0.02
18	$CdCl_2$	$LiBEt_3H$	2	23	99.4	0.0
19	$ReCl_3$	$LiBEt_3H$	2	23	95.4	0.0
20	$OsCl_3$	$NaBEt_3H$	2	23	95.8	0.0
21	$IrCl_3 \times 4\,H_2O$	$NaBEt_3H$	216	23	77.1	0.16
22	$IrCl_3$	$KBPr_3H$	2	65	94.7	0.08
23	$PtCl_2$	$NaBEt_3H$	5	23	98.2	0.21
24	$PtCl_2$	LiH + 10% BEt_3	12	65	98.8	0.0
25	$PtCl_2$	$LiBEt_3H$	4	65	99.0	0.0
26	$PtCl_2$	$LiBEt_3H$	2	0	99.0	0.0
27	$SnCl_2$	$LiBEt_3H$	2	23	96.7	0.0
28	$SnBr_2$	$LiBEt_3H$	2	23	87.1	0.0
29	$PdCl_2$	$Na(GaEt_2OEt)H$	2	40	92.7	Ga: 0.25
30	$Pt(NH_3)_2Cl_2$	$NaBEt_3H$	2	23	97.1	0.32
31	$Pt(Py)_2Cl_2$	$LiBEt_3H$	2	23	97.1	0.02
32	$Pt(Py)_4Cl_2$	$LiBEt_3H$	2	23	97.5	0.01
33	COD $PtCl_2$	$NaBEt_3H$	2	60	97.9	0.58

the MX_v according to Equation (9.4), ready to react with the next $M'H_u$ regenerating the hydrotriorganoborate complex [Eq. (9.5); Table 9-2, Nos. 6 and 24].

$$uMX_v + vM'H_u \xrightarrow{BR_3,\ THF} uM\downarrow + vM'X_u + \frac{uv}{2}\,H_2\uparrow \qquad (9.5)$$

X = halogen

Because ammonium halides are completely soluble in THF, the application of, for example, tetrabutylammonium hydrotriorganoborates, which are readily accessible from ammonium halides and alkali metal hydrotriorganoborates [13], may be advantageous for the reduction of metal halides in THF. The ammonium halide by-product formed during the reaction given in Eq. (9.6) remains completely dissolved, thus allowing the easy isolation of the precipitated metal powder in pure form. Some typical examples of

Table 9-3 Preparation of nanocrystalline metal powders in THF [Eq. (9.6)].

No.	Metal salt	Reducing agent	Conditions		Product	
			t (h)	T (°C)	Metal content (%)	Boron content (%)
1	$CoCl_2$	NBu_4BEt_3H	3	23	93.6	0.3
2	$PdCl_2$	NBu_4BEt_3H	3	23	96.2	0.47
3	$PtCl_2$	NBu_4BEt_3H	3	23	97.8	0.2
4	AgCl	NBu_4BEt_3H	3	23	98.0	0.2
5	$AuCl_3$	NBu_4BEt_3H	3	23	97.5	0.2
6	COD $PdCl_2$	NBu_4BEt_3H	3	23	86.7	0.2

the preparation of nanopowders from metal halides according to Eq. (9.6) are summarized in Table 9-3.

$$MX_v + vNR_4(BEt_3H) \xrightarrow{THF} M\downarrow + vNR_4X + vBEt_3 + \frac{v}{2} H_2\uparrow \tag{9.6}$$

X = halogen
v = 1, 2, 3
R = n-butyl

The isolated metal powders were characterized by electron microscopy, X-ray diffraction, and differential scanning calorimetry. A DSC diagram (Fig. 9-1) of a boron-free cobalt sample, prepared from $CoCl_2$ according to Eq. (9.4) (Table 9-2, No. 5), shows the endothermic recrystallization of the powder between 230 °C and 400 °C with a heat flow minimum at 298 °C. An enthalpy difference $\Delta H_R(Co) = 24.78$ J g^{-1} was calculated for the recrystallization. By comparing the melting and recrystallization enthalpy values, the degree of crystallinity of a given sample can be estimated. The melting enthalpy value $\Delta H_F(Co)$ is 259.63 J g^{-1} [20]. Since the value of $\Delta H_R(Co)$ was found to be only ca. 9.5% of the quoted melting enthalpy $\Delta H_F(Co)$, it can be concluded that the cobalt powder measured is nanocrystalline.

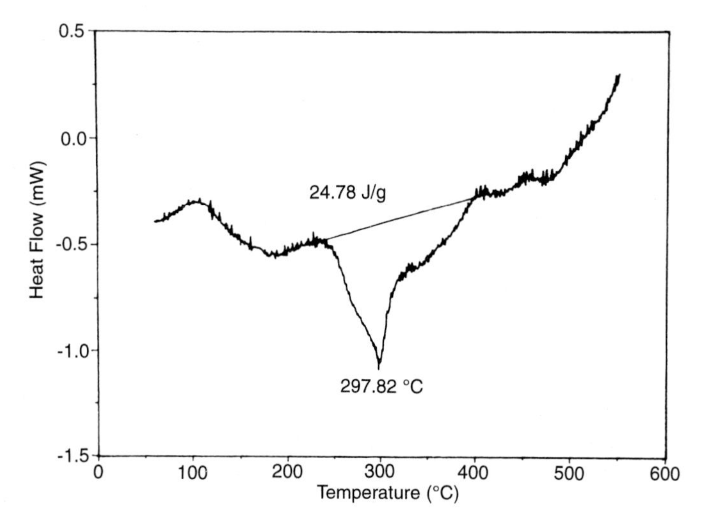

Fig. 9-1. DSC diagram of nanocrystalline cobalt powder (Table 9-2, No. 5); heating rate: 5 K min^{-1}. (Reprinted with the kind permission of Elsevier Science B.V. [42c]).

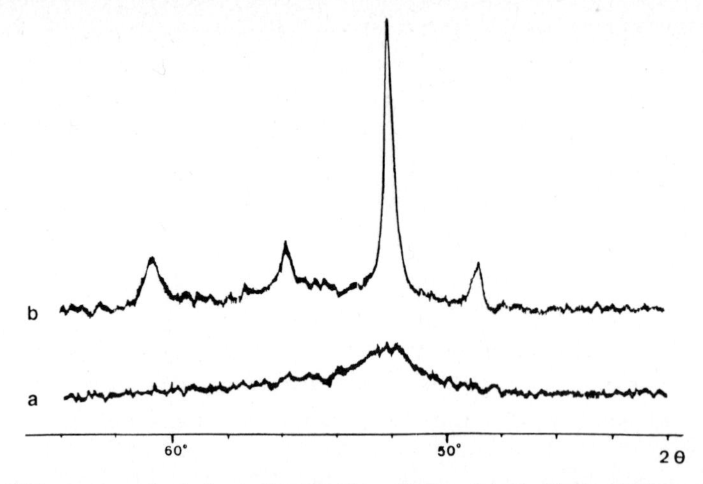

Fig. 9-2. X-ray diffractogram of cobalt powder (Co-K_α radiation with Fe filter). (a) Untreated original sample after reduction (nanocrystalline to X-ray amorphous); (b) after 3 hours at 450 °C; Co (hcp and/or fcc). (Reprinted with the kind permission of Elsevier Science B.V. [42e]).

The line widths observed in the X-ray diffraction patterns of the metal powders (Tables 9-1 to 9-3) indicate that the materials are generally nanocrystalline (X-ray amorphous). For example, X-ray powder patterns of a cobalt powder (Table 9-2, No. 5) are shown in Fig. 9-2, before (a) and after (b) annealing at 450 °C. The original sample (a) shows a single, very diffuse line which is typical for an amorphous or nanocrystalline material [21]. After annealing four sharp lines (b) are recorded, the strongest of which has a *d*-spacing of 2.04 Å corresponding to the (111) line of metallic cobalt. Since the mean particle size of the cobalt powder (100 nm) remains the same after annealing [as confirmed by scanning electron microscopy (SEM)] the sharp diffraction pattern given in Fig. 9-2 probably originates from the growth of the grains (nanocrystallites) inside the individual cobalt particles.

Table 9-4 summarizes the grain sizes of selected metal powders prepared by chemical reduction with hydrotriorganoborates, as determined by TEM (transmission electron microscopy). The observed grain size as well as the statistical distribution of the various metal powders depends strongly on the particular elements and not on the synthesis route. Metal powders having uniform size distributions may be isolated from colloidal metal solutions (see below).

In 1986 van Wontherghem et al. reported the preparation of nanoscale X-ray amorphous alloy particles by the co-reduction of $FeSO_4$ and $CoCl_2$ (molar ratio 7:3) with an excess of $K[BH_4]$ in aqueous solution [Eq. (9.7)] [22]. As determined by elemental analysis, the particles reveal the composition $Fe_{44}Co_{19}B_{37}$. The low saturation magnetization values measured

Table 9-4. Typical grain sizes of metal powders measured by TEM.

No.	Metal	Starting materials		Preparation	Grain size (nm)
1	platinum	$PtCl_2$	$LiBEt_3H$	Table 9-2, No. 25	2–5
2	palladium	$PdCl_2$	$LiBEt_3H$	analog. Table 9-2, No. 15	12–28
3	rhodium	$RhCl_3$	$LiBEt_3H$	Table 9-2, No. 14	1–4
4	copper	$CuCl_2$	$LiBEt_3H$	analog. Table 9-2, No. 9	25–90
5	nickel	$Ni(OH)_2$	$NaBEt_3H$	Table 9-1, No. 4	5–15
6	cobalt	$CoCl_2$	$NaBEt_3H$	Table 9-2, No. 5	3–5

for the sample before and after annealing (89 and 166 J T^{-1} kg^{-1} respectively) indicate the presence of boron, either in the alloy or as a separate phase. As mentioned earlier, since all four hydrogen atoms in BH_4^- take part in the reduction process described by Equation (9.7), the incorporation of boron (here 7.85%) in the targeted $Fe_{70}Co_{30}$ alloy is unavoidable.

$$\underset{7\quad:\quad3}{FeSO_4 + CoCl_2} \xrightarrow[H_2O/RT]{KBH_4} Fe_{44}Co_{19}B_{37} \qquad (9.7)$$

This problem can be circumvented by using hydrotriorganoborates for the co-reduction of mixtures of different metal salts, dissolved or suspended in THF, according to Equations (9.2) and (9.4). This method affords practically boron-free nanocrystalline alloys of two or more metals (Table 9-5). Within the known miscibility limits of a particular metal combination (taken from the phase diagram), the atomic ratio of the metals in the

Table 9-5. Preparation of nanocrystalline alloys by co-reduction of metal salts in THF.

No.	Metal salt	Reducing agent	Conditions		Product		DIF[a]		Comments
			t (h)	T (°C)	Metal content (%)	Boron content (%)	2θ[b] (°)	D[c] (Å)	
1	$Co(OH)_2$ $Ni(OH)_2$	$NaBEt_3H$	7	65	Co: 48.3 Ni: 45.9	0.25	51.7	2.05	Single phase nanocrystalline
2	$FeCl_3$ $CoCl_2$	LiH + 10% BEt_3	6	65	Fe: 47.0 Co: 47.1	0.0	52.7	2.02	Single phase grain size: 1–5 nm
3	$FeCl_3$ $CoCl_2$	$LiBEt_3H$	5	23	Fe: 54.8 Co: 24.5	0.0	52.5 99.9	2.02 1.17	Single phase nanocrystalline
4	$CoCl_2$ $PtCl_2$	$LiBEt_3H$	7	65	Co: 21.6 Pt: 76.3	0.0	55.4 47.4	1.93 2.23	Single phase
5	$RhCl_3$ $PtCl_2$	$LiBEt_3H$	5	65	Rh: 26.5 Pt: 65.5	0.04	40.2 46.3	2.24 1.96	Single phase grain size: 1–4 nm[d]
6	$RhCl_3$ $IrCl_3$	$LiBEt_3H$	5	65	Rh: 33.5 Ir: 62.5	0.15	42.3	2.14	Single phase + traces $IrCl_3$
7	$PdCl_2$ $PtCl_2$	$LiBEt_3H$	5	65	Pd: 33.6 Pt: 63.4	0.04	40.1 46.3	2.25 1.96	Single phase grain size: 2–6 nm[d]
8	$PtCl_2$ $IrCl_2$	$NaBEt_3H$	12	65	Pt: 50.2 Ir: 48.7	0.15	40.0 46.5	2.25 1.95	Single phase nanocrystalline
9	$CuCl_2$ $SnCl_2$	$LiBEt_3H$	4	65	Cu: 49.6 Sn: 47.6	0.0	30.2 53.5	2.96 1.80	Cu_6Sn_5 + Cu + Sn
10	$FeCl_3$ $CoCl_2$ $NiCl_2$	$LiBEt_3H$	1.5	23	Fe: 30.1 Co: 31.4 Ni: 30.9	0.0	52.7 60.8 77.7 100.3	2.02 1.77 1.43 1.17	Single phases nanocrystalline

a) X-ray diffractogram, CoK_α radiation, Fe filter; b) Strongest reflections; c) Reciprocal lattice distance; d) Measured by TEM.

resulting alloy may be adjusted by altering the proportion of the starting materials. For example, the co-reduction of $FeCl_3$ and $CoCl_2$ in the molar ratio 7:3 according to Equation (9.4) yields a boron-free $Fe_{70}Co_{30}$ powder (Table 9-5, No. 3). In the X-ray diffraction pattern only a single diffuse reflection with a *d*-spacing of 2.02 Å can be discerned (Fig. 9-3a). Measurement of the line width indicates that Fe/Co powder consists of a weakly crystalline to amorphous phase(s). After annealing (2 h, 400 °C) the diffraction line becomes sharper (Fig. 9-3b). Since the particle size remains unchanged after annealing (as shown by SEM), the sharpening of the reflection suggests that the grain size within the particles increases during thermal treatment.

A comparison of the X-ray diffraction patterns obtained for the alloy (Fig. 9-3) with that observed for a mixture of amorphous iron and cobalt powders (Fig. 9-4) suggests that the $Fe_{70}Co_{30}$ alloy is formed during the co-reduction step under thermodynamic control, and not as a result of subsequent heat treatment; the diffraction pattern of the annealed Fe/Co mixture (Fig. 9-4b) is, in contrast to that of the annealed alloy (Fig. 9-3b), a superimposition of the two set of lines for Fe and Co.

Unambiguous proof that an Fe/Co alloy is formed directly during co-reduction in solution was provided by Mössbauer spectroscopy. The Mössbauer spectrum (source: ^{57}Co/Pt, 27 °C) of an $Fe_{64}Co_{36}$ sample prepared by the coreduction of $FeCl_3$ and $CoCl_2$ is

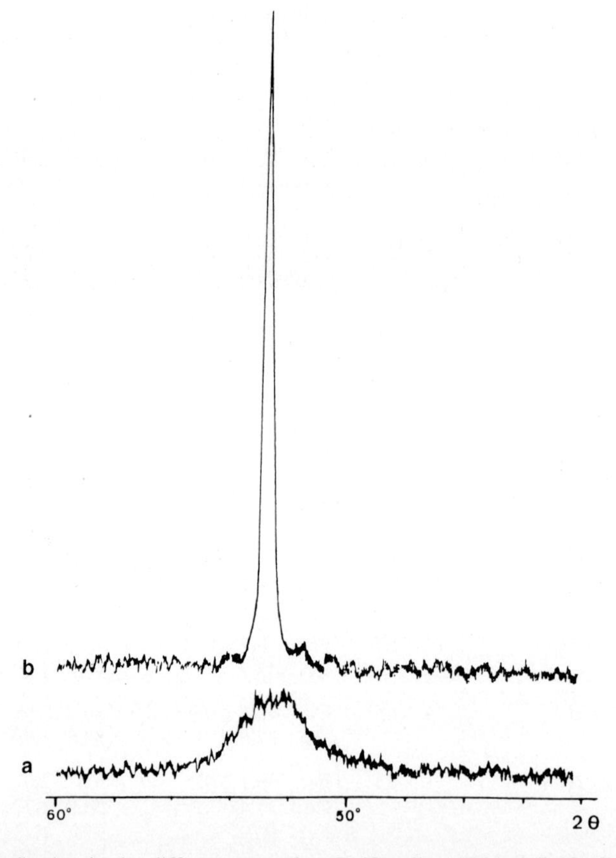

Fig. 9-3. Strongest reflection in the diffractogram of an Fe/Co alloy (Table 9-5, No 3). (a) Untreated original sample; (b) after 2 h at 400 °C. (Co-K_α radiation with Fe filter). (Reprinted with the kind permission of Elsevier Science B.V. [42e]).

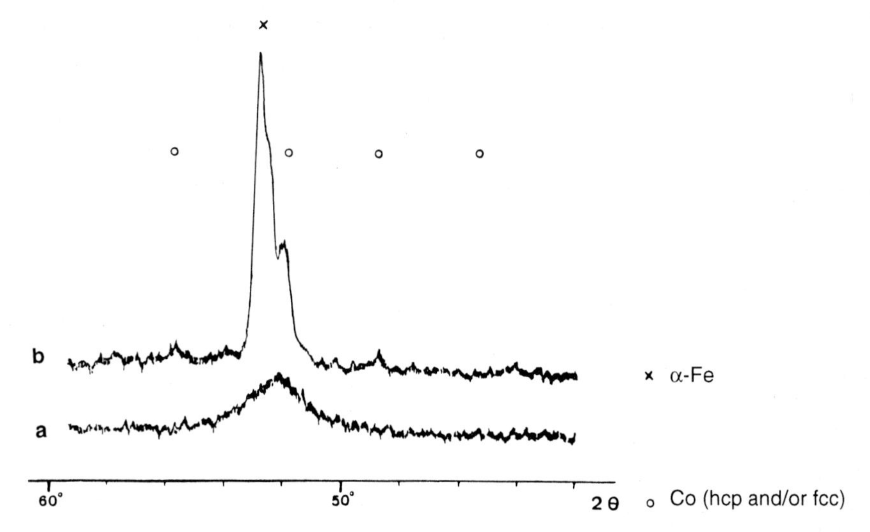

Fig. 9-4. Strongest reflection in the diffractogram of an Fe/Co mixture. (a) Untreated original sample; (b) after 2 h at 400 °C. (Co-K_α radiation with Fe filter). (Reprinted with the kind permission of Elsevier Science B.V. [42c]).

shown in Fig. 9-5 and compared with the spectrum of a well-characterized sample of the same composition [23] isolated by melt metallurgy techniques. The spectrum of the "co-reduction product" $Fe_{64}Co_{36}$ (Fig. 9-5a) reveals six resonances for metallic iron (but not elemental iron) with a hyperfine-field splitting of 36 T. The reference metallurgical

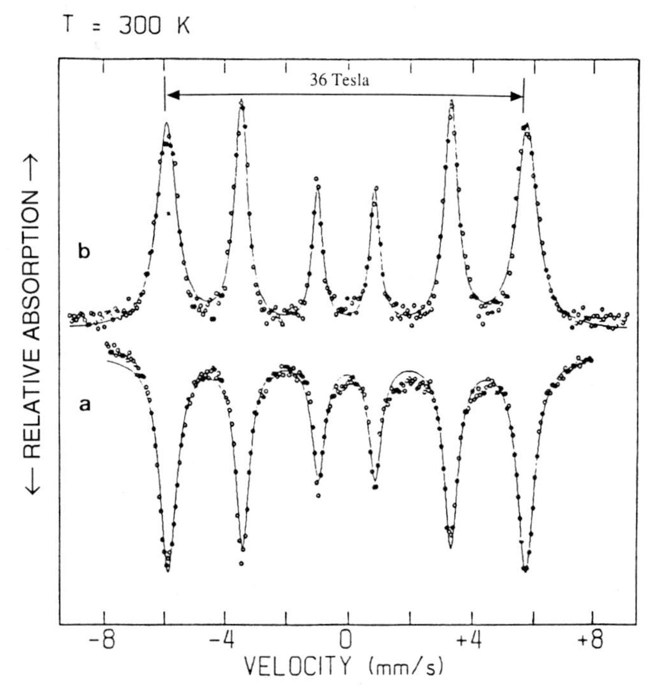

Fig. 9-5. Mössbauer spectra of (a) $Fe_{64}Co_{36}$ alloy obtained by coreduction, (b) $Fe_{60}Co_{40}$ alloy obtained by melting. Source: ^{57}Co/Pt, 27 °C. (Reprinted with the kind permission of Elsevier Science B.V. [42c]).

$Fe_{60}Co_{40}$ sample gives an identical spectrum (Fig. 9-5b), which confirms that iron and cobalt are in fact completely alloyed in the sample prepared by co-reduction.

Hence, the co-reduction of metal salts with hydrotriorganoborates in organic solvents provides a general route for the preparation of single-phase, X-ray amorphous alloys of two or more components (Table 9-5). Recently this method has been extended by Zeng and Hampden-Smith for the room temperature synthesis of Mo_2C and W_2C by the reduction of THF suspensions of $MoCl_4 \cdot 2\,THF$, $MoCl_3 \cdot 3\,THF$, and WCl_4 using $Li[BEt_3H]$ [24].

9.2.2 Metal and Alloy Powders from Metal Oxides

In analogy to Equation (9.2), hydrotriorganoborates can also be employed for the reduction of transition metal oxides under mild conditions in organic solvents to give practically boron-free metal powders. The deoxygenation of HgO, PtO_2, Ag_2O, and PdO in various organic media takes place smoothly even at moderate temperatures (20–130 °C; Table 9-6, Nos. 1–5). The course of the reaction has not yet been elucidated in full detail; however, the first step of the deoxygenation presumably entails the transformation of the metal oxo-groups into hydroxy-groups. This assumption is supported by the fact that the application of an excess amount of the hydrotriorganoborate is advantageous in order to afford a clean and complete deoxygenation of the metal oxide. Since the hydrotriorganoborate by-products readily dissolve in organic solvents, the resulting metal powders can be isolated by simple filtration. The overall reaction is represented by Equation (9.8), specifically for the deoxygenation of PtO_2 (Table 9-6, No. 2).

$$PtO_2 + NaBEt_3H\ (excess) \xrightarrow{4\,h,\ 20\,°C,\ THF} Pt\downarrow + NaBEt_3OH \tag{9.8}$$

The reduction of less noble metal oxides, such as CuO, is improved by adding hydrogen to the reaction system. The purity of the resulting copper powder is significantly

Table 9-6. Preparation of nanocrystalline metal powders from metal oxides in organic solvents.

No.	Metal oxide	Reducing agent	Solvent	Conditions			Product	
				H_2 pressure (bar)	t (h)	T (°C)	Metal content (%)	Boron content (%)
1	HgO	$NaBEt_3H$	Toluene	–	16	20 → 90	97.72	1.21
2	PtO_2	$NaBEt_3H$	THF	–	4	20	97.54	0.55
3	Ag_2O	$NaBEt_3H$	Toluene	–	16	20	97.66	0.10
4	Ag_2O	$NaBEt_3H$	Triglyme	–	16	20	97.95	0.13
5	PdO	$NaBEt_3H$	Toluene	–	16	80	98.22	0.03
6	CuO	$NaBEt_3H$	Toluene	–	16	20	89.8	0.0
7	CuO	$NaBEt_3H$	Toluene	100	16	130	96.3	0.0
8	NiO	$NaBEt_3H$	Toluene	100	16	130	94.05	0.0
9	NiO	$NaBEt_3H$	THF	100	16	130	98.4	0.13
10	CoO	$NaBEt_3H$	Toluene	3	16	130	98.1	–
11	CoO	$NaBEt_3H$	Toluene	100	4	130	97.96	–
12	CoO	$KBEt_3H$	THF	100	16	130	97.2	0.28
13	FeOOH	$NaBEt_3H$	Toluene	100	16	80	87.87	–
14	FeOOH	$NaBEt_3H$	THF	100	16	80	87.5	–
15	Fe_2O_3	$NaBEt_3H$	Toluene	100	16	80	88.3	–

higher (Table 9-6, No. 7) than that produced without additional hydrogen (No. 6). In contrast, the deoxygeneration of non-noble metal oxides with hydrotriorganoborates requires the presence of compressed hydrogen (see Table 9-6, Nos. 8–15).

X-ray powder diffraction measurements of the metal powders obtained by the deoxygenation of metal oxides show that the products are generally nanocrystalline. The grain size of the nickel and cobalt powders, as determined by TEM and quoted in Table 9-6, ranges from 6–10 nm (Nos. 8 and 9) to 1–5 nm (Nos. 10–12).

An interesting application of this deoxygenation reaction involves the transformation of finely-divided, needle-shaped iron oxides into acicular iron magnet pigments for magnetic signal recording [25]. The conventional industrial process for the production of such acicular iron pigments consists of reducing the corresponding iron oxides or oxide-hydroxides with excess gaseous hydrogen [Eq. (9.9)] at temperature generally above 250 °C in a fluid bed reactor [26].

$$\alpha\text{-}Fe_2O_3 + 3\,H_2 \xrightleftharpoons{400\,°C,\ 16\,h} 2\,Fe_{acicular} + 3\,H_2O \tag{9.9}$$

This process possesses the inherent disadvantage that the very labile acicular iron pigments break in the fluid bed reactor and agglomerate at the high reaction temperatures employed. Using hydrotriorganoborates in the presence of compressed hydrogen (20–100 bar), we were able, for the first time, to reduce iron oxides at only 80 °C to generate well-shaped acicular iron particles [Eqs. (9.10), (9.11)]. The metal oxide is suspended in an organic solvent and reduced using a mixture of hydrogen gas and a soluble hydrotriorganoborate. Surprisingly, under these mild reaction conditions oxygen is quantitatively removed from the iron oxide lattice. The pyrophoric iron pigments can be isolated from the clear organic solution by filtering. The trialkylborane acts as a complex carrier for both the metal hydride and the resulting metal hydroxide in the organic medium, and may be re-used after hydrolysis of the triorganohydroxoborate [25]. Fig. 9-6 shows a TEM image of the α-FeOOH used as the starting material for the deoxygenation reaction, Equation (9.10) (Table 9-6, No. 13). Well-shaped small needles about 350 nm long appear at a magnification of 50000:1. Even though the volume is observed to decrease by about 60% relative to the starting material during the preparation of the metal pigment, the needle shape is retained according to TEM. The product, shown in Fig. 9-7 on the same scale, consists of acicular iron pigment particles approximately 250 nm long with a diameter of about 20 nm. No evidence of sintering was noted.

$$\alpha\text{-}FeOOH + 2\,NaBEt_3H \xrightarrow{20\,bar\,H_2,\ 80\,°C,\ 4\,h} Fe_{acicular} + 2\,NaBEt_3OH \tag{9.10}$$

$$\alpha\text{-}Fe_2O_3 + 3\,NaBEt_3H \xrightarrow{20\,bar\,H_2,\ 80\,°C,\ 4\,h} 2\,Fe_{acicular} + 3\,NaBEt_3OH \tag{9.11}$$

The acicular iron pigments prepared according to Equations (9.10) and (9.11) reveal superior magnetic properties. The crucial characteristics of magnetic materials for recording applications include the saturation magnetization M_S, remanence M_R, and coercive field strength H_C. In practice a tuning of the coercive field force H_C of a magnet pigment is desirable. The magnetic properties of the iron pigments from the high temperature reduction process, Equation (9.9), can only be adjusted by varying the particle size and shape of the initial iron oxide reactant. In contrast, the low temperature deoxygenation according to Equations (9.10) and (9.11) allows the preparation of acicular iron pigments which exhibit high values of the saturation magnetization M_S and remanence M_R along

Fig. 9-6. TEM image of α-FeOOH (Table 9-6, No. 13). (Reprinted with the kind permission of Elsevier Science B.V. [42c]).

Fig. 9-7. TEM image of Fe (Table 9-6, No. 13). (Reprinted with the kind permission of Elsevier Science B.V. [42c]).

with a coercive force value which can be simply tuned for a particular application by thermal after-treatment (without causing undesirable sintering of the metal needles). Table 9-7 lists the H_C-values of the iron pigment quoted in Table 9-6 (No. 13) obtained after various thermal treatments. It can be increased from 82 kA m^{-1} in the initial product to over 100 kA m^{-1} by heating at temperatures above 200 °C in hydrogen or argon atmosphere or in vacuo. This change in H_C is probably the result of an augmentation of the grains (nanocrystallites) present in the acicular particles during the annealing.

By using hydrotriorganoborates and compressed hydrogen in combination, the deoxygenation of ternary oxides can also be achieved under mild conditions, giving nanocrystalline alloy powders (Table 9-8). For example, the spinel $NiFe_2O_4$ can be easily transformed into $NiFe_2$, as a single phase, at 130 °C in toluene [Eq. (9.12); Table 9-8, No. 2].

$$NiFe_2O_4 + 4\ NaBEt_3H \xrightarrow[\text{toluene}]{100\text{ bar } H_2,\ 130\,°C} NiFe_2 + 4\ NaBEt_3OH \tag{9.12}$$

Table 9-7. Adjustment of the coercive field force H_C of Fe powder (Table 9-6, No. 13) by thermal after-treatment (4 h).

After-treatment	M_S [mT cm^3 g^{-1}]	M_R [mT cm^3 g^{-1}]	H_C [kA m^{-1}]
none	193.0	101.3	82.2
150 °C (H_2 atmosphere)	–	–	83.0
200 °C (H_2 atmosphere)	–	–	99.8
250 °C (H_2 atmosphere)	199.0	111.6	101.6
250 °C (Ar atmosphere)	204.0	114.0	100.4
250 °C (vacuum)	203.0	113.0	101.4
300 °C (H_2 atmosphere)	–	–	98.2

Table 9-8. Preparation of alloys from spinels and metal oxides in organic solvents.

No.	Metal oxides	Reducing agent	Solvent	Conditions		
				H_2 pressure (bar)	t (h)	T (°C)
1	$CoNiO_2$	$NaBEt_3H$	Toluene	100	16	130
2	$NiFe_2O_4$	$NaBEt_3H$	Toluene	100	16	130
3	$CoFe_2O_4$	$NaBEt_3H$	THF	100	16	120
4	CoO NiO	$NaBEt_3H$	Toluene	100	16	130
5	NiO Fe_2O_3	$NaBEt_3H$	Toluene	100	16	130
6	CoO Fe_2O_3	$NaBEt_3H$	Toluene	100	16	130

No.	Product		DIF[a]		Comments
	Metal content (%)	Boron content (%)	$2\theta^b$ (°)	D^c (Å)	
1	Co: 45.3 Ni: 45.9	0.0	52.1	2.04	Single phase (fcc)
2	Ni: 29.0 Fe: 63.67	0.0	52.3	2.03	Single phase (bcc)
3	Co: 30.3 Fe: 61.1	0.45	52.5	2.02	Single phase (bcc)
4	Co: 43.7 Ni: 47.9	0.48	52.1	2.04	Single phase (fcc)
5	Ni: 48.5 Fe: 47.2	0.27	52.2	2.03	Single phase (fcc)
6	Co: 48.7 Fe: 48.4	0.36	52.2	2.03	Single phase (fcc)

a) X-ray diffraction, CoK_α radiation, Fe filter; b) Strongest reflection; c) Reciprocal lattice distance.

Surprisingly, the preparation of nanocrystalline metal alloys can also be effected through the deoxygenation of mixtures of binary metal oxides suspended in organic media, under reaction conditions analogous to those given for Equation (9.12). The advantage of this second route is that the ratio of the metals in the product alloy can be adjusted by varying the ratio of the metal oxides in the starting mixture. X-ray diffraction confirms the formation of the binary alloys and shows that there is no contamination by non-alloyed metals in their elemental form (see Table 9-8).

9.3 Formation of Colloidal Transition Metals and Alloys in Organic Solvents

Due to their catalytic potential, colloidal mono- and bimetallic materials have, after a period of hibernation, recently attracted more attention, notably through the contributions of Boutonnet [27], Bradley [28], Braunstein [29], Esumi [30], Evans [31], Heaton [32], Henglein [33], Klabunde [34], Knözinger [35], Larpent and Patin [36], Lewis [37], Moiseev [38], Schmid [39], and Toshima [40]. The first nanosized metals stabilized by tetraalkylammonium surfactants were reported in 1979 by Grätzel [41]. In the course of our research in this area, we have developed a new method for the preparation of stable and very soluble metal colloids in organic solvents having a narrow particle size distribution. The metallic core, derived from elements of groups 6–11 of the periodic table, is protected by tetraalkylammonium or -phosphonium halides [13]. However, in the case of titanium, metal stabilization is achieved by the solvent THF to give an ether-soluble compound $[Ti^0 \cdot 0.5\,THF]_x$ [12]. These novel colloidal systems have a narrow particle size distribution, are very stable, and may be handled at large concentrations [42]. Furthermore, electron microscopy confirmed that the discrete metal particles may be deposited onto supports without any unwanted agglomeration. So these materials fulfil all the major prerequisites necessary for precursors employed in heterogeneous catalysis.

9.3.1 Ether-Stabilized Ti^0, Zr^0, V^0, Nb^0, and Mn^0

The reduction of $TiCl_4 \cdot 2\,THF$ or $TiCl_3 \cdot 3\,THF$ in THF using $K[BEt_3H]$ gives, with the evolution of H_2, within 2 h a brown-black solution from which about 90% of the precipitated KCl can be removed by filtration. After evaporating the solvent and BEt_3 (identified by ^{11}B NMR spectroscopy) under vacuum, a black residue is obtained which is subsequently extracted with THF. The resulting THF solution is treated with pentane to generate the brown-black precipitate [**2** in Eq. (9-13)]. After thoroughly drying in vacuo a pyrophoric powder (**3**) containing small amounts of KCl (as identified by X-ray diffraction) is obtained. Compound **3** readily dissolves in THF and ether but is insoluble in hydrocarbons such as pentane.

$$x[TiCl_4 \cdot 2\,THF] + 4\,K[BEt_3H] \xrightarrow[2\,h,\,40\,°C]{THF}$$

$$\underset{\mathbf{2}}{[TiH_m \cdot 0.5\,THF]_x} + 4x\,BEt_3 + 4x\,KCl + x(2 - m/2)\,H_2\uparrow \qquad (9.13)$$

$$\Big\downarrow \text{vacuum} \quad \begin{matrix} -BEt_3 \\ -THF \\ -H_2 \end{matrix}$$

$$\underset{\mathbf{3}}{[Ti \cdot 0.5\,THF]_x}$$

The volume of gas evolved during the reduction was quantitatively measured and amounted to only 1 mol H_2 per mol Ti. This observation, together with results from the protonolysis of **2** and cross experiments using $K[BEt_3D]$ as the reducing agent, indicates

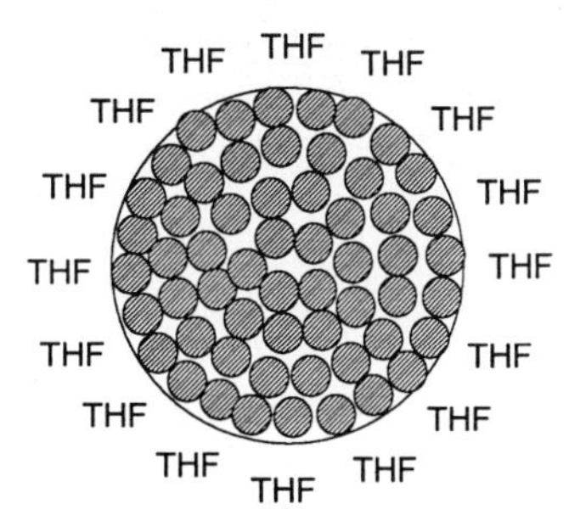

Fig. 9-8. Ether-soluble $[Ti^0 \cdot 0.5\ THF]_x$. (Reprinted with permission of Elsevier Science B.V. [42c]).

that **2** contains residual hydrogen in variable quantities. The IR spectrum of **2** shows that THF is coordinated to the metal core. Further characterization of **2** by X-ray diffraction, XPS, and EXAFS [12] supports its description as a colloidal titanium system stabilized by complexed THF and containing (interstitial) hydrogen (Fig. 9-8). Dehydrogenation of **2** under vacuum at room temperature gives **3** [Eq. (9-13)]. For experimental details see [12].

The determination of the particle size of this extremely oxophilic titanium colloid by HRTEM (high resolution transmission electron microscopy) failed. No particles could be detected. But the EDX analysis clearly shows titanium and smaller amounts of potassium and chlorine.

The corresponding Zr colloid was isolated by slowly adding the THF solution, after filtration from KCl, to pentane, in which the Zr colloid precipitates. The work-up of the V and Nb colloids was performed similarly. The reduction of the THF adduct of $MnBr_2$ at 40 °C yielded a stable, isolated colloid $[Mn \cdot 0.3\ THF)_x$ containing typically 60% Mn [Eq. (9-14)]. IR and NMR data for the colloid show intact THF coordinated to Mn. There is no evidence for ether cleavage. HRTEM showed the fringes of Mn particles of size 1–2.5 nm. An EDX analysis of the Mn nanoparticles showed no bromine to be present. This result confirms the assumption that colloidal Mn^0 was formed [43].

$$x[MnBr_2 \cdot 2\ THF] + 2x\ K[BEt_3H] \xrightarrow[2\ h,\ 40\ °C]{THF}$$

$$[Mn \cdot 0.3\ THF]_x + 2x\ BEt_3 + 2x\ KBr + x\ H_2\uparrow \qquad (9.14)$$

9.3.1.1 Hydrogenation of Ti and Zr Sponge with $[Ti^0 \cdot 0.5\ THF]_x$

In addition to the use of **2** or **3** as a dopant for noble metal hydrogenation catalysts (see below), the colloidal $[Ti^0 \cdot 0.5\ THF]_x$ material is a very efficient catalyst for the hydrogenation of titanium and zirconium sponges as well as for a nickel hydride battery alloy. The hydrogenation of titanium or zirconium sponges without a catalyst requires hydrogen pressures above 100 bar and a minimum reaction temperature of 150 °C. The hydrogenation of these metals under such drastic conditions leads to an undesired sintering so that the materials can only be used after additional grinding. Deposition of catalytic amounts of **3** (1 wt.-% Ti) onto the surface of the metal sponges allows for hydrogen absorption at much lower temperatures, now between 60 and 90 °C. Hydrogenation can be performed using the Ti and Zr sponges in the solid state or suspended in THF or toluene. The mass-

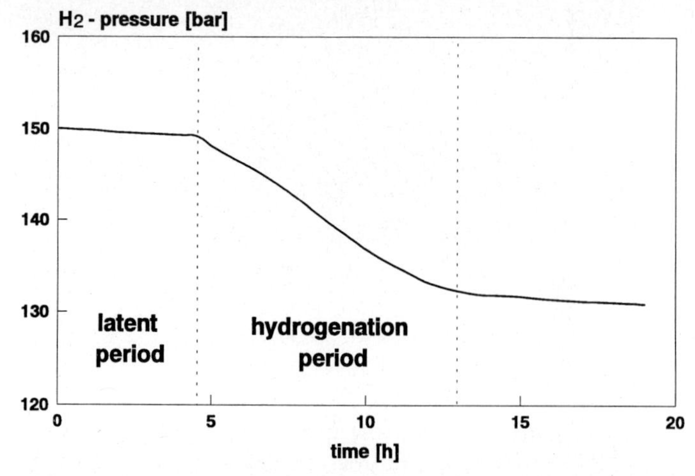

Fig. 9-9. Hydrogenation of Ti sponge catalyzed by **3**. (Reprinted with the kind permission of Elsevier Science B.V. [42c]).

specific uptake of hydrogen in relation to pressure, temperature, and time was monitored automatically using an apparatus especially designed for this purpose [44]. Fig. 9-9 shows the hydrogen uptake during the hydrogenation of a titanium sponge at 60 °C catalyzed by **3** (1 wt.-% Ti). After a latent period with only negligible uptake of H_2, the catalytic hydrogenation starts abruptly and is followed by a period of rather constant H_2 uptake (denoted as the "hydrogenation period"). After 13 h the H_2 uptake slowly tails off.

The mass-specific hydrogenation time [min g^{-1}] measured for a 1 g sample of titanium sponge at 100 bar H_2 in the presence of the catalyst is constant over reaction temperatures between 60 and 90 °C (Fig. 9-10). The hydrogenation period is also practically the same in THF and toluene and even in the absence of a solvent. However, significant differences are noted for the latent period. Performing the catalytic hydrogenation of titanium sponge

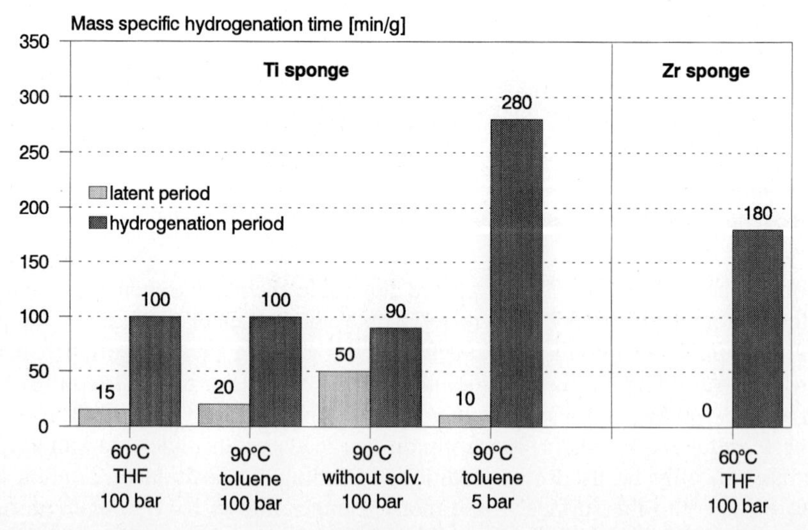

Fig. 9-10. Influence of temperature, solvent, and pressure in the hydrogenation of Ti and Zr sponges catalyzed by **3**. (Reprinted with the kind permission of Elsevier Science B.V. [42c]).

Fig. 9-11. SEM image of ZrH_2 powder. (Reprinted with permission of Elsevier Science B.V. [42c]).

at a very low hydrogen pressure (5 bar) prolongs the reaction time as expected. In the case of the Ti^0-catalyzed hydrogenation of zirconium sponge at 60 °C in THF no latent period is observed.

A welcome benefit of the Ti^0-catalyzed hydrogenation process is that the compact titanium and zirconium sponges break down during hydrogenation into finely-divided hydride powders. Fig. 9-11 shows a SEM image of a ZrH_2 sample produced by the catalytic hydrogenation of a zirconium sponge. The resulting 20 µm powder reveals a very uniform size distribution. X-ray diffraction patterns of the products verify the presence of TiH_2 and ZrH_2. Protonolysis of the solid-state hydrides with 5 M HCl yields 98% of the theoretically expected amount of H_2.

An interesting application of Ti^0-catalyzed metal hydrogenation is the hydrogenation of a Ni/Zr/V/Ti/Cr alloy at atmospheric pressure and room temperature. The hydride of this special alloy (38.2 wt.-% Ni, 24.3 wt.-% Zr, 12.8 wt.-% Ti, 18.7 wt.-% V, 6.0 wt.-% Cr) is of current interest for electrochemical applications, i.e. nickel hydride batteries. However, the hydrogenation at room temperature requires 5 bar of hydrogen pressure, and even under these conditions there is an unwanted latent period. After doping the surface of the alloy with 1 wt.-% of Ti using **3** as the precursor, spontaneous uptake of 0.6–0.7 wt.-% hydrogen is observed at room temperature without applying an H_2 over-pressure.

9.3.2 Colloidal Transition Metals and Alloys Protected by Tetraalkylammonium Halides via Reduction of Metal Salts with Tetraalkylammonium Hydrotriorganoborates

The application of hydrotriorganoborates for the production of colloidal metals is the result of an originally puzzling observation whereby the reduction of noble metal chlorides, e.g. $PtCl_2$, by $NBu_4[BEt_3H]$ in THF, gave dark red solutions besides the expected metal powders [Eq. (9.6); Table 9-3]. We were subsequently able to show that these THF solutions contain colloidal metals. Based on this finding, we developed a general method for the preparation of metal colloids of elements of groups 6–11 in organic solvents [13, 17].

Table 9-9. Preparation of metal colloids in THF solution or as an isolated powder.

No.	Metal salt	Reducing agent	Conditions t (h)	Conditions T (°C)	Product colloid solution color	Work-up solvent	Solvent added for precipitation	Metal content in isolated colloid (%)	Mean particle size (nm)
1	MnI_2	$N(octyl)_4BEt_3H$	1	23	Dark brown Mn completely dissolved				
2[a]	$FeBr_2$	$N(octyl)_4BEt_3H$	18	90	Dark brown to black Fe almost completely dissolved	Ethanol	Ether	11.34	3.0
3	$RuCl_3$	$N(octyl)_4BEt_3H$	2	50	Dark red-brown to black Ru almost completely dissolved	Ethanol	Pentane	68.72	1.3
4	$OsCl_3$	NBu_4BEt_3H	1	23	Deep red to black				
5	$CoBr_2$	$N(octyl)_4BEt_3H$	16	23	Dark brown to black Co completely dissolved	Ethanol	Ether	37.45	2.8
6	$RhCl_3$	$N(octyl)_4BEt_3H$	18	65	Deep red to black Rh completely dissolved	Ether	Ethanol	62.49	2.1
7	$IrCl_3$	$N(octyl)_4BEt_3H$	1	50	Dark red to black Ir almost completely dissolved	Ethanol	Ether	65.55	1.5
8	$NiBr_2$	$N(octyl)_4BEt_3H$	16	23	Dark red to black Ni completely dissolved	Ethanol	Ether	66.13	2.8
9	$NiBr_2$	$N(octyl)_3MeBEt_3H$	16	23	Dark red to black Ni completely dissolved	Ethanol	Ether	68.29	2.8
10	$PdCl_2$	$N(octyl)_4BEt_3H$	18	25	Dark brown to black Pd completely dissolved	Ether	Ethanol	32.39	2.5
11	$PtCl_2$	$N(hexyl)_4BEt_3H$	2	23	Dark brown to black Pt up to 80% dissolved				
12	$PtCl_2$	$N(octyl)_4BEt_3H$	18	23	Dark brown to black Pt completely dissolved	Ether	Ethanol	85.13	2.8
13	$PtCl_2$	$N(decyl)_4BEt_3H$	2	23	Dark brown to black Pt up to 80% dissolved				
14	$CuCl_2$	$N(octyl)_4BEt_3H$	2	23	Deep red to black Cu almost completely dissolved	Ether	Ethanol	77.04	
15	$CuBr_2$	$N(octyl)_4BEt_3H$	2	23	Deep red to black Cu completely dissolved	Toluene	Ethanol	52.15	

a) Solvent toluene.

THF suspensions of metal salts are treated with tetraalkylammonium hydrotriorganoborates in which the alkyl groups are preferably C_6–C_{20} chains. During reduction, H_2 is liberated and a brown-red metal colloidal solution is generated, from which only a small proportion of the reduced metal precipitates [Eq. (9.15)].

$$MX_v + vNR_4(BEt_3H) \xrightarrow{THF} M_{colloid} + vNR_4X + vBEt_3 + \frac{v}{2}H_2\uparrow \qquad (9.15)$$

M = metal of groups 6–11
X = Cl, Br
v = 1, 2, 3
R = alkyl, C_6–C_{20}

The long-chain tetraalkylammonium salts concentrate directly at the reduction center and act as effective complexing and protecting agents, keeping the freshly reduced metal particles in solution and at the same time preventing agglomeration. The filtered metal colloid solutions in THF are stable for months without any metal precipitation. In many cases the metal colloids can also be isolated from solution in the form of redispersible powders. The THF solvent is removed under vacuum, the waxy residue is dissolved in ether, toluene, or ethanol, and the precipitation of the metal colloid is achieved by adding a third solvent (for details see Table 9-9). The resulting grey-black metal colloid powders can be redissolved in various organic solvents (e.g. ethers, hydrocarbons, esters) with solubilities up to 1 g-at of metal per liter. Elemental analyses of the colloid powders as well as the results from mass spectrometry indicate that NR_4X is present. Presumably, the metal core of the colloid is protected against agglomeration by the surrounding tetraalkylammonium ions (Fig. 9-12). This screening of the metal particles by large lipophilic alkyl groups explains the remarkable solubility of the colloids in organic solvents and their extraordinary stability.

The isolated metal colloids were examined by TEM. The mean particle sizes are 3.0 nm and below (see Table 9-9) which indicates that the metals form relatively small metal aggregates. In general the colloids prepared according to Equation (9.15) exhibit a narrow size distribution. Figure 9-13 shows a typical particle size distribution determined from a TEM image of the Ir colloid (Table 9-9, No. 7). HRTEM (high resolution transmission electron microscopy) of the Ir colloid has further revealed that these small metal particles are crystalline (see the fringes in Fig. 9-14).

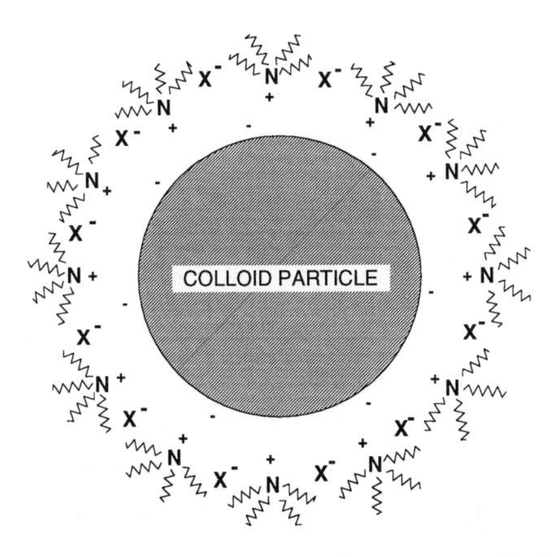

Fig. 9-12. Stabilization of the metal core by NR_4X. (Reprinted with the kind permission of Elsevier Science B.V. [42c]).

Table 9-10. Preparation of colloidal alloys.

No.	Metal salt	Reducing agent	Conditions		Colloidal alloy solution color	Work-up solvent	Solvent added for precipitation	Metal content in isolated colloid (%)	Mean particle size (nm)
			t (h)	T (°C)					
1	0.25 $RhCl_3$ 0.75 $PtCl_2$	$N(octyl)_4BEt_3H$	18	40	Deep red to black Rh and Pt completely dissolved	Ether	Ethanol	Rh: 14.2 Pt: 52.0	2.2
2	0.5 $RhCl_3$ 0.5 $PtCl_2$	$N(octyl)_4BEt_3H$	18	50	Deep red to black Rh and Pt completely dissolved	Ether	Ethanol	Rh: 34.7 Pt: 56.3	2.3
3	0.8 $RhCl_3$ 0.2 $PtCl_2$	$N(octyl)_4BEt_3H$	18	50	Deep red to black Rh and Pt completely dissolved	Ether	Ethanol	Rh: 61.2 Pt: 14.8	2.3
4	0.75 $PdCl_2$ 0.25 $PtCl_2$	$N(octyl)_4BEt_3H$	18	25	Deep brown to black Pd and Pt completely dissolved	Ether	Ethanol	Pd: 40.2 Pt: 18.0	–
5	0.5 $PdCl_2$ 0.5 $PtCl_2$	$N(octyl)_4BEt_3H$	18	25	Deep brown to black Pd and Pt completely dissolved	Ether	Ethanol	Pd: 25.9 Pt: 33.6	2.8
6	0.25 $PdCl_2$ 0.75 $PtCl_2$	$N(octyl)_4BEt_3H$	18	25	Deep brown to black Pd and Pt completely dissolved	Ether	Ethanol	Pd: 9.6 Pt: 34.7	–
7	0.5 $CuCl_2$ 0.5 $PtCl_2$	$N(octyl)_4BEt_3H$	16	23	Deep red to black Cu and Pt completely dissolved	Ether	Ethanol	Cu: 15.6 Pt: 55.4	2.3
8	0.5 $CuCl_2$ 0.5 $PdCl_2$	$N(octyl)_4BEt_3H$	18	25	Deep red to black Cu and Pd completely dissolved	Ether	Ethanol	Cu: 20.2 Pt: 38.5	2.6
9	0.8 $RhCl_3$ 0.2 $SnCl_2$	$N(octyl)_4BEt_3H$	18	50	Deep brown to black Rh and Sn completely dissolved	Ether	Ethanol	Rh: 12.5 Sn: 3.3	–
10	0.6 $RhCl_3$ 0.4 $SnCl_2$	$N(octyl)_4BEt_3H$	18	25	Deep brown to black Rh and Sn completely dissolved	Ether	Ethanol	Rh: 14.9 Sn: 15.3	2.1
11	0.44 $RhCl_3$ 0.56 $SnCl_2$	$N(octyl)_4BEt_3H$	18	40	Deep brown to black Rh and Sn completely dissolved	Toluene	Pentane	Rh: 5.9 Sn: 7.1	2.9
12	0.5 $PtCl_2$ 0.5 $CoBr_2$	$N(octyl)_4BEt_3H$	18	23	Deep redbrown to black Pt and Co completely dissolved	Toluene	Pentane/ Ethanol (25:1)	Pt: 25.4 Co: 6.5	–
13	0.5 $NiBr_2$ 0.5 $CoBr_2$	$N(octyl)_4BEt_3H$	16	25	Deep red to black Ni and Co completely dissolved	Ethanol	Ether	Ni: 23.8 Co: 23.8	2.8
14	0.5 $FeBr_2$ 0.5 $CoBr_2$	$N(octyl)_4BEt_3H$	18	50	Dark brown to black Fe and Co completely dissolved	Ethanol	Ether	Fe: 13.4 Co: 14.4	3.2

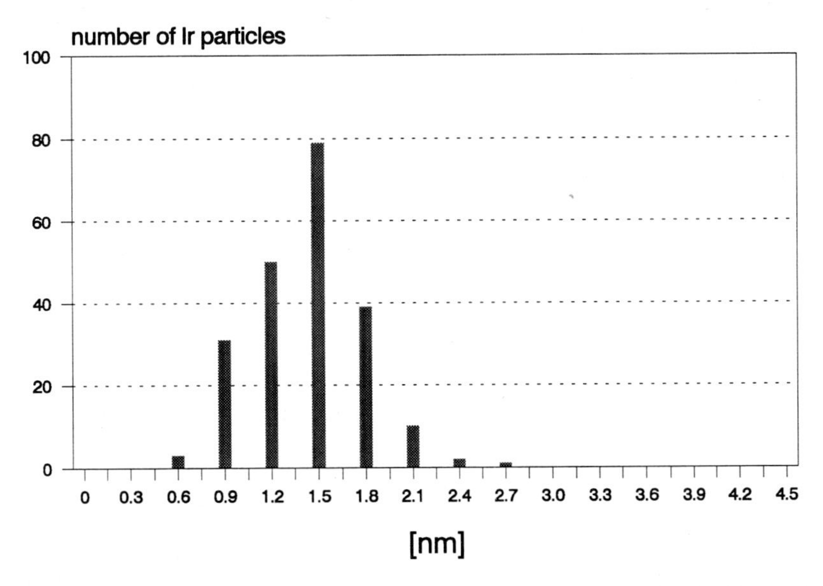

Fig. 9-13. Distribution of the particle size of Ir colloid (Table 9-9, No. 7). (Reprinted with the kind permission of Elsevier Science B.V. [42c]).

By exploiting the reaction given in Equation (9.15) for the coreduction of two different metal salts, bimetallic colloids (listed in Table 9-10) can be prepared, and subsequently isolated as redispersible powders. TEM images of the co-reduction products indicate that these colloids also have narrow size distributions (Figs. 9-15 and 9-16). Fig. 9-17 shows the HRTEM image of the Pt/Rh co-reduction product (Table 9-10, No. 1) prepared according to Equation (9.16). Metal aggregates with a diameter of about 2.3 nm can be identified, showing net plane distances of 0.25 nm in the crystal. Under the microscope 70 particles were analyzed by EDX with a point resolution of 1 nm. In every particle examined both Rh and Pt were found to be present, which indicates that

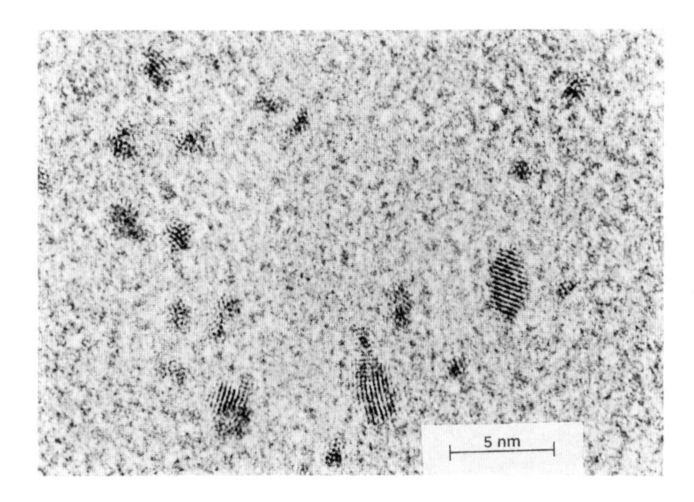

Fig. 9-14. HRTEM image of Ir colloid (Table 9-9, No. 7). (Reprinted with the kind permission of Elsevier Science B.V. [42c]).

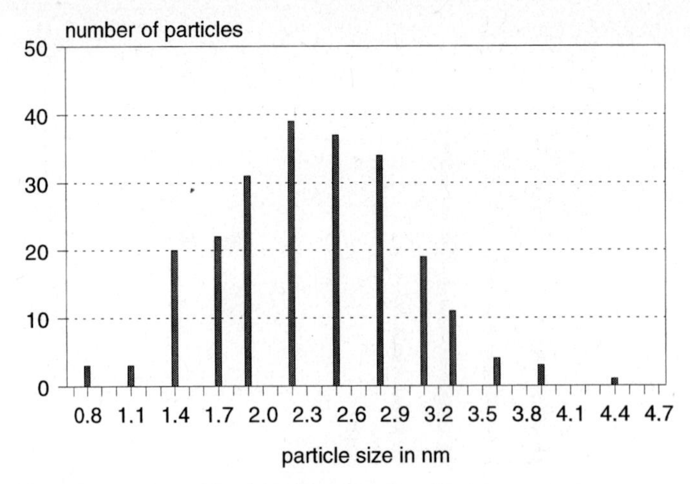

Fig. 9-15. Distribution of the particle size of $Pt_{50}Rh_{50}$ colloid (Table 9-10, No. 2).

in fact a colloidal Pt/Rh alloy is generated during co-reduction. An EXAFS study of a $Pt_{56}Rh_{44}$ colloid at both the Rh K-edge and the Pt L_3-edge verified the formation of a bimetallic alloy nanoparticulate system, in which the surface of the intermetallic aggregates is relatively richer in rhodium [45]. In subsequent X-ray absorption spectroscopy studies, the structural characterization of a series of Pt/Rh colloids with varying stoichiometries was achieved. The results of these investigations, together with others from electrochemical studies, XPS, and X-ray diffraction will be the subject of forthcoming papers [46].

$$m\mathrm{PtCl_2} + n\mathrm{RhCl_3} + (2m + 3n)\mathrm{N(octyl)_4BEt_3H} \xrightarrow{\mathrm{THF}}$$

$$\mathrm{Pt}_m\mathrm{Rh}_n\text{-colloid} + (2m + 3n)\mathrm{N(octyl)_4\,Cl} + (2m + 3n)\mathrm{BEt_3} + (2m + 3n)/2\ \mathrm{H_2}\uparrow \quad (9.16)$$

$$0 \leq m \leq 1;\ n = 1 - m$$

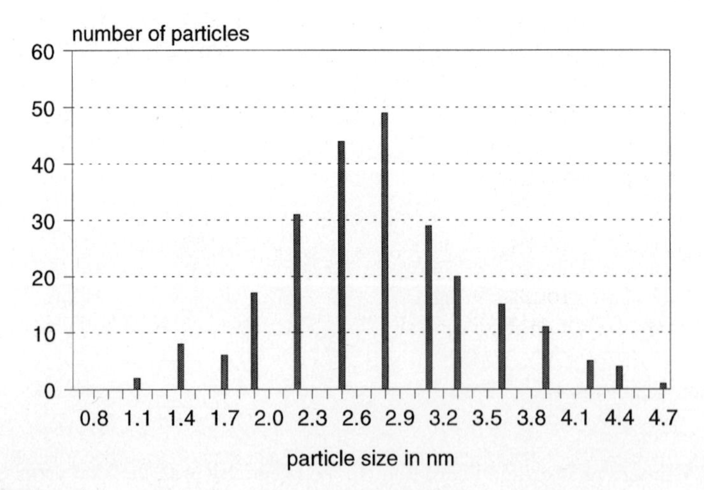

Fig. 9-16. Distribution of the particle size of $Pd_{50}Pt_{50}$ colloid (Table 9-10, No. 5).

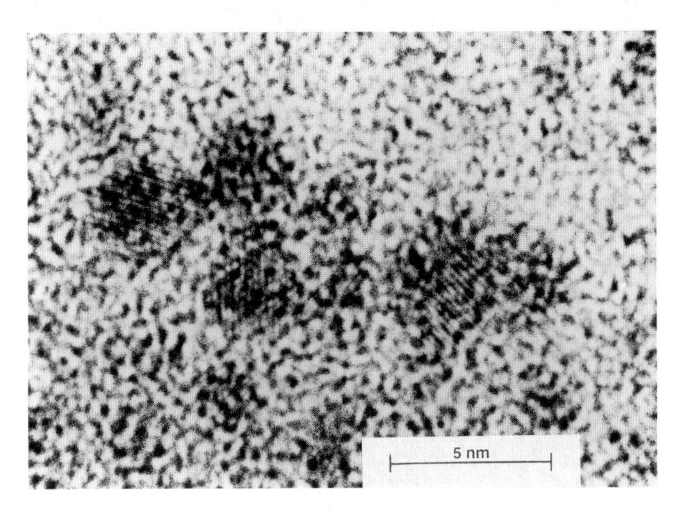

Fig. 9-17. HRTEM image of colloidal Pt/Rh alloy (Table 9-10, No. 2). (Reprinted with the kind permission of Elsevier Science B.V. [42c]).

9.3.3 General Approach to NR_4^+-Stabilized Metal and Alloy Colloids Using Conventional Reducing Agents [14]

In the syntheses of colloidal tansition metals according to Equation (9.15) as discussed above, the reducing agent (BEt_3H^-) is combined with a stabilizing agent (NR_4^+) in order to generate a high local concentration of NR_4^+ groups at the reduction center, thereby stabilizing the metal particles. This effect can also be achieved by first coupling the stabilizing agent to the metal salt to be reduced and then reacting further with common inorganic and organic reducing agents [Eq. (9.17)]. Employing this method, the disadvantages of using a special reducing agent and stoichiometric amounts of the stabilizing agent in the original synthetic route [Eq. (9-15)] can be avoided. The practicality of coupling the stabilizing agent to the metal center is exemplified for palladium: $(octyl)_4NBr$ or $(dodecyl)_4NBr$ is added to a solution of $Pd(Ac)_2$ in THF. A strong color intensification is observed, indicating that the desired interaction between the Pd salt and the stabilizing agent has been effected. Now by simply bubbling hydrogen through the solution at room temperature, the NR_4^+-stabilized Pd colloid is formed. The optimal ratio of stabilizing agent to palladium is 1:2, which represents only a quarter of the amount applied in the reduction procedure with $(octyl)_4NBEt_3H$. This decrease in the amount of stabilizing agent facilitates the work-up considerably.

$$(NR_4)_w MX_v Y_w + v\text{Red} \rightarrow M_{\text{colloid}} + v\text{Red X} + wNR_4Y \tag{9.17}$$

M = Metal of groups 6–11
Red = H_2, HCOOH, K, Zn, LiH, $LiBEt_3H$, $NaBEt_3H$, $KBEt_3H$, $Al(octyl)_3$
X, Y = Cl, Br, I
v, w = 1–3; R = C_6–C_{12}

Reduction of the NR_4^+-stabilized $Pd(Ac)_2$ by hydrogen, formic acid, or even thermal decomposition leads to stable palladium organo-sols. An XPS study of the Pd colloid

(No. 9 in Table 9-11) indicates that the binding energies for the Pd $3d_{5/2}$ and $3d_{3/2}$ electrons are 334.3 eV and 339.3 eV respectively. Since the values expected for zerovalent Pd are 334.9 eV and 340.1 eV, this result confirms that colloidal Pd^0 is produced in the reduction [42(b)]. Using this simplified preparation method [Eq. (9.17)] a number of noble and non-noble metal colloids have been obtained, which can be isolated in the form of redispersible powders by adding protic solvents such as water or ethanol to the organic solution (Table 9-11). The mean particle size of the resulting colloids, as analyzed by TEM, ranges from 1.5 nm in the case of ruthenium (Table 9-11, No. 4) to 4 nm for palladium (Table 9-11, No. 13).

The stabilizing agent [e.g. $(octyl)_4NBr$] can also be attached to the metal as a defined tetraalkylammonium metalate. Esumi [30], Blum and others [47] have introduced solvent–solvent extractions using tetraalkylammonium salts to transfer noble metal halides from aqueous HCl into organic solvents. However, we have found that these compounds can be more easily prepared by heating the metal halide (e.g. palladium, platinum, and nickel halides) with the tetraalkylammonium halide in THF [Eq. (9.18)]. The selection of the anion is critical for the success of the reductive colloid synthesis from the corresponding metalates prepared as in Equation (9.18). The reduction of $[(octyl)_4N]_2PdCl_4$ with hydrogen gives a metallic precipitate, whereas $[(octyl)_4N]_2PdBr_4$ cannot be reduced even under 50 bar of hydrogen atmosphere. And from $[(octyl)_4N]_2PdCl_2Br_2$ the desired palladium colloid (having a narrow size distribution) [14, 42(b)] is isolated upon reduction (Table 9-11, No. 13).

$$PdX_2 + 2\,N(octyl)_4Y \xrightarrow[\text{THF}]{\text{16 h, reflux}} [N(octyl)_4]_2PdX_2Y_2 \qquad (9.18)$$

$$X, Y = Cl, Br$$

The preparation of bimetallic colloids via the co-reduction of different metal salts according to this method is also successful [48], as exemplified for the Rh/Pt system in Equation (9.19).

Table 9-11. Preparation of NR_4^+-stabilized metal colloids using conventional reducing agents.

No.	Precursor	Reducing agent	Conditions t	Conditions T (°C)	Solvent	Metal in isolated colloid (%)	Mean particle size (nm)
1	$[N(octyl)_4]_2NiBr_4$	K	72 h	22	THF	57	2.8
2	$CoBr_2 \cdot 2N(octyl)_4Br$	$NaBEt_3H$	72 h	22	THF	31	2.8
3	$CoBr_2 \cdot 2N(octyl)_4Br$	K	72 h	22	THF	32	2.8
4	$RuCl_3 \cdot 3N(octyl)_4Br$	$KBEt_3H$	16 h	22	THF	–	1.5
5	$RhCl_3 \cdot 3$ Aliquat 336	H_2 (50 bar)	4 h	22	THF	74	3.0
6	$RhCl_3 \cdot 3\,N(octyl)_4Br$	$LiBEt_3H$	16 h	50	THF	–	2.1
7	$RhCl_3 \cdot 3\,N(octyl)_4Br$	LiH	16 h	60	THF	–	2.1
8	$RhCl_3 \cdot 3\,N(octyl)_4Br$	Zn	16 h	60	THF	–	2.2
9	$PdAc_2/N(dodecyl)_4Br$	H_2	16 h	22	THF	77	1.8
10	$PdAc_2/N(dodecyl)_4Br$	HCOOH	16 h	60	THF	73	1.8
11	$PdAc_2/NMe_2(dodecyl)_2Br$	H_2	16 h	22	THF	71	2.2
12	$PdAc_2/N(dodecyl)_4Br$	–	3 h	125	*p*-xylene	86	2.1
13	$[N(octyl)_4]_2PdCl_2Br_2$	H_2	14 d	22	THF	–	4.0
14	$[N(octyl)_4]_2PtCl_2Br_2$	$LiBEt_3H$	1 h	22	THF	–	1.9

Aliquat 336 = trioctylmethylammonium chloride.

$$[N(octyl)_4]_2PtCl_2Br_2 + [N(octyl)_4]_3RhCl_3Br_3 + 5\ red \xrightarrow{THF} Pt/Rh\text{-}colloid + 5\ N(octyl)_4Br + 5\ red\ Cl \quad (9.19)$$

red = $N(octyl)_4BEt_3H$, $LiBEt_3H$, $NaBEt_3H$, $KBEt_3H$, LiH, NaH, MgH_2, $LiAlH_4$, $(C_4H_9)_2AlH$, $Al(octyl)_3$, Zn, Li, HCOOLi, H_2

The production of Rh/Ti colloids can be achieved using a variation of the synthetic route given above, whereby $TiCl_4$ acts as the titanium source [Eq. (9.20)]. The colloid $Rh_{90}Ti_{10}$ can be isolated by evaporating the solvent and working up the bimetallic solid residue in Et_2O using a THF/EtOH mixture. The metal content of the purified, air-sensitive product amounts to 18%. Table 9-12 reviews the various bimetallic colloids obtained according the syntheses described here.

$$0.9\ [N(octyl)_4]_3RhCl_3Br_3 + 0.1\ TiCl_4 + 3.1\ LiBEt_3H \xrightarrow{THF} Rh_{0.9}Ti_{0.1}\text{-}colloid + 3.1\ BEt_3 + 2.7\ N(octyl)_4Br + 3.1\ LiCl + 1.65\ H_2\uparrow \quad (9.20)$$

Table 9-12. Preparation of NR_4^+-stabilized colloidal alloys using conventional reducing agents.

No.	Precursor	Reducing agent	Conditions *t* (h)	Conditions *T* (°C)	Solvent	Metal content in isolated colloid (%)	Mean particle size (nm)
1	0.5 $[N(octyl)_4]_2PtCl_2Br_2$ 0.5 $[N(octyl)_4]_3RhCl_3Br_3$	$N(octyl)_4BEt_3H$	18	50	THF	Pt: 16.0 Rh: 9.9	–
2	0.5 $[N(octyl)_4]_2PtCl_2Br_2$ 0.5 $[N(octyl)_4]_3RhCl_3Br_3$	$LiAlH_4$	24	50	THF	Pt: 46.5 Rh: 21.0	2.1
3	0.5 $[N(octyl)_4]_2PtCl_2Br_2$ 0.5 $[N(octyl)_4]_3RhCl_3Br_3$	Zn	24	50	THF	Pt: 52.7 Rh: 23.2	–
4	0.5 $[N(octyl)_4]_2PtCl_2Br_2$ 0.5 $[N(octyl)_4]_3RhCl_3Br_3$	Li	24	50	THF	Pt: 33.5 Rh: 17.9	–
5	0.5 $[N(octyl)_4]_2PtCl_2Br_2$ 0.5 $[N(octyl)_4]_3RhCl_3Br_3$	H_2	23 days	25	THF	Pt: 28.4 Rh: 15.2	–
6	0.5 $[N(octyl)_4]_2PtCl_2Br_2$ 0.5 $[N(octyl)_4]_3RhCl_3Br_3$	$N(octyl)_4BEt_3H$	18	25	THF	Pt: 13.0 Rh: 8.4	1.9
7	0.5 $[N(octyl)_4]_2PtCl_2Br_2$ 0.5 $TiCl_4$	$LiBEt_3BH$	18	50	THF	Pt: 16.8 Ti: 1.1	–

9.3.4 Synthesis of $PEt(octyl)_3^+$- and $SMe(octyl)_2^+$-Stabilized Metal and Alloy Colloids

The preparation of colloidal metals and alloys in organic solvents is not limited to the use of NR_4^+ as the stabilizing agent; PR_4^+ and SR_3^+ groups can also function as protective groups [48].

By reacting potassium hydrotriethylborate with $PEt(octyl)_3Br$, the preparation of $PEt(octyl)_3BEt_3H$ can be accomplished according to Equation (9.21). The solid KBr by-product is separated simply by filtering. In a manner similar to the reduction of metal halides by tetraalkylammonium hydrotriorganoborates [Eq. (9.15)], $PEt(octyl)_3BEt_3H$ can be used to reduce metal halides, generating the corresponding metal colloids. Equation (9.22) typifies such a reaction producing a Pt colloid.

$$PEt(octyl)_3Br + KBEt_3H \xrightarrow{THF} PEt(octyl)_3BEt_3H + KBr \downarrow \qquad (9.21)$$

$$PtCl_2 + 2\ PEt(octyl)_3BEt_3H \xrightarrow{THF} Pt\text{-colloid} + 2\ PEt(octyl)_3Cl + 2\ BEt_3 + H_2 \uparrow \qquad (9.22)$$

Again in analogy to the NR_4^+-protected metallic colloids, $PEt(octyl)_3Br$ can first be attached to the metal forming a metalate, which in a second synthetic step is reduced to give the $PEt(octyl)_3^+$-stabilized metal colloid. This synthetic route is illustrated in Equations (9.23) and (9.24), specifically for the production of a Rh colloid. As shown in Equation (9.23), the preparation of the intermediate rhodate (THF solution, red-colored) requires two equivalents of $PEt(octyl)_3Br$.

$$RhCl_3 \cdot 3\ H_2O + m\ PEt(octyl)_3Br \xrightarrow{THF} [PEt(octyl)_3]_mRhCl_3Br_m + 3\ H_2O \qquad (9.23)$$

$m = 2, 3$

$$[PEt(octyl)_3]_mRhCl_3Br_m + 3\ Al(octyl)_3 \xrightarrow{THF} Rh\text{-colloid} + m\ PEt(octyl)_3Br + 3\ Al(octyl)_2Cl + 1.5\ \text{octane} + 1.5\ \text{octene} \qquad (9.24)$$

$m = 2, 3$

The two preparative strategies described for $PEt(octyl)_3^+$-stabilized metal colloids are also applicable for the production of bimetallic colloids, as shown in Equations (9.25) to (9.28). The synthesis of Rh/Pt bimetallic colloids [Eq. (9.25)] proceeds under the same conditions as used for the preparation of Rh/Pt colloids with tetraalkylammonium hydrotriorganoborates according to Equation (9.16). During reduction with $PEt(octyl)_3BEt_3H$ metal precipitation was not observed. The solubility of the $PEt(octyl)_3^+$-stabilized $Pt_{50}Rh_{50}$ colloid is identical to that of the corresponding NR_4^+-stabilized $Pt_{50}Rh_{50}$ colloid. Thus, the purification of the former is performed in the same manner as for the NR_4^+-stabilized colloid, i.e. using $Et_2O/EtOH$ as a precipitant. Evaluation of TEM photographs of the $Pt_{50}Rh_{50}$ colloid reveals a mean particle size of 2.5 nm (Fig. 9-18), which corresponds well to that observed for the NR_4^+-stabilized $Pt_{50}Rh_{50}$ colloid. Thus, it can be concluded that changing the cation from nitrogen to phosphorus does not have a significant effect on the particle diameter.

$$PtCl_2 + RhCl_3 + 5\ PEt(octyl)_3BEt_3H \xrightarrow{THF} Rh/Pt\text{-colloid} + 5\ PEt(octyl)_3Cl + 5\ BEt_3 + 2.5\ H_2 \uparrow \qquad (9.25)$$

The synthesis of a platinate requires two equivalents of $PEt(octyl)_3Br$, as shown in Equation (9.26). The THF solution of the platinate is orange-colored. Employing the platinate together with the corresponding rhodate in an equimolar ratio [Eq. (9.23)], the generation of a $Pt_{50}Rh_{50}$ bimetallic colloid is possible using a simple reducing agent, such as zinc, as shown in Equation (9.27). The co-reduction product from the method of Equation (9.27) reveals a particle size of 2.4 nm (Fig. 9-19), which is comparable to that

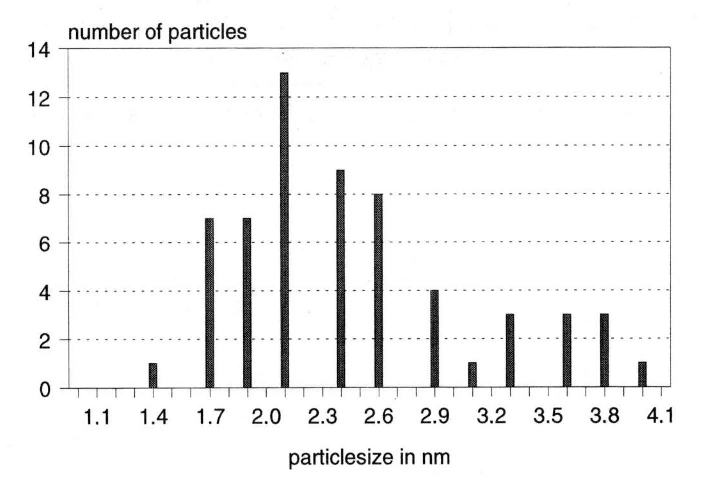

Fig. 9-18. Distribution of the particle size of $Pt_{50}Rh_{50}$ colloid (Table 9-13, No. 1).

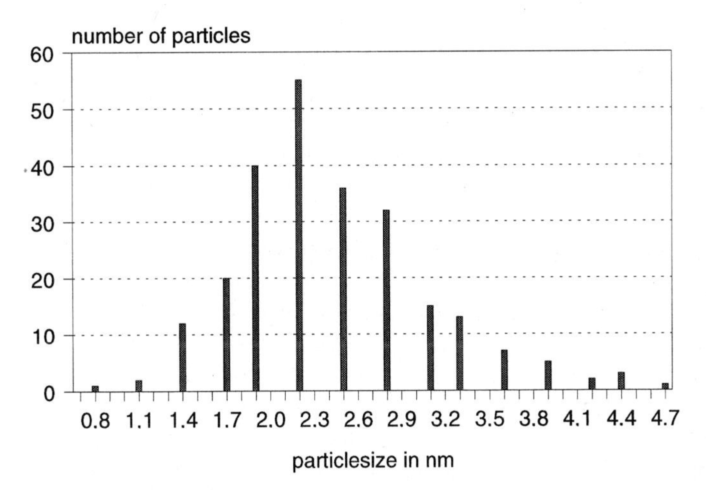

Fig. 9-19. Distribution of the particle size of $Pt_{50}Rh_{50}$ colloid (Table 9-13, No. 3).

of the bimetallic colloids prepared according the first strategy, i.e. co-reduction using $PEt(octyl)_3BEt_3H$ [Eq. (9.25)]. The co-reduction of platinate and rhodate by hydrogen gas [Eq. (9.28)] proceeds rapidly (Table 9-13, No. 2). Table 9-13 reviews the various $PEt(octyl)_3^+$-stabilized $Pt_{50}Rh_{50}$ bimetallic colloids prepared according to the synthetic routes discussed here.

$$PtCl_2 + 2\,PEt(octyl)_3Br \xrightarrow{THF} [PEt(octyl)_3]_2PtCl_2Br_2 \tag{9.26}$$

$$[PEt(octyl)_3]_2PtCl_2Br_2 + [PEt(octyl)_3]_2RhCl_3Br_2 + 2.5\,Zn \xrightarrow{THF} Pt/Rh\text{-colloid} + 4\,PEt(octyl)_3Br + 2.5\,ZnCl_2 \tag{9.27}$$

$$[PEt(octyl)_3]_2PtCl_2Br_2 + [PEt(octyl)_3]_3RhCl_3Br_3 + 2.5\,H_2 \xrightarrow{THF} Pt/Rh\text{-colloid} + 5\,PEt(octyl)_3Br + 5\,HCl \tag{9.28}$$

Table 9-13. Preparation of PR_4^+-stabilized colloidal alloys.

No.	Precursor	Reducing agent	Conditions *t* (h)	Conditions *T* (°C)	Solvent	Metal content in isolated colloid (%)	Mean particle size (nm)
1	0.5 $PtCl_2$ 0.5 $RhCl_3$	$[P(Et)(octyl)_3]BEt_3H$	18	50	THF	Pt: 34.0 Rh: 16.7	2.5
2	0.5 $[P(Et)(octyl)_3]_2PtCl_2Br_2$ 0.5 $[P(Et)(octyl)_3]_3RhCl_3Br_3$	H_2	9 days	25	THF	Pt: 23.9 Rh: 26.9	–
3	0.5 $[P(Et)(octyl)_3]_2PtCl_2Br_2$ 0.5 $[P(Et)(octyl)_3]_2RhCl_3Br_2$	Zn	20	50	THF	–	2.4

In contrast to the ammonium or phosphonium salts, long-chain sulfonium halides are not readily available commercially or synthetically in high yields. For the synthesis of $SR'R_2^+$-stabilized metal colloids, dioctylmethylsulfonium methylsulfate, prepared according to Equation (9.29) was used. During the preparation of the sulfonium as given in Equation (9.29), it is essential that no residue of the disulfide starting material is present in the final product, in order to avoid the formation of undesired sulfide complexes in the subsequent chemistry. The sulfonium methylsulfat was then coupled to the metal salt, as illustrated for $RhCl_3 \cdot 3\,H_2O$ in Equation (9.30). In a second reduction step using hydrogen, the $SMe(octyl)_2^+$-stabilized Rh colloid is then formed [Eq. (9.31)]. The purified Rh colloid contains 43.18 wt.-% Rh and reveals a mean particle diameter of 3.6 nm.

$$S(octyl)_2 + (CH_3O)_2SO_2 \xrightarrow{\text{THF}} (octyl)_2S(CH_3)^+ + (CH_3O)SO_3^- \tag{9.29}$$

$$RhCl_3 \cdot 3\,H_2O + 3\,(octyl)_2S(CH_3)(CH_3O)SO_3 \xrightarrow{\text{THF}} [(octyl)_2S(CH_3)]_3RhCl_3[(CH_3O)SO_3]_3 + 3\,H_2O \tag{9.30}$$

$$[(octyl)_2S(CH_3)]_3RhCl_3[(CH_3O)SO_3]_3 + 1.5\,H_2 \xrightarrow{\text{THF, 25 °C, 4 h}} \text{Rh-colloid} + 3\,(octyl)_2S(CH_3)ClSO_3 + 3\,CH_3OH \tag{9.31}$$

9.4 Nanoscale Metal Powders via Metal Colloids

During the physical characterization of the colloidal materials by electron microscopy (specifically EDX) and XPS, both of which employed ultra-high vacuum conditions (10^{-7} to 10^{-8} Pa), neither the halogen nor the nitrogen of the NR_4X groups could be detected. This clearly indicates that the protecting tetraalkylammonium halide may be removed under certain conditions leaving the bare metal core behind. This observation prompted us to develop a chemical procedure for the extraction of the protecting shell from the colloids (in Tables 9-9 and 9-10) at room temperature to produce nanoscale metal powders (listed in Table 9-14). The preparation of nanoscale platinum powder from the corresponding platinum colloid (Table 9-9, No. 12) represents a typical example. The grey-brown colloid powder is treated with an excess of ethanol, after which the supernatant solution containing the tetraalkylammonium halide is siphoned off. This procedure is repeated sev-

Table 9-14. Preparation of nanoscale metal powders via metal colloids.

No.	Starting material colloid		Mean particle size (nm)	Solvent for extraction of NR_4X	Product metal content after extraction (%)	Mean particle size	
						After extraction (nm)	After heat treatment (700 °C, 4 h, 10^{-3} mbar) (nm)
1	Co	Table 9-9, No. 5	2.8	Ethanol	82.84	3.8	–
2	Ni	Table 9-9, No. 8	2.8	Ethanol	88.18	3.0	–
3	Rh	Table 9-9, No. 6	2.1	Ethanol	82.76	2.7	2.9
4	Pd	Table 9-9, No. 10	2.5	Ethanol	98.13	5.8	6.0
5	Pt	Table 9-9, No. 12	2.8	Ethanol	92.90	2.8	2.8
6	Rh/Pt	Table 9-10, No. 1	2.3	Ether/Ethanol 1:10	–	2.7	3.0
7	Pd/Pt	Table 9-10, No. 4	2.8	Ether/Ethanol 1:10	–	2.8	–

eral times, during which a continuous darkening of the product color is observed. The resulting black pyrophoric metal powder (93 wt.-% Pt) is no longer redispersible in THF (Table 9-14, No. 5). The TEM image of this product (Fig. 9-20) shows that the mean particle size of the platinum powder after the extraction corresponds exactly to that of the initial colloidal platinum sample. As determined by EDX analysis, the extracted sample still contains traces of the protecting shell, which can be completely removed by heating the metal powder at 700 °C under vacuum (0.1 Pa). According to the TEM image of the

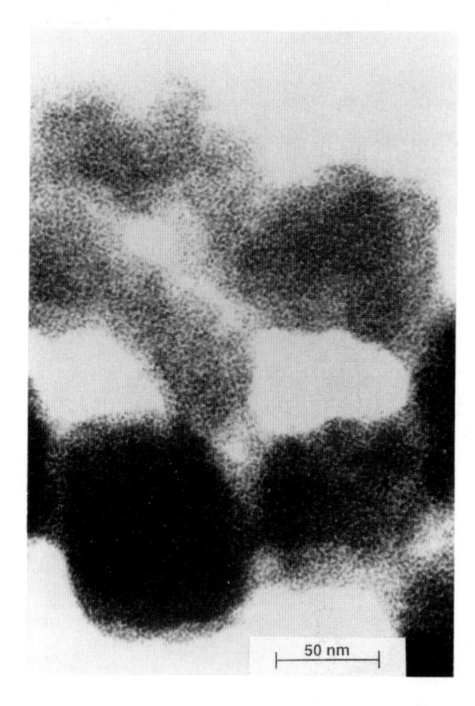

Fig. 9.20. TEM image of Pt powder obtained from Pt colloid (Table 9-14, No. 5). (Reprinted with the kind permission of Elsevier Science B.V. [42c]).

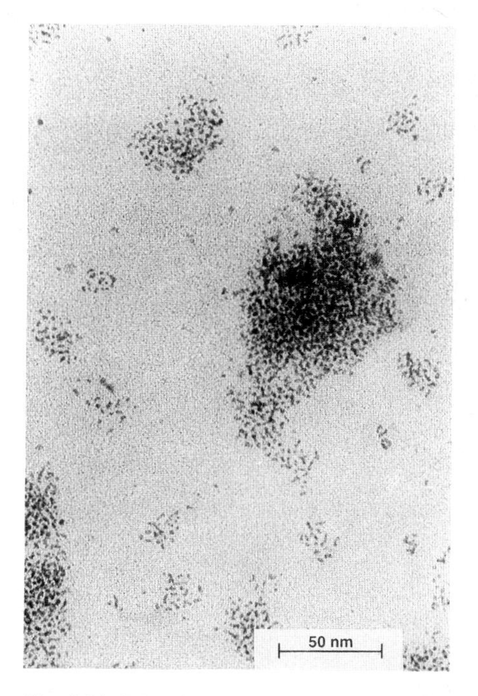

Fig. 9-21. TEM image of Pt powder obtained from Pt colloid (Table 9-14, No. 5) after thermal treatment (4 h, 700 °C, 0.1 Pa). (Reprinted with the kind permission of Elsevier Science B.V. [42c]).

heat-treated product (Fig. 9-21) the mean particle size is 2.8 nm, the same value as found for the original colloidal platinum. Thus, no agglomeration of the metallic nanoparticles as a consequence of heat treatment is observed. This finding basically holds true in almost every preparation of a metallic or intermetallic powder from the corresponding colloidal material (compare the particle sizes listed in Table 9-14). The only exception found so far involves palladium (Table 9-14, No. 4), where a particle growth from 2.5 to 5.8 nm is seen (by TEM) upon removing the protecting shell. After the thermal treatment no further particle augmentation is detected.

9.5 Catalytic Applications

One of the goals in the preparation of novel metal and alloy colloids is the isolation of promising new homogeneous and heterogeneous hydrogenation catalysts. The catalytic potential of the colloidal materials described here has been explored, both in the homogeneous phase and as supported heterogeneous catalysts. Regarding the former, the activities of a series of palladium colloids for the hydrogenation of cyclohexene in THF were measured under normalized conditions [14]. The catalytic activity, selectivity, and stability of noble metal colloids adsorbed on charcoal, specifically 5 wt.-% Pd [14], 5 wt.-% Rh [42], and 5 wt.-% Pt [42], were tested in the hydrogenation of cinnamic acid, butyronitrile, or crotonic acid, respectively. The experimental apparatus and conditions employed in the investigations of these heterogeneous charcoal-supported catalysts are described in detail in [42].

9.5.1 Free and Supported Metal and Alloy Colloid Catalysts in Liquid-Phase Hydrogenation

Non-supported, i.e. free nickel and palladium colloids are suitable for the selective hydrogenation of naturally occurring products such as soya bean oil. Table 9-15 summarizes the activities of a number of palladium colloids in the homogeneous hydrogenation of cyclohexene. The results reveal that neither the method used to prepare the colloid [Eqs. (9.15) or (9.17)] nor its particle size has a systematic influence on the resulting catalytic activity. Surprisingly however, the choice of the anion is important: a marked decrease in the catalytic activity of the colloidal palladium in the order Br > Cl > Ac > I is oberved [42b].

Table 9-15. Anion-dependent activity of colloidal palladium in the homogeneous cyclohexene hydrogenation test.

No.	Catalyst	Preparation method for the colloid	Particle size (nm)	Activity (N mL/g min)
1	$PdBr_2/N(octyl)_4BEt_3H$	Analogous Table 9-9, No. 10	–	1169
2	$PdCl_2/N(octyl)_4BEt_3H$	Table 9-9, No. 10	2.5	479
3	$PdAc_2/N(octyl)_4BEt_3H$	Analogous Table 9-9, No. 10	2.9	241
4	$PdI_2/N(octyl)_4BEt_3H$	Analogous Table 9-9, No. 10	–	0
5	$N(octyl)_4PdCl_2Br_2/H_2$	Table 9-11, No.13	4.0	762
6	$PdAc_2 \cdot 2\,N(dodecyl)_4Br/H_2$	Table 9-11, No. 9	1.8	377

Table 9-16. Comparison of the catalytic activity of a palladium colloid free and supported (5% on C)[a]

No.	Catalyst	Preparation method for the colloid	Support	Sol-vent	Substrate	Activity (N mL/g min)
1	$PdCl_2/N(octyl)_4BEt_3H$	Table 9-9, No. 10	–	THF	Cyclohexene	479
2	$PdCl_2/N(octyl)_4BEt_3H$	Table 9-9, No. 10	Charcoal	THF	Cyclohexene	566
3	$PdCl_2/N(octyl)_4BEt_3H$	Table 9-9, No. 10	–	THF	Cinnamic acid	329
4	$PdCl_2/N(octyl)_4BEt_3H$	Table 9-9, No. 10	Charcoal	THF	Cinnamic acid	207

a) Based on the weight of Pd used.

The major disadvantage of the free colloids lies in the tendency of the non-supported hydrogenation catalysts to agglomerate in solution, generating metallic precipitates and leading to a drastic decrease in activity. In the case of the free colloidal palladium catalysts, metal precipitation is already occurring after 6–7 minutes into the reaction. Adsorption of the metal colloids onto a carrier such as charcoal (a technically important support material) substantially improves the stability of the resulting catalysts. In contrast to the widespread prejudice that the catalytic activity of metal particles is reduced by fixing them onto a support, we found no deterioration in activity. Comparing the measured catalytic activities for free and charcoal-supported colloidal palladium (5 wt.-% Pd/C) listed in Table 9-16, only minor differences were noted in both the cyclohexene test (Table 9-16, Nos. 1 and 2) and the cinnamic acid test (Table 9-16, Nos. 3 and 4). Since the palladium colloid is insoluble in ethanol, the cinnamic acid test was performed in THF for both the free and supported catalyst. The standard activity of the supported 5 wt.-% Pd/C catalyst in ethanol is listed in Table 9-17, No. 2 [42(b)].

The non-supported and supported colloidal palladium catalysts (Table 9-11, No. 13) were examined by TEM (Fig. 9-22). The mean particle size in both samples is virtually the same. In other words, during and after the adsorption procedure particle agglomeration

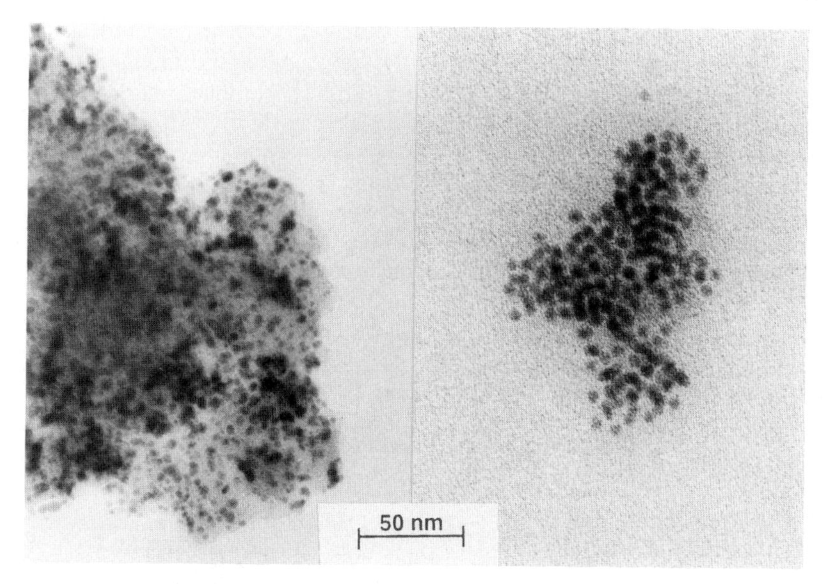

Fig. 9-22. TEM image of Pd colloid (Table 9-11, No. 13), free (left) and supported on charcoal (right). (Reprinted with the kind permission of Elsevier Science B.V. [42c]).

Table 9-17. Catalytic activity of supported colloidal palladium (5% Pd on C) in the cinnamic acid test.

No.	Catalyst	Colloid precursor	Activity (N mL/g/min)
1	Pd on C (Degussa E 10 R/D)[a]	–	356
2	$PdCl_2/N(octyl)_4BEt_3H$ on C	Table 9-9, No. 10	406
3	$N(octyl)_4PdCl_2Br_2/H_2$ on C	Table 9-11, No. 13	826
4	$PdAc_2/N(dodecyl)_4Br/H_2$ on C	Table 9-11, No. 9	586
5	$PdAc_2/N(dodecyl)_4Br/HCOOH$ on C	Table 9-11, No. 10	401
6	$PdAc_2/NMe_2(dodecyl)_2Br/H_2$ on C	Table 9-11, No. 11	389
7	$PdAc_2/N(dodecyl)_4Br$[b] on C	Table 9-11, No. 12	310

a) Precipitation of Pd on C; b) thermal decomposition at 125 °C.

on the support surface is not observed. Furthermore, the dispersion of the palladium colloid on the charcoal appears to be remarkably uniform. Both of these characteristics are probably due to the fact that the discrete particles in the free colloid are well shielded from each other by the long-chain hydrocarbon groups in the tetraalkylammonium protecting shell. A TEM study of the analogous colloidal Rh catalysts also reveals a conservation of the particle size of the metal particles after the adsorption onto charcoal [42a].

The measured activities (cinnamic acid test) of supported Pd/C catalysts (Table 9-17, Nos. 2, 5, 6, 7) prepared from NR_4^+-stabilized Pd colloids (Table 9-9, No. 10 and Table 9-11, Nos. 10, 11, and 12) are comparable to that of the industrial catalyst (Table 9-17, No. 1). Substantially higher activities are observed for the Pd/C catalysts (Table 9-17, Nos. 3 and 4) prepared from colloids Nos. 13 and 9 in Table 9-11 (a factor of 2.3 and 1.6, respectively, higher than for the conventional catalyst) [42b].

The activities of charcoal-supported colloidal platinum catalysts (5 wt.-% Pt) in the crotonic acid test are listed in Table 9-18 and compared with the activity of a conventional 5 wt.-% Pt/C catalyst. The Pt/C catalyst No. 3 was prepared from an isolated platinum colloid sample (Table 9-11, No. 14) and its activity exceeds that of the conventional catalyst. The catalysts Nos. 2 and 4 in Table 9-18 were produced using crude product solutions containing colloidal platinum. As shown by their measured catalytic activities, which are equal to or better than that of the industrial catalyst, complete work-up of the metal colloid before fixing onto a support is not necessary.

The catalytic activity of the charcoal-supported rhodium colloid catalyst in the butyronitrile test exceeds that of the conventional Rh/C catalyst (Table 9-18, No. 6 versus

Table 9-18. Catalytic activity of supported colloidal platinum and rhodium (5% on C).

No.	Catalyst	Colloid precursor	Substrate	Activity (N mL/g min)
1	Pt on C (Degussa F 103 R/D 314)[a]	–	Crotonic acid	254
2	$PtCl_2/N(octyl)_4BEt_3H$ on C	Table 9-9, No. 12 reaction solution	Crotonic acid	387
3	$PtCl_2/N(octyl)_4BEt_3H$ on C	Table 9-9, No. 12 isolated colloid	Crotonic acid	415
4	$[N(octyl)_4]_2PtCl_2Br_2/LiBEt_3H$ on C	Table 9-11, No. 14	Crotonic acid	258
5	Rh on C (Degussa G 10 S/W)[a]	–	Butyronitrile	71
6	$RhCl_3/N(octyl)_4BEt_3H$ on C	Table 9-9, No. 6	Butyronitrile	95

a) Precipitation on C.

No. 5). The industrial Rh/C catalyst is produced by precipitating rhodium metal from an aqueous solution of $RhCl_3$ onto the charcoal surface. According to a TEM study of this catalyst, only a minor proportion of the metal is well dispersed on the surface [42]. In contrast, the adsorption of the isolated NR_4^+-stabilized Rh colloid (Table 9-9, No. 6) on charcoal yields a uniform metal dispersion on the support surface without an increase in particle size from that of the colloidal precursor [42].

Both the conventional Rh/C and colloidal Rh/C catalysts were prepared under argon. In order to obtain air-stable products, the dry catalysts were carefully treated for 2 h with argon that contained 0.2 vol.-% molecular oxygen [42]. We were surprised to find that this careful oxygenation of the surface not only stabilizes both types of catalysts to air, but also considerably enhances their catalytic activity. In the case of the conventional Rh/C catalyst, oxygenation improves its activity by 17% (Table 9-19, Nos. 1 and 2). The more active rhodium colloid/C catalyst (Table 9-19, No. 4) almost doubles its activity after oxygenation (Table 9-19, No. 6). These findings correspond well with a report published by Willstätter as long ago as 1921 [49], concluding that the activity of noble-metal hydrogenation catalysts depends strongly on oxygen.

The most effective "strong metal–support interactions" (SMSIs) are those reported for noble metal hydrogenation catalysts using TiO_2 as a carrier [50, 51]. Ryndin et al. [52] systematically investigated the effect of early transition metal ions as dopants for noble metal hydrogenation catalysts on neutral supports; in this study, deposited tetravalent titanium, zirconium, and hafnium ions were reduced by thermal treatment under a hydrogen atmosphere. Reworking this doping procedure, we deposited zero-valent titanium in the

Table 9-19. Effect of dopant and oxygenation on the catalytic activity of precipitated and colloidal rhodium (5%) on C in the butyronitrile test.

No.	Catalyst	Preparation method	Dopant	Oxygenated	Activity (N mL/g min)	Enhancement (%)
1	Rh on C	Conventional precipitation under Ar	–	no	71	Standard
2	Rh on C	Conventional precipitation under Ar	–	yes	83	17
3	Rh on C	Conventional precipitation under Ar	0.2% $Ti(toluene)_2$	yes	124	75
4	$RhCl_3/N(octyl)_4BEt_3H$ on C	Colloid (Table 9-9, No. 6) adsorbed	–	no	95	34
5	$RhCl_3/N(octyl)_4BEt_3H$ on C	Colloid (Table 9-9, No. 6) adsorbed	0.2% $Ti(toluene)_2$	no	123	73
6	$RhCl_3/N(octyl)_4BEt_3H$ on C	Colloid (Table 9-9, No. 6) adsorbed	–	yes	203	186
7	$RhCl_3/N(octyl)_4BEt_3H$ on C	Colloid (Table 9-9, No. 6) adsorbed	0.2% $Ti(toluene)_2$	yes	262	269
8	$RhCl_3/N(octyl)_4BEt_3H$ on C	Colloid (Table 9-9, No. 6) adsorbed	0.2% $[Ti \cdot 0.5\,THF]_x$	yes	263	270

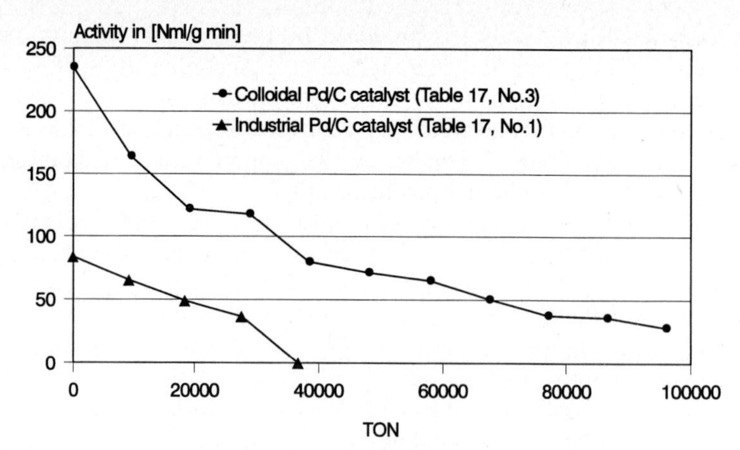

Fig. 9-23. Stability of Pd/C catalysts in the hydrogenation of cyclooctene.

form of bis(arene)-Ti^0 compounds or $[Ti^0 \cdot 0.5\ THF]_x$ at room temperature, without further thermal treatment or hydrolysis, onto Rh/C catalysts. The catalytic activity of the Ti^0-doped Rh/C catalysts in the butyronitrile test is considerably enhanced [42, 53]. A comparison of the activity figures quoted in Table 9-19 shows that the doping of the supported rhodium colloid catalyst with 0.2 wt.-% Ti^0 in combination with a subsequent oxygenation treatment provides the most effective catalyst.

The long-term stability of the conventional and colloidal noble metal catalysts supported on charcoal was tested by measuring their total turnover number in the hydrogenation of cyclooctene in ethanol at 40 °C [42b]. The results for an industrial Pd/C catalyst (Table 9-17, No. 1) are shown in Fig. 9-23 and compared with a supported palladium colloid catalyst (Table 9-17, No. 3). The activity of the conventional precipitation catalyst ceases after the performance of 38×10^3 catalytic cycles, whereas the colloidal Pd/C catalyst still shows a residual activity of 30 N mL H_2/(g min) after 96×10^3 catalytic turnovers!

The hydrogenation activity of a series of charcoal-supported bimetallic Pt/Rh colloids (Table 9-10, Nos. 1, 2, 3; 5 wt.-% metal on C) was investigated using the crotonic acid test [42(c)]. In order to elucidate the co-catalytic influence of rhodium, the activity of the bimetallic systems was plotted against increasing amount of rhodium present in various Pt/Rh colloids adsorbed on charcoal (Fig. 9-24, curve 1). In order to compare the observed activities to those for a corresponding mixture of the two metals on a charcoal surface, samples of platinum colloid (Table 9-9, No. 12) were combined in solution with increasing amounts of rhodium colloid (Table 9-9, No. 6) and then adsorbed on charcoal (5 wt.-% metal). The activities of the resulting catalysts in the crotonic acid test are also plotted in Fig. 9-24 (curve 3). Finally, mixtures consisting of the ready-made Pt/C catalyst (Table 9-18, No. 2) and increasing amounts of the corresponding Rh/C catalyst (Table 9-18, No. 6) were tested; the resulting activity values plotted against the Pt/Rh ratio are shown in Figure 9-24 (curve 2).

The catalysts prepared by mixing Pt/C and Rh/C (curve 2) or by supporting mixtures of colloidal platinum and rhodium samples on charcoal (curve 3) clearly show equivalent catalytic activities, as well as a linear increase of the activity with increasing content of rhodium (additive effect of platinum and rhodium). In contrast the activity plot of the bimetallic Pt/Rh catalysts (curve 1) exhibits a maximum at $Pt_{20}Rh_{80}$. This may be due to a synergistic effect of the two metals in the bimetallic colloid, which could also be inter-

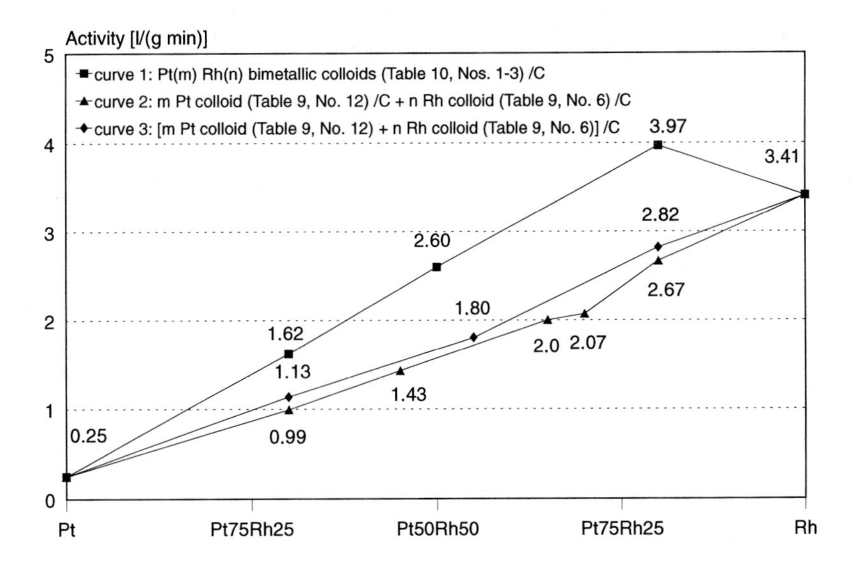

Fig. 9-24. Activity plot of the Pt_m/Rh_n bimetallic catalysts compared with mixed Pt and Rh catalysts (crotonic acid test).

pretated as a further indication that a Pt/Rh colloidal alloy exists in the precursor generated by coreduction (cf. Table 9-10, No. 2 and Fig. 9-17).

In Fig. 9-24 the reported catalytic activity is given in terms of catalyst mass, i.e. 5 wt.-% metal deposited on charcoal. However, to discuss a synergistic effect between platinum and rhodium, it is necessary to plot the activity in terms of metal molar mass, as in Fig. 9-25. The change in catalytic activity over the series of samples is not greatly different from that given in Fig. 9-24. The bimetallic Pt/Rh colloidal catalysts are more active than the mixed metal catalyst systems. The $Pt_{20}Rh_{80}$ bimetallic catalyst shows a

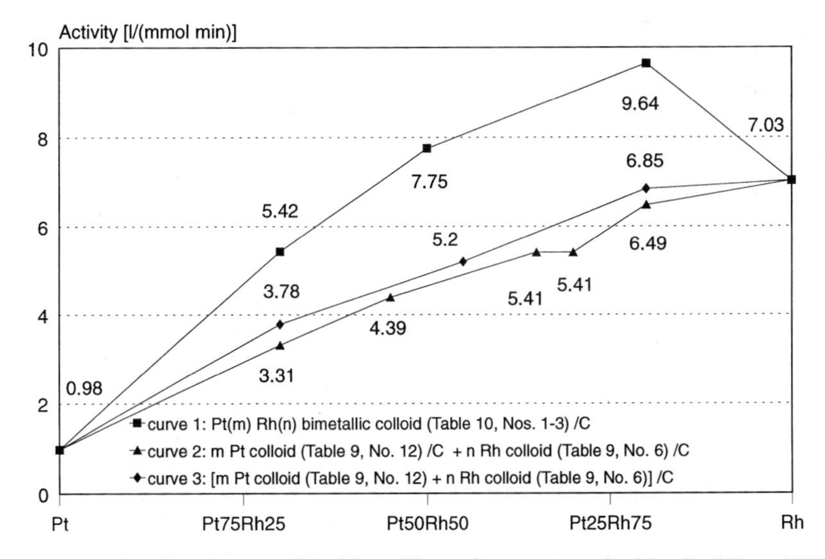

Fig. 9-25. Molar activity plot of the Pt_m/Rh_n bimetallic catalysts compared with mixed Pt and Rh catalysts (crotonic acid test).

significantly greater catalytic activity than the Rh metal catalyst. Is this increase in catalytic activity due to a synergistic effect between platinum and rhodium in the bimetallic catalysts, or is this apparent increase an artifact? In other words, per atom on the colloidal particle surface, i.e. those atoms available and active in catalysis, does the $Pt_{20}Rh_{80}$ catalyst show a higher catalytic activity than the Rh catalyst?

The mean particle diameter of the Rh clusters in the Rh colloidal catalyst is 2.1 nm. An octahedral fcc rhodium particle (diameter = 2.3 nm) has in total 489 atoms, 258 on the surface and 231 in the core. Thus, there are 258 Rh atoms accessible for catalysis. The activity observed for the Rh catalyst is 7025 mL/(mmol min). Hence, per Rh atom (on the surface) the activity equals 27.2 mL/(mmol min per atom), (7025 mL/(mmol min)/ 258 atoms). The mean particle size of the bimetallic clusters in $Pt_{20}Rh_{80}$ catalysts, 2.3 nm, is nearly the same as that of the rhodium colloidal particles. Thus, the octahedral particle now consists of 98 platinum atoms and 391 rhodium atoms. Since rhodium is segregated onto the particle surface in the Pt/Rh bimetallic clusters, there are again 258 Rh atoms available for catalysis. The activity of the $Pt_{20}Rh_{80}$ catalyst was found to be 9638 mL/ (mmol min), thus giving an activity per surface Rh atom of 37.4 mL/(mmol min). This value is clearly greater than the activity observed for the Rh colloidal catalysts per surface Rh atom. This increase in catalytic activity is the result of a synergistic effect of platinum and rhodium in the alloy colloid [48].

To further verify the presence of a synergistic effect in the bimetallic Pt/Rh catalysts, the activity of the $Pt_{50}Rh_{50}$ bimetallic catalysts was compared with that of a mixed Pt and Rh (1:1) colloidal catalyst using the butyronitrile test. As shown in Fig. 9-26, the $Pt_{50}Rh_{50}$ bimetallic catalyst is twice as active as the mixed catalyst. The synergistic effect between platinum and rhodium in the Pt/Rh bimetallic catalysts is the result of alloy formation.

Since the platinum–rhodium interaction in the $Pt_{50}Rh_{50}$ bimetallic catalyst results in an increase in catalytic activity (Fig. 9-26), the complete series of Pt/Rh bimetallic catalysts was tested. As with the results of the crotonic acid test (Fig. 9-24), the $Pt_{20}Rh_{80}$ bimetallic catalyst also gave the highest hydrogenation activity in the butyronitrile test (Fig. 9-27). The $Pt_{50}Rh_{50}$ bimetallic catalyst shows almost the same activity as the Rh metal catalyst, even though 50 mol-% of the more active rhodium is replaced by platinum [48].

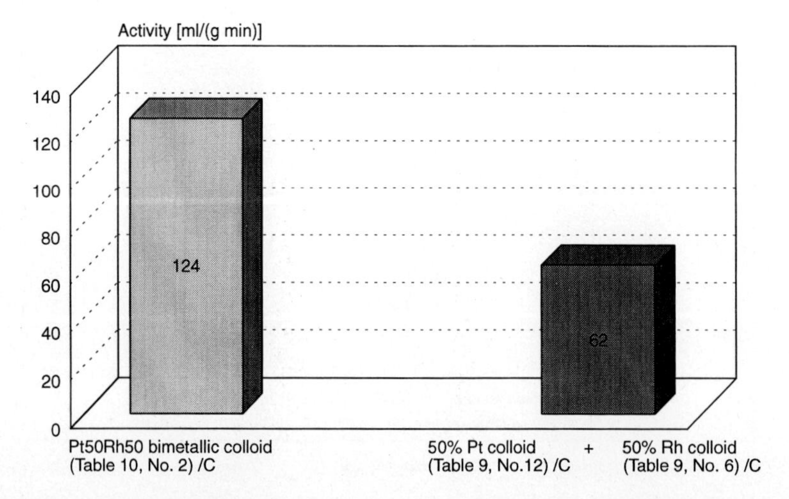

Fig. 9-26. Activity of the $Pt_{50}Rh_{50}$ bimetallic catalyst compared with the mixed Pt and Rh catalyst (butyronitrile test).

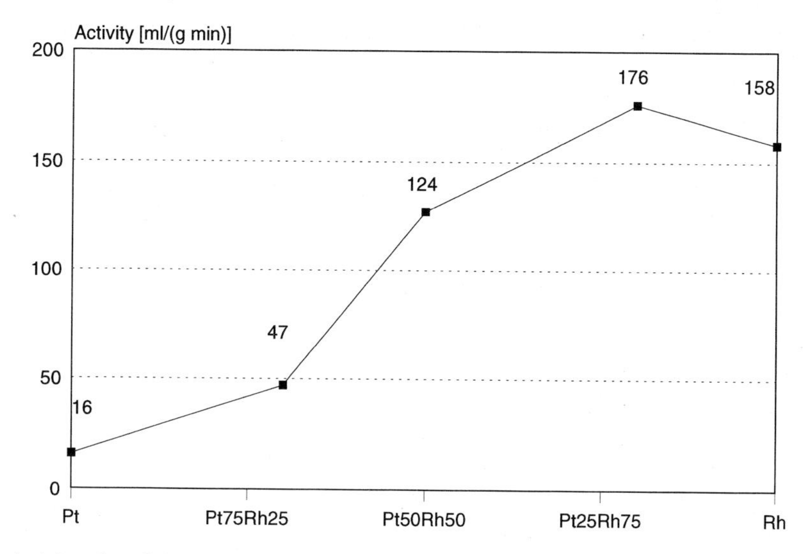

Fig. 9-27. Activity plot of the Pt_m/Rh_n bimetallic catalysts (butyronitrile test).

To determine whether the protective shell has an effect on catalytic activity, two $PEt(octyl)_3^+$-stabilized $Pt_{50}Rh_{50}$ bimetallic catalysts (colloids No. 1 and No. 3 of Table 9-13 deposited on charcoal) were also examined using the butyronitrile test. The activities of these catalysts (Fig. 9-28) are lower than those measured for the $N(octyl)_4^+$-stabilized $Pt_{50}Rh_{50}$ bimetallic catalysts (see Fig. 9-26) [48].

The catalytic activities of supported bimetallic Pd/Pt colloidal catalysts were also investigated, employing the cinnamic acid test. For comparison, Pd/Pt mixed catalyst systems were prepared in the same manner as described for the Pt/Rh mixed catalysts, i.e. mixtures of the two colloids deposited on charcoal or mixtures of the two indivdual supported metal catalysts (Fig. 9-29). The Pd/Pt bimetallic catalysts differ from the two types of mixed metal catalysts in their hydrogenation activity. The activity of the mixed Pd/Pt

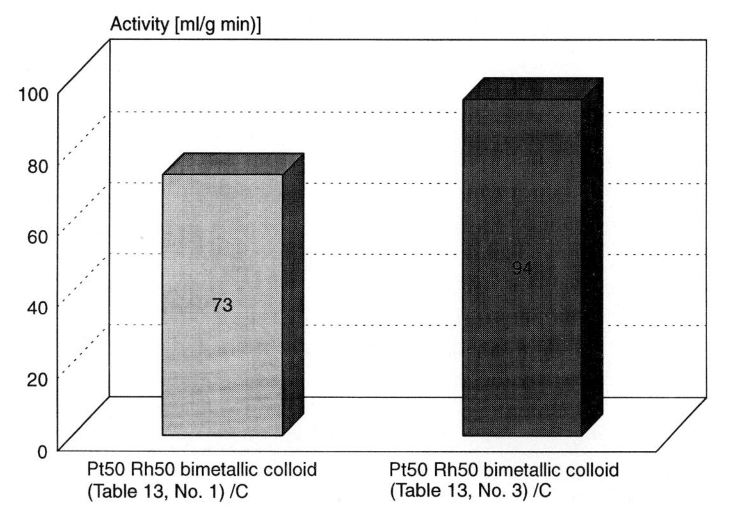

Fig. 9-28. Activity of the $PEt(Octyl)_3^+$-stabilized Pt_{50}/Rh_{50} bimetallic catalysts (butyronitrile test).

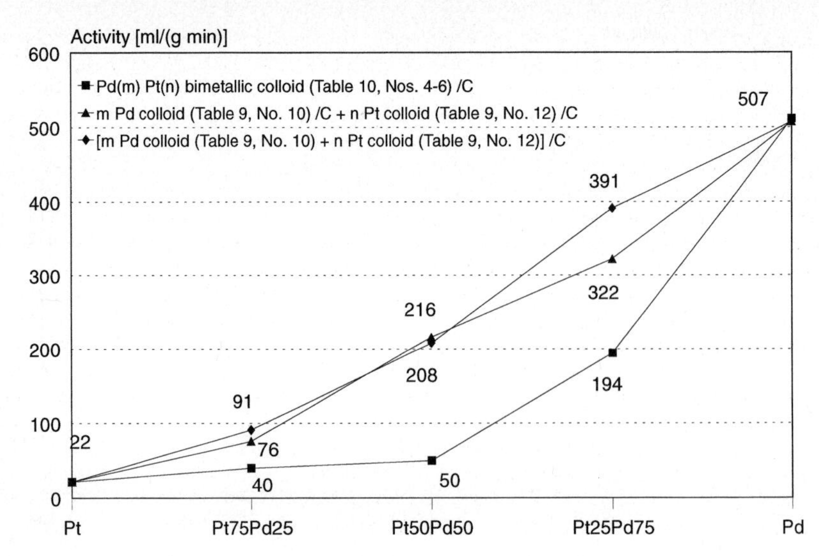

Fig. 9-29. Activity plot of the Pd_m/Pt_n bimetallic catalysts compared with mixed Pd and Pt catalysts (cinnamic acid test).

catalysts increases with increasing palladium content, resulting from an additive effect of the Pd- and Pt-metal catalysts. In contrast, over the series of Pd/Pt bimetallic catalysts a concave ascent in activity is noted. In particular, the activity measured for the $Pd_{50}Pt_{50}$ catalysts is twice that for the Pt metal catalyst and only a quarter of that measured for the corresponding mixed catalysts. The palladium–platinum interaction in the bimetallic catalysts leads to a deactivation in hydrogenation of cinnamic acid, which again points to a synergistic effect between the two metals resulting from alloy formation. Furthermore, these results also suggest the presence of alloyed nanoparticles in which platinum is now concentrated onto the cluster surface [48].

9.6 Conclusions

The results presented here establish that metals and metal alloys of 10–100 nm particle size are easily accessible through the chemical reduction of metal salts or metal oxides with hydrotriorganoborates. Further transition metal colloids (1–10 nm), including bimetallic colloids stabilized by NR_4X or related surfactants, may be prepared similarly. These particles are of special interest with respect to alloy formation and segregation processes. Furthermore, they have been shown to be active hydrogenation catalysts in both the homogeneous and heterogeneous phases. The noble metal colloids can be deposited on surfaces with an unusually high degree of dispersion. The pretreament of the support with low-valent titanium considerably enhances the catalytic activity in liquid-phase hydrogenations. Bimetallic Pt/Rh colloid catalysts show a higher activity compared with a combination of the individual metals. These results lead us to the conclusion that the concept of applying well characterized nanometal colloids as precursors for homogeneous and heterogeneous catalysts opens promising perspectives for rational catalyst design.

Acknowledgements. The authors gratefully acknowledge the valuable support of the following scientists, companies, and institutions: Dr. B. Tesche, Fritz-Haber-Institut der Max-Planck-Gesellschaft, Berlin (FRG), for numerous TEM images taken with a Siemens Elmiskop 102 and a DEEKO 100 at 100 kV and for many helpful discussions; Mr. T. Kamino, Hitachi Instruments Engineering Co. Ltd., 882 Ichige, Katsuta-shi (Japan), and Mr. B. Spliethoff, MPI für Kohlenforschung, Mülheim (FRG), for HRTEM images performed with a Hitachi HF 2000 at 200 kV including EDX point analyses; Prof. Dr. R. Courths and Dipl.-Phys. B. Heise, Universität-Gesamthochschule Duisburg (FRG), for XPS spectra obtained using an ESCALAB Mark II; Prof. Dr. W. Keune and Dipl.-Ing. U. von Hörsten, Universität-Gesamthochschule Duisburg (FRG), for the measurement and interpretation of the Mössbauer spectrum; Prof. Dr. J. Rozière and Dr. D. J. Jones, Université Montpellier (France), for the X-ray absorption measurements recorded on the TiK-edge (77 K) with an EXAFS 3 spectrometer in DC1 (French Synchrotron Facility in Lure); Prof. Dr. J. Hormes, Dipl.-Phys. Rothe, and Cand.-Phys. Becker, Universität Bonn (FRG), for the XANES and EXAFS measurements on the Pt/Rh colloid samples using synchrotron radiation at beamline BN3 of the storage ring ELSA at Bonn university. We are also indebted to Prof. P. Kleinschmit and Drs. P. Panster and A. Freund, Degussa AG, ZN Wolfgang, Hanau (FRG), for a gift of commercial noble metal catalysts and the test procedures for noble metal charcoal catalysts. The support of this work by Fonds der Chemischen Industrie, Frankfurt (FRG) and further, the financial support of this research by a grant (FKZ 03D0007A2) of the Bundesministerium für Forschung und Technologie, Bonn, is gratefully acknowledged. The English manuscript was revised by Dr. Lorraine Aleandri, MPI für Kohlenforschung, Mülheim (FRG), to whom the authors wish to express their sincere thanks.

References

[1] (a) R. Kieffer, F. Benesovsky, *Ullmanns Encykl. Techn. Chem., Vol. 19.* 4th Edition, VCH, Weinheim, **1980**, p. 563; (b) F. E. Luborsky (ed.), *Amorphous Metallic Alloys*, Butterworth, London, **1983**; (c) K. H. Roll, *Kirk-Othmer Encycl. Chem. Technol., Vol. 19.* 3rd Edition, Wiley, New York, **1982**, p. 28.

[2] (a) R. Krabetz, W. D. Mross, *Ullmanns Encykl. Techn. Chem., Vol. 13.* 4th Edition, VCH Weinheim, **1977**, p. 517; (b) G. Schmid (ed.), *Clusters and Colloids*, VCH, Weinheim, **1994**.

[3] S. C. Davies, K. J. Klabunde, *Chem. Rev.* **1982**, *82*, 152.

[4] N. Ibl, *Chem. Ing-Tech.*, **1964**, *36*, 601.

[5] R. D. Rieke, *Organometallics*, **1983**, *2*, 377.

[6] H. Bönnemann, B. Bogdanovic, R. Brinkmann, D.-W. He, B. Spliethoff, *Angew. Chem. Int. Ed. Engl.* **1983**, *22*, 728.

[7] H. Bönnemann, B. Bogdanovic, R. Brinkmann, B. Spliethoff, D.-W. He, *J. Organomet. Chem.*, **1993**, *451*, 23.

[8] (a) H. I. Schlesinger, H. C. Brown, A. E. Finholt, J. R. Gilbreath, H. R. Hoekstra, E. K. Hyde, *J. Am. Chem. Soc.* **1953**, *75*, 215; (b) H. C. Brown, C. A. Brown, *J. Am. Chem. Soc.* **1962**, *84*, 1492.

[9] N. N. Greenwood, A. Earnshaw, *Chemistry of the Elements*, Pergamon, Oxford, **1986**, p. 189.

[10] H. Bönnemann, W. Brijoux, Th. Joussen, *Angew. Chem. Int. Ed. Engl.* **1990**, *29*, 273.

[11] H. Bönnemann, W. Brijoux, R. Brinkmann, US Patent 5,053,075 (Oct. 1, 1991) to Studiengesellschaft Kohle mbH.

[12] H. Bönnemann, B. Korall, *Angew. Chem.* **1992**, *104*, 1506; *Angew. Chem. Int. Ed. Engl.* **1992**, *31*, 1490.

[13] H. Bönnemann, W. Brijoux, R. Brinkmann, E. Dinjus, Th. Joussen, B. Korall, *Angew. Chem. Int. Ed. Engl.* **1991**, *30*, 1312.

[14] H. Bönnemann, R. Brinkmann, R. Köppler, P. Neiteler, J. Richter, *Advanced Materials*, **1992**, *4*, 804.

[15] A. Corrias, G. Ennas, G. Licheri, G. Marongin, G. Paschina, *Chem. Mater.* **1990**, *2*, 363.

[16] (a) P. Binger, G. Benedikt, W. Rotermund, R. Köster, *Liebigs Ann. Chem.* **1968**, *717*, 21; (b) R. Köster, *Methoden Org. Chem. (Houben-Weyl) Bd. XIII, 13b.* 4th edition, Thieme, Stuttgart, **1983**, pp. 798ff.

[17] (a) H. Bönnemann, W. Brijoux, Th. Joussen, DE OS 3934351 (18. 04. 91), to Studiengesellschaft Kohle mbH; (b) H. Bönnemann, W. Brijoux, Th. Joussen, EP OS 0423627 (24. 04. 91), to Studiengesellschaft Kohle mbH.

[18] R. G. Pearson, *Surv. Prog. Chem.* **1969**, *5*, 1.

[19] L. Kolditz (ed.): *Anorganikum Teil 1*, VEB Deutscher Verlag der Wissenschaften, Berlin, **1981**, p. 474.

[20] *D'Ans Lax, Vol. 1*, 3rd edition. p. 1. Springer, Berlin, **1967**.
[21] H. P. Klug, L. E. Alexander, *X-Ray Diffraction Procedures for Polycrystalline and Amorphous Materials*, 2nd edition, Wiley, New York, **1974**.
[22] (a) J. van Wonterghem, S. Mørup, C. J. W. Koch, S. W. Charles, S. Wells, *Nature*, **1986**, *322*, 622; (b) J. van Wonterghem, S. Mørup, *Hyperfine Interact.* **1988**, *42 (14)*, 959.
[23] C. E. Johnson, M. S. Ridout, T. E. Cranshaw, *Proc. Phys. Soc.* **1963**, 1079.
[24] D. Zeng, M. J. Hampden-Smith, *Chem. Mater.* **1992**, *4*, 968.
[25] H. Bönnemann, W. Brijoux, R. Brinkmann, US Patent 5,053,075 (Oct. 1, 1991) to Studiengesellschaft Kohle mbH.
[26] (a) G. Schroeder, G. Buxbaum, F. Hund, H.-C. Schopper, R. Naumann, EP 0015 485 (26. 02. 1980) to Bayer AG; (b) H. C. Miller, A. L. Oppegard US Patent 2,900,246 (Sept. 14, 1965) to E. I. du Pont.
[27] (a) M. Boutonnet, J. Kizling, P. Stenius, G. Maire, *Coll. and Sulf.* **1982**, *5*, 209; (b) M. Boutonnet, J. Kizling, R. Touroude, G. Maire, P. Stenius, *Appl. Catal.* **1986**, *20*, 163.
[28] (a) J. S. Bradley, *Adv. Organomet. Chem.* **1983**, *22*, 1; (b) J. S. Bradley, E. Hill, M. E. Leonowicz, H. J. Witzke, *J. Mol. Catal.* **1987**, *41*, 59; (c) J. S. Bradley, J. M. Millar, E. W. Hill, S. Behal, B. Chaudret, A. Duteil, *Faraday Discuss.* **1991**, 92, 225.
[29] P. Braunstein, *Nouv. J. Chim.* **1986**, *10(7)*, 365.
[30] (a) K. Esumi, M. Shiratori, H. Ishizuka, T. Tano, K. Torigoe, K. Meguro, *Langmuir* **1991**, *7*, 457; (b) K. Meguro, M. Torizuka, K. Esumi, *Bull. Chem. Soc. Jpn.* **1988**, *61*, 341; (c) K. Esumi, M. Suzuki, T. Tano, K. Torigoe, K. Meguro, *Colloids and Surfaces* **1991**, *55*, 9; (d) T. Tano, K. Esumi, K. Meguro, *J. Colloid Interface Sci.* **1989**, *133*, 530.
[31] (a) J. Evans, *NATO ASI Ser., Ser. C (Surf. Organomet. Chem., Mol. Approaches Surf. Catal.)* **1988**, *231*, 47; (b) J. Evans, B. Hayden, F. Mosselmans, A. Murray, *J. Am. Chem. Soc.* **1992**, *114*, 6912; (c) J. Evans, B. Hayden, F. Mosselmans, A. Murray, *Surf. Sci.*, **1992**, *279(1–2)*, 159.
[32] B. T. Heaton, *Pure Appl. Chem.* **1988**, *60(12)*, 1757.
[33] (a) A. Henglein, *J. Phys. Chem.*, **1993**, *97*, 5457; (b) A. Henglein, P. Mulvaney, A. Holzwarth, T. E. Sosebee, A. Fojtik, *Ber. Bunsenges. Phys. Chem.* **1992**, *98*, 754.
[34] (a) K. J. Klabunde, Y.-X. Li, B. J. Tan, *Chem. Mater.* **1991**, *3*, 30; (b) K. J. Klabunde, *Science* **1984**, *224*, 1329; (c) K. J. Klabunde, Y. Imizu, *J. Am. Chem. Soc.* **1984**, *106*, 2721; (d) K. J. Klabunde, Y. Tanaka, *J. Mol. Catal.* **1982**, *47*, 843.
[35] H. Knözinger in *Cluster Models for Surface and Bulk Phenomena* (eds.: G. Paccioni, P. S. Bagus, F. Parmigiami) Plenum, New York, **1992**, p. 148.
[36] C. Larpent, H. Patin, *J. Mol. Catal.* **1988**, *44*, 191.
[37] (a) L. N. Lewis, N. Lewis, *Chem. Mater.* **1989**, *1*, 106; (b) L. N. Lewis, N. Lewis, *J. Am. Chem. Soc.* **1986**, *108*, 7228; (c) L. N. Lewis, R. Uriarte, N. Lewis, *J. Catal.* **1991**, *127*, 67.
[38] (a) T. A. Stromnova, I. N. Busygina, M. N. Vargaftik, I. I. Moiseev, *Metalloorg. Khim.* **1990**, *3*, 803; (b) I. I. Moiseev, *Pure Appl. Chem.* **1989**, *61*, 1755; (c) I. I. Moiseev, *Mekhanizm Kataliza* (Novosibirsk) **1984**, 21; from *Ref. Zh. Khim.* **1985**, Abstr. No. 4B4121; (d) I. I. Moiseev, *Itogi Nauki Tekh. VINITI. Kinet. Kataliz* **1984**, 13; from *Ref. Zh. Khim.* **1984**, Abstr. No. 16B4121.
[39] (a) G. Schmid, *Endeavour* **1990**, *14*, 172; (b) G. Schmid, *Aspects Homogen. Catal.* **1990**, *7*, 1; (c) G. Schmid, H. H. A. Smit, M. P. J. Staveren, R. C. Thiel, *New J. Chem.* **1990**, *14*, 559; (d) G. Schmid, N. Klein, B. Morun, A. Lehnert, J. O. Malm, *Pure Appl. Chem.* **1990**, *62*, 1175; (e) L. J. De Jongh, J. A. O. DeAguiar, H. B. Brom, G. Longoni, J. M. Van Ruitenbeek, G. Schmid, H. H. A. Smit, M. P. J. Van Staveren, R. C. Thiel, "Physical Properties of High Metal Compounds", in: *Z. Phys. D-Atoms, Molecules and Clusters*, **1989**, *12*, 455.
[40] (a) N. Toshima, M. Harada, T. Yonezawa, K. Kushihashi, K. Asakura, *J. Phys. Chem.* **1991**, *95*, 7448; (b) N. Toshima, T. Takahashi, *Bull. Chem. Soc. Jpn.* **1992**, *65*, 400; (c) N. Toshima, T. Takahashi, H. Hirai, *J. Macromol. Sci. Chem.* **1988**, *A25*, 669; (d) N. Toshima, *J. Macromol. Sci. Chem.* **1990**, *A 27 (9–11)*, 1225; (e) Zhao Bin, N. Toshima, *Chemistry Express* **1990**, *5, No. 10*, 721.
[41] J. Kiwi, M. Grätzel, *J. Am. Chem. Soc.* **1979**, *101*, 7214.
[42] (a) H. Bönnemann, W. Brijoux, R. Brinkmann, E. Dinjus, R. Fretzen, Th. Joussen, B. Korall, *J. Mol. Catal.* **1992**, *74*, 323; (b) H. Bönnemann, R. Brinkmann, P. Neiteler, *Appl. Organomet. Chem.*, **1994**, *8(4)*, 361; (c) H. Bönnemann, W. Brijoux, R. Brinkmann, R. Fretzen, Th. Joussen, R. Köppler, B. Korall, P. Neiteler, J. Richter, *J. Mol. Catal.* **1994**, *86*, 129.
[43] H. Bönnemann, W. Brijoux in: *MRS Symposium Proceedings Series, Vol. 351*, (eds: K. E. Gonsalves, G.-M. Chow, T. D. Xiao, R. C. Cammarata), Materials Research Society, Pittsburgh, **1994**, p. 3.
[44] B. Bogdanovic, T. Hartwig, A. Ritter, K. Straßburger, B. Spliethoff, *Abschlußbericht zum BMFT-Forschungsvorhaben* Nr. 0328939 C, **1991**.
[45] L. E. Aleandri, A. Baiker, H. Bönnemann, R. Brinkmann, W. Brijoux, R. Courths, B. Heise, D. J. Jones, T. Mallat, J. Rozière, J. Bradley, G. Via, *Chem. Mater.*, to be published.

[46] H. Bönnemann, W. Brijoux, J. Richter, R. Becker, J. Hormes, J. Rothe, *Z. Naturforsch.* **1995**, *50B*, 333.

[47] (a) J. Blum, Y. Sasson, A. Zoran, *J. Mol. Catal.* **1981**, *11*, 293; (b) J. Blum, I. Amer. K. P. C. Vollhardt, H. Schwarz, G. Höhne, *J. Org. Chem.* **1987**, *52*, 2804; (c) I. Amer. V. Orshaw, J. Blum, *J. Mol. Catal.* **1988**, *45*, 207; (d) I. Amer, T. Bernstein, M. Eisen, J. Blum, K. P. C. Vollhardt, *J. Mol. Catal.* **1990**, *60*, 313; (e) I. Amer. J. Blum, K. P. C. Vollhardt, *J. Mol. Catal.* **1990**, *60*, 323; (f) Y. Badrieh, J. Blum, I. Amer. K. P. C. Vollhardt, *J. Mol. Catal.* **1991**, *66*, 295; (g) J. Blum, H. Huminer, H. Alper, *J. Mol. Catal.* **1992**, *75*, 153; (h) Y. Badrieh, A. Kayyal, J. Blum, *J. Mol. Catal.* **1992**, *75*, 161; (i) A. Laufenberg, A. Behr, W. Keim, DOS 3841698 (1989) to Henkel KG. a. A; (j) N. Satoh, K. Kimura, *Bull. Chem. Soc. Jpn.* **1988**, *62*, 1758.

[48] J. Richter, Ph. D. Thesis, RWTH Aachen (1994).

[49] R. Willstätter, E. Waldschmidt-Leitz, *Ber. Dtsch. Chem. Ges.* **1921**, *54*, 113.

[50] L. L. Hegedus (ed.), *Catalyst Design*, Wiley, New York **1987**.

[51] S. A. Stevenson (ed.), *Metal–Support Interactions in Catalysis, Sintering and Redispersion*, Van Nostrand, Reinhold, New York, **1987**.

[52] (a) O. S. Alekseev, V. I. Zaikovskii, Y. A. Ryndin, *J. Appl. Catal.* **1990**, *63*, 37; (b) Y. A. Ryndin, O. S. Alekseev, P. A. Simonov, V. A. Likholobov, *J. Mol. Catal.* **1989**, *55*, 109; (c) Y. A. Ryndin, O. S. Alekseev, E. A. Paukshtis, A. V. Kalinkin, D. I. Kochubey, *Appl. Catal.* **1990**, *63*, 51.

[53] H. Bönnemann, W. Brijoux, R. Brinkmann, E. Dinjus, R. Fretzen, B. Korall, DE OS 4111719 (15. 12. 92) to Studiengesellschaft Kohle mbH.

10 Supported Metals

Alois Fürstner

10.1 Introduction

Supported reagents in general are rapidly gaining importance, despite the traditional preference of preparative chemists for homogeneous reactions [1]. Many examples in the literature and in industrial practice illustrate conclusively that the adsorption of a reagent on a carrier may significantly up-grade a chemical transformation into practicability, selectivity, and efficiency, not to mention the advantages in work-up and the options for the recovery of the reactant. The progress achieved by the Merrifield resin technique for peptide synthesis, by the use of supported enzymes for the conversion of porcine into human insulin, or by adsorbing TNT on kieselguhr is obvious [1]. These and other examples are encouraging precedents for the subject dealt with in this chapter. In the first part we will try to delineate the arguments for and against the immobilization of an activated metal on an inert support and will summarize the criteria that have to be considered in choosing the proper carrier for a particular metal. The second part of this review will then describe the scope, performance and limitations of the metal/carrier systems nowadays most frequently employed. However, this chapter will not cover applications of traditional, commercially available catalysts such as palladium on charcoal or other supports.

10.2 General Features of Supported Metals

By loading a reagent onto a carrier the site of reaction becomes clearly defined by the reduced (fractal) dimensionality of the surface of the support. Such a compartmentalization necessarily implies a local accumulation of the immobilized agent. In addition to this purely geometrical argument, the chemical constitution of a given support material may also contribute to improve the overall reaction rate. Many of the commonly used inorganic support materials have polar properties or are polyelectrolytes. Any charged or polarizable substrate in solution will accumulate near to their surface due to dipole–dipole and/or electrostatic interactions. This may increase the effective concentration of the dissolved reaction partner near the interface by several orders of magnitude which may result in a substantial acceleration of the reaction [1].

Synergisms between the supported reagent and its carrier, however, are not restricted to improved statistics of reactive encounters; the support may also directly interfere in the chemical reaction. Thus, the residual –OH groups on Al_2O_3 e.g. may well serve as proton sources or may become an internal source of H_2 in combination with a strongly reducing metal [2]. The decyanation of alkyl nitriles by K/Al_2O_3 in hexane and the isomerization of cycloalkadienes followed by their partial hydrogenation induced by the same reagent (see below) nicely illustrate these two possibilities.

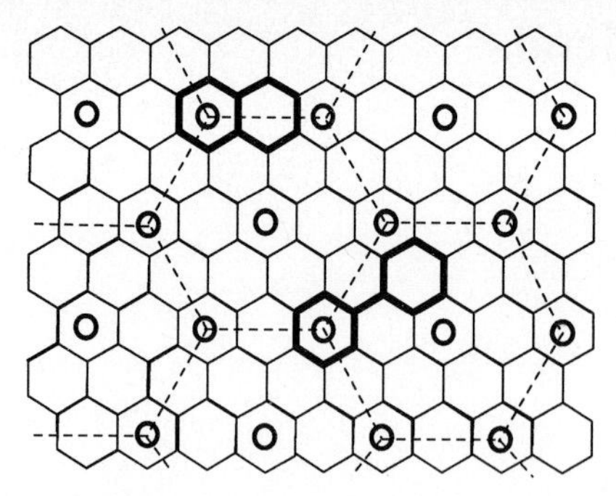

Fig. 10-1. Structure of C_8K. The highlighted segments emphasize that this graphite intercalation compound may be regarded as a polymeric array of alkali naphthalenide or alkali biphenyl radical anions. ○ = K

A further striking example of joint action of a metal/carrier couple is the chemistry of the potassium–graphite laminate C_8K (Fig. 10-1) [3], which will be discussed in detail below. In this compound the metal is intercalated between the graphite layers and hence not directly exposed to the substrate in solution. Because the 4s-electrons of the potassium are coupled with the valence electrons of the graphite, the "carrier" is not merely inert but plays an active role in electron transfer processes. By this interplay of the metal with its polyaromatic host, the electrons are rendered accessible at any site of the extended surface area, even though the metal itself is "hidden" within the graphite galleries. C_8K can thus be considered as a polymeric array of naphthalenide or biphenyl anions. This particular synergism boosts the reducing properties of C_8K as compared with that of lumps of metallic potassium [3].

Several practical advantages may stem from the immobilization of a metal on a support. Preparative chemists often find that the isolation of the product from the reaction mixture is the "rate-determining step" in the laboratory. By using supported reagents this becomes trivially easy and essentially consists of a simple filtration. Filtration also "switches off" the reaction instantaneously without the need for any external additive. This option of a rapid, non-aqueous work-up is of eminent importance for the isolation of hydrolytically and/or thermally labile products. It also offers the chance of recovering and/or regenerating the carrier-bound reagent or catalyst. Although many of the examples described in Section 10.3 of this chapter may well feature these advantages, the C_8K-induced one-pot formation of furanoid glycals from furanosyl chlorides is especially illustrative (Scheme 10-1) [4]. Both the substrates and the products formed are rather sensitive compounds and lengthy manipulations are not advisable. Treatment of a furanosyl chloride such as **1** with C_8K in THF at low temperatures leads almost instantaneously to the formation of the potassium alcoholate **2**, which may be trapped in situ upon addition of an appropriate electrophile RX. The inorganic by-products formed are graphite and KCl, both of which can be stripped off by filtration. The resulting filtrate contains a solution of the desired furanoid glycal **3** which is pure enough for direct further elaboration. Because of the exceptional performance of the C_8K-induced reductive scission together with the straightforward non-aqueous work-up, this cascade compares favourably with all other literature procedures for the formation of furanoid enol ethers of the sugar series in terms of yield, reaction rate, flexibility, ease of manipulation and purity of the products [4].

Yet another practical aspect of metal/carrier-induced reactions can be demonstrated by this particular example. C_8K is a brightly bronze colored and hence self-indicating reagent. The progress of the reaction is proportional to the conversion of C_8K and can be

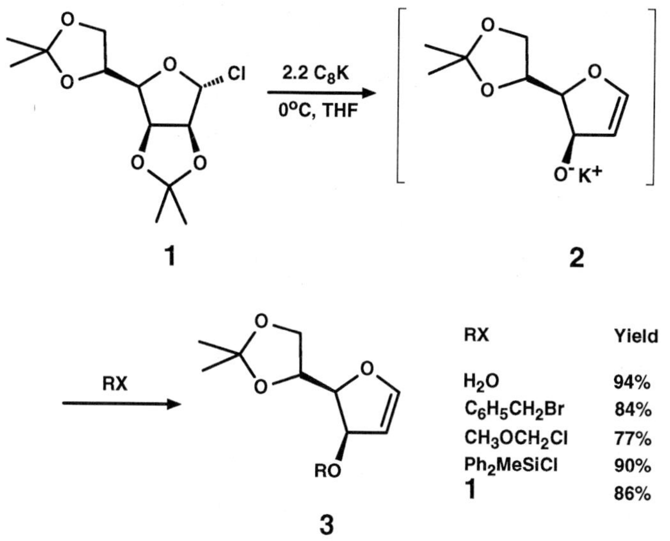

2.2 C_8K

0°C, THF

1 **2**

RX

3

RX	Yield
H_2O	94%
$C_6H_5CH_2Br$	84%
CH_3OCH_2Cl	77%
$Ph_2MeSiCl$	90%
1	86%

Scheme 10-1. One-pot conversion of a glycosyl chloride **1** into differently protected furanoid glycals **3** by means of C_8K [4].

easily monitored by the color change of the reaction mixture from bronze (C_8K) to black (graphite). This feature is not unique to C_8K; for example, K/Al_2O_3 (2–15 mol% metal content) is distinguished by a beautiful metallic-blue color.

High activation of the metal often stems from an exceptional degree of dispersion. However, particles with an average diameter of only a few nanometers show a strong tendency towards aggregation, recrystallization, and sintering [5]. All these phenomena are detrimental to activity. Adsorption of the clusters on an inert support may be a simple but efficient device to stabilize them or, at least, to retard such undesirable morphological changes. This may well have been one of the reasons for the attempts to fix Rieke metals on polystyrene matrices, and is also one of the major arguments in favor of the metal–graphite combinations, in which the highly dispersed metal particles get physisorbed at the graphite surface immediately upon formation. The repeated use of platinum–graphite e.g. as a selective catalyst for the hydrogenation of 4-iodonitrobenzene, did not change its activity, and no significant morphological deteriorations of the Pt particles were detectable by electron microscopic analysis when compared with a freshly prepared sample [6].

An interesting yet largely unexplored option for synergism of an activated metal with a carrier stems from the porosity of the latter. Shape selectivity as a result of the geometrical constraints of the inner surface of the support material is well documented in heterogeneous catalysis in general, but has not yet been exploited to a significant extent in the context of metal activation. This even holds true for the metal–graphite reagents. Although it has been proved that many organic molecules can penetrate into the interlayer galleries, most chemical reactions seem to take place at the surface [3]. An exception might be C_8K-induced anionic polymerizations, which probably occur in the interior of the laminate. Moreover, some indications that the Lewis-basicity exhibited by C_8K might be related to the size of the acidic reaction partner have been published (cf. Section 10.3.3).

The chosen support material should be a cheap, easily accessible fine powder which resists mechanical stress (grinding etc.) to some extent. This is why Al_2O_3, TiO_2, SiO_2, kieselguhr, graphite, charcoal etc. are widely used. Some fine-tuning is possible e.g. by choosing either an acidic, neutral or basic variety of Al_2O_3. The differences are due to the relative numbers of basic and acidic sites at the alumina surface [1]. Care must be taken that both the substrate and the product resist the catalytical properties exhibited by these different types of Al_2O_3 [1, 2]. Although only a few applications of metals deposited on NaCl have been described so far, this cheap support material might deserve more attention as it provides two options for product isolation [7], either by filtration or by dissolving the carrier during an extractive aqueous work-up. Organic polymers constitute another class of support materials. Although a laborious preparation may be necessary, they can be tailor-made for a given application. Successful examples of metal/polymer combinations are described in Section 10.3.6.

In most instances a pre-treatment of the chosen support material will be necessary in order to remove moisture and/or oxygen. This is usually achieved by heating it for several hours under reduced pressure. The progress of the drying can be followed either by thermogravimetry [1] or by titration. In the latter case, aliquots of the carrier powder are treated with a given amount of *n*BuLi. The proportion of the organometallic reagent that has not been hydrolysed by residual water can then be titrated with, for example, 2-propanol and one of the usual indicators. This method, however, is not applicable to TiO_2 as this oxide reacts with *n*BuLi.

Only in the case of the metal–graphite combinations is the loading predetermined by the stoichiometry of the reduction process that leads to their formation [3]. In most other cases the chemist is free to vary this parameter at will. Care taken in the choice of loading is often rewarded with superior performance of the supported system [1]. As a rule, a monolayer dispersion should be aimed for, but this might be difficult to accomplish in practice. Although the surface area of the support may be determined by PED and pore-size distribution analyses, the area occupied by the adsorbed reagent is difficult to estimate [1]. Its coagulation may further complicate matters, so that an empirical optimization of the ratio of loaded metal to amount of carrier is still the best means to address this point. In any case, severe overloading should be avoided as it will hide some of the adsorbed metal and thus render it unreactive. For potassium on alumina, for example, a metal content in the range 2–15% (w/w) was reported to lead to the highest catalytic activities in positional and configurational alkene isomerizations [2, 8].

Although the use of carrier-bound metals is still in its infancy, and the reasons for their special properties remain to be further elucidated on a molecular level, much favorable practical experience has already been gained. Chemistry on surfaces in general tends to be efficient and selective [1]. The paradox of an increased reactivity *and* an increased selectivity has often been encountered. This remarkable feature is nicely illustrated by several of the applications of supported metals summarized below.

10.3 Applications of Supported Metals in Synthesis

10.3.1 "High Surface Area Alkali Metals" [2]

By adding potassium to thoroughly dried basic or neutral alumina under argon at temperatures above 100 °C and stirring the mixture until a homogeneous appearance is reached, one can easily obtain "high surface area potassium" as a storable, moisture sensitive, *non-pyrophoric* powder. It has a shiny blue color if the metal content is in the range 2–15% (w/w). At higher loadings the reagent is grey to black and exhibits diminished activ-

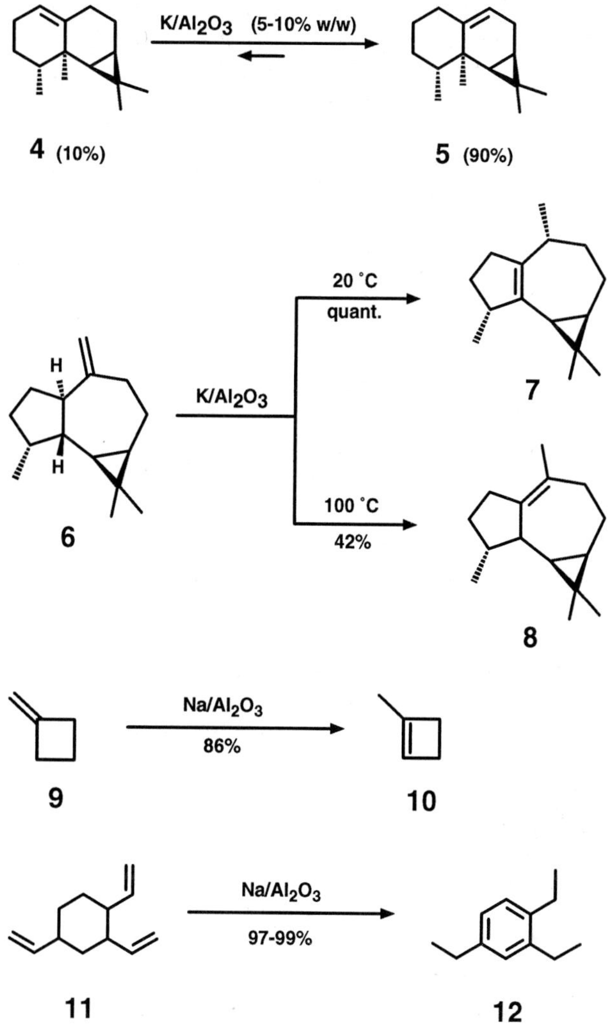

K/Al2O3 (5-10% w/w)

4 (10%) 5 (90%)

20 °C quant.

K/Al2O3

100 °C 42%

6 7 8

Na/Al2O3 86%

9 10

Na/Al2O3 97-99%

11 12

Scheme 10-2. Selected examples of M/Al_2O_3-catalyzed isomerization reactions (M = Na, K) [11, 12, 15].

ity [8]. In the X-ray diffraction pattern of a reagent consisting of 14% K on Al_2O_3, no reflections due to metallic potassium could be observed [9]. The same mixing/grinding procedure can also be applied for preparing high surface area variants of the other alkali metals, except lithium [8]. A more even distribution of the alkali metal on the surface of the inorganic carrier can be obtained by melting it and depositing it by means of a high-speed stirrer on a suspension of the support in boiling toluene (b.p. 110.6 °C, compared with m.p. of 97.8, 63.3, 38.9, and 28.4 °C for Na, K, Rb, and Cs respectively) [7]. This "wet" procedure is also applicable to supports other than Al_2O_3, such as TiO_2, charcoal, NaCl, etc. Alcoholysis of samples of high surface area sodium on different support materials with measurement of the amount of H_2 evolved has shown that in all cases the active reduction equivalent was at least 94% of the Na input [7]. Furthermore, Na/Al_2O_3 in a wax-coated form can be conveniently handled and used as an "off-the-shelf" reagent [10].

M/Al_2O_3 reagents efficiently catalyze configurational and/or positional alkene isomerizations, most likely via allyl anion intermediates (Scheme 10-2) [2, 8, 11–16]. The relative activities observed were in the order K $\geq$ Rb $\gg$ Cs $\approx$ Na [8]. The double bonds are usually shifted towards positions of higher substitution, or towards products with a thermodynamically favorable conformation. This can be used beneficially, e.g. in the terpene series. Thus, (+)-calarene (**4**) will equilibrate to (−)-aristolene (**5**), and the *exo*-methylene group of **6** may be isomerized to the tetrasubstituted alkenes **7** or **8** depending on the reaction conditions [12]. The alkene entities of dienes or polyenes are shifted towards conjugation, independently of their original position in the starting material. This is evidenced by the formation of 1,2,4-triethylbenzene **12** from 1,2,4-trivinylcyclohexane **11**, as well as by the isomerization of several macrocyclic 1,*n*-cycloalkadienes to the respective 1,3-isomers [8, 12, 13, 15]. After being formed by such double bond isomerizations, macrocyclic 1,3-cycloalkadienes may get slowly reduced to the corresponding cycloalkenes even in the absence of external hydrogen [8]. These selective hydrogenations are most likely due to a joint function of the alkali metal and the residual –OH groups on the alumina, which provide an internal H_2 source. Although these reductions are accelerated by running the reaction under a hydrogen atmosphere, even then no overreduction to the corresponding cycloalkanes takes place [8].

Surprisingly, M/Al_2O_3 reagents not only catalyze such hydrogenation reactions but may effect dehydrogenations as well. Thus, high surface area alkali metals exhibit a strong propensity to aromatize six membered ring systems bearing two neighboring alkene or cyclopropyl groups via a sequence of isomerization and dehydrogenation processes (Scheme 10-3) [12, 15].

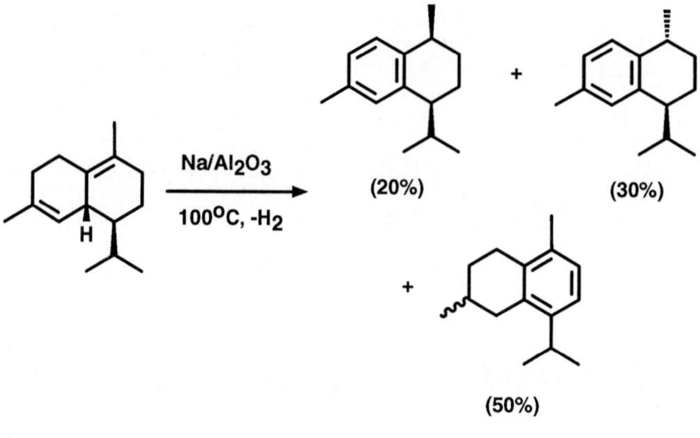

Na/Al2O3
100°C, -H2
H
(20%)
+
(30%)
+
(50%)

Scheme 10-3. Isomerization/dehydration sequence catalyzed by high surface area sodium [12].

In addition to the alkene isomerizations mentioned above, skeletal rearrangements and transannular reactions of macrocyclic compounds are also induced by M/Al_2O_3 [8, 12, 16]. Several catalytic processes induced by M/Al_2O_3 were used together in a reaction cascade, which made it possible to transform (*Z*,*E*,*E*)-cyclododeca-1,5,9-triene in a one-pot procedure into cyclododeca-1,7-dione (Scheme 10-4) [16].

High-surface alkali metals can also serve as bases and as electron transfer agents. Thus, K/Al_2O_3 can deprotonate ketones, alkyl nitriles, ethyl phenylacetate, *N*-cyclohexyl ketimines, and *N*,*N*-dimethylhydrazones of aliphatic aldehydes [17]. However, a large excess of the reagent turned out to be necessary, and the yields of the products obtained upon alkylation of the intermediate potassium anions were only moderate. Na on charcoal, graphite, or Al_2O_3 has also been used as a base for the monoalkylation of cyclic ketones. The different supports do essentially the same job, and the reaction is again not comple-

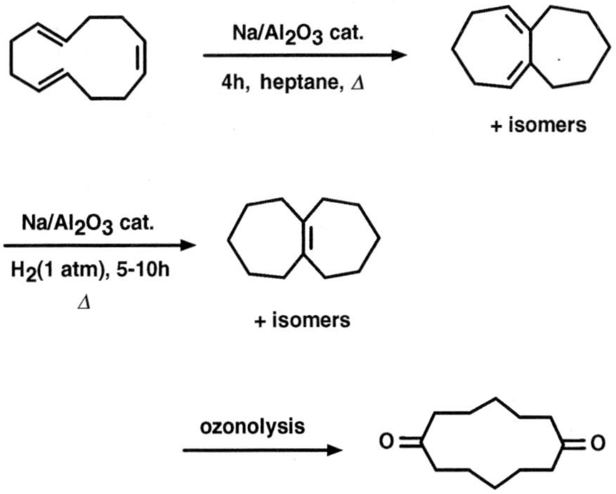

Na/Al2O3 cat.
4h, heptane, Δ
+ isomers
Na/Al2O3 cat.
H2(1 atm), 5-10h
Δ
+ isomers
ozonolysis

Scheme 10-4. Cascade comprising Na/Al_2O_3-catalyzed isomerization, transannular reaction, and hydrogenation reactions followed by an ozonolysis of the crude product mixture [16].

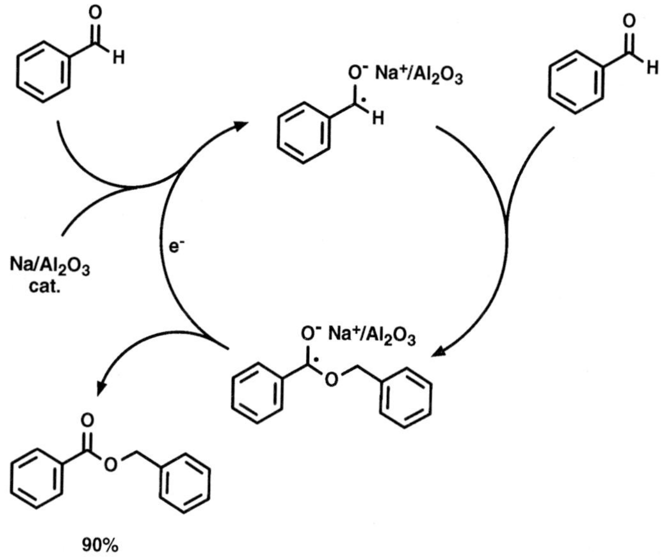

Scheme 10-5. Tishchenko coupling of benzaldehyde to benzyl benzoate catalyzed by high surface area sodium [18].

tely selective [17b]. In contrast, Na/Al_2O_3 in mixtures of THF with alcohols shows interesting electron transfer properties. It reduces different types of ketones to the corresponding alcohols [10]. In bicyclic systems the thermodynamically more stable isomer is formed preferentially. Oximes and esters may also be reduced by Na/Al_2O_3 in 2-propanol or *t*-butanol as solvent. The reagent also catalyzes the formation of benzyl benzoate from benzaldehyde via a single electron transfer process (Scheme 10-5), which triggers a subsequent Tishchenko coupling reaction [18].

How sensitive M/Al_2O_3-induced processes can be to the experimental conditions is illustrated by their reactions with alkyl nitriles. In THF as solvent, the deprotonation of these substrates is the dominant reaction, whereas in hexane a smooth decyanation reaction can be achieved (Scheme 10-6) [9, 17a]. In the latter case the residual –OH groups of the carrier once again play an important role as system-inherent proton sources, trapping the reactive intermediates during this reductive C–C bond scission. Disubstituted double bonds and acetal functions turned out to be compatible, while terminal alkene groups are simultaneously isomerized to internal ones during this decyanation process [9].

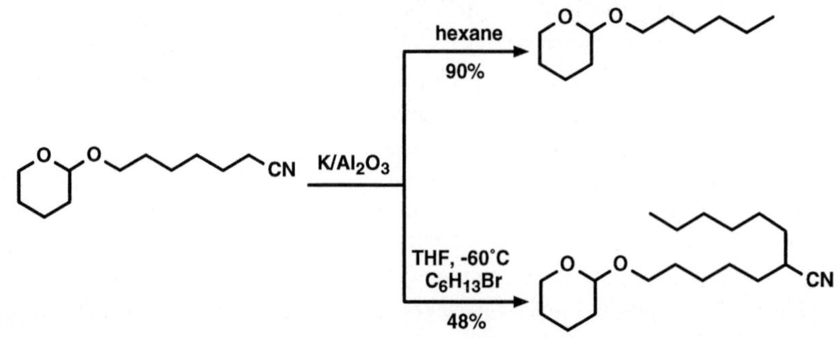

CN K/Al_2O_3 hexane 90% THF, -60°C $C_6H_{13}Br$ 48% CN

Scheme 10-6. Representative example of the solvent dependence of K/Al_2O_3-induced reactions of alkyl nitriles [9, 17a].

10.3.2 Activated Metals by Reduction of Metal Salts with "High Surface Area Sodium"

The essence of the Rieke method for metal activation [19] consists of preparing an active slurry by reduction of an appropriate metal salt by means of strong reducing agents such as K, Li, or Li-naphthalenide etc. in an inert solvent. Surprisingly, sodium has hardly ever been used for this purpose, despite its reasonably low melting point and much lower price per mole compared with K or Li [19].

The use of high-surface sodium should allow one to combine the favourable features of the Rieke activation procedure with the advantages of carrier supported systems. In fact, the reduction of $TiCl_3$ with Na on different carriers afforded a highly active low-valent titanium species which efficiently promoted different types of McMurry reactions (Scheme 10-7) [7]. However, the proper choice of the reaction conditions was crucial. The highest activity for the reductive coupling of carbonyl compounds to the respective alkenes was exhibited by a titanium species obtained from $TiCl_3$ with only two equivalents of Na/carrier after 1–2 h reduction time. Rather than being metallic titanium, the active species (represented here by [Ti]) consists of titanium in the formal oxidation state +1. Its

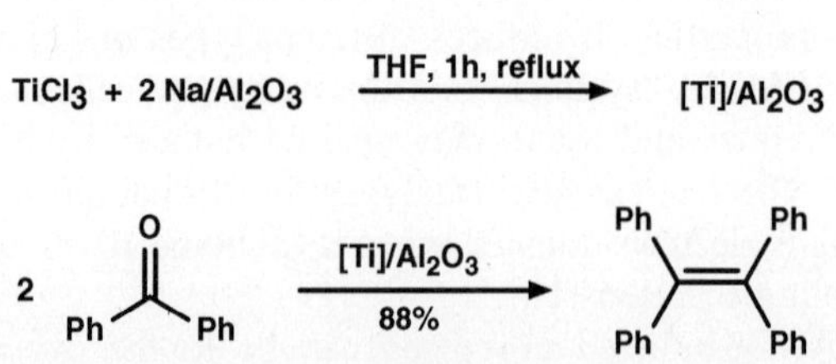

$$Al_2O_3 + Na\ (10\ \%\ w/w) \longrightarrow Na/Al_2O_3$$

$$TiCl_3 + 2\ Na/Al_2O_3 \xrightarrow{\text{THF, 1h, reflux}} [Ti]/Al_2O_3$$

2 Ph(C=O)Ph $\xrightarrow[88\%]{[Ti]/Al_2O_3}$ $Ph_2C=CPh_2$

Scheme 10-7. Preparation of low-valent titanium [Ti] on alumina. Application to the McMurry coupling of benzophenone [7].

exact nature has not yet been elucidated. With respect to the choice of support, Al_2O_3 and NaCl gave comparable results, while TiO_2 gave somewhat lower yields of coupling products [7].

[Ti]/Al_2O_3 readily dimerized aromatic aldehydes and ketones to the corresponding olefins. The lower yields observed in the aliphatic series were due to an incomplete deoxygenation of the intermediate pinacols and to concomitant two-electron reductions of the substrates. However, the most useful preparative feature of the reagent was its pronounced template effect. Dicarbonyl compounds undergo coupling to give the corresponding cycloalkenes independently of the resulting ring size, without recourse to high dilution techniques. Even for the formation of a 36-membered ring the substrate could be added to the [Ti] slurry in 2 h or less using a simple dropping funnel. Likewise, 2-acylamido benzophenone derivatives were readily coupled to substituted indoles according to a previously developed approach. Some representative examples are given in Table 10-1 [7].

A reaction of particular interest is the formation of (*E*)-1,2-diphenyl-1,2-bis(trimethylsilyl)ethene (**14**) from benzoyltrimethylsilane, (**13**) (Scheme 10-8). This is the first example of a successful McMurry-type coupling of an acylsilane. While [Ti]/NaCl afforded reasonable yields of **14** as a single diastereoisomer, other low-valent titanium species gave only poor results [21].

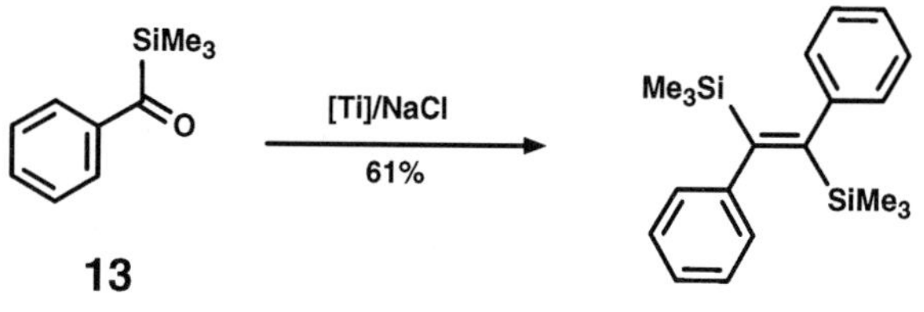

SiMe3
O
13
[Ti]/NaCl
61%
Me3Si
SiMe3
14

Scheme 10-8. The first successful McMurry coupling of an acylsilane [21].

The Na/support combination can also be used advantageously for the reduction of metal salts other than $TiCl_3$. Thus, Knochel et al. have used it as reducing agent for the preparation of carrier-supported Zn from $ZnCl_2$ [22, 23]. In contrast to the preparation of activated titanium described above, TiO_2 turned out to be the support material of choice in this particular application. Zn/TiO_2 smoothly inserts into secondary alkyl- and benzyl bromides. With these latter substrates Wurtz coupling was negligible, although it constitutes a major side reaction if other zinc samples are employed. The organozinc halides obtained can be trapped with different electrophiles such as acyl chlorides, allyl halides, tosyl cyanide, or nitroolefins, after transmetallation with $CuCN \cdot 2\,LiCl$. The examples shown in Table 10-2 illustrate the preparative possibilities [22].

10.3.3 Potassium–Graphite Laminate (C_8K)

It has been known for several decades that a variety of atoms and molecules are able to penetrate into the layered crystal lattice of graphite, thus forming "laminar" or "intercalation" compounds [3, 24–31]. The most prominent among them is the potassium–graphite laminate C_8K, which is readily obtained within a few minutes by adding the appropriate amount of potassium to a previously degassed graphite powder with vigorous stirring under Ar at ≥150 °C [29]. In C_8K, which is a "first stage" intercalation compound, each

Table 10-1. [Ti]/Al_2O_3-promoted cyclization reactions of aromatic diketones and of oxoamides [7].

Substrate	Product	Yield
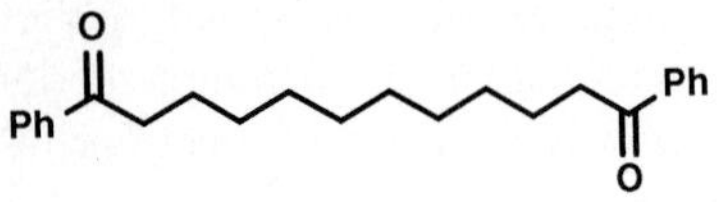 O, Ph, Ph, O	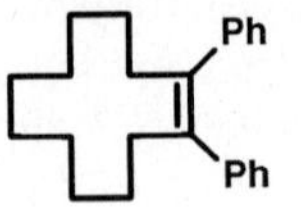 Ph, Ph	72%
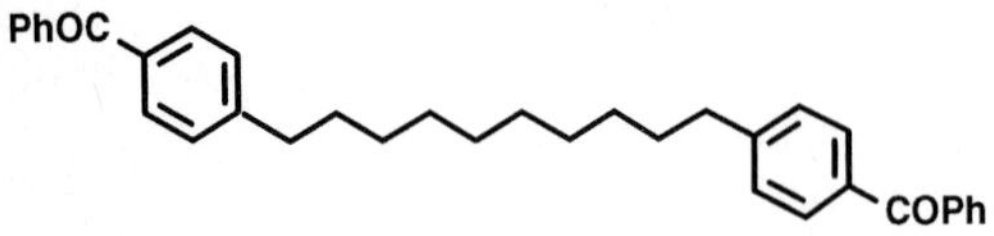 PhOC, COPh	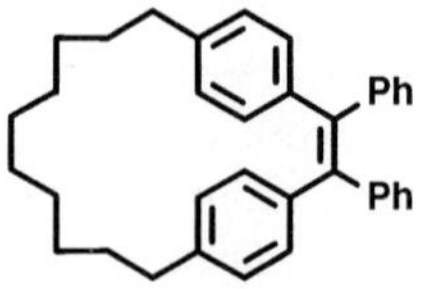 Ph, Ph	86%
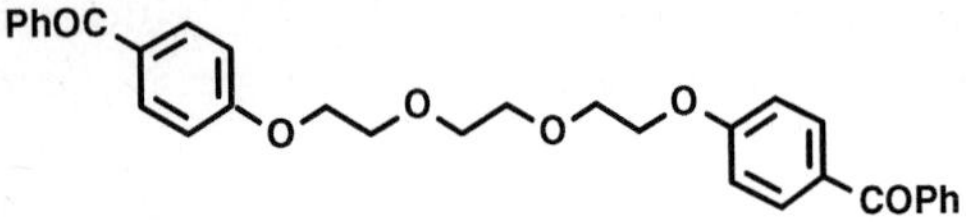 PhOC, O, O, O, O, COPh	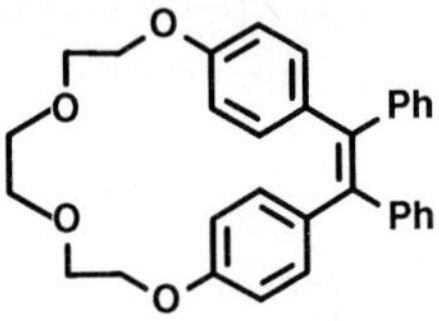 O, O, O, O, Ph, Ph	83%
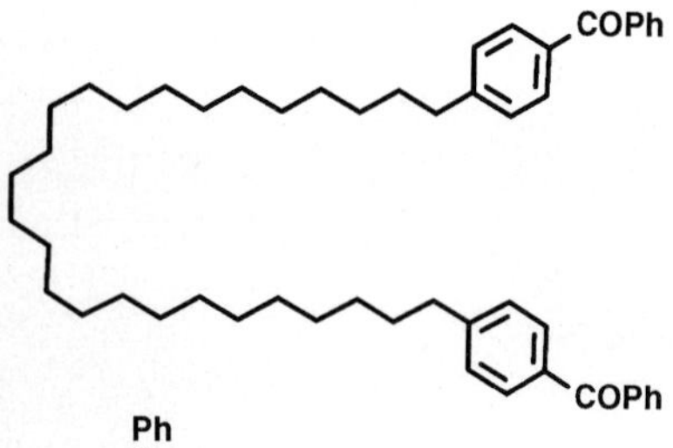 COPh, COPh	 Ph, Ph	83%
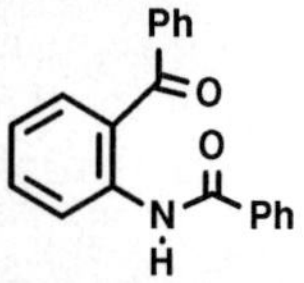 Ph, O, O, N, H, Ph	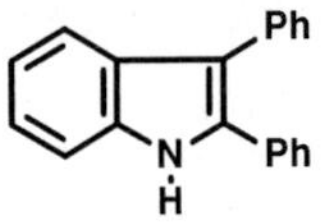 Ph, N, H, Ph	75%
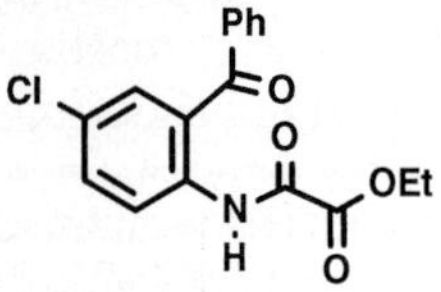 Ph, Cl, O, O, N, H, OEt, O	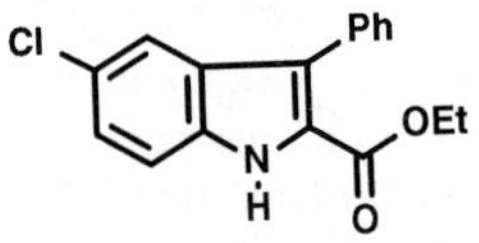 Cl, Ph, N, H, OEt, O	72%

layer of graphite is followed by a layer of potassium, with the metal atoms being highly mobile within the rafts (Fig. 10-2) [30, 32, 35]. "Higher-stage" laminar graphite–potassium compounds (more than one graphite layer separating those of potassium) of the stoichiometries $C_{24}K$, $C_{36}K$, $C_{48}K$, and $C_{60}K$ as well as intercalation compounds of Li, Rb, and Cs have also been described. However, from crystallographic investigations it must be concluded that they have less ordered structures [30].

C_8K is a paramagnetic, nicely bronze-colored, pyrophoric powder, which can be stored under argon for extended periods of time without any significant loss in reactivity.

Table 10-2. Zn/TiO_2-promoted metallation of secondary alkyl or benzyl bromides. The organozinc species formed are trapped by various electrophiles after transmetallation with CuCN · 2 LiCl. Newly formed bonds are highlighted [22].

Substrate	Electrophile	Product	Yield
Bromocyclohexane	Ethyl 2-bromomethyl propenoate	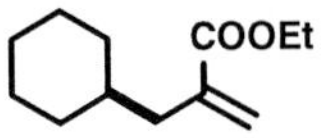	88%
Bromocyclohexane	4-Chlorobutanoyl chloride	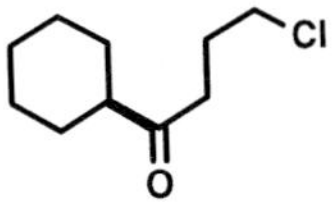	93%
Bromocycloheptane	Ethyl 2-bromomethyl propenoate	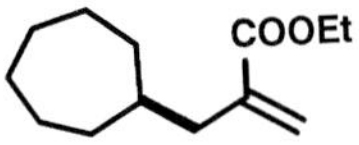	96%
Bromocycloheptane	Benzoyl chloride	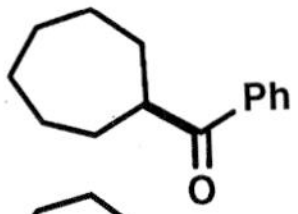	72%
Bromocycloheptane	Nitrostyrene	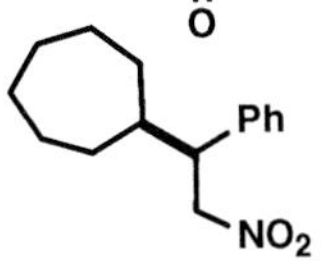	57%
1-Bromo-1-phenylpentane	3-Iodo-2-cyclohexenone	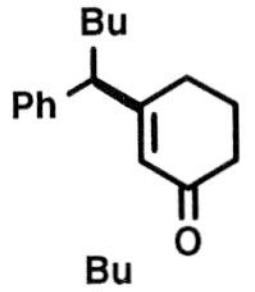	83%
1-Bromo-1-phenylpentane	Nitrostyrene	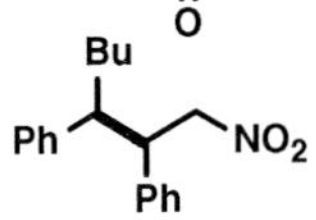	76%
1-Bromo-1-phenylpentane	Diethyl benzylidene malonate	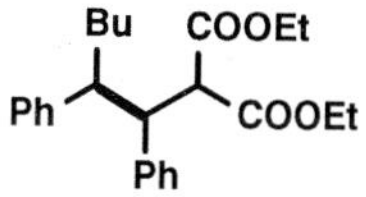	71%
1-Bromo-1-phenylpentane	Tosylcyanide	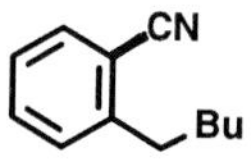	78%[a)]

a) Without transmetallation.

It reacts violently with water [32, 33], alcohols, and ammonia, but may be used in hydrocarbon or ethereal solvents without problems. Both σ- and π-electron donors such as THF, 1,4-dioxane, DME, benzene, toluene, furan, and pyridine easily enter into the interlayer galleries of C_8K with formation of ternary compounds, thereby causing considerable swelling of the graphite flakes [24–27, 30]. Some aromatic guest molecules, however, may undergo irreversible reactions. This ability to transfer electrons to aromatic compounds makes C_8K a Birch-type reducing agent, although in practice large excesses are necessary in order to achieve preparatively useful conversions of, for example, substituted naphthalenes [34].

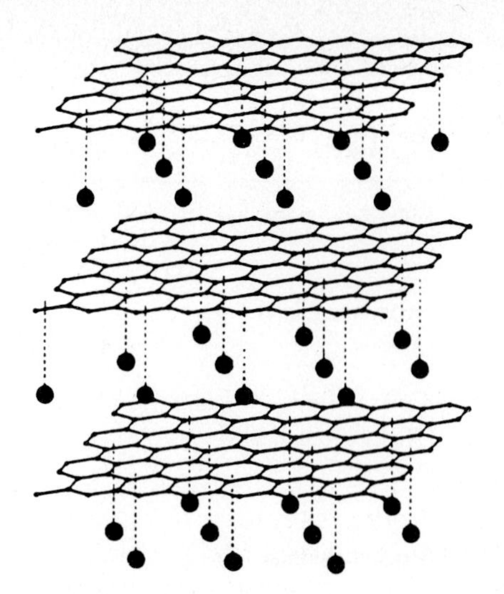

Fig. 10-2. Perspective drawing of the structure of C_8K.

While the physical properties and structural peculiarities of C_8K (Table 10-3) and its higher homologues soon evoked considerable interest [30], their preparative potential has been largely underestimated for decades. This may be due in part to the pyrophoric character of C_8K, which might intimidate potential users. However, more recent investigations are likely to correct the previous image of C_8K as a mere laboratory curiosity.

Despite some dispute on the best representation of the particular bonding situation in C_8K, there is a general consensus that the 4s-electrons of the potassium are strongly coupled with the valence band of the graphite layers. C_8K has a translationally periodic structure. Therefore band analysis rather than a simple chemical description is necessary in order to understand its properties adequately [35, 36]. For preparative purposes, however, it is of minor importance whether the electrons at the Fermi level are in graphite or in potassium orbitals. Chemically speaking, the intercalation of potassium is a redox process that leads to the formation of an infinite aromatic polyanion. Thus, C_8K behaves much like a polymeric array of alkali metal naphthalenide units, combining an extraordinary electron transfer ability with the advantages of an easy work-up by filtration [3, 24–31, 36]. Moreover, any C_8K-promoted transformation is easy to monitor by the characteristic color change from the bronze of the reagent to the black of the liberated graphite.

This close relationship between the chemistry of C_8K and that of alkali metal naphthalenides ($M^+C_{10}H_8^-$) becomes evident from a recent study of the reduction of $Cr(CO)_6$ to $M_2Cr(CO)_5$ (M = Na, K) (Scheme 10-9) [37]. Although both $Na^+C_{10}H_8^-$ and C_8K will induce this conversion, the latter emerged as the reagent of choice, being clearly superior

Table 10-3. Some physical properties of C_8K [30].

Interlayer distance	5.39 Å
Density	0.713–0.730 g cm^{-3}
Space group	D_2^6-$C222$
Electrical conductivity (288 K)	980 Ω^{-1} cm^{-1}
Superconductivity transition temperature	0.55 K

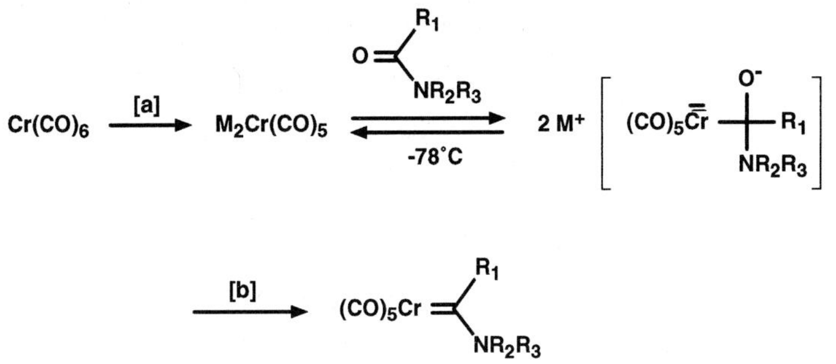

Scheme 10-9. Preparation of chromium carbenes by reduction of $Cr(CO)_6$; (a) 2.2 equiv. C_8K or Na-naphthalenide; (b) Me_3SiCl [37].

from a practical point of view. The work-up of the reaction mixture consists of a simple filtration only, thus avoiding the tedious removal of two equivalents of naphthalene from the desired product. $M_2Cr(CO)_5$ reacts with esters or amides in the presence of TMSCl to afford chromium carbenes, which have found many interesting applications in advanced organic chemistry [37]. Table 10-4 summarizes the results of the comparative study mentioned above. Similar reductions of $M(CO)_6$ (M = Cr, Mo, W) and $Fe(CO)_5$ to dinuclear carbonyl complexes [38, 39] as well as the formation of $Ni[P(OR)_3]_4$ from $NiX_2[P(OR)_3]_2$ precursors [40] by means of C_8K have also been successfully carried out.

Another striking example is the reduction of $[(\eta^5\text{-}C_5H_5)(CO)Ni]_2$ with C_8K to the nickelate $[(\eta^5\text{-}C_5H_5)(CO)Ni]^-K^+$, followed by an in-situ trapping of this labile but highly nucleophilic species [41]. Alkali metals, alkali metal amalgams or naphthalenides, as well as trialkylhydrido borates lead to the formation of the thermodynamically more stable $(\mu^3\text{-}CO)_2\ [(\eta^5\text{-}C_5H_5)Ni]_3$ complex in > 80% yield. However, C_8K is the only reagent rapidly affording the desired nickelate (Scheme 10-10), which can subsequently be used, for example, to prepare π-allylnickel complexes upon addition of allyl bromide [41].

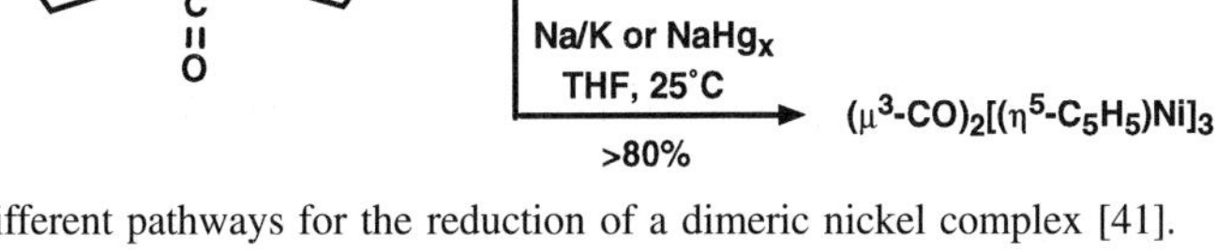

Scheme 10-10. Different pathways for the reduction of a dimeric nickel complex [41].

Other applications of C_8K in inorganic and coordination chemistry include a new route to η^4-nickel anthracene complexes with fluxional behaviour [42], the rapid reduction of 9-ferrocenylfluorenium tetrafluoroborate to 9-ferrocenylfluorene [43], and the likely conversion of $MeBBr_2$ into highly reactive methylborylene, which can be trapped by cycloalkenes or by alkynes (Scheme 10-11). Again, C_8K was reported to be by far superior to lithium, lithium naphthalenide, or Na/K alloy [44].

C_8K efficiently promotes the reduction of chlorosilanes and -stannanes to the corresponding disilanes or distannanes, respectively [45, 46]. Disilanes carrying at least one

Table 10-4. Comparative investigation of the synthesis of chromium carbenes according to Scheme 10-9 [37b].

Product		C_8K	$Na^+C_{10}H_8^-$
$Cr(CO)_5$, H, NMe_2		93%	84%
$Cr(CO)_5$, Ph, NMe_2		87%	93%
$Cr(CO)_5$, NBn_2		50%	44%
Cl, $Cr(CO)_5$, NEt_2		50%	a)
$Cr(CO)_5$, NEt_2, O		63%	a)
$Cr(CO)_5$, $CH_3(CH_2)_{10}$, N		74%	a)
$Cr(CO)_5$, N, Bn		78%	8%
$Cr(CO)_5$, N, O		54%	78%
$(CO)_5Cr$, H, N, R, R, O, Ph	R = H	63%	a)
	R = Me	96%	93%

a) Not reported.

phenyl substituent may be subsequently cleaved in one pot by an excess of C_8K (Scheme 10-12). The silylpotassium reagents formed undergo cross-coupling reactions or afford highly selective nucleophilic silylations of an array of organic electrophiles after transmetallation with an appropriate transition metal salt [45]. Representative examples are shown in Table 10-5. Although metallic lithium is also suited for these purposes, the C_8K-based methodology is distinguished by its considerably increased reaction rate. While the cleavage of $Me_2PhSiSiPhMe_2$, e.g. with lithium powder in THF, usually takes

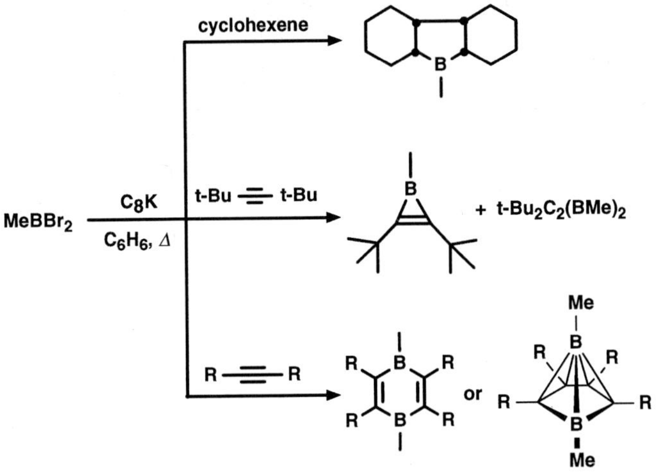

Scheme 10-11. Formation and interception of methylborylene [44].

$$2\ R_3SiX \xrightarrow{>2\ C_8K} R_3Si\text{-}SiR_3 \xrightarrow{2\ C_8K} 2\ R_3SiK$$

Scheme 10-12. C_8K-induced formation and cleavage of disilanes [45].

Table 10-5. Cross-coupling of Ph_2MeSiK with various electrophiles after transmetallation with MX: selected examples [45].

Electrophile	MX	Product	Yield
Benzylbromide	–	$C_6H_5CH_2SiPh_2Me$	90%
2-Cyclohexenone	CuCN	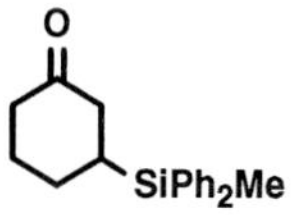 3-($SiPh_2Me$)cyclohexanone	70%
Ethyl cinnamate	CuCN	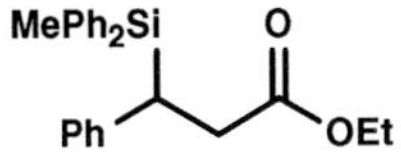 $PhCH(SiPh_2Me)CH_2CO_2Et$	74%
Benzoyl chloride	MnI_2	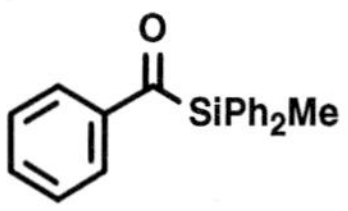 $PhCOSiPh_2Me$	74%
Cyclohexanecarboxylic acid chloride	VCl_3	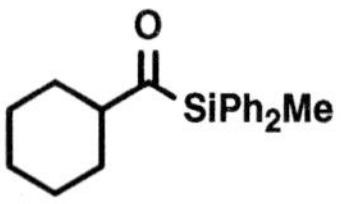 $C_6H_{11}COSiPh_2Me$	80%
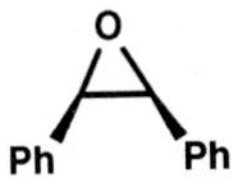 Stilbene oxide (Ph/Ph epoxide)	–	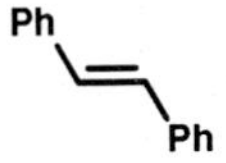 PhCH=CHPh	90%[a]

a) Using $PhMe_2SiK$ as nucleophile.

about 18 h at room temperature [47], the same disilane is reduced by C_8K within minutes at 0 °C. Here again the completion of the reaction can be easily determined by the disappearance of the characteristic bronze color of the reagent.

Disilanes bearing only alkyl substituents on silicon, however, cannot be cleaved to the corresponding silylpotassium reagents by means of C_8K [45]. This resistance is also seen, for example, in the behavior of *gem*-bis(pentamethyldisilyl)cyclopentadiene derivatives, which suffer either deprotonation or C–Si bond cleavage rather than Si–Si bond scission upon exposure to potassium–graphite laminate (Scheme 10-13) [48].

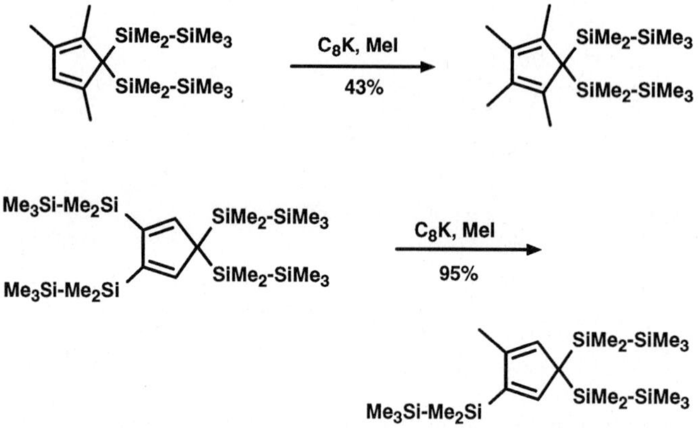

$SiMe_2$-$SiMe_3$
$SiMe_2$-$SiMe_3$
C_8K, MeI
43%
$SiMe_2$-$SiMe_3$
$SiMe_2$-$SiMe_3$

Me_3Si-Me_2Si
Me_3Si-Me_2Si
$SiMe_2$-$SiMe_3$
$SiMe_2$-$SiMe_3$
C_8K, MeI
95%
$SiMe_2$-$SiMe_3$
$SiMe_2$-$SiMe_3$
Me_3Si-Me_2Si

Scheme 10-13. C_8K-induced transformations of *gem*-bis(pentamethyldisilyl)cyclopentadiene derivatives [48].

Reactions of C_8K with purely organic substrates have long been considered to be of limited value [49]. In early screenings with a set of simple compounds no striking advantages have been found. Its reactions with different Brönsted acids seem to depend on the size of the substrate [49]. C_8K was also used as a polymeric Lewis base for selective mono-alkylations of alkyl nitriles and phenylacetates, although dialkylation of these substrates could not be completely suppressed [50]. Imines and dihydro-1,3-oxazines have also been deprotonated by C_8K (Scheme 10-14) [51].

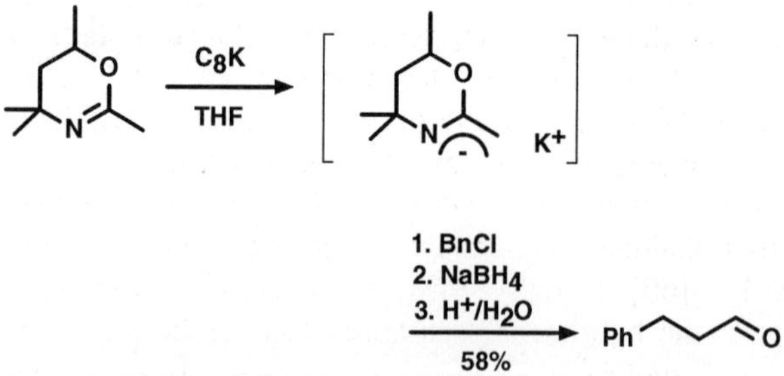

C_8K
THF
K^+
1. BnCl
2. $NaBH_4$
3. H^+/H_2O
58%
Ph

Scheme 10-14. C_8K as polymeric base: deprotonation of a dihydro-1,3-oxazine [51].

The strong electron donor capacity of C_8K is exemplified by the reactions with different types of carbonyl compounds. Ketones usually afford the respective alcohol and/or pinacol [52, 53]. However, the double bonds of enones and α,β-unsaturated carboxylic acids are selectively reduced by C_8K in THF/hexamethyldisilazane as solvent system in a kind of heterogeneous Birch reaction [54]. Alkyl enoates undergo dimeriza-

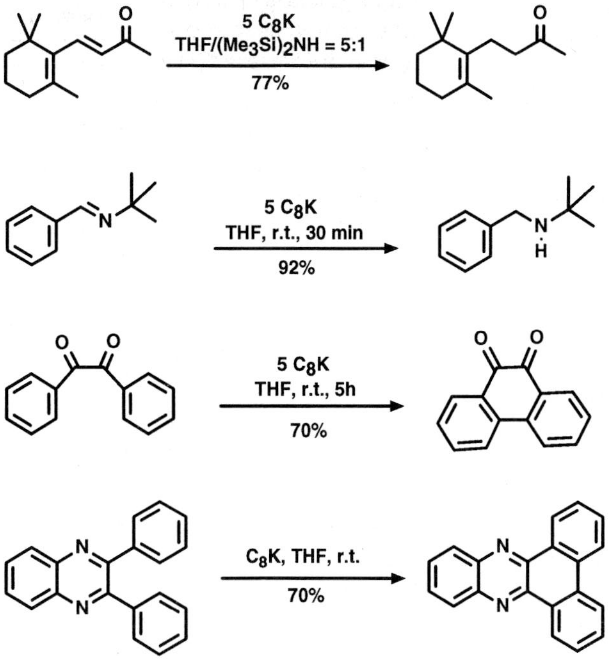

Scheme 10-15. C_8K in heterogeneous Birch-type reductions and arene coupling reactions [54–57].

tion under the same conditions. Schiff bases afford the respective secondary amines in good yield. Benzil and derivatives thereof undergo coupling of the phenyl rings, thus providing a convenient route into the phenanthren-9,10-quinone series (Scheme 10-15) [55–57].

Whereas aryl halides and alkyl chlorides afford the corresponding hydrocarbons on exposure to C_8K, alkyl iodides undergo Wurtz coupling and alkyl bromides generally follow both pathways giving product mixtures [49]. By using structural probes in the molecules it has been concluded that these reductions proceed via single electron transfer processes. *vic*-Dibromides afford alkenes in good yields [49, 58]. C_8K cleaves the C–O bond of esters, carbonates, xanthates, and aryl ethers, but the S–O rather than the C–O bond of sulfonates (Scheme 10-16) [49, 64]. The propensity of C_8K to cause bond cleavage of both aryl ethers and halides can be used cooperatively for efficiently destroying highly toxic polyhalo-1,4-dibenzodioxins and related poisons [59].

Of particular interest are the applications of C_8K to deoxyhalogeno sugar derivatives. In some cases dehydrohalogenation reactions due to its basic character have been observed (Scheme 10-17) [60]. Usually, however, reductive elimination reactions triggered by electron transfer to the halide function take place, followed by scission of the vicinal C–O bond in the sugar molecule (dealkoxyhalogenation process). This made the smooth conversion of glycosyl halides to the corresponding glycal derivatives possible. The effectiveness of this approach turned out to be independent of the ring size, thus providing ready access to the precious furanoid series [4, 61]. Taking into consideration the lability of such carbohydrate-derived five-membered enol ethers, as well as the option of preparing either 3-*O*-unprotected or 3-*O*-substituted derivatives in one pot (cf. Scheme 10-1), it is the high performance of C_8K together with the straightforward non-aqueous work-up that are crucial. This particular application illustrated for the first time that a polyfunc-

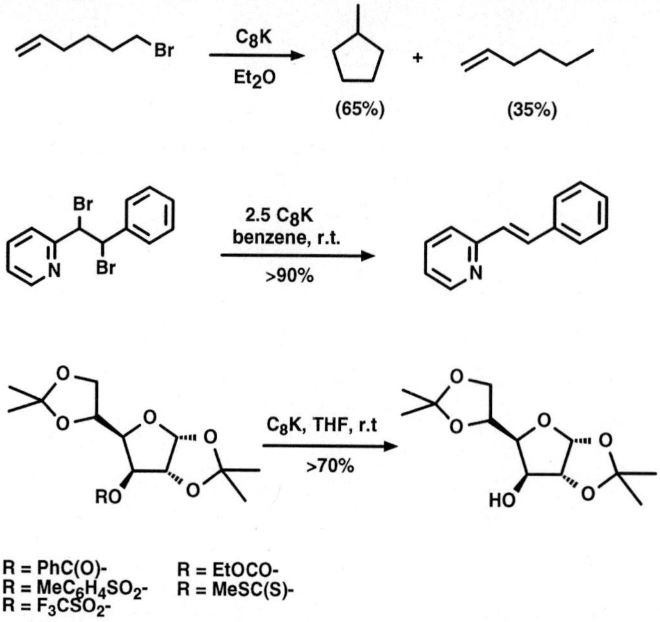

Scheme 10-16. Representative examples of C_8K-induced reactions of halides and aprotic ester cleavages [49, 58, 64].

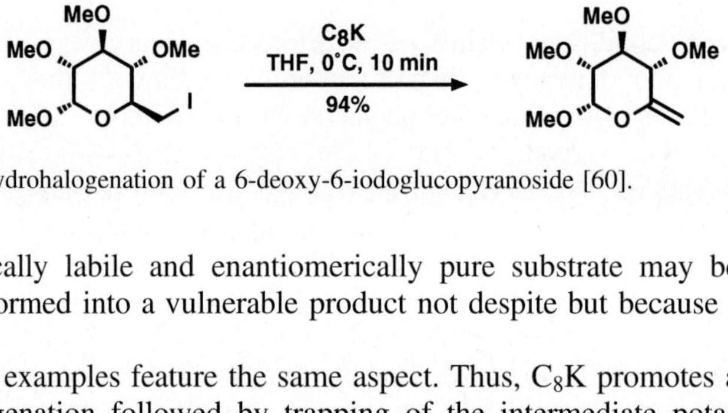

Scheme 10-17. Dehydrohalogenation of a 6-deoxy-6-iodoglucopyranoside [60].

tional, hydrolytically labile and enantiomerically pure substrate may be efficiently and selectively transformed into a vulnerable product not despite but because of the high reactivity of C_8K [4].

Several other examples feature the same aspect. Thus, C_8K promotes a similar cascade of dealkoxyhalogenation followed by trapping of the intermediate potassium alcoholate in the one-pot conversion of 1-deoxy-1-iodofructopyranose or -sorbofuranose derivatives to rather sensitive exocyclic enol ethers (e.g. **15** → **16**, Scheme 10-18) [62]. 1-Deoxy-1-iodopentitol derivatives were likewise cleaved giving access to the full set of stereoisomeric

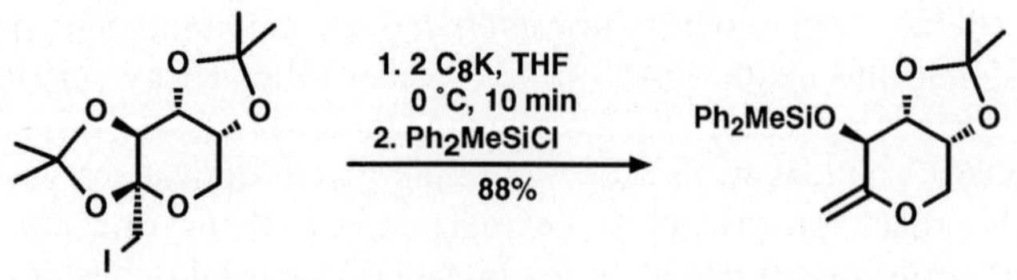

Scheme 10-18. Formation of an exocyclic enol-ether by dealkoxyhalogenation of 1-deoxy-1-iodo-2,3;4,5-di-*O*-isopropylidene-D-fructopyranose **15**. The intermediate potassium alcoholate was trapped in situ with $Ph_2MeSiCl$ [62].

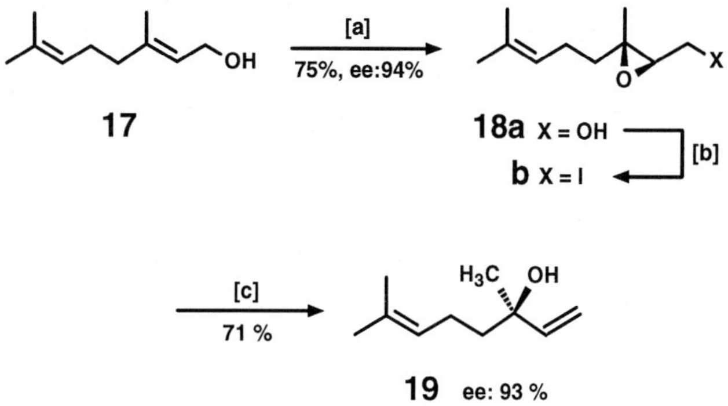

Scheme 10-19. Synthesis of (*R*)-linalool from geraniol; (a) *t*-BuOOH, (D)-(−)-diethyl tartrate, $Ti(O\text{-}iPr)_4$, MS 4 Å, CH_2Cl_2; (b) PPh_3, I_2, imidazole, toluene, 85%; (c) C_8K (2.5 equiv.), THF [64].

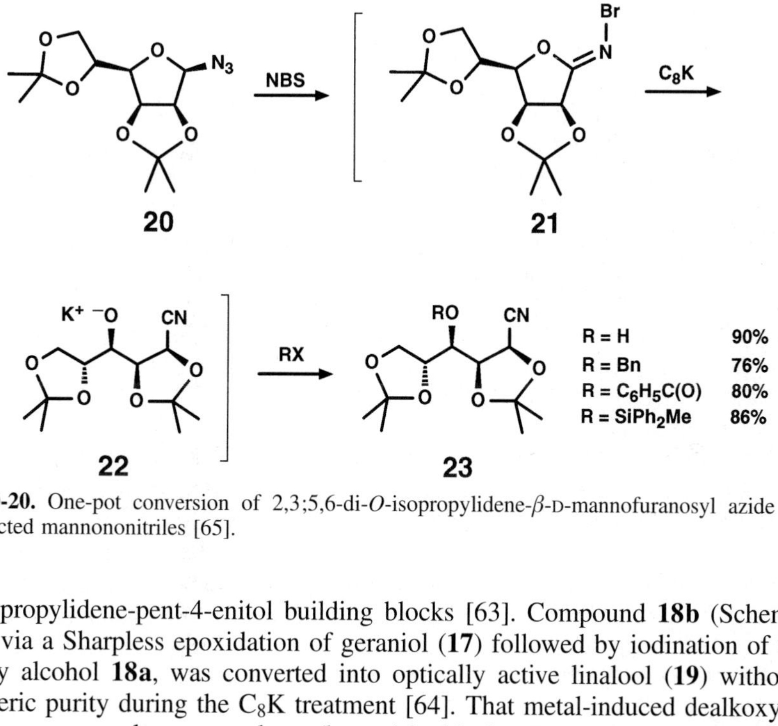

Scheme 10-20. One-pot conversion of 2,3;5,6-di-*O*-isopropylidene-*β*-D-mannofuranosyl azide into differently protected mannononitriles [65].

1,2-*O*-isopropylidene-pent-4-enitol building blocks [63]. Compound **18b** (Scheme 10-19), prepared via a Sharpless epoxidation of geraniol (**17**) followed by iodination of the resulting epoxy alcohol **18a**, was converted into optically active linalool (**19**) without loss of enantiomeric purity during the C_8K treatment [64]. That metal-induced dealkoxyhalogenation processes may also occur along (hetero)double bond systems was illustrated by the formation of aldononitriles (**23**) from *N*-bromoglycosyl imines (**21**) (Scheme 10-20) [65]. This is the first successful chemical transformation of this thermally and hydrolytically labile class of monosaccaride derivatives. When exposed to C_8K in THF they readily become *N*-metallated, causing the formation of the nitrile group by the extrusion of the ring oxygen atom. The resulting acyclic alcoholates **22** may be protonated, alkylated, silylated, or acylated in situ upon addition of the appropriate electrophile [65].

C_8K-induced chemo- and regioselective transformations of complex substrates may also be triggered at C–S bonds. For example, phenylthio glycosides such as **24** (Scheme 10-21) may be used as substitutes for glycosyl chlorides in glycal formation

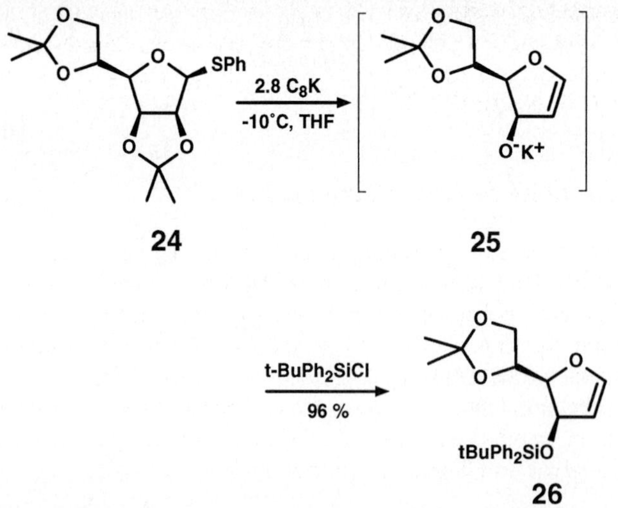

2.8 C_8K
-10°C, THF
24
25
t-BuPh$_2$SiCl
96 %
26

Scheme 10-21. Representative example of the formation of glycals from thioglycosides [66].

[66]. They are metallated by C_8K at the anomeric center and undergo immediate dealkoxysulfuration with scission of the 2,3-*O*-alkylidene group. The 3-*O*-potassium alcoholate **25** thus formed may once again be trapped in situ. Because phenylthio glycosides are much more stable than glycosyl halides and can be stored for extended periods of time, this approach to furanoid glycals (**26**) is particularly convenient [66].

Selective cleavage of C–S bonds has also been observed with allyl- and vinyl sulfones as substrates, which are reduced by C_8K to the corresponding alkenes with moderate (*E*)-selectivity [67, 68]. In the case of allyl sulfones this desulfuration reaction is accompanied by a rearrangement of the double bond.

All the above-mentioned reactions are stoichiometric in C_8K. In addition, there exist many reports in the literature concerning the use of potassium–graphite laminates as catalysts [24–31, 36, 69]. C_8K reacts with H_2 to give a dark-blue ternary compound with a stoichiometry close to $C_8KH_{2/3}$ [70]. This compound is a "solid solution" of hydrogen in the graphite-potassium double layers, albeit the precise positions of the hydrogen atoms are not yet known. C_8K and its higher homologues show catalytic properties for the hydrogenation of simple alkenes, alkadienes, acetylenes, and under more drastic conditions, benzene derivatives. In the absence of hydrogen they are able to induce geometrical and positional isomerizations of alkenes. The catalytic activity observed is, however, lower than that of K/Al_2O_3 or K/SiO_2 [24, 36, 69, 71].

Graphite intercalation compounds have also been used as initiators for anionic polymerizations of a variety of monomers. As observed in many catalytic processes, $C_{24}K$ exhibits higher activities than C_8K for these purposes. Since the polymerization reactions seem to proceed between the graphite galleries, the observed activity may in fact reflect the different diffusion rate of a given monomer into the interlayer space of the laminar reagents [24, 36, 72]. In this respect the more closely packed structure of C_8K compared with that of $C_{24}K$ represents a kinetic obstacle for the propagation of the poly-reaction. Due to the interlayer character of potassium–graphite-induced polymerizations, the reaction rates are in general lower than for polymerizations in homogeneous solutions. This was reconfirmed by a recent comparative study of the polymerization of ε-caprolactam [72a].

Graphite–alkali metal intercalation compounds also exhibit catalytic activities for the Fischer–Tropsch synthesis as well as for the formation of ammonia from the elements. In both cases, however, their efficiency is markedly improved by doping them with different transition metals prior to use [24, 36, 69].

10.3.4 Metal–Graphite Surface Compounds [3]

C_8K may also be used as a reducing agent for metal salts. Reductions as shown in Scheme 10-22 readily lead to so-called "metal–graphite combinations", which consist of finely dispersed metal particles physisorbed on the surface of the graphite flakes [3, 24, 25]. Detailed electron-microscopic investigations gave *no indication* of any intercalation of these metals between the graphite layers [6, 73, 74], thus refuting previous assumptions of a potassium–metal exchange with formation of laminar metal–graphite reagents [113]. An in depth treatise on the morphology of this valuable class of highly activated metals is given in Chapter 11 of this book.

$$MX_n + n\, C_8K \xrightarrow{THF,\ \Delta} M^* \text{ on graphite} + n\, KX$$

(MX_n = i.a. $ZnCl_2(AgOAc)$, $TiCl_3$, $TiCl_4$, $SnCl_2$, $PdCl_2$, $PtCl_2$, $NiBr_2$, $MgCl_2$, $FeCl_3$)

Scheme 10-22. Preparation of metal–graphite surface compounds [3].

This approach to active metals combines some favorable properties. Its principle is inspired by Rieke's pioneering work [19]; however, instead of occurring on a melting potassium lump, the reduction of the salt can now take place simultaneously at any site on the extended surface area of the graphite, whose π-system mediates the electron transfer process [3]. This leads to the formation of highly dispersed and hence very reactive metal slurries. As they immediately get adsorbed on the inert support, the detrimental effects of aggregation, sintering, and recrystallization of the nanoparticles are suppressed or at least retarded. Thus, graphite reagents combine an exceptional performance with the convenience of supported reagents in terms of stability, handling, and work-up. Needless to say, this approach is rather general in scope, since C_8K is capable of reducing almost any metal salt that is partly soluble in ethereal solvents [75]. This is why metal–graphite reagents have attracted considerable attention in recent years and found many applications in organic and organometallic chemistry [3, 24, 25]. The most important examples are reviewed below.

10.3.4.1 Zinc- and Zinc/Silver–Graphite

Zn–graphite is obtained within a few minutes by adding $ZnCl_2$ to a suspension of C_8K in THF or other ethereal solvents [76]. Its morphology has been studied in detail [73]. Doping with about 10 mol% of silver by co-reduction of $ZnCl_2$ and AgOAc yields an even more active variant of this reagent [77].

Zn– and Zn(Ag)–graphite readily insert into the activated C–X (X = Cl, Br) bond of α-haloesters and α-bromolactones with formation of the respective zinc enolates [76, 77]. Such insertions may occur at unprecedentedly low temperatures and without any Wurtz coupling of the substrates. This led to considerable improvements of the Reformatsky reaction [78] by avoiding any problems with entrainment, by extending its scope to less reactive donors such as α-chloroalkanoates, and by greatly increasing the diastereoselectivity. Whereas Reformatsky reactions are conventionally carried out at reflux temperatures in ethereal or hydrocarbon solvents, Zn(Ag)–graphite promotes them in excellent yields and with high reaction rates even at –78 °C [77]. In addition to reactions with aldehydes and ketones, the zinc ester enolates formed have also been trapped with other electrophiles such as acyl chlorides [25], Eschenmoser salt [76], allyl palladium complexes [79], or aldonolactones [80]. α,α-Dihalo esters, which tend to polymerize in a conventional set-up, react with Zn(Ag)–graphite and carbonyl compounds under exception-

Table 10-6. Examples of Zn– or Zn(Ag)–graphite-promoted Reformatsky-type reactions [76, 77, 81–83].

Substrates	Method [a)]	Temp.	Product	Yield
Cyclohexanone Ethyl bromoacetate	A B	0 °C –78 °C	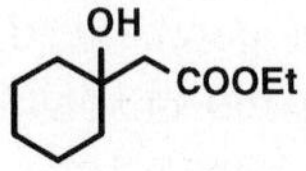	88% 92%
Cyclopentanone Ethyl 2-bromobutanoate	B	–10 °C	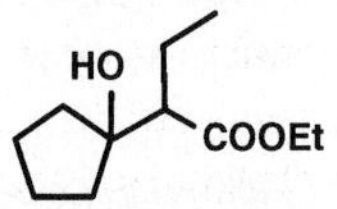	76%
9-Fluorenone 2-Bromobutyrolactone	A	0 °C	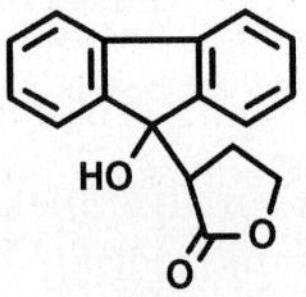	88%
Benzaldehyde Methyl 4-bromo-2-butenoate	A	0 °C	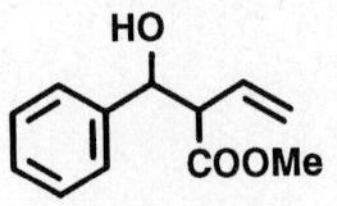	85%
Benzophenone Ethyl dibromoacetate	B	–78 °C	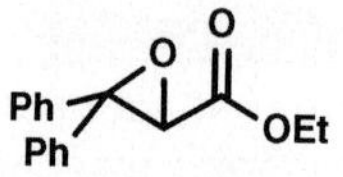	84% [b)]
Cyclohexanone Ethyl bromo(trimethylsilyl)acetate	B	–78 °C	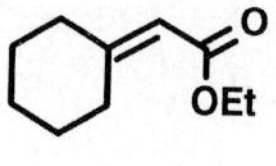	88%
Methyl 4,6-O-benzylidene- α-D-erythro-hex-2-ulopyranoside Ethyl 2-bromomethyl-2-propenoate	B	–78 °C	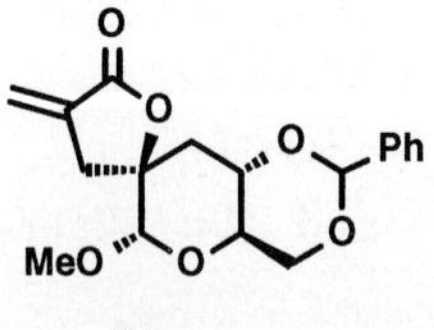	92% [c)]
2,3-O-Isopropylidene-D-erythrono- lactone Ethyl 2-bromomethyl-2-propenoate	B	–78 °C	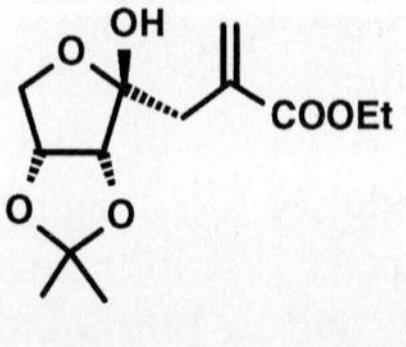	92% [c)]

a) Method A: Zn-graphite, THF; Method B: Zn/Ag-graphite, THF. b) After basic work-up. c) One diastereoisomer only.

ally mild conditions to give α-halo-β-hydroxyalkanoates [81]. A basic work-up will transform the latter into glycidates. Furthermore, with ethyl bromo(trimethylsilyl)acetate as donor, a Reformatsky reaction/Peterson elimination tandem process leads regioselectively to 2-alkenoates [81]. Some representative examples of Reformatsky-type reactions using Zn– or Zn(Ag)–graphite are given in Table 10-6.

The very mild conditions of Zn(Ag)–graphite-induced transformations guarantee kinetic control and hence high selectivity in reactions of 2-bromomethyl-2-alkenoates with diastereotopic carbonyl compounds. This transformation which involves an allylic rearrangement of the double bond has found applications in, for example, the diastereoselective formation of sugar-derived α-methylene-γ-butyrolactones (Table 10-6) [82, 83] and in a total synthesis of *N*-acetylneuraminic acid [84].

Zn–graphite has also been used for the preparation of allylzinc reagents in THF at low temperature. These react with carbonyl compounds to give homoallyl alcohols [76], with nitrones to give homoallyl hydroxylamines [85], and with nitriles to give β,γ-unsaturated ketones after hydrolysis of the intermediate metalloaldimines (Table 10-7) [86]. Allyl chlorides and bromides seem to be metallated with comparable rates.

Aryl halides are known to be difficult to zincate [23]. Conventional zinc dust reacts only at high temperatures and/or in aprotic dipolar solvents such as DMF. Zn(Ag)–graphite, in contrast, turned out to be capable of metallating aryl- and heteroaryl iodides and even an activated heteroaryl bromide in THF at or slightly above room temperature (Scheme 10-23, Table 10-8) [87]. The arylzinc reagents formed may bear electrophilic sites such as keto-, cyano-, ester-, alkoxy-, or trifluoromethyl groups. These functionalized nucleophiles can be smoothly acylated or allylated after transmetallation with $CuCN \cdot 2\,LiCl$ or cross-coupled with iodoalkenes or bromoarenes in the presence of catalytic amounts of Pd^0. This allows particularly flexible reaction planning, since the functional groups can be introduced into the target via the electrophile and/or the nucleophile [87].

Zn(Ag)–graphite in THF transforms glycosyl halides to the corresponding glycals, presumably via metallation at the anomeric center followed by reductive elimination

Table 10-7. Examples of Zn–graphite induced reactions of allyl halides with various electrophiles [76, 85, 86].

Substrates		Product	Yield
2-Octanone	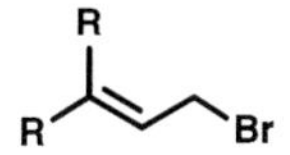 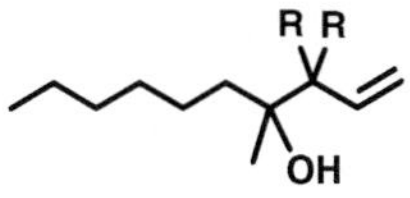 R, R, Br	R R, OH; R = H R = Me	94% 94%
p-RC_6H_4CN	Br	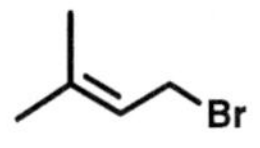 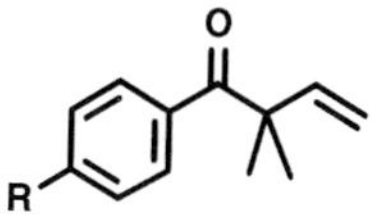 O, R; R = H R = Me R = Cl R = COOEt	83% 81% 80% 81%
1-Cyanoheptane 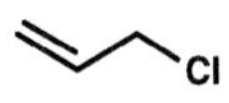 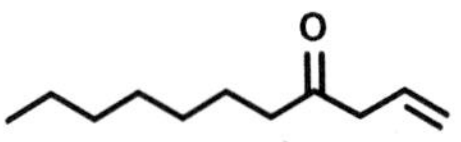	Cl	O	61%
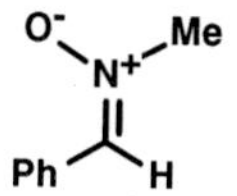 O^-, N^+, Me, Ph, H	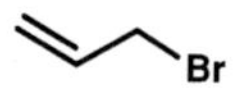 Br	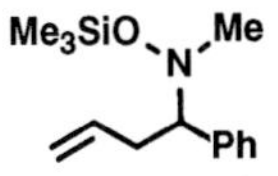 Me_3SiO, N, Me, Ph	92%[a)]

a) After silylation of the crude hydroxylamine.

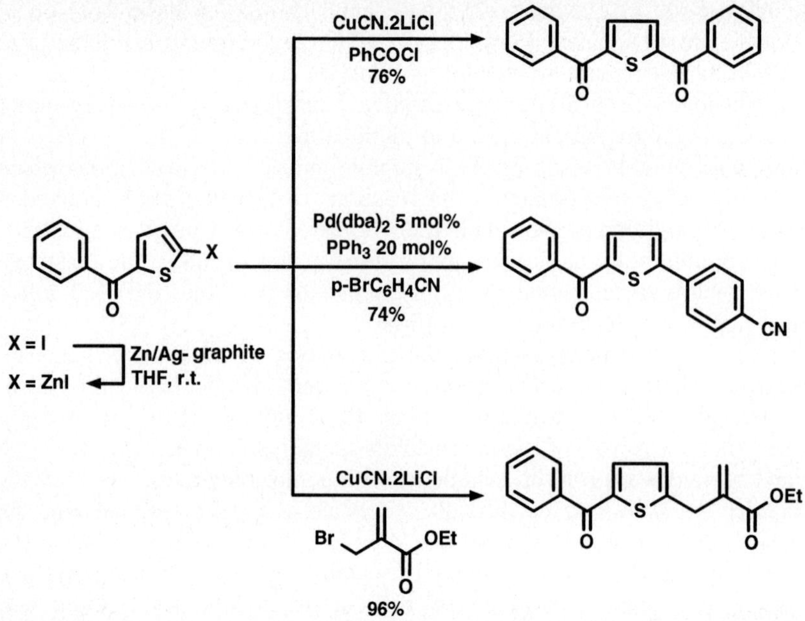

Scheme 10-23. Zincation of 5-benzoyl-2-iodo-thiophene and subsequent cross-coupling reactions [87].

(Scheme 10-24) [88]. Because of the aprotic conditions and the low reaction temperatures this synthesis is not only compatible with different protecting groups in the substrates but, more importantly, is also applicable to the furanoid series despite the pronounced tendency of furanoid glycals to irreversible allylic rearrangements which hampered previous approaches. Less activated zinc reagents are unsuitable for this purpose. The high performance of this method is also evidenced by the significantly improved yield reported in a synthesis of the labile compound L-fucal [89], and by the formation of a carbohydrate-derived bridgehead enol ether [90].

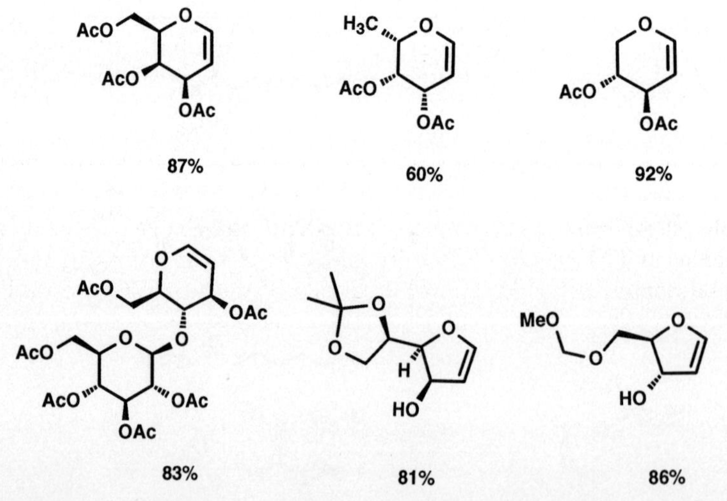

Scheme 10-24. Representative examples of glycals obtained by Zn(Ag)–graphite-promoted reduction of glycosyl halides [88, 89].

Table 10-8. Examples of Zn(Ag)–graphite-induced metallation of aromatic or heteroaromatic halides. The organozinc species formed are cross-coupled with the given electrophiles in the presence of Cu^{I} salts or catalytic amounts of Pd^{0}. Newly formed bonds are highlighted [87].

Substrate	Electrophile	Product	Yield
2-Iodobenzophenone	PhCOCl	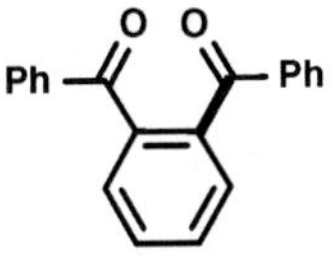	80%
2-Iodobenzothiazole	Ethyl 2-bromomethyl propenoate	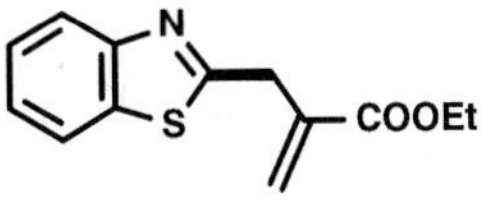	81%
Iodo-2-trifluoromethylbenzene	PhCOCl	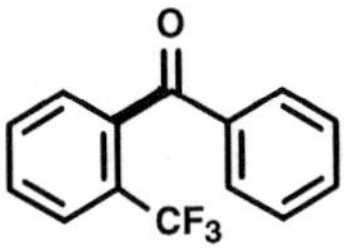	80%
Ethyl 4-iodobenzoate	Acetyl chloride	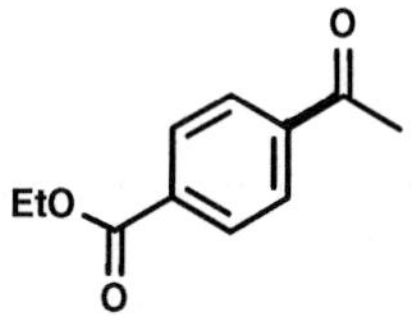	68%
2-Bromo-4,5-dicyano-1-methylimidazole	1-Iodo-1-octene	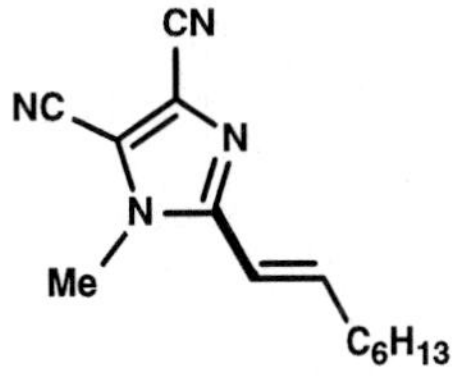	41%
2-Iodo-1-methylpyrrole	Ethyl 2-bromomethyl propenoate	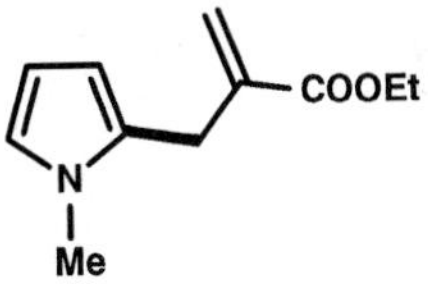	85%

Deoxyhalo sugar derivatives undergo reductive ring-opening upon treatment with Zn(Ag)–graphite in THF or DME, with formation of enantiomerically pure enal building blocks of wide applicability (Scheme 10-25, Table 10-9) [60, 62, 63, 91–94]. This is essentially the only reagent which can achieve this conversion under strictly anhydrous con-

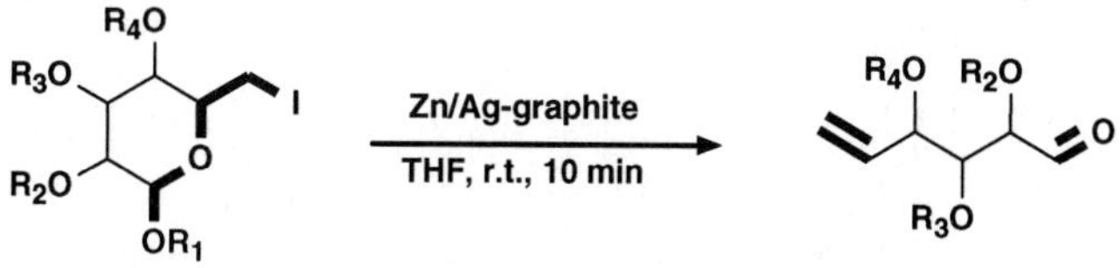

Zn/Ag-graphite

THF, r.t., 10 min

Scheme 10-25. Zn(Ag)–graphite-induced reductive ring-opening of deoxyhalo sugars.

Table 10-9. Examples of Zn(Ag)–graphite-induced dealkoxyhalogenation reactions of various types of deoxyhalo sugars [60, 63, 65, 91, 93].

Substrate	Product		Yield
		R = Me	93%
		R = Ac	87%
		R = Bz	81%
			76%
			90%[a)]
			64%
			96%[b)]
			92%
			71%[c)]

a) After $NaBH_4$ reduction of the crude aldehyde. b) After acidic work-up. c) After acetylation.

ditions. This transformation offers excellent compatibility with different functional and protecting groups in the substrates. Iodides react more smoothly than bromides, and pyranosides more readily than furanosides [91]. The ring cleavage is by and large independent of the position of the halogen atom, of the conformation and configuration of the substrate, of the leaving group at the anomeric center, and proceeds at comparable rates with either primary or secondary halides. The dealkoxyhalogenation reaction can be applied to aldose [60, 91, 92, 94], alditol [63], and ketose derivatives [62], and may also

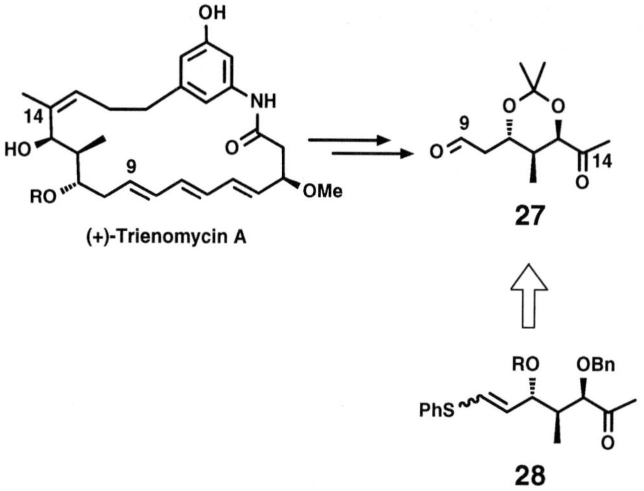

Scheme 10-26. Degradation of (+)-trienomycin A. All members of the family of trienomycin, mycotrienin, and ansatrienin antibiotics share the same C9–C14 segment except for the substituent R. The synthetic equivalent **28** allows the proper differentiation of the two carbonyl groups and the two secondary alcohols of the target [93].

occur along hetero-double bond systems, as exemplified by the formation of aldononitriles from *N*-bromo glycosylimines [65]. Sugar-derived α-chloro-phenylthio ethers are another class of valuable substrates, which lead to building blocks with a carbonyl group on one end and a vinylsulfide group on the other end upon treatment with Zn/Ag–graphite [93].

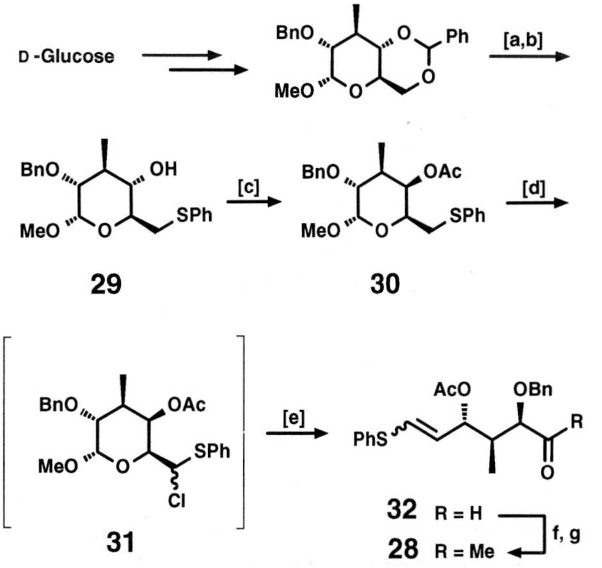

Scheme 10-27. Synthesis of the chiral core region of the trienomycin antibiotics; (a) *p*-TsOH · H_2O, MeOH, 92%; (b) PhSSPh, Bu_3P, pyridine, 83%; (c) 1. triflic anhydride, pyridine/CH_2Cl_2; 2. Bu_4NOAc, MeCN, 78%; (d) NCS, CCl_4; (e) Zn(Ag)–graphite, DME, 66% (both steps); (f) $MeTi(O\text{-}iPr)_3$, THF; (g) PCC, CH_2Cl_2, 72% (both steps) [93].

This particularly dense functionalization has recently been used for the synthesis of compound **28**, which is a synthetic equivalent of the C9–C14 fragment **27** common to all macrolides of the trienomycin family (Scheme 10-26) [93]. After the three contiguous chiral centers of this target molecule had been set up with the right absolute and relative stereochemistry on conformationally rigid pyranoid precursors, the substrate **30** (Scheme 10-27) was treated with NCS and the resulting crude α-chloro-phenylthio ether **31** was reductively ring-opened to give the 6-phenylthio-hex-5-enal **32**. In view of the thermal and hydrolytic lability of **31**, the mild and strictly anhydrous conditions in the Zn/Ag–graphite-induced dealkoxyhalogenation are crucial for success. Transformation of the aldehyde group into a methyl ketone provided **28**, which is suitable for further elaboration to the trienomycin antibiotics because of the proper differentiation of the two carbonyl groups and the two secondary alcohols [93].

Conformationally biased installation of stereogenic centers on a pyranose template followed by a reductive ring scission were also used for a convergent synthesis of the C1–C13 segment of amphoteronolide B [93]. The approach to this polyhydroxylated target molecule was based upon its hidden C_2-symmetry. We have prepared both enantiomers of a properly functionalized hex-5-enal building block from D-glucose, with a Zn(Ag)–graphite-induced ring-opening of a 6-deoxy-6-iodopyranoside and a subsequent transposition sequence [94] as the key steps (Scheme 10-28).

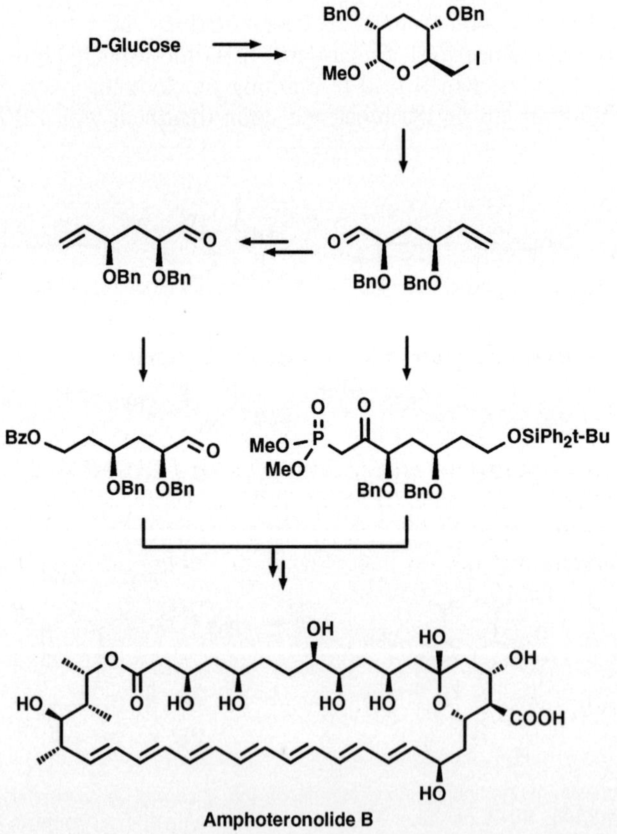

Scheme 10-28. Convergent approach to the C1–C13 fragment of amphoteronolide B [93].

Azides exhibit a unique influence on the course of zinc-induced reactions. In their presence all deoxyhalo sugars undergo only simple reduction of the C–X bond instead of the dealkoxyhalogenations otherwise induced by Zn(Ag)–graphite (Schemes 10-29 and 10-30) [92]. This unprecedented azide switch from one reaction pathway to another is effective in an intra- as well as in an intermolecular way, and afforded some insight into the mechanism of zinc-induced reductions. The azide group itself is recovered unchanged except under forcing conditions, although it obviously becomes involved in the electron transfer process [92].

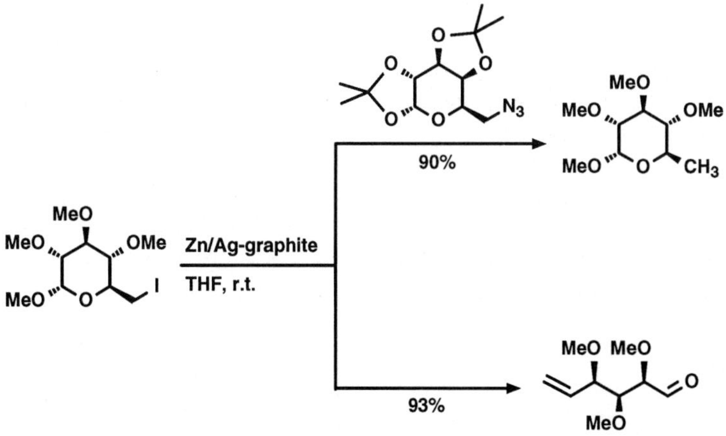

Zn/Ag-graphite
THF, r.t.
90%
93%

Scheme 10-29. Representative example of an intermolecular "azide switch" between two reaction pathways [92].

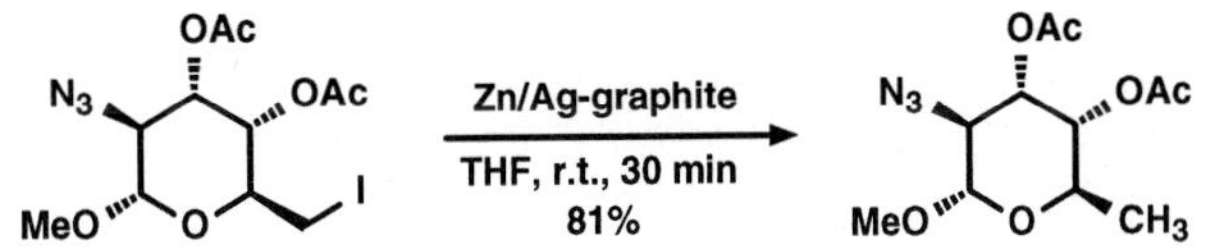

Zn/Ag-graphite
THF, r.t., 30 min
81%

Scheme 10-30. Zn(Ag)–graphite induced reduction of a C–I bond enforced by the presence of an azido group [92].

The favourable profile of the Zn(Ag)–graphite promoted dealkoxyhalogenation reactions also becomes evident from applications in the non-carbohydrate series. Thus, a new approach to enantiomerically pure aldols by scission of appropriate spiroketals has been developed (Scheme 10-31) [95]. This methodology was then applied to the total synthesis of the strongly immunosuppressing macrocyclic antibiotic 9-dihydro-FK-506 [96]. Treat-

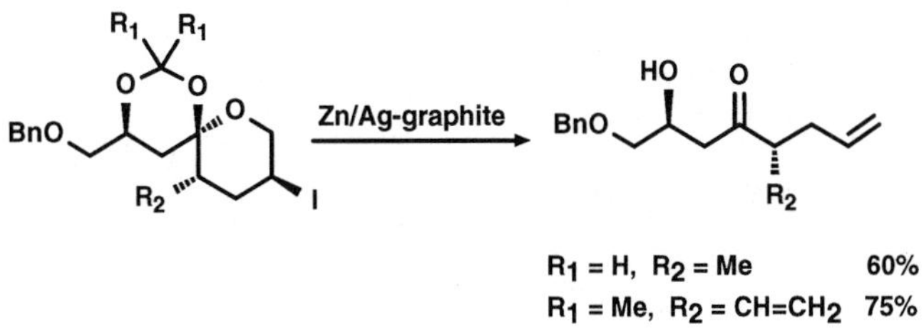

Zn/Ag-graphite

R_1 = H, R_2 = Me 60%
R_1 = Me, R_2 = CH=CH$_2$ 75%

Scheme 10-31. Approach to aldols by reductive cleavage of spiroketals [95].

ment of **33** with Zn(Ag)–graphite in THF (Scheme 10-32) gave the α-allyl aldol entity of the macrolide in 84% yield. This example nicely illustrates the high performance and the compatibility of the reagent, and shows conclusively that structural complexity of the target certainly does not exclude the use of a highly active metal reagent.

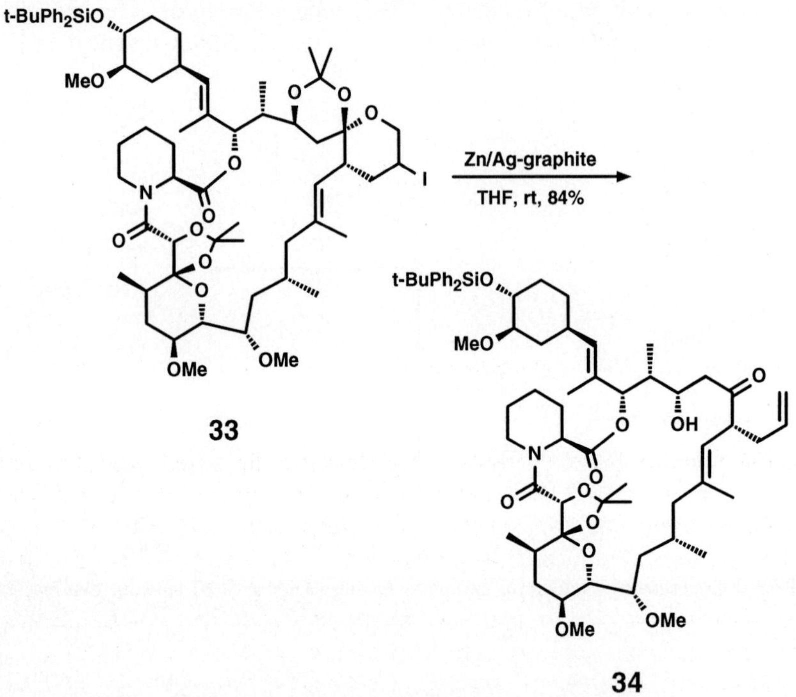

t-BuPh2SiO, MeO, I, OMe, OMe, **33**, Zn/Ag-graphite, THF, rt, 84%, OH, **34**

Scheme 10-32. Penultimate step of a total synthesis of 9-dihydro-FK 506 [96].

10.3.4.2 Titanium–Graphite

Low-valent titanium reagents effect the reductive coupling of carbonyl compounds to alkenes (McMurry reaction, Scheme 10-33) [20]. "Titanium–graphite" combinations serve this purpose very well, although distinct differences in reactivity were encountered depending on the mode of preparation. As we still lack a deeper insight into the actual nature of these species, this chapter can only provide a phenomenological rather than a morphologically based description of the results obtained with different titanium–graphite samples.

If $TiCl_4$ is reduced with 4 equivalents of C_8K in THF a chemoselective reagent is formed, which converts aromatic carbonyl compounds to alkenes, but affords pinacols as the major products with aliphatic substrates [97]. In contrast, $TiCl_3$-derived Ti–graphite samples

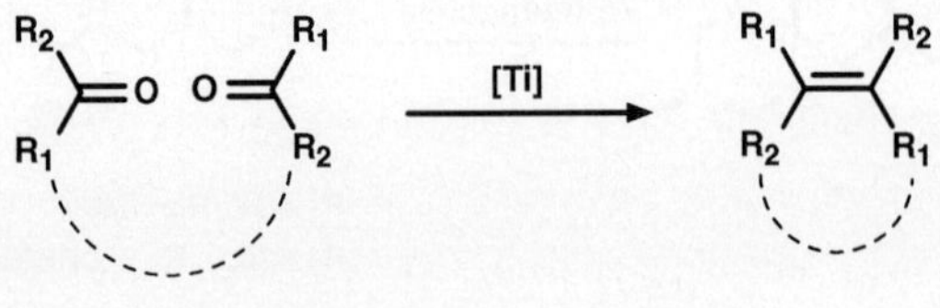

R2, R1, R1, R2, [Ti], R1, R2, R2, R1

Scheme 10-33. The McMurry reaction; R_1, R_2 = H, alkyl, aryl [20].

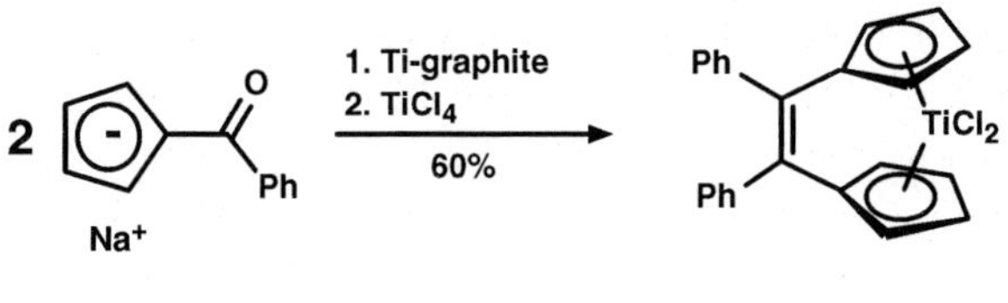

Scheme 10-34. Reductive coupling of acylated cyclopentadienyl complexes by Ti–graphite ($TiCl_3$:C_8K = 1:3) [100, 101].

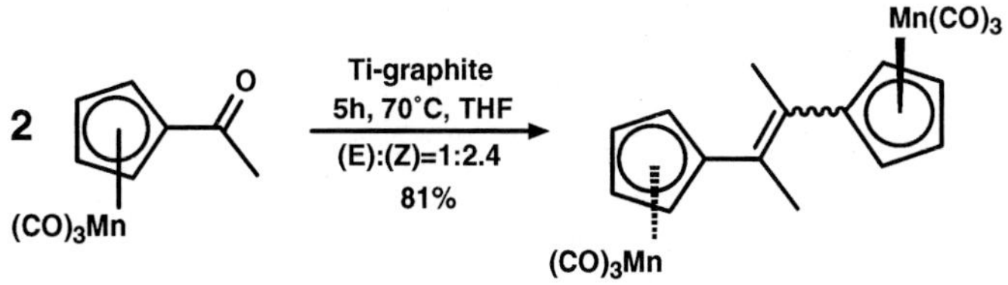

are unique McMurry coupling agents of high performance for all types of substrates [97–105]. They promote inter- as well as intramolecular coupling reactions and have found applications to different organometallic substrates. For example, Ti–graphite proved to be very efficient in the synthesis of alkene bridged *ansa*-titanocene derivatives (**36**) via dimerization of acylated cyclopentadienyl anions, **35** (Scheme 10-34) [100]. Similarly, $TiCl_3$/3 C_8K turned out to be advantageous for the reductive coupling of [η^5-C_5H_4-C(O)CH_3]$Mn(CO)_3$ (**37**) to **38**, avoiding undesirable reductions of this ketone to monomeric by-products as observed with other McMurry coupling agents [101]. Ti–graphite was also proposed for sealing the "slit" of the fullerene-derived diketone $C_{60}O_2$, obtained by light-induced incision of excited molecular oxygen into the fullerene cage [102]. Ti–graphite reagents have been widely used for intramolecular coupling reactions affording cycloalkenes independently of the ring size formed [97, 98], and turned out to be well suited for annulation reactions [103, 104]. Based on this methodology, elegant total syntheses of (+)-compactin and (+)-mevinolin were achieved with a McMurry cyclization reaction as the key step (Scheme 10-35) [105]. Con-

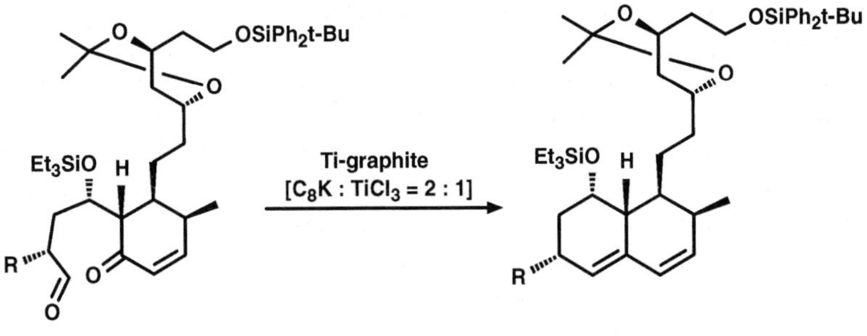

Scheme 10-35. Total syntheses of (+)-mevinolin and (+)-compactin: titanium–graphite-promoted closure of the A-ring [105].

sidering the dense functionalization of the precursors (**39**) and the high oxophilicity of low-valent titanium, this conversion to **40** is by no means trivial. Among all titanium reagents tested, Ti–graphite was the only one to promote the conversion **39** → **40**. The stoichiometry chosen in its preparation, however, was crucial: only the reagent obtained by reducing $TiCl_3$ with 2 equivalents of C_8K (thus formally a Ti(+1) species) turned out to guarantee both high yield and good reproducibility [103–105]. The validity of this observation was later confirmed in heterocycle syntheses (see below) in which Ti–graphite prepared from $TiCl_3$ + 2 C_8K proved to be more reliable than Ti–graphite from $TiCl_3$ + 3 C_8K as initially proposed. Based on extensive comparative studies, we strongly recommend the use of this formally univalent titanium reagent [106].

McMurry reactions were for a long time essentially confined to aldehydes and ketones as substrates [20]. One exception was the cyclization of oxo-esters to enolethers, which afford cyclanones as final products on hydrolytic work-up of the reaction mixtures. Ti–graphite turned out to efficiently promote such intramolecular *Type I* alkylidenations of esters (Scheme 10-36) [97, 98]. Of even wider scope are *Type II* cyclizations of oxo-alkanoates, which open up a new entry into the furan- and benzo[*b*]furan series starting from simple precursors (Scheme 10-37) [107, 108]. More importantly, it was demonstrated that acylamido carbonyl compounds of the general constitution **41** are also readily

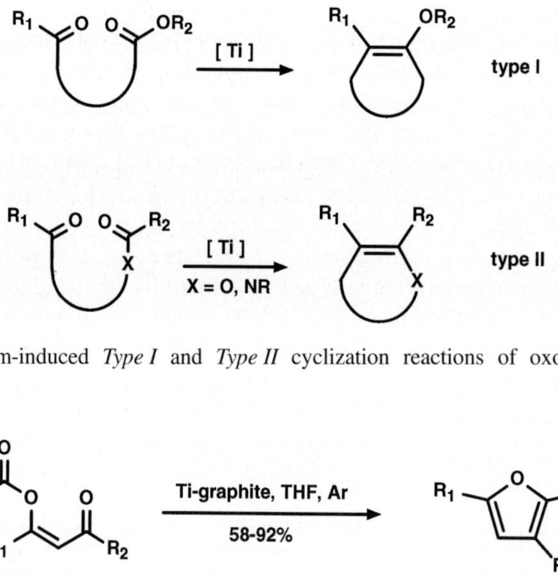

R_1 O O OR_2 [Ti] R_1 OR_2 type I

R_1 O O R_2 X [Ti] X = O, NR R_1 R_2 X type II

Scheme 10-36. Titanium-induced *Type I* and *Type II* cyclization reactions of oxoesters and oxoamides [106, 108].

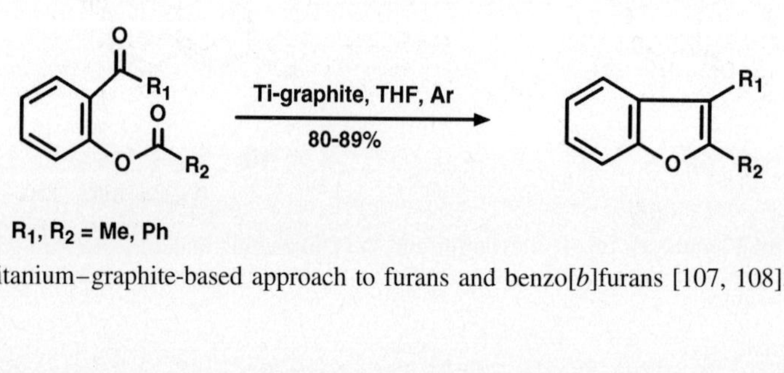

O Ph O O R_1 R_2 Ti-graphite, THF, Ar 58-92% R_1 O Ph R_2

R_1, R_2 = Me, Ph

O R_1 O O R_2 Ti-graphite, THF, Ar 80-89% R_1 O R_2

R_1, R_2 = Me, Ph

Scheme 10-37. Titanium–graphite-based approach to furans and benzo[*b*]furans [107, 108].

cyclized to the corresponding indole derivatives (**42**) upon treatment with Ti–graphite (Scheme 10-38). Although originally effected by Ti–graphite with $TiCl_3 : C_8K = 1:3$ [107–109] a re-investigation showed the stoichiometry $TiCl_3 : C_8K = 1:2$, as proposed by Clive et al., to be preferable in all cases [106]. Besides Ti–graphite a variety of other low-valent titanium samples may effect such oxo-amide coupling reactions, with a most convenient alternative procedure having been worked out recently [106a, c].

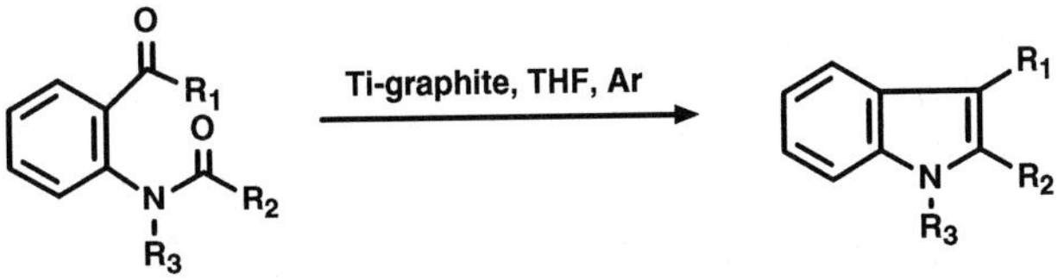

Ti-graphite, THF, Ar

41 **42**

Scheme 10-38. Indoles by titanium–graphite-induced *Type II* coupling reactions of acylamido carbonyl compounds; R_1 = alkyl, aryl; R_2 = H, alkyl, aryl, heteroaryl; R_3 = H, alkyl, tosyl [106–110].

The scope of such reductive indole syntheses seems to be wide, as all types of amides of constitution **41** undergo reductive cyclization. *N*-Substitution does not affect the yield, and a variety of reducible functional groups in the substrates turned out to be compatible [106–110]. The most striking feature, however, is the pronounced chemo- and regioselectivity of this new approach to heterocycles. Indole formation is favored to such an extent over other possible titanium-promoted processes that residual ester, amide, or even ketone groups are tolerated. This is illustrated by the examples shown in Schemes 10-39 to 10-41. Only one of the amide groups of the imide **43** will react, giving access to the tricyclic skeleton **44** [108]. Compound **45** undergoes exclusively oxo-amide coupling without oxoester cyclization interfering; the indole **46** thus obtained is a known precursor for diazepam and for the synthesis of serotonin and histamine antagonists [109]. Even more sur-

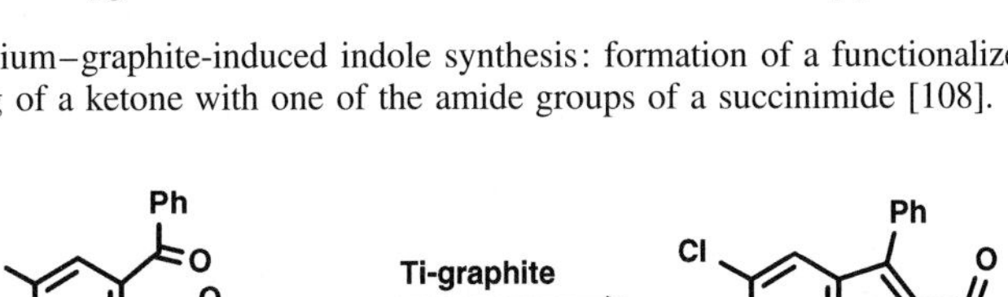

Ti-graphite, THF, Ar
61%

43 **44**

Scheme 10-39. Titanium–graphite-induced indole synthesis: formation of a functionalized tricyclic skeleton by selective coupling of a ketone with one of the amide groups of a succinimide [108].

Ti-graphite
DME, reflux

45 **46** R = H 93%, R = Me 94%

Scheme 10-40. Regio- and chemoselective oxoamide cyclizations by titanium-graphite [109].

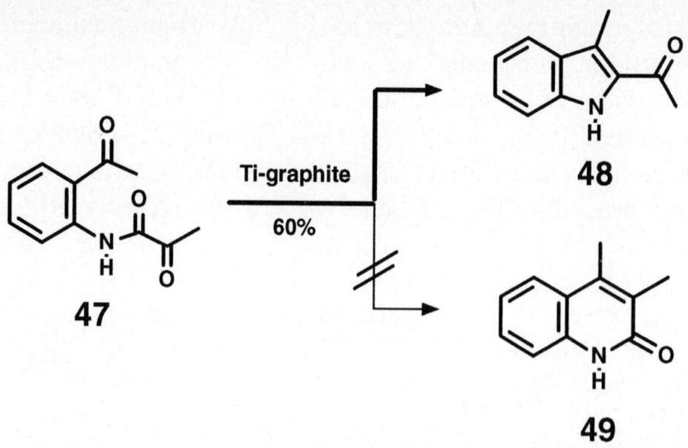

Ti-graphite

60%

47

48

49

Scheme 10-41. Synthesis of salvadoricine **48** by a chemo- and regioselective oxoamide cyclization reaction. The reductive coupling of the two ketone groups with formation of **49** does not intervene [109].

prisingly, the substrate **47** is selectively cyclized to salvadoricine, **48**, although a conventional McMurry reaction of the two ketone groups with formation of 2-quinolone (**49**) can also be envisaged. Not only is this preference of a ketoamide over a diketone coupling unprecedented, this is also one of the rare examples in the literature in which an unprotected keto group resists exposure to activated titanium [109].

Ti–graphite was used for the synthesis of zindoxifene (**52c**) and a variety of analogues (e.g. **52a, b, d, e**) of this tumor-growth inhibiting drug, from simple precursors (Scheme 10-42) [110]. It has also been successfully applied in the key steps of recent total syntheses of the indole alkaloids flavopereirine and the closely related indolopyridocoline (Scheme 10-43) [106b]. Congested indole derivatives bearing bulky groups at both C-2 and C-3 can also be prepared by this approach [106, 108]. Concerning the mechanism of such *Type II* cyclizations, a keto- rather than an amide-triggered process must be assumed.

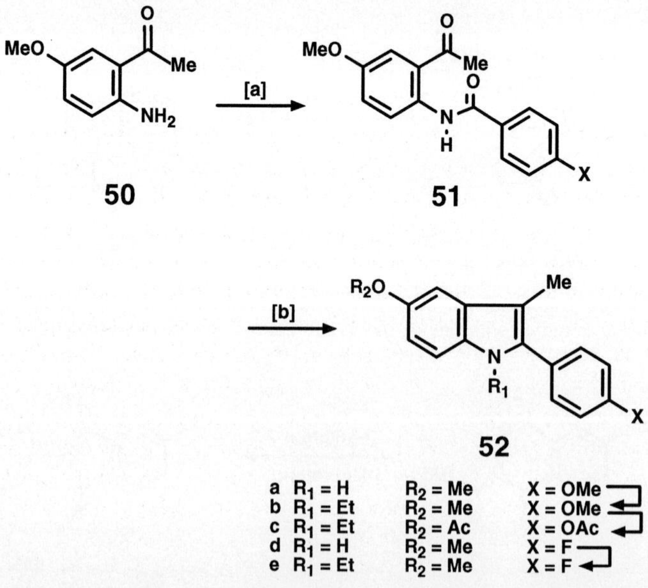

50

[a]

51

[b]

52

	R_1	R_2	X
a	R_1 = H	R_2 = Me	X = OMe
b	R_1 = Et	R_2 = Me	X = OMe
c	R_1 = Et	R_2 = Ac	X = OAc
d	R_1 = H	R_2 = Me	X = F
e	R_1 = Et	R_2 = Me	X = F

Scheme 10-42. Synthesis of the tumor-growth inhibitor zindoxifene, **52c**, and analogues; (a) *p*-$MeOC_6H_4COCl$ or *p*-FC_6H_4COCl, pyridine, CH_2Cl_2; (b) Ti–graphite, THF, reflux [110].

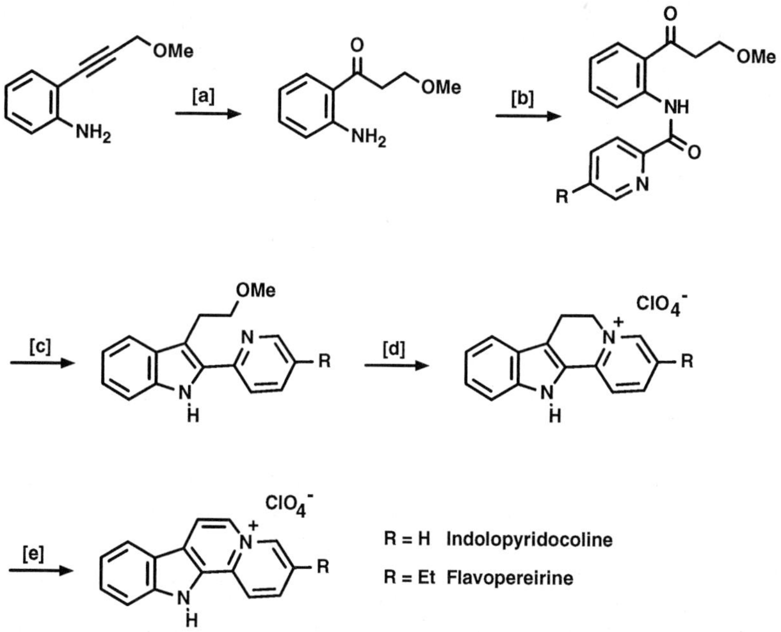

Scheme 10-43. Syntheses of Indolopyridocoline and Flavopereirine. (a) $HgSO_4$ (10 mol%), $MeOH/H_2O$, 48 h, slow addition, reflux, 53%; (b) pyridine-2-carboxylic acid chloride · HCl or 5-ethyl-pyridine-2-carboxylic acid chloride · HCl, pyridine, CH_2Cl_2, DMAP cat., 89% (R = H), 85% (R = Et); (c) Ti–graphite, THF, reflux, 62% (R = H), 52% (R = Et); (d) 1. BBr_3, CH_2Cl_2, –78 °C to –15 °C; 2. HCl, $CHCl_3$; 3. $NaClO_4$, H_2O, 79% (R = H); 67% (R = Et); (e) 1. DDQ, HOAc, reflux; 2. NaOH, $CHCl_3$; 3. $NaClO_4$, H_2O, 74% (R = H); 78% (R = Et).

Titanium dianions are the most likely intermediates, whereas no evidence for a radical scenario was found [106].

10.3.4.3 Magnesium–Graphite

Magnesium–graphite is obtained by reducing MgI_2 in diethyl ether or $MgCl_2$ in THF with two equivalents of C_8K. Although it was reported to be a true intercalation compound [111], more recent electron microscopic investigations on related systems render this original report doubtful [6, 73, 74]. Anyhow, Mg–graphite reacts readily with alkyl- and aryl halides to give the corresponding Grignard reagents, which seem to be included into the graphite host to some extend. Thus, filtration left a solid which could be stored for months without affecting its activity towards electrophiles such as ketones, CO_2, or water [111].

Mg–graphite is a strong single electron transfer (SET) agent readily effecting inter- and intramolecular pinacol coupling reactions of carbonyl compounds (Scheme 10-44) [98, 112]. Despite the obvious relationship to classical reagents such as Mg/Hg or Al/Hg, amalgamation was unnecessary for increasing the SET capacity of Mg–graphite. Its performance compares favorably with other methods in the literature, although some shortcomings due to the incompatibility with $-NO_2$ and halogen groups except –F (cf. Grignard formation) must be considered [98].

Its SET ability is also evident from the efficient Wurtz-type coupling of a 6-deoxy-6-iodopyranoside derivative as depicted in Scheme 10-45. Mg–graphite affords the dimer in 68% yield, whereas ordinary Mg turnings gave only 6.5% [60].

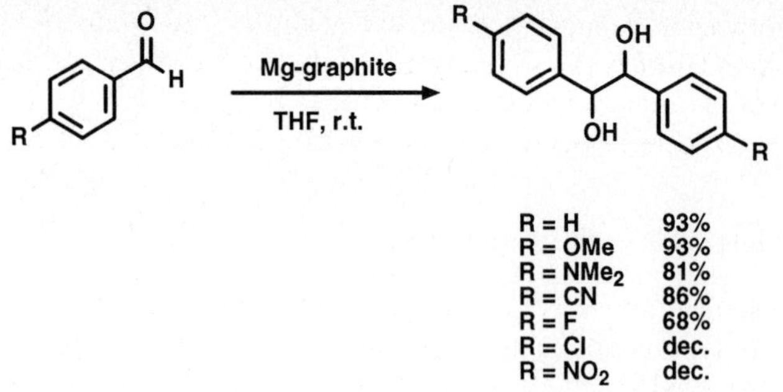

R	Yield
R = H	93%
R = OMe	93%
R = NMe_2	81%
R = CN	86%
R = F	68%
R = Cl	dec.
R = NO_2	dec.

Scheme 10-44. Mg–graphite promoted pinacol coupling of substituted benzaldehyde derivatives [98].

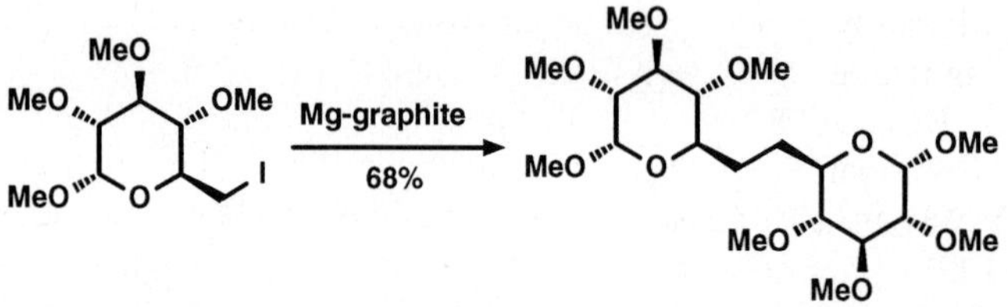

Scheme 10-45. Mg–graphite induced Wurtz coupling of methyl 6-deoxy-6-iodo-2,3,4-tri-*O*-methyl-α-D-glucopyranoside [60].

10.3.4.4 Group-10-Metal–Graphite Combinations as Catalysts

The graphite combinations of nickel, palladium and platinum prepared by the C_8K route exhibit interesting catalytic properties. Once again they seem to consist of finely dispersed metals, with an average particle size of only a few nanometers, adsorbed on the graphite which merely acts as an inert support [6, 74]. Previous assumptions of intercalation of these metals between the graphite layers [113] are incorrect.

Ni–graphite, as obtained by reduction of $NiBr_2 \cdot 2\,DME$ with $2\,C_8K$ in THF/HMPA, is an interesting alternative to the conventional Lindlar catalyst for the semihydrogenation of alkynes to (*Z*)-alkenes at ambient temperature and 1 atm H_2 [114]. The stereoselectivity observed is good to excellent. An inconvenience, however, might be that the catalyst is best prepared in situ for each run. Upon exposure to air and washing of the catalyst, the finely dispersed Ni particles get coated by a hydroxide layer [74, 115]. Although this treatment reduces the activity, the selectivity of the catalyst is increased and it may be stored on the shelf for months without deterioration [115].

Closely tied to these results, palladium–graphite doped with ethylenediamine turned out to catalyze semihydrogenations of alkynes to alkenes with preparatively useful selectivity in favor of the (*Z*)-isomer [116]. The same reagent was also employed as a substitute for soluble Pd complexes in Heck reactions (arylation or vinylation of alkenes) [117] as well as for the preparation of π-allyl palladium species from allyl acetates, carbonates, or phosphates [118]. Since the leaching of the Pd from the support seems to be low and the catalyst can be used repeatedly, Pd–graphite might offer interesting options for large-scale applications.

Pt–graphite showed a promising selectivity profile in the reduction of $X\text{-}C_6H_4\text{-}NO_2$ (X = Cl, Br, I) to $X\text{-}C_6H_4\text{-}NH_2$ [6]. Dehalogenation was low even in the case of X = I, as was the hydrogenation of the arene nucleus. Moreover, the catalyst showed neither significant morphological changes on AEM inspection nor a noticeable decrease in activity after 20 cycles.

10.3.4.5 Other Metal–Graphite Combinations

Ni–graphite as described above is not only a useful hydrogenation catalyst, but efficiently promotes the Wurtz-type coupling of allylic and benzylic halides as a stoichiometric reagent (Scheme 10-46) [119]. A variety of functional groups such as –Cl, –Br, –OMe, $-CF_3$, $-NO_2$, –CN, and –COOMe are compatible. It is interesting to note that co-reduction of $NiBr_2/CuCl_2$ (10/1) leads to a more active reagent. An incomplete reduction of $CuCl_2$ to Cu^I, which is known for its effect on Wurtz-type reactions, accounts for the observed synergism [74]. Together with Zn(Ag)–graphite [77] this is a second example in which doping significantly improved the preparative performance of a supported metal [120].

Sn–graphite undergoes oxidative insertion into allylic bromides at ambient temperature in THF [99]. Both allyl groups of the diallyltin dibromides obtained are transferred to aldehydes with formation of homoallyl alcohols. The reaction proceeds with allylic inversion and has been used, for example, for the synthesis of yomogi alcohol, **55** (Scheme 10-47) [99].

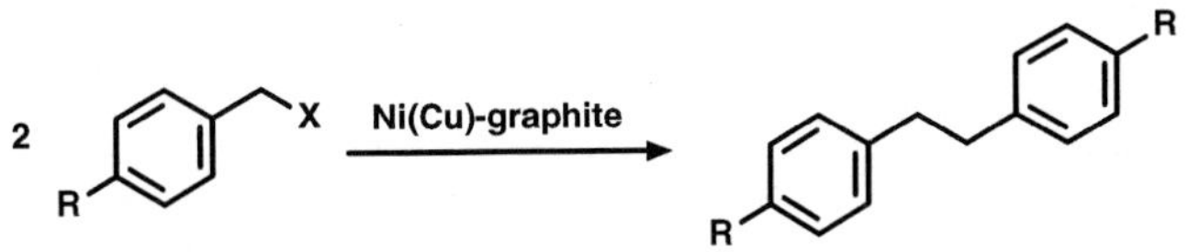

X	R	Yield
Cl	H	87%
Br	H	99%
Cl	Cl	85%
Cl	NO_2	54%
Br	CN	62%
Br	Br	90%
Cl	Me	90%

Scheme 10-46. Ni(Cu)–graphite-induced Wurtz coupling of substituted benzyl halides [119].

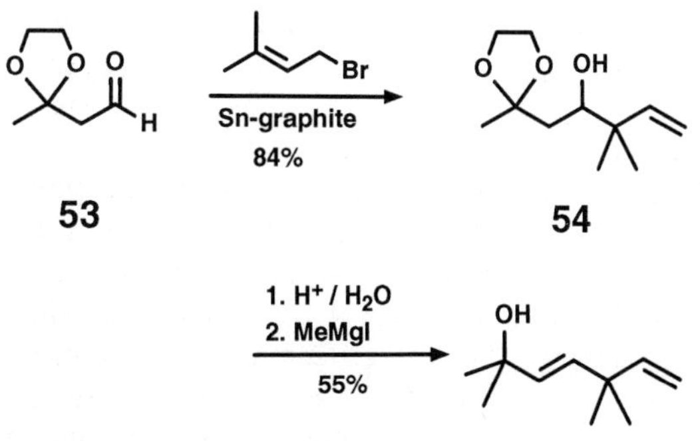

Scheme 10-47. Synthesis of yomogi alcohol **55** [99].

Fe–graphite has been used to reduce *vic*-dibromides to alkenes [121]. This reaction is greatly accelerated when a small amount of de-aerated water is added to the reaction mixture. The *anti*-debromination pathway translates configurational integrity of the dibromide into an alkene of high isomeric purity. Thus, *erythro*-1,2-diphenylethanedibromide forms (*E*)-stilbene selectively, while *threo*-5,6-dibromodecane affords 5-(*Z*)-decene. Fe–graphite has also been employed for the dehalogenation of α-bromo ketones [121]. This latter process passes through Fe(+2)-enolates, which have been trapped with either D_2O or trimethylsilylchloride. α,α'-Dibromoketones on treatment with Fe–graphite form 2-oxoallyl cations, which react with enamines or furan to give the corresponding [3+2]- and [3+4]-cycloaddition products (Scheme 10-48).

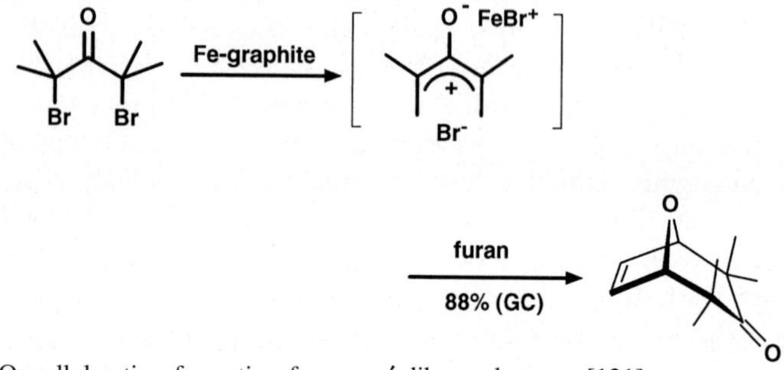

Fe-graphite

O^- $FeBr^+$

Br^-

furan

88% (GC)

Scheme 10-48. 2-Oxoallyl cation formation from α,α'-dibromoketones [121].

10.3.5 Graphimets

Besides the C_8K route to metal–graphite combinations outlined above, there exists yet another method for preparing such reagents [3]. Whereas the C_8K-based approach takes advantage of the ease of intercalation of potassium into the graphite matrix and the favorable reducing properties of C_8K for metal salts in solution, a reversed strategy can also be envisaged. Several metal salts are able to intercalate into graphite [122]. The laminates formed may then be treated with an appropriate reducing agent (such as H_2, Li-biphenyl, alkali metal naphthalenides, potassium vapor, sodium benzophenone ketyl, sodium in liquid ammonia, $NaBH_4$, or $LiAlH_4$) to give metal–graphite combinations. The products derived from the reduction with lithium-biphenyl are marketed under the tradename "Graphimets" and are claimed to consist of metals being intercalated into graphite [123].

There is some controversy about the actual location of the metals in these reagents. AEM investigations and X-ray diffraction studies of several commercially available samples of this type showed the presence of small islands of either metals or metal oxides *on* the graphite surface [6, 74, 124]. However, some authors maintain that a mixture of adsorbed particles (major) and encapsulated rafts (minor) is present [125–127]. As these latter tend to migrate to the surface, differences in the pre-treatment and aging of such polyphasic catalysts may explain the different morphological conclusions obtained so far. It is, however, beyond doubt that under reaction conditions there will remain no significant amount of metal intercalated into the graphite. For this reason Graphimets were found unsuitable for shape-selective catalysis [127b].

Furthermore, their initial composition depends critically on the reducing agent chosen. Potassium vapor is able to form a bi-intercalation compound in which potassium and the intercalated metal salt seem to coexist [125, 126]. Above 100 °C they react with one another, initially forming two-dimensional metal layers within the graphite galleries. A further increase in temperature or the duration of the reduction process leads to the exfoliation of the metal with formation of merely adsorbed clusters [125].

Some catalytic properties of Graphimets deserve mentioning. Cu-Graphimet exhibits a significantly higher catalytic activity than other Cu catalysts for the conversion of tetramethyloxirane to *tert*-butylmethylketone via 1,2-methyl group migration and to 2,3-dimethylbut-3-en-2-ol via a 1,3-hydrogen shift, respectively [128]. Fe-Graphimet proved to be a remarkably efficient, air-stable, and long-lived catalyst for the synthesis of acetylene from syngas (H_2:CO = 3:1) [129]. A strong interaction between metal and support may be responsible for this behavior. Pt-Graphimet was used as an efficient catalyst for the selective reduction of aromatic nitro groups in halonitrobenzene derivatives [6]. The catalytic activity of Graphimets of Pt, Pd, and Ni for the hydrogenation of olefins was reported to surpass that of supported metal catalysts prepared by traditional methods [127].

10.3.6 Supported Rieke Metals

Early attempts to immobilize Rieke metal slurries on alumina met with little success. Thus, the lifetime of a Rieke Co/Al_2O_3 catalyst for the Fischer–Tropsch process was not significantly increased compared with the unsupported sample [130]. Likewise, Rieke Ni on Al_2O_3 obtained by reduction of a nickel salt with lithium and naphthalene in the presence of alumina showed properties similar to ordinary Rieke Ni slurries in the reaction with allyl halides [131].

However, a promising new development stems from the immobilization of Rieke Cu on phosphane-bearing polystyrene matrices [132, 133]. Rieke Cu has great importance for synthesis, since it is not only capable of oxidative insertion into a wide variety of alkyl, aryl, allyl and benzyl halides, and even into allyl acetates, but is compatible with many electrophilic groups within these substrates. This allows the preparation of *functionalized* organocopper intermediates which are beyond the scope of traditional organocopper chemistry [19].

Rieke Cu is prepared by low-temperature reduction of CuI · PR_3 with lithium naphthalenide. Removal of the stoichiometric amounts of naphthalene and phosphane from the products formed may, however, be tedious. These shortcomings can be solved by using polymer supported Rieke copper variants [132, 133]. A polystyrene derivative bearing $-PPh_2$ groups was loaded with CuI, which was then reduced with lithium naphthalenide at 0 °C. The naphthalene formed was conveniently removed by filtration and washing of the black solid with THF. The active copper species retained inside the matrix of the resin [133] will metallate alkyl and aryl bromides or iodides at low temperature without difficulty. Various functional groups in the substrates, such as –Cl, –COOEt, or –CN, are compatible with this process. After cross-coupling of the functionalized organocopper species thus obtained, e.g. with acylhalides, epoxides, aldehydes, or enones, the work-up is straightforward as the phosphane-carrying support is easily separated by a simple filtration [132].

10.3.7 Miscellaneous Applications

Several other ways of preparing carrier supported active metal reagents have been reported in the literature. Thus, a rather particular kind of fixation has been developed for Mg–anthracene [134] by using a tailor-made polystyrene which bears anthracene residues

[135]. Recently, the immobilization of Mg-anthracene on silica via propylsilyl tethers has been found to be more practical. Many benzyl-Grignard reagents have been prepared with this supported Mg source in excellent yields [135b].

The thermodynamically most powerful reducing agent in a given solvent is the solvated electron ($e^-_{solv.}$) [136]. Alkali metal anions (M^-) are almost as effective. Based on the use of such electrides or alkalides, respectively, a whole branch of metal activation has been developed and used in particular for studying the physico-chemical aspects of highly dispersed systems. It has been reported that electride-promoted reductions have been carried out in the presence of neutral alumina, leading in the case of Au to particles of about 6 nm average diameter randomly dispersed on the surface of the admixed support [136].

The nanoscale metals obtained via the triethylhydroborate route have also been immobilized on inert carriers without significant changes in the average particle size [137]. A fairly homogeneous distribution on the surface of the support can be achieved. These materials show promising activity as catalysts in different types of hydrogenation reactions, clearly surpassing the performance of technical standards. Doping with, for example, bis(toluene)titanium, further enhances their performance [137].

10.4 Some Representative Procedures

10.4.1 "High Surface Area Sodium" on NaCl [7]

100 g pre-dried NaCl (200 °C, ≥ 24 h, 10^{-3} torr) was suspended in boiling toluene by means of an Ultra-turrax high-speed stirrer. Sodium sand (10 g, 435 mmol, ∅ 1–2 mm) was added to the boiling suspension under Ar over a period of 20 min. After complete addition reflux was continued for another 15 min, then the mixture was allowed to cool to ambient temperature and filtered under Ar. The crude Na/NaCl was washed with pentane (300 mL) in several portions and dried in vacuo (0.1 torr). This homogeneously grey, *non-pyrophoric* material can be stored under inert atmosphere for extended periods of time without loss of activity. Na/Al_2O_3 and Na/TiO_2 have also been prepared according to this general procedure.

In alcoholyses of sample aliquots with 2-ethyl-1-hexanol, 96% of the theoretical amount of H_2 was collected in a gas burette.

10.4.2 Example of a [Ti]/Al_2O_3-Induced Macrocyclization of a Diketone: 1,2-Diphenyl(2,26)[36]paracyclophane [7] (cf. Table 10-1)

A 100 mL two-necked flask equipped with a Teflon-coated stirring bar and a reflux condenser was flame-dried under reduced pressure (1 torr) and connected to the Ar-line after cooling to ambient temperature. It was charged with Na/Al_2O_3 (3.22 g, 10% w/w, ≈ 12.2 mmol Na), which was suspended in anhydrous THF (60 mL). $TiCl_3$ (940 mg, 6.1 mmol) was added, the mixture was refluxed for 1 h, and a solution of 1,26-bis(4-benzoylphenyl)hexacosane (727 mg, 1.0 mmol) in THF (20 mL) was added to the refluxing slurry over a period of 1.5 h. The mixture was filtered through a pad of silica, the inorganic residues were washed with THF (approx. 100 mL in several portions), the solvent evaporated, and the pale yellow residue purified by flash chromatography on silica with hexane/ethyl acetate (10/1) as eluent. Thus, the title compound was obtained as a colorless, waxy solid (575 mg, 83%). All analytical and spectroscopic data (^1H and ^{13}C NMR, IR, MS, elemental analysis) were in full agreement with the expected structure.

10.4.3 Large-Scale Preparation of Potassium–Graphite Laminate (C_8K)

Caution: *C_8K is pyrophoric and reacts violently with water, ammonia, low molecular weight alcohols, etc. It must be handled with great care and used in freshly distilled anhydrous solvents. Excess C_8K can be safely destroyed by suspending it in anhydrous THF and slowly adding an excess of isopropanol (technical grade).*

A 500 mL three-necked flask equipped with a gas-tight Hershberg stirrer and connected to the vacuum line was charged with graphite (Lonza KS 5-44, 63.52 g, 5.288 mol). The graphite powder was degassed and dried by heating it to 140–160 °C at reduced pressure (approx. 5 torr) with occasional stirring for 30 min. The flask was flushed with argon and a positive argon pressure was maintained throughout the following manipulations. One of the necks was connected to another Schlenk tube charged with potassium sand (∅ ≈ 2–3 mm, 25.85 g, 0.661 mol), which was then added in portions over a period of 30 min to the vigorously stirred graphite powder at 150 °C. After stirring for another 30 min at that temperature the bronze colored C_8K obtained was allowed to cool and transferred into a flame-dried 250 mL Schlenk tube via an argon-flushed polyethylene tube. It can be stored under inert atmosphere at ambient temperature for extended periods of time without any loss of activity.

Other samples of synthetic graphite with different specifications supplied by Lonza AG, Switzerland (HSAG 9, KS 5-75) as well as natural graphite have also been used for C_8K preparations. No significant differences have been noticed.

10.4.4 C_8K-Induced Synthesis of a Furanoid Glycal from a Thioglycoside: 1,4-Anhydro-3-*O*-(*tert*-butyldiphenylsilyl)-2-deoxy-5,6-*O*-isopropylidene-D-arabino-hex-1-enitol [66] (26, cf. Scheme 10-21)

A pre-dried two-necked 50 mL flask equipped with a Teflon-coated magnetic stirring bar and connected to an Ar line was charged with C_8K (575 mg, 4.26 mmol). After suspension in anhydrous THF (10 mL) a solution of phenyl 2,3;5,6-di-*O*-isopropylidene-1-thio-*β*-D-mannofuranoside, **24** (500 mg, 1.42 mmol) in THF (2 mL) was added via syringe at −10 °C, causing an immediate color change of the slurry from bronze to black. After 10 min *tert*-butylchlorodiphenylsilane (780 mg, 2.84 mmol) dissolved in the minimum amount of THF (≈1 mL) was injected, and the reaction mixture was allowed to stir at ambient temperature for another 30 min, then diluted with THF (30 mL) and filtered through silica. The colorless filtrate was evaporated and the residue purified by flash chromatography (SiO_2) with hexane/ethyl acetate (15/1) as eluent. The product was obtained as colorless syrup (580 mg, 96%) showing satisfactory analytical and spectroscopic data.

10.4.5 Zn(Ag)–Graphite-Induced Reductive Ring-Opening of a Deoxyiodo Sugar: (2*R*,4*S*)-Bis(benzyloxy)hex-5-enal [93]

A 250 mL two-necked flask equipped with a magnetic stirring bar and a reflux condenser was charged with graphite powder (6.0 g, 500 mmol), which was degassed and dried under reduced pressure at 150 °C for 15 min. After connecting the flask to the argon line, potassium (2.40 g, 61.4 mmol) was added in pieces with vigorous stirring. The bronze colored C_8K obtained within a few minutes was allowed to cool to ambient temperature and suspended in THF (80 mL), and a mixture of anhydrous $ZnCl_2$ (4.00 g, 29.4 mmol) and AgOAc (0.40 g, 2.4 mmol) was added causing the solvent to boil. After the exothermic reaction had subsided the mixture was refluxed for 20 min to ensure complete reduction. After cooling to room temperature a solution of methyl 2,4-di-*O*-benzyl-3,6-dideoxy-6-iodo-α-D-*ribo*-hexopyranoside [93] (3.00 g, 6.41 mmol) in THF (≈5 mL) was added via syringe and the mixture stirred for another 10 min, then filtered through a short pad of silica. The graphite residue was washed with THF in several portions, the combined filtrates were evaporated, and the crude product was purified by column chromatography with toluene/ethyl acetate (25/1) as eluent, affording the title compound as a colorless oil (1.64 g, 83%).

All spectroscopic and analytical data were in accordance with the expected structure. This compound has been further elaborated to form both halves of the C1–C13 segment of the fungicide amphoteronolide B (cf. Scheme 10-28) [93].

10.4.6 Titanium–Graphite-Induced Cyclization of Oxoamides: Synthesis of 2-(2′-Furanyl)-3-phenylindole [106a] (**42**, R_1 = Ph, R_2 = 2-furanyl, R_3 = H, cf. Scheme 10-38)

Graphite powder (1.97 g, 164 mmol) was degassed and dried by heating it to 150 °C for 15 min under vacuum (15 torr) in a 50 mL two-necked flask equipped with a Teflon-coated stirring bar and a reflux condenser. The flask was then connected to the argon line and pieces of potassium (802 mg, 20.5 mmol) were added to the graphite with vigorous stirring while still hot, affording the bronze-colored C_8K within a few minutes. After cooling it to ambient temperature and suspending in anhydrous THF (30 ml), $TiCl_3$ (1.58 g, 10.3 mmol; Aldrich, 99% purity) was added and the slurry refluxed for 1.5 h. After addition of N-(2′-benzoylphenyl)-furan-2-carboxylic acid amide (500 mg, 1.71 mmol) in THF (5 ml) reflux was continued for 10 min, then the mixture was cooled and filtered through a pad of silica. The residue was washed with THF (50 ml in several portions), the combined filtrates were evaporated, and the crude product was purified by flash chromatography on silica (Merck, 230–400 mesh) with hexane/ethyl acetate (10/1) as eluent affording the title compound as a pale yellow syrup (429 mg, 97%). All spectroscopic and analytical data (^1H and ^{13}C, NMR, IR, MS, elemental analysis) were in full agreement with the expected structure.

Acknowledgements. I sincerely thank all my colleagues and coworkers who contributed to the success of our metal activation project. Their names appear in the references. Financial support by the Fonds zur Förderung der Wissenschaftlichen Forschung (Vienna), the Max-Planck-Institut für Kohlenforschung, the Volkswagen Stiftung (Hannover), and the Fonds der Chemischen Industrie are gratefully acknowledged.

References

[1] (a) *Preparative Chemistry using Supported Reagents* (ed.: P. Laszlo), Academic Press, New York, **1987**; (b) J. H. Clark, A. P. Kybett, D. J. Macquarrie, *Supported Reagents. Preparation, Analysis, and Applications*, VCH, Weinheim, **1992**; (c) *Solid Supports and Catalysts in Organic Synthesis* (ed.: K. Smith), Ellis Horwood, Chichester, **1992.**

[2] A. Fürstner in *Encyclopedia of Reagents for Organic Synthesis* (ed.: L. A. Paquette), Wiley, New York, in press.

[3] A. Fürstner, *Angew. Chem.*, **1993**, *105*, 171; *Angew. Chem. Int. Ed. Engl.* **1993**, *32*, 164.

[4] A. Fürstner, H. Weidmann, *J. Carbohydr. Chem.* **1988**, *7*, 773.

[5] For an excellent review of the chemistry and properties of nanomaterials see: *Clusters and Colloids. From Theory to Applications* (ed.: G. Schmid), VCH, Weinheim, **1994**. The fundamental aspects of particle formation and growth are discussed in depth by Klabunde and Cardenas-Trivino in Chapter 6 of this book.

[6] A. Fürstner, F. Hofer, H. Weidmann, *J. Catalysis* **1989**, *118*, 502.

[7] A. Fürstner, G. Seidel, *Synthesis*, **1995**, 63.

[8] A. J. Hubert, *J. Chem. Soc. (C)* **1967**, 2149.

[9] D. Savoia, E. Tagliavini, C. Trombini, A. Umani-Ronchi, *J. Org. Chem.* **1980**, *45*, 3227.

[10] S. Singh, S. Dev, *Tetrahedron* **1993**, *49*, 10959.

[11] J. Shabtai, E. Gil-Av, *J. Org. Chem.* **1963**, *28*, 2893.

[12] R. Rienäcker, J. Graefe, *Angew. Chem.* **1985**, *97*, 348; *Angew. Chem. Int. Ed. Engl.* **1985**, *24*, 320.

[13] A. J. Hubert, J. Dale, *J. Chem. Soc. (C)* **1968**, 188.

[14] G. Suzukamo, M. Fukao, M. Minobe, *Chem. Lett.* **1987**, 585.

[15] G. Ruckelshau*β*, K. Kosswig, *Chem.-Zeitung* **1977**, *101*, 103.

[16] T. Alvik, J. Dale, *Acta Chem. Scand.* **1971**, *25*, 1153.

[17] (a) D. Savoia, E. Tagliavini, C. Trombini, A. Umani-Ronchi, *J. Organomet. Chem.* **1981**, *204*, 281; (b) H. Hart, B. L. Chen, C. T. Cheng, *Tetrahedron Lett.* **1977**, 3121.

[18] F. Scott, F. R. van Heerden, H. G. Raubenheimer, *J. Chem. Res. (S)*, **1994**, 144.

[19] R. D. Rieke, *Science* **1989**, 246, 1260. For a comprehensive review see Chapter 1 of this book.

[20] J. E. McMurry, *Chem. Rev.* **1989**, *89*, 1513. For a comprehensive review see also Chapter 3 of this book.

[21] A. Fürstner, G. Seidel, B. Gabor, C. Kopiske, C. Krüger, R. Mynott, *Tetrahedron* **1995**, *51*, 8875.

[22] H. Stadtmüller, B. Greve, K. Lennick, A. Chair, P. Knochel, *Synthesis* **1995**, 69.

[23] P. Knochel, R. Singer, *Chem. Rev.* **1993**, *93*, 2117. For another comprehensive treatise see Chapter 5 of this book.

[24] R. Csuk, B. I. Glänzer, A. Fürstner, *Adv. Organomet. Chem.* **1988**, *28*, 85.

[25] D. Savoia, C. Trombini, A. Umani-Ronchi, *Pure Appl. Chem.* **1985**, *57*, 1887.

[26] H. F. Klein, *Kontakte (Merck)* **1982**, 3.

[27] R. Setton in *Preparative Chemistry Using Supported Reagents* (ed.: P. Laszlo), Academic Press, New York, **1987**, p. 255.

[28] H. B. Kagan, *Pure Appl. Chem.* **1976**, *46*, 177.

[29] Originally reported in: (a) E. Weintraub, U.S. Patent 922.645, Chem. Abstr. **1909**, 3, 2040; (b) K. Fredenhagen, G. Cadenbach, *Z. Allg. Anorg. Chem.* **1926**, *158*, 249.

[30] For a compilation of the physical properties of C_8K and related intercalation compounds see: *Gmelins Handbuch der Anorganischen Chemie*, Kohlenstoff, Teil B, Lieferung 3, System Nr. 14, VCH, Weinheim, **1968**, pp. 881–892.

[31] R. Setton, F. Beguin, S. Piroelle, *Synth. Met.* **1982**, *4*, 299.

[32] H. P. Böhm, R. Schlögl, *Carbon* **1987**, *25*, 583.

[33] D. E. Bergbreiter, J. M. Killough, *J. Chem. Soc. Chem. Commun.* **1976**, 913.

[34] I. S. Weitz, M. Rabinovitz, *J. Chem. Soc. Perkin Trans.* 1, **1993**, 117.

[35] (a) A detailed discussion and evaluation of knowledge about the bonding situation in C_8K and its chemical implications is given in ref [32] and references therein. Further reading: (b) S. P. Kelty, C. M. Lieber, *J. Phys. Chem.* **1989**, *93*, 5983; (c) M. F. Quinton, C. Fretigny, A. P. Legrand, *Synth. Met.* **1989**, *34*, 569; (d) D. Anselmetti, R. Wiesendanger, V. Geiser, H. R. Hidber, H. J. Güntherodt, *J. Microscopy*, **1988**, *152*, 509; (e) M. H. Whangbo, W. Liang, J. Ren, S. N. Magonov, A. Wakuschewski, *J. Phys. Chem.* **1994**, *98*, 7602 and references therein.

[36] L. B. Ebert, *J. Mol. Catal.* **1982**, *15*, 275.

[37] (a) L. S. Hegedus, *Pure Appl. Chem.* **1990**, *62*, 691; (b) M. A. Schwindt, T. Lejon, L. S. Hegedus, *Organometallics*, **1990**, *9*, 2814.

[38] C. Ungurenasu, M. Palie, *J. Chem. Soc. Chem. Commun.* **1975**, 388.
[39] G. P. Boldrini, A. Umani-Ronchi, *Synthesis*, **1976**, 596.
[40] K. A. Jensen, B. Nygaard, G. Elisson, P. H. Nielsen, *Acta Chem. Scand.* **1965**, *19*, 768.
[41] R. A. Fischer, J. Behm, E. Herdtweck, C. Kronseder, *J. Organomet. Chem.* **1992**, *437*, C29.
[42] A. Stanger, K. P. C. Vollhardt, *Organometallics*, **1992**, *11*, 317.
[43] M. Buchmeiser, H. Schottenberger, *Organometallics*, **1993**, *12*, 2472.
[44] (a) S. M. van der Kerk, J. C. Roos-Venekamp, A. J. M. van Beijnen, G. J. M. van der Kerk, *Polyhedron*, **1983**, *2*, 1337; (b) S. M. van der Kerk, P. H. M. Budzelaar, A. L. M. van Eekeren, G. J. M. van der Kerk, *Polyhedron* **1984**, *3*, 271. There is, however, some dispute on the interpretation of the reaction mechanism and the structure of the products formed, cf.: (c) R. Schlögl, B. Wrackmeyer, *Polyhedron*, **1985**, *4*, 885.
[45] A. Fürstner, H. Weidmann, *J. Organomet. Chem.* **1988**, *354*, 15.
[46] (a) H. Müller, U. Weinzierl, W. Seidel, *Z. Anorg. Allg. Chem.* **1991**, *603*, 15; (b) For an application to the synthesis of a functionalized distannane see: F. Richter, H. Weichmann, *J. Organomet. Chem.* **1994**, *466*, 77.
[47] D. J. Ager, I. Fleming, S. K. Patel, *J. Chem. Soc. Perkin Trans.* 1, **1981**, 2520.
[48] P. Jutzi, J. Kleimeier, R. Krallmann, H. G. Stammler, B. Neumann, *J. Organomet. Chem.* **1993**, *462*, 57.
[49] D. E. Bergbreiter, J. M. Killough, *J. Am. Chem. Soc.* **1978**, *100*, 2126.
[50] D. Savoia, C. Trombini, A. Umani-Ronchi, *Tetrahedron Lett.* **1977**, 653.
[51] D. Savoia, C. Trombini, A. Umani-Ronchi, *J. Org. Chem.* **1978**, *43*, 2907.
[52] J. M. Lalancette, G. Rollin, P. Dumas, *Can. J. Chem.* **1972**, *50*, 3058.
[53] D. Tamarkin, M. Rabinovitz, *Synth. Met.* **1984**, *9*, 125.
[54] M. Contento, D. Savoia, C. Trombini, A. Umani-Ronchi, *Synthesis*, **1979**, 30.
[55] D. Tamarkin, D. Benny, M. Rabinovitz, *Angew. Chem.* **1984**, *96*, 594; *Angew. Chem. Int. Ed. Engl.* **1984**, *23*, 642.
[56] D. Tamarkin, M. Rabinovitz, *J. Org. Chem.* **1987**, *52*, 3472.
[57] D. Tamarkin, Y. Cohen, M. Rabinovitz, *Synthesis*, **1987**, 196.
[58] M. Rabinovitz, D. Tamarkin, *Synth. Commun.* **1984**, *14*, 377.
[59] (a) M. Lissel, J. Kottmann, D. Tamarkin, M. Rabinovitz, *Z. Naturforsch.* **1988**, *43b*, 1211; (b) M. Lissel, J. Kottmann, D. Lenoir, *Chemosphere*, **1989**, *19*, 1499.
[60] A. Fürstner, H. Weidmann, *J. Org. Chem.* **1989**, *54*, 2307.
[61] W. Abramski, K. Badowska-Roslonek, M. Chmielewski, *Tetrahedron Lett.* **1993**, 2403.
[62] A. Fürstner, U. Koglbauer, H. Weidmann, *J. Carbohydr. Chem.* **1990**, *9*, 561.
[63] A. Fürstner, *Tetrahedron Lett.* **1990**, *31*, 3735.
[64] A. Fürstner, unpublished results.
[65] A. Fürstner, J. P. Praly, *Angew. Chem.* **1994**, *106*, 779; *Angew. Chem. Int. Ed. Engl.* **1994**, *33*, 751.
[66] A. Fürstner, *Liebigs Ann. Chem.* **1993**, 1211.
[67] D. Savoia, C. Trombini, A. Umani-Ronchi, *J. Chem. Soc. Perkin Trans.* 1, **1977**, 123.
[68] P. O. Ellingsen, K. Undheim, *Acta Chem. Scand.* **1979**, *B 33*, 528.
[69] W. Jones, J. M. Thomas, D. Tilak, B. Tennakoon, R. Schlögl, P. Diddams, *ACS Symp. Series*, **1985**, *288*, 472.
[70] P. Lagrange, A. Metrot, A. Herold, *Compt. Rend. Acad. Sci.* **1974**, *278 C*, 701.
[71] For leading references see: (a) S. Tsuchiya, K. Konishi, T. Kawashita, H. Imamura, *J. Mol. Catal.* **1989**, *57*, 13; (b) J. M. Lalancette, R. Roussel, *Can. J. Chem.* **1976**, *54*, 2110; (c) M. Ichikawa, Y. Inoue, K. Tamaru, *J. Chem. Soc. Chem. Commun.* **1972**, 928; (d) M. Ishikawa, M. Soma, T. Onishi, K. Tamaru, *J. Catalysis* **1967**, *9*, 418; (e) M. P. Rosynek, Y. P. Wang, *J. Mol. Catal.* **1984**, *27*, 277; (f) S. Tsuchiya, T. Misumi, N. Ohuye, H. Imamura, *Bull. Chem. Soc. Jpn.* **1982**, *55*, 3089.
[72] For leading references see: (a) R. Puffr, N. Vladimirov, *Makromol. Chem.* **1993**, *194*, 1765; (b) I. B. Rashkov, S. L. Spassov, I. M. Panayotov, *Makromol. Chem.* **1973**, *170*, 39; (c) H. Podall, W. E. Foster, A. P. Giraitis, *J. Org. Chem.* **1958**, *23*, 82; (d) J. Parrod, G. Beinert, *J. Polymer Sci.* **1961**, *53*, 99; (e) I. B. Rashkov, I. Gitsov, I. Panayotov, J. P. Pascault, *J. Polymer Sci.: Polymer Chem.*, **1983**, *21*, 923; (f) I. B. Rashkov, I. Gitsov, *J. Polymer Sci.: Polymer Chem.* **1984**, *22*, 905.
[73] A. Fürstner, F. Hofer, H. Weidmann, *J. Chem. Soc. Dalton Trans.* **1988**, 2023.
[74] A. Fürstner, F. Hofer, H. Weidmann, *Carbon*, **1991**, *29*, 915.
[75] We failed, however, to reduce $AlCl_3$ to activated Al by means of C_8K, most likely because of concomitant cleavage of the ethereal solvent, cf. [112].
[76] G. P. Boldrini, D. Savoia, E. Tagliavini, C. Trombini, A. Umani-Ronchi, *J. Org. Chem.* **1983**, *48*, 4108.
[77] R. Csuk, A. Fürstner, H. Weidmann, *J. Chem. Soc. Chem. Commun.* **1986**, 775.

[78] A. Fürstner, *Synthesis* **1989**, 571.
[79] G. P. Boldrini, M. Mengoli, E. Tagliavini, C. Trombini, A. Umani-Ronchi, *Tetrahedron Lett.* **1986**, *27*, 4223.
[80] R. Csuk, B. I. Glänzer, *J. Carbohydr. Chem.* **1990**, *9*, 797.
[81] A. Fürstner, *J. Organomet. Chem.* **1987**, *336*, C33.
[82] R. Csuk, A. Fürstner, H. Sterk, H. Weidmann, *J. Carbohydr. Chem.* **1986**, *5*, 459.
[83] R. Csuk, B. I. Glänzer, Z. Hu, R. Boese, *Tetrahedron* **1994**, *50*, 1111.
[84] R. Csuk, M. Hugener, A. Vasella, *Helv. Chim. Acta* **1988**, *71*, 609.
[85] F. Mancini, M. G. Piazza, C. Trombini, *J. Org. Chem.* **1991**, *56*, 4246.
[86] P. Marceau, L. Gautreau, F. Beguin, G. Guillaumet, *J. Organomet. Chem.* **1991**, *403*, 21.
[87] A. Fürstner, R. Singer, P. Knochel, *Tetrahedron Lett.* **1994**, *35*, 1047.
[88] R. Csuk, A. Fürstner, B. I. Glänzer, H. Weidmann, *J. Chem. Soc. Chem. Commun.* **1986**, 1149.
[89] P. Pudlo, J. Thiem, V. Vill, *Chem. Ber.* **1990**, *123*, 1129.
[90] R. Csuk, B. I. Glänzer, A. Fürstner, H. Weidmann, V. Formacek, *Carbohydr. Res.* **1986**, *157*, 235.
[91] A. Fürstner, D. Jumbam, J. Teslic, H. Weidmann, *J. Org. Chem.* **1991**, *56*, 2213.
[92] A. Fürstner, J. Baumgartner, D. N. Jumbam, *J. Chem. Soc. Perkin Trans.* 1, **1993**, 131.
[93] A. Fürstner, J. Baumgartner, *Tetrahedron* **1993**, *49*, 8541.
[94] A. Fürstner, H. Weidmann, *J. Org. Chem.* **1990**, *55*, 1363.
[95] R. E. Ireland, P. Wipf, W. Miltz, B. Vanasse, *J. Org. Chem.* **1990**, *55*, 1423.
[96] R. E. Ireland, T. K. Highsmith, L. D. Gegnas, J. L. Gleason, *J. Org. Chem.* **1992**, *57*, 5071.
[97] A. Fürstner, H. Weidmann, *Synthesis*, **1987**, 1071.
[98] A. Fürstner, R. Csuk, C. Rohrer, H. Weidmann, *J. Chem. Soc. Perkin Trans.* 1, **1988**, 1729.
[99] G. P. Boldrini, D. Savoia, E. Tagliavini, C. Trombini, A. Umani-Ronchi, *J. Organomet. Chem.* **1985**, *280*, 307.
[100] P. Burger, H. H. Brintzinger, *J. Organomet. Chem.* **1991**, *407*, 207.
[101] S. Pittner, G. Huttner, O. Walter, L. Zsolnai, *J. Organomet. Chem.* **1993**, *454*, 183.
[102] G. Taliani, G. Ruani, R. Zamboni, R. Danieli, S. Rossini, V. N. Denisov, V. M. Burlakow, F. Negri, G. Orlandi, F. Zerbetto, *J. Chem. Soc. Chem. Commun.* **1993**, 220.
[103] D. L. J. Clive, C. Zhang, K. S. K. Murthy, W. D. Hayward, S. Daigneault, *J. Org. Chem.* **1991**, *56*, 6447.
[104] D. L. J. Clive, K. S. K. Murthy, C. Zhang, W. D. Hayward, S. Daigneault, *J. Chem. Soc. Chem. Commun.* **1990**, 509.
[105] D. L. J. Clive, K. S. K. Murthy, A. G. H. Wee, J. S. Prasad, G. V. J. da Silva, M. Majewski, P. C. Anderson, C. F. Evans, R. D. Haugen, L. D. Heerze, J. R. Barrie, *J. Am. Chem. Soc.* **1990**, *112*, 3018.
[106] (a) A. Fürstner, A. Hupperts, A. Ptock, E. Janssen, *J. Org. Chem.* **1994**, *59*, 5215; (b) A. Fürstner, A. Ernst, *Tetrahedron*, **1995**, *51*, 773; (c) A. Fürstner, A. Hupperts, *J. Am. Chem. Soc.* **1995**, *117*, 4468.
[107] A. Fürstner, D. N. Jumbam, H. Weidmann, *Tetrahedron Lett.* **1991**, *32*, 6695.
[108] A. Fürstner, D. N. Jumbam, *Tetrahedron* **1992**, *48*, 5991.
[109] A. Fürstner, D. N. Jumbam, *J. Chem. Soc. Chem. Commun.* **1993**, 211.
[110] A. Fürstner, D. N. Jumbam, G. Seidel, *Chem. Ber.* **1994**, *127*, 1125.
[111] C. Ungurenasu, M. Palie, *Synth. React. Inorg. Met.-Org. Chem.* **1977**, *7*, 581.
[112] R. Csuk, A. Fürstner, H. Weidmann, *J. Chem. Soc. Chem. Commun.* **1986**, 1802.
[113] D. Braga, A. Ripamonti, D. Savoia, C. Trombini, A. Umani-Ronchi, *J. Chem. Soc. Dalton Trans.* **1979**, 2026.
[114] D. Savoia, E. Tagliavini, C. Trombini, A. Umani-Ronchi, *J. Org. Chem.* **1981**, *46*, 5340.
[115] D. Savoia, E. Tagliavini, C. Trombini, A. Umani-Ronchi, *J. Org. Chem.* **1981**, *46*, 5344.
[116] D. Savoia, C. Trombini, A. Umani-Ronchi, G. Verardo, *J. Chem. Soc. Chem. Commun.* **1981**, 540.
[117] D. Savoia, C. Trombini, A. Umani-Ronchi, G. Verardo, *J. Chem. Soc. Chem. Commun.* **1981**, 541.
[118] G. P. Boldrini, D. Savoia, E. Tagliavini, C. Trombini, A. Umani-Ronchi, *J. Organomet. Chem.* **1984**, *268*, 97.
[119] P. Marceau, F. Beguin, G. Guillaumet, *J. Organomet. Chem.* **1988**, *342*, 137.
[120] The morphological reasons for the observed synergisms however, seem to be different. In the case of Zn(Ag)-graphite an electrochemical half-element is the most likely explanation for the increased reactivity; this is formed by the intimate contact of fine silver and zinc particles in the conglomerate deposited on the graphite surface. A. Fürstner, F. Hofer, unpublished results. See also Chapter 11 of this book.
[121] D. Savoia, E. Tagliavini, C. Trombini, A. Umani-Ronchi, *J. Org. Chem.* **1982**, *47*, 876.
[122] W. Rüdorff, E. Stumpp, W. Spriessler, F. W. Siecke, *Angew. Chem.* **1963**, *75*, 130.
[123] J. M. Lalancette, U.S. Patent 3,847,963 (November 12, **1974**). Commercially available from Alfa Division, Ventron Corp.

[124] D. J. Smith, R. M. Fisher, L. A. Freeman, *J. Catalysis* **1981**, *72*, 51.
[125] (a) A. Messaoudi, M. Inagaki, F. Beguin, *J. Mater. Chem.* **1991**, *1*, 735; (b) M. Ohira, A. Messaoudi, M. Inagaki, F. Beguin, *Carbon*, **1991**, *29*, 1233.
[126] C. Herold, J. F. Mareche, G. Furdin, *Microsc. Microanal. Microstruct.* **1991**, *2*, 589.
[127] (a) F. Notheisz, A. Mastalir, M. Bartok, *J. Catalysis* **1992**, *134*, 608; (b) G. Sirokman, A. Mastalir, A. Molnar, M. Bartok, Z. Schay, L. Guczi, *Carbon* **1990**, *28*, 35.
[128] A. Molnar, A. Mastalir, M. Bartok, *J. Chem. Soc. Chem. Commun.* **1989**, 124.
[129] W. Jones, R. Schlögl, J. M. Thomas, *J. Chem. Soc. Chem. Commun.* **1984**, 464.
[130] G. L. Rochfort, R. D. Rieke, *Inorg. Chem.* **1986**, *25*, 348.
[131] R. D. Rieke, A. V. Kavaliunas, L. D. Rhyne, D. J. J. Fraser, *J. Am. Chem. Soc.* **1979**, *101*, 246.
[132] R. A. O'Brien, R. D. Rieke, *J. Org. Chem.* **1990**, *55*, 788.
[133] R. A. O'Brien, A. K. Gupta, R. D. Rieke, R. K. Shoemaker, *Magn. Reson. Chem.* **1992**, *30*, 398.
[134] For a comprehensive review on Mg-anthracene see Chapter 8 of this book.
[135] (a) S. Harvey, C. L. Raston, *J. Chem. Soc. Chem. Commun.* **1988**, 652; (b) T. R. van den Ancker, C. L. Raston, *Organometallics*, **1995**, *14*, 584.
[136] K. L. Tsai, J. L. Dye, *J. Am. Chem. Soc.* **1991**, *113*, 1650.
[137] For a comprehensive treatise see Chapter 9 of this book.

11 Morphological Considerations on Metal–Graphite Combinations

Ferdinand Hofer

11.1 Introduction

The efficiency of chemical reactions involving metals as catalysts or reducing agents for various transformations largely depends on the active surface area, which is generally a direct function of the degree of metal dispersion. The activity of the metals is greatly improved when they are finely and homogeneously distributed on an appropriate support. Due to the continuing interest in metal activation, a considerable variety of activating procedures has been developed. These methods invariably consist of two main strategies, either aiming at the effective removal of the deactivating oxide layers from the metal surface by chemical or mechanical means, or at achieving a fine distribution of the metal in an appropriate solvent. Many activation procedures have been developed, such as simple grinding, dispersion to form blacks and colloids [1, 2], formation of amalgams [3, 4] and metal couples [5–7], ultrasonic irradiation [8, 9], condensation with an inert solvent [10], reduction of metal salts in solution [11, 12], and lastly fine dispersions on various supports [13–16].

Compared with previous metal activation methods, the reduction of metal halides by potassium in solution was a first milestone in the history of metal activation (Rieke metals) [11, 12]. This method was further extended by using sodium or lithium naphthalenide as reducing agents [17, 18]. However, the most outstanding metal activating procedure turned out to be the reduction of metal halides by intercalated potassium–graphite (C_8K) [16, 19]. This quite universally applicable and efficient method has recently led to considerable scope for extending the chemistry of zinc [5, 20], magnesium [21], titanium [22] and other metals [23].

According to Braga and Volpin [24–26] the excellent activity of the metal salt reduced by C_8K originates from the intercalation of the metal atoms into graphite. Although some ambiguous evidence for metal intercalated graphite has been found, a lamellar structure for reagents such as zinc–graphite obtained by C_8K reduction appeared rather doubtful [27, 28]. Therefore, several groups have tried to characterize these interesting metal-graphite compounds during the last few years, and in every case very finely dispersed metal particles distributed over the graphite support have been found. No intercalation of metals into graphite at all was noted [29, 30], and therefore we will use the terms "metal–graphite combinations" or "metal–graphites" instead of metal–graphite "compounds".

In this contribution we review these investigations to reach a conclusive view of the actual structure of metal–graphite combinations and related reducing agents such as the Rieke metals. After an introduction to the most important methods for characterizing these materials, we look at the structure of graphite and the phenomenon of intercalation. However, the emphasis in this chapter is on summarizing investigations of the morphology and structure of metal–graphite combinations and comparing them with similar agents such as the commercial 'Graphimets' and the Rieke metals.

11.2 Characterization Methods

For the conclusive elucidation of the morphology of metal–graphite compounds (or heterogeneous catalysts), which is a prerequisite for understanding the relationship between morphology and activity, a combination of various analytical methods is called for [31]. X-ray diffraction and X-ray absorption are powerful tools that provide structural information on the atomic scale, spatially averaged over a total of, generally, not less than some 10^{18} unit cells. However, due to the heterogeneity of catalysts and metal–graphite combinations, a complete knowledge of the morphology of these materials can only be obtained by also using microscopic methods.

11.2.1 X-Ray Methods

X-ray diffraction (XRD), in particular X-ray powder diffraction, has been extensively used to identify and characterize intercalated graphites and metal–graphite combinations. Single crystal and, more recently, powder XRD methods can be used to find the atomic positions within the crystal structures of new and as yet uncharacterized materials. In particular, powder XRD is useful for identification of crystalline phases, determination of stage fidelity in intercalated compounds, and estimates of metal crystallite sizes [32].

A more recent addition to the diverse armory of X-ray based methods is X-ray absorption spectroscopy (XAS) [33]. In contrast to X-ray diffraction methods, which derive their utility from the properties of well-defined crystallites, X-ray absorption methods are atomic probes, capable of obtaining both electronic and structural information about a specific type of atom. The study of absorption edge fine structures can provide information about the symmetry of an atom's environment, its oxidation state, and bond lengths and bond angles. However, X-ray absorption has only rarely been used for investigations of metal–graphite combinations.

11.2.2 Electron Microscopy

In the development of catalyst systems the electron microscope is nowadays being used increasingly as an interactive tool. For example, the first clear insight into the morphology of metal–graphite combinations was obtained by using electron microscopy [29]. Many reviews have been written on the use of electron microscopy in characterizing catalysts [34, 35]. Clearly, electron microscopy is well suited for studying the distribution of metals. At a resolution of 2 to 10 nm the scanning electron microscope (SEM) provides morphological information over a large lateral range, and by using energy-dispersive or wavelength-dispersive X-ray spectrometry in the SEM, the elemental composition in micrometer regions can also be determined. Since the samples investigated in the SEM are usually thick, electron beam broadening limits the analytical resolution to 1 to 5 μm [36]. However, in catalyst research smaller particles are important and therefore the analytical transmission electron microscope (AEM, TEM) is the method of choice [37]. This microscope allows sample characterization on a nanometer scale, and to gain the full benefits from it, very thin samples have to be investigated in high vacuum, with high-voltage, high-intensity electron beams. Sample thicknesses for AEM are typically only 10 to 200 nm, but nevertheless such a thickness contains many unit cells, in fact many more than are probed by surface sensitive techniques (see Section 10.2.3).

A modern AEM capable of performing high resolution work (0.1–0.2 nm), to which has been attached an electron energy-loss spectrometer (EELS) and an energy-dispersive X-ray spectrometer (EDXS), can give the following information for materials ranging upwards in mass from ca. 10^{-19} g and in volume from 10^{-19} cm^3:

a) the crystallographic phase and its space group from electron diffraction techniques (ED), with a lateral resolution down to some nm;
b) the projected image and in favourable circumstances the crystal structure from high resolution electron microscopy (HREM) with a resolution of ca. 0.2 nm;
c) the elemental composition from X-ray emission (EDXS) and electron energy-loss spectrometry (EELS), with nanometer resolution in very thin samples;
d) the valence state, and coordination of specific (especially light) atoms from EELS.

11.2.3 Surface-Sensitive Methods

X-ray photoelectron spectroscopy (XPS) is among the most frequently used techniques in catalysis, and has often been used for the characterization of metal–graphite compounds [38, 39]. It provides information about the elemental composition, the oxidation state of the elements, and in favourable cases the dispersion of one phase over another [40]. Similar information can be obtained by Auger electron spectroscopy [41]. Electron energy-loss spectroscopy (EELS) employing low-energy electrons has become an important technique for studying the electronic structure of graphite compounds, and has been applied to several intercalated graphites [42]. In recent years a variety of new high resolution microscopic probes has become available, such as scanning tunneling microscopy (STM) [43] and atomic force microscopy (AFM) [44], which reveal the surface topology on an atomic scale, but which have no element-specific image with which to work. In particular, STM has been applied in several cases for investigating the topology of graphite compounds [45]. These surface sensitive methods have to be performed in ultrahigh vacuum and the metal–graphite combinations are often covered by potassium halide from the dry residue of the solvent thus hindering surface analysis. Therefore these methods are not always well suited for heterogeneous systems such as we are concerned with in this chapter.

11.2.4 In-situ Investigations

However, of the techniques available for the characterization of catalysts, only a few are applicable under in-situ conditions. Many of the powerful methods that can reveal so much about the nature of a catalyst in the ex-situ state are inapplicable for in-situ studies. In surface investigations and most microscopy studies the catalyst has to be examined under vacuum. For example, all those techniques that depend on either primary or liberated electrons to probe the properties of surfaces are ruled out for investigations in the gas phase or in solvents. To prevent formation of artifacts, the metal–graphite samples need to be examined under reaction conditions, which means they are covered by the solvent. This can essentially only be achieved using the X-ray methods, but these have only rarely been used for in-situ investigation of metal–graphite combinations [30]. In all other cases, sample preparation for characterization requires careful operations, excluding any influence of oxygen or water.

11.3 Intercalation in Graphite and the Structure of C_8K

Graphite is the most thermodynamically stable form of carbon at room temperature, and has a simple layered structure as shown in Fig. 11-1a. The carbon-carbon distance in the planar hexagonal sheets is 0.142 nm, compared with the value of 0.135 nm found in benzene. Adjacent layers are separated by 0.335 nm, which is approximately double the van der Waals' radius of carbon. Successive layers are displaced so that the centroids of the hexagons of one layer lie above the carbon atoms in the next layer. An overlap of the valence and conduction energy levels results in graphite being described as a semimetal rather than a semiconductor [46]. It also explains one of the unique features of graphite chemistry in that the lattice can intercalate both electron acceptors and electron donors. The term intercalation describes the reversible insertion of mobile guest species (atoms, molecules, or ions) into a crystalline host lattice that contains an interconnected system of empty lattice sites (□) of appropriate size. Generally the intercalation reaction can be described by the Equation (11.1).

$$x_{\text{Guest}} + \square_x[\text{Host}] \leftrightarrow [\text{Guest}]_x[\text{Host}] \tag{11.1}$$

The first intercalated graphite compound reported in the scientific literature was graphite sulphate which was discovered by Schafhäutl in 1840 [47]. Fredenhagen and Cadenbach in 1926 [48] were the first to report the intercalation of alkali metals in graphite by exposure to the metal vapor at 400 °C. Since then graphite intercalation chemistry has been extensively investigated, and many review articles have appeared through the years [16, 49–52]. Graphite intercalation compounds of potassium can be easily prepared by heating the metal with finely divided graphite in vacuum at 300–400 °C. The resulting intercalation compounds can have the stoichiometric compositions C_8K, $C_{24}K$, $C_{36}K$, $C_{48}K$, and $C_{60}K$. X-ray studies carried out by Schleede and Wellmann [53] on C_8K have revealed that in this first stage compound every interlayer space is filled with a layer of potassium (Fig. 11-1b). Because of this intercalation, the interplanar distance is increased from 0.335 nm in graphite to 0.54 nm. In stages 2–5, intercalation of metal atoms takes place in each second, third, fourth, and fifth interlayer space.

It is well known that C_8K reacts with molecules that do not intercalate (for instance, CO, O_2), leading to high stage $C_{12s}K$ (stage $s > 2$) binary compounds [54, 55]. The alkali

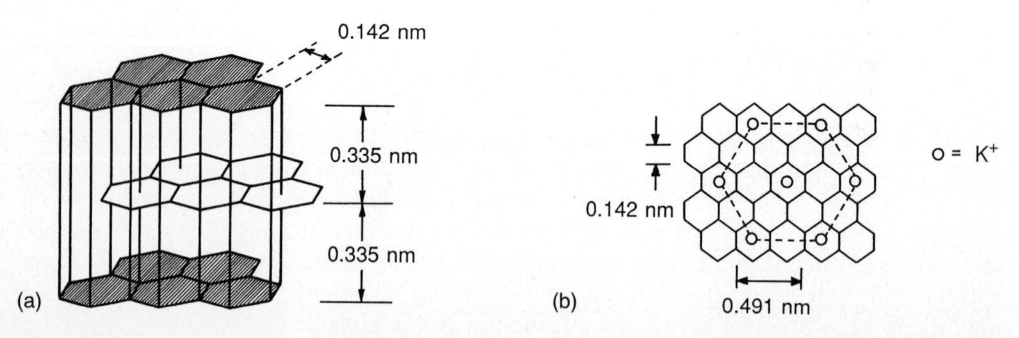

Fig. 11-1. (a) Layer structure of graphite. (b) Schematic diagram of the packing of the potassium ions in the basal planes of C_8K.

metal intercalated graphites of higher stages readily take up polar molecules such as ammonia, tetrahydrofuran, and heterocyclic molecules giving ternary intercalation compounds [56, 57]. In contrast to this feature, the C_8K intercalate is non-sorptive. This behavior can be explained, however, by the fact that in C_8K the potassium layer is packed so densely that the gas molecules are sterically hindered from entering the interplanar space [58].

11.4 The Structures of the Metal–Graphite Combinations

The metal–graphite combinations are prepared by adding anhydrous metal halides MX_n to a stirred suspension of C_8K in anhydrous tetrahydrofuran. Usually the reaction has to be performed under a stream of argon. After typically 30 min at reflux the black suspension is cooled. The following reaction shown in Equation (11.2) occurs.

An in-situ X-ray investigation of this reaction showed that intermediate products such as first-stage graphite intercalation compounds, $K(THF)_{2.5}C_{24}$ and $K(THF)_{1.7}C_{24}$, are formed initially and then transformed to higher stages [30]. Therefore, one must be very careful in the interpretation of the diffraction patterns of the compounds obtained after reaction of C_8K with metal halides in THF or similar solvating agents. The product is the metal–graphite combination accompanied by KX, and sometimes small amounts of complex potassium metal halides have also been found [29]. If the metal halide MX_n exhibits low solubility in Lewis-base solvents (as, for example, with Cu and Ni salts), much longer reaction times are necessary, and THF has sometimes been replaced by diglyme as the preferred solvent, which allows higher reaction temperatures [59].

$$nC_8K + MX_n \xrightarrow{\text{THF}} \text{M–C} + n\text{KX} \tag{11.2}$$

Since C_8K and the reaction products – the activated metals – are extremely sensitive to oxidizing agents, the entire experimental process must be carried out under an inert atmosphere. Thus, special techniques are necessary to introduce the samples into the analytical instrument without deterioration. This is an especially important consideration, because the first X-ray investigations in the seventies were performed on water treated samples, thus leading to oxidized samples. Consequently, erroneous conclusions were drawn about the nature of metal-graphites.

11.4.1 The Graphite Support

If a metal is intercalated into the graphite lattice, in-plane and *c*-axis diffraction patterns obtained with both X-rays (bulk samples) and electrons (thin samples) reveal the type and extent of intercalation. Additionally, microanalysis of thin samples in the AEM gives elemental information about the intercalated species. In an early investigation using X-ray powder diffraction, Braga et al. [24, 25] found diffraction lines that were additional to the lines of the graphite lattice, and were attributed to a first stage structure. The first arguments against the intercalation hypothesis were launched by Schäfer-Stahl [27], who showed in the case of Fe–graphite that an oxidation product of Fe, namely iron oxyhydroxide (FeOOH), yields diffraction lines at the same positions as those attributed to a hypothetical first stage Fe–graphite. Detailed investigations [29] employing a combination of XRD and AEM showed that the metal is not intercalated into the graphite lattice, but unevenly distributed on the surface of the graphite flakes. Information on intercalation, and therefore on staging, would best be obtained by using the (00l) X-ray reflections. However, in no case were any reflections other than the (002) ones, which are characteristic of the graphite lattice, observed. Furthermore, X-ray diffraction work showed that diffraction lines which were found in addition to the reflections due to graphite, KCl, and metal originate from small amounts of complex potassium transition metal salts (e.g. K_2ZnCl_4). These lines could well have been erroneously interpreted as evidence of intercalated metals. Intercalation has never been observed using X-ray diffraction in any metal-salt-reduced C_8K system [29, 59–61] and this was later confirmed by other investigations [30, 39]. In two cases, intercalated transition metals in graphite have been reported [62, 63], but one has to consider that these compounds were prepared in a different way from graphites intercalated with metal halides (like the Graphimets, see Section 5) and that these investigations were performed with the scanning electron microscope.

On the other hand, intercalated graphites have been observed in AEM investigations of metal-halide-reduced C_8K [64]. In these cases the graphite flakes still contained a small amount of potassium, but no other metal. Additionally, the in-plane electron diffraction patterns showed extraneous diffraction spots which could be attributed to $C_{24}K$ and $C_{48}K$. However, it is difficult to explain how the potassium could remain in the graphite host despite the aggressive metal salt reduction; perhaps these compounds correspond to the intermediate reaction products which have been observed in an in situ investigation employing X-ray diffraction [30].

Intercalation reactions are usually reversible [as indicated by Equation (11.1)], and they may also be characterized as topochemical processes, since the structural integrity of the host lattice is formally conserved during the forward and reverse reactions. Therefore, the graphite lattice should not be altered during the reduction of metal salts by C_8K. However, detailed investigations by XRD and HREM showed that the graphite lattice can be disturbed to some extent, depending on the vigor of the reaction. To demonstrate this effect the structure of synthetic graphite was determined before C_8K preparation and after metal salt reduction [61]. The XRD results for synthetic graphite are shown in Fig. 11-2 (top) and exhibit a clear broadening of the (002) reflection obviously associated with the reduction of the metal salt by C_8K. As has been demonstrated by Thomas and Millward [65], lattice fringe imaging by HREM provides a complementary means of examining the *c*-axis structure of graphite on a non-statistical, microscopic basis. Typical lattice images of graphite with the basic layers parallel to the electron beam are shown in Fig. 11-2 (bottom), clearly revealing that the reduction of the metal salt by C_8K introduces structural inhomogeneities in the graphite lattice. The broadening of the (002) reflection in XRD and the irregularities in the HREM-images may be a consequence of the intercalation-deintercalation process, because residues of potassium have also been found in the graphite (Fig. 11-3) [61]. Intercalation of transition metals, as was previously reported [66], could not be confirmed.

11.4.2 Platinum- and Palladium-Graphites

Platinum and palladium supported on graphite have been successfully used as selective catalysts for hydrogenations [67, 68]. Recently [60] it has been shown that C_8K-derived catalysts are of comparable activity and sometimes superior to other supported noble me-

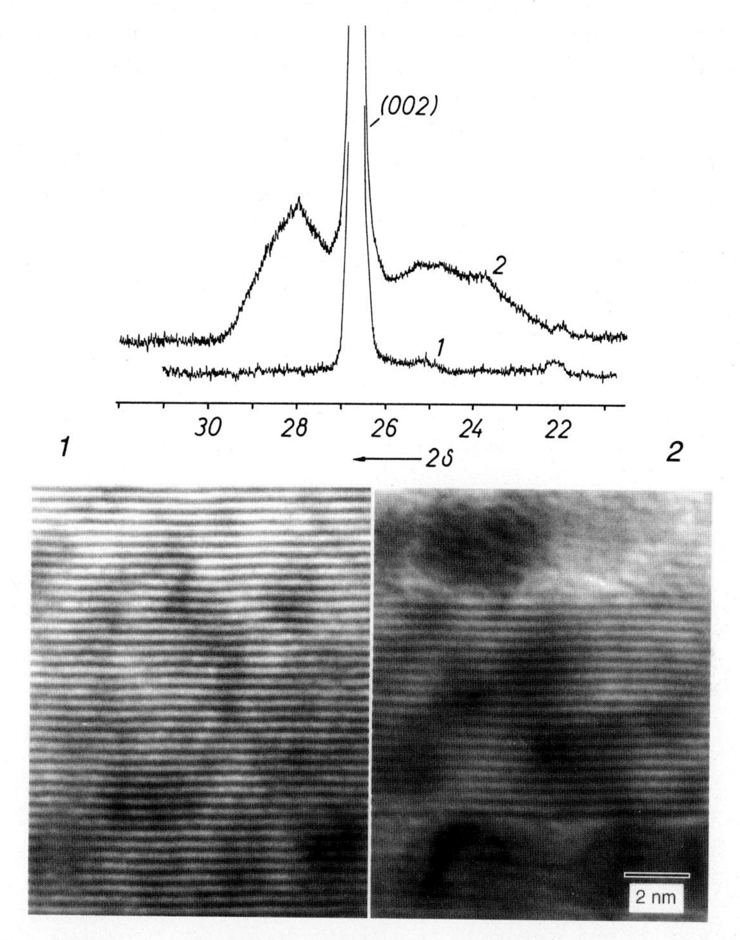

Fig. 11-2. X-Ray diffraction patterns [(002) reflection] (top) and TEM bright field images with lattice fringes (HREM, graphite flakes orientated with their layer planes parallel to the electron beam) for synthetic graphite. In both cases 1 indicates results before intercalation of potassium and 2 indicates results after reduction of a metal salt by C_8K.

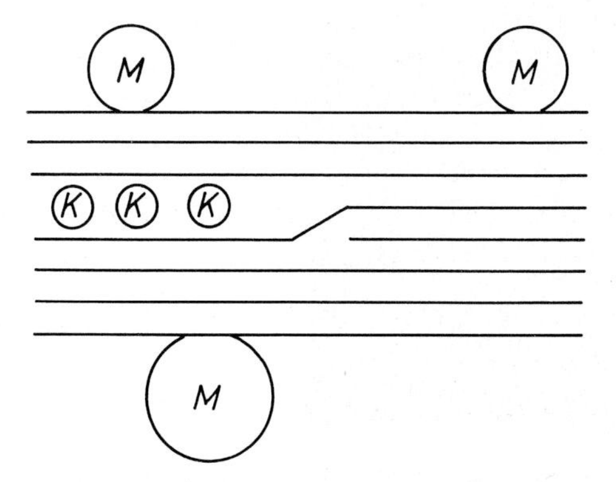

Fig. 11-3. Model of the structure of a metal-graphite combination which has been prepared by reduction of a metal halide by C_8K, with the metal particles dispersed on graphite, and intercalated residues of potassium in the graphite lattice.

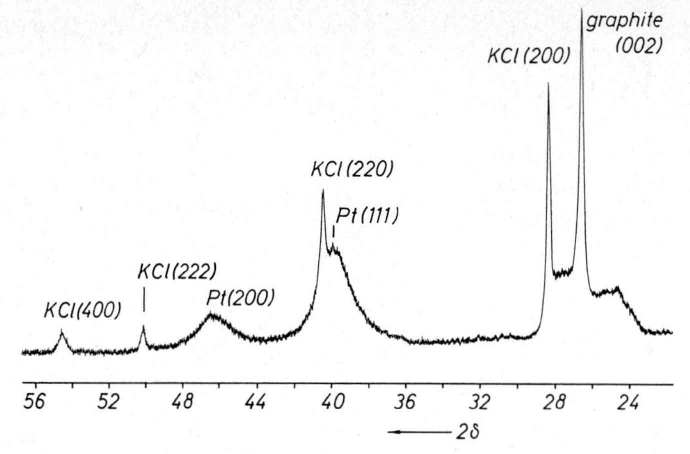

Fig. 11-4. X-Ray diffraction pattern of Pt-graphite prepared from C_8K and $PtCl_2$ in THF; besides the graphite and KCl reflections, broad Pt reflections are visible.

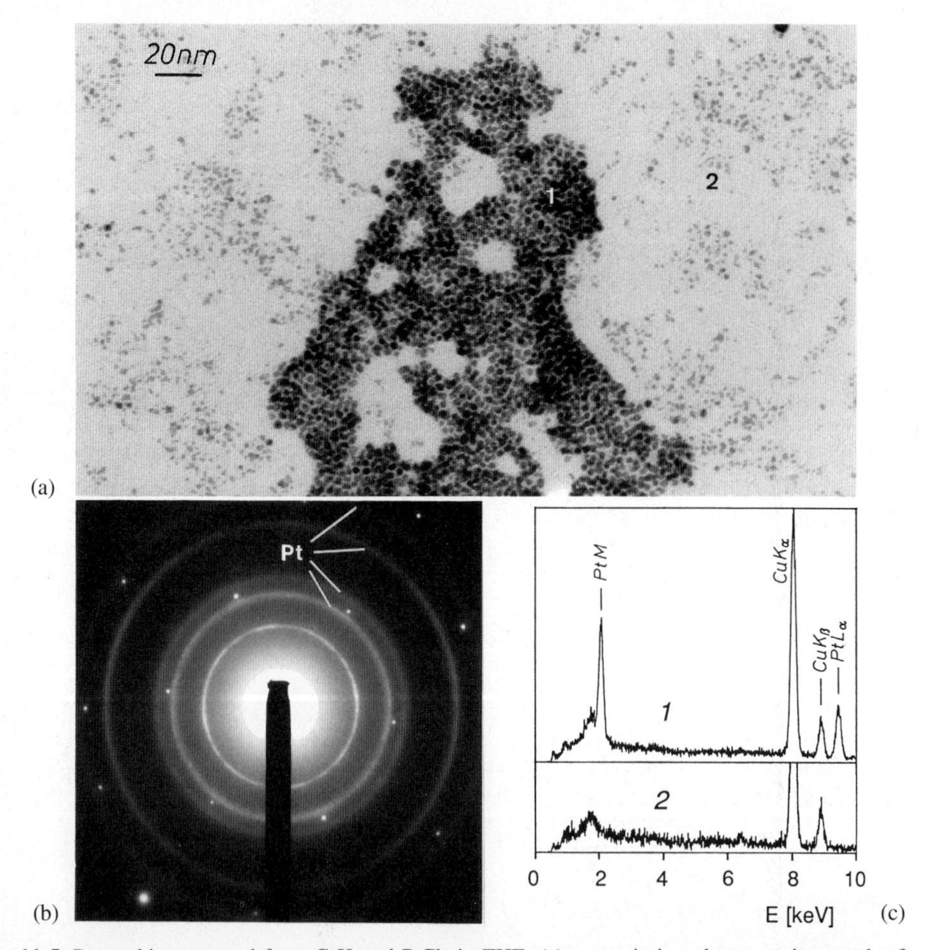

Fig. 11-5. Pt-graphite prepared from C_8K and $PtCl_2$ in THF; (a) transmission electron micrograph of a thin graphite crystal with Pt particles; (b) electron diffraction pattern of specimen region shown in Fig. 11-5a. Bright spots are the (hk0) reflections and rings are due to KCl and Pt particles (as indicated); (c) EDX analyses of (1) graphite covered by Pt and (2) graphite without Pt particles, clearly showing that Pt is not intercalated.

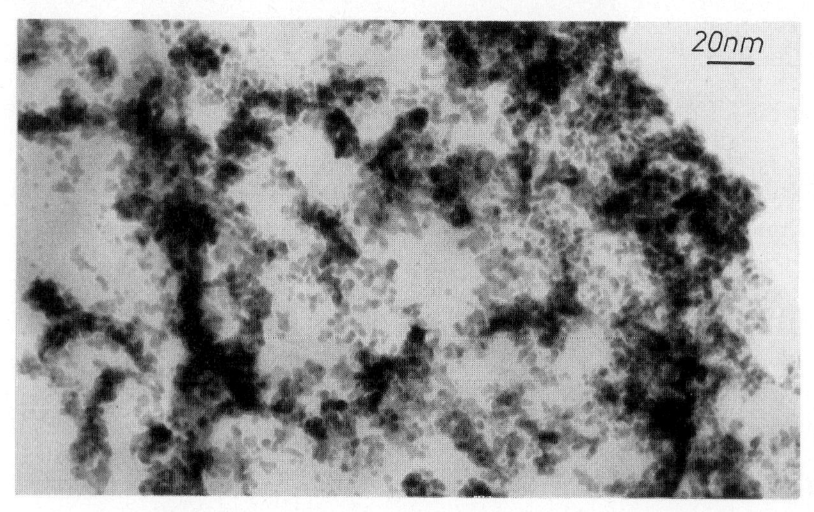

Fig. 11-6. Transmission electron micrograph of Pd-graphite prepared from C_8K and $PdCl_2$ in THF.

tal catalysts [69, 70]. It has long been known that the adsorptive and catalytic properties of metals on supports depend on the particle size and on the mode of catalyst preparation [71]. Therefore, much attention has been paid to the morphology of these catalysts. Since the heavy element Pt gives good contrast relative to graphite in electron microscopy, we will look at this material in more detail than the other metal–graphite combinations, in order to outline the principles for the characterization of such systems.
Fig. 11-4 shows the X-ray diffraction pattern of Pt–graphite prepared from C_8K and $PtCl_2$ in THF, clearly revealing the (002) reflection of graphite, the KCl reflections, and some broadened peaks which can be attributed to Pt particles of very small size. No sign of intercalated Pt is visible. The TEM bright field image of Pt–graphite (Fig. 11-5a) shows uniformly and finely distributed Pt particles on graphite with an average particle size of 3 nm, although smaller particles down to 1 nm are also observed. In the EDX spectra (Fig. 11-5c) taken from regions of covered and uncovered graphite, Pt is only visible in the former, a clear indication against intercalation. This result is supported by electron diffraction (Fig. 11-5b) exhibiting no detectable alterations to the graphite lattice. Besides the (hk0) reflections of the graphite crystal lying on the basal plane, diffuse rings caused by KCl and Pt can be observed. In contrast to other Pt/C catalysts, which sometimes have a cube-octahedral or hexagonal outline [72–74], we have only found spherical particle shapes without any sign of twinning. Similar results have been obtained for Pd–graphite [61] which has been prepared from C_8K and $PdCl_2$ in THF. A typical TEM image (Fig. 11-6) shows small Pd particles on the graphite flake with an average size of about 3 nm. If the catalyst is washed with water to remove the KCl, the Pd particles become coarser and form large aggregates on the graphite crystals [61].

11.4.3 Silver– and Copper–Graphites

Copper plays an important role in the classical Ullmann biaryl synthesis [75], and the highly active Rieke copper has often been successfully used [76, 77]. In contrast to this, Ag–graphite is mainly of interest for a comparative study, and no applications have yet

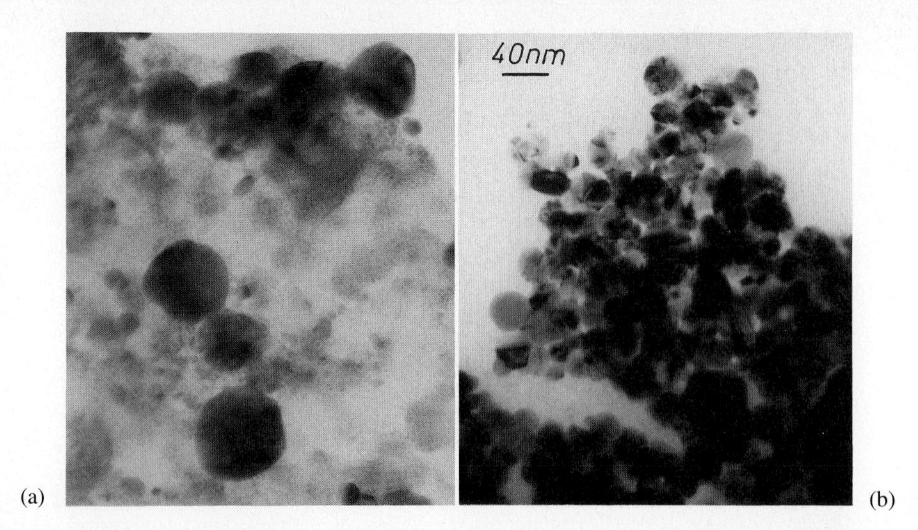

Fig. 11-7. Transmission electron micrographs of (a) Cu-graphite and (b) Ag-graphite, prepared from C_8K with $CuCl_2$ and AgOAc respectively.

been reported in organometallic reactions. However, the morphology of Ag– and Cu–graphites is relevant, because both metals are used to enhance the reactivity of other metal–graphites (Ag for Zn and Cu for Ni, see Sections 10.4.4 and 10.4.5). Cu–graphite can be prepared by reducing $CuCl_2$ with C_8K in diglyme. In view of the high reactivity of the C_8K it is noteworthy that very long reaction times have been observed (up to 17 hours, similar to the Ni–graphites, see Section 10.4.5) [59]. Contrary to first assumptions by Braga et al. [24, 25], intercalation of copper into the graphite could not be observed, but due to the slowness of the reaction (especially in THF at lower temperatures) large amounts of potassium remain intercalated. X-ray diffraction shows lines which are indicative of potassium intercalation after short reaction times [61]. These lines may have been erroneously interpreted as caused by Cu intercalation [24, 25]. Fig. 11-7a shows a typical TEM bright field image of Cu–graphite with both small particles and large Cu-crystals on a graphite flake. Ag–graphite is obtained after reaction of C_8K with AgOAc in THF, and the product has been investigated in the AEM (Fig. 11-7b) [61]. It consists of Ag-particles on graphite and large needle shaped KOAc crystals. The Ag particles show an inhomogeneous surface distribution and tend to form spongelike agglomerations; they are often twinned, which is very seldomly observed with other metal–graphite combinations. In contrast to all other metal–graphite combinations, both Ag– and Cu–graphites have large crystals of particle sizes ranging from 30 to 100 nm.

11.4.4 Zinc–Graphite and Zinc/Metal–Graphites

Finely dispersed zinc plays a central role in the Reformatsky reaction, which consists of the zinc-induced formation of β-hydroxy alkanoates from ethyl haloacetates and aldehydes or ketones [20, 78, 79]. In extending the scope of the Reformatsky reaction, various parameters such as metal activation, solvent and reaction temperature as well as appropriately designed reagents and substrates have been investigated. As a result, it was found that considerable progress could be achieved by steadily increasing the reactivity of the zinc and by combining the zinc with various other metals. The use of highly activated

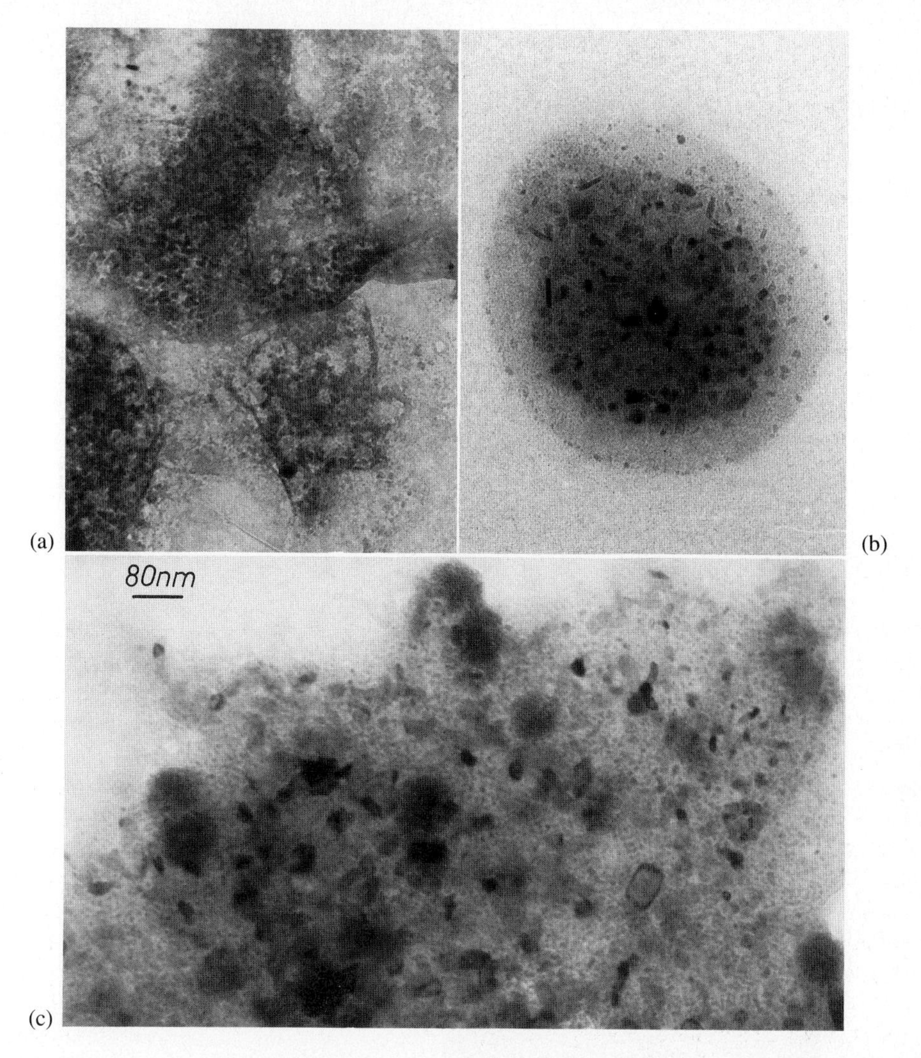

Fig. 11-8. Transmission electron micrographs of Zn–graphite and Rieke zinc; (a) and (b) Zn–graphite prepared from $ZnCl_2$ and C_8K in THF, showing the small Zn particles, (a) Zn and KCl on a graphite crystal; (b) dry residue of the THF solution consisting of KCl and clearly visible Zn crystals; (c) Rieke zinc consisting of KCl and Zn particles on the carbon support film.

zinc and/or the proper selection of the solvent help to suppress various long-known side reactions and to improve selectivities. Detailed investigations by XRD and AEM revealed that the zinc particles in Zn-graphite are dispersed on the graphite support [29]. A typical example is shown in Fig. 11-8, although here the Zn-particles are difficult to distinguish from KCl. Two groups of particles could be detected: small crystals with diameters ranging from 5 to 30 nm (Fig. 11-8a), and very small clusters with a diameter of about 4 nm (Fig. 11-8b), explaining the unprecedented high reactivity. In addition, admixed K_2ZnCl_4 could often be detected, thus complicating the interpretation of the electron and X-ray diffraction patterns [29]. X-ray diffraction showed relatively weak and broadened Zn reflections, thus confirming the small particle size of the metal. If this material is washed with water as recommended in an earlier study [80], the zinc metal is completely oxidized [29].

At this stage it is useful to compare the Zn–graphite with the similar Rieke zinc, resulting from reduction of anhydrous zinc chloride by potassium [81,12]. Although the development of Rieke zinc was a milestone in the history of zinc activation, there are noteworthy shortcomings that can detract from the utility of the Rieke metals [20]. There are only a few investigations concerning the structure of Rieke metals, mainly employing scanning electron microscopy [82]. However, a detailed insight into the morphology is again only possible by using both X-ray diffraction and analytical electron microscopy, and therefore we include here our results obtained on Rieke zinc [61]. Besides KCl and K_2ZnCl_4 peaks, X-ray diffraction showed relatively sharp and intense Zn reflections,

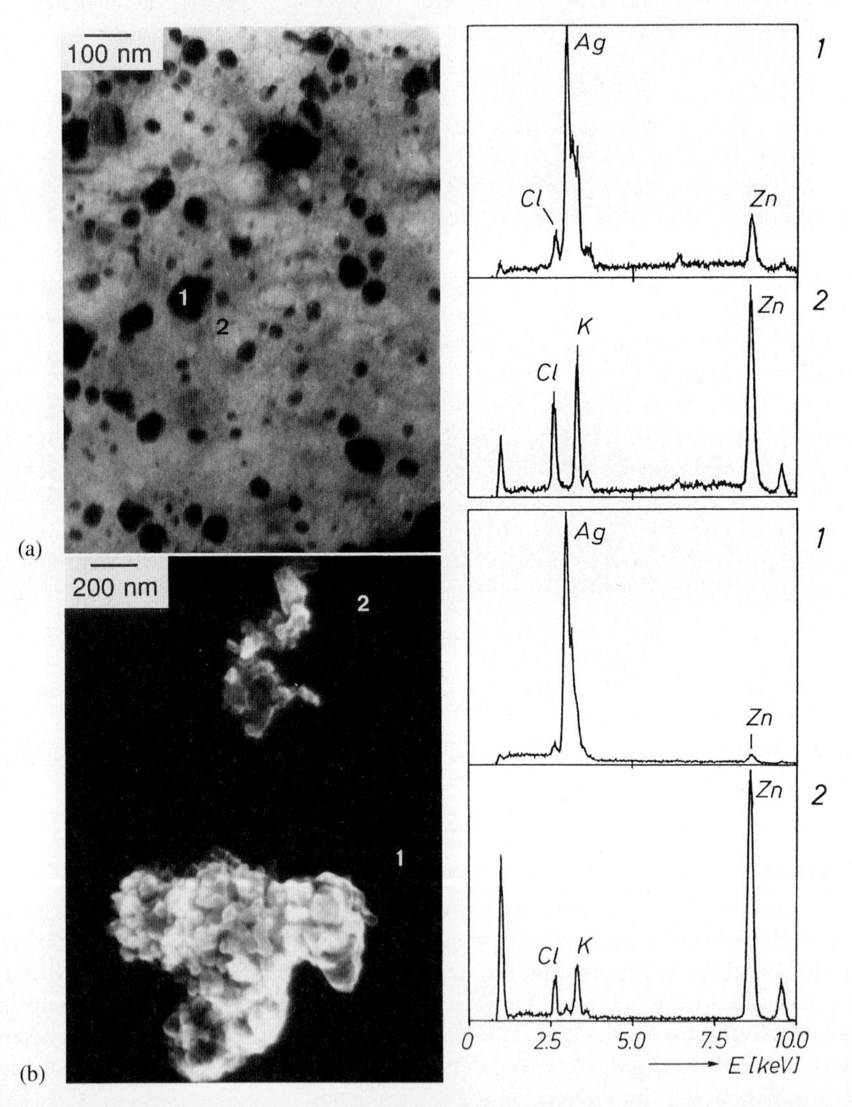

Fig. 11-9. Left: electron micrographs of (a) Zn/Ag–graphite and (b) Rieke Zn/Ag; right: EDX spectra of marked regions 1 and 2 of the two specimens. In (a), TEM image of Zn/Ag–graphite prepared from $ZnCl_2$, AgOAc, and C_8K in THF shows that the Ag and Zn particles are uniformly distributed on graphite; in (b), scanning electron micrograph of Rieke Zn/Ag shows the agglomeration of the metal particles; Ag and Zn are not uniformly mixed.

which are an indication of larger particles than those observed with Zn–graphites. AEM work confirmed these findings (Fig. 11-8c). The sample region shown in this TEM bright field image reveals relatively large Zn particles with sizes ranging from 10 to 100 nm, which are embedded in a matrix of very small Zn clusters and KCl. Zn crystals and small Zn clusters can be found in both reagents, but the Zn–graphite sample shows considerably smaller Zn crystals. This may be one main reason for the higher reactivity of zinc–graphite (besides the uniform distribution of the metal species on the graphite support in this case).

A special type of zinc activation, distinctly improving the classical Reformatsky reaction, was accomplished by using a zinc–silver couple in the conventional manner [83]. If the zinc-graphite combinations are also doped with silver [5], the reactivity is considerably enhanced over all other zinc–metal couples and zinc samples. Zinc/silver–graphite of about 10:1 molar ratio in THF combines several favorable properties. Reactions proceed rapidly at very low temperatures, and troublesome side-reactions are almost completely suppressed [20]. The important question of the origin of the high reactivity can be clearly answered by electron microscopy. Fig.11-9a shows a TEM bright field image of zinc/silver–graphite obtained using equimolar amounts of C_8K and $ZnCl_2$/AgOAc (0.1 molar ratio). EDX spectrometry of the various particles revealed that the dark particles are pure silver embedded in Zn and KCl. These results were confirmed by X-ray diffraction, which showed reflections due to Zn and Ag, whereas no hint of a Zn/Ag alloy could be detected.

If we look at the morphology of a comparable Rieke-zinc/silver couple we do not find such a uniform mixture of the two metal species, as is evident from Fig. 11-9b. The SEM image of the Rieke reagent shows only pure Ag and pure Zn particles attached to large aggregates. Mixtures of Ag and Zn like those found in the Zn/Ag graphite occur only occasionally. Furthermore, in Rieke-Zn/Ag a high proportion of the silver is transformed into AgCl, as was shown by XRD [61], and is therefore ineffective for enhancing the activity of the zinc.

The unprecedented high reactivity of the Zn/Ag–graphite combination may be explained by the intimate mixing of Ag and Zn particles, which is only possible by direct reduction of the metals on the graphite surface and not by the formation of Zn–Ag alloys. We may expect this to result in many active sites where the more noble metal Ag greatly enhances the reactivity of the Zn metal. Similarly, it is possible to enhance the activity of zinc–graphite by doping with Hg or Cu. The Zn/Ag–graphite combination clearly demonstrates the advantage of metal couples which are distributed and immobilized on a support.

11.4.5 Nickel–Graphite and Nickel/Metal–Graphites

Active nickel and nickel/copper reagents have various applications as hydrogenation catalysts [67] and as reagents for organometallic reactions [84]. Besides the Rieke nickel which has often been applied in synthesis [85–87], the Ni–graphite combination offers high reactivity combined with high efficiency and general applicability. Compared with these reagents the related Ni-Graphimet (discussed in Section 10.5) shows lower activity.

The Ni-graphite combination is usually prepared by reducing $NiBr_2$ with C_8K in THF [39] or diglyme [59]. Due to the comparatively low solubilities of nickel and copper salts in Lewis-base solvents, rather low rates of reduction have been found (when compared with the other metal salt reductions). To ensure complete reduction, both nickel and copper salts needed to be treated with C_8K for up to 17 hours in diglyme, as the preferred solvent allowing higher temperatures.

In contrast to the Zn–graphites, the morphology of the Ni-graphites has been characterized in several investigations. Braga et al. reported that the nickel is intercalated into graphite [24, 25]; however, in all subsequent analyses (using XRD [30, 59], SEM [62], AEM [59], and XPS [30]) no sign of metal intercalation has been found (see Section 10.4.1). A typical TEM bright field image of Ni–graphite (Fig. 11-10a) shows uniformly and finely distributed nickel particles on graphite, with an average size of 5 nm and a range of 2–7 nm. In EELS and EDXS analyses taken from regions of uncovered and covered graphite, nickel is only observed in the latter, a clear indication against intercalation of nickel [59]. However, small amounts of potassium can be found in the graphite crystals – obviously residues of the intercalation compound C_8K. Electron diffraction and X-ray diffraction work confirmed these findings, because only reflections due to graphite, potassium bromide and nickel could be found.

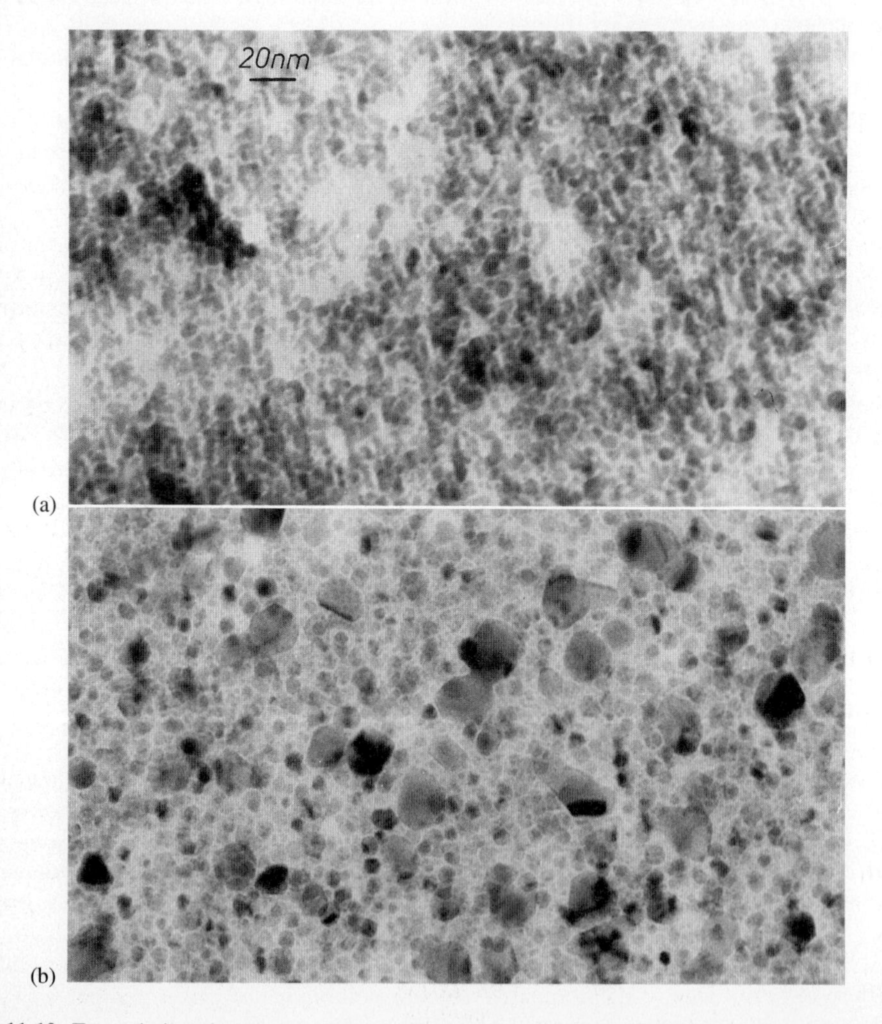

Fig. 11-10. Transmission electron micrographs of (a) Ni–graphite and (b) Ni/Cu–graphite, prepared from $NiBr_2$, $CuCl_2$, and C_8K in diglyme.

If Ni–graphite is washed with water, the fine Ni particles are oxidized to $Ni(OH)_2$ and NiO while the bigger ones are covered by a 1–2 nm thin Ni-hydroxide layer [59], thus reducing the catalytic activity, but improving the selectivity for particular hydrogenations [67]. The washed Ni–graphite can be stored over months without further loss of activity, and it exhibits a morphology quite similar to that of the parent Ni-Graphimet.

Based on the pronounced activation of zinc by the Zn/Ag graphite combination (see Section 10.4.4), a Ni/Cu–graphite reagent has also been proposed [84] and recently characterized [59]. A mixture of $NiBr_2$ and $CuCl_2$ (molar ratio 8:1) was reduced by C_8K in diglyme forming a complex system which essentially consisted of metallic nickel, α-CuBr, potassium halide, graphite, and some unreacted C_8K. The TEM image (Fig. 11-10b) shows nickel particles and CuBr crystals distributed on graphite which have been identified by EDX spectroscopy. The reduction of mixtures of nickel and copper salts, unlike that of zinc and silver, does not lead to a nickel–copper couple with enhanced reactivity. Alloy formation between Ni and Cu could not be observed. A possible mechanism for the coupling process described [84] is a Cu^{I}-salt-assisted, nickel induced Wurtz-type reaction.

11.4.6 Titanium–Graphite Combinations

The Ti–graphite combination is a highly reactive and universally applicable McMurry reagent [88–91], which efficiently couples aryl and alkyl carbonyl and dicarbonyl compounds to alkenes and cycloalkenes, respectively. The McMurry reaction was first favored by the development of Rieke Ti [11], and numerous recipes for the preparation of Rieke Ti have been proposed; e.g. reduction of $TiCl_3$ by Li [92], Na [93], K [94], Mg [90], a Zn/Cu couple [94], or $LiAlH_4$ [91] in THF with various additives such as pyridine or triethylamine. Likewise, the McMurry reaction can be promoted by the Ti-graphite combinations, which are prepared either from titanium trichloride and lamellar potassium graphite (C_8K) [22] or from titanium tetrachloride and C_8K [21].

It is well known that commercial titanium powder cannot be used for the McMurry reaction, because the Ti particles are covered by a very thin oxide layer due to the high affinity of titanium for oxygen. Furthermore the activity of the titanium largely depends on the preparation conditions. Although much effort has been put into the characterization of these reagents, there is as yet no clear understanding of their morphology [23]. The characterization of Rieke Ti reagents showed that the active species is not necessarily titanium in the metallic state [95], and that Ti compounds of low valency could also be active [22]. For example, it is already known that $TiCl_4$ cannot be reduced to Ti metal by common reducting agents [22, 96, 97], but nevertheless the resulting reagent can be used for carbonyl coupling reactions.

Recent AEM investigations of Ti–graphite prepared from $TiCl_3$ and C_8K [61] showed that this combination consists of very small TiO particles (nanometer-sized) which are embedded in potassium chloride. However, the TiO particles are very difficult to distinguish from the KCl, because their scattering power for the transmitted electrons is quite similar. The existence of TiO was proved by EELS spectrometry, and a typical spectrum is shown in Fig. 11-11. It is possible to see the Ti L_{23}- and O K-edges of the Ti-oxide phase and additionally the Cl L_{23}- and K L_{23}-edges of potassium chloride. The atomic ratio of Ti and O has been derived from this spectrum using previously described methods [102] and corresponds to that of TiO. To confirm that the near-edge fine structure of the O K-edge is characteristic of TiO, we have subtracted the background beyond the edge.

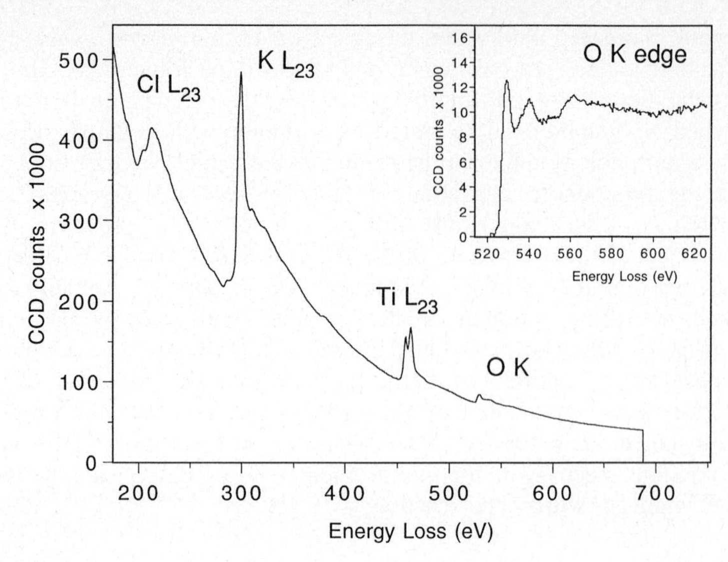

Fig. 11-11. EELS spectrum of Ti-graphite prepared from $TiCl_3$ and C_8K in THF; TiO particles are embedded in KCl. The inset shows the near-edge fine structure of the K-edge of oxygen which is similar to that of a TiO standard.

In the inset of Fig. 11-11 the O K-edge is shown, and it exhibits all the near-edge fine structure typical of TiO.

Summarizing these results, one can say that McMurry reagents prepared by different routes exhibit different morphologies and consequently different chemical behavior.

11.5 The Structure of Graphimets

Although graphite cannot be directly intercalated with transition metals, there have been other attempts to produce lamellar compounds by the reduction of intercalated metal salts. A wide range of "metallic intercalates" is commercially available from the Ventron Corporation under the general name "Graphimets" [98]. These compounds are prepared by the reduction of a graphite-metal chloride intercalation compound with lithium biphenyl at −50 °C under a helium atmosphere. After washing with tetrahydrofuran, acetone, and water, the Graphimets are vacuum dried at 140 °C. Despite their numerous applications for catalytic transformations [16, 99] contradictory results have been reported with respect to their structures [60, 66, 100]. In view of the exceptional catalytic activity of these materials, it is of considerable interest to investigate their structure and to compare it with the C_8K-derived metal-graphite combinations.

Electron microscopic investigations [59, 60, 66] have shown that the reduction of the metal-halide intercalated graphite has primarily resulted in the formation of small metal and metal oxide particles ranging from 1 to 10 nm in diameter. The morphology of Pt-Graphimet and Ni-Graphimet is shown by the TEM bright field images in Fig. 11-12. Whereas Pt-Graphimet consists of uniformly and finely distributed platinum particles of 1–7 nm diameter on graphite (Fig. 11-12a) [60], Ni-Graphimet shows a less homogeneous distribution of particles with diameters ranging from 1 to 20 nm [Fig. 11-12(b)]. In the Ni, Co, and Fe samples two kinds of particles could be identified: some larger metallic

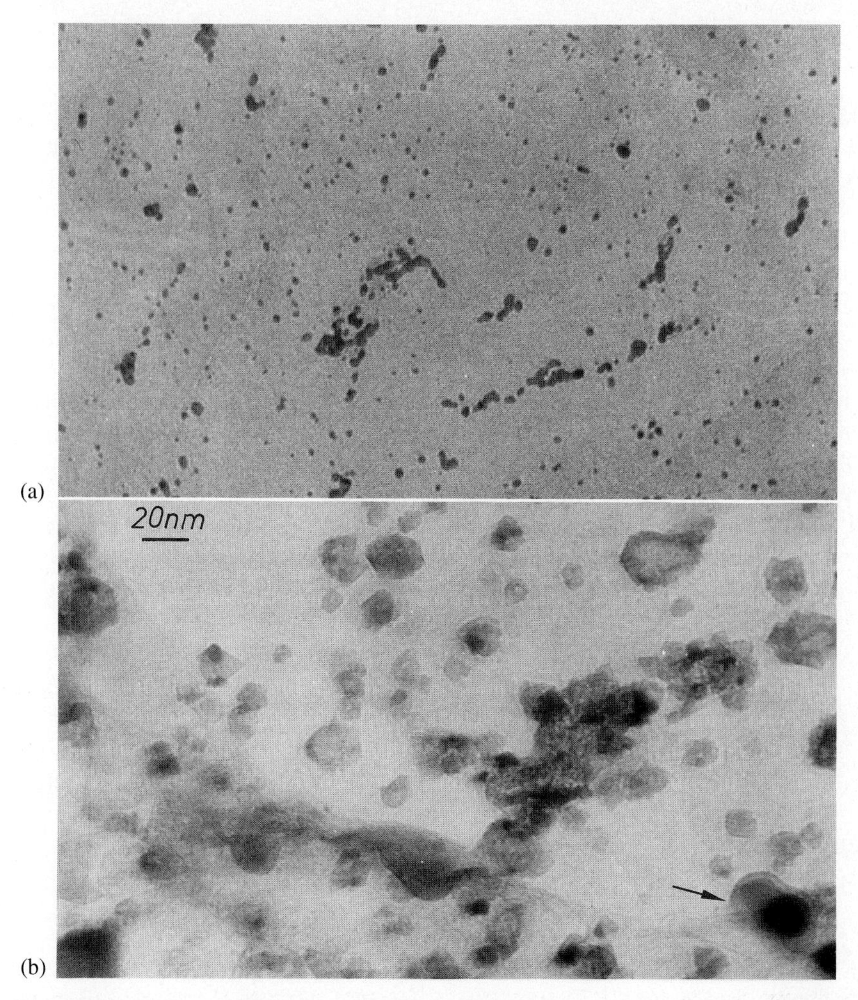

Fig. 11-12. TEM images of "Graphimet" samples; (a) "Pt-Graphimet" with Pt particles on a graphite flake, 1–7 nm diameter; (b) "Ni-Graphimet" with larger Ni particles covered by Ni-oxide (see arrow) and with small fine-grained NiO particles which have agglomerated.

particles that are uniformly covered by oxide films (see also Fig. 11-12b), and some very-fine-grained material which has been found to consist of oxides [59, 100]. This is obviously caused by oxidation which occurs when the respective Graphimet powders are exposed to water and air after vacuum drying. It should be noted that the oxides observed are in keeping with earlier observations on various chloride intercalates [101]. Obviously most of the metal is distributed on the graphite surface and the catalytic activity has to be associated with the surface of the metal particles (or oxide particles). It should however, be mentioned that there have been arguments for residual intercalation of transition metals in Graphimets. Lattice images of the graphite perpendicular to the base plane revealed that the lattice planes are sometimes wider apart than in pure graphite [66] (an effect that may be similar to the occurrence of potassium in reduced graphites, see Section 10.4.1). However, it is obvious that only a small part of the metal remains intercalated, most of the metal species being distributed on the surface of the graphite support.

11.6 Conclusions and Outlook

The metal-graphite combinations which are generally prepared by the reduction of metal halides by C_8K in THF solution essentially consist of fine metal particles more or less evenly distributed on a graphite support. There has been no indication of active metals intercalated in the graphite lattice, but some residual potassium has been found intercalated, even after reduction and washing with water. In this review we have also compared metal–graphite combinations with the well-known Rieke metals and Graphimets, both of which have been found to exhibit a similar morphology. However, sometimes (e.g. in the Zn/Ag system) differences occur, explaining the unprecedented activity of the metal-graphite combinations.

We have shown that a detailed knowledge of the structure and morphology of heterogeneous catalysts and metal-graphite systems is necessary for further development and optimization of these systems in synthetic applications. However, techniques of investigation are still under development, for example the various scanning probe microscopies [43] and energy-selective transmission electron microscopy [103, 104]. Particularly from the latter we can expect important advances in the characterization of these materials, since elemental mapping on a nanometer scale is now possible almost on a routine basis [105].

Acknowledgements. I would like to thank Dr. Alois Fürstner for providing the samples and for numerous stimulating discussions.

References

[1] A. Mendel, *J. Organomet. Chem.* **1966**, *6,* 97.
[2] G. Atnes, G. P. Chiusoli, A. Marracchini, *J. Organomet. Chem.* **1973,** *49,* 239.
[3] A. A. P. Schreibmann, *Tetrahedron Lett.* **1970**, 4271.
[4] C. S. Chao, C. H. Cheng, C. T. Chang, *J. Org. Chem.* **1983**, *48,* 4904.
[5] R. Csuk, A. Fürstner, H. Weidmann, *J. Chem. Soc., Chem. Commun.* **1986**, 775.
[6] E. Santaniello, A. Manzocchi, *Synthesis* **1977**, 698.
[7] R. D. Clark, C. H. Heathcock, *J. Org. Chem.* **1973**, *38,* 3658.
[8] P. Boudjouk, *Nachr. Chem. Tech.* **1983**, *31,* 798.
[9] K. S. Suslick, S. J. Doktycs, *J. Am. Chem. Soc.* **1989**, *111,* 2342.
[10] K. J. Klabunde, *Acc. Chem. Res.* **1975**, *8,* 393.
[11] R. D. Rieke, *Top. Curr. Chem.* **1975**, *59,* 1.
[12] R. D. Rieke, *Acc. Chem. Res.* **1977**, *10,* 301.
[13] D. Savoia, E. Tagliavini, C. Trombini, A. Umani-Ronchi, *J. Org. Chem.* **1980**, *45,* 3227.
[14] H. Hart, B. Chen, C. Peng, *Tetrahedron Lett.* **1977**, 3121.
[15] A. B. McEven, M. J. Guttieri, W. F. Maier, R. M. Laine, Y. Shvo, *J. Org. Chem.* **1983**, *48,* 4436.
[16] R. Csuk, B. I. Glänzer, A. Fürstner, *Adv. Organomet. Chem.* **1988**, *28,* 85.
[17] R. T. Arnold, S. T. Kulenovic, *Synth Commun.* **1977**, *7,* 223.
[18] R. D. Rieke, P. T. J. Li, T. P. Burns, S. T. Uhm, *J. Org. Chem.* **1981**, *46,* 4323.
[19] D. Savoia, C. Trombini, A. Umani-Ronchi, *Pure Appl. Chem.* **1985**, *57,* 1887.
[20] A. Fürstner, *Synthesis* **1989**, *8,* 571.
[21] A. Fürstner, R. Csuk, C. Rohrer, H. Weidmann, *J. Chem. Soc. Perkin Trans.* **1988,** *1,* 1729.
[22] A. Fürstner, H. Weidmann, *Synthesis* **1987**, *12,* 1071.
[23] A. Fürstner, *Angew. Chem.* **1993**, *105,* 171; *Angew. Chem. Int. Ed. Engl.* **1993**, *32,* 164.
[24] D. Braga, A. Ripamonti, D. Savoia, C. Trombini, A. Umani-Ronchi, *J. Chem. Soc. Chem. Commun.* **1978**, 927.

[25] D. Braga, A. Ripamonti, D. Savoia, C. Trombini, A. Umani-Ronchi, *J. Chem. Soc. Dalton Trans.* **1979**, 2026.
[26] M. E. Volpin, Y. N. Novikov, N. D. Lapkina, V. I. Kasatochkin, Y. T. Struchkov, M. E. Kazakov, R. A. Stulkan, V. A. Povitskij, Y. S. Karimov, V. A. Zvarikina, *J. Am. Chem. Soc.* **1975** *97,* 3366.
[27] H. Schäfer-Stahl, *J. Chem. Soc. Dalton Trans.* **1981**, 328.
[28] D. Braga, A. Ripamonti, D. Savoia, C. Trombini, A. Umani-Ronchi, *J. Chem. Soc. Dalton Trans.* **1981**, 329.
[29] A. Fürstner, F. Hofer, H. Weidmann, *J. Chem. Soc. Dalton Trans.* **1988**, 2023.
[30] A. Messaoudi, F. Beguin, *J. Mater. Res.* **1992**, *7,* 418.
[31] J. W. Niemantsverdriet, *Spectroscopy in Catalysis,* VCH, Weinheim, **1993**.
[32] H. Krischner, *Einführung in die Röntgenfeinstrukturanalyse,* 2nd ed., Vieweg, Braunschweig, **1980**.
[33] D. C. Koningsberger, R. Prins, *X-ray absorption. Principles, Applications, Techniques of EXAFS, SEXAFS and XANES,* Wiley, New York, **1988**.
[34] M. S. Dresselhaus, J. S. Speck, in *Intercalation in Layered Materials*, (ed.: M. S. Dresselhaus), Plenum, New York, **1986**, p. 213.
[35] A. K. Datye, D. J. Smith, *Catalysis Rev.-Sci. Eng.* **1992**, *34,* 129.
[36] L. Reimer, *Scanning Electron Microscopy,* Springer, Berlin, **1986**.
[37] L. Reimer, *Transmission Electron Microscopy,* Springer, Berlin, **1984**.
[38] G. Dresselhaus, M. Lagues, in *Intercalation in Layered Materials,* (ed.: M. S. Dresselhaus), Plenum, New York, **1986**, p.271.
[39] R. Erre, A. Messaoudi, F. Beguin, *Synthetic Metals* **1988**, *23,* 493.
[40] P. K. Glosh, *Introduction to Photoelectron Spectroscopy,* Wiley, New York, **1983**.
[41] M. Grasserbauer, H. J. Dudek, M. F. Ebel, *Angewandte Oberflächenanalyse,* Springer, Berlin, **1986**.
[42] L. A. Grunes, I. P. Gates, J. J. Ritsko, E. J. Mele, D. P. DiVincenzo, M. E. Preil, J. E. Fischer, *Phys. Rev.* **1983**, *B28,* 6681.
[43] G. Binning, H. Rohrer, *Helv. Phys. Acta* **1982**, *55,* 726.
[44] G. Binning, C. F. Quate, C. Gerber, *Phys. Rev. Lett.* **1986**, *56,* 930.
[45] S. P. Kelly, C. M. Lieber, *J. Phys. Chem.* **1989**, *93,* 5983.
[46] D. E. Soule, J. W. McClure, L. B. Smith, *Phys. Rev.* **1964**, *134A,* 453.
[47] C. Schafhäutl, *J. Prakt. Chem.* **1840**, *3,* 129.
[48] K. Fredenhagen, G. Cadenbach, *Z. Anorgan. Allg. Chem.* **1926**, *158,* 249.
[49] G. R. Henning, *Progr. Inorg. Chem.* **1959**, *1,* 125.
[50] W. Rüdorff, E. Stumpp, W. Spriessler, F. W. Siecke, *Angew. Chem.* **1963**, *75,* 130.
[51] H. Selig, L. B. Ebert, *Adv. Inorg. Radiochem.* **1980**, *3,* 281.
[52] M. S. Dresselhaus (ed.), *Intercalation in Layered Materials*, Plenum, New York, **1986**.
[53] A. Schleede, M. Wellmann, *Z. Phys. Chem.* **1932**, *B18,* 1.
[54] N. Daumas, A. Herold, *C.R. Acad. Sci. Paris* **1969**, *268,* 373.
[55] N. Akuzawa, T. Fujizawa, T. Amemiya, *Synthetic Metals* **1983**, *7,* 57.
[56] F. Beguin, R. Setton, *Carbon* **1975**,*13,* 293.
[57] F. Beguin, R. Setton, A. Hamwi, P. Touzain, *Mater. Sci. Eng.* **1979**, *40,* 167.
[58] M. A. M. Boersma, in *Advanced Materials in Catalysis,* (eds.: J. J. Burton, R. L. Garten), Academic Press, New York, **1977**, p. 67.
[59] A. Fürstner, F. Hofer, H. Weidmann, *Carbon* **1991**, *29,* 915.
[60] A. Fürstner, F. Hofer, H. Weidmann, *J. Catalysis* **1989**, *118,* 502.
[61] F. Hofer, A. Fürstner, to be published.
[62] M. Ohira, A. Messaoudi, M. Inagaki, *Carbon* **1991**, *29,* 1233.
[63] C. Herold, J. F. Mareche, G. Furdin, *Microsc. Microanal. Microstruct.* **1991**, *2,* 589.
[64] F. Hofer, unpublished results, **1994**.
[65] J. M. Thomas, G. R. Millward, N. C. Davies, E. L. Evans, *J. Chem. Soc. Dalton Trans.* **1976**, 2443.
[66] D. J. Smith, R. M. Fisher, L. A. Freeman, *J. Catalysis* **1981**, *72,* 51.
[67] D. Savoia, E. Tagliavini, C. Trombini, A. Umani-Ronchi, *J. Org. Chem.* **1981**, *46,* 5340.
[68] D. Savoia, C. Trombini, A. Umani-Ronchi, G. Verdardo, *J. Chem. Soc. Chem. Commun.* **1981**, 540.
[69] P. N. Rylander, *Hydrogenation Methods,* Academic Press, New York, **1985**.
[70] J. Lehmann, in *Methoden Org. Chem. (Houben-Weyl), Vol. 4/1c,* (eds.: E. Müller, O. Bayer), Thieme, Stuttgart, **1980**, p. 520.
[71] M. Boudart, A. W. Aldag. L. D. Ptak, J. E. Benson, *J. Catalysis* **1968**, *11,* 35.
[72] M. Chen, L. D. Schmidt, *J. Catalysis* **1978**, *55,* 348.
[73] M. Jose-Yacaman, J. E. Dominguez, *J. Catalysis* **1980**, *64,* 213.
[74] M. Jose-Yacaman, M. Avalos-Borja, *Catal. Rev.-Sci. Eng.* **1992**, *34,* 55.
[75] P. E. Fanta, *Chem. Rev.* **1964**, *64,* 613.

[76] R. D. Rieke, L. D. Rhyne, *J. Org. Chem.* **1979**, *44*, 3445.
[77] P. Boudjouk, D. P. Thompson, W. H. Ohrbom, B. H. Han, *Organometallics* **1986**, *5*, 1257.
[78] S. Reformatsky, *Ber. Dtsch. Chem. Ges.* **1887**, *20*, 1210.
[79] K. Nützel, in: *Methoden Org. Chem. (Houben-Weyl), Vol. XIII/2a,* Thieme, Stuttgart, **1973**, p. 805.
[80] G. P. Boldrini, D. Savoia, E. Tagliavini, C. Trombini, A. Umani-Ronchi, *J. Org. Chem.* **1983**, *48* 4108.
[81] R. D. Rieke, S. J. Uhm, *Synthesis,* **1975**, 452.
[82] Chao Li-Chung, R. D. Rieke, *J. Org. Chem.* **1975**, *40*, 2253.
[83] K. Maruoka, S. Hashimoto, Y. Kitagawa, H. Yamamoto, H. Nozaki, *J. Am. Chem. Soc.* **1977** *99*, 7705.
[84] P. Marceau, F. Beguin, G. Guillaumet, *J. Organomet. Chem.* **1988**, *342*, 137.
[85] H. Matsumoto, S. I. Inaba, R. D. Rieke, *J. Org. Chem.* **1983**, *48*, 840.
[86] S. I. Inaba, R. D. Rieke, *Synthesis* **1984**, *842.*
[87] S. I. Inaba, R. D. Rieke, *J. Org. Chem.* **1985**, *50*, 1373.
[88] J. E. McMurry, *Acc. Chem. Res.* **1983**, *16*, 405.
[89] T. Mukaiyama, T. Sato, J. Hanna, *Chem. Lett.* **1973**, 1041.
[90] S. Tyrlik, I. Wolochowicz, *Bull. Soc. Chim. France* **1973**, 2147.
[91] J. E. McMurry, M. P. Fleming, *J. Am. Chem. Soc.* **1974**, *96*, 4708.
[92] J. E. Pauw, A. C. Weedon, *Tetrahedron Lett.* **1982**, 5485.
[93] R. Dams, M. Malinowski, I. Westdorp, H. Y. Geise, *J. Org. Chem.* **1982**, *47*, 248.
[94] J. E. McMurry, M. P. Fleming, K. L. Kees, L. R. Krepski, *J. Org. Chem.* **1978**, *43*, 3255.
[95] a) H. Idriss, K. Pierce, M. A. Barteau, *J. Am. Chem. Soc.* **1991**, *113*, 715; b) L. E. Aleandri, S. Becke, B. Bogdanovic, D. J. Jones, J. Roziere, *J. Organomet. Chem.* **1994**, *472*, 97; c) L. E. Aleandri, B. Bogdanovic, A. Gaidies, D. J. Jones, S. Liao, A. Michalowicz, J. Roziere, A. Schott, *ibid.* **1993**, *459*, 87.
[96] D. Lenoir, *Synthesis* **1977**, 553.
[97] E. J. Corey, R. L. Danheiser, S. Chandrasekaran, *J. Org. Chem.* **1976**, *41*, 260.
[98] J. M. Lalancette, U.S. Patent No. 3,847,963, dated November 12, 1974.
[99] K. Otto, M. Shelef, *Carbon* **1977**, *15*, 117.
[100] G. Sirokman, A. Mastalir, A. Molnar, M. Bartok, Z. Schay, L. Guczi, *Carbon* **1990**, *28*, 35.
[101] E. L. Evans, J. M. Thomas, *J. Solid State Chem.* **1975**, *14*, 99.
[102] F. Hofer, *Microsc. Microanal. Microstruct.* **1991**, *2*, 215.
[103] D. Krahl, *Mat.-wiss. Werkstofftech.* **1990**, *21*, 84.
[104] A. J. Gubbens, O. L. Krivanek, *Ultramicroscopy* **1993**, *51*, 146.
[105] F. Hofer, P. Warbichler, *Ultramicroscopy* **1995**, in press.

Experimental Procedures

alkylation of α,β-unsaturated ketones, one-pot double 82
allenylmagnesium chloride 334
allylation of a carbonyl compound 71
2-amino alcohol 120
1,4-anhydro-3-*O*-(*tert*-butyldiphenylsilyl)-2-deoxy-5,6-*O*-isopropylidene-D-arabino-hex-1-enitol 421

Barbier reaction with 1-bromo-4-chlorobutane 184
barium, reactive (Ba*) 4, 65
– homocoupling reactions of allylic halides using 77
barium reagent
– allylation of a carbonyl compound 71
– carboxylation of 72
– protonation of 65
– silylation of 66
barium, Rieke 4, 65
in-bicyclo[4.4.4]-1-tetradecene 105
(2*R*,4*S*)-bis(benzyloxy)hex-5-enal 421

calcium, Rieke 4
carboxylation of an allylic barium reagent 72
(+)-casbene 101
C_8K *see* potassium–graphite laminate
conjugate addition of 6-bromo-1-hexene to 1-buten-3-one 184
crassin intermediates 113
cyclohexylidenecyclohexane 93
cyclotetradecene 100

deoxyiodo sugar, ring-opening 421
dimerization of Boc-L-phenylalaninal 116
(*E*)-4,8-dimethyl-3,7-nonadienoic acid 72
(*E*)-4,8-dimethyl-1-phenyl-3,7-nonadien-1-ol 71
1,2-diphenyl(2,26)[36]paracyclophane 420

ethylcadmium iodide, nonsolvated 242

furanoid glycal 421
2-(2′-furanyl)-3-phenylindole 422

geranyl barium reagent, protonation of 65
graphite *see also* potassium–graphite laminate
graphite, titanium on 123

“high surface area sodium” 420
homocoupling reactions of allylic halides using reactive barium 77
hydrogenation of 4,4-dimethylcyclopent-2-en-1-one 184

lithium diisopropylamide 184

magnesium
– manipulation of, under inert atmosphere 243
– Rieke 4
– solvated metal atom dispersed (SMAD Mg) 243
magnesium–anthracene · 3 THF 334
magnesium–platinum intermetallic t-$MgPtC_xH_y$ 334
McMurry ketoester coupling 122
McMurry-type reaction
– of acylamidocarbonyl compounds 124, 422
– of acyloxycarbonyl compounds 124
trans-2-(3-methyl-2-butenyl)-3-(2-propenyl)cyclopentanone 82
2-methylcyclotridecanone 122
methyltin iodides [CH_3SnI_3, $(CH_3)_2SnI_2$, $(CH_3)_3SnI$] 242

nickel carbide (Ni_3C) from Ni-FOP 242
nickel–fragments-of-pentane (Ni-FOP) 241
niobium trichloride/DME [$NbCl_3$(DME)] 119
nonsolvated ethylcadmium iodide 242

Boc-L-phenylalaninal, dimerization of 116
pinacol coupling 111–113, 116
platinum–iron intermetallic 335
potassium–graphite laminate C_8K
– large scale preparation of 421
– synthesis of a furanoid glycal 421

Rieke metals
– barium 4, 65
– calcium 4
– magnesium 4
– strontium 4
– zinc 24

sarcophytol B 112
silylation of (*E*)-2-decenylbarium chloride 66
sodium, “high surface area” 420

solvated metal atom dispersed magnesium (SMAD Mg) 243
squalene 77
strontium, Rieke 4

tetraisopropylethylene 94
titanium–alumina-induced macrocyclization of a diketone 420
titanium–graphite-induced cyclization of oxoamides 124, 422
titanium Grignard reagent $[Ti(MgCl)_2(THF)_x]_2$ 335
titanium-mediated dicarbonyl coupling, Clive procedure 104
titanium on graphite 104, 123, 124, 422
titanium trichloride/DME $[TiCl_3(DME)_{1.5}]$ 92

zinc–copper couple 91
zinc(silver)–graphite-induced reductive ring-opening 421
zinc, Rieke 24

Subject Index

acenaphtylene, lithium radical anion of 142
acetals
– halo-, ring closure of 200
– spiro-, scission of 409f
acid chlorides, reaction with
– benzylnickel reagents 49
– dien–magnesium complexes 8
– organocopper reagents 33, 36, 38, 419
– organozinc reagents 25ff, 156, 158, 215f, 224f, 389, 391
– zinc enolates 402
acoustic waves 133f, 182
acylamidocarbonyl compounds, reductive cyclization 124, 389, 412ff, 422
acyloin condensation 148f, 151
acyloxycarbonyl compounds, reductive cyclization 123f, 412
acylsilanes
– in McMurry couplings 108, 389
– reaction with allylic barium reagents 68
adatoms 284
1,2-addition 38, 68, 169, 217
1,4-additions
– of allylic barium reagents 77ff
– of organocopper reagents 36, 38, 40, 45ff
– of organozinc compounds 25f, 156, 184, 214f, 216ff, 222f
– under Barbier conditions 174f, 184
agglomeration 137, 280, 285, 352, 357, 368, 383
aggregation 339, 383, 401
aldehydes
– chiral, diastereoselective additions 217
– enantioselective additions to 226ff
aldol condensation 80
aldols 117, 209, 221, 227f, 409f
aldononitriles, formation of 399, 407
alkali amides, preparation of 145, 154, 184
alkalides 1, 420
alkali metals
– effect of ultrasound on 136, 139
– ketyls 147
alkenes
– bridgehead 104ff, 404
– by desulfuration of sulfones 400
– by reductive carbonyl coupling, *see* McMurry reaction
– complexation with Ag(1) 106
– cyclopropanation of 166ff
– from vic-dibromides 397, 418
– from vic-dinitriles 152
– hydrogenations under sonication 181, 184
– hydrogenations with SMAD metal powders 255ff
– hydrosilylations 182
– isomerizations 122, 384ff, 400
– nitro-, formation of 221
– nitro-, reaction with organozinc reagents 214f, 220f, 389, 391
– reaction with dichloroketene 167
– selective reductions 146
– strained 94, 104ff
alkenyl halides
– formation of 222
– in intramolecular Barbier reaction 170
– reaction with organozinc reagents 214, 224ff, 405
alkenyl triflates 176, 224
alkylidenemalonates 214, 220
alkynes
– as ligands 313
– carbometallation of 165
– cyclotrimerization of 314
– functionalized, formation of 214
– hydrosilylation of 182
– reaction with methylborylene 393, 395
– reaction with organozincs 222f
– semihydrogenation of 319, 416
– terminal, deprotonation by sonicated Li 151, 160
alkynoates, 1,4-addition of organozinc compounds 222f
alkynyl halides 78, 214
allenes 167, 213, 216
allenyl magnesium chloride 301, 306, 334
alloys, *see also* intermetallics
– by deoxygenation of ternary oxides 350f
– by reduction of metallates 361ff
– electrochemically prepared 279
– Fe–Co 344f, 346ff
– Na–K 3, 165, 393
– of immiscible metals, by SMAD approach 272ff
– phosphonium ion stabilized 363ff, 375
– Pt–Rh and Pt–Pd 359f, 362, 364, 372, 375
– sulfonium ion stabilized 363ff
– surface versus bulk composition 360, 374, 376

– synergistic effects in 372ff
– via hydrotriorganoborate route 339, 345ff, 355ff, 372ff
– via inorganic Grignard reagents 330ff, 335f
– via Mg route 299, 315ff
allyl acetates
– reaction with Rieke Cu 42, 419
– with activated palladium 158, 416
allyl chlorides, *see also* Grignard reagents, allyl
– formation of tetraallylstannane 165
– reaction with activated Ba 61ff
– reaction with organozinc reagents 212f
– reaction with Rieke Cu 42f, 47
allylic alcohols
– cyclopropanation of 167
– reductive coupling of 125f
allylic barium reagents 61ff
– carboxylation of 71ff
– couplings of 73ff
– functional group compatibility 69
– in Michael additions 77ff
– in synthesis of homoallylic alcohols 82
– secondary 69
– silylation of 64
π allyl metal complexes
– in preparation of intermetallics 315ff, 323ff
– preparation of 158, 306, 313f, 393, 416
– reaction with Et_2Mg or MgH_2 299, 316ff
allyl metal reagents 61ff, 169f, 209, 300f, 306, 385
allyl phosphates
– in Barbier reactions 170
– reaction with organozinc reagents 212
– reaction with Pd–graphite 416
– zinc insertion into 195
– stereoretention 63ff, 70
– stereospecific formation of 61ff
Al_2O_3, *see also* high surface area sodium *or* high surface area potassium
– alkali metals on, relative activity 386
– as support 253, 383f, 389, 419ff
– residual –OH groups, chemical functions 381, 386, 388
– varieties of 384
aluminum
– alloy with Ti, industrial application 279
– amalgam 152
– clusters
– – attempted electrochemical synthesis 285
– – by SMAD 238, 249f, 253
– in liquid ammonia 148
– sonication 136, 158
amalgams 152, 415, 427
amides
– chromium carbenes derived of 393f
– keto-, reductive coupling, *see* acylamidocarbonyl compounds
amines
– deprotonation of 145, 160, 184
– secondary, formation of 23
amino acids 158, 224f
amino alcohols, by reductive coupling 116, 118ff, 121
amino aldehydes
– pinacol coupling of 116, 121
– reaction with organozinc reagents 158, 216
ammonia, formation over graphite catalysts 401
amphoteronolide B, synthesis of 408
anhydrous salts, preparation of 2
annelations, *see* carbocycles
anthracene, *see also* magnesium–anthracene
– anthracene radical anion 142, 300, 302
– as electron carrier 24
– cycloaddition with 2-hydroxyallyl cation 177
– nickel complex of 393
anthracyclinones, synthesis via xylylenes 171
(–)-aristolene 386
aryl fluorides, reaction with Rieke Mg 5
aryl chlorides
aryl bromides 42, 48
– reaction with activated Ni 304
– reaction with C_8K 397
– reaction with copper anion 46
– reaction with sonicated sodium 155
aryl halides, cross coupling with organozinc reagents 224ff, 403ff
aryl lithium, functionalized 37, 206
arylcalcium reagents 5
aryl triflates 162, 224
asymmetric synthesis
– ligands for 312f
– with organozinc reagents 198, 203, 226ff
atomic force microscopy, AFM 429
Auger spectroscopy 136f, 261ff, 429
azides, effect on zinc insertion 206, 409

Baldwin rules 36
Barbier reaction 137, 163, 168ff, 184, 216
barium–diene complexes 9
barium iodide, *see also* Rieke barium, allylic barium reagents
– transmetallation of allyllithium reagents 72, 82
Barton olefin synthesis 107
benzaldehyde 5, 38f, 66f, 69, 71, 118, 161, 387
benzil, reaction with C_8K 397
benzofurans, by reductive coupling 123f, 412
benzyl barium reagents 78
benzylic alcohols, reductive coupling of 125f
benzyl nickel halides 49f
BET 3, 304f
$BF_3 \cdot Et_2O$ 27, 38, 48, 157, 217, 219
biaryls
– 4,4′-di(*tert*-butyl)-, radical anions of 142, 144
– as electron carrier 4, 24, 32, 65
– functionalized, formation of 49, 195, 243
bibenzyl, *see* 1,2-diarylethane
bicyclogermacene, synthesis of 101f
bicyclo[4.4.4]-1-tetradecene and homologes 104ff
bilobalide, synthesis of 222
bimetallic compounds 209ff

Birch-type reductions 142, 391, 396f
bismuth, intermetallic with Pt 332
Blaise reaction, sonochemical 178
Blake treshold 134
borates, *see* hydrotriorganoborates *or* borohydrides
borides, as impurities in metal powders 280, 339f, 345
borohydride, as reducing agents for metal salts 280, 299, 339f, 344, 418, 420
boron–zinc exchange 202ff, 226ff
borylene, formation and trapping of methyl- 393, 395
Bouveault aldehyde synthesis, sonochemical 176
2-bromomethyl-2-alkenoates 195ff, 403
building blocks, sugar derived 397ff, 405ff, 421
butyllithium, in situ generation 160
2-butyne, cyclotrimerization of 313
butyronitrile test, hydrogenation 368, 370, 372, 374f

C1 fragments, *see* solvents, fragments of cadmium 167, 242, 252
C_8K
– as catalyst 400f
– electron transfer by –, synergistic effects 382
– in anionic polymerizations 383, 400
– in Birch-type reductions 391, 396f
– in cascade reactions 382, 397ff
– in synthesis of metal–graphite surface compounds 401ff, 427ff
– Lewis basicity of 383, 396f
– preparation and handling of 389, 421
– reaction with organosilicon compounds 393ff
– reactions with sugar derivatives 382f, 397ff, 421
– reduction of metal complexes by 392f
– self-indicative character of 382f, 392, 396
– structure and properties 382, 390ff, 427, 430f
– ternary compounds of 391, 400, 431
$C_{24}K$ and higher homologes 390, 400, 430, 432
(+)-calarene 386
calcium, *see also* Rieke calcium
– reactivity, cluster size dependence 250
camphor, reduction of 147f
cannithrene II, synthesis of 101
carbenoid organometallics 160, 166f, 191, 206ff, 213
carbides
– formation of, via SMAD method 242, 259ff
– of Mo and W, via hydrotriorganoborate route 348
– via Mg route 299, 315ff, 319ff, 324
carbocations
– 2-hydroxyallyl-, from dibromoketones 177, 418
– non-classical, from strained alkenes 105f
carbocycles
– annelated 12, 150, 411
– by cycloaddition of 2-hydroxyallyl cations 177f, 418
– by McMurry couplings, *see* McMurry reaction, intramolecular
– from ω-chloro carboxylic acids 145
– from magnesium–diene complexes 8ff
– macrocyclic, *see* macrocycles
– spiro 12ff, 167
– via organozinc reagents 29f, 203, 211, 216, 222ff
carbohydrates
– allylation of, in water 173
– building blocks, derived from 397ff, 405ff, 421
carbometallations 165f, 209f, 223
carbon, interstitial 258ff, 317, 320, 324
carbonates
– allyl-, reaction with Pd–graphite 416
– cleavage of, by C_8K 397
carbon dioxide 5, 16ff, 71ff, 415
carbon monoxide 140, 218, 238, 249, 260, 430
– β,γ-unsaturated, stereoselective synthesis 71f
carbonyl compounds
– reductions 145, 147
– sonochemical effect, in Grignard additions 155
carboxylic acid derivatives, *see also* acid chlorides
– dianions of 145, 160f
– from organomagnesium reagents and CO_2 5, 16ff, 415
– unsaturated, reaction with C_8K 396
β-carotene, by McMurry coupling 95
carvone, selective reductions of 146
casbene, synthesis of 101
cascade reactions 382, 386, 397ff, 403
catalysts, *see also* hydrogenations
– activity, comparative studies of 368ff
– bimetallic, synergistic effects in 372ff
– controlled oxidation of 371f
– doping of 339, 352ff, 371f, 416, 420
– electrochemically prepared clusters as 279, 295f
– graphimets as 419, 439, 442
– hydrotriorganoborate route to 339, 352ff, 368ff, 420
– strong metal–support interactions 371f
– ultrafine powders from SMAD as 255ff
cavitation 134ff, 138f
cembrol A, synthesis of 77
cementation 140, 146, 167, 184, 427
charcoal, as support 283, 296, 368ff, 381, 383, 385f
characterization methods, for heterogeneous reagents 428ff
(–)-chinchonidine, zinc reagent of 204
chlorosilanes, *see also* trimethylchlorosilane
– homocouplings 163, 393ff
– in hydrosilylations 182
– oligomerizations 163
– silyl-Grignard formation 312
chlorostannanes 165, 212, 301, 393ff
chromium
– carbenes 219, 393f
– complexes, via Mg route 313f

– pentacarbonyl–THF complex, reaction with organozincs 218f
cinnamic acid test, for hydrogenation catalysts 368, 370, 375
cleaning bath 182
cleavage
– of C–S bonds 152f, 399f
– of C–Se bonds 154
– of C–Si bonds 396
– of Se–Se bonds 154
Clemmensen reduction, sonochemical 150
clusters 237ff, 279ff, *see also* respective metal
– bimetallic 285f, 299
– by electrochemical salt reduction 286, 295
– combined TEM/STM study of 287ff, 295
– electrochemical preparation of 279ff, 299
– gas phase 237f, 249, 264, 271, 294
– growth process 237ff, 244f, 246ff, 266
– industrial applications 279
– magnetic properties 270ff, 294
– physical properties 249f, 264, 268, 286ff, 292f
– reactivity 249ff, 257f
– shape 237f, 245, 255, 287ff
– size control 244f, 281ff
– size dependent reactivity 250
– solubility of 286f, 295, 357
cobalt
– alloys with Fe 344ff
– clusters, electrochemically prepared 285, 294
– magnetic properties 270ff, 279, 294
– powder, via hydrotriorganoborate route 343f
coercivity 271f, 275, 349f
colloids
– as catalysts 339ff, 368ff
– bimetallic 359ff, 363ff, 372ff
– by reduction of metallates 361ff
– by SMAD method 263ff
– ether stabilized 352ff
– in polymers 263ff
– "living" 265
– physical properties of 357
– removal of surfactant shell 366ff
– silver –, by sonication 141
– tetraalkylammonium stabilized 355ff, 368ff
– tetraalkylphosphonium stabilized 363ff, 375
– trialkylsulfonium stabilized 363ff, 366
– via hydrotriorganoborate route 339, 352ff, 361ff, 368ff
(+)-compactin, synthesis of 411
concentration, increased effective 381
conjugate additions, *see* 1,4-additions
copper, *see also* organocopper reagents, Rieke Cu
– alloy with Pb, industrial use 279
– anion 44ff
– bronze 33f
– clusters 238, 285
– formation of 2-hydroxyallyl cations 177
– graphimet 419
– intermetallics of 332
– powder, from CuO 348f
– reactive powder, via Mg route 304
– soluble complexes 34, 212ff
– sonicated 162
copper–graphite 435f
co-reduction, *see also* alloys and intermetallics
– of Fe and Co salts 344ff
– of $NiBr_2/CuCl_2$ 417, 441
– of $ZnCl_2/AgOAc$ 401, 417, 439
– using hydrotriorganoborates 339ff
crassin 112
$CrCl_3$ 1, 302f
$Cr(CO)_6$ reduction with C_8K 392f
cross coupling
– of allylic barium reagents 73ff
– of benzyl glyoxalate 117
– of benzyl halides and acyl chlorides 49
– of imines 117ff
– of silylhalides 162f
– of vinyl (allyl) halides and perfluoroalkylhalides 162
crotonic acid test, for hydrogenation catalysts 368, 370, 372, 374f
crucibles 239ff, 247
crystallinity of metal powders, estimation of 343f
crystallite sizes 241, 246, 260ff, 264 *see also* grain size
$CuCN \cdot n$ LiBr ($n = 1, 2$)
– as catalyst 25ff, 198, 201, 206, 209, 212ff, 216, 389, 391, 403f
– catalytic effect on zinc insertion into diiodomethane 195
– reduction of 34, 41ff
$CuI \cdot PR_3$ 34ff, 46, 55f, 419
cuprates, *see* organocopper reagents
cyanide, adsorbed on zinc surface 32
cyanohydrines, *O*-silylated 178
cycloalkenes
– Ag(+1) complex 106f
– as test substrates in hydrogenations 368ff
– by McMurry, *see* McMurry reaction, intramolecular
– from cycloalkadienes 386
– large, *see also* macrocycles
– reaction with methylborylene 393, 395
cyclodimerization of carbonyls 124f
cyclohexene
– hydrogenation with electrochemically prepared Pd 279
– test for hydrogenation catalysts, comparative study 368ff
cyclooctene, in stability test of catalysts 372
cyclopentadiene
– acylated, McMurry coupling 411
– silylated, reaction with C_8K 396
cyclopentenols 15
cyclopropanes 14, 30, 150, 166f, 207, 252
– sila-, 163
cyclotrimerization, of alkynes 131f

DAIB, as catalyst in reactions of organozinc reagents 203
DBB, *see* biphenyl, 4,4′-di-(*tert*-butyl)-
dealkoxyhalogenation reactions 397ff, 405ff
decomposition, thermal – of Pd at-complexes 361
dehalogenation
– of deoxyhalo sugars in presence of azides 409
– of organoboron compounds 311
– of organophosphorous compounds 312f
dehydrogenations 318, 386
dehydrohalogenations, induced by C_8K 397
DeMayo reaction 102
deoxygenation
– Clemmensen, sonochemical 150
– of diols 87, 89f
– of transition metal oxides 339, 348ff
deoxyhalogeno sugars
– dehalogenation of 409
– reactions with C_8K 397ff
– reductive ring-opening by Zn(Ag)–graphite 405ff, 421
– Wurtz coupling of 415f
deprotonations
– by C_8K 396
– under sonication 145, 160f, 184
– using high surface area alkali metals 386f
diamines, vicinal, by reductive coupling 118ff
1,2-diarylethanes, synthesis of 49, 125, 243
diazepam 413
vic-dibromides 397, 418
1,2-dibromoethane, activation of metals by 193, 195
dichloroketene, sonochemical preparation 167
dichloropropene, double substitution of 213
Dieckmann condensations, with sonciated potassium 151, 183
Diels–Alder reactions 50, 171
1,3-dienes, *see also* barium–diene-, magnesium–diene complexes
– as ligands 313f
– cycloadditions with 2-hydroxyallyl cations 177f, 418
– for synthesis of functionalized organozincs 205
– from sulfolenes and UDP 152
– hydrogenation with metal SMAD 256ff
– hydrovinylation of 313
– partial hydrogenation by M/Al_2O_3 381, 386f
– reaction with Rieke metals 7ff
– radical anions of 145
– synthesis of 28f
1,4-dienes, preparation of 165
1,5-dienes 74, 125
diethylborane 203ff
diethylmagnesium 299, 316f, 319f, 328ff, 335
diethylzinc 191, 195ff, 202ff, 226ff
differential scanning calorimetry (DSC) 247f, 326, 332f, 343
1,*n*-dihaloalkanes, reaction with magnesium–diene complexes 8ff
dihaloarenes
– in Grignard reactions 5, 310
– polymerization of 50ff
diiodomethane 195f, 207, 254
(dimethyldodecylammonio)propane sulfonate 287
1,2-dimethylenecycloalkanes, reaction with Rieke Mg 11ff
dimethylpropylene urea (DMPU) 195, 202, 214f
dimethylsilylacetals, as models for pinacol couplings 110f, 115
1,2-diols
– by reductive coupling, *see* pinacol coupling
– deoxygenation of 89f
– from magnesium–diene complexes 16, 20
– stereochemical prediction 110ff, 115
dioxins, polyhalo-, destruction by C_8K 397
diphenyldiselenide, reductive cleavage of 154
1,3-dithiane, in situ deprotonation 160
dummy ligands, in organocopper reagents 42, 47

electrocatalysis 280, 285, 295
electrochemistry
– activation of zinc 195
– dibromoketone reduction 177
– for preparation of clusters 279ff
– sono 158f
electron carriers
– for amine deprotonations 145
– for metal salt reductions 1, 24, 32f, 49
electron diffraction 429, 431, 434f, 437ff, 440
electron energy loss spectroscopy (EELS) 429, 440ff
electrons, solvated 420
electron spin resonance (ESR, EPR) 88, 152
electron transfer, *see also* single-electron *or* two-electron transfer
– by C_8K 382
– correlation to cluster ionization energy 249f
– effect of azides on zinc-induced 409
– sonochemical acceleration 137ff
– to arenes 142
– to multiple bonds 145ff
– to single bonds 151ff
electrophoretic studies, on colloidal metals 264, 269
elmene, synthesis of 100f
enals
– 1,2-attack on 68, 217
– enantioselective additions to 226ff
– sugar derived 405ff, 421
enantioselective synthesis, organozinc compounds in 226ff
energy dispersive X-ray analysis (EDX)
– as analytical technique 429
– of bimetallic clusters 286
– of ether soluble colloids 353, 359
– of metal–graphite reagents 429, 434f, 438f, 440f
– of Pt powder from Pt colloid by removal of surfactant shell 367
– of Rieke Zn 438

enolates 78, 207, 218, 418, *see also* Reformatsky reaction
enol ethers, carbohydrate derived 398, 404, *see also* glycals
enones, *see also* 1,2- and 1,4-additions
– β-iodo- 193, 215
– double alkylation of 80f
– reaction with C_8K 396
enthalpy of recrystallization and melting of clusters 343f
epoxides
– epoxyalkylcopper reagents 36f
– epoxyesters, *see* glycidate
– epoxyketones, reduction of 152
– from ketones and bromochloromethane 170
– halogenated 176, 399
– opening 33, 35ff, 41ff, 419
– reaction catalyzed by Cu graphimet 419
– reaction with magnesium–diene complexes 17ff
Eschenmoser salt 402
esters
– α,β-unsaturated, by Reformatsky/Peterson tandem 403
– α-halo- 24, 49, 158, 178f, 401ff, 436
– α-silylated 222, 403
– β-bromo-, cyclopropanes from 150
– chromium carbenes derived of 393f
– cleavage of aprotic – 397f
– epoxy-, *see* glycidates
– reaction with arene radical anions 143
– reaction with C_8K 396f
– reaction with organozincs 216, 220
– reduction by high surface area alkali metals 387
ethanol as reducing agent for metal salts 280
ethers, cleavage of 1, 48, 253, 310f, 335, 397
extended X-ray absorption fine structure (EXAFS)
– of bimetallic clusters 286
– of inorganic Grignard reagents 326ff
– of McMurry coupling agents 321
– of Mg–Pd intermetallics 317
– of nanosized Pt–Rh alloy 360

fatty acid derivatives 167, 180, 218
ferromagnetism 239, 245, 270ff, 279, 294
Fischer–Tropsch reaction 141, 401
FK-506, total synthesis of 9-dihydro- 409f
flavopereirine, synthesis of 414f
flexibilene, synthesis of 101f
fluoride, as leaving group 150
fly-over ligands 314
force field calculations on ammonium stabilized clusters 290f
formaldehyde as reducing agent for metal salts 280
formylations with DMF 176
fragments of solvents, *see* solvents, fragments of
frequency effects in sonochemistry 135, 139, 176, 180, 182
furans
– by reductive coupling 123, 412
– cycloadditions of 418
– via Simmons–Smith reactions 167

gallates, *see* hydrotriorganogallates
gallium 325
geranylgeraniol, synthesis of 77
D-glucose, as starting material for natural product synthesis 407f
glycals
– from glycosyl halides 382f, 397f, 403f
– from thioglycosides 399f, 421
glycidates 403
glycosyl azides 399
glycosyl halides 382f, 397f, 400, 403f
gold
– bimetallic clusters with tin 246ff
– clusters, electrochemically prepared 285
– colloids 263f, 266ff, 280
grain size
– effect of annealing on 346
– effect on coercivity of acicular iron 350
– of powders from hydrotriorganoborate route 344f
graphimets 418f, 427, 432, 439, 442f
graphite
– as additive in formation of intermetallics 316
– binary intercalation into 418, 430f
– effect of intercalation/deintercalation on 431, 433
– intercalation into 389, 418f, 427, 430ff
– metal adsorbed on, *see* metal–graphite reagents
– metal salts intercalated into 418f, 442f
– on surface of SMAD powders 263
– potassium laminate, *see* C_8K
– residual potassium in 432f, 436, 440, 444
– sodium on 386f
– structure of 427, 430
Grignard reaction/reagents 3ff, 61, 191, 216
– alkenyl 4ff, 209
– allenyl 301, 306, 334
– allyl 66, 244, 251, 300, 306, 310
– aryl 4ff, 244, 251, 310, 415
– benzocyclobutenylmethyl 251f
– benzyl 300f, 306, 420
– bimetallic 209ff
– butyl, kinetics of formation 155
– cyclopropylmethyl 251
– di- 5, 7ff, 300, 310,
– in hydrocarbon solvents 310f
– "inorganic", *see* inorganic Grignard reagents
– mono-, of dihalides 5, 310
– of polymeric halides 300f
– of trimethylsilylbromide 312
– one pot version, *see* Barbier reaction
– role of Mg surface 135f
– using Mg–anthracene 300ff, 306f, 420
– with active Mg from Mg–anthracene or MgH_2 307ff

– with Mg–graphite 415
– with Mg SMAD 243f, 250ff, 306
– with Rieke Mg 5ff, 306, 316
– with sonicated Mg 155f
grinding 427

halide–zinc exchange 195ff
– mechanism of 198
heat storage systems, chemical 304, 309, 315
Heck reaction 296, 416
helminthogermacene, synthesis of 101f
high dilution technique 99ff, 109ff, 130, 389
high resolution electron microscopy (HR(T)EM)
– as analytical technique 280, 429
– of graphite, effect of intercalation/deintercalation 432f
– of transition metal colloids 280, 353, 357, 359, 361
"high surface area potassium" 381, 383, 384ff, 400
"high surface area sodium"
– alcoholysis of 385, 420
– for reduction of metal salts 193, 388f, 420
– preparation of 420
– solvent effect on reaction path 387f
– synthetic applications of 385ff
– wax coated 385
hirsutene, synthesis of 102
histamin antagonist 413
HIV-protease inhibitors, synthesis of 116
$[HMgCl \cdot THF]_2$ 303, 315
HMPA 36, 222, 252, 416
homoallylic alcohols 82, 207, 403, 417
homocoupling, *see* Wurtz coupling
homoenolates 82, 207, 217, 224
„hot spot" 134f, 139
HSAB concept 341
$HTiCl(THF)_{0.5}$ in McMurry couplings 87, 320ff
humulene, synthesis of 101f
hydrazines 181, 280
hydrazones, deprotonation by high surface area alkali metals 386
hydrides, *see also* magnesium hydride *or* HTiCl *or* HMgCl *or* borohydrides *or* hydrotriorganoborates
– as reagents and catalysts 319
– by hydrogenation of Ti and Zr sponges 353ff
– nickel hydride battery alloy 353, 355
– of titanium via hydrotriorganoborate route 352f
– of transition metals via Mg route 299, 315ff, 319ff, 323ff
hydroboration-transmetallation, *see* boron–zinc exchange
hydrogenation, *see also* butyronitrile, cyclohexene, cinnamic acid *or* crotonic acid test
– alloys for, synergistic effects 372ff
– comparative studies 368ff
– of intermetallics 316ff
– of metallates 361f, 365f
– of metal oxides in presence of hydrotriorganoborides 348ff
– of Mg 254f, 299, 302f, 307, 309f
– of Ni/Zr/V/Ti/Cr alloy 355
– of Ti and Zr sponges 353ff
– selective, with intermetallics 319
– semihydrogenation of alkynes 319, 416
– using active powders from Mg route 304
– with C_8K as catalyst 400
– with electrochemically prepared clusters 279, 295f
– with graphimets 419, 439
– with high surface area alkali metals 386f
– with metal–graphite catalysts 383, 416f, 432, 441
– with SMAD metal powders 255ff
hydrogen/deuterium exchange, catalyzed by SMAD metal powders 255ff
hydrogen storage 254f, 304, 309, 315ff
hydrophopic effect 172
hydrotriorganoborates
– alkali metal –, as reducing agents 340ff, 420
– alkali metal –, in carbide formation 348
– in formation of ether stabilized transition metals 352f
– redox potential of 340
– tetraalkylammonium –, as reducing agents for metal salts 339ff, 342f, 355ff
– tetraalkylammonium –, formation and solubility of 342
– tetraalkylphosphonium –, as reducing agents for metal salts 339, 363ff
hydrotriorganogallates, as reducing agents for metal salts 340
hydrovinylation, with chiral Ni catalyst 313
hydroxylamines 403

icosahedra, McKay 237
imidazole, 2,4,4-triphenyl –, via Rieke Mg 6
imides 148, 413
imines
– couplings of 117ff, 127
– dealkoxyhalogenation of *N*-bromoglycosyl- 399, 407
– deprotonations of 386, 396
– reaction with C_8K 396f
– reaction with magnesium–diene complexes 22ff
– reaction with organozincs 216, 218
indium, ultrafine powder by SMAD 253, 265
indoles
– by reductive coupling 123f, 389, 412ff, 422
– indolyl zinc reagents 31
indollactam V, synthesis of 146
indolopyridocolin, synthesis of 414f
infrared spectroscopy 266, 353
inorganic Grignard reagents
– formation of 299, 310, 319, 326ff, 333, 335
– in inorganic synthesis 330ff
– silyl- 312
intercalation 389ff, 418f, 427, 430ff, *see also* C_8K

intermetallics
– by electrochemical methods 286
– by SMAD approach 246ff, 259
– formation via inorganic Grignard reagents 330ff
– metastable 272ff, 315ff, 319ff, 323f, 333
– t-$MgPtC_xH_y$, preparation of 334
– Mg route to 299, 303, 315ff, 319ff, 323ff, 330ff
– Mg–Ti, in McMurry reactions 87, 320ff, 326f
– Pt–Fe, preparation of 335f
– thermal treatment of 316ff, 323, 332f, 335
– via hydrotriorganoborate route 339ff
– Zintl-type 325f
iodine
– reaction with sonicated samarium 139
– reaction with vinyl copper 44
iodomethylzinc iodide 207f, 213
iridium
– colloid, HREM study of 357, 359
– inorganic Grignard of 329f, 333
– intermetallics with Sn 333
– powder, finely divided 330
iron
– acicular magnet pigments 339, 349f
– alloy
– – with Co 344ff
– – with Li, Mg, Ag 272ff
– – with Ni 350f
– catalysts, for zinc–halide exchange 198
– clusters
– – annealing of 237
– – electrochemically prepared 285
– complexes via Mg route 313
– encapsulated nanoparticles of 273ff
– ferromagnetic –, electrochemically prepared 279
– graphimet 419, 442
– inorganic Grignard of 328
– intermetallics of 331f
– magnetic properties of SMAD Fe 270ff
– metastable alloys with Li, Mg, Ag 272ff
– oleic acid stabilized powder 271f
– oxide, deoxygenation of 339
– pentacarbonyl 141, 177
– powder by sonolysis of $Fe(CO)_5$ 141
– reactive powder via Mg route 304
iron–graphite 418, 431f
isocarbocyclin, synthesis of 213
isocarophyllene, synthesis of 122f
isocyanates, in Barbier reactions 176
isokhusimone, synthesis of 102f
isolobophytolide, synthesis of 112f
isomerization
– of alkenes 256ff, 385ff, 400
– of allylic metal reagents 62ff, 70
isoprene 7, 15, 145, 165

KAPA, *see* 1,3-propanediamine
ketoester coupling 85, 91, 121ff, 412
ketones
– α,α'-dibromo 177f, 418
– α-bromo, reaction with Fe–graphite 418
– β,γ-unsaturated 12, 14f, 403
– benzyl- 49f
– Clemmensen reduction of 150
– cyclic, by ketoester coupling 122, 412
– deprotonation by high surface area alkali metals 386f
– dianions of 97, 143, 415
– 1,2-diketones 50, 143, 147
– formation of bicyclic 220
– functionalized, synthesis of 36, 50, 215f, 221, 225
– monoalkylation of 386f
– reduction 147, 387, 396
– trichloromethyl 156
ketyl radicals 86, 88, 118, 120, 143, 146ff
kieselguhr 381, 383
K_2ZnCl_4 432, 437f

lactams
– β-, by Reformatsky reaction 178
– ε-capro-, polymerization by C_8K 400
– γ- 20ff
lactones
– α-bromo- 401f
– aldono- 402
– butyro- 178, 200ff, 207, 402f
– spiro- 16ff, 402f
latent period in hydrogenations of metal sponges 354f
lead 253f, 279
lepidozene 101f
(R)-linalool, synthesis of 399
lithium
– activation by ultrasound 136
– as reducing agent 1f, 24, 33, 49, 65, 87, 91, 97, 140
– cleavage of disilanes by 394, 396
– in Barbier reactions 137, 168ff, 184
– in homocoupling of chlorosilanes/stannanes 163ff
– metastable alloy with Fe by SMAD 272ff
– proton abstraction by 151, 160, 184
– Wurtz couplings by 157, 161
lithium aluminum hydride as reducing agent 86f, 91f, 122, 321f, 418, 441
lithium biphenylide 4, 61f, 418, 442
lithium naphtalenide
– as reducing agent 2, 4, 24, 34ff, 37, 42, 44f, 55, 193, 388, 392f, 419, 427
– formation of, under sonication 142
– in Graphimet formation 418
lithium 2-thienylcyanocuprate 27, 34, 37ff

macrocycles 89, 99f, 104ff, 109ff, 122, 124, 203, 386, 389f, 420
magic numbers 237
magnesium, *see also* Grignard reagents
– for reduction of $TiCl_3$ 87, 91, 323, 326ff, 335, 441

– for reduction of transition metal salts 303 ff, 319
– formation of inorganic Grignard reagents 326 ff
– highly reactive
– – from Mg–anthracene 302, 308 ff
– – from MgH_2 307 ff
– hydrogenation of 254 f, 299, 309
– in Barbier reactions 168 ff
– intermetallics of 315 ff, 319 ff, 323 ff
– metastable alloy with Fe by SMAD 272 ff
– phase transfer by Mg–anthracene systems 302
– powders of, by SMAD 243 ff, 250 ff
– reaction with trimethylsilylbromide 312
– reactivity
– – comparative study 306, 309 f
– – to alkyl halides, cluster size dependence 250
– ultrasonic activation 135 f
magnesium–anthracene 1, 299
– as source of soluble Mg 300 ff
– decomposition to active Mg 302, 308 ff
– formation under sonication 144, 302
– in formation of intermetallics from metal salts 319 ff, 325 f
– in Grignard formations 300 f, 306 f
– polymer bound 301, 419 f
– preparation of, procedure 334
– role in phase transfer of conventional Mg 302
– solubility of 300
magnesium dichloride for solubilizing Mg–anthracene 302 ff
magnesium–diene complexes 7 ff, 12 ff, 16 ff, 20 ff
magnesium–ene reaction 144, 306
magnesium–graphite 415 f, 427
magnesium hydride 299
– catalytic synthesis 302 f
– dehydrogenation of 307 ff
– from Mg SMAD 254 f
– in preparation of intermetallics 315 ff, 319 ff, 323 ff, 334
– in preparation of McMurry coupling agents 320 ff
– preparation of t-$MgPtC_xH_y$ 334
– solubility of 302 f
magnetic properties
– of acicular iron 339, 349 f
– of electrochemically prepared clusters 280, 294 f
– of electrochemically prepared pigments 279 f
– of Fe–Co alloy, from borohydride reduction 344 f
– of SMAD metal particles 270 ff
manganese
– catalysts, for halide–zinc exchange 198, 202
– colloid, ether stabilized 352 f
mass spectrometry for cluster analysis 237, 266, 282
matrix effects on clustering 238 f, 242 ff
McMurry reaction 85 ff, 320 ff, 327 f, 388 f, 410 ff
– cyclodimerization 124 f
– functional group compatibility 126 f, 412 ff
– in natural products syntheses 100 ff, 411, 414 f
– in polymer syntheses 128 f
– intramolecular 88, 99 ff, 104 ff, 389, 411 ff
– mixed couplings 96 ff
– of acylated cyclopentadienyl complexes 411
– of acylsilanes 108, 389
– of allyl and benzyl alcohols 85, 125 f
– of ketoesters, *see* ketoester coupling
– optimized procedures 90
– pinacol formation, *see* pinacols
– prototype couplings 93
– selectivity in 413 f
– strained alkene synthesis 94 ff, 104 ff
– with Ti–graphite 410 ff, 441
– with titanium on supports 388 f, 420
MeLi in at-complex formation 42 f, 47
mercury, ultrasonically dispersed 177
Merrifield resin technique 381
metal atom/vapor approach 237 ff, 280, 427
metal complexes
– formation via activated metals 140 f, 279, 304 ff, 313 f
– reduction by C_8K 392 f
metal films
– chemical liquid deposition technique 268
– conductivity of 265 f
– formation of 263 ff, 267 ff
metal–graphite reagents 1, 383, 401 ff, 418 f, *see also* C_8K
– in situ investigations on 429, 431
– morphology of 427 ff, 431 ff
metallacycles
– during alkyne cyclotrimerization 314
– in Pederson cross couplings 119
– structure of magnesium–diene complexes 7
metallocenes 306, 411
metal oxides, *see* oxides
metals with formal negative charge 44 ff, 326 ff, 420
(–)-methylenolactocin, synthesis of 200 ff
(+)-mevinolin, synthesis of 103, 411
Michael reactions, *see* 1,4-additions
molybdenum carbide 348
Mössbauer spectroscopy
– of annealed Pt–Fe intermetallics 332
– of Au–Sn intermetallics 246 ff
– of Fe–Co alloy from hydrotriorganoborate route 346 f
– of Fe encapsulated in Li or Mg particles 273 f
– of Ti–Fe intermetallics 331
(*R*)-(–)-muscone, asymmetric synthesis of 203

NaCl, supported metals on 384 f, 389, 420
naphthalene 1, 24, 32 f, 49, 393, 419
neuraminic acid, *N*-acetyl 403
Ni_3B 340
$Ni(COD)_2$ 140, 279
nickel
– activation by ultrasound 136 ff, 180 ff
– alloy with La by SMAD 259
– alloy with Fe 350 f
– allyl complexes in preparation of intermetallics 315 ff
– anthracene complex 393

– ate-complex formation with C_8K 393
– carbide by SMAD 242, 260ff
– catalysts 51, 195, 198, 200, 213, 313
– chiral catalyst for hydrovinylation 313
– clusters, electrochemically generated 285f
– inorganic Grignard reagent of 328, 332
– intermetallics with Fe or Cu 332
– magnetic properties of Ni SMAD 271
– powder, by reduction with BH_4 340
– reaction with aryl halides 254
– reactive powder via Mg route 304
– sacrificial anode 285
– sonicated, in hydrosilylations 182
– ultrafine powder by SMAD 239, 241f, 245f, 249, 254, 255ff, 260ff
nickel Graphimet 439, 441, 442f
nickel–graphite 416f, 439ff
niobium, ether stabilized colloid 352f
niobium halides 118f
nitriles
– α-halo-, Reformatsky reaction of 50
– aldono-, formation of 399, 407
– aromatic, reaction with Rieke Mg 6
– cross coupling with imines 117f
– decyanation of 152, 381, 387f
– deprotonation of 386f, 396
– from organozinc reagents 211, 391
– reaction with organozinc reagents 216, 403
– solvent effect in reactions with high surface area alkali metals 387f
nitroaldol reaction 221f
nitrone, reaction with allyl zinc reagents 403
nitroolefins, *see* alkenes, nitro-
NMR, cluster-shell analysis by 266, 282
10-nonacosanol, enantioselective synthesis of 228
nonlinear optical properties 270
norcaran, synthesis of 252

oleic acid, as surfactant for Fe SMAD 271f
organoaluminum halides 48, 158
organoboron compounds, *see* diethylborane *or* triethylborane
– by sonication 157
– non-classical 311
– transmetallation to organozinc reagents 202ff
– triorganoboranes, as complexing agents for metal hydrides 340
organocadmium compounds 207, 242
organocalcium reagents 5, 54f, 61
organocopper reagents, *see also* acid chlorides
– allylic 42, 44, 46f, 77, 419
– aryl 36, 42, 46, 56, 419
– benzyl 419
– by transmetallation of organozinc reagents 24ff, 212ff
– di-, 44
– functionalized 33ff, 37ff, 42ff, 419
– higher order 27
organolithium reagents 33, 155f, 160, 169, 176, 206, 216, 220
organomagnesium compounds, *see* Grignard
organomercury compounds 49, 154, 206, 253
organophosphorous compounds, reaction with activated Mg 312f
organosilicon compounds
– by sonochemical heterocouplings 162
– reaction with activated Mg 312
– reaction with C_8K 393ff
organotin compounds 165, 242, 253f, 417
organozinc compounds, *see also* alkenes *or* acid chlorides *or* imines *or* diethylzinc
– α-boronylated 225
– alkenyl 24ff, 156, 193, 206, 216
– alkyl 32, 156, 191ff, 252, 389, 391
– allyl 173, 191ff, 195, 203, 207, 216, 403
– aryl 24ff, 50ff, 156f, 193ff, 206, 403ff
– benzyl 156, 191ff, 201, 203, 209, 389, 391
– bis 25, 195ff, 205
– by boron–zinc exchange 202ff, 226ff
– by halogen–zinc exchange 195ff, 226ff
– by homologations 207ff
– by mercury–zinc exchange 206
– by oxidative addition 24ff, 157f, 191ff, 403
– by transmetallation from organozirconium reagents 220
– by transmetallation of organolithium reagents 156, 206
– carbometallations of 165f, 209ff, 223
– copper-mediated reactions 25ff, 212ff
– di- 202ff, 315, 318
– diallylzinc in formation of Mg intermetallics 315, 318
– dimetallic 209ff
– from 1,*n*-bimetallic compounds 209ff
– from functionalized dienes 205
– heteroaryl 31, 193ff, 403ff
– monoorganozinc compounds 50ff
– multiple substitutions with 213
– oxidation of 200, 211
– perfluoro 195, 206, 207
– propargylic 207, 216
– reaction
– – with alkenyl halides 214, 224ff, 405
– – with alkynoates 222f
– – with metal complexes 213f, 218f
– reactivity of 211ff
– synthetic applications 211ff
– with enones 25ff, 156, 184, 215, 219f
– zinc malonates 223
– zincates 207, 220
osmium clusters, electrochemically prepared 286
overpotential 283f, 293
1,3-oxazines, dihydro-, deprotonation of 396
oxazols, 2-bromomethyl-, zincation 218
oxidative addition, effect of azides 206, 409
oxidation
– controlled, of activated metals 265, 371f
– of nitronates by ozone 221
– of organozinc reagents 200, 211
oxide shells on activated metals 271, 416, 441

oxides, reductive deoxygenation of 339, 348ff
oxims 387

palladium
– alloy with Pt 360, 375
– allyl complex in preparation of intermetallics 315ff
– bimetallic with Ni 286
– carbide from SMAD 260ff
– clusters
– – cyclic voltammetry 292f
– – electrochemically prepared 279ff, 286ff
– – solubility of 283, 286
– – TEM/STM study 287ff
– – tetraalkylammonium stabilized 282ff, 361
– – UV/vis spectra 291f
– colloids
– – by reduction of metallates 361f
– – effect of counterion on catalytic activity 368
– – on charcoal, as hydrogenation catalysts 368ff
– – via SMAD 263ff
– inorganic Grignard reagent of 328f, 332
– intermetallics with Fe or Te 332
– reactive powder via Mg route 304
– sacrificial anode 282, 285
– sonicated 140, 158, 179, 181
palladium catalysts, homogeneous
– for dialkylzinc formation 198
– for reactions of organozinc reagents 28ff, 51, 162, 165, 191, 211, 213, 224ff, 403ff
palladium–graphite 416, 432ff
passivation layer, effect of sonication 135f, 155
particle size
– control of 281ff
– distribution, of electrochemically prepared Pd clusters 283
– effect on electrochemical properties 292ff
– estimation by UV/vis 291f
– in metal Graphimets 442
– metal core versus ligand shell, by TEM/STM 287ff
– of activated Mg 309
– of bimetallic Rh–Pt catalysts 374
– of colloidal sodium 139
– of colloids obtained by reduction of metallates 362, 364f
– of Co–Fe alloy 346
– of Co powder from hydrotriorganoborate route 344
– of free versus supported Pd colloid 369f
– of metal–graphite reagents 435ff
– of powder versus colloid 367
– of Rieke metals 3, 34, 439
– of SMAD metals 246, 264, 271
– of sonicated zinc 139
– of transition metal colloids from hydrotriorganoborate route 356, 358ff
– of trialkylsulfonium stabilized Rh colloid 366
– of ZrH_2 355
Pederson pinacol coupling 117ff
Peterson elimination 403
pentane, matrix properties and cleavage 241, 258f
perfluoro-*n*-tributylamine, solvent for Au colloids 266f
periplanone C, synthesis of 114f
phenanthroline as stabilizer for clusters 287
phenols
– as proton donors 152
– Birch reductions of 142
phosphines
– chiral, synthesis of 312f
– in formation of Rieke Cu 34f f, 46, 55, 419
– transition metal complexes in Mg route 306, 313f
phospholenes 312f
pinacol coupling
– by HTiCl, kinetic study 321f
– cross couplings 117ff
– in natural products syntheses 111ff
– of amino aldehydes 116, 121
– Pederson procedure 117ff
– prototype reactions 109
– with Al/NH_3 148
– with C_8K 396
– with Mg–graphite 415f
– with SMAD indium 265
– with sonicated zinc 148
– with Ti 85, 89, 109ff, 115ff, 148, 321f, 410
pinacolones 148
platinum
– active powder, formation and dissolution of 329
– alloying, synergistic effects by 372ff, 375f
– colloidal alloys 359ff, 364f, 372, 375
– colloids
– – by SMAD 263ff
– – on charcoal as catalysts 368ff
– – tetraalkylphosphonium stabilized 364, 375
– electrochemically prepared clusters 280, 286
– inorganic Grignard reagent of 328f, 332, 335f
– intermetallics of 315, 317f, 323ff, 332f
– powder
– – by removal of surfactants from colloid 366ff
– – from PtO_2 348
– preparation
– – of t-$MgPtC_xH_y$ 334f
– – of PtFe 335f
– sonicated 140, 179
platinum Graphimet 442f
platinum–graphite 383, 416f, 432ff
1,3-propanediamine, deprotonation by potassium 154
protecting shell, removal from colloids 366ff
polyarylenes 50ff
polyenes 106f, 128f, 213, 386
polymers
– by C_8K-induced anionic polymerizations 383, 400
– coating for metal powders 261, 280
– conducting 50ff, 128f
– metal colloids in 269f

– metal films on 268f
– metallation of 53f, 300f
– Mg–anthracene, bound to 301, 419
– organometallic 128
– polymer precursor for diamond 52
poly(methyl methacrylate) coating of SMAD powders 261f
poly(phenylcarbyne) 52f
polyphenylenesulfide, gold film on 268f
polysilanes 163ff
polystyrenes
– as support for Rieke metals 55f, 383f, 419
– Au colloid in 269
– by tribochemical polymerization 138
– halogenated, metallation of 53
porphycenes, by cyclodimerization 124
potassium
– as reducing agent for metal salts 1, 24, 33, 48, 49, 91, 102, 401, 418, 427, 438, 441
– –graphite, *see* C_8K
– in reductions of tetraalkylammonium metallate complexes 361f
– iodide, as additive to active metals 5, 253, 310
– KAPA formation 154
– on inorganic supports, *see* high surface area potassium
– radical anion with DBB 144
– residual, in graphite 432f, 436, 440, 444
– ultrasonically dispersed 136, 139, 151f, 183
propargylic halides in Grignard reactions 300f, 306f, 310, 334
prostaglandin derivatives 213, 220
pyridine
– as additive in McMurry reactions 91, 441
– pyridinyl zinc reagents 31

quarternary centers, formation of 16, 19f, 23
quinones
– phenanthren-9,10-, formation of 397
– reaction with Zn/TMSCl 147
quinuclidine 303

radical anions, *see also* ketyl radical anions
– as intermediates in Grignard formation 301
– by sonication 141ff, 145, 151ff, 169
– from Mg–anthracene 300ff
radicals
– allyl, benzyl 126, 301
– as intermediates 137, 161, 172, 175
– chain mechanism 198
– effect of ultrasound on 139
Raney nickel
– activity, comparative studies 255, 263, 319
– enantioselective catalysis with modified 180
– hydrogenolysis of hydrazines 181
– sonicated 180f
rare-earth metals, ultrafine powders by SMAD 259
reaction cascades, *see* cascade reactions
rearrangement
– anionic oxy-Cope 80
– Claisen 100
– Cope 100, 108
– cyclopropylidenes to allenes 167
– during desulfuration of allylsulfones 400
– metallotropic 62, 70
– of cyclopropylmethylmagnesium bromide 251
– skeletal, by high surface area alkali metals 386
Reformatsky reaction 191
– under sonication 157f, 168, 178f
– with Rieke metals 24, 49, 50
– with Zn–graphite 401ff, 436f
remanence, of acicular iron pigments 349
retinal, McMurry coupling of 96
rhodium
– alloying with Pt, synergistic effects 372ff
– alloy with Pt 359ff, 363f, 372
– ammonium stabilized clusters 280, 285, 286
– colloids
– – doped with Ti(0) 371f
– – effect of oxygen on activity 371f
– – on charcoal as catalysts 368ff
– – tetraalkylphosphonium stabilized 364
– – trialkylsulfonium stabilized 366
– industrial Rh–C catalyst 371
– inorganic Grignard of 329, 333
– intermetallics 325, 332f
– sonicated 140, 179
Rieke aluminum 1, 48
Rieke barium, *see also* allylic barium reagents
– in polymer synthesis 52
– preparation of 3ff, 61ff, 65
– reaction with dienes 9
Rieke cadmium 2
Rieke calcium 3f, 5ff, 52ff
Rieke chromium 1
Rieke cobalt 419
Rieke copper 33ff, 435, *see also* organocopper reagents
– copper anion 44ff
– cyanide based 34, 41ff, 47
– in polymer synthesis 53f
– phosphine based 34ff, 42, 55f, 419
– physical properties of 3, 34
– polymer bound 55, 419
– thienylcyanocuprate based 37ff, 47
– Wurtz couplings 33f, 161
Rieke indium 1, 48
Rieke magnesium
– ether cleavage with 311
– in Grignard reactions 3ff, 306, 310
– reaction with 1,3-dienes 7ff
Rieke metals 1ff, 61ff, 138, 161, 299, 383, 388, 401, 419, 427
– annealing of 138
– morphology of 3, 437ff
– sonochemical preparation 140
Rieke nickel 3, 49f, 263, 419
Rieke palladium 49
Rieke platinum 49

Rieke strontium 3ff, 52
Rieke uranium 2
Rieke zinc 24ff
– from zinc cyanide 32
– in polymer synthesis 50ff
– morphology of 437ff
– preparation and applications of 24ff, 50ff, 161, 193, 223
ring expansion 15
ruthenium
– electrochemically prepared clusters 286
– finely divided powder 330
– inorganic Grignard of 329f, 332
– intermetallics with Pt 332

sacrifical electrodes 159, 195, 281ff, 285, 295
salvadoricin, synthesis of 414
samarium 139, 259
sarcophytol B, synthesis of 111f
saturation magnetization 271, 349
scanning electron microscopy (SEM)
– as analytical method 428
– of activated titanium 87
– of Co powder from hydrotriorganoborate route 344
– of metal–graphite reagents 432, 440
– of Rieke metals 3, 438f
– of sonicated lithium 136
– of ultrafine SMAD powders 245, 265
– of ZrH_2 powder 355
scanning tunneling microscopy (STM) 280, 283, 287ff, 295, 429
segregation of metals in alloy clusters 360, 374, 376
serotonin antagonists 413
Sharpless epoxidation of geraniol 399
silacyclopropanes 163
silanes
– allyl 64f, 77
– di- 163, 393ff
– dichloro- 10f, 14, 163
– chloro- *see* chlorosilanes
– silacyclopropane 163
silver
– as doping agent for Zn 401, 417, 439
– chloride 439
– colloidal 141, 266
– electrochemically prepared clusters 285
– metastable alloy with Fe by SMAD 272ff
silver–graphite 435f
silylenes 163
silyl-Grignard reagent 312
silyloxyallylbarium reagents 82
silylpotassium reagents 394ff
Simmons–Smith reaction 166f, 252
single-electron transfer (SET) 53, 88, 152, 167, 169
– by Mg–anthracene 300
– by Mg–graphite 415
– effect of ultrasound on 137ff
– sonochemical switching 139, 150
sintering 1ff, 33, 353, 383, 401
SiO_2 as support 259, 383, 400, 420
SMAD, *see* solvated metal atom dispersion
S_N2'-attack 28, 212f
sodium
– arene radical anions 142
– as reducing agent for metal salts 1, 24, 48
– in liquid ammonia 418
– Na–K alloy 3, 165, 393
– on inorganic supports, *see* "high surface area sodium"
– polymerization of dichlorosilanes 164
– sodium phenylselenide 154
– ultrasonically dispersed 136, 139, 150, 152, 154f, 167, 171
solvated metal atom dispersion (SMAD) 239ff
solvents, *see also* pentane *or* tetrahydrofuran
– DMF, in Bouveault formylations 176
– effect on catalytic activity of Ni SMAD 255ff
– effect on reactions with high surface area alkali metals 387f
– electron abstraction from –, by clusters 264
– for SMAD approach 239, 242, 244f, 247f, 255ff, 263f
– for zinc insertions 193ff
– fragments of –, on ultrafine powders 241ff, 257ff, 267
– hydrocarbon –, in Grignard synthesis 310f
– in electrochemical preparation of Pd clusters 282, 295
– in McMurry reactions 88, 92
– non-coordinating, in diastereoselective additions 217
– perfluorohydrocarbon –, for SMAD metal colloids 266f
– polymerizable monomers as 269f
– siloxanes, for metal vapor deposition into 269
– solvent effects 148, 157, 387f
– ternary compounds with C_8K 391, 431
– undried, in Grignard reactions 155
– water
– – for clusters 286f, 296
– – in Barbier reactions 172ff
– – in reactions with Fe–graphite 418
– – in reductions of metal salts 340, 344
sonochemical activation, *see* ultrasound
sonochemical switching 139, 150
sonoelectrochemistry 158
spirocycles 12ff, 16ff, 167
squalene
– (3S)-2,3-oxido 76f
– synthesis of 75f, 77
SQUID, Superconducting Quantum Interference Device 294
squaric acid derivatives 167f, 215
stannanes
– allyl 48, 165, 212
– by homocoupling 165, 393
– di-, polymer bound 301
– chloro-, *see* chlorostannanes

stereochemistry, *see also* asymmetric synthesis
– control in cyclizations of organozinc reagents 200ff, 211
– control in reactions of bimetallic reagents 211
– diastereoselectivity, in additions of organozinc reagents to chiral aldehydes 217
– in Simmons–Smith cyclopropanation 167
– loss, in 1,4-additions to alkynoates 222
– of secondary allylbarium reagents 69
– prediction, in pinacol couplings 110ff, 115
– retention of, by allylic metal reagents 62, 70ff
steroids, oxo-
– Clemmensen reduction of 150f
– homodimer 96
– mixed McMurry couplings 97
strain 85, 94ff, 104ff
strigol, synthesis of 103
strong metal–support interactions (SMSI) 371f
styrene
– as electron carrier 145
– as solvent 269
– asymmetric hydrovinylation of 313
– hydrosilylation of 182
– tribochemical polymerization of 138
sublimation energy of metals 138
sulfolenes 152f, 171
sulfones
– 1,4-addition to alkenyl- 214
– reductive C–S scission 142, 152f, 400
superparamagnetism 271ff, 294f
support
– loading on 381, 384
– pre-treatment of 384
– strong metal–support interactions (SMSI) 371f
– synergism with loaded metals 381f, 386, 388
supported metals, *see also* "high surface area potassium" *or* "high surface area sodium" *or* metal–graphite reagents
– preparation and applications 55f, 253, 283, 296, 352, 368ff, 381ff
surface area, *see also* BET
– of Ni SMAD, influence of heat treatment 260
– of reactive powders via Mg route 304f, 307ff
– ultimate 237
surface composition,
– in alloy colloids 360, 374, 376
– of Graphimets 443
– of oxidized Ni–graphite 416, 441
surfactants, *see also* tetraalkylammonium, tetra-alkylphosphonium *or* trialkylsulfonium salts
– oleic acid on SMAD Fe 271f
– removal of –, from colloids 366ff
– size of surfactant shell on clusters by TEM/STM 287ff, 295
synergistic effects
– between metal and carrier 381ff, 386, 388
– in bimetallic alloys 372ff

TADDOL in asymmetric synthesis 226ff
tamoxifene, synthesis of, by mixed coupling 98
taxol 115f
tellurium, intermetallics with Pd 332
template effect of low-valent titanium 89, 109, 126
terpenes 100ff, 386
tetraalkylammonium salts 280ff
– as surfactants for clusters 281ff, 352ff, 355ff, 361ff, 366ff
– chiral, as surfactants 283
– cyclic voltammetry of 292f
– effect of concentration on cluster size 283
– electrostatic interaction with metals 280, 287ff, 357
– hydrotriorganoborates, *see* hydrotriorganoborates
– metallate complexes, formation and reduction of 361ff
– removal from colloids, 366ff
tetralkylphosphonium salts
– as surfactants for clusters 283, 285, 295, 363ff, 375
– hydrotriorganoborates, *see* hydrotriorganoborates
tetrahydroborates, *see* borohydrides
tetrahydrofuran (THF)
– as hydrogen radical donor 92
– as ligand to colloidal transition metals 339, 352ff
– cleavage of 1, 48, 311, 335
– formation of substituted 200ff
– reaction with magnesium and anthracene 299
– solubility of metal halides in 341f
– ternary compounds with C_8K 391, 431
tetraisopropylethene 93, 95
tetrakis(*tert*-butyl)ethene, failed synthesis 95
thermogravimetry 384
thioethers
– α-chloro-, reaction with Zn(Ag)–graphite 407f
– alkynyl- 222
– vinyl-, derived from carbohydrates 407f
thioglycosides in glycal formation 399f, 421
thioketones in McMurry reactions 126
Thorpe–Ziegler cyclization with sonicated potassium 151
$TiCl_3$, reduction
– under sonication 140
– with C_8K 86, 91, 103, 123, 410ff, 441
– with high surface area sodium 388f, 420
– with $LiAlH_4$ 86f, 91, 92, 122, 321, 441
– with Mg 87, 91, 323, 326ff, 335, 441
– with MgH_2 320ff
– with potassium 91, 102, 441
– with Zn/Cu 86, 88, 91f, 104, 441
$TiCl_3(DME)_{1.5}$ 92ff, 109ff
$TiCl_3(THF)_3$ 87, 320, 323, 352
$TiCl_4$
– in hydrogenations of magnesium 302f
– in McMurry reactions 91, 327, 410, 441
– reduction with borates 339, 352
– in Rh–Ti alloy formation 363
TiH_2 323
$[TiMgCl_2 \cdot x THF]$ 323, 327, 331f

$[Ti(MgCl)_2 \cdot x THF]$ 323, 326ff, 331f, 335
tin
– intermetallics 246ff, 325, 332f
– powder by SMAD 242, 246ff, 253f
– sonicated –, in Barbier reactions 173
tin–graphite 417
$Ti(0) \cdot 0.5\ THF$ 339, 352ff, 371f, 420
TiO_2, as support material 193, 371, 383, 385, 389, 391
$Ti(silox)_3$ 88
Tishchenko coupling 387
titanium catalysts, soluble
– as doping agent 339, 353ff, 371f
– in reactions of organozinc reagents 191, 211, 217, 226ff
titanium alloys 279, 363
titanium, *see also* McMurry reaction, titanium–graphite
– ether soluble *see* $Ti(0) \cdot 0.5\ THF$
– hydrogenation of metallic 353ff
– interstitial hydrogen 353
– nature of low valent 87, 320ff, 327f, 441f
– on inorganic supports 388f, 410ff, 420
– template, *see* template effect
titanium–graphite 86, 91, 103f, 123, 410ff, 422, 427, 441f
titration, of water on supports 384
TMEDA (*N,N,N′,N′*-tetramethylethylenediamine) 2, 312
tosyl cyanide, reaction with organozinc reagents 211, 389, 391
transannullar reactions 386f
transition state
– anomeric stabilization in ring closures 200
– for α-selective allylation 71
– for stereoisomerizations of allylmetal reagents 70
– in Pederson pinacolizations 120
– oxidative additions of clusters 250
transmission electron microscopy (TEM)
– as analytical method 428
– of acicular Fe pigments 349f
– of Au colloids in polystyrene 269
– of colloids from hydrotriorganoborate route 357, 359
– of Co powder from hydrotriorganoborate route 344
– of electrochemically prepared Pd clusters 282f, 287ff, 295
– of ferromagnetic Fe powder 273
– of free and supported Pd colloids 369f
– of Graphimets 418, 427, 442f
– of industrial Rh/C catalysts 371
– of metal–graphite combinations 383, 401, 415f, 417, 427ff
– of Pd powder, effect of heat treatment 368
– of Pt powder from Pt colloid by removal of surfactant shell 367
– of Rh colloid on charcoal 371
– of Rieke Zn 437ff
– of Ru and Pd colloids from metallates 362
– of Ru–Pt intermetallics 332
– of tetraalkylphosphonium stabilized Pt–Rh alloys 364
trialkylsulfonium halides as surfactants for colloids 363ff
triazines by trimerization of aromatic nitriles 6
tribochemistry 137
trienomycin, synthesis of 407f
triethylamine as additive in McMurry reactions 122, 441
triethylbenzene, catalytic formation 386
triethylborane 203ff, 340, 352
trimethylchlorosilane (TMSCl)
– activation of Zn dust 193ff
– homocoupling of 163, 393f
– in acyloin condensations 148
– in additions of homoenolates 217
– in Birch reductions 142, 147
– in conjugate additions 27, 30, 36, 38, 48, 217, 219, 222
– in oxidations of organozinc reagents 200
– in syntheses of chromium carbenes 393f
– reaction with allylbarium reagents 64f
– synthesis of silylated esters 222
– trapping of Fe enolates 418
– with Na, for SET 152
trioctylaluminum as reducing agent 361
trivinylcyclohexane, isomerization of 386
tungsten carbide 348
two-electron transfer
– by HTiCl 322
– by Mg–anthracene 300
– in reductive indole synthesis 415
– to benzyl glyoxalate 117
– to diarylketones 97

UDP, *see* potassium, ultrasonically dispersed
Ullmann reaction 33f, 161f, 435
ultrasound 133ff
– chemical effects of 137f, 161
– formation of inorganic Grignard of Ti 335
– frequency, *see* frequency effects
– general approach to sonochemistry 139
– in decomposition of Mg–anthracene 304, 308
– in preparation of metal complexes 306
– morphological changes by 135ff
– physical effects of 133ff, 164, 185
– pulsed-wave technique 162
– technical aspects of 182f
uranium 2, 318
UV/vis spectroscopy 280, 291f

$[V_2Cl_3(THF)_6]_2[Zn_2Cl_6]$ 120ff
vanadium, ether stabilized colloid 352f
vaporization of metals 237ff
– prototype examples 241ff
verticillene, synthesis of 101f
vindolinine, synthesis of 171
vinyl halides, *see* alkenyl halides

vinyl silane via McMurry coupling 108, 389
vitamine D3, analogues of 175
voltammetry, cyclic 280, 292f

water, *see* solvents
Wurtz coupling
– as side reaction in Grignard formation 306
– by activated Cu 33ff
– by C_8K 397
– by lithium 157, 161f
– by Ni–graphite 417, 441
– of allylic barium reagents 73ff
– of aryl triflates 162
– of benzylic halides 49, 243, 254, 417
– of deoxyhalosugar by Mg–graphite 415f
– stereoselective –, metal effect on 73f
– suppression of 156f, 193, 389, 401
– under sonication 161f

xanthates, cleavage by C_8K 397
xanthene-9-one, umpolung of 143
X-ray absorption spectroscopy (XAS) 428
X-ray photoelectron spectroscopy (XPS)
– as analytical technique 429
– of Au colloids 266f
– of ether soluble colloids 353
– of metal–graphite reagents 429, 440
– of Ni–La intermetallics from SMAD 259
– of Pd colloids 361f
– of Rieke metals 3
X-ray powder diffraction (XRD)
– as analytical method 428
– of activated Mg samples 308f
– of Au–Sn intermetallics 246
– of C_8K 430
– of ether soluble colloids 353
– of Graphimets 418, 442f
– of graphite, effect of intercalation/deintercalation 432f
– of high surface area potassium 385
– of intermetallics of transition metals 332f, 335
– of magnetic SMAD powders 271ff
– of metal–graphite reagents 428ff
– of Mg intermetallics 316f, 323, 326, 329
– of *t*-$MgPtC_xH_y$ 335
– of nickel carbide from SMAD 262
– of Ni-FOP 241, 260
– of powders and alloys from hydrotriorganoborate route 344f, 349, 351
– of Rieke metals 3
– of ZrH_2 355
ortho-xylylenes 50, 171

yomogi alcohol, synthesis of 417

zinc, *see also* acid chlorides *or* Rieke zinc *or* organozinc compounds *or* zinc–graphite
– activation
– – by TMSCl/dibromoethane 193ff
– – by ultrasound 136f, 139
– activity, catalytic effect of Pb on 195
– allyl complex in preparation of intermetallics 315, 318
– bis(iodomethyl) – 195, 207
– cementation of 140, 146, 167, 184
– dichloroketene formation 167f
– in Barbier reactions 171ff, 184
– in reductions of metallate complexes 361f, 364
– in Simmons–Smith reactions 166f
– in *ortho*-xylylene formation 171
– on inorganic supports 193, 389, 391
– powder, electrochemical formation of 285
– sacrificial anode 195
– sonicated, in acetic acid 146, 151
– ultrafine powder by SMAD 252
zinc cyanide 32
zinc–graphite 193, 401ff, 421, 427, 436ff
zindoxifene, synthesis of 414
Zintl compounds by wet chemical synthesis 325f
zirconium
– ether stabilized colloids 352f
– sponge, hydrogenation of 353ff
Zn–Cu couple
– applications of 157, 174, 177, 184
– for reduction of $TiCl_3$ 86, 88, 92, 104, 441
– preparation of 91, 140